Climate Action and Sustainable Development Goals

This book investigates the critical relationship and interactions between two major environmental agendas of this century: the Paris Climate Agreement and the UN 2030 Agenda for Sustainable Development. Through numerous global case studies, this book provides extensive evidence explaining the synergies and trade-offs that exist between these two agendas and illustrates the varying approaches employed by governments and institutions while implementing them. It highlights essential lessons learned, key policy and regulatory measures focused on potential mitigation and adaptation solutions, and the pivotal role played by a diverse range of stakeholders in achieving the goals of both agendas.

Features

- Explains how to create synergies among the SDGs and the Climate Goals for the betterment of the planet.
- Raises awareness on urgent global issues through comprehensive analysis of policy measures and decision-making.
- Provides in-depth case studies from various regions with diverse perspectives and successful implementation strategies.
- Discusses financing gaps and solutions to guide financial planning and resource allocation.
- Addresses monitoring and verification mechanisms that will enhance accountability and transparency.

This book is an insightful, comprehensive reference for professionals, researchers, senior undergraduates, and graduate students working on projects in support of climate change and sustainable development. It's also an excellent book for policymakers and NGOs involved with UN SDGs.

Environmental and Societal Dimensions of Sustainable Development Goals

There is a global need for a structured examination of a number of environmental science projects that are producing innovative approaches to providing governments, communities, and people with outcomes from the specific themes of the United Nations Sustainable Development Goals (UN SDG) 2030 initiative. This book series takes a new and comprehensive look at interdisciplinary ways of applying environmental science, technology, and science communication to explain how environmental science serves the societal needs of specific communities around the world to achieve Sustainable Development Goals. Each of the books in this series directly explores the use of environmental science, technology, and associated approaches to environmental management in a range of biophysical, societal, and theoretical-scientific topics, specifically related to energy, science and other partnerships, food security, societal productivity and infrastructure, water resources, land resources, climate, ecology, and sustainable practices. This series provides a multi-volume resource for ecologists, engineers, natural resource planners, and professionals in other related disciplines at the local, regional, national, and global levels of decision making.

Series Editor: Ricardo D. Lopez

Climate Actions
Local Applications and Practical Solutions
Brenda Groskinsky

Rangeland Sustainability
Social, Ecological, and Economic Assessments
Kristie Maczko, John Tanaka, Aaron Harp, and Matthew Clark Reeves

Climate Action and Sustainable Development Goals
Implementation and Integration
Gowhar Meraj and Ahmad Mohd Khalid

Climate Action and Sustainable Development Goals

Implementation and Integration

Edited by
Gowhar Meraj and Ahmad Mohd Khalid

CRC Press
Taylor & Francis Group
Boca Raton London New York

CRC Press is an imprint of the
Taylor & Francis Group, an **informa** business

Designed cover image: © Getty Images

First edition published 2026
by CRC Press
2385 NW Executive Center Drive, Suite 320, Boca Raton FL 33431

and by CRC Press
4 Park Square, Milton Park, Abingdon, Oxon, OX14 4RN

CRC Press is an imprint of Taylor & Francis Group, LLC

ISBN: 978-1-032-98718-7 (hbk)
ISBN: 978-1-032-98721-7 (pbk)
ISBN: 978-1-003-60015-2 (ebk)

DOI: 10.1201/9781003600152

Typeset in Times
by Apex CoVantage, LLC

Contents

Series Editor Foreword

As humankind has moved fully into the 21st century, societies have been confronted with the urgency of environmental challenges and indeed the immediate and practical interconnections between our climate system and how disruptions to that system impact fundamental human needs. These needs, which at their core are societal imperatives, span the full spectrum, from breathable air and clean water to safe shelter and secure food supplies, from physical infrastructure to the sociological and cultural components that define and shape our communities and their sense of place.

This book series, and this particular volume, seeks to answer such issues of sustainability, specifically focusing on the fundamental needs of humanity, which increasingly pose new challenges and opportunities for humanity. You will notice a common theme throughout these chapters: humanity's long-standing desire for a livable, comfortable, and enjoyable life for those living today, and for future generations, which is the fundamental desire generally referred to as human well-being; humanity shares this value as varied as human societies themselves.

From an institutional perspective, humanity has come together in numerous ways to find solutions to sustaining ourselves. One such structure representing most societies is the United Nations (UN). In recent years, the UN has undertaken a deliberate focus on the key elements of human well-being through the UN Sustainable Development Goals (UN SDGs). Among the 17 articulated UN SDGs, the following tie directly to environmental sciences and geography, the primary focus of this series:

- Climate Action (UN SDG 13)
- No Poverty (UN SDG 1) and Zero Hunger (UN SDG 2)
- Good Health and Well-being (UN SDG 3)
- Ensuring Clean Water and Sanitation (UN SDG 6)
- Developing Affordable and Clean Energy (UN SDG 7)
- Decent Work and Economic Growth (UN SDG 8)
- Stimulating Industry, Innovation, and Infrastructure (UN SDG 9)
- Reduced Inequalities (UN SDG 10)
- Supporting Sustainable Cities and Communities (UN SDG 11)
- Responsible Consumption and Production (UN SDG 12)
- Protecting Life below Water (UN SDG 14) and Life on Land (UN SDG 15)
- Peace, Justice and Strong Institutions (UN SDG 16)
- Developing Partnerships to Achieve the Goals (UN SDG 17)

This volume strategically examines these interlinkages through diverse contexts and scales, from corporate environmental disclosure and family firm leadership in India to coal transition finance in Indonesia and Vietnam; from AI-enabled resilience frameworks and digital twins to district cooling and thermal energy storage in the United Arab Emirates and Middle East and North Africa region; from community-led water systems and nature-based solutions in South Africa to glacier grafting and agroforestry in Himalayan communities; and through critical governance and data monitoring challenges across the Gulf Cooperation Council and BRICS+ countries. Each contribution recognizes that effective environmental science must serve the communities and places it aims to support, integrating diverse ways of knowing, from Indigenous practices and local knowledge to satellite observations and quantitative modeling.

Contributors address four interconnected dimensions that shape effective climate action. First, integration: embedding climate considerations within broader development planning while actively navigating synergies and trade-offs across sectors and geographies. Second, governance: achieving policy coherence, establishing transparent monitoring systems using earth observations and ground-based data, and ensuring inclusive decision-making that genuinely empowers, rather than tokenizes, Indigenous peoples, women, and marginalized communities. Third, implementation: closing persistent gaps in climate finance mobilization, technology transfer, institutional capacity, and verification systems. Fourth, opportunity: strategically leveraging falling renewable energy costs, scaling nature-based solutions anchored in specific landscapes, and deploying blended finance mechanisms to enable just transitions that respect the cultural and environmental values of affected communities.

While this volume demonstrates that climate action and development goals can reinforce one another, contributors also critically examine where tensions and trade-offs emerge, where climate mitigation may conflict with immediate development needs, where technology transfer faces barriers, and where Indigenous knowledge systems and modern science offer different solutions. Effective action requires navigating these tensions transparently rather than assuming harmony.

As the 2030 SDG target year and the 1.5°C climate window narrow rapidly, this volume charts realistic pathways forward. The evidence presented here demonstrates that when climate action embraces equity principles, integrates local and Indigenous knowledge with environmental sciences (an integration that remains incompletely understood), and pursues co-benefits across development goals, it can simultaneously address the climate emergency and accelerate progress toward the societal imperatives that all communities share. This volume offers policymakers evidence-based frameworks for consideration and reform, researchers' identification of critical knowledge gaps across spatial and temporal scales, practitioners' evidence-based replicable models adaptable to diverse geographies, and communities' validation that place-based participatory approaches are essential to effective and durable climate action.

Ricardo D. Lopez, PhD
Series Editor
Environmental and Societal Dimensions of Sustainable Development Goals

Preface and Acknowledgements

The Paris Agreement and the 2030 Agenda for Sustainable Development together define the most consequential policy horizon of the 21st century. The climate and development crises are interconnected and require synergistic approaches and solutions. The United Nations' Third Report on Climate and Sustainable Development Goals (SDG) Synergies 2025 shows that 80% of the SDG targets are linked to climate, and synergy can boost efficiency and cut government crisis costs by 40%, easing massive funding gaps, which together surpass 10 trillion dollars. These crucial frameworks alone do not align incentives, mobilize finance, or deliver implementation at scale. It requires a participatory approach from multi-stakeholders and tailored dual-focused strategies aligning investments with specific country development and climate goals.

This volume, Achieving Climate Action and SDG, was conceived to close that gap by tracing, with precision, how climate mitigation and adaptation can be designed, financed, and governed to accelerate progress across the SDGs, while acknowledging trade-offs, capacity constraints, and the political economy that shapes real choices. The book progresses from core principles to actionable tools, like standards, finance, and digital AI solutions, and community-centered processes so that readers can move deliberately from first principles to execution.

The opening chapter establishes a common lexicon and situates climate action within the broader governance architecture comprising the United Nations Framework Convention on Climate Change, the Paris Agreement, and linked processes such as the Global Stocktake and Sendai Framework for Disaster Risk Reduction. It clarifies the aims and targets of SDG 13 and maps interlinkages to poverty, food and water security, health, cities, oceans, terrestrial ecosystems, equity, and partnerships. Rather than treating "synergy" as a slogan, the chapter treats it as a design problem: it specifies where climate measures amplify SDG progress through co-benefits and where poorly designed interventions can undermine affordability, labor markets, and land use. This framing recurs throughout the volume, which treats implementation gaps in ambition, finance, technology, and capacity not as generic obstacles but as points for specific interventions.

Subsequent chapters examine the institutions and actors that determine whether integrated strategies are adopted. The discussion of family-owned firms and the wider corporate sector analyzes environmental leadership, disclosure practices, and the credibility of transition plans, recognizing the dual role of businesses as both sources of innovation and subjects of accountability. A complementary treatment of higher-education institutions shows how governance, diversified finance, and technology adoption can make universities effective engines of sustainability, research translation, and workforce development. Legal and policy frameworks are then considered as coherence devices: where statutes, regulations, and standards are properly aligned, they shape procurement, tariffs, investment mandates, and data regimes in ways that lower risk and enable long-term capital flows.

Regional realities are treated directly rather than by inference. A critical review of MENA's progress assesses why several SDG targets are not on track and identifies the geopolitical and socio-economic mechanisms that impede delivery. Health-climate dynamics in BRICS countries, analyzed using a pooled mean group approach, illuminate how temperature and precipitation extremes, health expenditure, and energy structure interact to shape life expectancy and youth mortality. Transition finance receives granular attention in the study of coal retirement initiatives in Indonesia and Vietnam, where the feasibility of mechanisms such as Just Energy Transition Partnerships and Energy Transition Mechanisms ultimately turns on coordination across donors, domestic institutions, and private capital. A complementary review of sustainable aviation across BRICS+ charts research trajectories, technology readiness, and fuel pathways, relating sectoral decarbonization to SDG 12 and SDG 13, while underscoring cost, infrastructure, and scaling constraints.

The book also addresses decision-dense systems, where early choices lock in long-term outcomes. In hot-arid regions, the chapter on urban heat, district cooling, and thermal storage argues that treating cooling as core infrastructure allows countries to reconcile load growth with COP28 ambitions to triple renewable power and double efficiency. Here, execution detail matters: harmonized appliance and refrigerant standards, stronger building envelopes, time-of-use tariffs and demand response, district energy and thermal storage, desalination–power coupling, and open-data monitoring, reporting, and verification are considered together with blended finance ladders that match risk to capital. The data chapter takes a similarly practical stance, diagnosing current reporting gaps for SDG 13 and proposing statistical and institutional reforms that can raise data quality, periodicity, and comparability. This shows why finance follows standards and data, not the reverse. A policy framework for AI-centric climate resilience then demonstrates how sensor networks, digital twins, and large language models can support early warning, resource allocation, and urban planning if governance, privacy, and capacity are explicitly designed for deployment. Community-led adaptation and indigenous knowledge are treated as determinants of durability rather than "soft" add-ons, with South African participatory models and Himalayan water-system practices showing how co-creation, transparent budgeting, citizen science, and knowledge rights produce lasting results. A gender-and-technology perspective closes this arc by illustrating how differentiated vulnerabilities and financial constraints shape the uptake and effectiveness of climate technologies and how culturally appropriate design improves adoption and equity.

Methodologically, contributors were asked to make claims falsifiable and bounded, to name trade-offs, to link analysis to instruments that shift real decisions, and to attend to inclusion by specifying who decides, who benefits, who bears cost, and how accountability is maintained. Chapters draw on econometric analysis, comparative policy studies, geospatial methods, bibliometric mapping, and case-based inference. Where administrative, corporate, or community data were used, the authors obtained the necessary permissions or relied on de-identified public sources under institutional ethics guidelines. When Indigenous knowledge was cited, informed consent and recognition of knowledge rights were observed. To aid cumulative learning, acronyms are expanded at first use and standardized across chapters, SI units are used throughout, currency values are reported in USD unless otherwise indicated, and where code or datasets support figures, repositories are referenced in endnotes.

The opening chapter establishes the scientific basis for rapid mitigation and adaptation and situates SDG 13 within the wider governance architecture, including the UNFCCC, the Paris Agreement, the Global Stocktake, and the Sendai Framework. It maps interlinkages with goals on poverty, food, health, water, energy, cities, oceans, land, equity, and partnerships, and introduces climate-resilient development as an integrating concept. Implementation gaps in ambition, finance, technology, and capacity are treated alongside opportunities for innovation, public health co-benefits, nature-based solutions, and multi-level governance. The chapter concludes with actionable implications for policy coherence and international cooperation.

Focusing on Business Responsibility and Sustainability Reporting in India, Chapter 2 assesses environmental leadership among top family-owned firms. It documents practices on disclosure, stewardship, and long-term sustainability commitments and examines how family governance interacts with regulatory pressure and stakeholder expectations. The analysis highlights credible target setting, investment in cleaner processes, and supply chain engagement while noting constraints in financing, indicator design, and verification. Lessons are framed to inform both family and non-family firms in India and comparable economies.

Chapter 3 reviews the structural barriers to SDG delivery in the NENA region. It synthesizes evidence on conflict, macroeconomic volatility, inequality, gender gaps, water scarcity, and environmental stress and links them to indicator trends and policy trajectories. The chapter argues for targeted interventions that move beyond generic templates and emphasizes regional collaboration, shock-responsive systems, and fit-for-purpose finance. It proposes a set of priority reforms that recognize geopolitical realities, while safeguarding social outcomes.

In Chapter 4, cooling is treated as the core infrastructure for hot-arid cities. It aligns COP28 goals to triple renewables and double efficiency with measures that curb peak demand: district cooling, thermal storage, stronger appliance standards, and better building envelopes. A 2030 implementation spine is proposed that connects remote sensing, National Cooling Action Plans, demand response, and a blended finance ladder. The G7/G20 levers include harmonized efficiency and refrigerant standards, green procurement, and open-data MRV. The UAE case studies illustrate the integrated gains in reliability, emissions, and health.

Using TVP VAR and transfer entropy, Chapter 5 examines the time-varying connectivity among equity, ESG, climate, and geopolitical risk indices. It identifies directional information flows and nonlinear dependencies with implications for portfolio risk and policy signaling. The results show a distinctive transmission from geopolitical shocks to key indices. This chapter discusses how investors and regulators can use connectivity diagnostics to anticipate volatility, design hedges, and interpret disclosure-based market movements.

Applying a pooled mean group approach for 2000 to 2022, Chapter 6 estimates the long-run relationship between precipitation, temperature, health expenditure, renewable energy shares, and population health outcomes. It finds heterogeneous effects by age cohort and climate variable and links pathways to water-related hazards, heat stress, and food security. Policy implications include targeted health investments, faster renewable adoption, and locally tuned mitigation and adaptation packages to protect adolescent and young adult health.

Chapter 7 analyzes how businesses drive or impede SDG and climate outcomes through financial mobilization, technology deployment, and policy advocacy. It assesses the promise and limits of ESG disclosure as a monitoring tool, identifies credible elements of corporate transition plans, and flags the risks of goal fragmentation and greenwashing. Practical guidance is offered on indicators, governance, and partnerships that translate corporate commitment into measurable social and environmental gains.

Focusing on SDG 13 measurement, Chapter 8 reviews data coverage, periodicity, and quality across income groups and regions. It explains why few indicators reach global reporting and proposes reforms for national statistical plans, legal mandates, technology-enabled collection, and interoperable MRV systems. Case contrasts show how consistently reporters built capacity, while non-reporting contexts faced resource and coordination gaps. A staged roadmap links better data to better policies.

Chapter 9 presents a scalable policy framework for integrating sensor networks, advanced communications, digital twins, large language models, and agentic decision support into climate resilience. Governance, ethics, and privacy are treated as first-order design constraints. The framework targets systemic barriers, such as data scarcity, infrastructure gaps, and limited computing, and illustrates how modular rules and regional collaboration can enable low-cost pilots. Use cases from Latin America and the Caribbean demonstrate improvements in disaster preparedness, resource allocation, and urban planning.

Through a qualitative case study of a Turkish private university, Chapter 10 examines the diversification of funding, digital transformation, and environmental programs aligned with the SDGs. It identifies how innovation hubs and technology-integrated teaching can strengthen academic and financial resilience. This chapter offers policy suggestions for ministries and boards seeking to strengthen university systems against fiscal and demographic shocks while advancing sustainability.

Chapter 11 argues that legal and policy frameworks are decisive for integration in the EU. It reviews domestic legislation and international commitments, including the Paris Agreement, and analyzes national case studies to show how mandates, procedures, and multi-level governance can deliver or derail coherence. Environmental justice and community participation are placed at its center. The chapter concludes with reforms that strengthen alignment while managing trade-offs transparently.

Using qualitative comparative analysis, Chapter 12 examines the design features and outcomes of coal transition platforms such as ETM, JETP, and GFANZ in two industrializing countries. It identifies the conditions for systemic impact across social, environmental, and financial outcomes and proposes governance mechanisms for coordinating initiatives. The analysis speaks directly to the timing, sequencing, and verification of coal-fired power plant retirement and replacement.

Chapter 13 reviews how GCC governance shapes climate and SDG delivery under conditions of hydrocarbon dependence and a high climate risk. It maps institutional architecture, subsidy reform, renewable procurement, and adaptation in water, cities, and coasts. Horizontal and vertical coordination, regional platforms, and grid interconnections are assessed alongside the roles of sovereign funds, the private sector, academia, and youth. Priorities for 2030 include mandate clarity, budget tagging, open-data MRV, common regional standards, and structured stakeholder engagement.

Comparing urban Tshwane and rural Limpopo, Chapter 14 shows how co-creation, citizen science, and nature-based solutions translate climate goals into local gains. An iterative six-step participatory framework enables joint problem definition, implementation, and skills transfer. Enablers and barriers are distilled into policy guidance for institutionalizing community-led mechanisms, scaling multiple-use water services where appropriate, investing in capabilities, and designing finances that match local delivery capacity.

Chapter 15, centering Himalayan communities, documents practices such as glacier grafting and agroforestry that enhance water security, biodiversity, and adaptive capacity. It identifies risks from marginalization and policy neglect and proposes legal recognition, co-production of evidence, knowledge rights, and community-led financing and monitoring. The argument is pragmatic: Indigenous institutions are critical partners for equitable adaptation and measurable SDG progress.

Chapter 16 shows why gender-responsive design and recognition of indigenous knowledge are not add-ons but core to effective climate technology adoption. It traces differentiated impacts and access constraints, using a case from rice farming systems to illustrate how finances, capabilities, and cultural fit shape uptake. This chapter offers design principles for inclusive governance and technology transfer that protect rights and improve resilience.

Through a bibliometric analysis of the literature from 1995 to 2024, Chapter 17 maps research patterns and collaboration networks in sustainable aviation across BRICS+. It links pathways such as sustainable aviation fuels, electric propulsion, and hydrogen systems to SDG 13 and SDG 12 and contrasts policy and infrastructure readiness across countries. The persistent constraints include production costs, supply chains, and airport infrastructure. The chapter calls for coordinated investment, standards, and partnerships to accelerate carbon-neutral aviation.

The scope of this volume is intentionally selective. It concentrates on arenas where near-term choices create long path dependencies, such as energy and cooling systems, urban design, data and MRV, finance architecture, institutional capacity, and social inclusion. Some sectors and geographies are addressed through case studies rather than exhaustive surveys, and where the evidence base is emergent, the text marks uncertainty and delineates research needs. The intended audience includes public officials designing bankable, standards-aligned programs, corporate leaders and investors seeking credible transition pathways, researchers and students building bridges between scholarship and practice, and civil society organizations embedding justice and participation in delivery. The book can be read linearly or modularly: readers focused on governance may begin with the legal-policy and GCC chapters; those concerned with implementation can pair the cooling, data, and AI chapters; and those seeking sectoral decarbonization debates may turn to coal transitions and aviation.

Two cross-cutting conclusions emerge from this study. First, synergistic integration requires coherence, transparent standards, disclosure, and MRV systems to facilitate capital flow from pilots to portfolios. Second, community and justice-centered design is pivotal to both agendas. Without participation, benefits are fragile and costs are poorly distributed, eroding political support over time. This volume posits that aligning climate action with sustainable development hinges less on identifying new principles and more on rigorously implementing established ones with discipline, evidence, and accountability.

The editors are grateful to the authors, reviewers, institutions, and communities whose work and experiences enrich the chapters. Interpretation errors remain our own. The decisive decade is already underway; the task now is to convert global aspirations into durable systems that compound gains rather than dissipate them. This book is offered in that spirit, as a resource for synergistic design and execution, and as an invitation to test, adapt, and openly share the standards, data, and practices that make climate-compatible developmental progress replicable across contexts. The authors would also like to acknowledge the Japan Society for the Promotion of Science (JSPS), as the initial idea for this book was conceptualized through a series of discussions that took place during both authors' postdoctoral fellowships in Japan. Gowhar Meraj is grateful to the American University of Sharjah for its full support in carrying out this project.

Finally, we owe a special debt to our families and parents, whose patience, quiet sacrifices, and unwavering faith sustained the long hours behind this work and made its completion possible.

Editors
Gowhar Meraj
Sharjah, United Arab Emirates
Ahmad Mohd Khalid
Mumbai, India

About the Editors

Dr. Gowhar Meraj is a research fellow at the American University of Sharjah (UAE), where he develops geospatial AI, geospatial analytics, and remote-sensing workflows to support SDG-aligned climate action across the GCC and beyond. He works on global issues such as climate risk, land–ocean interactions, ecosystem services, blue carbon, and sustainable urbanization, translating machine-learning and Earth-observation insights into policy-ready decision support. His research integrates artificial intelligence with climate and ocean drivers and socio-environmental metrics to strengthen carbon accounting, nature-based solutions, and climate-resilient planning. Previously, he served as a postdoctoral fellow at the American University of Sharjah (AUS). Before that, he held the highly competitive JSPS Postdoctoral Fellowship at the University of Tokyo, where he advanced transboundary ecosystem-services modelling and decision-support tools for river-basin governance, and he was also a Young Scientist Fellow under the Scheme for Young Scientists and Technologists, Government of India (SYST-GoI).

Working at the science-to-policy interface, Dr. Meraj links climate change mitigation and adaptation to delivery instruments such as MRV systems, standards, and implementation pathways, with recent work also covering energy and urban systems, including cooling demand, urban heat risk, and efficiency-focused transitions relevant to the GCC and MENA. He has authored 150+ peer-reviewed publications, produces policy-facing outputs (briefs, toolkits, and decision-support applications), and has edited more than 10 academic books and volumes, coordinating multi-author teams across regions and disciplines. He has delivered invited and keynote presentations internationally, including at UNCCD COP16 (Riyadh) and Big 5 Global (Dubai).

Dr. Ahmad Mohd Khalid's research examines how developing economies can align growth with sustainability—linking climate action, corporate disclosure, and urban governance to the Sustainable Development Goals. At NMIMS Mumbai (PDSE & FBM), he teaches Microeconomics, Business Statistics, Business Sustainability, Business & Society, and Policy & Regulation, integrating empirical methods with real-world cases from India and abroad. His recent work investigates city-scale climate action planning, data assurance in corporate environmental reporting, and strategies to enhance SDG–climate co-benefits. Before joining NMIMS, Khalid was awarded Japan's highly competitive JSPS Postdoctoral Fellowship, during which he conducted research at Keio University and the United Nations University (Tokyo). His earlier industry-policy experience includes over two years at Adelphi (New Delhi), focusing on research, project management, and training for sustainability initiatives.

Khalid's publications reflect both methodological rigor and policy relevance. He has also authored policy briefs (UNU-IAS) and book chapters on sustainable consumption (EV adoption) and tourism-SDG linkages. His conference record includes invited and peer-reviewed presentations in South Korea, Japan, the USA, and India. Across teaching, research, and policy outreach, Dr. Khalid's editorial focus emphasizes actionable evidence and clear communication for decision-makers. He brings to the volume a cross-disciplinary perspective that connects quantitative analysis, institutional design, and stakeholder engagement, aimed at accelerating climate-aligned, SDG-consistent development pathways in emerging economies.

Contributors

Letetia Addison
The University of the West Indies
St. Augustine Campus
St Augustine, Trinidad and Tobago

S. Aiswarya
Department of Commerce
School of Social Sciences and Languages
Vellore Institute of Technology
Vellore

Abdullah Ali
The University of the West Indies
St. Augustine Campus
St Augustine, Trinidad and Tobago

C. Ayas
Monash University
Melbourne, Australia

Tarek Ben Hassen
Program of Policy, Planning, and Development
Department of International Affairs
College of Arts and Sciences
Qatar University
Doha, Qatar

Mehmet Akın Bulut
Department of Management
Ibn Haldun University
Başakşehir, Istanbul

Hamid El Bilali
International Centre for Advanced Mediterranean Agronomic Studies (CIHEAM-Bari)
Valenzano, Italy

Lesego Gaegane
Water Research Commission
South Africa

P. Raja Lingam Goud
Central University of Karnataka
Kalaburagi, Karnataka, India

E.M. Gui
Monash University
Melbourne, Australia

Yedhu Harikumar
CFAI Foundation for Higher Education
Off Campus Center Bangalore (Deemed to be University)
Karnataka, India

Abdo Hassoun
Sustainable AgriFoodtech Innovation & Research (SAFIR)
Arras, France

Patrick Hosein
The University of the West Indies
St. Augustine Campus
St Augustine, Trinidad and Tobago

Siti Hafsyah Idris
Universiti Teknologi MARA (UiTM)
Malaysia

Ahmad Mohd Khalid
Pravin Dalal School of Entrepreneurship and Family Business Management
NMIMS
Vile Parle West, Mumbai, Maharashtra

M.R. Kresnawan
Monash University
Melbourne, Australia

Karnati Kiran Kumar
Central University of Punjab
Bathinda, Punjab, India

Naveen Kumar
Central University of Punjab
Bathinda, Punjab, India

Lone Rafiya Majeed
Department of Life Sciences
Vivekananda Global University
Jaipur, Rajasthan, India

Gowhar Meraj
Department of Civil Engineering
College of Engineering
American University of Sharjah
Sharjah, United Arab Emirates

V. Molose
Water Research Commission
South Africa

M. Muthumeenakshi
Department of Commerce
School of Social Sciences and Languages
Vellore Institute of Technology
Vellore

Mahfoozur Rahman
Department of Management
Ibn Haldun University
Başakşehir, Istanbul

Tamika Ramkissoon
The University of the West Indies
St. Augustine Campus
St Augustine, Trinidad and Tobago

Deepak Ramsubhag
The University of the West Indies
St. Augustine Campus
St Augustine, Trinidad and Tobago

Pankaj Singh Rawat
Jindal Global Business School
OP Jindal Global University
Sonipat, Haryana, India

Charlotte Remteng
Independent Researcher
Douala, Cameroon

Shailendra Singh
IILM University Gurugram
Gurugram, Haryana, India

Reshma Sucheran
Department of Hospitality and Tourism
Faculty of Management Sciences
Durban University of Technology
South Africa

Adem Yurdunkulu
Department of Management
Ibn Haldun University
Başakşehir, Istanbul

1 Climate Action and Sustainable Development Goals (SDGs)

Gowhar Meraj

1.1 INTRODUCTION

Climate change has emerged as one of the defining global challenges of our time, with profound implications for society, economies, and ecosystems worldwide (Richardson et al., 2011). The warming of the planet, driven primarily by human activities, threatens to undermine decades of developmental progress and poses risks to the well-being of current and future generations. Recognizing these dangers, the international community has prioritized climate action and embedded it within the broader framework of sustainable development. In 2015, the United Nations adopted the 2030 Agenda for Sustainable Development, featuring 17 interlinked Sustainable Development Goals (SDGs)[1] that collectively provide a roadmap for global prosperity, social inclusion, and environmental sustainability (Arora and Mishra, 2019). Among these goals is SDG 13: Climate Action, a dedicated commitment to take urgent action to combat climate change and its impacts. This chapter provides a comprehensive introduction to climate action and the SDGs in a global context. It outlines the scientific basis and urgency of climate action, the origins and objectives of the SDGs, the integration of climate goals within the sustainable development agenda, and the key linkages between climate change and other development objectives. Additionally, we examine major policy frameworks, such as the Paris Agreement under the UN Climate Convention, discuss challenges and opportunities in aligning climate action with sustainable development, and highlight the crucial role of evidence-based policy and international cooperation in achieving these intertwined goals. Figure 1.1 illustrates the conceptual pathway linking climate action with the SDGs, moving from climate change threat to a sustainable global future.

1.2 CLIMATE CHANGE: A GLOBAL CHALLENGE AND THE IMPERATIVE FOR ACTION

Climate change refers to long-term shifts in temperature and weather patterns, primarily caused in the modern era by increased concentrations of greenhouse gases (GHGs), such as carbon dioxide, methane, and nitrous oxide, in the atmosphere due to human activities, such as fossil fuel combustion, deforestation, and industrial processes. The IPCC's Sixth Assessment Report states plainly: it is unequivocal that human influence has warmed the atmosphere, ocean, and land (IPCC, 2023). The global average surface temperature in the decade 2011–2020 was approximately 1.09°C higher than that in pre-industrial times (1850–1900 average) (Omali, 2022). Each of the last four decades has been successively warmer than any decade since at least 1850, indicating an accelerating trend of global warming. This warming has already led to observable changes in the Earth's system, including rising sea levels, retreating glaciers, diminishing Arctic sea ice, more frequent and intense heatwaves, shifting precipitation patterns, and other climate-related extreme events. Without significant reductions in GHG emissions, global temperatures will continue to rise, leading to increasingly dangerous impacts.

The consequences of unmitigated climate change are far-reaching and pose severe threats to human societies and natural ecosystems. Higher global temperatures intensify the hydrological

DOI: 10.1201/9781003600152-1

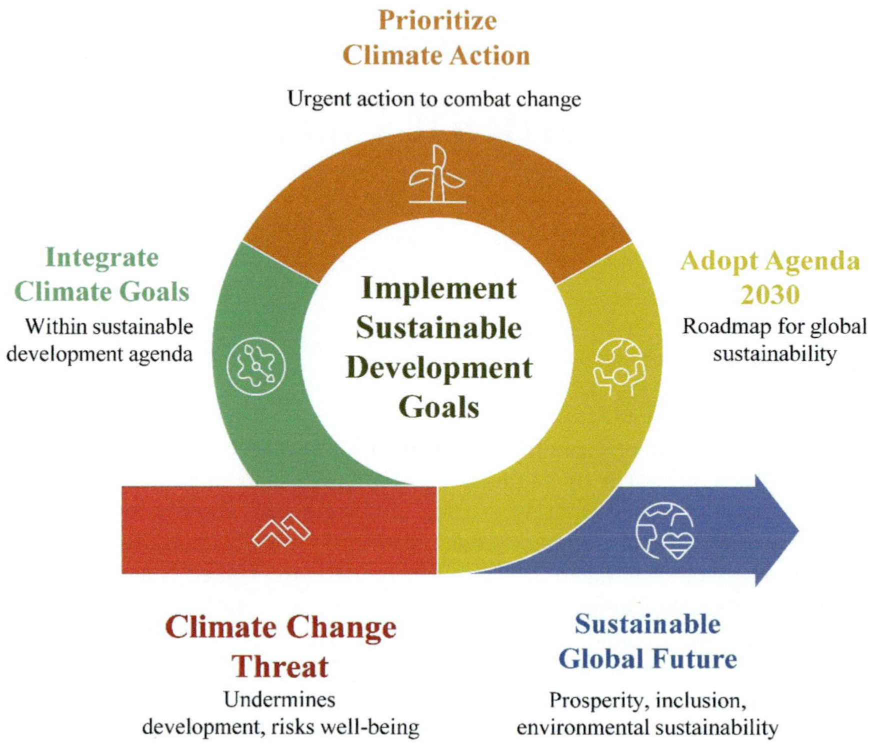

FIGURE 1.1 Climate–SDG integration pathway. The diagram frames climate action as central to implementing the Sustainable Development Goals (SDGs): prioritizing climate action and adopting Agenda 2030 feed into integrating climate goals within the broader development agenda, steering progress from a climate change threat toward a sustainable global future (prosperity, inclusion, environmental sustainability).

Source: Author's illustration from literature sources discussed in this chapter.

cycle, contributing to more extreme weather events, such as heavy rainfall and floods, as well as longer or more intense droughts in some regions of the world. Ocean warming and acidification, caused by CO_2 absorption, harm marine life and threaten fishery and coral reef health. Sea-level rise, which has accelerated in recent decades as ice sheets and glaciers have melted, endangers low-lying coastal zones and small island nations worldwide. From a human standpoint, climate change exacerbates existing vulnerabilities and inequalities, and its impacts disproportionately affect the poorest and most marginalized communities, who often have the least resources to cope with disasters or adapt to changing conditions. The United Nations highlights that "climate change is one of the greatest challenges of our time and its adverse impacts undermine the ability of all countries to achieve sustainable development."[2] In other words, climate change is not just an environmental issue but also a developmental and humanitarian one; left unchecked, it could reverse progress in poverty reduction, food security, health, and many other areas.

The urgency of climate action arises from the narrowing window of opportunity to limit warming to levels deemed relatively safe. Through the Paris Agreement of 2015, an international treaty under the UN Framework Convention on Climate Change (UNFCCC), nations have agreed to pursue efforts to hold the increase in the global average temperature to "well below 2°C above pre-industrial levels" and to "pursue efforts to limit the temperature increase to 1.5°C."[3] These targets reflect scientific findings that warming beyond 1.5–2°C would lead to exponentially higher risks of severe and potentially irreversible impacts, such as widespread coral reef die-offs, inundation of coastal cities, more extreme heat stress on human health, and destabilization of critical Earth systems. Indeed, the IPCC's Special Report on Global Warming of 1.5°C concluded that every fraction of a degree of further warming matters, and the difference between 1.5°C and 2°C of warming could mean substantially more frequent extreme heatwaves, greater declines in crop yields, and an increased likelihood of tipping points in the climate system.[4] This has galvanized a worldwide call for rapid and deep reductions in GHG emissions over the next few decades to achieve net-zero emissions. To meet the 1.5°C goal, scientists project that global CO_2 emissions must be halved by 2030 (from 2010 levels) and reach net zero around mid-century. Such a trajectory requires transformational changes in energy systems, transportation, industry, land use, and consumption patterns on a global scale, essentially a comprehensive shift toward low-carbon and sustainable practices. Figure 1.2 summarizes the post-Paris signal, temperature limits, scientific assessment of risks, and the resulting call for rapid, transformational emissions cuts.

However, despite growing awareness and political commitment, the world's current actions remain inadequate. The IPCC notes that the national pledges (Nationally Determined Contributions, or NDCs) made under the Paris Agreement, as of the early 2020s, are not ambitious enough to meet the agreed-upon temperature limits. In fact, the projected emissions under present policies and pledges would likely lead to warming well above 2°C by 2100 unless significantly strengthened. As one IPCC assessment states, the aggregate reductions implied by current NDCs "would still make it impossible to limit warming to 1.5°C with no or limited overshoot," and even limiting warming below 2°C would require much steeper emission cuts after 2030 than those currently planned (Grubb et al., 2022). Recent analyses by the UN likewise warn that the world is on the brink of a climate catastrophe if transformational action is not taken. For instance, the UN Department of Economic and Social Affairs reported in 2024 that without drastic measures to reduce emissions across all sectors this decade, the 1.5°C target will be missed,[5] putting the lives of more than 3 billion people at risk from intensifying heatwaves, droughts, floods, wildfires, and other disasters (IPCC, 2023). The gravity of the situation is perhaps best captured by the IPCC's 2022 report on

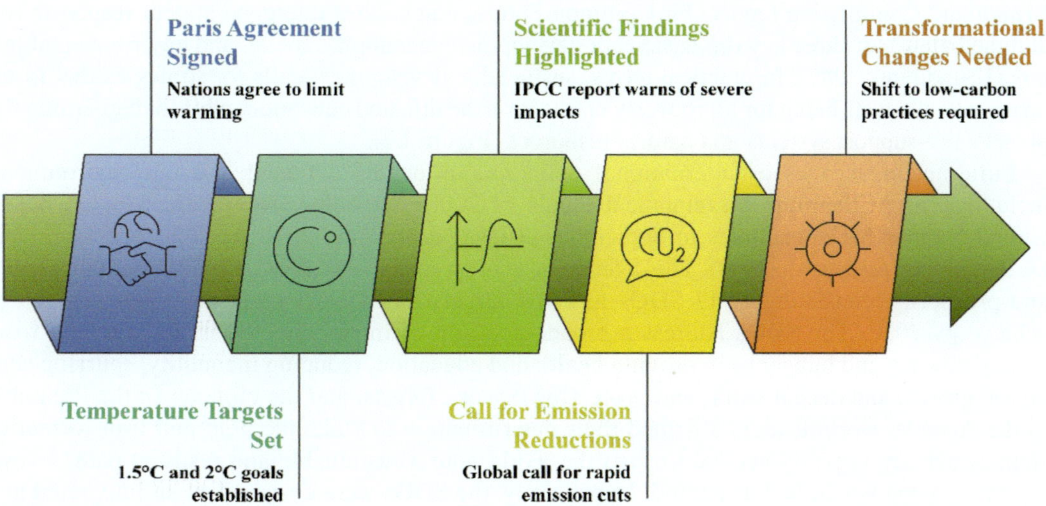

FIGURE 1.2 Post-Paris pathway from targets to transformation. Stylized policy–science timeline highlighting: the Paris Agreement commitment to limit warming (2015), establishment of 1.5°C/2°C temperature goals, subsequent IPCC findings on escalating risks, the global call for rapid emission reductions, and the imperative for transformational, low-carbon change.

Source: Author's illustration from literature sources discussed in this chapter.

impacts and adaptation, which states, "The scientific evidence is unequivocal: climate change is a threat to human well-being and the health of the planet. Any further delay in concerted global action will miss a brief and rapidly closing window to secure a liveable future."[6] In summary, the scientific and international policy consensus is that immediate and ambitious climate action is imperative to avert the worst outcomes and ensure a sustainable future.

Crucially, climate action is two-pronged and involves both mitigation and adaptation. Mitigation refers to efforts to reduce or prevent GHG emissions, for example, by transitioning to renewable energy sources (solar and wind), enhancing energy efficiency, protecting and expanding forests, which absorb CO_2, and developing low-carbon technologies and practices in sectors such as transportation, agriculture, and industry. Mitigation tackles the root cause of climate change by stabilizing atmospheric GHG concentrations. Adaptation, on the other hand, involves adjusting our societies and ecosystems to climate changes that are already happening or cannot be avoided. Adaptation measures include building resilient infrastructure, such as sea walls or climate-proof buildings, improving water management to deal with droughts or floods, developing climate-smart agriculture, such as drought-resistant crops, changing planting schedules, strengthening early warning systems for extreme weather, and ensuring public health systems can handle new disease patterns or heat-related illnesses. Both mitigation and adaptation are essential: mitigation can limit the magnitude of long-term climate change, whereas adaptation reduces vulnerability to changes that occur. The concept of "climate action" as embodied in SDG 13 and the UNFCCC encompasses both these aspects, reducing emissions on one side and enhancing resilience on the other, reflecting a comprehensive approach to addressing climate risks.

1.3 THE SDGS FRAMEWORK

To place climate action in context, it is important to understand the broader framework of sustainable development, as articulated by the United Nations. Sustainable development is a paradigm that seeks to harmonize economic growth, social inclusion, and environmental protection, ensuring that the needs of the present generation are met without compromising the ability of future generations to meet their own needs. This widely cited definition, originating from the 1987 Brundtland Commission report Our Common Future, emphasizes intergenerational responsibility and the balance of three key dimensions of development: economic, social, and environmental pillars (Brundtland, 1987). In practical terms, sustainable development calls for strategies that foster prosperity and well-being for all (poverty eradication, health, and education) while safeguarding the planet's life-support systems and natural resources (Figure 1.3).

Building on decades of international deliberations on development and the environment, including the Millennium Development Goals of 2000–2015, the SDGs were adopted by all United Nations Member States in September 2015 as part of the 2030 Agenda for Sustainable Development (Aziz et al., 2021). This agenda is a comprehensive plan of action for people, planet, and prosperity, consisting of 17 SDGs and 169 targets that UN member countries pledged to achieve by 2030. The SDGs address a broad spectrum of interconnected issues, ranging from ending poverty and hunger to improving health and education, reducing inequality, spurring economic growth and decent work, and protecting oceans, forests, and the climate. In the Preamble to the Agenda, world leaders affirmed their determination to "take the bold and transformative steps which are urgently needed to shift the world onto a sustainable and resilient path,"[7] vowing that no one would be left behind. Importantly, the SDGs were designed as an integrated and indivisible set; they are deliberately interlinked and meant to balance the three dimensions of sustainable development, including economic, social, and environmental, in an integrated way. This means that progress on one goal should not undermine another, and, in many cases, one goal's advancement can have co-benefits.

The SDGs have a global scope is another defining feature. Unlike some previous development frameworks, the SDGs are universal; they apply to all countries, rich and poor, recognizing that

FIGURE 1.3 The 17 Sustainable Development Goals (SDGs) of the UN's 2030 Agenda. SDG 13 (Climate Action) focuses specifically on combating climate change, but climate action also supports many other goals (poverty alleviation, food security, health), reflecting the integrated nature of sustainable development.

Source: https://sdgs.un.org/goals.

sustainable development is a shared challenge for all humanity. Each country is expected to work toward all 17 goals, taking into account its own context and priorities, and report on its progress. Goals are supported by a set of targets and indicators to guide the implementation and monitoring of the same. For example, SDG 1 (No Poverty) has targets such as eradicating extreme poverty (<$1.90 a day)[8] and implementing social protection systems, while SDG 7 (Affordable and Clean Energy) includes targets on ensuring universal access to modern energy and substantially increasing the share of renewables in the energy mix. Achieving the SDGs requires action across all levels of governance from local, national, regional, and global, as well as engagement from a wide range of stakeholders, including governments, international organizations, civil society, the private sector, academia, and communities. Agenda 2030 emphasizes a revitalized global partnership for sustainable development, acknowledging that mobilizing the means of implementation in finance, technology, capacity building, trade, and systemic reforms is critical (Yiu and Saner, 2017).

One of the key motivations behind the SDGs is to address the emerging global challenges that threaten development gains. In the negotiating outcome document, member states explicitly recognized issues such as climate change, environmental degradation, and inequality as fundamental challenges. Paragraph 14 of the Agenda's Declaration[9] makes a striking statement: "Climate change is one of the greatest challenges of our time and its adverse impacts undermine the ability of all countries to achieve sustainable development." It further notes that increases in global temperature, sea-level rise, and other climate impacts are seriously affecting vulnerable countries, including least developed countries and small island developing states, and that the very survival of some societies and the natural ecosystems on which they depend are at risk. This acknowledgement within the core development agenda marks a significant convergence of climate and development discourses. This highlights that combating climate change is not a standalone environmental goal but is integral to achieving all aspects of sustainable development. Many SDGs have environmental dimensions, such as Goal 6 on water, Goal 7 on energy, Goal 14 on oceans, and Goal 15 on terrestrial ecosystems, reflecting the understanding that a healthy planet underpins long-term development and human well-being.

1.4 CLIMATE ACTION AS A SDG 13

Climate Action is a dedicated goal in the 2030 Agenda that focuses on climate change challenges. The goal is phrased succinctly as: "Take urgent action to combat climate change and its impacts." By elevating climate action to one of the 17 global goals, the international community affirmed that responding to climate change is central to achieving sustainable development. However, it is worth noting that SDG 13 is somewhat unique among the goals. While other SDGs have specific quantitative targets, for instance, SDG 1 aims to eradicate extreme poverty, and SDG 2 aims to end hunger, the ultimate objectives of climate action, such as limiting global temperature rise, are addressed through the separate but complementary processes of the UNFCCC and the Paris Agreement. Thus, SDG 13 serves to bridge and integrate the climate agenda with the development agenda, ensuring that climate considerations are embedded across all sectors and levels of planning by 2030 (Babacan, 2025). SDG 13 comes with a set of targets and indicators that provide concrete focus areas for countries' efforts to combat climate change. These targets highlight both mitigation and adaptation, as well as means of implementation, aligning with the pillars of the UN climate regime.

1.4.1 Target 13.1

Strengthen resilience and adaptive capacity to climate-related hazards and natural disasters in all countries. This emphasizes disaster risk reduction and adaptation efforts, acknowledging that countries must prepare for and reduce their vulnerability to the impacts of climate change, such as extreme weather events. Progress on this target can be measured by indicators such as the number of deaths and economic losses from disasters and the number of countries with national disaster risk reduction strategies.

1.4.2 Target 13.2

Integrate climate change measures into national policies, strategies, and planning. This calls for mainstreaming climate action across governmental planning, essentially ensuring that climate mitigation and adaptation are not confined to environment ministries alone but are part of economic planning, development strategies, and urban planning. This reflects the need for holistic policy coherence so that development plans are climate-informed and do not inadvertently worsen emissions or vulnerabilities.

1.4.3 Target 13.3

Improve education, awareness, and human and institutional capacity for climate change mitigation, adaptation, impact reduction, and early warning systems. This target recognizes that knowledge and capacity are critical for effective climate action in the health sector. It covers climate change education at various levels, such as through public awareness campaigns, training for officials or communities, and the development of institutions that can manage climate risks, such as meteorological services for early warning.

1.4.3.1 Target 13.a

Implement the commitment by developed countries that are parties to the UNFCCC to mobilize jointly $100 billion annually by 2020 from all sources to address the needs of developing countries in the context of meaningful mitigation actions and transparency, and fully operationalize the Green

Climate Fund (GCF) (Datta and Jena, 2025). This target refers to the international climate finance pledge first made in 2009, and reiterated in the Paris Agreement, that wealthy countries would, by 2020, mobilize \$100 billion per year in climate finance to support developing countries' mitigation and adaptation efforts. It also mentions GCF, a key multilateral fund established under the UNFCCC to channel climate finance. The inclusion of this target in SDG 13 highlights the critical importance of financial support and equity in the climate-development nexus. Developing nations, which often contribute the least to emissions but are the most vulnerable, require funding and technology transfers to leapfrog to cleaner development paths and build their resilience.

1.4.3.2 Target 13.b

Promote mechanisms for raising capacity for effective climate change-related planning and management in least developed countries and small island developing states, including focusing on women, youth, and local and marginalized communities. This target is about ensuring that support in terms of financial, technical, and educational means reaches the most vulnerable countries and communities and that those groups, often on the frontline of climate impacts, such as women, youth, and Indigenous and local communities, are empowered to participate in climate action. This highlights the importance of inclusive approaches and equity within countries, not just between them.

Together, these targets provide a comprehensive view of what climate action entails under the SDG framework, from national policy integration and education to specific concerns regarding finance and vulnerable groups' capacity. They are intended to be pursued in conjunction with legally binding climate agreements under the UNFCCC. In fact, a footnote in the SDG 13 description explicitly acknowledges the UNFCCC as "the primary international, intergovernmental forum for negotiating the global response to climate change" (Lofts et al., 2017). This indicates that SDG 13 does not replace the Paris Agreement or other climate treaties but rather complements them by ensuring that climate change is addressed in the context of development planning by all development actors. Figure 1.4 summarizes the SDG 13 governance linkages, UNFCCC/Paris processes, targets 13.1/13.2/13.3/13.b, and means of implementation, highlighting mid-course gaps and DRR/SDG coherence.

Progress on SDG 13 is tracked using indicators such as the number of countries with NDCs, national adaptation plans, disaster risk reduction strategies aligned with the Sendai Framework, climate education incorporated in curricula, and the volume of climate finance provided. As of the midpoint toward 2030, the assessments have been sober, and the UN's annual SDG progress reports indicate that the world is not on track to meet SDG 13 by 2030; in fact, GHG emissions have continued to increase, and climate impacts are accelerating. For instance, global CO_2 emissions rebounded sharply after a brief pandemic-related dip in 2020, and the year 2021 saw a 6% rise in energy-related CO_2 emissions, reaching a new record (Aljohani et al., 2025). Based on current trends and pledges, global emissions in 2030 are projected to be significantly higher than the level required for a 1.5°C pathway. The \$100 billion climate finance goal, originally promised for 2020, has not yet been fully met. About \$83–89 billion per year was being mobilized by 2019–2021, leaving a shortfall and delays in scaling up support (Tyson and Weiss, 2025). Adaptation finance, in particular, remains far below estimated needs, and only a fraction of the required funding is flowing to help developing countries adapt to climate impacts (Meraj and Hashimoto, 2025). These gaps highlight that much stronger action and international solidarity are required to fulfill the ambitions of SDG 13. They also clarify that SDG 13 is tightly interwoven with the success of other goals; if climate action falters, many other SDGs will also be in jeopardy.

1.5 INTERLINKAGES BETWEEN CLIMATE ACTION AND OTHER SDGS

One of the fundamental features of the SDGs is their integrated nature, the recognition that progress in one area often depends on progress in others, and that development objectives can reinforce

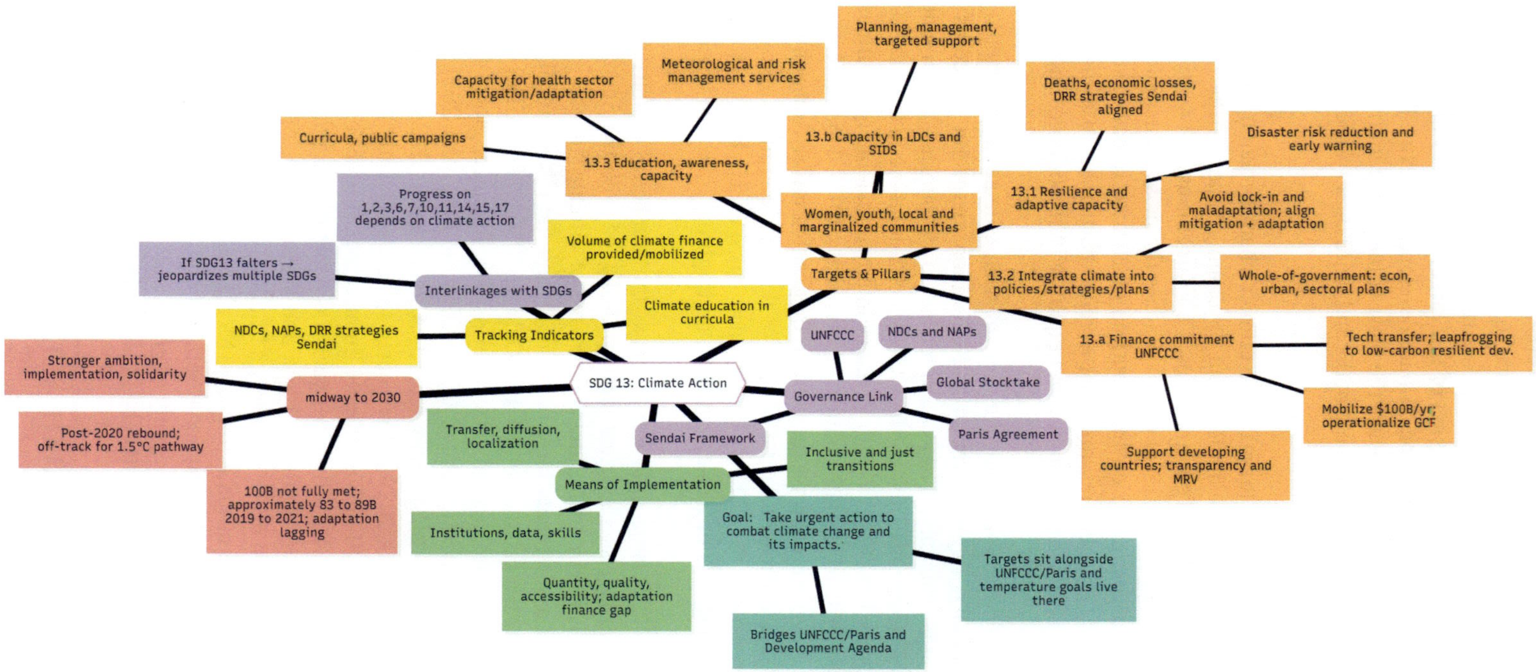

FIGURE 1.4 SDG 13 governance and implementation map. Concept map linking SDG 13: Climate Action to the UNFCCC/Paris architecture (NDCs, Global Stocktake), the targets/pillars (13.1 resilience, 13.2 policy integration, 13.3 education/awareness, 13.b capacity in LDCs/SIDS), and the means of implementation (finance, technology transfer, capacity, inclusive/just transitions). Tracking indicators, Sendai DRR coherence, and SDG interlinkages are shown alongside status notes at the 2030 midpoint (ambition/implementation and finance gaps; risks of lock-in and maladaptation).

Source: Author's synthesis, adapted from literature sources discussed in this chapter.

each other. Climate action (SDG 13) is a prime example of this interdependence. Achieving climate stability will greatly influence the world's ability to meet targets for poverty, hunger, health, water, and energy. Conversely, pursuing multiple SDGs can aid climate mitigation and adaptation. In this section, we explore how climate change and actions to address it intersect with other key SDGs, creating both synergies and trade-offs that policymakers and practitioners must navigate to achieve the SDGs. Figure 1.5 maps the SDG 13 hub of risks and levers across other goals, including cross-cutting synergies and trade-offs.

1.5.1 Climate and Poverty (SDG 1: No Poverty)

Climate change poses a serious threat to poverty eradication in developing countries. The poorest populations are often the most vulnerable to climate shocks; they tend to live in high-exposure areas, such as flood plains, drought-prone regions, and informal settlements, and have fewer resources to cope. Climate-related disasters, such as floods, storms, and droughts, can destroy homes, assets, and livelihoods, pushing people into extreme poverty or trapping them in a state of chronic poverty. The World Bank and other analyses have warned that without climate action, tens of millions of people could be pushed back into poverty by 2030 owing to climate impacts, such as reduced crop yields, higher food prices, lost income, and health costs. Moreover, slow-onset changes, such as sea-level rise or desertification, can erode economic opportunities over time, leading to displacement and loss of livelihoods for communities. On the positive side, climate action through investments in resilience, such as climate-smart agriculture, disaster-proof infrastructure, or social safety nets that help people recover from climate shocks, can directly support poverty reduction by protecting vulnerable communities. Mitigation actions, such as expanding clean energy access, for example in the form of solar panels in remote areas and decentralized renewables, can also provide affordable energy to poor households and spur local development, again contributing to poverty alleviation. In summary, ending poverty and tackling climate change are mutually reinforcing goals: development gains reduce vulnerability to climate shocks, and climate stabilization prevents future poverty caused by unrestrained warming.

1.5.2 Climate and Hunger/Food Security (SDG 2: Zero Hunger)

Agriculture and food systems are highly sensitive to climate variability. Changes in temperature and precipitation patterns, along with more frequent extreme events, directly affect crop productivity, livestock health, and fishery yields. Climate change is already linked to increased crop failures in some regions and the degradation of arable land through processes such as drought and salinization. This threatens to exacerbate hunger and malnutrition, especially in regions that are heavily reliant on rain-fed agriculture, such as sub-Saharan Africa and South Asia. For instance, more frequent droughts can devastate harvests, leading to food shortages and spikes in food prices that affect the poorest consumers the most. Similarly, warming and acidifying oceans disrupt fish stocks that coastal communities depend on for their protein intake. Addressing these challenges requires integrating climate adaptation into food security strategies, such as developing drought-resistant crop varieties, improving irrigation efficiency, diversifying crops, strengthening agricultural extension services to help farmers adjust planting times or farming techniques, and building food storage and distribution systems that can buffer against adverse seasons. Simultaneously, the agricultural sector is a significant source of GHG emissions, such as methane from rice paddies and ruminant livestock, nitrous oxide from fertilizers, and CO_2 from deforestation for farmland. Thus, mitigation measures such as sustainable land management, reducing food waste, and shifting toward climate-smart agriculture can contribute to climate goals while ensuring long-term food production. These are win–win solutions in which climate action bolsters food security. Conversely, failing to limit climate change could severely jeopardize the world's ability to eliminate hunger. As one study succinctly puts it, climate change "threatens SDG 2 by affecting crop growth, food security, and

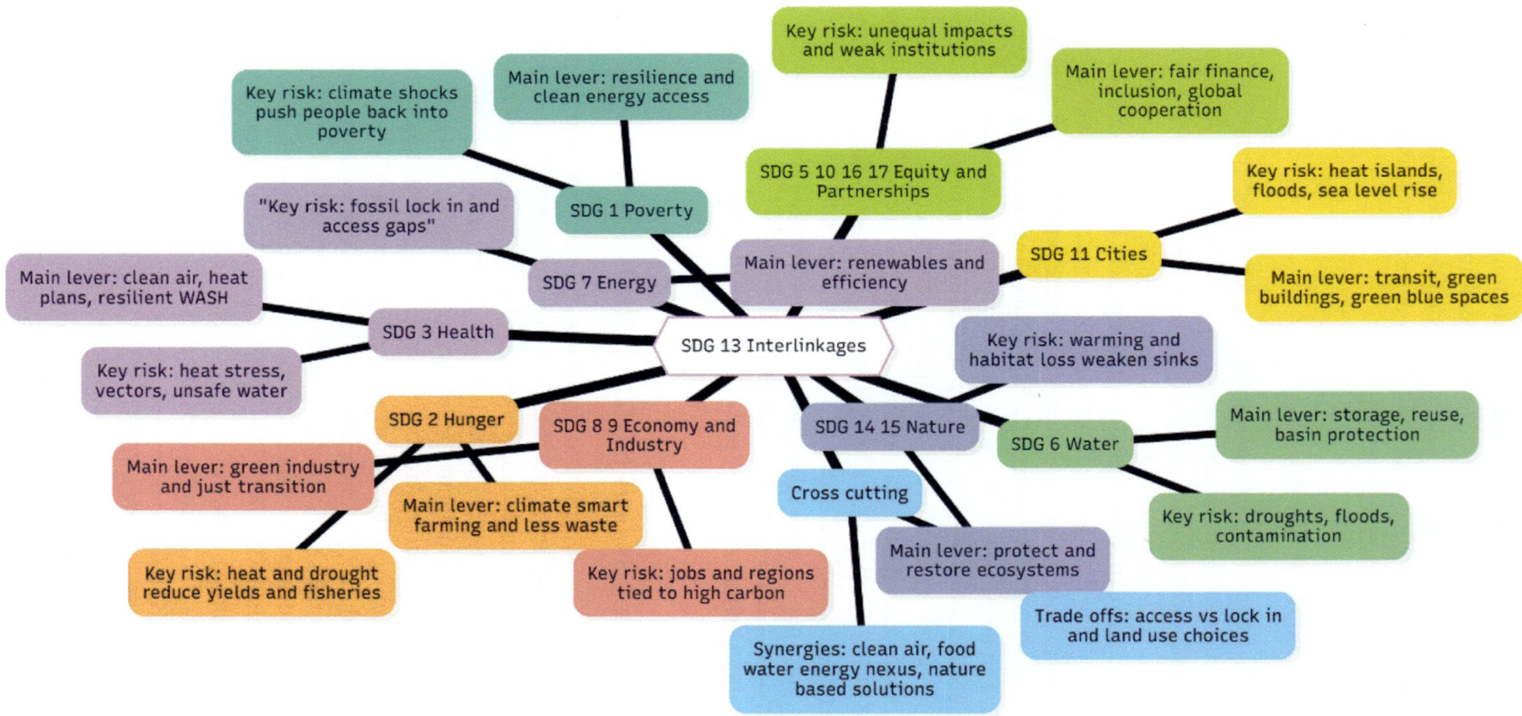

FIGURE 1.5 SDG 13 interlinkages, risks, and levers across the SDGs. Mind-map showing SDG 13 at the hub, with core risks (e.g., poverty from climate shocks, yield losses, heat/health burdens, droughts/floods, fossil lock-in, urban heat/flooding, ecosystem degradation, unequal impacts) and main levers (resilience and clean energy access, climate-smart farming and waste reduction, clean air and heat plans, storage/reuse/basin protection, renewables and efficiency, green industry and just transition, transit/green buildings/green-blue spaces, protect, and restore ecosystems). Cross-cutting nodes flag synergies (air-quality co-benefits, food-water-energy nexus, nature-based solutions) and trade-offs (access vs. lock-in; land-use choices).

Source: Author's synthesis, adapted from literature sources discussed in this chapter.

increasing hunger and malnutrition (Mugambiwa and Tirivangasi, 2017)," making it difficult to achieve zero hunger.

1.5.3 Climate and Health (SDG 3: Good Health and Well-Being)

The connections between climate and health are complex and interrelated. Climate change affects human health directly and indirectly. Directly, extreme heat can cause heatstroke and cardiovascular strain, particularly among older adults and outdoor workers. Higher temperatures also expand the range of vector-borne diseases, such as malaria, dengue, and Zika, as mosquitoes and other carriers thrive in warmer conditions and new regions. Changing rainfall patterns and flooding can contaminate water supplies, leading to diarrheal diseases. Extreme weather events cause injuries, death, and mental health impacts such as trauma from disasters. Indirectly, climate impacts on food production and nutrition can lead to health problems, such as malnutrition and stunting in children, and air pollution from fossil fuels, while technically not climate change itself, often shares sources with GHG emissions and contributes to respiratory and cardiovascular diseases. Therefore, climate action can yield major health co-benefits. For example, reducing air pollution by transitioning to clean energy and transport reduces carbon emissions and prevents millions of premature deaths due to cleaner air. The IPCC and health research have highlighted such co-benefits: climate mitigation aligned with the Paris Agreement could save many lives through improved air quality (Markandya et al., 2018). Adaptation measures, such as heat health warning systems in cities, improved vector control, and resilient water and sanitation infrastructure, directly protect health in a warming climate. Thus, SDG 13 (climate action) and SDG 3 (health) are closely linked, and ambitious climate policies can be justified not only on environmental grounds but also as public health interventions.

1.5.4 Climate and Clean Water (SDG 6: Clean Water and Sanitation)

Water is the primary medium through which climate change influences the Earth's ecosystems and human societies. Warmer temperatures alter the hydrologic cycle, leading to an increased occurrence of droughts and heavy precipitation. Water scarcity is projected to worsen in many areas owing to reduced rainfall and higher evaporation, thereby impacting drinking water supplies, agriculture, and hydropower generation. Simultaneously, more intense downpours and floods can damage water infrastructure and contaminate water sources. Ensuring the availability and sustainable management of water (SDG 6) in the face of these challenges requires climate-informed water resource management strategies. This includes investing in water storage and conservation techniques, improving wastewater treatment and reuse, especially during droughts, protecting watersheds and wetlands that regulate water flows, and extending safe water and sanitation services that are resilient to climate extremes, such as flood-proof sanitation. Climate action, in terms of reducing emissions, can limit long-term changes in precipitation patterns, potentially avoiding the worst-case scenarios of water stress that would occur with higher levels of warming. In turn, attaining SDG 6 helps adaptation: communities with reliable water and sanitation systems are better equipped to handle climate variability without experiencing waterborne diseases or conflicts over water access.

1.5.5 Climate and Energy (SDG 7: Affordable and Clean Energy)

Energy production and use are at the heart of the climate change problem, and roughly three-quarters of global GHG emissions come from the energy sector, including electricity generation, heating, transportation, and industry. Therefore, SDG 7, which seeks universal access to affordable, reliable, sustainable, and modern energy, significantly overlaps with the climate mitigation objectives. The transition to clean energy sources, such as solar, wind, hydro, geothermal, and possibly advanced nuclear energy, is one of the most critical strategies for reducing emissions. Expanding renewable energy and improving energy efficiency can dramatically reduce carbon output while supporting other developmental goals. For instance, providing solar power or off-grid renewable solutions to rural communities can improve their livelihoods and reduce the use of polluting fuels such as kerosene or diesel. Achieving SDG 7 in a climate-friendly way means focusing on sustainable energy systems, phasing out unabated coal-fired power, limiting investments in new fossil fuel infrastructure that would lock-in emissions, and leapfrogging to cleaner technologies. Fortunately, the costs of renewable energy have fallen rapidly over the past decade, making them increasingly competitive with those of fossil fuels. Many countries have already integrated significant shares of wind and solar energy into their respective grids.

Energy efficiency measures, such as in appliances, buildings, and industrial processes, represent another synergy; using less energy to achieve the same services saves money, reduces pollution, and lowers emissions. There can also be trade-offs to manage: ensuring energy access, especially for the poorest, sometimes leads governments to invest in the cheapest short-term options, which might be fossil-based fuels. However, with international support and falling clean tech costs, SDG 7 can be achieved through green pathways, aligning with climate goals. By doing so, progress on SDG 7 directly advances SDG 13, as decarbonizing energy is indispensable for stabilizing the climate.

1.5.6 Climate and Economic Growth/Industry (SDG 8 and 9)

Sustainable economic growth (SDG 8) and industrialization/innovation (SDG 9) must be reimagined in the context of a carbon-constrained world. Historically, industrial growth has been strongly coupled with increased emissions; industrialized nations have grown their economies using fossil fuels, and many developing countries are following a similar trajectory. However, there is a growing recognition that decoupling economic growth from emissions is possible and necessary for sustainable development. The IPCC has documented that some countries have managed to grow their GDP while stabilizing or even lowering their emissions in recent years. This is often due to a combination of factors, including shifting to service-based economies, improvements in energy efficiency, deployment of renewables, and market or policy measures that incentivize the use of lower-carbon technologies. Effective climate action encourages innovation in clean technologies and can drive new industries, such as renewable energy technology manufacturing, electric vehicles, energy-efficient appliances, carbon capture, and storage, which create jobs and economic opportunities, a concept often referred to as the "green economy." Thus, climate and SDG 8/9 can be synergistic: a transition to sustainable infrastructure and industries can boost economic development and employment; for example, jobs in renewable energy now number in the millions globally. The key concept here is the just transition, which ensures that workers and communities currently dependent on high-carbon industries, such as coal mining and oil extraction, are not left behind. Climate policies must incorporate measures for retraining, social protection, and regional development to support those affected, thereby aligning with SDG 8's emphasis on decent work and inclusive growth. Technological and industrial innovation (SDG 9) is critical for climate solutions, whether it involves advancements in battery storage, hydrogen fuel, climate-resilient construction, or smart grid systems. Investment in research and development and building robust, sustainable infrastructure, including transport, energy, and buildings, directly links climate goals to SDG 9.

1.5.7 Climate and Cities (SDG 11: Sustainable Cities and Communities)

Over half of the world's population lives in urban areas, a proportion expected to rise to approximately two-thirds by 2050.[10] Cities are often hotspots for emissions and vulnerability. Urban areas account for a significant share of energy consumption and carbon output through transport, buildings, and industry; however, they are also uniquely positioned to implement innovative climate solutions and improve their living standards. Many large cities are already experiencing climate impacts: coastal cities face flooding from sea-level rise and storm surges; many cities struggle with heatwaves exacerbated by the urban heat island effect; and others contend with water scarcity or extreme rainfall that overwhelms drainage systems. SDG 11's vision of inclusive, safe, resilient, and sustainable cities inherently involves climate action. Urban climate mitigation might include developing efficient public transport to cut vehicle emissions, adopting green building codes, and retrofitting for energy efficiency, expanding green spaces, which help sequester carbon and reduce heat, and moving toward zero-waste and circular economy models within cities. Urban adaptation involves improving disaster preparedness, such as early warning and evacuation plans, upgrading slum housing and infrastructure to withstand floods or heat, and investing in resilient infrastructure such as flood barriers and drought-resistant water supply. Some of the most ambitious climate actions are being led by city governments, showing that municipal policies on zoning, transport, waste, and energy can significantly contribute to achieving the national climate targets. In turn, effective climate action ensures that cities remain livable and thriving, supporting SDG 11 outcomes such as reduced disaster losses and improved urban air quality. This is another win–win area: a low-carbon, climate-resilient city is typically cleaner, healthier, and more efficient for its residents.

1.5.8 Climate and Oceans (SDG 14) and Ecosystems (SDG 15)

The environment-focused SDGs, Life Below Water and Life on Land, have direct climate linkages. Healthy oceans and terrestrial ecosystems are crucial for absorbing CO_2; for instance, forests act as carbon sinks, and oceans absorb a large proportion of anthropogenic CO_2 emissions. Climate change is already causing ocean warming, acidification, and deoxygenation, which puts marine biodiversity at risk and undermines fisheries (affecting SDG 14). Likewise, on land, changing climate patterns can shift habitats and contribute to biodiversity loss, and extreme events, such as wildfires, which are intensified by heat and drought, devastate forests and wildlife (affecting SDG 15). Conversely, efforts to conserve and sustainably manage ecosystems, such as protecting forests, restoring wetlands, and establishing marine protected areas, have climate mitigation benefits by enhancing carbon sequestration and building natural buffers against climate impacts; for example, mangroves protect the coasts from storm surges. There is an increasing emphasis on nature-based solutions for climate, utilizing ecosystems for mitigation and adaptation, which simultaneously advances SDGs 13, 14, and 15. For example, reforestation or stopping deforestation, a target under SDG 15, is one of the most cost-effective climate mitigation strategies, and it preserves biodiversity and water regulation services. Similarly, conserving coral reefs and mangroves, targets under SDG 14, can help shield communities from storms, provide fishery income, and store carbon. However, climate change can undermine these conservation efforts; if warming exceeds certain thresholds, such as 1.5–2°C, many ecosystems will face limits in their ability to adapt. For instance, coral reefs have bleak prospects beyond 1.5°C of warming.

Thus, achieving SDG 13 is essential to allow the possibility of achieving SDG 14 and 15, and vice versa, as protecting nature is part of climate action.

1.5.9 Equity, Justice, and Partnerships (SDGs 5, 10, 16, and 17)

Finally, climate action intersects with goals related to equity and global cooperation. Climate change has often been described as a "threat multiplier" that can exacerbate inequalities (SDG 5 on gender equality and SDG 10 on reduced inequalities) and even contribute to conflict or instability if not managed (SDG 16 for peaceful and just institutions). For example, women frequently face higher risks and burdens in climate-affected regions due to gender roles, such as responsibilities for water and food provision and systemic inequalities; empowering women and ensuring their participation in climate planning improves outcomes, for example, studies have shown that women's leadership in disaster risk management or resource management can strengthen community resilience (Alam and Rahman, 2018). Similarly, Indigenous peoples and local communities often possess traditional knowledge for managing ecosystems sustainably, and including them, and respecting their rights, is vital for successful adaptation and mitigation on their lands. Addressing climate change in a manner consistent with SDG 10 means recognizing and rectifying the disparities in who is most affected and who has the resources to respond, which leads to calls for climate justice, such as financial mechanisms for "loss and damage" to support those who suffer irreversible climate harm (Boyd et al., 2021). Internationally, climate action demands a strong global partnership (SDG 17) because no country can solve climate change alone. Sharing technology, expertise, and finance, as well as coordinating policies, is crucial. The Paris Agreement is a model of global partnership in climate action, and its success relies on trust and collaboration between developed and developing countries. Indeed, the IPCC highlights that international cooperation enhances efforts for ambitious climate mitigation in the context of sustainable development. The principle of "common but differentiated responsibilities and respective capabilities" under the UNFCCC (Brunnée and Streck, 2013) embodies the need for partnership tempered by equity: all countries are responsible for addressing climate change, but wealthier nations are expected to lead in emissions reduction and support others, given their greater historical emissions and financial and technological capacity. This principle aligns with the ethos of SDG 17 and the broader spirit of solidarity.

1.6 GLOBAL POLICY RESPONSES AND FRAMEWORKS FOR CLIMATE ACTION

The global recognition of the crisis has led to the development of a robust international governance framework to coordinate climate actions. This framework operates in parallel and in collaboration with the SDG. Here, we outline the key milestones and instruments of international climate policy and how they integrate with the development goals. Figure 1.6 synthesizes the climate governance architecture, UNFCCC foundations, Paris mechanisms, SDG/Sendai coherence, non-state and finance roles, and IPCC science, plus the emerging metrics agenda.

The foundation was laid by the creation of the UNFCCC at the 1992 Earth Summit in Rio de Janeiro.[11] The UNFCCC, which nearly all countries have joined, set out the objective of stabilizing GHG concentrations "at a level that would prevent dangerous anthropogenic interference with the climate system" and did so in a way that would allow ecosystems to adapt naturally, ensure food production is not threatened, and enable sustainable economic development. The Convention established principles such as common but differentiated responsibilities and required developed countries to take the lead in cutting emissions and providing finance and technology. Under the UNFCCC, parties meet annually at the Conference of the Parties (COPs) to negotiate and review progress.

In 1997, the Kyoto Protocol was adopted as the first legally binding climate treaty under the UNFCCC. It set quantified emission reduction targets for industrialized countries (Annex I parties) for the period 2008–2012, averaging about a 5% reduction below 1990 levels (Mor et al., 2024). While Kyoto was a landmark, as it introduced emissions trading and a clean development mechanism for offset projects in developing countries, its scope was limited; however, developing countries had no targets, and the US never ratified it. Emissions continued to rise globally, and it became clear that a more inclusive approach was needed to address the issue.

This led to negotiations that culminated in the 2015 Paris Agreement (COP21). The Paris Agreement is considered a transformative milestone in climate governance and a triumph of multilateral diplomacy in this regard. As noted earlier, its central aim is to limit global warming to well below 2°C and pursue efforts to achieve 1.5°C. Under Paris, every country, developed and developing alike, has to define and submit its own NDC, which is essentially a climate action plan with targets for reducing emissions, and often includes adaptation goals. The Paris Framework emphasizes a "bottom-up" approach where national pledges are brought to the table but with a mechanism to ratchet up ambition over time. Every five years, countries are expected to update their NDCs with more ambitious efforts than the last (the so-called ambition cycle). This is paired with a Global Stocktake (Northrop et al., 2018), occurring every five years starting in 2023, to assess collective progress toward the Paris goals and inform the next round of NDC improvements.

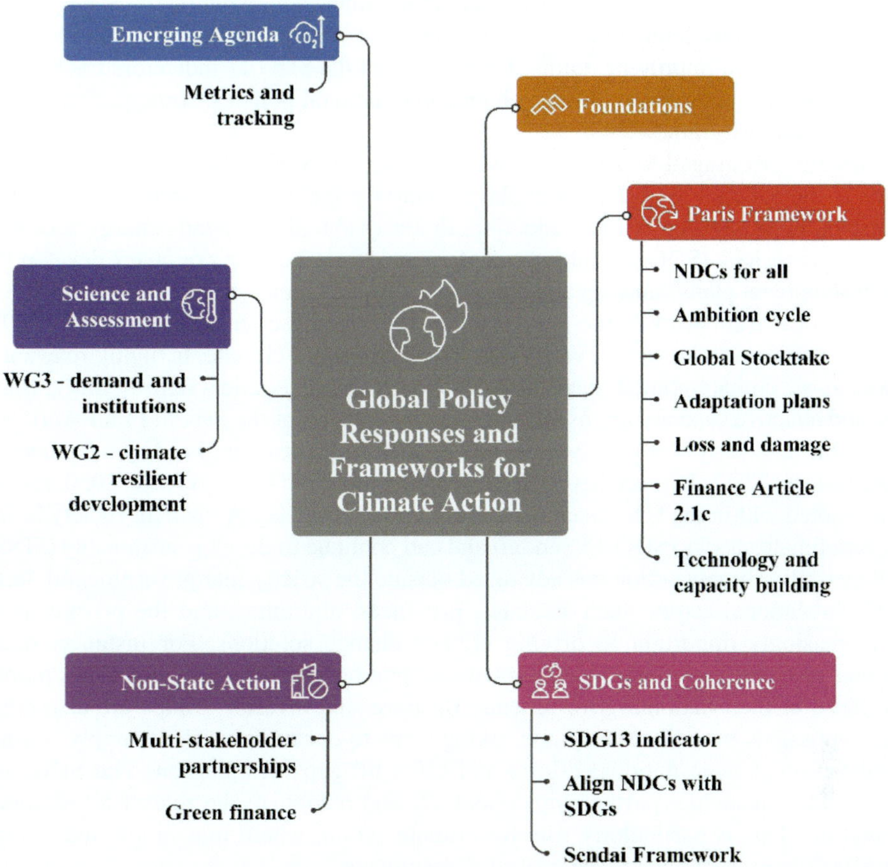

FIGURE 1.6 Global policy architecture for climate action. Central schematic linking the governance core to six pillars: Foundations (UNFCCC); Paris Framework (NDCs for all, ambition cycle, Global Stocktake, adaptation plans, loss and damage, finance under Article 2.1c, technology, and capacity building); SDGs and coherence (SDG 13 indicator, aligning NDCs with SDGs, Sendai Framework); Non-state action (multi-stakeholder partnerships, green finance); Science and assessment (IPCC WG3 on demand/institutions; WG2 on climate-resilient development); and an emerging agenda (metrics and tracking, e.g., adaptation goal).

Source: Author's synthesis, adapted from literature sources discussed in this chapter.

The Paris Agreement also contains important provisions for adaptation. Each country should plan and take measures to adapt to climate impacts and communicate their efforts through Adaptation Communications on finance, reiterating the $100 billion commitment and beyond, and encouraging financial flows consistent with low-emission, climate-resilient development (LECRD), technology transfer, and capacity building for developing countries.[12] Furthermore, it created a framework to discuss and address loss and damage, that is, the harms from climate impacts that countries cannot adapt to, such as extreme events or slow-onset changes destroying infrastructure or ecosystems, although this area is still evolving in terms of support mechanisms.

From a sustainable development perspective, the timing of the Paris Agreement was significant as it was adopted a few months after the SDGs in 2015. This synchronicity signaled a convergence: the world's climate change strategy and its development strategy were now being set together. Many have noted that the Paris Agreement and Agenda 2030 are complementary: Paris provides the temperature and emissions goals and a process for climate action, while the SDGs integrate that action into a holistic development agenda, ensuring that climate measures go hand-in-hand with eradicating poverty and improving equity. Indeed, one of the SDG 13 indicators tracks the number of countries that have "integrated climate change into national policies, strategies and planning," a bridge between the two frameworks.

Countries are encouraged to align their NDCs with their SDG strategies to operationalize this integration at the national level. For example, a country might design its renewable energy expansion, contributing to NDC emissions reduction, in a way that also expands energy access (SDG 7) and creates green jobs (SDG 8). International agencies are helping countries develop "climate-sensitive development plans" and tag and track how climate finance contributes to the SDGs.

Another critical framework is the Sendai Framework for Disaster Risk Reduction (2015–2030) (Pearson and Pelling, 2015), which, while not exclusively about climate, is highly relevant because many disasters are climate-related, such as storms, floods, droughts, and wildfires. SDG target 13.1 on resilience and adaptive capacity aligns with the implementation of the Sendai Framework. Reducing disaster risk through early warning systems and resilient infrastructure is a key adaptation strategy for climate change. The coherence between these agreements and Paris, Agenda 2030, and Sendai is actively promoted within the UN system to avoid siloed approaches. A resilient society is one that is prepared for climate shocks (SDG 13/Sendai) and can continue to develop sustainably (SDGs).

Simultaneously, climate action has advanced outside the strictly intergovernmental arena of the UNFCCC. Subnational actors, such as states, provinces, and cities, and the private sector, have become increasingly important in driving climate change solutions. For instance, many cities have set their own emissions targets and climate adaptation plans that are sometimes more aggressive than those of their national governments. Business and investor groups are also committing to net-zero emissions by mid-century and taking steps to decarbonize their supply chains. These developments complement national efforts and often fill gaps in ambition. The SDGs explicitly encourage multi-stakeholder partnerships (Goal 17) and recognize the role of all sectors in their implementation. This is particularly true for climate action, where innovation and scaling up of solutions often come from outside the central governments.

One notable policy trend is the growth of green finance and the incorporation of climate risk into financial decision-making. Global institutions, such as the World Bank and International Monetary Fund (IMF), have integrated climate considerations into their development finance portfolios to ensure that development projects are climate-resilient and contribute to mitigation (Lucy, 2025). Many countries and blocs, such as the EU, are establishing green taxonomies to define what qualifies as a sustainable investment and guide capital toward climate-positive projects. This responds toward Article 2.1c of the Paris Agreement, which calls for "making finance flows consistent with a pathway toward low greenhouse gas emissions and toward climate resilient development" (Pauw et al., 2021). It is increasingly evident that achieving climate and SDG goals will require trillions of dollars of investment in infrastructure, energy, and adaptation, much of which must come from private sources leveraged by public policy.

In terms of global stocktaking and accountability, the IPCC remains a cornerstone, providing scientific assessments that inform policies. The Sixth Assessment Reports (AR6, 2021–2022) were fed into the first Global Stocktake under Paris and highlighted the urgency and feasibility of achieving climate targets (IPCC, 2023). The IPCC has also deepened its analysis of mitigation pathways consistent with sustainable development, including issues of equity and of justice. For example, AR6 Working Group III devoted chapters to "Demand, Services and Social Aspects of Mitigation" and "Institutional and Economic Aspects," exploring how lifestyle changes, behavioral aspects, and just transitions could contribute to mitigation. The AR6 Working Group II introduced the concept of CRD, which is defined as a process that integrates mitigation and adaptation to advance sustainable development for all (Schipper et al., 2022). According to the IPCC, CRD is already challenging at current warming levels and will become progressively harder if warming exceeds 1.5°C or 2°C, highlighting that delaying action undermines our ability to achieve any development goals.[13] This concept essentially merges the climate and development agendas: it is development done in a way that is resilient to climate impacts and contributes to climate stability.

Finally, it is important to mention the Global Goal on Adaptation, a concept from the Paris Agreement that is still being developed. While mitigation has clear global goals, that is, temperature limits and net-zero by mid-century, adaptation is context-specific and more difficult to aggregate and compare. The international community is working on ways to track and enhance adaptation progress globally, which will feed into both SDG 13 and the Paris Agreement. Adaptation communications and metrics, such as the number of people protected by adaptation investments and the percentage of climate-proofed infrastructure, are being refined.

1.7 CHALLENGES AND OPPORTUNITIES IN ACHIEVING CLIMATE AND SDGS

Despite the clear frameworks and mutual benefits, aligning climate action with sustainable development faces several challenges worldwide. These hurdles are technical, financial, political, and social in nature. Simultaneously, there are significant opportunities to be seized, co-benefits, innovations, and synergies that can accelerate progress if properly leveraged.

1.7.1 Challenges

1.7.1.1 Insufficient Ambition and Implementation Gaps

As discussed, the current commitments by countries fall short of what is needed to meet the Paris climate goals. There is an ambition gap; many governments are cautious in setting targets owing to perceived economic costs or political constraints. Even where targets exist, there is often an implementation gap; policies on paper may not be fully implemented or enforced. For instance, some countries have renewable energy goals but still subsidize fossil fuels or have adaptation plans that are not funded by the government. Overcoming this requires raising awareness that the long-term costs of inaction, such as economic damages and disaster losses, far outweigh the upfront costs of action and the building of political coalitions that support climate-friendly policies.

1.7.1.2 Financial Constraints and Unequal Capacity

Transitioning to a low-carbon, resilient economy requires massive investments, estimated to run in the trillions of dollars globally for clean energy infrastructure, efficiency upgrades, public transportation, and adaptation measures such as flood defenses and resilient agriculture. Developing countries, in particular, face financial and capacity constraints. Many low-income countries have pressing development needs and limited fiscal space; asking them to invest heavily in climate mitigation or adaptation can be unrealistic without external support from high-income countries (HICs). The promised international climate finance, the $100 billion/year and more beyond 2025, is, therefore, critical, but as noted it has not fully materialized by the target date, and even $100 billion is considered insufficient relative to needs.[14] Adaptation finance is especially lagging, yet adaptation mostly concerns developing countries that have done the least to cause problems. This raises issues of equity and social justice. If climate finance remains under-delivered, it could breed mistrust and hinder future cooperation. Additionally, capacity issues mean that, even when funding is available, some countries struggle with project design, implementation, and absorption of funds, necessitating capacity-building efforts, as stated in SDG 13. b emphasizes.

1.7.1.3 Technology and Infrastructure Lock-In

A challenge for mitigation is the long lifespan of energy infrastructure. Power plants, vehicles, and buildings, once constructed, may lock in emissions for decades. Many developing countries are still building their infrastructure, which is an opportunity to build sustainably but also a risk if they adopt outdated, high-carbon models. There is a tension between the immediate needs, for example, providing electricity to those without it, and doing so in a clean way, if the latter has a higher upfront cost or requires new technologies. Transitioning away from existing fossil fuel-based systems, such as coal power or internal combustion vehicles, can face resistance from industries and workers in those sectors, as well as concerns about the reliability and cost of alternatives. Rapid technological advancements are needed in areas such as energy storage, grid management, and industrial processes, such as green steel production or cement, to achieve full decarbonization. Although technologies such as solar and wind are now mature and cost-competitive for electricity, other sectors, such as long-distance transport, heavy industry, and agriculture, still require innovation to achieve zero emissions. International collaboration on research and development and sharing of technology, including addressing intellectual property barriers, is therefore an important challenge that must be addressed.

1.7.1.4 Social and Political Resistance

Climate policies can be politically sensitive because they often require changes that affect industries, jobs, and consumers' daily lives. For example, raising carbon prices or removing fossil fuel subsidies to discourage emissions can lead to higher fuel or electricity costs in the short term, which may provoke public pushback or yellow vest-type protests (Patterson et al., 2025), as seen in some countries. Phasing out coal mining or fossil fuel extraction can threaten jobs and regional economies if done without proper planning and support. Hence, even when leaders acknowledge the need for climate action, they may face internal opposition or fear of economic disruption or both. Misinformation and politicization of climate science, although less widespread now than in previous decades, still persist and can delay consensus on action plans. Overcoming this requires proactive measures to ensure just transitions, for instance, using revenues from carbon pricing to compensate and retrain workers or subsidize energy bills for low-income households. It also requires strong communication and stakeholder engagement to build public support, emphasizing the health and economic co-benefits of climate action, such as clean air, modern jobs, and energy security. In many cases, climate action can align with national priorities, such as reducing air pollution or achieving energy independence, which can help depoliticize climate action.

1.7.1.5 Adaptation Limits and Residual Losses

Even with robust adaptation efforts, there are limits to how much we can adapt to certain impacts, especially at higher levels of warming. For example, beyond a certain point, low-lying islands cannot adapt to sea-level rise other than by relocation, coral reef ecosystems cannot survive beyond certain temperature or acidity thresholds, and outdoor work becomes dangerous under extreme heat and humidity. This implies that if mitigation fails and the world warms substantially, losses are inevitable. The possibility of loss and damage on a large scale, such as the permanent loss of territory, cultural heritage, and livelihoods, is a looming challenge that raises questions of compensation and support for affected populations. The 2022 COP27 made progress by establishing a loss and damage fund to assist vulnerable nations (Naylor and Ford, 2023), but operationalizing it and providing adequate resources will be a challenge going forward. This intersects with SDGs, such as poverty and inequality, because the poorest will bear disproportionate losses. Ethically, the global community faces the challenge of addressing these harms cooperatively, which tests the solidarity envisioned in the SDGs' principles.

1.7.1.6 Timeline and Short-Termism

Climate problems are characterized by misalignment of timescales. The benefits of strong climate action, avoiding catastrophe by 2050 or 2100, are long-term and global, whereas the costs or disruptions are immediate and local. Political and business cycles tend to prioritize near-term gains; it is challenging to mobilize action today for benefits that accrue mainly in the future. This is sometimes referred to as a discount rate or an incentive problem. Sustainable development frameworks attempt to counter this by emphasizing intergenerational justice and the well-being of future generations; however, in practice, leaders often grapple with more immediate pressures such as economic downturns and elections. Overcoming short-termism might require institutional innovations, such as mandating climate risk disclosure so businesses think long-term, setting up independent climate commissions that keep governments on track, or enshrining climate commitments into law to survive electoral changes. The SDGs' 2030 deadline has helped concentrate minds to some extent, but climate change will require an even longer view, net zero by 2050, and climate stabilization through 2100 and beyond. Figure 1.7 maps the major barriers to climate action and paired remedies, synthesizing insights from IPCC AR6 and the Paris Agreement architecture.

1.7.2 Opportunities

Despite these challenges, there are powerful reasons for optimism and many opportunities that can be leveraged, such as the following:

1.7.2.1 Economic Opportunities in Green Growth

Far from being merely a cost, the transition to a low-carbon economy is stimulating the development of new industries and markets. The renewable energy sector has become a major job creator. According to the International Renewable Energy Agency, the renewable energy sector has employed over 12 million people globally in recent years, and this number could multiply with further investments.[15] Energy efficiency retrofits, sustainable transport, such as electric vehicles and public transit systems, and nature restoration, such as tree planting and landscape rehabilitation, create employment opportunities. Many studies suggest that green investments, such as those in renewable energy, energy efficiency, and sustainable transport, often generate more jobs per dollar than traditional fossil fuel investments. For developing countries, pursuing a green growth pathway can attract international financing and technologies. There is an opportunity to leapfrog: rather than follow the pollute-then-cleanup model of industrialization, emerging economies can go straight to cleaner technologies. This can avoid the costs of retrofitting later and can produce cleaner air and water immediately, yielding health and productivity benefits for residents. For example, expanding solar mini-grids in Africa can provide electricity to underserved areas faster than waiting to extend the national grid, and it does so without emissions (Saleh, 2024). Similarly, sustainable public transportation can improve urban mobility and air quality. These opportunities can be framed as part of a new development paradigm that decouples emissions from economic growth and promotes sustainable development.

1.7.2.2 Technological Innovation and Falling Costs

Rapid technological advancements and deployment are key reasons for hope. Solar and wind power are among the cheapest sources of electricity in many countries. Battery costs for energy storage and electric vehicles have fallen dramatically, enabling a growing market for EVs and making it feasible to integrate a high share of renewable energy into power grids. Breakthroughs are emerging in areas such as green hydrogen, which can decarbonize industries such as steel and long-haul shipping, advanced nuclear designs, carbon capture and storage, and alternative proteins, to reduce emissions from livestock. Digital technologies, such as big data, AI, and satellite monitoring, are also being applied to improve climate modeling, precision agriculture, and efficient energy management. As these technologies mature, the costs of climate action decline and the feasibility increases. For adaptation, innovations include drought-tolerant crop varieties developed through biotechnology and new materials for flood defense. Supporting research and innovation is a clear opportunity area linked to SDG 9 (innovation and infrastructure). International initiatives, such as Mission Innovation,[16] a coalition to accelerate clean energy innovation, and various public–private partnerships, aim to push these frontiers. Countries that invest in R&D and cultivate innovative industries can gain competitive advantages, which is another incentive, the first-mover advantage in future industries.

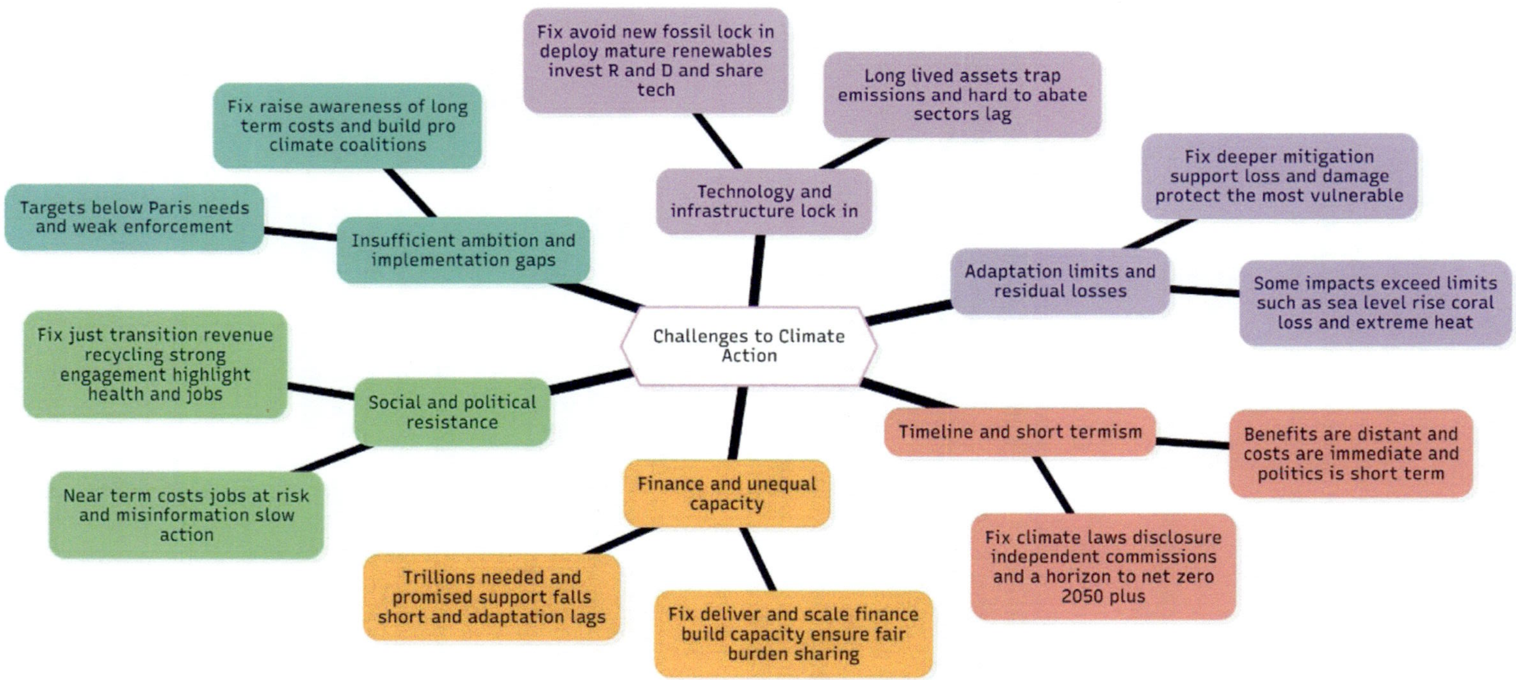

FIGURE 1.7 Challenges to climate action, risks, and fixes. Mind-map summarizing six recurring barriers, insufficient ambition/implementation, finance and unequal capacity, technology and infrastructure lock-in, social and political resistance, adaptation limits and residual losses, and timeline/short-termism, with corresponding remedy cues (e.g., stronger coalitions and awareness, scaled and fair finance, R&D and tech sharing, just-transition policies, support for loss and damage, and long-horizon governance).

Source: Author's synthesis, adapted from literature sources discussed in this chapter.

1.7.2.3 Co-Benefits and Public Health

Some of the strongest arguments for climate action are the immediate co-benefits, especially in terms of public health. For instance, air pollution from burning fossil fuels currently causes millions of premature deaths annually owing to respiratory and cardiovascular diseases. By aggressively reducing fossil fuel use to mitigate climate change, countries can save on healthcare costs and have healthier and more productive populations. The World Health Organization has highlighted how climate action aligns with health benefits, from promoting active transport (walking and cycling) that reduces emissions and improves fitness to shifting diets toward more plant-based options, reducing agricultural emissions, and improving health outcomes. These co-benefits can often justify climate policies on their own merits, even before climate benefits are considered. Recognizing and quantifying co-benefits helps build broader coalitions for action, including among ministries of health, finance, and planning, which might otherwise see climate change as outside their remit.

1.7.2.4 Youth Movements and Public Awareness

There has been a significant surge in public awareness and concern about climate change in recent years, largely driven by youth-led movements and civil society activism. This societal momentum is an opportunity because it creates political space for more ambitious actions. Many young people view climate action and sustainable development as defining issues for their future and are increasingly holding their leaders accountable. This can counteract some short-termism in politics. Additionally, as awareness grows, consumer and investor preferences are shifting, there is a rising demand for sustainable products and services, and investors are paying more attention to companies' environmental, social, and governance (ESG) performance. If this trend continues, markets will drive change, for instance, making it harder for carbon-intensive businesses to obtain loans or investments while rewarding green enterprises.

1.7.2.5 Integration and Holistic Planning

The integrated nature of the SDGs provides a framework for addressing multiple problems with a single policy solution. For example, well-designed urban development can simultaneously address housing (SDG 11), energy (SDG 7), and climate (SDG 13) by constructing energy-efficient and affordable housing near mass transit. Another example: restoring a mangrove forest can count toward climate mitigation (carbon sink), climate adaptation (storm surge protection), biodiversity (SDG 15), and livelihoods for local communities (fisheries improvement, SDG 1 and 14). Recognizing these overlaps means funding and efforts can achieve more results, essentially killing two (or more) birds with one stone. Governments and donors are increasingly interested in such integrated solutions because they maximize the impact of limited resources. It is an opportunity to promote cross-sectoral collaboration by breaking silos in government ministries or international agencies. The concept of nexus approaches, like water-energy-food nexus and policy coherence for sustainable development, is gaining traction, which is encouraging because climate action can benefit from being mainstreamed into all sectors (Hejnowicz et al., 2022).

1.7.2.6 Climate-Resilient Development Pathways

The IPCC's emphasis on climate-resilient development is itself an opportunity; it provides a conceptual roadmap that combines mitigation and adaptation with development, yielding better outcomes than pursuing them separately. For instance, sustainable urban planning that reduces emissions, through compact, transit-oriented cities, can also enhance resilience by avoiding building in flood plains, including green spaces to absorb heat and water, and simultaneously improving the quality of life. Such holistic approaches mean that countries do not have to choose between development and climate; they can design strategies that accomplish both goals. Pilot projects and case studies of CRD are emerging; for example, communities adopting agroforestry techniques that increase crop yields, diversify income, and sequester carbon, or small islands developing renewable energy microgrids that provide reliable power and replace expensive imported diesel while also being more robust against hurricanes. Scaling up these successes is a promising approach. Figure 1.8 summarizes near-term opportunity spaces for climate action, economic, technological, health, social, integration, and CRD pathways

1.8 TOWARD CLIMATE-RESILIENT SUSTAINABLE DEVELOPMENT

Given the interplay between climate and development, the concept of CRD has emerged as a unifying vision. CRD involves integrating climate change adaptation and mitigation measures with the pursuit of sustainable development to enhance well-being for all and reduce vulnerability to the impacts of climate change. It recognizes that efforts to alleviate poverty, improve health, and expand economic opportunities must now be undertaken in the context of a changing climate; conversely, climate strategies must contribute to, or at least not impede, broader development goals. Achieving CRD entails several strategic shifts. Figure 1.9 depicts the CRD pathway from drivers through shifts and enablers to practices and outcomes.

1.8.1 Long-Term Planning

Development planning must incorporate long-term climate risks. This means that, for example, when building infrastructure such as roads, bridges, or power plants, one must consider how future climate, higher temperatures, and more extreme rainfall may affect them. Similarly, urban expansion plans must consider climate projections to avoid putting more people at risk. Long-term planning also applies to the economic structure, diversifying economies that are heavily reliant on climate-sensitive sectors or fossil fuels to ensure their future resilience.

1.8.2 Inclusive and Participatory Governance

CRD emphasizes the inclusion of all stakeholders in planning and decision-making. As climate impacts and responses affect various groups differently, a broad range of voices, including marginalized communities, Indigenous peoples, women and youth, the private sector, and local governments should shape policies. This inclusive approach aligns with SDG 16 (inclusive societies) and helps ensure that the measures are culturally appropriate and locally accepted. It also helps identify innovative solutions that top-down approaches might overlook.

1.8.3 Adaptive Management

Given the uncertainties in how climate change will unfold locally, development processes must be flexible and adaptive. Policies might start with no-regret options, such as measures that make sense under a range of scenarios, such as improving water efficiency, and then adjust as new information comes in. Monitoring and learning mechanisms are crucial, basically applying the scientific method to policy: implement, monitor outcomes, learn, and revise. This is somewhat new to development practices, which have traditionally relied on static planning.

1.8.4 Linking Mitigation and Adaptation

Traditionally, climate mitigation (emissions reduction) and adaptation efforts have been pursued separately, sometimes by different agencies or funding streams. The CRD calls for linking them where possible. One example is nature-based solutions, such as restoring ecosystems like mangroves or forests, which can simultaneously sequester carbon (mitigation) and protect communities from storms or floods (adaptation) while also supporting livelihoods (development). Another example is

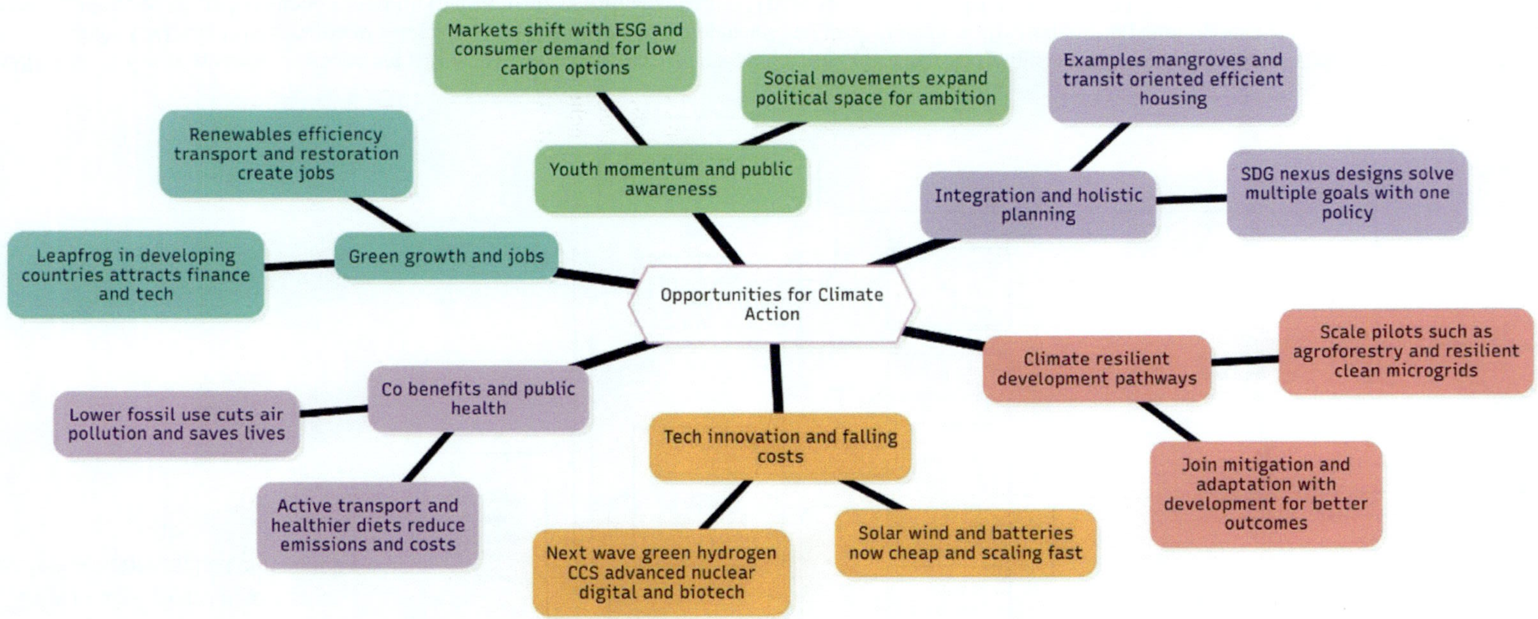

FIGURE 1.8 Opportunities for climate action, growth, technology, co-benefits, and governance. Mind-map highlighting six opportunity spaces: green growth and jobs (renewables, efficiency, transport, restoration, leapfrogging in developing countries), tech innovation and falling costs (solar/wind/batteries now cheap, next-wave options such as green hydrogen, CCS, advanced nuclear, digital/biotech), co-benefits and public health (clean air, active mobility, and healthier diets), youth momentum and public awareness (ESG/consumer shifts widening political space), integration and holistic planning (SDG-nexus designs, examples like mangroves and transit-oriented efficient housing), and climate-resilient development pathways (joining mitigation and adaptation with development, scaling pilots such as agroforestry and resilient clean microgrids).

Source: Author's synthesis, adapted from literature sources discussed in this chapter.

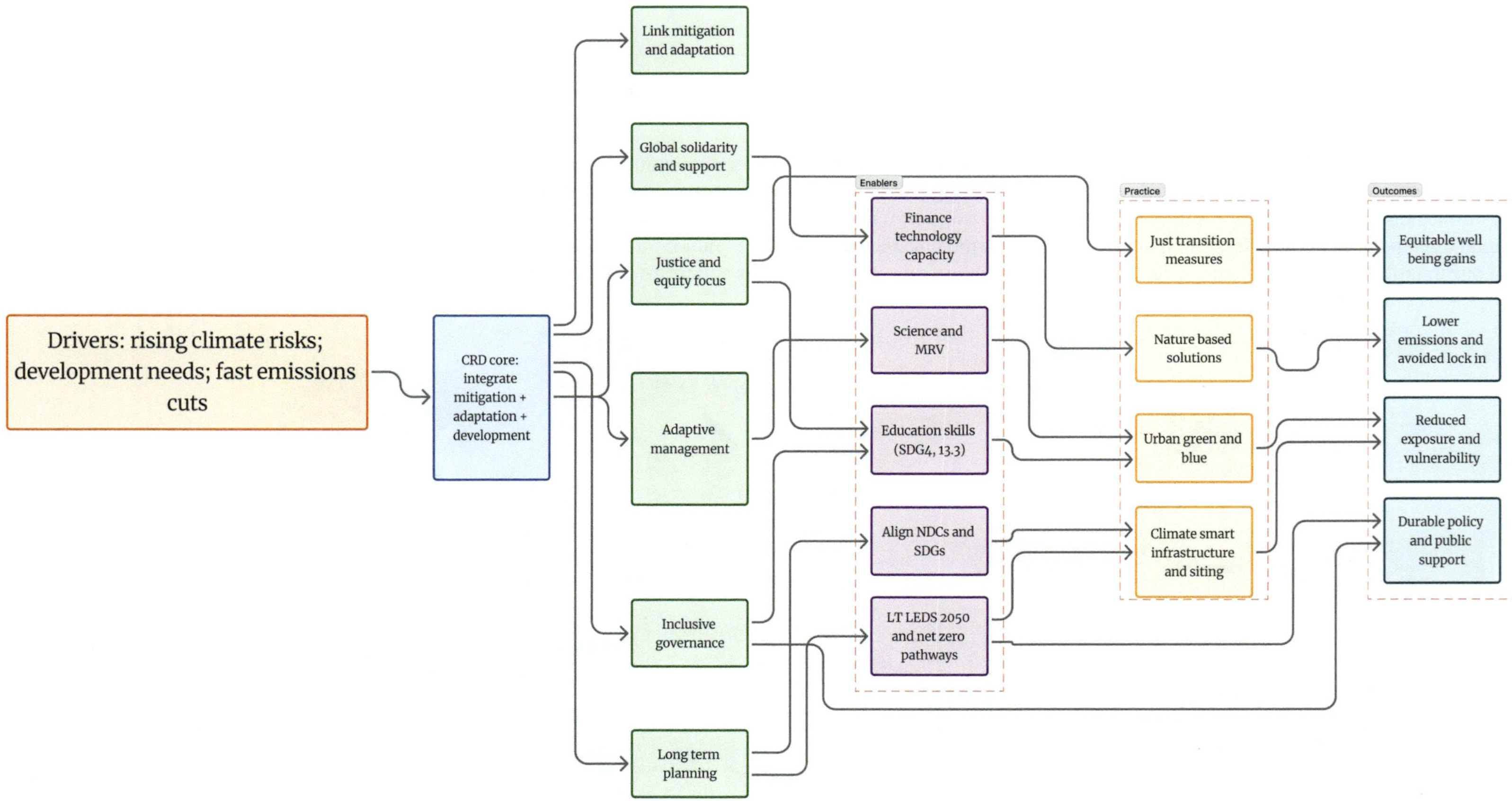

FIGURE 1.9 Climate-resilient development pathway (CRD). Schematic showing the flow from drivers (rising climate risks, development needs, need for rapid emissions cuts) to the CRD core (integration of mitigation, adaptation, and development) and then strategic shifts (long-term planning, inclusive governance, adaptive management, justice/equity, global support, and linking mitigation with adaptation). These feed enablers (LT-LEDS and NDC–SDG alignment; finance, technology, and capacity; education/skills; science and MRV) translate into practice (climate-smart infrastructure, urban green-blue, nature-based solutions, just-transition measures) and ultimately outcomes (lower emissions/avoided lock-in, reduced exposure and vulnerability, equitable well-being gains, durable policy, and public support).

Source: Author's synthesis, adapted from literature sources discussed in this chapter.

that urban green infrastructure (parks, green roofs) can reduce heat islands (adaptation) and absorb CO_2 (mitigation) while improving the urban living conditions. By finding such integrated solutions, resources can provide multiple benefits. However, there can also be tensions; for example, large-scale biofuel cultivation could mitigate emissions from fossil fuels, but if done poorly, it might compete with food production and thus harm development or adaptation. Therefore, careful assessment is required to avoid maladaptations or unintended harm.

1.8.5 Justice and Equity Focus

CRD places justice at its center. This translates to prioritizing the needs of the most vulnerable, socially and economically, in climate and development policies, for example, building sea walls to protect poor coastal neighborhoods as well as affluent city centers, or ensuring that low-income households can access clean energy, perhaps through subsidies or targeted programs. It also means intergenerational justice, not achieving present development at the cost of future generations' well-being. Equity considerations can increase public support for climate policies, thereby making them more robust. When people see policies as fair and beneficial to the many, not just elites, there is a stronger constituency for sustained actions. The IPCC notes that attention to questions of equity and justice in climate policy can "help ameliorate distributional impacts across income groups, countries and generations" and failing to do so can undermine social cohesion (Grubb et al., 2022).

1.8.6 Global Solidarity and Support

CRD at a global scale requires richer nations to support poorer ones through financial, technological, and capacity-building assistance. This is partly an ethical commitment and partly pragmatic: the SDGs and Paris Agreement both rely on mutual trust and cooperation, and global climate stability cannot be achieved if large parts of the world lack the means to act on it. Mechanisms such as the Green Climate Fund, technology-sharing platforms, and capacity-building networks are instrumental. The opportunity here is for renewed multilateralism, where common goals, such as the SDGs, drive collaboration rather than competition. As climate impacts worsen, it becomes increasingly evident that no country is immune and that collective action is in everyone's interest; for example, climate-driven migration or supply chain disruptions can affect global stability and the economy.

Although climate-resilient sustainable development is an aspirational concept, there are concrete signs of its implementation. Many countries are preparing long-term low emissions development strategies (LT-LEDS) looking out to 2050, as encouraged by the Paris Agreement, which often discusses how to reach net-zero emissions while also achieving development goals (Buckle et al., 2020). For instance, a country might plan how to restructure its energy, transport, and agriculture in a way that by 2050, it has near-zero emissions and improved quality-of-life metrics. When linked to shorter-term NDCs and SDG plans, these strategies can guide consistent action.

At the city level, "resilient city" strategies are being developed that integrate climate adaptation, such as flood control and heat action plans, with low-carbon infrastructure, such as public transit and renewable energy districts. Some cities, such as Copenhagen, have set targets to be carbon neutral by a certain date while also climate-proofing themselves; they aim to be net zero by 2025 and are investing in stormwater management due to heavy rain.[17] These examples provide learning opportunities for others.

Another emerging area is the emphasis on education and capacity building for CRD (linked to SDG 4 on education). Universities and training institutes are increasingly offering interdisciplinary programs on sustainability and climate, recognizing that tomorrow's leaders and professionals, including engineers, urban planners, and economists, need climate change competencies. Educating younger generations empowers them to adapt to their communities and innovate new solutions to problems. Target 13.3 of SDG 13 stresses the importance of improving education and awareness, which is vital for enabling all other actions.

It is also worth noting the ethical or philosophical shift underpinning CRD, which calls for a rethinking of our relationship with nature and recognition of limits. The old model of development was often about dominating nature for human progress, while the new model sees human progress as intertwined with respect for planetary boundaries.

Concepts such as the planetary health or one health approach unify human health and environmental health (Lerner and Berg, 2017), and doughnut economics frameworks suggest that development should occur within a safe and just space (Raworth, 2018), bounded by social foundations and ecological ceilings. These ideas are permeating policy discussions, indicating a cultural change that values sustainability inherently rather than as an add-on to the curriculum.

1.9 CONCLUSION

Climate action and sustainable development are the two pillars of the same edifice: a viable future for humanity on Earth. In this chapter, we examined how climate change, a global phenomenon predominantly caused by human activities, poses an existential threat to ecosystems and human societies, and why urgent action to mitigate and adapt to climate change is indispensable. We also explored the SDGs as the international community's blueprint for achieving social, economic, and environmental objectives in tandem, highlighting that SDG 13 (Climate Action) is both a goal in itself and a linchpin for the success of many other objectives. The interdependencies are clear: without stabilizing the climate, efforts to end poverty, ensure food security, improve health, and conserve nature will be severely undermined. Conversely, a development path that prioritizes equity, education, health, and environmental stewardship enhances our capacity to address climate change.

This chapter describes the major frameworks guiding the global response, notably the Paris Agreement under the UNFCCC for climate change and Agenda 2030 for sustainable development, and shows that they are mutually reinforcing. The Paris Agreement's temperature goals of limiting warming to well below 2°C, aiming for 1.5°C, set the benchmark for mitigation ambition, while its provisions for adaptation and finance stress the need for resilience and support from vulnerable nations. These have been integrated into SDG targets, such as SDG 13.1 on resilience and 13.a on finance, to ensure that climate action permeates development planning. Achieving the SDGs will require a sharp downward bend in the curve of global emissions within this decade and building adaptive capacity at every level of society. In essence, meeting the SDGs and the Paris goals is part of the same endeavor of sustainable development in the age of climate change.

We also discussed the challenges that remain, from inadequate current commitments and financial shortfalls to technology gaps and social resistance, which explain why, at the midpoint to 2030, the world is not yet on track to meet these lofty objectives. However, we also highlighted opportunities: the economic benefits of a green transition, the falling costs of renewable energy and other clean technologies, the public health gains from reducing pollution, and the rise of global awareness and youth advocacy that demands action. The required transformation is unparalleled in speed and scope, often described as akin to a new industrial revolution that must unfold within a few decades and reach all countries. This transformation must be managed justly, ensuring that it leaves no one behind, the sole motto of the SDGs, and, ideally, leaves everyone better off through sustainable growth and improved welfare.

A recurring theme is the importance of equity and inclusion in the workplace. Climate change has ethical dimensions; those least responsible are often the most affected. Thus, climate action must include mechanisms to support the vulnerable, whether farmers facing crop failures, coastal communities confronting sea-level rise, or workers transitioning from high-carbon industries to new livelihoods are concerned. International equity, between developed and developing countries, and intra-society equity, that is, protecting the poor and marginalized, are crucial for both moral and practical reasons: policies seen as fair tend to garner broader support, and global solidarity is necessary for global problems. The SDGs encapsulate this through their universal and inclusive approach.

We also observed that scientific research plays a critical role in guiding policy by informing us of the risks of various warming levels, the efficacy of solutions, and the urgency of timelines. Science tells us that the window for limiting warming to 1.5°C is rapidly closing and that action in the next few years will be pivotal. However, it also tells us that we have many tools and pathways available to avert the worst outcomes and create a sustainable society if we choose to employ them. The notion of CRD is effectively a call to action: an evidence-based appeal to integrate strategies, innovate, cooperate, and act decisively.

From an academic standpoint, we highlighted that addressing climate change is not an impediment to development but rather a precondition for sustainable development in the 21st century. Forward-thinking policies that reduce emissions and strengthen resilience can drive economic modernization, reduce inequality, and protect the natural systems that sustain life. There are already signs of momentum: renewable energy capacity worldwide has been growing exponentially, more countries are legislating net-zero emissions targets, and the concepts of a circular economy and regenerative development are gaining traction. The challenge is to accelerate and scale up these positive trends while phasing out the old, unsustainable practices.

In concluding, we can reflect on the words of the 2030 Agenda's preamble: "We are determined to protect the planet from degradation, including through sustainable consumption and production, sustainably managing its natural resources and taking urgent action on climate change, so that it can support the needs of the present and future generations"[18] This encapsulates the essence of climate action within sustainable development; it is about safeguarding the planetary home we all share through urgency and collaboration for the sake of both current inhabitants and future generations. The tasks ahead are daunting, but the stakes could not be higher: achieving the SDGs and climate goals is fundamentally about ensuring a livable, equitable, and prosperous world. With strong leadership, inclusive engagement, and adherence to scientific guidance, the global community can overcome this challenge. The years up to 2030 and beyond will test our collective resolve, but they also hold the promise of laying the foundation for a new era, one defined by harmony between humanity and the environment and by development that truly meets the needs of all within the limits of our precious planet.

The challenges ahead are considerable, but the knowledge and tools to address them are at our disposal, and the imperative to use them cannot be overstated. As we advance in this book and beyond, understanding these concepts equips us to contribute meaningfully to one of the most important undertakings of our time: securing a sustainable and climate-safe future for all.

DECLARATION STATEMENT

The author declares that this research work is original and has duly acknowledged all the sources of information used in the paper. There are no conflicts of interest.

ACKNOWLEDGMENTS

The author gratefully acknowledge the organisations that produced and maintain the data reports and supporting datasets used in this chapter. All such sources are cited in the chapter text and reference list. The author also acknowledge the American University of Sharjah for providing full institutional support that enabled the completion of this book contribution. Appreciation is extended to the Japan Society for the Promotion of Science (JSPS) for JSPS fellowship support and for providing a formative opportunity to broaden academic and international perspectives. The author further acknowledge the support and professional exposure provided through The University of Tokyo (UTokyo), UNU-IAS, Tokyo University of Agriculture and Technology (TUAT), Hokkaido University, the Department of Science and Technology (Government of India), and the World Bank, which collectively contributed to the author's academic development and professional trajectory. The views expressed in this chapter are those of the author and do not necessarily reflect the views of the above organisations.

ETHICAL APPROVAL

Not applicable.

CONSENT TO PARTICIPATE

Agree to participate.

CONSENT TO PUBLISH

Agree to publish.

AUTHOR'S CONTRIBUTIONS

Conceptualization, Methodology, Software, Formal analysis, Investigation, Resources, Writing—original draft preparation, Writing—review and editing: Gowhar Meraj.

FUNDING

This work received no external funding.

AVAILABILITY OF DATA AND MATERIALS

All the data is presented in the paper.

DECLARATION OF GENERATIVE AI AND AI-ASSISTED TECHNOLOGIES IN THE WRITING PROCESS

During the preparation of this work, the author used an AI service only (PaperPal/Grammarly) to correct grammar and expression errors. After using this service, the author reviewed and edited the content as needed and takes full responsibility for the content of the publication.

NOTES

1 https://sdgs.un.org/goals
2 https://libraryresources.unog.ch/climatechange/UN
3 https://unfccc.int/most-requested/key-aspects-of-the-paris-agreement
4 https://www.un.org/en/climatechange/science/climate-issues/degrees-matter
5 https://sdgs.un.org/goals/goal13
6 https://news.un.org/en/story/2022/02/1112852
7 https://sdgs.un.org/2030agenda
8 https://unitedarabemirates.un.org/en/sdgs/1
9 https://sustainabledevelopment.un.org/topics/climatechange
10 https://www.un.org/uk/desa/68-world-population-projected-live-urban-areas-2050-says-un
11 https://unfccc.int/process/the-convention/history-of-the-convention
12 https://www.orfonline.org/research/towards-a-low-carbon-and-climate-resilient-world-expectations-from-cop26
13 https://www.ipcc.ch/2023/03/20/press-release-ar6-synthesis-report
14 https://unctad.org/news/countries-agree-300-billion-2035-new-climate-finance-goal-what-next
15 https://www.irena.org/Digital-Report/World-Energy-Transitions-Outlook-2023
16 https://mission-innovation.net/
17 https://www.c40knowledgehub.org/s/article/Copenhagen-2025-Climate-Plan-Roadmap-2017-2020
18 https://sdgs.un.org/2030agenda

REFERENCES

Alam, K. and Rahman, M.H., 2018. Chapter 29: The role of women in disaster resilience. In C.N. Madu and C. Kuel (eds.) Handbook of disaster risk reduction & management (pp.697–719). Scientific Publishing, Singapore. https://doi.org/10.1142/9789813207950_0029.

Aljohani, T.M., Assolami, Y.O., Alrumayh, O., Mohamed, M.A. and Almutairi, A., 2025. Sustainable energy systems in a post-pandemic world: A taxonomy-based analysis of global energy-related markets responses and strategies following COVID-19. Sustainability, 17(5), p.2307.

Arora, N.K. and Mishra, I., 2019. United Nations Sustainable Development Goals 2030 and environmental sustainability: Race against time. Environmental Sustainability, 2(4), pp.339–342.

Aziz, F., Tahir, F. and Qureshi, N.A., 2021. Millennium development goals (MDGs-2000–2015) to sustainable development goals (SDGs-2030): A chronological landscape of public sector health care segment of Pakistan. Journal of the Pakistan Medical Association, 71(2), pp.596–601.

Babacan, H., 2025. From crisis to policy action: Tackling climate change through SDG 13. The International Journal of Community and Social Development, p.25166026251351370.

Boyd, E., Chaffin, B.C., Dorkenoo, K., Jackson, G., Harrington, L., N'guetta, A., Johansson, E.L., Nordlander, L., De Rosa, S.P., Raju, E. and Scown, M., 2021. Loss and damage from climate change: A new climate justice agenda. One Earth, 4(10), pp.1365–1370.

Brundtland, G.H., 1987. Our common future world commission on environment and development. Oxford University Press, Oxford, UK. http://ir.harambeeuniversity.edu.et/bitstream/handle/123456789/604/Our%20Common%20Future%20World%20Commission%20on%20Environment%20andDevelopement.pdf?sequence=1&isAllowed=y.

Brunnée, J. and Streck, C., 2013. The UNFCCC as a negotiation forum: Towards common but more differentiated responsibilities. Climate Policy, 13(5), pp.589–607.

Buckle, S., Ellis, J., Jaber, A.A., Rocha, M., Anderson, B. and Bjersér, P., 2020. Addressing the COVID-19 and climate crises: Potential economic recovery pathways and their implications for climate change mitigation, NDCs and broader socio-economic goals. OECD and IEA. https://www.oecd-ilibrary.org/environment/addressing-the-covid-19-and-climate-crises_50abd39c-en.

Datta, K. and Jena, P.R., 2025. Its implications on sustainable development goals for developing. In Proceedings of the 3rd international symposium on disaster resilience and sustainable development—Volume 2: Resilient systems for climate change adaptation (Vol. 624, p.225). Springer, Singapore.

Grubb, M., Okereke, C., Arima, J., Bosetti, V., Chen, Y., Edmonds, J., Gupta, S., Köberle, A., Kverndokk, S., Malik, A. and Sulistiawati, L., 2022. Introduction and framing. In IPCC, 2022: Climate change 2022: Mitigation of climate change. Contribution of Working Group III to the Sixth Assessment Report of the Intergovernmental Panel on Climate Change [P.R. Shukla, J. Skea, R. Slade, A. Al Khourdajie, R. van Diemen, D. McCollum, M. Pathak, S. Some, P. Vyas, R. Fradera, M. Belkacemi, A. Hasija, G. Lisboa, S. Luz and J. Malley (eds.)]. Cambridge University Press, Cambridge, UK and New York, NY, USA. https://doi.org/10.1017/9781009157926.003.

Hejnowicz, A.P., Thorn, J.P.R., Giraudo, M.E., Sallach, J.B., Hartley, S.E., Grugel, J., Pueppke, S.G. and Emberson, L., 2022. Appraising the water-energy-food nexus from a sustainable development perspective: A maturing paradigm? Earth's Future, 10(12), p.e2021EF002622.

IPCC, 2023. Summary for policymakers. In Climate change 2023: Synthesis report. Contribution of Working Groups I, II and III Tothe Sixth Assessment Report of the Intergovernmental Panel on Climate Change [Core Writing Team, H. Lee and J. Romero (eds.)]. IPCC, Geneva, Switzerland, pp.35–115. https://doi.org/10.59327/IPCC/AR6-9789291691647.001.

Lerner, H. and Berg, C., 2017. A comparison of three holistic approaches to health: One health, ecohealth, and planetary health. Frontiers in Veterinary Science, 4, p.163.

Lofts, K., Shamin, S., Zaman, S.T. and Kibugi, R., 2017. Brief on Sustainable Development Goal 13 on taking action on climate change and its impacts: Contributions of international law, policy and governance. McGill Journal of Sustainable Development Law, 13, p.183.

Lucy, 2025. Climate action and resilience for a sustainable future: The roles of the UN, WTO, and IMF. In R. Singh and D. Crowther (eds.) Climate neutrality and sustainable ecosystems: Practical approaches to addressing global warming (pp.177–198). Springer Nature, Singapore. https://doi.org/10.1007/978-981-96-7181-6_10.

Markandya, A., Sampedro, J., Smith, S.J., Van Dingenen, R., Pizarro-Irizar, C., Arto, I. and González-Eguino, M., 2018. Health co-benefits from air pollution and mitigation costs of the Paris Agreement: A modelling study. The Lancet Planetary Health, 2(3), pp.e126–e133.

Meraj, G. and Hashimoto, S., 2025. Bridging the adaptation finance gap: The role of nature-based solutions for climate resilience. Sustainability Science, 20, pp.1093–1107. https://doi.org/10.1007/s11625-025-01655-1.

Mor, S., Aneja, R., Madan, S. and Ghimire, M., 2024. Kyoto protocol and Paris agreement: Transition from bindings to pledges—a review. Millennial Asia, 15(4), pp.690–711.

Mugambiwa, S.S. and Tirivangasi, H.M., 2017. Climate change: A threat towards achieving 'Sustainable Development Goal number two' (end hunger, achieve food security and improved nutrition and promote sustainable agriculture) in South Africa. Jàmbá: Journal of Disaster Risk Studies, 9(1), pp.1–6.

Naylor, A.W. and Ford, J., 2023. Vulnerability and loss and damage following the COP27 of the UN Framework Convention on Climate Change. Regional Environmental Change, 23(1), p.38.

Northrop, E., Dagnet, Y., Höhne, N., Thwaites, J. and Mogelgaard, K., 2018. Achieving the ambition of Paris: Designing the global Stocktake. World Resources Institute (WRI), Washington, DC, USA.

Omali, T., 2022. Monitoring climate change using Satellite-observed earth's surface temperature: A review. International Journal of Scientific Research in Physics and Applied Sciences, 10(4), pp.15–22.

Patterson, J., Anisimova, K., Logg-Scarvell, J. and Kaiser, C., 2025. Reactions to policy action: Socio-political conditions of backlash to climate change policy. Policy Sciences, pp.1–34.

Pauw, W.P., König, M., Sadikhova, K. and Stutzmann, T., 2021. Financing low-carbon and climate resilient development: Do countries integrate article 2.1 (c) of the Paris agreement in their long-term strategies? Utrecht University Repository (Report). https://dspace.library.uu.nl/handle/1874/417005.

Pearson, L. and Pelling, M., 2015. The UN Sendai framework for disaster risk reduction 2015–2030: Negotiation process and prospects for science and practice. Journal of Extreme Events, 2(01), p.1571001.

Raworth, K., 2018. Doughnut economics: Seven ways to think like a 21st century economist. Chelsea Green Publishing. https://www.google.co.in/books/edition/Doughnut_Economics/.

Richardson, K., Steffen, W. and Liverman, D. eds., 2011. Climate change: Global risks, challenges and decisions. Cambridge University Press.

Saleh, M.T., 2024. Mini-grids: Empowering Africa's sustainable energy transition. Advances in Science and Technology, 142, pp.67–77.

Schipper, E.L.F., Revi, A., Preston, B.L., Carr, E.R., Eriksen, S.H., Fernandez-Carril, L.R., Glavovic, B.C., Hilmi, N.J.M., Ley, D., Mukerji, R., Muylaert de Araujo, M.S., Perez, R., Rose, S.K. and Singh, P.K., 2022. Climate resilient development pathways. In Climate change 2022: Impacts, adaptation and vulnerability. Contribution of Working Group II to the Sixth Assessment Report of the Intergovernmental Panel on Climate Change [H.-O. Pörtner, D.C. Roberts, M. Tignor, E.S. Poloczanska, K. Mintenbeck, A. Alegría, M. Craig, S. Langsdorf, S. Löschke, V. Möller, A. Okem and B. Rama (eds.)]. Cambridge University Press, Cambridge, UK and New York, NY, USA, pp.2655–2807. https://doi.org/10.1017/9781009325844.027.

Tyson, L.D. and Weiss, D., 2025. Climate finance: Progress, challenges, and opportunities on the path to a sustainable planet. California Management Review, 67(2), pp.5–39.

Yiu, L. and Saner, R., 2017. Business diplomacy in implementing the global 2030 development agenda: Core competencies needed at the corporate and managerial level. In International business diplomacy (Vol. 18, pp.33–58). Emerald Publishing Limited.

2 Environmental Leadership
Insights from India's Most Valuable Family Firms

Ahmad Mohd Khalid

2.1 INTRODUCTION

Family firms have a unique contribution to "business, society, and nature" (Ramírez-Pasillas et al. 2023). Family firms are one of the most common types of ownership prevailing globally (Aminadav and Papaioannou 2020), especially in developing countries (Lins 2003), and this also makes them a key contributor to climate change (Dyck et al. 2024). Research shows that the share of developing countries in global emissions is significant and rising (Copeland et al. 2021; Khalid and Rawat 2025).

Literature shows that firms globally have become more concerned about the environment and increasingly are embedding sustainability initiatives in their businesses because of rising climate risks, inequalities, environmental degradation and biodiversity loss (Rockström et al. 2009; Ripple et al. 2019). Family-owned firms have played a crucial role in promoting and shaping CSR activities at the community and national level in India (Panicker 2017). Several of these family firms are major players globally in industries ranging from manufacturing, automobile, mining and energy, oil and gas, IT and financial services, among others. India is among the top 5 global GHG emitters, and businesses, including family businesses, are key contributors and thus present an important case for exploring their environmental action and disclosures that contribute toward global environmental protection agendas such as the Sustainable Development Goals (SDGs) and the Paris Climate Agreement (Khalid and Rawat 2025). There are several SDGs that align with environmental issues and aspects, such as SDG 6—clean water and sanitation, SDG 7—affordable and clean energy, SDG 11—sustainable cities and communities, SDG 12—sustainable consumption and production, SDG 13—climate action, SDG 14—life below water and SDG 15—life on land. Firms, whether family or non-family, have to deal with these aspects in their day-to-day operation, planning and meeting stakeholder aspirations.

This study focuses on India's top 10 most valuable family businesses (see Table 2.1). Many of these family firms have a global presence and feature in the Forbes 500 list. These firms are also among the top 100 listed firms on the Bombay Stock Exchange (BSE) and the National Stock Exchange (NSE). These firms hold high market valuations, rank among the top employers, have significant ecological footprints and play a key role in both Indian and global supply chains.

The key objective of this study is to present a simple comparative analysis of the environmental performance leadership of the top family business firms in India to reflect their environmental commitment and long-term sustainability goals. In India, "small and medium enterprises" is a generic term used to describe small-scale industrial units and medium-scale industrial units. This study has not explored the factors that help us explain why and how the family business firms behave or disclose data with respect to environmental issues, goals and commitments.

2.2 BACKGROUND LITERATURE

Family ownership may influence environmental investments differently under existential environmental risks, as families aim to protect long-term private benefits and outperform other firms

DOI: 10.1201/9781003600152-2

TABLE 2.1
Top 10 Family Business Firms in India

Rank	Family	Flagship Company	Value (INR Crore)	Person Leading the Company	Industry	HQ City[a]
1	Ambani family	Reliance Industries	28,23,100	Mukesh Ambani	Energy	Mumbai
2	Bajaj family	Bajaj Financial Services	8,02,000	Sanjiv Bajaj	Financial Services	Pune
3	Kumar Mangalam Birla family	Ultratech Cement	6,47,700	Kumar Mangalam Birla	Cement & Cement Products	Mumbai
4	Jindal family	JSW Steel	5,70,900	Sajjan Jindal	Metals & Mining	Mumbai
5	Mahindra family	Mahindra & Mahindra	5,43,800	Anand Mahindra	Automobile & Auto Components	Mumbai
6	Nadar family	HCL Technologies	4,68,900	Roshni Nadar Malhotra	Software & Services	National Capital Region
7	Murugappa family	Cholamandalam Investment & Finance Company	2,92,400	Vellayan Subbiah	Financial Services	Chennai
8	Premji family	Wipro	2,78,600	Rishad Premji	Software & Services	Bengaluru
9	Anil Agarwal family	Hindustan Zinc	2,55,000	Anil Agarwal	Metals & Mining	Udaipur
10	Dani family, Choksi family and Vakil family	Asian Paints	2,20,900	R Seshasayee[b]	Chemicals & Petrochem	Mumbai

[a] HQ—Head Quarter City of the Flagship Company.
[b] Person heading is not a family member.
Source: 2025 Barclays Private Clients Hurun India Most Valuable Family Businesses List.

(Dyck et al. 2024). Based on the literature extracted by Ramírez-Pasillas et al. (2023), key aspects that affect a family's engagement with social or environmental agenda and sustainability practices within their family firm include family values, socioemotional wealth, the CEO's choices and family involvement, among others. Several factors may influence a firm's environmental disclosure, including its age, size, profitability, industry type, gender diversity, female leadership, regional context and regulatory norms (Driss et al. 2024; Agyemang et al. 2025; Khalid and Rawat 2025).

Disclosures are costly, and family members in leadership roles are already well aware of the firm's environmental decisions and plans (Dyck et al. 2024). Many firms, including family business firms, have established sustainability or environmental committees to improve the firm's environmental performance, focus on long-term strategies and thereby also address increasing stakeholder pressure (Khalid and Rawat 2025).

Typically, firms may be reluctant to disclose their environmental decisions to external stakeholders, such as investors. However, when facing environmental risks, they are obligated to provide at least the minimum required information and mitigation plans to regulators to prevent penalties or operational disruptions and are expected to follow disclosure norms similar in the industry (Dyck et al. 2024).

A family firm's business orientation is concerned about the community in which it operates, and its socioemotional values shape its plans and strategies (Baù et al. 2021). Family firms have huge potential to contribute toward the global agendas, such as the SDGs and the Paris Climate Agreement, because of their long-term commitment to sustainability and stakeholder concerns

(Myslivcova et al. 2023). This is further prioritized considering some of the corporate environmental scandals that shook the world, such as the Exxon Valdez, BP Oil Spill, Volkswagen's Emissionsgate Scandal has tightened vigil by government, investors and the concerned regulatory authorities (Khalid and Rawat 2025).

There is a lack of literature that reflects on the ecological or environmental sustainability, disclosures and best practices of family business (Ramírez-Pasillas et al. 2023). Differences have been noted in the case of social and environmental practice of family firms (Miroshnychenko and De Massis 2022) because of differences in the family's involvement in ownership and management of the firm (Berrone et al. 2010; Memili et al. 2018; Dyck et al. 2024). It is also not clear what drives family firms while implementing environmental concerns in their firm strategies (Myslivcova et al. 2023). Such evidence is also lacking in the case of developing countries, including India.

2.3 METHODOLOGY

The environmental leadership performance (ELP) of India's top 10 family businesses is assessed using content analysis. Content analysis has been widely used in environmental or sustainability assessment studies (Raimo et al. 2021; Maji and Kalita 2022; Mou and Ma 2023; Khalid and Rawat 2025).

BRSR reports of the top 10 family firms for the fiscal year 2024–25 have been referred to. Principle 6 under the BRSR focuses specifically on businesses' efforts to protect and restore the environment. It lists various indicators to assess firms' environmental performance. These indicators are classified into essential indicators and leadership indicators. There are 13 essential indicators and eight leadership indicators under Principle 6 of the BRSR. Disclosure of environmental information on essential indicators is mandatory, while leadership indicators are voluntary and encouraged (SEBI 2023). This study focuses on the eight leadership indicators to assess family businesses' environmental leadership and assess their performance with each other. These indicators also align well with the global targets under SDG 13 that focus on climate action (see Table 2.2).

In case of Bajaj Group, Aditya Birla Group and Murugappa Group, for ease of comparison only the top listed firm of the groups: Bajaj Finance Limited, Ultratech Cements and Cholamandalam Investment & Finance Co. Ltd (CIFCL) are covered based on family firm's official financial reports, and this captures nearly 48.9%, 44% and 38.1%, respectively, of the group's total market capitalization for the fiscal year 2024–25.

Table 2.3 shows the systematic approach used to assign scores based on different levels of environmental disclosure under the leadership indicators of the BRSR reports. In addition to BRSR reports, family firm's Integrated Annual Reports (IAR), Annual Reports, Sustainability Reports (SR) and the ESG Reports, based on availability, were referred to as filling data gaps or missing information. For example, in the case of two firms, a few leadership indicators were missing because of missing page numbers in BRSR reports but were present in the IAR. In a few cases, details on the indicators were present in their SR and ESG reports. It should be noted that data availability issues and data gaps are common in developing countries, including India (Khalid et al. 2020).

As shown in Table 2.3, the maximum score that can be achieved by any family business on the environmental leadership performance across eight indicators is 25. So, the higher the ELP score, the better the environmental leadership of the firm.

The firms are also categorized into three groups: top performers (ELP score > 0.75), medium performers (0.40–0.75) and low performers (<0.40). Family business companies and their ELP are also compared with respect to their capital valuation to reflect the seriousness and resource allocation of companies on environmental disclosure and leadership.

It should be noted that the environmental leadership scores based on the eight BRSR indicators may be termed as an "Indicative" or "firsthand" environmental leadership performance tool, as the actual environmental performance may require finer details on an enlarged set of indicators, ambitious resource allocation on climate activities and initiatives of the firm that are in tune with international standards and benchmarks. However, as environmental disclosure is evolving in India and

TABLE 2.2
Leadership Indicators and Their Alignment with SDG 13 Global Targets

Leadership Indicator BRSR	Alignment with SDG 13 Target and its Number
Indicator 1 Water withdrawal, consumption and discharge in areas of water stress	**Target 13.2** Integrate climate change measures into policies and planning **Target 13.3** Build knowledge and capacity to meet climate change
Indicator 2 Scope 3 emissions	**Target 13.2** Integrate climate change measures into policies and planning **Target 13.3** Build knowledge and capacity to meet climate change
Indicator 3 Impact on ecologically sensitive areas	**Target 13.2** Integrate climate change measures into policies and planning **Target 13.3** Build knowledge and capacity to meet climate change
Indicator 4 Initiatives for innovation, resource efficiency or impact reduction	**Target 13.2** Integrate climate change measures into policies and planning **Target 13.3** Build knowledge and capacity to meet climate change
Indicator 5 Business continuity and disaster management plan	**Target 13.1** Strengthen resilience and adaptive capacity to climate-related disasters **Target 13.3** Build knowledge and capacity to meet climate change
Indicator 6 Value chain environmental impact, and mitigation or adaptation measures taken	**Target 13.2** Integrate climate change measures into policies and planning **Target 13.5** Promote mechanisms for raising capacity for effective climate change-related planning and management
Indicator 7 Percentage of value chain partners that were assessed for environmental impacts	**Target 13.2** Integrate climate change measures into policies and planning **Target 13.5** Promote mechanisms for raising capacity for effective climate change-related planning and management
Indicator 8 Green credits generated or procured	**Target 13.3** Build knowledge and capacity to meet climate change

Source: SDG 13 Targets available at https://globalgoals.org/goals/13-climate-action/

the leadership indicators are still voluntary, the ELP based on the BRSR methodology may provide a simple and robust way to compare the ELP of firms in India. It may be noted here that indicators selected under the BRSR also align with existing reporting standards such as GRI, TCFD, SASB and global agendas such as the SDGs and Paris Climate Agreement (Khalid and Rawat 2025).

2.4 RESULTS AND DISCUSSION

Based on the indicator scoring criteria, Table 2.4 shows the environmental performance of India's top 10 business firms. It may be observed that high-performing firms (EPL score > 0.75) comprise the Major IT (Wipro, HCL) and energy-intensive firms like Hindustan Zinc and Ultratech Cements. It is established from literature that energy-intensive firms are very cautious of their image, and hence the impact they are putting on ecology is also serious in disclosing their environmental

TABLE 2.3
Scoring System for Leadership Indicators

Indicator	Scoring System for Disclosure	Maximum Score
Indicator 1 Water withdrawal, consumption and discharge in areas of water stress	**Data disclosed** (0–3 points) • Detailed quantitative data for withdrawal, consumption and discharge provided (3) • Some quantitative data provided or qualitative mention (1) • No data/not applicable (0) **Water intensity reported** (0/1 points) Yes (1), No (0) **Independent assurance** (0/1 points) Yes (1), No (0)	5 points
Indicator 2 Scope 3 emissions	**Scope 3 emissions and intensity** (0–3 points) • Quantified emissions and intensity provided (3) • Mentioned but not quantified/not currently tracking (1) • Not applicable/No data (0) **Independent assurance** (0/1 points) Yes (1), No (0)	4 points
Indicator 3 Impact on ecologically sensitive areas	**Discloser on impact, prevention or remediation** (0–3 points) • Detailed impacts, prevention and remediation activities reported (3) • Acknowledged but stated "not applicable" or minimal detail (1) • No mention or blank (0)	3 points
Indicator 4 Initiatives for innovation, resource efficiency or impact reduction	**Disclosure quality and quantity** (0–3 points) • Detailed and outcome-oriented initiatives (preferably quantified outcomes) (3) • Some initiatives with stated outcomes (2) • Few or vague initiatives (1) • No initiative reported (0)	3 points
Indicator 5 Business continuity and disaster management plan	**Plan detail and quality** (0–3 points) • Detailed plan with specific examples (e.g., ISO alignment, regular drills, climate risk integration, IT/DR specifics) (3) • Plan mentioned with some details but not comprehensive (2) • Plan mentioned, but vague or limited details (e.g., just "Yes") (1) • No plan details (0)	3 points
Indicator 6 Value chain environmental impact and mitigation or adaptation measures taken	**Details on impacts and responsive measures** (0–3 points) • Identified significant impacts and described robust mitigation or adaptation measures (3) • Stated "no adverse impact" but explained proactive measures to prevent such impacts (2) • Identified potential impacts but with vague mitigation or stated "no adverse impact" without explanation (1) • No information (0)	3 points
Indicator 7 Percentage of value chain partners that were assessed for environmental impacts	**Quantified assessment percentage** (0–3 points) • >50% of partners assessed by value of business, with quantitative data provided (3) • 26–50% of partners assessed by value of business, with quantitative data provided (2) • 1–25% of partners assessed value of business, with quantitative data provided (1) • No assessment or tracking or "Nil" (0)	3 points
Indicator 8 Green credits generated or procured	**Details** (0/1 points) • Quantified Green Credits generated/procured with details (1) • "Nil," "None," or no information provided (0)	1 point

TABLE 2.4
ELP Indicator Scores and Categories for Top 10 Family Business Firms in India

Company	I1	I2	I3	I4	I5	I6	I7	I8	ELP Score	ELP Score Ratio	Company Valuation Rank	ELP Category
Wipro	5	4	1	3	3	3	3	0	22	0.88	8	High
HCL Tech	5	4	1	3	3	2	3	0	21	0.84	6	High
Hindustan Zinc	5	4	1	3	3	2	3	0	21	0.84	9	High
Ultratech Cement	4	4	3	3	2	3	2	0	21	0.84	3	High
Mahindra & Mahindra (M&M)	5	4	1	2	3	2	2	0	19	0.76	5	High
Asian Paints	4	4	1	3	2	2	3	0	19	0.76	10	High
Reliance Industries Limited (RIL)	5	0	1	2	3	3	2	0	16	0.64	1	Medium
Bajaj Finance	0	4	1	3	3	2	0	0	13	0.52	2	Medium
CIFCL	0	3	1	1	2	1	0	0	8	0.32	7	Low
Jindal Steel	0	0	1	2	3	0	0	0	6	0.24	4	Low

initiatives and data metrics to win stakeholder trust, as stressed by Driss et al. (2024) and Maji and Kalita (2022). These firms demonstrate exemplary performance and overall leadership. These firms demonstrate a proactive and mature approach to environmental management, not just in their direct operations but also in understanding and addressing broader environmental impacts and risks. Their transparency and the tangible outcomes of their initiatives set them apart.

It is also worth noting that Reliance Industries, which is India's top firm, an energy-intensive firm in the oil and gas sector, is not placed among the leading performers and has a medium ELP score of 0.64. One of the major factors that impacts its performance is no disclosure on Scope 3 emissions (Indicator 2). It also reflects a very casual and non-serious attitude of the top management at India's top firm on valuation, which is supposed to inspire other industry peers. Similarly, a few other leading firms, such as Bajaj Finance and Jindal Steel, placed at second- and fourth-largest valuation among the top 10 firms, are placed in medium and low categories in environmental performance. This also provides a clear message that firms which are big, make more profit and are ranked high on valuation may not necessarily also be the firms which are most sustainable or sensitive toward environmental leadership.

The medium-performing firms with ELP scores between 0.4 and 0.75, such as Reliance Industries and Bajaj Finance, show varying levels of engagement and transparency. While they may excel in specific indicators, they often have significant gaps in others, such as missing quantitative data, undeveloped reporting processes for certain metrics (e.g., Scope 3 emissions) or less mature value chain assessments. They are actively engaged in some aspects of environmental leadership but have considerable room for improvement in holistic reporting and performance across all indicators.

The firms rated low on ELP leadership, such as CIFCL and Jindal Steel, either provide very little to no information across the leadership indicators or explicitly state that certain critical environmental metrics are not applicable or not currently being tracked. This indicates a significant lack of transparency or a very early stage or absence of structured reporting on these key environmental leadership aspects.

Another critical aspect is with respect to Indicator 8 on Green Credits, where none of the top 10 firms have any initiative or clear details.

2.4.1 Beyond Compliance: Innovation and Key Strategy

Beyond their environmental performance, the "leadership indicators" under Principle 6 of BRSR offer a fascinating glimpse into how companies are pushing the boundaries of environmental responsibility, moving beyond compliance to embrace innovation and strategic foresight. These indicators are a testament to their ambition to contribute meaningfully to a sustainable future, addressing challenges from climate change to resource scarcity. A few of the key innovations and strategies being undertaken by the leading business firms to showcase their environmental prowess are as under:

2.4.2 Moving Beyond Scope 1 and 2 Emissions: Quest for Decarbonization and Net Zero

A standout theme among the leaders is the aggressive pursuit of decarbonization, extending beyond direct operational emissions (Scope 1 and 2) to tackle the more complex Scope 3 emissions. Companies like Wipro, HCLTech, Hindustan Zinc and Ultratech Cements are actively measuring and reporting their Scope 3 emissions, acknowledging their significant environmental footprint embedded within their value chains. Ultratech Cement reports substantial Scope 3 emissions (1,27,11,286 tCO2e) for FY24–25, compared to HCLTech and Wipro, showcasing their energy-intensive nature compared to IT firms. Wipro and HCL Tech highlight that the majority (>80%) of their environmental impact from GHG emissions originates from their extended value chains, including purchased goods and services, business travel and employee commute. This transparency is crucial for a holistic approach to climate action.

Beyond reporting, companies are setting ambitious targets and implementing innovative solutions. RIL is a frontrunner with its net carbon zero target by 2035, a bold commitment supported by a multi-pronged strategy encompassing energy conservation, renewable energy transition, carbon capture and utilization and industrial decarbonization. Their initiatives, like the development of a heterogeneous catalyst for CO_2 utilization to produce sustainable chemicals and fuels, demonstrate a commitment to deep technological innovation in carbon management.

With regard to energy transition initiatives, Hindustan Zinc aims to add 500 GW of non-fossil fuel-based energy by 2030, reduce its Scope 3 emissions by 25% by 2030 and become net zero by 2070. The company has also launched "EcoZen" Asia's first low-carbon zinc with 75% lower carbon footprint. HCLTech aims to source 80% of its electricity from renewable sources by 2030, having already reached 34.38% in FY24–25. Mahindra & Mahindra (M&M) are investing in captive solar and windmill projects across its manufacturing sites to reduce GHG emissions and energy costs. Similarly, at Asian Paints, renewable electricity accounted for 57.6% of their total electricity consumption in the FY 24–25. These investments signal a fundamental reorientation of energy portfolios toward cleaner sources.

2.4.3 Advanced Water Management

Some of the manufacturing or energy and oil firms demonstrate proactive water management. Asian Paints exemplifies advanced water stewardship through the implementation of zero liquid discharge (ZLD) systems at all its manufacturing facilities. Similarly, Wipro has ZLD across most locations. Firms like RIL, Ultratech and Hindustan Zinc have implemented ZLD at key properties and plants, respectively. This signifies a commitment to minimizing freshwater withdrawal and treating wastewater for reuse within their operations.

Companies like HCLTech and M&M, despite not explicitly stating ZLD, *do* identify specific locations in water stress areas (e.g., Bengaluru, Chennai for HCLTech, Jaipur, Mohali, Pithampur, Chennai for M&M) and report their water withdrawal, consumption and discharge data for these regions. This granular reporting is a crucial leadership indicator, showing a detailed understanding of local environmental contexts and potential vulnerabilities.

2.4.4 Innovation for Circularity

Beyond basic waste management, several companies are embracing the principles of a circular economy. CIFCL highlights efforts in sludge drying, waste-to-energy initiatives such as converting bagasse to steam and ensuring waste segregation for compliant disposal. Ultratech Cement integrates circularity into its production practices by utilizing 44.15 million tons of recycled and alternative raw materials in cement production in 2024–25 and 100% fly ash utilization from their power plants for cement manufacturing to reduce embodied carbon and promote the circular use of industrial waste. This demonstrates a foresight in design and material selection to reduce waste generation.

Hindustan Zinc has achieved 100% fly ash usage and tripled the usage of smelting process waste (e.g., jarosite and jarofix) in the cement industry, road and railways. Similarly, the company has an innovative circular process to recover iron from jarosite.

M&M offers a unique example of resource circularity through initiatives like CERO Recycling for vehicle takebacks, onboarding of lithium-ion battery recyclers, authorized plastic recyclers and used oil collection. HCL Tech proudly reports that all its owned campuses in India have received TRUE Zero waste to landfill Platinum certification, setting a high benchmark for operational waste management and demonstrating exceptional performance in waste conversion and recycling.

2.4.5 Supply Chain Sustainability

A critical aspect of leadership in environmental responsibility is extending sustainable practices to the value chain. Both HCL Tech and Wipro acknowledge that a significant portion of their environmental impact (>80% of GHG emissions) originates from their extended value chains, including purchased goods and services, business travel and employee commute. This recognition is a crucial first step. Their mitigation strategies involve engagement and disclosures with suppliers, efforts to reduce business travel, promotion of electric vehicles and public transport for employee commute and renewable energy procurement for upstream energy emissions. This reflects a strategic understanding that true sustainability requires influencing partners beyond direct control.

RIL also addresses value chain risks, particularly concerning health, safety and environmental incidents and oil spills from crude or product transport. Their mitigation measures include strong vessel vetting, incident monitoring, emergency response systems, vendor management audits and comprehensive training programs for drivers. CIFCL assesses 73% of its upstream and downstream partners for social and environmental practices, indicating a proactive approach to understanding and managing its supply chain's sustainability performance. M&M, while encouraging sustainable practices through its Sustainable Supplier Impact Programme, has covered ~33% of its suppliers by value in FY2024–25 as part of its ESG onsite audit, with none of them having any adverse environmental impact. Three of the business firms, namely Bajaj Finance, CIFCL and Jindal Steel, mention that formal environmental impact assessment in their value chains is yet to be undertaken.

2.4.6 Building Resilience for Environmental Shocks

Having a robust business continuity and disaster management plan is a sign of responsible business conduct, especially today when the world is witnessing increasing environmental risks. While not all companies explicitly detail their plans with respect to this under leadership Indicator 7, a few like Wipro, HCL Tech, Hindustan Zinc and M&M provide comprehensive insights. All three Wipro, HCL Tech, Hindustan Zinc boast a robust business continuity plan aligned with ISO 22301 standards, incorporating a "Resilience-by-Design" philosophy that integrates technological, geopolitical and environmental risks, including climate change, into its planning across various dimensions like workplace, people management, supply chain and leadership. In addition, Hindustan Zinc keeps track of its sites with regular Business Continuity Testing & Disaster Recovery Drills. Similarly, Jindal Steel also emphasizes its adherence to the Information Technology General Controls framework, with a Disaster Recovery (DR) system and policies for data backup, recovery and retention, underpinned by routine assessments and risk management governance.

M&M, on the other hand, has an Onsite Emergency Plan and Disaster Control Plan Unit, which not only conducts mock drills from time to time but also assigns a high-risk priority number to increase effectiveness and reduce reaction time in case of an environmental or business emergency. These initiatives and plans clearly demonstrate that business firms are taking unique and sophisticated approaches to anticipate and mitigate environmental disruptions.

2.4.7 Unexplored Green Credits and Holistic Reporting

The concept of "Green Credits" is a newer leadership indicator, and most firms in the sample do not report any generated or procured Green Credits. HCL Tech explicitly states that while they engage in extensive tree plantation drives, these activities are not registered under any formal Green Credit Program. Wipro further clarifies that its focus remains on decarbonization efforts aligned with the Science Based Targets Initiative (SBTi), rather than on generating Green Credits. None of the firms formally reports Green Credits generated or procured, and no assessment of value chain partners has been undertaken in this regard. This suggests that while environmental initiatives are widespread, formal participation in emerging green credit mechanisms is still at a nascent stage in India. Also, it may be difficult to conclude that it is not a priority for business; rather, they are building capability and understanding the process.

2.5 CONCLUSION

This study assessed a qualitative overview of the environmental performance leadership of the top 10 family firms in India based on indicators prescribed under the Leadership Indicators of Principle 6 of the BRSR Report. It was found that firms are disclosing data and details on the leadership indicators, but the level of detail and transparency varies. The primary reason for this is that disclosure on the leadership indicator is currently not mandatory in India. The most environmentally proactive family business firms are IT firms and those in the energy-intensive sector. While analyzing the family firms, it was also evident that the highest valuation firms may not necessarily be among the top environmental performers. Some of the key initiatives, innovations and plans that have supported family firms in their environmental stewardship, pertain to their decarbonization initiatives and net-zero targets, advanced water management, innovation for circularity, building resilience and state-of-the-art disaster management plans and the firms try to align with international best practices and benchmarks and create a holistic supply chain assessment. It was also observed that none of the family firms in India have initiatives that are registered under the Green Credits scheme. As the disclosure norms are being tightened globally, it is expected that environmental leadership and its disclosure will also witness an improvement over the years in Indian firms.

This study's key objective was to provide an overview of the environmental performance leadership of the top 10 firms using the eight leadership indicators under Principle 6 of the BRSR Report. Such a small set of indicators may be a limitation to truly judge the environmental leadership of firms. Another aspect that needs attention is a deeper and detailed understanding of the factors that characterize or push family business firms to behave or act proactively toward environmental leadership. Future studies may think along these lines and try to analyze a larger set of business firms, say the Top 100 or Top 200 firms listed on the BSE or the NSE of India. The analysis may also consider comparing the family business firms with respect to non-family business firms in India and globally.

ACKNOWLEDGEMENT

This chapter has used Quillbot to enhance the fluency and readability of the text.

REFERENCES

Agyemang, A. O., K. Yusheng and A. Osei. 2025. Addressing sustainability footprint disclosure for high pollutant firms in China and the US: The roles of firms governance structure, financing decisions, and eco-technology. *Corporate Social Responsibility and Environmental Management* 32, no. 2: 2835–2858.

Aminadav, G. and E. Papaioannou. 2020. Corporate control around the world. *The Journal of Finance* 75, no. 3: 1191–1246.

Baù, M., J. Block, A. D. Cruz and L. Naldi. 2021. Bridging locality and internationalization—a research agenda on the sustainable development of family firms. *Entrepreneurship and Regional Development* 33, nos. 7–8: 477–492.

Berrone, P., C. Cruz, L. R. Gomez-Mejia and M. Larraza-Kintana. 2010. Socioemotional wealth and corporate responses to institutional pressures: Do family-controlled firms pollute less? *Administrative Science Quarterly* 55, no. 1: 82–113.

Copeland, B. R., J. S. Shapiro and M. S. Taylor. 2021. Globalization and the environment. Working Paper, National Bureau of Economic Research.

Driss, H., W. Drobetz, S. El Ghoul and O. Guedhami. 2024. The committee and environmental disclosure: International evidence. *Journal of Economic Behavior & Organization* 221: 602–625.

Dyck, I. J., K. V. Lins, L. Roth, M. Towner and H. F. Wagner. 2024. Family-controlled firms and environmental sustainability: All bite and no bark. HKU Jockey Club Enterprise Sustainability Global Research Institute-Archive, European Corporate Governance Institute–Finance Working Paper 983. https://dx.doi.org/10.2139/ssrn.4829162

Khalid, A. M. and P. S. Rawat. 2025. Corporate environmental disclosures and role of top management: Evidence based on the business responsibility and sustainability reporting in India. *Corporate Social Responsibility and Environmental Management* 32, no. 4: 4619–4635.

Khalid, A. M., S. Sharma and A. K. Dubey. 2020. Some methodological issues in India's SDG index report 2019. *Economic and Political Weekly* 55, nos. 32–33: 29–31.

Lins, K. V. 2003. Equity ownership and firm value in emerging markets. *Journal of Financial and Quantitative Analysis* 38: 159–184.

Maji, S. G. and N. Kalita. 2022. Climate change financial disclosure and firm performance: Empirical evidence from Indian energy sector based on TCFD recommendations. *Society and Business Review* 17, no. 4: 594–612.

Memili, E., H. C. Fang, B. Koç, Ö. Yildirim-Öktem and S. Sonmez. 2018. Sustainability practices of family firms: The interplay between family ownership and long-term orientation. *Journal of Sustainable Tourism* 26, no. 1: 9–28.

Miroshnychenko, I. and A. De Massis. 2022. Sustainability practices of family and nonfamily firms: A worldwide study. *Technological Forecasting and Social Change* 174: 121079.

Mou, R. and T. Ma. 2023. A study on the quality and determinants of climate information disclosure of a-share-listed banks. *Sustainability* 15, no. 10: 8072.

Myslivcova, K., N. Hausmannova and L. Zapletalová. 2023. Drivers of environmental responsibility in family firms: The role of an environmental manager. European Conference on Knowledge Management 1: 924-R29, Academic Conferences International Limited. https://papers.academic-conferences.org/index.php/eckm/issue/view/22

Panicker, V. S. 2017. Ownership and corporate social responsibility in Indian firms. *Social Responsibility Journal* 13, no. 4: 714–727.

Raimo, N., E. de Nuccio and F. Vitolla. 2021. Corporate governance and environmental disclosure through integrated reporting. *Measuring Business Excellence* 26, no. 4: 451–470.

Ramírez-Pasillas, M., U. A. Saari and H. Lundberg. 2023. Business groups owned by family and sustainability embeddedness: Understanding the family sustainability spectrum. In *The Palgrave Handbook of Managing Family Business Groups*, ed. M. Rautiainen, M. J. Parada, T. Pihkala, N. Akhter, A. Discua Cruz and K. Mukherjee, 429–457. Cham: Springer International Publishing.

Ripple, W., C. Wolf, T. Newsome, P. Barnard, W. Moomaw and P. Grandcolas. 2019. World scientists' warning of a climate emergency. *BioScience* 70, no. 1: 8–12.

Rockström, J., W. Steffen, K. Noone, Å. Persson, F. S. III Chapin, E. F. Lambin T. M. Lenton, M. Scheffer, C. Folke, H. J. Schellnhuber, B. Nykvist, C. A. de Wit, T. Hughes, S. van der Leeuw, H. Rodhe, S. Sörlin, P. K. Snyder, R. Costanza, U. Svedin, M. Falkenmark, L. Karlberg, R. W. Corell, V. J. Fabry, J. Hansen, B. Walker, D. Liverman, K. Richardson, P. Crutzen and J. A. Foley. 2009. A safe operating space for humanity. *Nature* 461: 472–475.

SEBI. 2023. BRSR core—framework for assurance and ESG disclosures for value vhain. *Securities Exchange Board of India*. https://www.sebi.gov.in/legal/circulars/jul-2023/brsr-core-framework-for-assurance-and-esg-disclosures-for-value-chain_73854.html

3 Time to Sound the Alarm
Why May the UN Sustainable Development Goals Not Be Achieved by 2030 in the NENA Region?

Tarek Ben Hassen, Hamid El Bilali, and Abdo Hassoun

3.1 INTRODUCTION

In 2016, the United Nations (UN) initiated the 2030 Agenda for Sustainable Development as a comprehensive and all-encompassing appeal to tackle urgent global issues and strive toward a future characterized by inclusivity, equity, and sustainability by 2030 (United Nations General Assembly, 2015). The core of this agenda is 17 Sustainable Development Goals (SDGs), including a wide variety of objectives such as poverty eradication, hunger alleviation, promotion of quality education, gender equality, clean water, and affordable clean energy, comprising five of the most critical aspects (viz. 5Ps: people, prosperity, planet, partnership, and peace) (UNDG, 2016). The SDGs include a comprehensive and interrelated framework designed to guide policy endeavors, mobilize resources, and facilitate international collaboration in addressing some of the most intricate global challenges (United Nations General Assembly, 2015). Early efforts after the SDGs were adopted produced some favorable trends. Extreme poverty and child mortality rates have continued to decline, while significant progress has been made in combating diseases such as HIV and hepatitis. Some gender equality objectives have been producing favorable outcomes. Electricity availability in the least developed countries, along with the percentage of renewables in their energy mix, has increased. Global unemployment rates have decreased to the levels of the pre-2008 financial crisis (United Nations, 2023b).

However, as we surpass the halfway point of the 15-year agenda and the 2030 deadline approaches, doubts regarding achieving these commitments continue to grow (Leal Filho et al., 2023). As highlighted by Malekpour et al. (2023), "The world is not on track to achieve any of the 17 SDGs." At the present pace of progress, the world will not be able to achieve some of the fundamental goals, such as eliminating poverty, ending hunger, or offering decent education for everyone by 2030. Instead, the world will have 575 million people living in severe poverty, 600 million people experiencing hunger, and 84 million children and young people out of school by the end of this decade (Malekpour et al., 2023). This is due to a combination of issues and obstacles. The achievement of the SDGs is dependent on three key areas: implementation, governance, and stability (European Union, 2022). Social, economic, political, and ecological stability is critical to achieving the SDGs. Historically, crises have triggered either beneficial or adverse social, political, and economic transformations in society. Global crises highlight the deep interconnection between the components of sustainable development, underscoring how these linkages can be leveraged to achieve the SDGs (Leal Filho et al., 2023). Multiple crises have arisen since 2019, including the ongoing pandemic, rising inflation and resulting cost-of-living increase, global environmental and economic instability, regional disturbances, wars, and natural catastrophes (United Nations, 2023a).

DOI: 10.1201/9781003600152-3

Further, many major regions are significantly off-track in achieving the SDG targets, especially those facing numerous challenges like the Near East and North Africa (NENA) region (ESCWA, 2022). The NENA region is one of the most diverse regions in the world, yet it shares several common challenges, including poverty, inequality, conflict, and environmental degradation. Indeed, the region is characterized by high population growth with a rate nearly twice the world average and low domestic agricultural productivity, leading to an increased reliance on global markets and trade to meet the demand for food. In this context, the Arab Forum for Environment and Development (AFED, 2023) indicates that pre-existing challenges have left Arab countries particularly vulnerable to the regional impacts of both the COVID-19 pandemic and the Ukraine conflict. The region was already grappling with food insecurity due to increasing populations, climate change-induced droughts, and internal strife. This situation has been exacerbated by the region's limited self-sufficiency in food and dependence on food imports. Additionally, Arab nations typically face significant water shortages, with the water demand consistently outpacing the growth in water resource availability (Djoundourian, 2021).

Consequently, NENA countries have been adversely impacted by recent global food shocks like the COVID-19 pandemic and the war in Ukraine due to their reliance on food imports (Al-Saidi, 2023; Ben Hassen & El Bilali, 2022b). Further, NENA is the most water-stressed region in the world. It will likely face the most significant economic losses due to climate-related water scarcity compared to other areas worldwide (World Bank, 2018). Moreover, the region has a multitude of persistent political instabilities and wars, as seen by the situations in Libya, Palestine, Iraq, Syria, Somalia, and Yemen, compromising human security and projections for sustainable development (United Nations, 2021). Additionally, the COVID-19 pandemic and the ongoing war in Ukraine and Gaza have impeded prospects for development in numerous countries, many of which were already dealing with violent conflicts or internal political turmoil (Bayoumi et al., 2022; Faris et al., 2025; Hassoun et al., 2024).

In general, the accomplishment of the SDGs in the NENA region partly depends on the level of previous successes connected to the UN Millennium Development Goals (MDGs), pursued between 2000 and 2015 (Göll et al., 2019). The region's progress on the MDGs has been uneven, mainly due to the region's diversity. Middle- and high-income countries scored exceptionally well in several areas, such as increasing education and improving health indices. However, countries most in need of development were left behind, and in several instances, indicators deteriorated over the MDG timeframe. Significant economic and political reforms did not take place. Numerous countries in the region lacked reliable data on most indicators for most years (Asi, 2021). Finally, the region failed to fulfill several MDG targets, particularly in conflict-affected countries and rural areas, as well as due to instability in the aftermath of the Arab Spring.

Regarding the SDGs, a recent report (Bayoumi et al., 2022) demonstrates encouraging advances in three critical areas: education, sustainable energy, and climate change. Furthermore, many countries in the region are on pace to meet SDG 4 (Quality Education), SDG 7 (Affordable and Clean Energy), and SDG 13 (Climate Action). Six countries have reached two-thirds of the way to achieving the SDGs, with a total index score of 66 or above. Nevertheless, the data indicates that many countries still face persistent obstacles in achieving gender equality, promoting decent jobs and economic development, and ensuring sustainable food production. SDG 3 (Good Health and Well-being), SDG 6 (Clean Water and Sanitation), SDG 9 (Industry, Innovation, and Infrastructure), SDG 14 (Life Below Water), and SDG 16 (Peace, Justice, and Strong Institutions) all as well face challenges (Bayoumi et al., 2022).

Accordingly, this chapter aims to examine the prospects for NENA countries to achieve the SDGs by critically evaluating progress and obstacles across key goals and representative country cases. It uncovers the systemic barriers undermining efforts to attain the SDGs through a comprehensive analysis of critical indicators, policy trends, and regional dynamics. From issues of conflict and instability to disparities in wealth distribution, gender inequality, and resource scarcity, the NENA region grapples with complex challenges that pose formidable obstacles to the SDG agenda.

By shedding light on these challenges, this chapter contributes to a deeper understanding of the barriers that must be addressed to pave the way for sustainable development in the NENA region.

This study utilized a broad literature review approach to gather references from various sources, including academic databases such as Scopus. Data collection was performed in September 2024 by using the combination of terms related to SDGs and barriers in the MENA region, and the Boolean operators "AND" and "OR." The best combination of search strings was the following: (TITLE-ABS-KEY: (SDGs OR "Sustainable Development Goals") AND (MENA OR "Middle East and North Africa" OR NENA OR "Near East and North Africa")). The data collection process followed a strategic search of peer-reviewed articles, policy reports, and UN documents to obtain comprehensive insights into SDG progress barriers. Grey literature was also included to ensure that recent and practical developments not yet published in academic journals were also considered. This combined approach of academic and grey literature provided a well-rounded perspective on the ongoing challenges to SDG implementation in the region. From this broad collection of literature, the study identified not only the challenges but also key recommendations for accelerating the SDGs progress in the NENA region. This methodology ensured a wide-ranging exploration of the factors affecting sustainable development across the region, offering actionable insights for policymakers and researchers.

In the sections that follow, this chapter is structured to provide a comprehensive exploration of the barriers impacting SDG progress in the NENA region. Section 3.2 examines the repercussions of the COVID-19 pandemic on SDG advancement, highlighting the pandemic's effect on the achievement of SDG 1 and SDG 2. Section 3.3 focuses on the consequences of the Ukraine conflict, emphasizing its disruption of essential food supplies in the region. In Section 3.4, the chapter delves into the core challenges impeding SDG achievement, including conflict, social and gender disparities, climate vulnerability, and weak governance. Finally, the discussion and conclusion synthesize the findings, proposing targeted policy recommendations aimed at overcoming these obstacles and advancing sustainable development in the NENA region.

3.2 IMPACT OF THE COVID-19 PANDEMIC ON THE PROGRESS OF THE SDGS IN THE NENA REGION

The COVID-19 pandemic has hampered progress toward the SDGs in the NENA region, significantly delaying initiatives to achieve the ambitious 2030 Agenda (Priyadarshini, 2022) (Figure 3.1). Indeed, the pandemic has impacted almost all of the SDGs substantially, with no country left unaffected (Martín-Blanco et al., 2022). Further, the pandemic has arrived at an inopportune moment for the area. Indeed, COVID-19 represents the region's fourth crisis, a decade after the Arab Spring, the 2014–16 oil price drops, and the 2019 protest upsurge (World Bank, 2021).

Since the outbreak began in March 2020, the pandemic has exposed and exacerbated pre-existing issues in poverty, hunger, health, education, inequality, and economic development. While governments have responded with massive stimulus and social protection initiatives, the SDGs are projected to suffer long-term effects for society and institutions. Progress from 2015 to 2019 has halted or reversed, indicating the need for more action during the remaining period. Looking at specific SDGs, the pandemic has made it much more difficult to achieve objectives linked to poverty (SDG 1), hunger (SDG 2), health (SDG 3), education (SDG 4), economic development (SDG 8), inequality (SDG 5), responsible consumption (SDG 12), climate action (SDG 13), and justice (SDG 16) (Shulla et al., 2021; Yuan et al., 2023).

The confirmed cases, as well as the deaths, have been relatively high in the NENA region (Table 3.1). The impact of the pandemic has been particularly severe in this region, with multiple countries, including Iraq, Jordan, Morocco, Lebanon, Tunisia, and the United Arab Emirates, reporting over a million confirmed cases (WHO, 2023). However, the vaccine distribution and immunization rate in the NENA countries have generally been substantial, except for nations facing political instability and conflict, such as Yemen, Iraq, Libya, and Syria (Table 3.1).

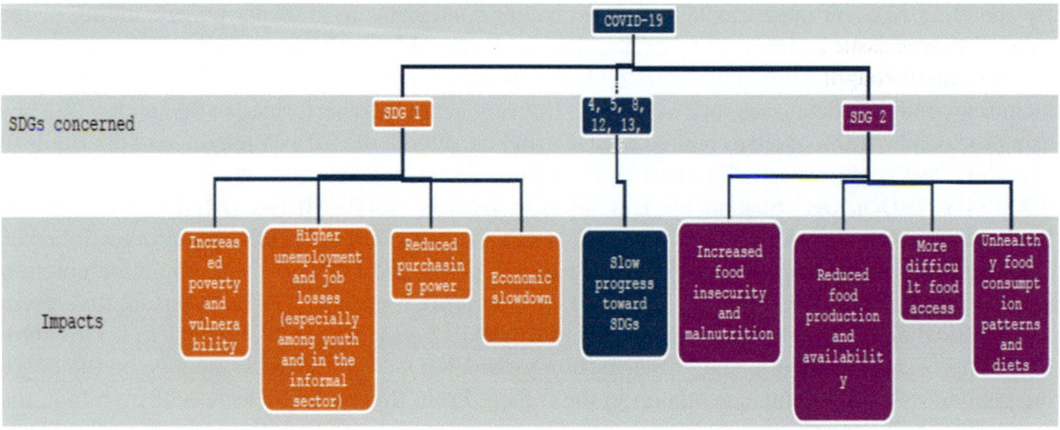

FIGURE 3.1 Impact of the COVID-19 pandemic on the progress of the SDGs in the NENA region.

TABLE 3.1
COVID-19 Pandemic in the NENA Region

Country	Confirmed Cases[a]	Deaths[b]	Vaccine Doses Administered[c]	Population (Thousands, 2022)
Algeria	271,945	6,881	15,267,442	44,903.22
Bahrain	696,614	1,536	3,476,633	1,472.23
Egypt	516,023	24,830	112,673,535	110,990.10
Iraq	2,465,545	2,537	19,557,364	44,496.12
Jordan	1,746,997	14,122	10,057,975	11,285.87
Kuwait	666,486	2,570	8,261,153	4,268.87
Lebanon	1,239,904	10,947	5,814,699	5,489.74
Libya	507,269	6,437	3,739,158	6,812.34
Mauritania	63,774	997	4,075,704	4,736.14
Morocco	1,277,342	16,297	55,389,602	37,457.97
Oman	399,449	4,628	7,086,050	4,576.30
Qatar	514,524	690	7,609,178	2,695.12
Saudi Arabia	841,469	9,646	68,534,631	36,408.82
Sudan	63,993	5,046	22,598,737	46,874.20
Syria	57,423	3,163	5,090,630	22,125.25
Tunisia	1,153,361	29,423	13,253,317	12,356.12
United Arab Emirates (UAE)	1,067,030	2,349	24,922,054	9,441.13
Yemen	11,945	2,159	1,287,68	33,696.61
Source	WHO (WHO, 2023)	WHO (WHO, 2023)	WHO (WHO, 2023)	World Bank (World Bank, 2023)

[a] As of November 4, 2023.

3.2.1 Impact of the Pandemic on SDG 1

The COVID-19 pandemic has hampered progress toward achieving all of the SDGs in the NENA area, with some goals, such as SDG 1 (No Poverty), experiencing a particularly severe impact. Indeed, the pandemic has intensified pre-existing development issues and introduced new ones

(ESCWA, 2023). Even before COVID-19 emerged in March 2020, the region had various major socioeconomic issues, further exacerbated as the pandemic progressed. NENA has a low yearly growth rate of roughly 1.4% per capita, which is lower than that of its peers. Extreme poverty has more than doubled, from 2.4% in 2011 to 4.2% in 2015, spurred in part by conflicts in Libya, Syria, and Yemen. The region also suffers from several issues, such as an unfavorable business climate and a lack of quality jobs, very high levels of unemployment (particularly among youth), and low female labor force participation, with many working in the informal sector. These circumstances restricted people's prospects and impacted life satisfaction, which had already dropped in the region before the pandemic, unlike in all other areas worldwide. Life satisfaction in 2019 was lower than in 2010 in 11 of the 14 NENA nations for which data is available (World Bank, 2021).

The COVID-19 crisis is distinct from previous crises due to its extensive ramifications and the resulting disparities, which have exhibited differences among countries. These differences can be attributed to the diverse economic composition of the region, encompassing high-, middle-, and low-income countries and economies, as well as those reliant on oil exports and those characterized by fragility, conflict, and vulnerabilities associated with violence (ESCWA, 2024).

SDG 1, which aims to eradicate poverty, has seen a concerning regression. Despite previous progress in reducing extreme poverty, the January 2021 Global Economic Prospects forecast estimates that between 7 million and 8 million people in the region will fall into poverty due to the pandemic (World Bank, 2021). Consequently, the pandemic resulted in a marked reduction in purchasing power, mainly because of extensive job losses and its adverse effects on different facets of living. This decline in purchasing power worsened due to the economic slowdown and interruptions in sources of income. This downturn in purchasing power was exacerbated by the economic recession and disruptions in income sources, which further strained the financial capabilities of individuals and families, making it more challenging for them to afford essential goods and services, including food (Rahmat et al., 2022).

Second, the COVID-19 pandemic had significant unemployment impacts, especially in the informal sectors, generally characterized by a lack of social protection coverage and low and unstable revenues (OECD, 2020). As COVID-19 spreads across regions with large informal economies, especially low-and middle-income countries, more informal workers lose their jobs, thus increasing extreme poverty and food insecurity (FAO, 2020). In fact, the COVID-19 pandemic exposed significant deficiencies within domestic social protection systems. Social protection systems in the Arab region persistently exhibit notable flaws in coverage, sufficiency, and comprehensiveness. Moreover, these systems confront significant governance, coordination, and funding challenges complicated by the economic, social, and demographic shifts occurring throughout the area (International Labour Organization, 2021). Indeed, even before the COVID-19 pandemic, a significant portion of the populations in most countries in the region lacked adequate and comprehensive social protection. A mere 40% of individuals residing in MENA possess adequate social protection coverage in at least one domain, notably lower than the global average of 46.7% (International Labour Organization, 2021).

Lebanon, for instance, started 2020 already in the grip of a catastrophic economic crisis, massive demonstrations, and political instability. The COVID-19 outbreak stressed Lebanon's already frail institutions even further, resulting in severe humanitarian consequences (Bizri et al., 2021). In this regard, CIHEAM (CIHEAM, 2020) observed that the combined effects of the financial crisis and COVID-19 increased the price of all food products in Lebanon. Before the pandemic, Lebanon's food security and nutrition situation was already dire. The pandemic struck at a time when Lebanon's food security was already under stress (Ben Hassen et al., 2021). Further, since 2011, the Syrian crisis, with a significant influx of refugees, has had a considerable impact on Lebanon's food security (ESCWA, 2016). Food has become costly for many Lebanese families, particularly those below the upper poverty level (45% of the population) and the severe poverty line (22%). In 2019, persons living in extreme poverty had to spend 51% of their income on food. In 2020, the same minimal necessities accounted for 79% of their income (Wood et al., 2020). Furthermore, according to Lebanon's Central Administration of Statistics (CAS), the Consumer Price Index increased by approximately

146% year on year between December 2019 and December 2020, the highest pace recorded since the country's current economic crisis began (Bahn et al., 2021).

Lebanon once appeared on track to lower its extreme poverty rate to under 3%. However, the country fell into political turmoil and financial crisis in 2019, reversing poverty reduction gains and plunging growing numbers of households into economic deprivation. Over the last five years, the Lebanese currency has experienced dramatic devaluation, causing prices of staple food and consumer goods to surge beyond the means of many citizens. An UN expert report labeled Lebanon as a "failing state" amid hyperinflation, monetary collapse, and an inability to provide basic public services—dynamics threatening to push nearly three-quarters of the population into poverty (UN, 2022). Consequently, poverty in Lebanon has more than quadrupled in the last decade (2014–2024), reaching 44% of the entire population (World Bank, 2024).

3.2.2 Impact of the Pandemic on SDG 2

Before the pandemic outbreak, the NENA region was confronted with significant challenges related to malnutrition and food insecurity (Table 3.2). Indeed, numerous NENA countries experienced a substantial prevalence of undernourishment. From 2017 to 2019, the rate of undernourishment varied considerably, from under 2.5% in Algeria, Kuwait, and Tunisia to 11.9% in Mauritania, 12.4% in Sudan, and 23.7% in Iraq. When considering the prevalence of moderate to severe food insecurity during the same period, which ranges from 12.3% in Kuwait to 44.8% in Mauritania and as high as

TABLE 3.2
Food (In)Security and Cereal Import Dependency in NENA Countries

	Prevalence of Undernourishment (% Population)		Prevalence of Moderate or Severe Food Insecurity (% Population)		Cereal Import Dependency Ratio 2016–2018 (%)
Country	2017–19	2010–22	2017–19	2020–22	
Algeria	2.8	<2.5	17.6	19.4	70.80
Bahrain	n.a.	n.a.	n.a.	n.r.	n.a.
Egypt	4.7	7.2	34.2	28.5	47.80
Iraq	23.7	16.3	n.a.	n.r.	56.80
Jordan	8.5	n.a.	n.a.	n.r.	100.0
Kuwait	<2.5	<2.5	12.3	10.9	98.80
Lebanon	5.7	n.a.	n.a.	36.5	99.00
Libya	n.a.	8.4	35.9	39.8	n.a.
Mauritania	11.9	8.7	44.8	53.7	n.a.
Morocco	4.3	6.3	25.9	n.r.	56.90
Oman	7.8	2.8	n.a.	n.a.	93.60
Qatar	n.a.	n.a.	n.a.	n.a.	n.a.
Saudi Arabia	4.8	3.8	n.a.	n.r.	95.60
Sudan	12.4	11.9	48.9	51.8	23.60
Syria	n.a.	27.8	n.a.	n.a.	n.a.
Tunisia	<2.5	3.0	20.0	28.5	65.50
UAE	3.1	<2.5	n.a.	9.8	100.00
Yemen	n.a.	34.5	n.a.	67.2	n.a.
Source	FAO et al. (2020)	FAO (2023)	FAO et al. (2020)	FAO (2023)	United Nations Statistics Division (2023)

n.a. no available data; n.r. not reported.

48.9% in Sudan, the scope of the issue becomes more evident and alarming. Several countries in the region experienced an increase in the prevalence of undernourishment from 2017 to 2019, including Egypt, Morocco, and Tunisia, where the prevalence increased during that period. During the same period, moderate or severe food insecurity increased in Tunisia, Libya, Mauritania, and Sudan. Moreover, the food security situation of nations within the NENA region is precarious due to their significant dependence on imports to fulfill their increasing food needs. For example, the percentage of grain import reliance is very high in all nations within the NENA region, as seen in Table 3.2.

The pandemic impacted all levels of the food supply chain, including production, processing, transportation, distribution, and consumption. Moreover, the pandemic affected all four pillars of food security (availability, access, use/utilization, and stability) (El Bilali & Ben Hassen, 2024b).

Concerning food availability, the pandemic impacted both local and international food production. The outbreak of COVID-19 led to a decline in domestic food production, which decreased the overall food supply. In certain instances, obtaining inputs like fertilizers, pesticides, and seeds became more challenging and costly, reducing their usage. This decrease in input usage contributed to lower yields and productivity. Indeed, the pandemic affected crop production and productivity in different ways. It made it more challenging for farmers, especially small-scale producers, to access inputs, which affected yield and production (Ali & Gad, 2022). For example, in Tunisia, Koussani and Khamassi (Koussani & Khamassi, 2022) point out that the pandemic led to significant disruptions in agricultural operations, especially in supply chains for raw materials (such as treatment products, fertilizers, seeds, and animal feed) and marketing channels. These issues were due to the closures of weekly markets selling perishable goods, like leafy vegetables, and small livestock. Further, disruptions of global supply chains and markets influenced the trade and import of some products and inputs and, consequently, their availability in the domestic markets. Referring to Yemen, Rahmat et al. (2022) observed that COVID-19 restrictions severely hinder food acquisition and the import of essential commodities.

Concerning food access, the COVID-19 pandemic influenced both economic and physical accessibility/affordability. Virus containment measures, such as lockdowns, impacted physical access in NENA countries. Nevertheless, these effects differed not only among NENA countries, based on the stringency and duration of the containment measures (cf. lockdown) but also among socioeconomic groups. In particular, the poor (McCall et al., 2022), women (Ben Hassen et al., 2022; Bossenbroek & Ftouhi, 2021; Bouzidi & Abdellaoui, 2022; Butt et al., 2022; Tayyem et al., 2022; Turki et al., 2022), and children (Alhaffar et al., 2022; Benmerzoug et al., 2022; Butt et al., 2022; Gedeon et al., 2022; Hoteit et al., 2022) seem to be more affected. Rahmat et al. (2022), in their study on Yemen, caution that the combination of intense economic instability, worsened by the COVID-19 pandemic, ongoing conflict, and soaring food prices, has brought the Yemeni population to the verge of famine, with women and children suffering particularly from malnutrition. Economic access was negatively affected due to the increase in food prices.

Numerous studies indicate that the COVID-19 pandemic adversely influenced dietary habits and food consumption patterns, particularly during lockdowns, leading to notable health consequences such as increased rates of overweight and obesity (Alghadir et al., 2021; Alkhalaf et al., 2022; Almousa & Alagal, 2022; Alouani et al., 2022; Bushnaq et al., 2022; El-Akabawy et al., 2022; Radwan et al., 2021; Turki et al., 2022). Alkhalaf et al. (2022) found a reduction in the consumption of vegetables and an increase in sweets consumption among Saudi Adults. Meanwhile, Braiji et al. (2022) in their study focusing on Jeddah, Saudi Arabia, observed a notable decrease in the consumption of fast foods like pizza and burgers during the quarantine period, while the intake of snacks, sugars, and pastries saw a significant increase during the same time.

3.3 IMPACT OF THE WAR IN UKRAINE ON THE PROGRESS OF THE SDGS IN THE NENA REGION

The Ukraine conflict caused global shockwaves, especially in the NENA region. The war has intensified global instability and uncertainty, compounding the challenges already heightened by the

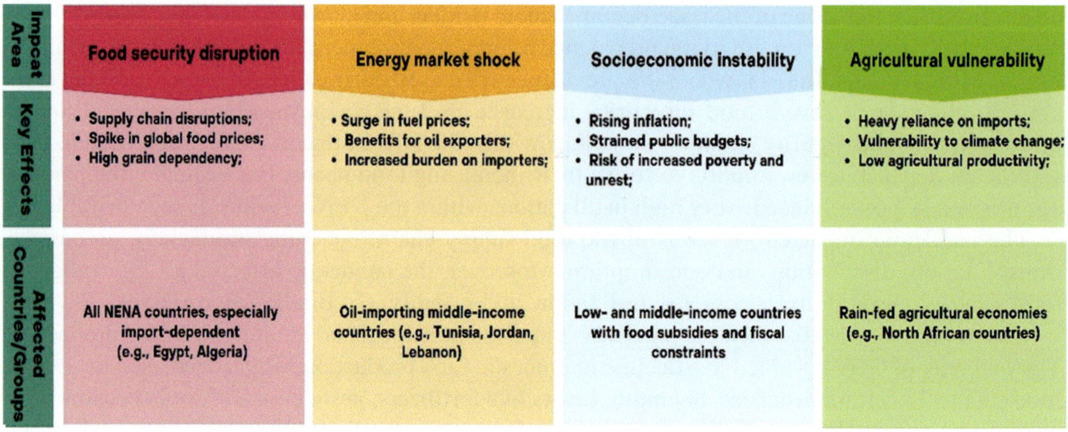

FIGURE 3.2 Impact of the Ukraine War on SDGs in the NENA Region.

Source: Authors' elaboration.

COVID-19 pandemic (El Bilali & Ben Hassen, 2024a). Aside from the loss of life, the current conflict has far-reaching consequences for the environment, economy, and society. Given that the world economy is still recovering from COVID-19, the sanctions on the Russian Federation and the disruption of several crucial global supply chains are already significantly impacting the world economy (ESCWA et al., 2022). The conflict triggered an international chain reaction, particularly in the energy and agricultural sectors. The escalation of this conflict poses severe challenges to achieving the UN SDGs, not only for the countries directly involved in the conflict but also for other countries, particularly developing countries that are more vulnerable to the economic crisis (Pereira et al., 2022). Accordingly, the conflict poses a significant risk to achieving the SDGs in the NENA region since it impacts several aspects, such as energy and food markets, trade relations, and investment conditions (Figure 3.2).

First, the conflict has limited crop production and trade in one of the world's most critical food-exporting areas, putting global food security at risk (Ben Hassen & El Bilali, 2022b; Jagtap et al., 2022). It has considerably increased global food prices, increasing the food system's vulnerability, which has already been compounded by climate change and the COVID-19 pandemic (Bloem & Farris, 2022; Rice et al., 2022). The conflict's full military consequences and outcomes are still unknown almost two years later, and most analysts foresee a lengthy conflict (Foreign Affairs, 2022). The implications for Ukrainian agricultural output and food security, on the other hand, are evident (Allam et al., 2022; Ben Hassen & El Bilali, 2022b; Jagtap et al., 2022; Zhou et al., 2023).

The primary outcome of the war was the impact on agricultural production in Ukraine. The conflict has already impacted agricultural production for three consecutive years: 2021, 2022, and 2023 (Glauber, 2023). Most countries in the NENA region are affected by price volatility. Their agricultural productivity is constrained by geography, climate, and growing populations, leaving them reliant on global markets (OECD & FAO, 2018). Already four months after the outbreak of the conflict, Ben Hassen and El Bilali (2022b) warned that "If the war deepens, the food crisis will worsen, posing a challenge to many countries, especially those that rely on food imports, such as those in the Middle East and North Africa (MENA) region." The NENA region constitutes a mere 4% of the global population, yet it remarkably accounts for 30% of the total grain consumption. Ukraine is a significant exporter of agricultural commodities, with more than 50% of its wheat and around one-third of its maize and sunflower oil being exported to nations in the MENA region (Abis & Bertin, 2022). The upward trajectory of commodity prices drives food inflation in many countries, escalating production costs and devaluing currencies (Knowledge Centre for Global Food and Nutrition Security, 2022). Inflation has exacerbated the issue of food insecurity and has also contributed to heightened social tensions. Additionally, it burdens governmental finances, which are already facing challenges in financing food imports and social welfare programs to support poor populations (IMF, 2022).

While the surge in energy prices has helped the oil-rich NENA countries reduce their vulnerability to food staples' historically high prices, Middle-income countries in the region that depend primarily on imported energy and food appear to be the most vulnerable to the economic fallout from Ukraine's conflict (ESCWA et al., 2022). According to predictions, supply constraints and price increases for key imports of wheat, fertilizers, oil, and gas from Russia and Ukraine would significantly affect middle-income MENA countries such as Egypt, Jordan, Lebanon, Morocco, and Tunisia (Ben Hassen & El Bilali, 2024).

Food inflation is a primary source of socioeconomic instability in NENA nations since most of their food, especially grains, is imported. Wheat accounts for roughly one-third of these nations' food supply. Food price increases have sparked worries about public budgets due to government subsidies. With many countries recovering from the COVID-19 pandemic, mounting public debt, and expenditure cuts, price increases have compounded the challenging financial situation (FAO, 2022a). Global grain prices have not affected local prices due to food subsidies, but MENA governments' long-term inflation control has been challenged. The structural fragility of regional food systems adds to socioeconomic stress, and the ongoing Russia-Ukraine war raises concerns about regional destabilization (Liga, 2022). Low productivity, heavy reliance on rainfall, and climate change characterize MENA agriculture. This is due to a lack of investment, markets, information, and training on new and better farming methods and technologies. For instance, in North Africa, the sector's reliance on rainwater makes it particularly vulnerable to variations in rainfall, which are expected to become more unpredictable due to climate change (OECD & FAO, 2018). As a result, the region imports a significant portion of its food, particularly grain, accounting for 16–17% of world wheat imports. Algeria, for example, imports 75% of its food, whereas Egypt is the world's largest wheat importer. Food imports expose the region to price swings and increase its reliance on international markets.

3.4 THE MAIN CHALLENGES TO ACHIEVING THE SDGS IN THE NENA REGION

The NENA region faces a complex set of structural challenges that continue to hinder progress toward achieving the SDGs. These challenges are deeply interconnected, often reinforcing one another and compounding their impacts. Figure 3.3 provides a visual summary of the four primary obstacles discussed in this section. Each will be examined in more detail in the subsections that follow.

3.4.1 Conflict and Instability

Conflict and instability pose significant obstacles to the development process in the MENA region. Several MENA countries are embroiled in protracted wars with global ramifications. The war in Syria and Gaza, Yemen's humanitarian catastrophe, civil war in Sudan, and Libya's instability after the fall of Muammar Gaddafi are just a few instances. Not only are these conflicts are military or political, but they often cross with ethnic, religious, and socioeconomic fault lines, complicating resolution attempts even more. Conflict results in the displacement of people, the destruction of physical infrastructure, and a decline in economic growth. Further, such armed conflicts and wars challenge the implementation of policies and programs. Moreover, the ensuing displacement issues, which have resulted in millions of people seeking asylum, have had a domino effect within the region and beyond. Consequently, the region has recently become home to many refugees. The number of forcibly displaced and stateless people in the MENA region increased from 15.8 million

FIGURE 3.3 Key structural challenges to achieving the SDGs in the MENA region.

Source: Authors' elaboration.

recorded in 2020, reaching 16 million by the end of 2021. Most displaced individuals currently reside in non-camp environments, specifically in urban areas.

This situation places additional strain on local governments responsible for addressing their daily needs (United Nations, 2020). The situation is particularly challenging given that most displaced people in the region live in extreme poverty and face significant barriers to accessing essential services. For instance, Lebanon, the world's top refugee-hosting country per capita, is facing a dire situation where nine out of every ten refugee households live in extreme poverty (UNHCR, 2021). This situation has increased tensions with host communities due to competition for scarce resources (Daher et al., 2021). In Iraq, many out-of-camp households have accumulated high levels of debt, facing difficulties in accessing adequate housing, purchasing food, ensuring their children attend school, and obtaining healthcare (UNHCR, 2021). Accordingly, the prevalence of acute food insecurity and malnutrition in Jordan and Lebanon is closely related to refugee hosting (Da Silva & Shenggen, 2017).

3.4.2 Social and Gender Disparities

Promoting social inclusion and gender equality is a fundamental aspect of the SDGs. The MENA region faces the world's most significant gender gap, measured across multiple dimensions, including women's participation in politics and the economy, access to education, and health outcomes. Progress toward gender equality in the region has been gradual and inconsistent. At the current pace, MENA is projected to take more than 140 years to achieve gender parity, notably lagging behind areas like South Asia and Sub-Saharan Africa, which are expected to reach this milestone about 20 years sooner (ESCWA, 2024; Khadduri et al., 2022). The MENA region is the most unequal globally, and inequality is rising. The COVID-19 pandemic, high interest rates, rising debt

El Bilali, H., & Ben Hassen, T. (2024b). Regional Agriculture and Food Systems Amid the COVID-19 Pandemic: The Case of the Near East and North Africa Region. *Foods*, *13*(2), 297. https://doi.org/10.3390/foods13020297

El Chami, D., Trabucco, A., Wong, T., Monem, M. A., & Mereu, V. (2022). Costs and Effectiveness of Climate Change Adaptation in Agriculture: A Systematic Review from the NENA Region. *Climate Policy*, *22*(4), 445–463. https://doi.org/10.1080/14693062.2021.1997703

El-Akabawy, G., Abukhaled, J. K., Alabdullah, D. W., Aleban, S. A., Almuqhim, S. A., & Assiri, R. A. (2022). Prevalence of Eating Disorders among Saudi Female University Students during the COVID-19 Outbreak. *Journal of Taibah University Medical Sciences*, *17*(3), 392–400. https://doi.org/10.1016/j.jtumed.2022.02.001

ESCWA. (2016). *Strategic Review of Food and Nutrition Security in Lebanon*. https://www.unescwa.org/sites/www.unescwa.org/files/page_attachments/escwa_food_and_nutrition_security_in_lebanon_final_version_high_res_en.pdf

ESCWA. (2022). *Annual SDG Review 2022*. https://www.unescwa.org/publications/annual-sdg-review-2022

ESCWA. (2023). *Development Challenges Index: Impact of the COVID-19 Pandemic*. https://www.unescwa.org/publications/development-challenges-index-impact-covid-19-pandemic

ESCWA. (2024). *Arab Sustainable Development Report 2024*. https://www.unescwa.org/publications/arab-sustainable-development-report-2024

ESCWA, FAO, UNEP, & WFP. (2022). *Impacts of the War in Ukraine on the Arab Region*. https://www.unescwa.org/news/war-ukraine-will-lead-record-about-126-million-poor-arab-region

European Union. (2022). *Sustainable Development in the European Union*. https://op.europa.eu/en/publication-detail/-/publication/0cf33c27-e706-11ec-a534-01aa75ed71a1/language-en

FAO. (2013). *First Regional Training Workshop "Launching Workshop of the Regional Water Scarcity Initiative."* https://openknowledge.fao.org/server/api/core/bitstreams/4f104780-63fa-4f4c-864e-c6dae956b838/content

FAO. (2020). *Impact of COVID-19 on Informal Workers*. http://www.fao.org/3/ca8560en/CA8560EN.pdf

FAO. (2022a). *The Importance of Ukraine and the Russian Federation for Global Agricultural Markets and the Risks Associated with the War in Ukraine. December 2022*. Information Note. https://www.fao.org/3/cc3317en/cc3317en.pdf

FAO. (2022b). *The State of Land and Water Resources for Food and Agriculture in the Near East and North Africa region*. FAO. https://doi.org/10.4060/CC0265EN

FAO. (2023). *The State of Food Security and Nutrition in the World 2023*. FAO; IFAD; UNICEF; WFP; WHO. https://doi.org/10.4060/cc3017en

FAO, IFAD, UNICEF, WFP, & WHO. (2020). *The State of Food Security and Nutrition in the World 2020*. FAO. https://doi.org/10.4060/ca9692en

Faris, M. E., Ben Hassen, T., Parra-López, C., & Hassoun, A. (2025). From Acute Food Insecurity to Famine: The Struggle of Gazans for Food Security. In A. Hassoun (Ed.), *War on Gaza* (pp. 29–42). Springer. https://doi.org/10.1007/978-3-031-88500-6_3

Foreign Affairs. (2022). *No One Would Win a Long War in Ukraine*. https://www.foreignaffairs.com/ukraine/no-one-would-win-long-war-ukraine

Gedeon, R., Hallit, S., & Wakim, L. H. (2022). Food Insecurity and Eating Habits of Lebanese Children Aged 5–11 Years during the COVID-19 Pandemic and the Socioeconomic Crisis: A National Study. *BMC Public Health*, *22*(1), 1982. https://doi.org/10.1186/s12889-022-14387-z

Glauber, J. (2023). *Ukraine One Year Later: Impacts on Global Food Security*. IFPRI: International Food Policy Research Institute. https://www.ifpri.org/blog/ukraine-one-year-later-impacts-global-food-security

Göll, E., Uhl, A., & Zwiers, J. (2019). *Sustainable Development in the MENA Region*. MENARA Future Notes. http://menara.iai.it/publications/

Hassoun, A., Al-Muhannadi, K., Hassan, H. F., Hamad, A., Khwaldia, K., Buheji, M., & Al Jawaldeh, A. (2024). From Acute Food Insecurity to Famine: How the 2023/2024 War on Gaza Has Dramatically Set Back Sustainable Development Goal 2 to End Hunger. *Frontiers in Sustainable Food Systems*, *8*. https://doi.org/10.3389/fsufs.2024.1402150

Hoteit, M., Ibrahim, C., Saadeh, D., Al-Jaafari, M., Atwi, M., Alasmar, S., Najm, J., Sacre, Y., Hanna-Wakim, L., & Al-Jawaldeh, A. (2022). Correlates of Sub-Optimal Feeding Practices among under-5 Children Amid Escalating Crises in Lebanon: A National Representative Cross-Sectional Study. *Children*, *9*(6), 817. https://doi.org/10.3390/children9060817

IMF. (2022). *Global Food Prices to Remain Elevated Amid War, Costly Energy, La Niña*. https://www.imf.org/en/Blogs/Articles/2022/12/09/global-food-prices-to-remain-elevated-amid-war-costly-energy-la-nina

International Labour Organization. (2021). *World Social Protection Report 2020–22: Regional Companion Report for the Middle East and North Africa (MENA) Region*. https://researchrepository.ilo.org/esploro/outputs/report/World-social-protection-report-2020-22-Regional/995219260102676

Jagtap, S., Trollman, H., Trollman, F., Garcia-Garcia, G., Parra-López, C., Duong, L., Martindale, W., Munekata, P. E. S., Lorenzo, J. M., Hdaifeh, A., Hassoun, A., Salonitis, K., & Afy-Shararah, M. (2022). The Russia-Ukraine Conflict: Its Implications for the Global Food Supply Chains. *Foods*, *11*(14), 2098. https://doi.org/10.3390/foods11142098

Khadduri, R., Haegeman, N., Briansó, A., & Tomlin, A. (2022). *MENA Regional Overview: Gender and Equality*. https://manniondaniels.com/wp-content/uploads/2022/12/MENA-overview-Gender-and-Equality.pdf

Knowledge Centre for Global Food and Nutrition Security. (2022). *The Impact of Russia's War against Ukraine on Global Food Security*. European Commission, December. https://knowledge4policy.ec.europa.eu/publication/impact-russia%E2%80%99s-war-against-ukraine-global-food-security-%E2%80%93-december-2022_en

Koussani, W., & Khamassi, F. (2022). Impact de la crise sanitaire Covid-19 sur les petites exploitations agricoles et perspectives pour un système alimentaire durable en Tunisie. *New Medit*, *21*(1). https://doi.org/10.30682/nm2201i

Leal Filho, W., Viera Trevisan, L., Simon Rampasso, I., Anholon, R., Pimenta Dinis, M. A., Londero Brandli, L., Sierra, J., Lange Salvia, A., Pretorius, R., Nicolau, M., Paulino Pires Eustachio, J. H., & Mazutti, J. (2023). When the Alarm Bells Ring: Why the UN Sustainable Development Goals May Not Be Achieved by 2030. *Journal of Cleaner Production*, *407*, 137108. https://doi.org/10.1016/j.jclepro.2023.137108

Liga, A. (2022). *Ukraine War Feeds Food Inflation in North Africa*. https://www.ispionline.it/it/pubblicazione/ukraine-war-feeds-food-inflation-north-africa-33973

Malekpour, S., Allen, C., Sagar, A., Scholz, I., Persson, Å., Miranda, J. J., Bennich, T., Dube, O. P., Kanie, N., Madise, N., Shackell, N., Montoya, J. C., Pan, J., Hathie, I., Bobylev, S. N., Agard, J., & Al-Ghanim, K. (2023). What Scientists Need to Do to Accelerate Progress on the SDGs. *Nature*, *621*(7978), 250–254. https://doi.org/10.1038/d41586-023-02808-x

Martín-Blanco, C., Zamorano, M., Lizárraga, C., & Molina-Moreno, V. (2022). The Impact of COVID-19 on the Sustainable Development Goals: Achievements and Expectations. *International Journal of Environmental Research and Public Health*, *19*(23), 16266. https://doi.org/10.3390/ijerph192316266

McCall, S. J., El Khoury, T., Salibi, N., Abi Zeid, B., El Haddad, M., Alawieh, M. F., Abdulrahim, S., Chaaya, M., Ghattas, H., & Sibai, A. M. (2022). Development of a Prediction Model for the Management of Noncommunicable Diseases among Older Syrian Refugees Amidst the COVID-19 Pandemic in Lebanon. *JAMA Network Open*, *5*(10), e2231633. https://doi.org/10.1001/jamanetworkopen.2022.31633

Nazemi, N., Foley, R. W., Louis, G., & Keeler, L. W. (2020). Divergent Agricultural Water Governance Scenarios: The Case of Zayanderud Basin, Iran. *Agricultural Water Management*, *229*, 105921. https://doi.org/10.1016/j.agwat.2019.105921

OECD. (2017). *The Pursuit of Gender Equality an Uphill Battle*. https://doi.org/10.1787/9789264281318-en

OECD. (2020). *COVID-19 and the Food and Agriculture Sector: Issues and Policy Responses*. https://www.oecd.org/coronavirus/policy-responses/covid-19-and-the-food-and-agriculture-sector-issues-and-policy-responses-a23f764b/

OECD, & FAO. (2018). *OECD-FAO Agricultural Outlook 2018–2027*. OECD. https://doi.org/https://doi.org/10.1787/agr_outlook-2018-en

Pereira, P., Zhao, W., Symochko, L., Inacio, M., Bogunovic, I., & Barcelo, D. (2022). The Russian-Ukrainian Armed Conflict Will Push Back the Sustainable Development Goals. *Geography and Sustainability*, *3*(3), 277–287. https://doi.org/10.1016/j.geosus.2022.09.003

Priyadarshini, P. (2022). The COVID-19 Pandemic Has Derailed the Progress of Sustainable Development Goals. *Anthropocene Science*, *1*(3), 410–412. https://doi.org/10.1007/s44177-022-00032-2

Radwan, H., Al Kitbi, M., Hasan, H., Al Hilali, M., Abbas, N., Hamadeh, R., Saif, E. R., & Naja, F. (2021). Indirect Health Effects of COVID-19: Unhealthy Lifestyle Behaviors during the Lockdown in the United Arab Emirates. *International Journal of Environmental Research and Public Health*, *18*(4), 1964. https://doi.org/10.3390/ijerph18041964

Rahmat, Z. S., Islam, Z., Mohanan, P., Kokash, D. M., Essar, M. Y., Hasan, M. M., Hashim, H. T., & Basalilah, A. F. M. (2022). Food Insecurity during COVID-19 in Yemen. *The American Journal of Tropical Medicine and Hygiene*, *106*(6), 1589–1592. https://doi.org/10.4269/ajtmh.22-0059

Rice, B., Hernández, M. A., Glauber, J., & Vos, R. (2022). *The Russia-Ukraine War Is Exacerbating International Food Price Volatility*. IFPRI Blog: Issue Post. https://www.ifpri.org/blog/russia-ukraine-war-exacerbating-international-food-price-volatility

Shulla, K., Voigt, B.-F., Cibian, S., Scandone, G., Martinez, E., Nelkovski, F., & Salehi, P. (2021). Effects of COVID-19 on the Sustainable Development Goals (SDGs). *Discover Sustainability*, *2*(1), 15. https://doi.org/10.1007/s43621-021-00026-x

Tayyem, R., Ibrahim, M. O., Mortada, H., AlKhalaf, M., Bookari, K., Al Sabbah, H., Qasrawi, R., Kamel, I., Dashti, S., Allehdan, S., Bawadi, H., Waly, M., Abuhijleh, H., Hammouh, F., Al-Awwad, N., Al-Bayyari, N., Cheikh Ismail, L., Abu Al-Halawa, D., Othman, M., & Hoteit, M. (2022). Sex Disparities in Food Consumption Patterns, Dietary Diversity and Determinants of Self-Reported Body Weight Changes before and Amid the COVID-19 Pandemic in 10 Arab Countries. *Frontiers in Public Health*, *10*. https://doi.org/10.3389/fpubh.2022.1029219

Turki, S., Bouzekri, K., Trabelsi, T., & El Ati, J. (2022). Impact of COVID-19 Lockdown on Food Habits, Appetite and Body Weight in Tunisian Adults. *Journal of Nutritional Science*, *11*, e60. https://doi.org/10.1017/jns.2022.58

UN. (2022). *Report of the Special Rapporteur on Extreme Poverty and Human Rights*. https://www.srpoverty.org/schedule-lebanon-mission/.

UNDG. (2016). *MAPS—A Common Approach to the UNDG's Policy Support to the SDGs—An Update on Implementation*, March.

UNDP. (2023). *Adapting to a New Climate in the MENA Region*. https://www.undp.org/arab-states/publications/adapting-new-climate-mena-region

UNHCR. (2021). *Global Report 2021*. https://reporting.unhcr.org/globalreport2021

United Nations. (2020). *Arab Sustainable Development Report 2020*. https://asdr.unescwa.org/sdgs/pdf/en/ASDR2020-Final-Online.pdf

United Nations. (2021). *The Attainment of SDGs in Conflict-Affected Countries in the Arab Region*. https://www.unescwa.org/sites/default/files/pubs/pdf/attainment-sdgs-conflict-affected-countries-arab-region-english.pdf

United Nations. (2023a). *Global Sustainable Development Report (GSDR) 2023*. https://sdgs.un.org/gsdr/gsdr2023?_gl=1*5e7b84*_ga*MTI0Nzk4NDI0OC4xNjg2NDY1MzUw*_ga_TK9BQL5X7Z*MTY5ODIyMjczMS43LjEuMTY5ODIyMjc2OC4wLjAuMA

United Nations. (2023b). *The Sustainable Development Goals Report 2023 Special Edition*. https://unstats.un.org/sdgs/report/2023/

United Nations General Assembly. (2015). *Transforming Our World: The 2030 Agenda for Sustainable Development*. https://sdgs.un.org/2030agenda

United Nations Statistics Division. (2023). *Cereal Import Dependency Ratio*. http://data.un.org/Data.aspx?q=dependency+ratio&d=FAO&f=itemCode%3A21035

WHO. (2023). *WHO Coronavirus Disease (COVID-19) Dashboard*. https://covid19.who.int/

Wood, D., Boswall, J., & Halabi, S. (2020). *The Empty Plates and Pockets of Lebanon: The Fight to Prevent Hunger Amidst Pandemic and Recession*. http://www.thinktriangle.net/wp-content/uploads/2020/05/Going-Hungry-The-Empty-Plates-and-Pockets-of-Lebanon.pdf

World Bank. (2018). *Beyond Scarcity: Water Security in the Middle East and North Africa*. The World Bank. https://doi.org/10.1596/978-1-4648-1144-9

World Bank. (2021). *Distributional Impacts of COVID-19 in the Middle East and North Africa*. https://blogs.worldbank.org/arabvoices/covid-19-sets-mena-back-poverty

World Bank. (2023). *The World Bank Data—Population, Total*. World Bank, December 9. https://data.worldbank.org/indicator/SP.POP.TOTL

World Bank. (2024). Lebanon Poverty and Equity Assessment 2024—Weathering a Protracted Crisis. In *Lebanon Poverty and Equity Assessment 2024—Weathering a Protracted Crisis*. World Bank. https://doi.org/10.1596/41619

Yuan, H., Wang, X., Gao, L., Wang, T., Liu, B., Fang, D., & Gao, Y. (2023). Progress towards the Sustainable Development Goals Has Been Slowed by Indirect Effects of the COVID-19 Pandemic. *Communications Earth & Environment*, *4*(1), 184. https://doi.org/10.1038/s43247-023-00846-x

Zhou, X.-Y., Lu, G., Xu, Z., Yan, X., Khu, S.-T., Yang, J., & Zhao, J. (2023). Influence of Russia-Ukraine War on the Global Energy and Food Security. *Resources, Conservation and Recycling*, *188*, 106657. https://doi.org/10.1016/j.resconrec.2022.106657

4 Urban Heat, District Cooling, and Thermal Storage

G7/G20 Tools to Triple Renewables and Cut Peak Demand in the UAE and MENA

Gowhar Meraj

4.1 INTRODUCTION

Climate change and sustainable development are now inseparable agendas. This is especially evident in the Middle East and North Africa (MENA), where surging electricity demand, largely for space cooling in heat-stressed cities, intersects with global mitigation goals. Conference of Parties (COP)-28 in the United Arab Emirates (UAE) marked a significant shift where nearly 200 parties endorsed targets to triple renewable power capacity and double the rate of energy efficiency improvements by 2030, a goal subsequently echoed by the group of seven countries (G7: Canada, France, Germany, Italy, Japan, the United Kingdom, and the United States) (Tenedios et al., 2018) and group of 20 countries (G20: Argentina, Australia, Brazil, Canada, China, France, Germany, India, Indonesia, Italy, Japan, Mexico, Russia, Saudi Arabia, South Africa, Republic of Korea, Türkiye, United Kingdom, United States, European Union, and African Union, permanent member since 2023) (Rwigema, 2023). The companion Global Renewables and Energy Efficiency Pledge links these goals to SDG outcomes via job creation, energy security, and improved livelihoods (UAE COP28). For the MENA region, and the water-scarce Gulf, in particular, the stakes are immediate (Bhattarai and Yousef, 2025). Regional electricity use has already tripled since 2000. Under current policies, it could rise by roughly half again by the mid-2030s, with air conditioning, desalination, and allied cooling loads driving a substantial share of the increase.[1] Without a rapid pivot in supply and end-use efficiency, cooling-driven peaks will entrench higher emissions and system costs through the 2030s (Pohl et al., 2023). Put simply, the development and climate trajectories of the MENA region will be shaped by how it meets its cooling needs in a warming world.

A different pathway is emerging. The UAE and several of its neighbours are seeking to electrify and decarbonise their supply via rapid solar and wind deployment, while structurally reducing cooling demand through efficiency standards, building design, district cooling, and urban planning. This chapter examines how the COP28 UAE Consensus, which calls for tripling renewables and doubling efficiency, can serve as an essential step for climate mitigation and SDG progress in the MENA region. It first profiles regional demand and the cooling challenge (Section 4.2). It then articulates a 2030 plan that integrates clean-supply expansion with end-use efficiency (Section 4.3). Persistent hindrances, such as financing, regulatory standards, technology diffusion, and Monitoring, Reporting, and Verification (MRV), are analysed with corresponding G7/G20 levers, including de-risked capital, harmonised efficiency and refrigerant standards, green procurement, and trade facilitation (Section 4.4). Implementation is organised around an integration spine that couples remote-sensing MRV, national cooling action plans, demand response,

 DOI: 10.1201/9781003600152-4

and a blended finance ladder capable of crowding in private investment at scale (Section 4.5). UAE-anchored case studies, district cooling and building efficiency (Dubai and Abu Dhabi) and blue carbon initiatives that link coastal restoration to urban cooling and resilience, illustrate the approach in practice (Section 4.6). As Figure 4.1 shows, treating cooling as core infrastructure, backed by clean-power expansion, G7/G20 levers, and an implementation spine, turns rising heat-driven demand into a virtuous pathway that advances SDGs across water, ecosystems, health, energy, climate, cities, and jobs.

Throughout this chapter, a policy-oriented frame is employed, drawing on recent international assessments and UAE and MENA policy sources. The central claim is that cooling must be treated as infrastructure, not merely as a series of appliances. Aligning efficient, system-level cooling with an accelerated clean-power buildout yields a climate–SDG multiplier based on cleaner air and lower bills (SDGs 3, 7, 11), more resilient water systems (SDG 6), green employment and innovation (SDG 8), and sustained mitigation (SDG 13), while safeguarding terrestrial and marine ecosystems (SDGs 14–15). The concluding sections situate these findings within the context of G7/G20 cooperation, arguing that the UAE's COP28 legacy offers a transferable template for the MENA region and other emerging regions (Sections 4.7 and 4.8).

4.2 MENA'S SURGING DEMAND AND THE COOLING CHALLENGE

The MENA energy landscape is characterised by rapid load growth under extreme climatic stress. Since 2000, regional electricity use has tripled (~1,440 TWh in 2023; ~5% per annum), driven by population and income growth, urban–industrial expansion, energy-intensive desalination, and, above all, space cooling in cities where summer maxima routinely exceed 45°C. The International Energy Agency (IEA) estimates that ~40% of the additional demand to 2030 will arise from cooling and desalination, highlighting how "heat- and water-stress" directly translates into increased electricity demand (IEA, 2025). As summarised in Table 4.1, ~40% of added demand to 2030 comes from cooling and desalination, with air conditioning dominating GCC peaks.

Cooling is reshaping hourly load profiles and straining system adequacy. In GCC states, cooling dominates peak demand. Saudi residential electricity use exceeds 70% for cooling in hot months (Howarth et al., 2020), the UAE's peak demand is approximately 60% AC-driven (DEWA, 2013), and Qatar's summer peak approaches 70%, with approximately 50% of annual demand driven by air conditioning (Darwish, 2013). Because peaks cluster in hot afternoons and evenings, utilities must finance capacity utilised for only a few hundred hours each year, an expensive equilibrium. Without intervention, climate-driven heat extremes will amplify these peaks through the 2030s; some projections indicate that GCC cooling demand could triple by 2030 compared to 2010 under business-as-usual scenarios (Eveloy and Ayou, 2019; IEA, 2025).

Compounding the challenge, more than 90% of MENA power is still oil- and gas-fired,[2] so incremental load raises greenhouse gas and local pollutant emissions while imposing an opportunity cost on domestic hydrocarbons. Even with a shift from oil to natural gas, methane leakage and fuel-supply constraints persist. In a delayed-transition scenario, IEA analysis suggests more than 25% growth in oil and gas used for power by 2035, eroding export revenues, which in turn raises the macroeconomic stakes (IEA, 2025).

The largest near-term relief lies on the demand side. The average AC sold in the MENA region operates at less than half the efficiency of its Japanese counterparts, indicating substantial gains from the tightening of Minimum Energy Performance Standards (MEPS), accelerated stock turnover, and improved building envelopes. Raising AC efficiency alone could avert peak additions on the order of 35 GW by 2035, equivalent to Iraq's current installed capacity, effectively creating a virtual power plant (IEA, 2025). Since 2013, the UAE, Saudi Arabia, Qatar, and Kuwait have introduced MEPS and updated building codes, including insulation, glazing, and solar water heating (Dubey, 2016). Nevertheless, enforcement gaps and a long tail of inefficient units persist.

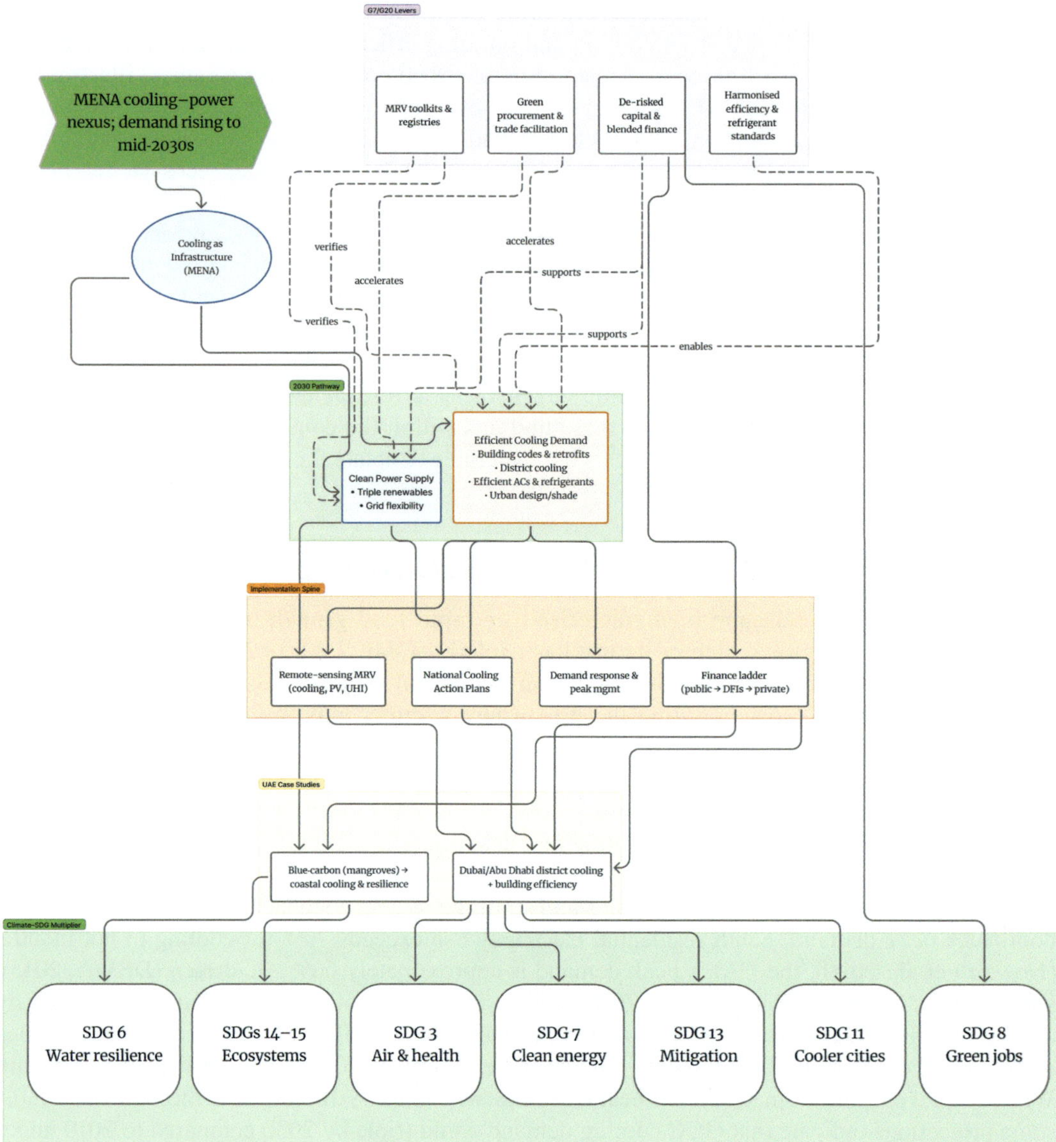

FIGURE 4.1 MENA cooling power nexus. The diagram summarises the chapter's logic: treat cooling as core infrastructure and couple it with rapid clean-power expansion, turning a heat-driven vicious cycle into a virtuous one aligned with the COP28 "tripling renewables/doubling efficiency" trajectory. From the top, G7/G20 levers, MRV toolkits and registries, green procurement and trade facilitation, de-risked/blended finance, and harmonised efficiency/refrigerant standards, enable a 2030 pathway that pairs clean-power supply (triple renewables, grid flexibility) with efficient cooling demand (building codes/retrofits, district cooling, efficient ACs/refrigerants, urban design/shade). An implementation spine, remote-sensing MRV, National Cooling Action Plans, demand response/peak management, and a public–DFI–private finance ladder, connects strategy to delivery. UAE case studies (Dubai/Abu Dhabi district cooling + building efficiency; blue carbon mangroves) illustrate practice. The result is a climate–SDG multiplier spanning SDG 6 (water), SDGs 14–15 (ecosystems), SDG 3 (health), SDG 7 (clean energy), SDG 13 (mitigation), SDG 11 (cooler cities), and SDG 8 (green jobs).

Source: Author's synthesis, adapted from literature sources discussed in this chapter.

TABLE 4.1
The Cooling Power Nexus: Rapid Demand Growth, AC-Driven Peaks, and >90% Fossil-Fired Generation, Alongside Efficiency Potential of ~35 GW Peak Avoidance by 2035

Metric	Value	Region	Source
Regional electricity use	Tripled since 2000; ~1,440 TWh (2023); ~5% per annum	MENA, 2000→2023	IEA (2025)
Added demand due to cooling + desalination	~40% of additional demand to 2030	MENA, to 2030	IEA (2025)
Saudi residential cooling share	>70% in hot months	Saudi Arabia	Howarth et al. (2020)
UAE peak that is AC-driven	~60%	United Arab Emirates	DEWA (2013)
Qatar peak and annual AC share	~70% summer peak; ~50% annual demand	Qatar	Darwish (2013)
GCC cooling demand projection	Could triple by 2030 vs 2010 (BAU)	GCC, 2010→2030	Eveloy and Ayou (2019); IEA (2025)
Fossil use in delayed transition	>25% growth in oil & gas for power by 2035	MENA, to 2035	IEA (2025)
Peak avoided via AC efficiency	~35 GW by 2035 (≈ Iraq's installed capacity)	MENA, to 2035	IEA (2025)
Policy steps	MEPS & building codes since 2013	UAE, Saudi Arabia, Qatar, Kuwait	Dubey (2016)

Source: Author's synthesis, adapted from literature sources discussed in this chapter.

A comprehensive approach should pair appliance standards with urban design, passive cooling, and district cooling with thermal storage, exploiting economies of scale and enabling load shifting.

In short, the MENA region stands at an inflexion point. Meeting cooling needs with fossil-centric, device-by-device responses will lock in higher emissions, fiscal exposure, and air-quality burdens. An integrated strategy, tripling renewables and doubling efficiency by 2030 (Section 4.3), targeted de-risking and standards harmonisation via G7/G20 (Section 4.4), and an implementation spine built on MRV, demand programmes, and blended finance (Section 4.5), can convert the cooling challenge into a triple dividend: reliable service, lower emissions, and system-wide cost savings. The UAE's emerging practices, such as district energy and blue carbon co-benefits, provide a concrete testbed and inform the forward policy agenda.

4.3 TRIPLING RENEWABLES AND DOUBLING EFFICIENCY: A 2030 GAME PLAN

Recognising the need to reduce emissions while sustaining growth, the COP28 UAE Consensus centres on two quantitative mid-term targets: tripling global renewable power capacity by 2030, from ~3.5 TW in 2022 to around 11 TW,[3] and doubling the global rate of energy efficiency improvement by 2030, that is, faster declines in energy intensity (UNFCCC, 2023; RENA et al., 2024). Although global, these targets are particularly salient for the MENA region, implying a rapid scaling up of solar and wind energy alongside comprehensive efficiency programmes in buildings and industry. The premise is dual: essential for 1.5°C alignment and catalytic for socio-economic gains (COP28 UAE, 2023). As shown in Figure 4.2, parties endorsed tripling renewables to ~11 TW and doubling efficiency to ~4%/yr by 2030.

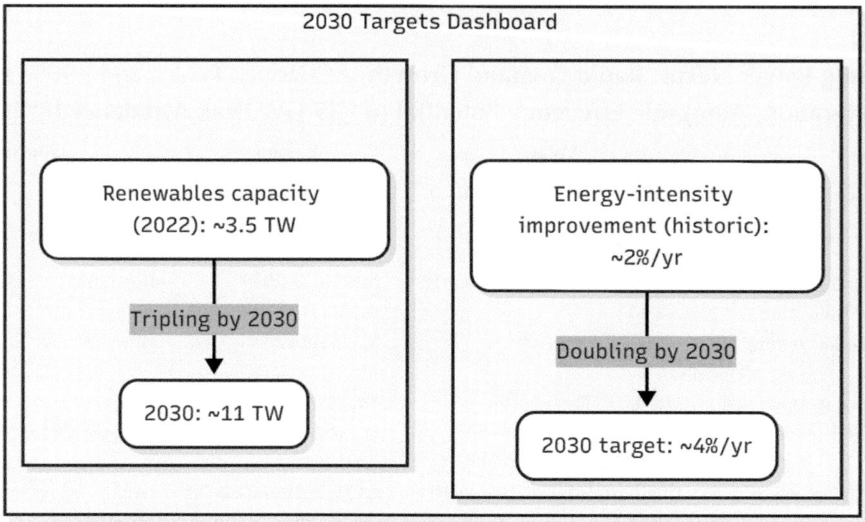

FIGURE 4.2 2030 targets dashboard. Global renewables capacity to be tripled from ~3.5 TW (2022) to ~11 TW (2030), and the global rate of energy-intensity improvement to be doubled from ~2%/yr (historic) to ~4%/yr (2030 target), as reflected in the COP28 UAE Consensus and related pledges.

Source: Author's synthesis, adapted from literature sources discussed in this chapter.

At COP28 in late 2023, Parties incorporated these goals into the global stocktake decision, marking the first collective, quantitative endorsement of renewables and efficiency by nearly 200 Parties. In parallel, the Global Renewables and Energy Efficiency Pledge, now with over 120 signatories, emphasises enabling conditions, including concessional and blended finance, accelerated permitting and grid buildout, technology diffusion, and public awareness (European Commission, 2023). The agenda also implies a managed transition from unabated fossil supply while preserving national technology choices such as renewables, efficiency, nuclear, and carbon abatement (UNFCCC, 2023).

For MENA, tripling by 2030 is ambitious yet increasingly plausible. The region's non-hydro renewable base remains modest, but deployment has accelerated, driven by the growth of utility-scale solar photovoltaic (PV). National plans envision a solar expansion of one order of magnitude by the mid-2030s. Flagship programmes, such as Saudi Arabia's increased 2030 renewables share (World Nuclear Association, 2024), Egypt's long-term targets (Ministry of Electricity & Renewable Energy (MoEE), 2022; UNDP, 2025), and the UAE's rising clean-energy capacity share (MOEI, 2024, 2025; DEWA, 2025a), are underpinned by excellent solar resources, declining costs, and the strategic aim to free hydrocarbons for export. Commissioned assets, such as Al Dhafra PV and Dumat Al Jandal wind, and expanding platforms, like the Mohammed bin Rashid Solar Park, demonstrate that where policy certainty meets investor confidence, multi-gigawatt (GW) pipelines materialise (Masdar, 2023a, 2023b; DEWA, 2024, 2025a). The UAE has paired targets with capital commitments, for example, a 2030 clean-energy investment envelope of AED 150–200 billion, signalling credibility to markets (UAE Government, 2024; U.S. Department of Commerce, 2025; ESG Today, 2024).

On efficiency, doubling implies a roughly 4% annual global energy-intensity improvement, compared to ~2% historically (COP28 UAE, 2023). For Gulf economies with high per-capita use, this is substantial but tractable given the prevailing inefficiency baseline. The UAE's Energy Strategy 2050 explicitly targets 42–45% improvements in individual and institutional energy-consumption efficiency by 2030 (vs 2019) (UAE Government, 2024); Saudi initiatives have already curbed peak growth by several gigawatts through appliance standards and industrial measures (Krarti and Howarth, 2020). Cooling is the linchpin: tighter MEPS and stock turnover for ACs, better envelopes, smart thermostats, and demand response, as well as district cooling where density permits (Eveloy and Ayou, 2019). National Cooling Action Plans (NCAPs) are emerging in the region (e.g., Egypt NCAP launch; Jordan's Cooling Strategy), supported by UNEP's Cool Coalition and the Kigali Cooling Efficiency Programme (now referred as, Clean Cooling Collaborative), with frameworks that extend beyond appliances to urban form, green infrastructure, cool roofs, and the cold-chain (Cool Up, 2022; UNEP/Cool Coalition, 2023; UNDP, 2023; K-CEP, 2019). Framing cooling as infrastructure enables cross-sector coordination.

The synergy is structural. Cleaner supply benefits from flattened load profiles; conversely, electrifying cooling, that is, retiring diesel chillers and legacy window units, delivers climate benefits only if the grid decarbonises apace. Daytime PV can directly drive chillers and charge thermal storage, such as ice or chilled water, for evening peak demand, thereby embedding flexibility into urban energy systems. SDG co-benefits are material. SDG 7 (affordability) thorugh stable solar tariffs and reduced consumption; SDG 3 through lower indoor/outdoor air pollution and reduced heat stress risk; SDG 11 through improved thermal comfort and reliability; SDG 13 through lower power-sector emissions and Kigali-aligned refrigerant transitions, based on transitions aligned with the Kigali Amendment to the Montreal Protocol (phasing down high, Global Warming Potential (GWP) hydrofluorocarbons (HFCs) and shifting to lower-GWP alternatives, with proper leak management, technician training, and end-of-life recovery); SDG 6 through lower water intensity of solar/wind relative to thermal plants; and SDG 8 via labour-intensive deployment and Operations and Maintenance (O&M) in PV, Heating, Ventilation, and Air Conditioning (HVAC), storage, and digital grids. Fiscal co-benefits accrue as efficiency dampens subsidised electricity outlays, creating space for social spending.

Hence, the tripling of renewables and the cooling cities framework underpins the COP28 UAE climate–SDG nexus. Nevertheless, delivery at scale faces financial, regulatory, technical, and data as well as MRV constraints. Sections 4.4 and 4.5 detail the G7/G20 levers, de-risked capital, standards harmonisation in efficiency and refrigerants, green procurement, trade facilitation, and open-data MRV, while Section 4.6 demonstrates operational pathways through UAE case studies.

4.4 BOTTLENECKS AND G7/G20 LEVERS FOR SCALING UP

Despite a rising commitment to accelerating renewables and sustainable cooling in the MENA region, progress is constrained by four interlinked bottlenecks: financing and investment risk, policy and standards, procurement, technology and knowledge transfer, and MRV. The G7/G20, as conveners of capital, standards, and technology, can align incentives and lower execution risk across all four (see Sections 4.5 and 4.6 for implementation pathways). As shown in Figure 4.3, the four bottlenecks align with specific G7/G20 levers: finance, standards/procurement, pilots/training, and open MRV, clarifying how each converts to tangible execution gains in the region.

4.4.1 Financing and Investment Risks

Outside the GCC, high sovereign and currency risks, shallow capital markets, and limited project track records elevate the cost of capital. Even within the GCC, grid, storage, and district cooling projects demand large upfront outlays and credible off-take. COP28 and subsequent G20 statements call for reform of the Multilateral Development Bank (MDB) and the adoption of blended finance to reduce hurdle rates. Vehicles such as the UAE-launched ALTÉRRA, a US$30 billion catalytic climate fund aiming to mobilise US$250 billion by 2030, illustrate this approach (ALTÉRRA, 2023; ADGM, 2023). G7 development finance institutions can add first-loss tranches, guarantees, and political risk insurance, while export credit agencies coordinate de-risking packages for

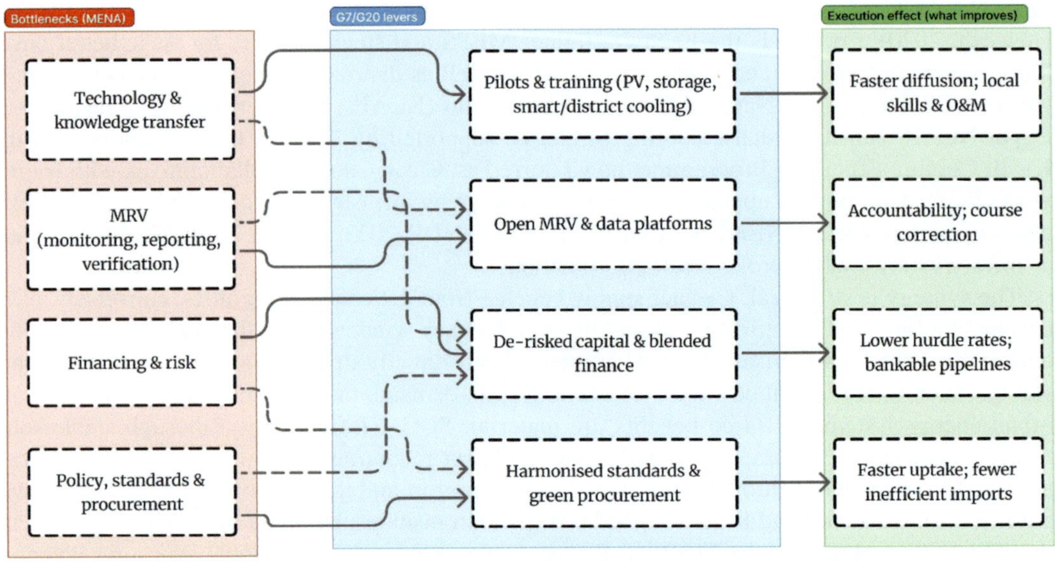

FIGURE 4.3 Bottlenecks and G7/G20 levers for scaling up. Four constraints in MENA, financing and risk, policy/standards and procurement, technology and knowledge transfer, and MRV, map to G7/G20 actions: de-risked capital and blended finance, harmonised efficiency/refrigerant standards with green procurement, city-scale pilots and training, and open MRV/data platforms. The paired levers drive execution effects: lower hurdle rates and bankable pipelines, faster uptake with fewer inefficient imports, quicker diffusion and local O&M skills, and accountability with course correction.

Source: Author's synthesis, adapted from literature sources discussed in this chapter.

clean-power and high-efficiency cooling. Climate funds, for example, the Green Climate Fund, must prioritise cooling and end-use efficiency but with simplified access. In brief, G7/G20 can lead finance innovation, stacking grants, concessional debt, and commercial capital to convert pipelines into bankable portfolios.

4.4.2 Policy Frameworks, Standards, and Procurement

Weak or uneven enforcement of renewable auctions, building codes, and MEPS slows uptake and allows inefficient imports. G7 communiqués (2024) endorse the Global Cooling Pledge; G20 can amplify this via harmonised efficiency classes for ACs and refrigeration, refrigerant management, through Kigali alignment, and urban-cooling strategies (Khosla, 2025). Green public procurement by G7 governments and MDB-funded projects can mandate top-tier cooling efficiency and require district cooling feasibility for large developments, creating a demand-pull for efficient technologies. Technical twinning, for example, with Japan and Germany, can accelerate code design, software tools, and enforcement protocols (Evans et al., 2017; Khosla 2025). On trade, reviving an Environmental Goods Agreement to lower tariffs on PV, wind, controls, and efficient chillers would reduce costs across MENA.

4.4.3 Technology and Knowledge Transfer

High-renewables, cooling-heavy systems need storage, flexible demand, and advanced HVAC. The G7 can fund city-scale demonstrations, such as PV, batteries, and smart cooling and thermal storage, in MENA, such as those in Masdar City, paired with training for technicians and energy managers. Industry platforms, for example, Oil and Gas Climate Initiative (OGCI), provide practice transfer for methane/flare reduction; the G7 can reinforce this by tying import standards to stringent methane MRV, thereby accelerating abatement in regional oil and gas operations (co-benefits for power-sector footprints). These partnerships build the human infrastructure to sustain gains.

4.4.4 Monitoring, Reporting, and Verification

Execution depends on granular, credible data on energy use, emissions, and realised savings. G7/G20 can support national energy data centres, open satellite-enabled MRV (such as methane super-emitters, plant heat signatures, and urban thermal maps), and standardised metrics via platforms like International Renewable Energy Agency (IRENA) (IRENA, 2025) and Climate-TRACE (Tracking Real-time Atmospheric Carbon Emissions, a coalition that compiles independent, near–real-time emissions estimates using satellites, AI, and other data) (Climate TRACE, 2024a, 2024b). This enables the transparent tracking of tripling or doubling in 2025 NDC updates and toward the 2028 stocktake and triggers corrective action, such as dispatching Leak Detection and Repair teams or targeting heat island retrofits (European Commission, 2025). The 2024 G7 emphasis on Oil and Gas Methane Partnership, OGMP 2.0 and Kigali ratification aligns refrigerants, efficiency, and MRV, key to sustainable cooling.

These bottlenecks are tractable if addressed in a coherent manner. The G7/G20 can (i) unlock affordable capital at scale; (ii) raise and harmonise standards for appliances, buildings, refrigerants, and leverage procurement; (iii) diffuse technologies and skills through joint pilots and training; and (iv) institutionalise MRV for accountability. As a COP28 host and G20 invitee, the UAE has advocated for transitioning from billions to trillions and for collective cooling targets.

4.5 IMPLEMENTATION: DATA, DEMAND, AND FINANCE—AN INTEGRATION SPINE

Delivering a tripling of renewables and a step-change in cooling efficiency by 2030 is primarily an execution problem. A practical spine comprises three pillars: (1) advanced MRV, (2) demand-side programmes and policies, and (3) a blended finance ladder to move projects from concept to scale. In short, precisely measuring demand, mobilising capital, and being underpinned by international cooperation and transparent accountability, is the key.

4.5.1 Remote Sensing and MRV for Climate and Cooling

Given compressed timelines, near-real-time data are indispensable.

4.5.1.1 Macro-MRV

Public and private satellites already identify methane super-emitters and observe power plant heat signatures; forthcoming CO_2 missions will refine the attribution of these plumes (Schuit et al., 2023; Wei et al., 2023). Integrating these streams with national inventories enables rapid abatement (for example, pipeline repairs) and cross-checks reported generation/emissions (Section 4.4 on MRV standards).

4.5.1.2 Micro-MRV

High-resolution thermal imagery (from satellites/UAVs) maps urban heat islands (UHIs) at the building scale, guiding cool-roof retrofits, shading/greening, or district cooling connections. Smartmeters/IoTs provide time-of-use data to verify precooling and demand-response outcomes. Dubai Electricity and Water Authority (DEWA)-style smart-meter rollouts illustrate the enabling data layer (DEWA, 2025b).

4.5.1.3 Institutionalisation

A joint Cooling and Renewables Watch (IEA–IRENA–UNEP) updated annually would benchmark progress toward tripling or doubling and the Global Cooling Pledge, identifying shortfalls early and spotlighting replicable success (cf. Section 4.3). The objective is adaptive management: detecting deviations, adjusting policies and investments, and verifying impacts.

4.5.2 Demand-Side Programmes: NCAPs, Efficiency, and Smarter Grids

Managing demand is coequal with adding supply.

4.5.2.1 NCAPs Are the Umbrella

NCAPs should set targets for cooling energy intensity (kWh/m²), AC stock turnover, district cooling penetration, and refrigerant management, integrated with power planning. Jordan's NCAP demonstrates the template; the UNEP Cool Coalition[4]/K-CEP supports scaling across the MENA region (MoEnv Jordan, 2024).

4.5.2.2 District Energy and Thermal Storage

In dense loads, district cooling performed by Dubai's Empower[5] and Abu Dhabi's Tabreed[6] substantially reduces cooling electricity use and enables load shifting: producing and storing coolth during midday, that is, PV surplus, or at night, and dispatching it at evening peaks (Yang, 2024). Planning levers include mandated feasibility assessments for large developments, pipe-corridor rights-of-way, and connection-friendly tariffs.

4.5.2.3 Codes, MEPS, and Incentives

Predictable MEPS ratchets for ACs and refrigeration, as well as strengthened building envelopes, remain foundational; G7 harmonisation reduces technology dumping. Utilities can accelerate uptake via rebates, on-bill finance, and swap programmes for legacy units.

4.5.2.4 Digital Demand Response

With subsidy reforms restoring price signals, time-of-use tariffs, automated demand response, and smart thermostats can choreograph cooling loads with variable renewables, precool when PV is abundant, and coast through peaks. Operator twinning, for example, between Italy and/or Canada with Gulf utilities, shall expedite the transfer of capability (Kaczmarski et al., 2022).

4.5.3 The Blended Finance Ladder

Finance must match project granularity and risk over time.

4.5.3.1 Aggregation and Super-Energy Service Company

Many cooling retrofits are sub-scale. Super-Energy Service Companies (ESCOs), for example, Dubai's Etihad ESCO, bundle projects with standardised contracts and M&V, seeded by concessional tranches/guarantees; once savings stabilise, portfolios can be refinanced commercially (Section 4.4 on de-risking).

4.5.3.2 Utility-Scale Renewables

Typical steps include converting feasibility grants into blended construction finance (MDB/Export Credit Agencies debt plus private equity), followed by post-COD refinancing via green bonds and YieldCos (Yield Companies, companies that own operating, cash-generating infrastructure assets and pay out dividends), to recycle public capital. Deepening local markets through green and sustainability sukuk[7] broadens investor pools.

4.5.3.3 Consumer Finance

Bill-neutral instalment plans or green micro-credit close affordability gaps for high-efficiency ACs/envelope upgrades; partial guarantees crowd in local banks.

4.5.3.4 Innovation Pipeline

Early-stage R&D, for example, non-HFC refrigerants, solar-thermal or advanced heat pumps, through Mission Innovation–style grants; pilots co-funded by impact and venture capital; manufacturing scale-up supported by procurement and blended capital. Prize mechanisms, for example, the Global Cooling Prize, should be followed by commercialisation finance.

4.5.3.5 Governance and Coordination Bind the Spine

The UAE's National Net Zero by 2050 Platform,[8] which is multi-agency, exemplifies whole-of-government alignment across solar parks, codes, EVs, and R & D; analogues in other MENA states, supported by G7/G20 expert secondments and institutional twinning, can accelerate execution.

With rigorous MRV, demand-side discipline, and fit-for-purpose finance, systems enter a virtuous cycle: moderated peak growth lowers integration costs; falling technology prices reinforce adoption; and fiscal savings from efficiency fund further measures.

4.6 UAE CASE STUDIES: COOLING AND RENEWABLES IN ACTION

To ground the discussion, we examine two UAE-anchored case studies that exemplify the climate–SDG integration approach: (a) urban district cooling coupled with efficiency in Dubai/Abu Dhabi, and (b) coastal blue carbon ecosystem restoration linked to cooling and resilience. These cases demonstrate how the principles discussed are electrifying and decarbonising supply, while aggressively reducing and managing demand come together in practice. The UAE's experience as a regional innovator offers a template that could be adapted across other MENA cities and indeed in any hot-climate urban centres globally.

4.6.1 Case 1: District Cooling and Building Efficiency—Dubai and Abu Dhabi

Traditional air conditioning in hot climates often means that each building runs its own chiller or package units, which is akin to every house having a small diesel generator for power, which is inefficient and duplicative. The UAE has instead increasingly treated cooling as shared infrastructure. In Dubai, the Emirates Central Cooling Systems Corporation, commonly known as Empower, was established in 2003 as a public–private partnership backed by the Dubai government to develop and operate district cooling systems.[9] Two decades on, Empower is the largest district cooling services provider in the world, supplying cooling to landmarks such as Downtown Dubai, Business Bay, and the Palm Jumeirah through a network of plants and chilled water pipes. Its Business Bay mega-project achieved two Guinness World Records for scale, with an ultimate capacity of over 450,000 refrigeration tonnes (RT) planned, which is roughly equivalent to replacing tens of thousands of individual building chillers with a far more efficient centralised system.[10] District cooling plants utilise large, industrial-grade chillers that have higher coefficients of performance than small units, and they can optimise load across multiple buildings (Eveloy and Ayou, 2019). According to an analysis by the Atlantic Council's Arsht-Rock Resilience Centre well-designed district cooling can cut electricity use and emissions by up to 50% compared to conventional decentralised AC.[11] It also reduces the UHI effect because heat is dissipated in centralised plants, often by cooling towers or even seawater in some coastal systems, rather than every building rejecting heat into the city air.

Abu Dhabi's Tabreed has a similar story. As the Middle East's first district cooling firm, founded in 1998, Tabreed now delivers about 1.5 million RT of cooling across 86 plants in the UAE and GCC, serving iconic developments like Abu Dhabi's Al Maryah Island and Yas Island, with its theme parks.[12] The government of Abu Dhabi has been a major shareholder and driver, seeing district cooling as both an energy efficiency measure and a business opportunity. Tabreed is publicly listed and profitable.[13] Policy has supported it. Abu Dhabi's Department of Energy set a goal for 30% of the Emirate's cooling demand to be met by district cooling by 2030, up from roughly 20% today.[14] New large developments are often required or incentivised to connect to a district cooling system rather than install standalone chillers.

The integration of renewables with district cooling is now on the agenda. One idea piloted in Dubai is using excess solar PV output during midday to produce chilled water and store it in thermal storage tanks for later use in peak times, effectively storing solar energy as cooling energy. Empower and DEWA have collaborated on such concepts, as Dubai aims to meet 100% of its desalinated water needs and a significant share of its cooling requirements via clean energy.[15,16] On the building side, the UAE has also pushed green building standards that complement district cooling. Dubai's Green Building Regulations (Al Sa'afat) and Abu Dhabi's Estidama Pearl Rating system both mandate improved insulation, efficient glazing, LED lighting, and low-power-density HVAC designs for new buildings.[17] These reduce the cooling load a building requires in the first place, by cutting waste and heat gains, making the district cooling network's job easier and allowing smaller pipe and plant sizes than otherwise. Retrofitting existing buildings is another key initiative. In Dubai, Etihad ESCO has retrofitted over 7,000 buildings by 2020,[18] implementing measures such as improved chilled water controls, variable-speed drives, and efficient lighting, which have yielded substantial electricity savings, often 20–30% per building. This not only reduces emissions but also saves money for building owners through lower utility bills, supporting SDG 7's affordability aspect.

From a policy integration perspective, the UAE case shows multiple agencies working in concert. Municipal planners ensure new urban developments consider district cooling at the design stage. Utility regulators structure tariffs to encourage efficiency. For example, Dubai has a slab tariff, where high consumption incurs a higher rate per kWh, encouraging consumers to adopt conservation measures. The federal government, through initiatives such as the UAE Energy Strategy and the Net Zero by 2050 pathway (UAE Embassy (USA), 2024), sets high-level targets that filter down, for example, the goal to reduce emissions from power/water by 100% (net zero) by 2050, along with interim goals for 2030, including tripling renewable energy and slashing power-sector emissions.[19] These targets implicitly rely on widespread cooling efficiency because one of the only ways to reduce power-sector emissions fast, aside from adding renewables, is to slow demand growth so that new clean supply is not outpaced by rising usage.

Finally, the SDG outcomes in this case are tangible. Residents in district-cooled buildings benefit from more reliable cooling and often lower long-term cooling costs, which contribute to SDG 11 (sustainable cities) by improving comfort and reducing strain on the power grid, resulting in fewer outages. By using electricity more efficiently, local air pollutant emissions from power plants are avoided, benefiting health (SDG 3). The utilities burn less gas or oil, lowering carbon emissions (SDG 13) and conserving resources. Local employment is also created, as Empower and Tabreed employ thousands of engineers, technicians, and operators, and have spurred the local manufacturing of equipment such as insulated pipes and chillers, supporting SDG 8 (decent work and economic growth). These companies are now exporting their expertise abroad. For example, Empower has consulted on projects in Egypt, Tabreed has concessions in Oman, Bahrain, and recently in India, showcasing a new high-tech service export for the region.[20] As shown in Figure 4.4, the UAE's district cooling model, anchored by Empower and Tabreed, combines high-COP plants, PV-thermal storage, and building-efficiency standards with supportive policy and delivery mechanisms, yielding reliability, affordability, and emissions gains.

Thus, the UAE's cooling sector transformation, treating it as core urban infrastructure to be optimised and supplied efficiently, provides a blueprint for other cities facing cooling demand explosions. The key is a mix of technology (such as district systems and efficient buildings), policy (in terms of standards and targets), and business models (such as public–private utilities and ESCOs), underpinned by the increasingly green electricity that feeds those cooling systems.

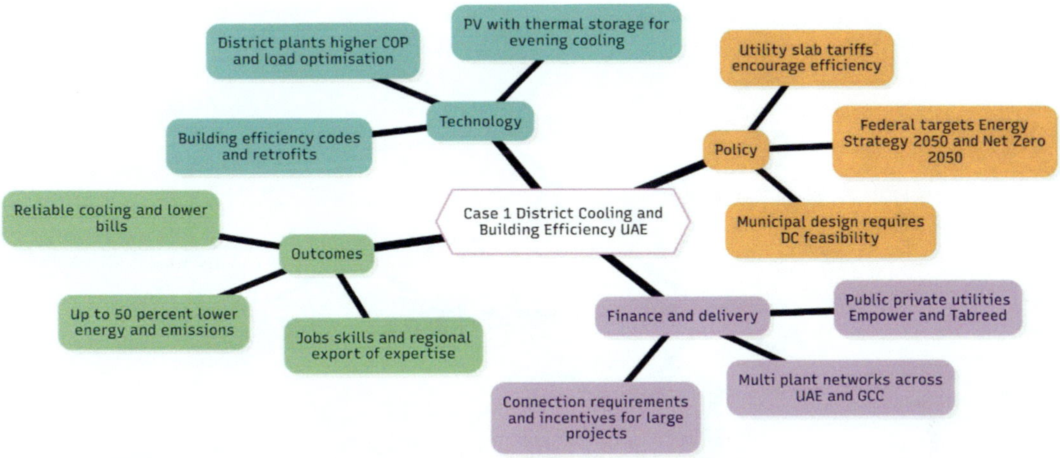

FIGURE 4.4 Mind map, Case 1: District cooling and building efficiency in Dubai and Abu Dhabi. Map groups the case study into four strands: Technology (high-COP district plants; PV-charged thermal storage for evening cooling; building-efficiency codes and retrofits), Policy (municipal DC-feasibility at design, utility slab tariffs, federal Energy Strategy 2050/Net Zero 2050), Finance & delivery (public–private utilities Empower and Tabreed; connection requirements; multi-plant networks across the UAE/GCC), and Outcomes (up to 50% lower energy/emissions, more reliable cooling and lower bills, jobs/skills and export of expertise).

Source: Author's synthesis, adapted from literature sources discussed in this chapter.

4.6.2 Case 2: Blue Carbon and Urban Cooling Synergies—Mangrove Coastal Ecosystems

While energy and technology solutions are critical, the UAE has also turned to nature-based solutions (NbS) to complement its climate strategy, notably through its championing of blue carbon ecosystems, such as mangrove forests (Ocean Conservancy, 2023). Mangroves in the UAE and the wider MENA coasts play multiple roles: they sequester significant amounts of carbon dioxide, protect shorelines from erosion and storm surges, and create microclimates that can have a cooling effect on surrounding areas. The UAE's leadership has committed to planting 100 million mangroves by 2030[21] as part of its second Nationally Determined Contribution, and launched the Abu Dhabi Mangrove Initiative in 2022 to be a global hub for mangrove research and innovation (Aljneibi, 2023). As shown in Figure 4.5, the UAE's blue carbon strategy couples large-scale mangrove restoration with MRV and finance mechanisms, delivering urban cooling and resilience co-benefits that advance SDGs 13/14/15/11.

Abu Dhabi, in particular, has a long history of mangrove conservation dating back to the country's founding, when Sheikh Zayed initiated efforts in the 1970s. In recent decades, the Emirate has expanded its mangrove cover by 6,400 hectares, planting over 20 million mangrove seedlings, including through innovative methods such as the drone-dispersed seeds.[22] Studies have demonstrated that Abu Dhabi's mangroves capture carbon at a rate up to four times higher than tropical rainforests (Alsumaiti and Shahid, 2019). This carbon bank locks CO_2 in biomass and waterlogged soils for millennia, directly contributing to climate mitigation (SDG 13) and making progress toward the UAE's net-zero and nature-positive commitments. Importantly, mangroves also cool their local environment. Al-Huqail et al. (2024) have found that expanding mangrove cover led to a significant decrease in land surface temperatures in those areas, demonstrating that increased mangrove coverage significantly contributes to local cooling. This cooling occurs through shading and evapotranspiration. Mangroves and vegetation in general reflect more sunlight than bare ground or concrete, and they release moisture that cools the air.

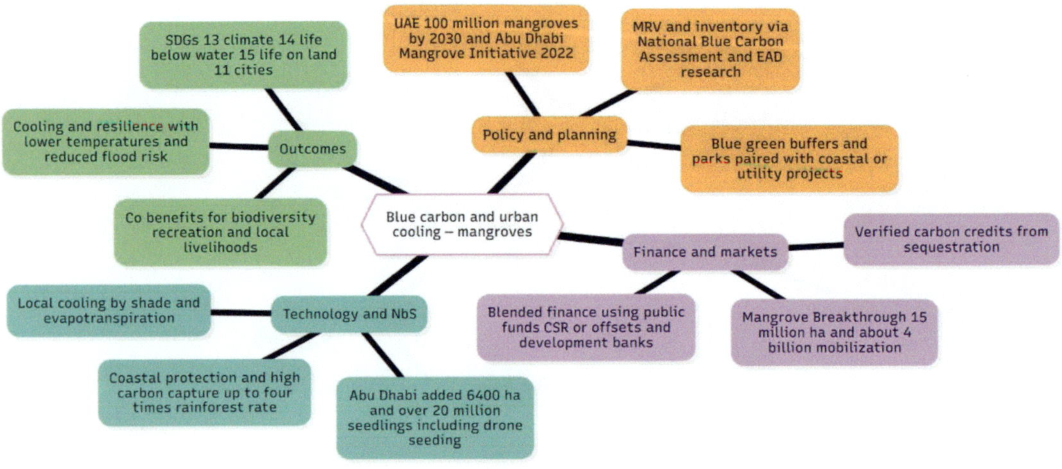

FIGURE 4.5 Mind map—Blue carbon and urban cooling (mangroves). Four strands link the UAE's mangrove programme to cooling and resilience: Technology/NbS (restoration and expansion near cities, incl. Abu Dhabi's +6,400 ha and >20 million seedlings with drone seeding; local cooling via shade/evapotranspiration; coastal protection and up to 4× tropical-rainforest carbon capture), Policy & planning (UAE pledge to plant 100 million mangroves by 2030; Abu Dhabi Mangrove Initiative 2022; MRV and inventory via the National Blue Carbon Assessment and EAD research), Finance & markets (Mangrove Breakthrough: ~15 million ha and ~US$4 billion mobilisation; verified carbon credits; blended finance), and Outcomes (lower local temperatures and flood risk; SDGs 13, 14, 15, 11; biodiversity/recreation/livelihood co-benefits).

Source: Author's synthesis, adapted from literature sources discussed in this chapter.

In urban contexts, strategic conservation or planting of mangroves near coastal developments can thus help moderate the heat island effect at the fringe of cities. For example, Abu Dhabi city has protected mangrove sanctuaries, such as the Eastern Mangrove Lagoon National Park, adjacent to the city's downtown area, which acts as a green (or blue-green) buffer that not only shields the coast from waves but also cools the adjacent urban neighbourhoods to some extent (Frey et al., 2009). A healthier mangrove belt can lower ambient temperatures, providing natural cooling that complements engineered solutions. Moreover, public access mangrove parks, such as Jubail Mangrove Park in Abu Dhabi, enhance urban quality of life and promote awareness of nature, aligning with SDG 11 (inclusive, green public spaces).

The resilience co-benefits are substantial. Mangroves are often referred to as the guardians of the coast for their ability to reduce flood risks. Their dense roots and biomass break wave energy, studies worldwide show that mangroves can significantly reduce the height of storm surges, protecting communities and infrastructure behind them (Van Hespen et al., 2023). In the UAE, where much critical infrastructure, such as cities, desalination plants, and ports, is at sea level, this is a necessity. It has been reported that coastal cities face dual threats of heat and rising seas, and cites mangrove restoration as a cost-effective defence that also provides habitat benefits. Indeed, nature-based approaches often have a better cost-benefit ratio than hard engineering for flood control, while delivering multi-fold co-benefits. From an SDG perspective, restoring mangroves contributes to achieving SDG 14 (Life Below Water) by nurturing fisheries and biodiversity, SDG 15 (Life on Land) through ecosystem conservation, and SDG 13 via climate mitigation/adaptation.

The UAE has taken this a step further by linking finance to blue carbon. It is helping spearhead the Mangrove Breakthrough with the UN Climate Champions and the Global Mangrove Alliance, an effort to secure 15 million hectares of mangroves by 2030 and mobilise ~US$4 billion (Global

Mangrove Alliance (2022). The idea is to generate verified carbon credits from mangrove projects, which can attract corporate or international finance (Choudhary et al., 2024). For instance, a mangrove restoration project can quantify the CO_2 sequestered and sell credits, providing revenue to sustain the project and local communities. While carbon markets are complex, the blending of climate finance with conservation here exemplifies the integrated approach: a blended finance ladder in a different sense, combining public conservation funds, private CSR or offset funds, and potentially development bank support, for example, the Global Environment Facility's Blue Forests project, to scale up mangrove protection.[23]

In an urban planning sense, coupling blue-green infrastructure with cooling targets is a novel approach. Imagine coastal cities like Dubai implementing blue carbon and cooling corridors, protected mangrove parks along the coast, connected to inland green parks, forming a network that cools the city and provides recreation. This is analogous to how some cities plan green belts; here, it is a hybrid green-blue belt. Such measures can reduce local temperatures by several degrees and lower building cooling requirements for nearby districts, allowing AC units to run slightly more efficiently with less heat stress. While quantifying this effect is complex, it reinforces our rationale that cooling is not just an appliance issue but an infrastructure and environmental design issue.

Abu Dhabi's experience offers a model: The Environment Agency, Abu Dhabi (EAD) has integrated mangrove expansion in its climate adaptation strategy (Zavadski, 2023). They are studying the precise impact of mangroves on carbon sequestration and other benefits by establishing a mangrove research facility in collaboration with global partners. Additionally, the UAE's Ministry of Climate Change and Environment, in collaboration with the UNEP, has completed a National Blue Carbon Assessment that catalogues carbon stocks in the UAE's mangroves, salt marshes, and seagrasses (IUCN and AGEDI, 2016). This data helps incorporate these ecosystems into the national greenhouse gas inventory, giving them proper due in climate plans.

The main takeaway from the blue carbon case is that NbS can amplify and complement technical solutions in the context of cooling and climate mitigation. They should be planned jointly. For example, when a coastal power plant or desalination plant is built, incorporating a mangrove restoration component nearby can help offset some of its emissions and provide cooling water with reduced thermal pollution, as mangroves improve water quality. Additionally, as cities consider protective measures against sea level rise, planting mangroves and restoring wetlands is often a more cost-effective and sustainable alternative to building concrete seawalls. G7/G20 support can encourage such approaches by counting NbS in climate finance commitments and sharing expertise on their implementation.

Looking beyond the UAE, other MENA countries with coastlines, such as Saudi Arabia, Qatar, Bahrain, Egypt, and Morocco, could initiate similar programmes. For instance, Saudi Arabia has a Vision 2030 goal to rehabilitate its Red Sea mangroves. Another study from Al-Wajh Lagoon, KSA, demonstrated significant cooling effects from mangrove expansion. So the science and policy stars are aligning to make blue carbon a part of climate–SDG integration (Al-Huqail et al., 2024).

In closing this case study section, the examples of district cooling and mangrove restoration may seem disparate; one is high-tech infrastructure, the other is ecosystem management. However, both are integral pieces of a holistic climate strategy. They demonstrate that mitigation and adaptation can be pursued simultaneously, yielding benefits across the entire SDG spectrum. The UAE's willingness to innovate in both realms and share its lessons (hosting global forums and investing abroad) is a positive sign. Through platforms like the G20, which, for the first time in 2023, formally discussed NbS and agreed to accelerate the adoption of renewables. A city in India or West Africa could look to Dubai's cooling systems or Abu Dhabi's mangroves and adapt the ideas to its local context. The next and final section will discuss the broader future scope and policy implications of embracing such integrated approaches, concluding with reflections on what is needed to generalise success across the MENA region and beyond.

4.7 FUTURE OUTLOOK AND POLICY IMPLICATIONS

The chapter's central proposition is infrastructural i.e., treat cooling as core infrastructure and couple it with a rapid transition to clean power. This reframes a potential vicious cycle (i.e., heat leading to more AC, which in turn leads to higher emissions) into a virtuous cycle (i.e., efficient cooling, lower demand, and easier decarbonisation), operationalised through the framework discussed in Section 4.3 and the integration in Section 4.5. As Figure 4.6 illustrates, treating cooling as core infrastructure and activating six policy levers, planning, international support, private innovation, regional cooperation, MRV/governance, and equity, turns a heat-driven vicious cycle into a virtuous one aligned with 2030 tripling/doubling targets and a mid-century net-zero power vision.

4.7.1 Mainstream Cooling in National and Urban Planning

Cooling should be considered alongside electricity, water, and transportation in planning hierarchies. NCAPs, standardised in hot-climate economies, ought to set targets, mandates, and finance pathways, coordinated with power-sector expansion. Urban design instruments can reduce structural

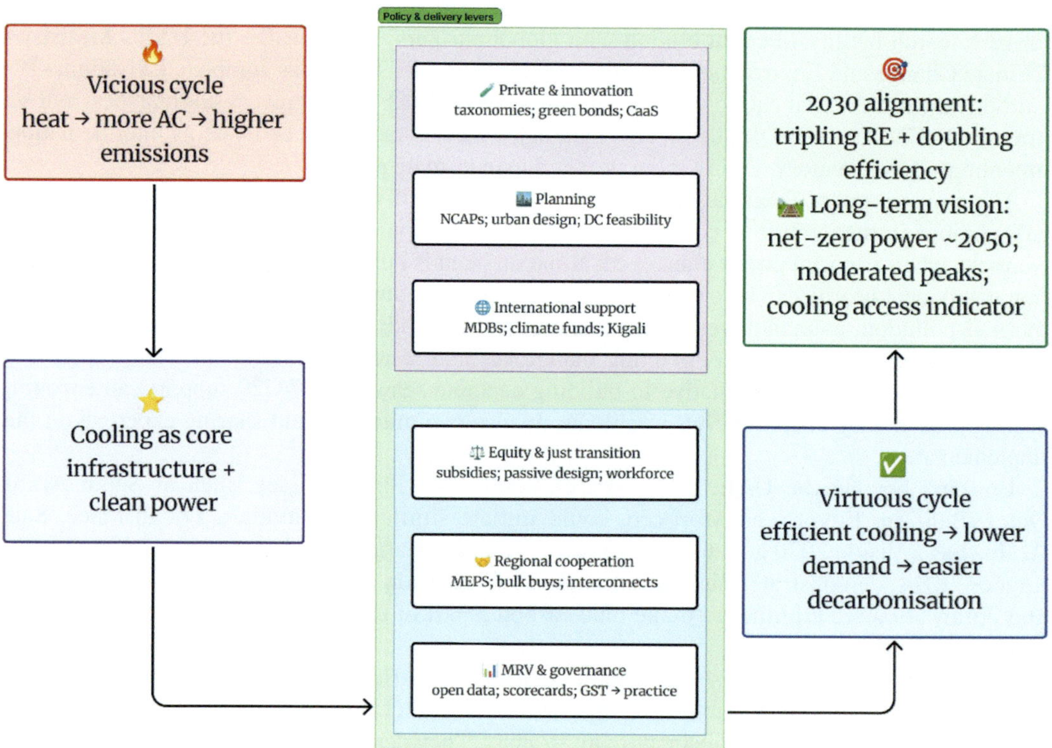

FIGURE 4.6 From vicious to virtuous: policy levers that make cooling an infrastructure win. The diagram shows the shift from a vicious cycle (heat more AC higher emissions) to a virtuous cycle (efficient cooling lower demand easier decarbonisation) when cooling is treated as core infrastructure coupled to clean power. The centre panel groups six concise levers: Planning, International support, Private & innovation, Regional cooperation, MRV & governance, and equity & just transition. The levers align 2030 goals with a long-term vision of near-net-zero power by mid-century, moderated peaks, and a cooling-access indicator.

Source: Author's synthesis, adapted from literature sources discussed in this chapter.

heat loads, including zoning for minimum green/reflective surfaces, mandatory district cooling feasibility for large developments, and long-horizon siting of plants, storage, and pipe corridors.

4.7.2 Align International Support With the Cooling–Renewables Agenda

Multilateral banks and climate funds should explicitly prioritise integrated proposals, for example, solar-linked district cooling, city-scale retrofit programmes with verifiable savings, and thermal storage. A post-COP28 global alliance on sustainable cooling can coordinate donor windows, standard-setting, and progress tracking. Kigali implementation is integral: refrigerant transitions, technician training, and updated codes complement energy efficiency. Blending climate finance with Montreal Protocol resources creates an efficient pooled channel.

4.7.3 Private Sector and Innovation

Delivery depends on market signals. Incorporate cooling into sustainable-finance taxonomies; recognise district energy and building retrofits as green-bond/sukuk-eligible assets; standardise performance contracts. Governments can co-fund R&D partnerships, such as advanced heat pumps, non-HFC refrigerants, and solid-state cooling, and catalyse cooling-as-a-service models that align provider incentives with efficiency. The UAE's innovation platforms, for example, Masdar, position the country as a regional hub.

4.7.4 Regional Cooperation

Regional standards for MEPS and joint bulk procurement can lift the efficiency floor and cut costs, for example, a GCC-wide tender for high-efficiency ACs targeting low-income households. To address supply, expand grid interconnections and flexibility markets to enable surplus solar energy to cross borders, thereby reducing reserves and integration costs.

4.7.5 Monitoring and Governance—Stocktake to Practice

The COP28 global stocktake shifts the agenda to implementation and heightened ambition for 2025 NDCs. Countries should set domestic 2030 targets consistent with tripling/doubling (Section 4.3) and embed annual scorecards/peer review (G7/G20) to track alignment. Open MRV, satellite methane/thermal imagery, standardised energy-intensity metrics, and IRENA/UNEP reporting underpin accountability and mid-course correction.

4.7.6 Equity and Just Transition

Sustainable cooling for all frames is a necessity for both health and productivity. Policy instruments include targeted subsidies/credit for efficient fans and ACs, passive-design standards for social housing, and tariff design that protects basic needs while preserving conservation signals. For hydrocarbon economies, workforce transition pathways from oil and gas to HVAC, renewables, storage, and grid digitalisation are essential; near-term methane abatement through OGCI-aligned practices delivers immediate climate benefits.

4.7.7 Long-Term Vision

Suppose the 2020s deliver on tripling renewable energy and doubling energy efficiency. In that case, the MENA region can approach a net-zero power system by mid-century, with moderated peak demand and high thermal comfort resilience. Efficient, electrified cooling powered by clean grids avoids heat-related health crises while lowering lifecycle costs (Sections 4.3 and 4.6). Planning should adopt cooling access as a development indicator (analogous to electricity access), with a G20 declaration linking sustainable cooling to SDG 3 (health) and SDG 7 (energy for all). Mainstreaming cooling, aligning international finance and standards, mobilising private innovation, and institutionalising MRV can turn the region's most challenging end-use into a lever for climate-aligned development, consistent with the UAE-anchored practice in Section 4.6 and the policy pathways synthesised in Sections 4.3–4.5.

4.8 CONCLUSION

The twin challenges of a warming climate and growing development needs demand integrated solutions. In examining the UAE and MENA context, this chapter has illustrated that cooling and renewables are not separate agendas but deeply interdependent ones. Tripling the renewable energy supply without addressing cooling demand would risk a constant race just to stand still, as unchecked demand growth could consume the new green capacity and necessitate continued fossil fuel backup. Conversely, improving cooling efficiency without clean power would only marginally reduce emissions if the electricity is still generated from fossil fuels. But, together, these strategies form a powerful synergy: electrify and decarbonise the supply, while bending down the demand curve through technology and design.

The UAE's experience, from pioneering massive solar projects and committing to net zero, to rolling out world-class district cooling and championing a global cooling pledge, demonstrates both the feasibility and benefits of this approach. It demonstrates that a hydrocarbon-based economy can revamp its energy system in a relatively short period with the right vision and investment. Regional initiatives, such as efficiency standards and emerging cooling action plans, indicate that momentum is spreading across the MENA region. The involvement of G7 and G20 nations amplifies this momentum by providing the tools, such as finance, standards, and innovation, to accelerate and scale up progress.

To return to the SDG approach which is the transformations described are not only about tonnes of CO_2 or megawatts of capacity, as important as those are but translate into better lives for people. Imagine a Gulf city in 2030 under this paradigm: the air is cleaner because power plants have cut oil use by 95% and eventually phased out unabated coal/oil entirely; the electric grid is stable and largely green, integrating storage and regional trade; on a sweltering August day, citizens still feel cool indoors, but their high-efficiency air conditioners sip electrons frugally, many offices and homes are linked to quiet district cooling systems, and peak power demand no longer spikes uncontrollably. The cityscape is dotted with greenery, including parks and green roofs, and along the coast, mangrove forests buffer the city from storm surges, offering recreational opportunities. Households pay less of their income on energy bills than they would have in a business-as-usual scenario, thanks to efficiency, and many locals have found new careers in the burgeoning clean-energy and sustainability sectors. That is a vision of a climate-resilient, sustainable city, aligning with the SDGs on energy, cities, climate, and ecosystems simultaneously.

Realising this vision will require sustained political will, public–private collaboration, and community engagement. There will be hurdles: incumbent interests, upfront costs, the need to train the workforce, and ensuring equity so that no one is left behind in the transition. However, the direction set by COP28 and related international efforts provides a clear mandate. As UN Secretary-General António Guterres told the G20, we must contribute to the global goals agreed upon at COP28 to triple renewable energy capacity, double energy efficiency, and halt deforestation, all by 2030. The inclusion of those points in the highest diplomatic communiqués is a signal that these are no longer niche issues, they are central to the global climate solution.

In closing, what is different in the approach advocated here is a holistic treatment of cooling as part of the energy system and urban fabric. It calls for breaking silos: energy ministries working with urban planners, environment agencies with finance ministries, international climate negotiators with industry alliances. The UAE's establishment of a high-level climate task force spanning various sectors is a good governance practice. The payoffs of getting this right are immense. By 2050, with continued effort, cooling could become a solved problem, providing comfort without climate guilt, and renewables could be the undisputed backbone of the MENA region's energy, powering not just cooling but a diversified, sustainable economy. This would exemplify the mantra of climate action as an SDG multiplier. The years to 2030 are thus a critical window. With the G7 and G20 utilising their influence to support countries like the UAE in implementing the outlined playbook, we can hope that by the next global stocktake, the world will be well on its way to a cooler, cleaner, and more prosperous future for all.

DECLARATION STATEMENT

The author declares that this research work is original and has duly acknowledged all the sources of information used in the chapter. There are no conflicts of interest.

ACKNOWLEDGEMENTS

The author gratefully acknowledge the organisations that produced and maintain the data reports and supporting datasets used in this chapter. All such sources are cited in the chapter text and reference list. The author also acknowledge the American University of Sharjah for providing full institutional support that enabled the completion of this book contribution. Appreciation is extended to the Japan Society for the Promotion of Science (JSPS) for JSPS fellowship support and for providing a formative opportunity to broaden academic and international perspectives. The author further acknowledge the support and professional exposure provided through The University of Tokyo (UTokyo), UNU-IAS, Tokyo University of Agriculture and Technology (TUAT), Hokkaido University, the Department of Science and Technology (Government of India), and the World Bank, which collectively contributed to the author's academic development and professional trajectory. The views expressed in this chapter are those of the author and do not necessarily reflect the views of the above organisations.

ETHICAL APPROVAL

Not applicable.

CONSENT TO PARTICIPATE

Agree to participate.

CONSENT TO PUBLISH

Agree to publish.

AUTHOR'S CONTRIBUTIONS

Conceptualisation, Methodology, Software, Formal analysis, Investigation, Resources, Writing—original draft preparation, Writing—review and editing: Gowhar Meraj.

FUNDING

This work received no external funding.

AVAILABILITY OF DATA AND MATERIALS

All the data is presented in the chapter.

DECLARATION OF GENERATIVE AI AND AI-ASSISTED TECHNOLOGIES IN THE WRITING PROCESS

During the preparation of this work, the author used an AI service only (PaperPal/Grammarly) to correct grammar and expression errors. After using this service, the author reviewed and edited the content as needed and takes full responsibility for the content of the publication.

NOTES

1 https://www.wam.ae/en/article/blrrd4b-electricity-demand-surges-across-mena-driven
2 https://www.iea.org/data-and-statistics/charts/share-of-fossil-fuels-in-electricity-generation-in-the-middle-east-and-north-africa-by-country-2023
3 https://www.energy-transitions.org/bitesize/cop28-assessment-mitigation-proposals
4 https://coolcoalition.org/about/what-we-do/national-cooling-action-plans
5 https://www.empower.ae/media/e1ullsih/empower_integrated-report_2024_e_08-03-2025.pdf
6 https://tabreed.ae/sites/default/files/2025-03/2020%20ESG%20Report_17%20MAR%202021.pdf
7 https://www.majidalfuttaim.com/en/media-centre/press-releases/detail/2019/05/majid-al-futtaim-lists-world-s-first-benchmark-corporate-green-sukuk-on-nasdaq-dubai
8 https://moccae.gov.ae/en/media-center/news/16/11/2023/climate-neutrality-a-national-priority-as-uae-net-zero-2050-strategy-accelerated-following-the-annou-d34d0d00
9 https://empower.ae/media/w5nidr2b/prospectus.pdf
10 https://www.guinnessworldrecords.com/world-records/671519-largest-district-cooling-plant
11 https://heatactionplatform.onebillionresilient.org/heatactionpolicy/district-cooling-other
12 https://www.tabreed.ae/en/news/tabreed-celebrates-year-collaboration-emaar-and-recent-connection-dunya-towers-downtown-dubai
13 https://www.tabreed.ae/sites/default/files/2025-03/2024_Annual_Integrated_Report_En.pdf
14 https://www.doe.gov.ae/-/media/Project/DOE/Department-Of-Energy/Media-Center-Publications/1_Abu-Dhabi-Energy-Outlook-2050_En.pdf
15 https://www.wam.ae/en/article/bl9axls-empower-signs-contract-design-new-district-cooling
16 https://www.dewa.gov.ae/en/about-us/media-publications/latest-news/2025/03/etihad-esco-completes-dubais-first
17 https://www.bdo.ae/en-gb/insights/shaping-a-sustainable-future-esg-and-sustainability-in-uae-real-estate-construction-sector
18 https://www.dipmf.ae/en/the-uae-has-effectively-implemented-sustainability-and-environmental-preservation
19 https://unfccc.int/sites/default/files/resource/UAE_LTLEDS.pdf
20 https://www.tabreed.ae/sites/default/files/2025-03/2024_Annual_Integrated_Report_En.pdf
21 https://www.wam.ae/en/article/hszrdhyi-uae-announces-enhanced-target-plant-100-million
22 https://www.weforum.org/stories/2023/07/how-abu-dhabi-is-using-the-carbon-capture-capabilities-of-its-mangrove-reserves-to-combat-climate-change
23 https://www.thegef.org/news/greening-blue-championing-coastal-climate-solutions

REFERENCES

Abu Dhabi Global Market (ADGM), 2023. *ADGM concludes a landmark participation at COP28*, 15 December. Available at: https://www.adgm.com/media/announcements/adgm-concludes-a-landmark-participation-at-cop28

Al-Huqail, A.A., Islam, Z. and Al-Harbi, H.F., 2024. Mangroves trend and their impact on surface temperature in Al-Wajh Lagoon: A study aligned with Saudi Arabia's vision 2030. *Frontiers in Environmental Science*, *12*, p.1439425.

Aljneibi, A.K., 2023. *Strategies to mitigate climate change impact in UAE through nationally determined contributions NDCs* (Doctoral dissertation, Khalifa University of Science).

Alsumaiti, T.S. and Shahid, S.A., 2019. Mangroves among most carbon-rich ecosystem living in hostile saline rich environment and mitigating climate change—a case of Abu Dhabi. *Journal of Agricultural and Crop Research*, *7*(1), pp.1–8.

ALTÉRRA, 2023. *UAE commits US$30 billion in catalytic capital to launch landmark climate-focused investment vehicle at COP28*, 1 December. Available at: https://www.alterra.ae/news/uae-commits-us-30-billion-in-catalytic-capital-to-launch-landmark-climate-focused-investment-vehicle-at-cop28

Bhattarai, K. and Yousef, M., 2025. Water scarcity and climate change in MENA: Challenges, innovations, and geopolitical impacts. In *The Middle East: Past, present, and future* (pp. 105–136). Cham, Switzerland: Springer Nature.

Choudhary, B., Dhar, V. and Pawase, A.S., 2024. Blue carbon and the role of mangroves in carbon sequestration: Its mechanisms, estimation, human impacts and conservation strategies for economic incentives. *Journal of Sea Research*, *199*, p.102504. https://doi.org/10.1016/j.seares.2024.102504

Climate TRACE, 2024a. *Approach*. Available at: https://climatetrace.org/approach

Climate TRACE, 2024b. *Data downloads*. Available at: https://climatetrace.org/data

Cool Up Programme, 2022. *Cooling sector status report: Egypt*. Frankfurt: Guidehouse for Cool Up. Available at: https://www.coolupprogramme.org/wp-content/uploads/2022/07/Cool-Up_Cooling-Sector-Status-Report_Egypt_2022.pdf

COP28 UAE, 2023. *Global renewables and energy efficiency pledge*. Dubai: COP28 Presidency. Available at: https://www.cop28.com/en/global-renewables-and-energy-efficiency-pledge

Darwish, M., 2013. Towards energy conservation in Qatar. *Open Journal of Energy Efficiency*, *2013*. http://dx.doi.org/10.4236/ojee.2013.24023

Dubai Electricity and Water Authority (DEWA), 2013. *DEWA recommends rational consumption during summer*. Available at: https://www.dewa.gov.ae/en/about-us/media-publications/latest-news/2013/07/dewa-recommends-rational-consumption-during-summer

Dubai Electricity and Water Authority (DEWA), 2024. *Mohammed bin Rashid Al Maktoum Solar Park*. Available at: https://www.dewa.gov.ae/en/about-us/strategic-initiatives/mbr-solar-park

Dubai Electricity and Water Authority (DEWA), 2025a. *DEWA adds 800MW of clean energy production capacity to its energy mix in 2025 through the MBR Solar Park's projects*, 10 June. Available at: https://www.dewa.gov.ae/en/about-us/media-publications/latest-news/2025/06/dewa-adds-800mw-of-clean-energy-production (Accessed 5 October 2025).

Dubai Electricity and Water Authority (DEWA), 2025b. *Smart applications via smart grid and meters*. Available at: https://www.dewa.gov.ae/en/about-us/dewa-digital-journey/smart-grid-and-meters-smart-applications

Dubey, K., 2016. *Buildings sector: Energy productivity in the GCC*. Presentation at the 7th International Forum on Energy for Sustainable Development, Baku, 18 October. Riyadh: KAPSARC.

ESG Today, 2024. *UAE plans to invest $54 billion by 2030 to meet sustainable energy demand*, 6 November. Available at: https://www.esgtoday.com/uae-plans-to-invest-54-billion-by-2030-to-meet-sustainable-energy-demand/ (Accessed 5 October 2025).

European Commission, 2023. *Global renewables and energy efficiency pledge*. Brussels: Directorate-General for Energy. Available at: https://energy.ec.europa.eu/publications/global-renewables-and-energy-efficiency-pledge_en

European Commission, 2025. *Methane regulation—import requirements Q&A*. Available at: https://energy.ec.europa.eu/document/download/b545c5a6-03c9-4cfa-805a-0411de927ce6_en

Evans, M., Roshchanka, V. and Graham, P., 2017. An international survey of building energy codes and their implementation. *Journal of Cleaner Production*, *158*, pp.382–389. https://doi.org/10.1016/j.jclepro.2017.01.007

Eveloy, V. and Ayou, D.S., 2019. Sustainable district cooling systems: Status, challenges, and future opportunities, with emphasis on cooling-dominated regions. *Energies*, *12*(2), p.235. https://doi.org/10.3390/en12020235

Frey, C.M., Rigo, G. and Parlow, E., 2009. Investigation of the daily Urban Cooling Island (UCI) in two coastal cities in an arid environment: Dubai and Abu Dhabi (UAE). *City*, *81*(2.06), pp.1–5.

Global Mangrove Alliance, 2022. *The mangrove breakthrough*. https://www.mangrovealliance.org/news/the-mangrove-breakthrough

Howarth, N., Odnoletkova, N., Alshehri, T., Almadani, A., Lanza, A. and Patzek, T., 2020. Staying cool in a warming climate: Temperature, electricity and air conditioning in Saudi Arabia. *Climate*, *8*(1), p.4. https://doi.org/10.3390/cli8010004

International Energy Agency (IEA), 2025. *The future of electricity in the Middle East and North Africa*. Paris: IEA. Available at: https://iea.blob.core.windows.net/assets/71d04046-37df-45fb-bb39-a4f818c9aaec/TheFutureofElectricityintheMiddleEastandNorthAfrica.pdf

IRENA, 2025. *Data hub & statistics*. Abu Dhabi: International Renewable Energy Agency. Available at: https://www.irena.org/Data

IUCN and AGEDI, 2016. *National blue carbon policy assessment*. UAE: Summary. IUCN, AGEDI.

Kaczmarski, J., Jones, B. and Chermak, J., 2022. Determinants of demand response program participation: Contingent valuation evidence from a smart thermostat program. *Energies*, *15*(2), p.590. https://doi.org/10.3390/en15020590

Khosla, R., 2025. *The role of cooling standards in climate adaptation to extreme heat*. World Bank Background Paper. Available at: https://thedocs.worldbank.org/en/doc/8dccee9af27ee9024d6b13251016de8b-0050062025/original/Radhika-Khosla-Climate-adaptation-cooling.pdf

Kigali Cooling Efficiency Program (K-CEP), 2019. *Guidance on incorporating efficient, clean cooling into the enhancement of nationally determined contributions*. London: K-CEP. Available at: https://www.k-cep.org/wp-content/uploads/2019/07/Guidance-on-Incorporating-Efficient-Clean-Cooling-into-the-Enhancement-of-Nationally-Determined-Contributions.pdf

Krarti, M. and Howarth, N., 2020. Transitioning to high efficiency air conditioning in Saudi Arabia: A benefit cost analysis for residential buildings. *Journal of Building Engineering*, *31*, p.101457. https://doi.org/10.1016/j.jobe.2020.101457

Masdar, 2023a. *World's largest single-site solar power plant inaugurated: Al Dhafra Solar PV (2GW)*, 16 November. Available at: https://masdar.ae/en/news/newsroom/worlds-largest-single-site-solar-power-plant-inaugurated

Masdar, 2023b. *Dumat Al Jandal wind farm (400 MW)*. Project Page. Available at: https://masdar.ae/en/renewables/our-projects/dumat-al-jandal (Accessed 5 October 2025).

Ministry of Electricity & Renewable Energy (MoEE), 2022. *Egypt energy transition* (presentation), 24 May. Available at: https://www.moee.gov.eg/english_new/Presentations/EGYPT_Energy_Transition.pdf

Ministry of Energy & Infrastructure (UAE), 2024. *UAE energy strategy 2050 (updated 2023/24)*. Official UAE Government Portal, 7 May. Available at: https://u.ae/en/about-the-uae/strategies-plans-and-visions/environment-and-energy/uae-energy-strategy-2050

Ministry of Energy & Infrastructure (UAE), 2025. *Energy strategies to achieve net zero*, 2 September. Available at: https://www.moei.gov.ae/en/about-ministry/energy-strategies-to-achieve-net-zero

Ministry of Environment (MoEnv) (Jordan), 2024. *National cooling strategy of Jordan*. Available at: https://www.moenv.gov.jo/ebv4.0/root_storage/en/eb_list_page/national_cooling_strategy_of_jordan_finalest.pdf

Ocean Conservancy, 2023. *Ocean-based climate solutions in Nationally Determined Contributions: June 2023 update*. Washington, DC: Ocean Conservancy. Available at: https://bvearmb.do/handle/123456789/5195

Pohl, A., Eisbrenner, K., Surmeli-Anac, N., Abdalla, N., Ozkan, C. and Asan, G., 2023. *MENA region cooling status report: Progress, opportunities, and insights (Issue 2)*. Berlin: Guidehouse Germany GmbH, December. Available at: https://www.coolupprogramme.org/wp-content/uploads/2023/12/Cool-Up_MENA-Region-Cooling-Status-Report_Issue-2.pdf

RENA, COP28, COP29, GRA, MoEA and Government of Brazil, 2024. *Delivering on the UAE Consensus: Tracking progress toward tripling renewable energy capacity and doubling energy efficiency by 2030*. International Renewable Energy Agency, COP28 Presidency, COP29 Presidency, Ministry of Energy of the Republic of Azerbaijan, and Government of Brazil, Abu Dhabi. Available at: https://www.irena.org/-/media/Files/IRENA/Agency/Publication/2024/Oct/IRENA_UAE_Consensus_2030_tripling_renewables_doubling_efficiency_2024.pdf

Rwigema, P.C., 2023. G20 and African Union development: Building alliances for sustainable development. *The Strategic Journal of Business & Change Management*, *10*(1), pp.38–79.

Schuit, B.J., Maasakkers, J.D., Bijl, P., Mahapatra, G., Van den Berg, A.W., Pandey, S., Lorente, A., Borsdorff, T., Houweling, S., Varon, D.J. and McKeever, J., 2023. Automated detection and monitoring of methane super-emitters using satellite data. *Atmospheric Chemistry and Physics Discussions*, *2023*, pp.1–47. https://doi.org/10.5194/acp-23-9071-2023

Tenedios, C., O'Leary, S., Capocci, M. and Desai, S.P., 2018. History of anaesthesia: Nurse anaesthesia practice in the G7 countries (Canada, France, Germany, Italy, Japan, the United Kingdom and the United States of America). *European Journal of Anaesthesiology | EJA*, *35*(3), pp.158–164.

UAE Embassy (USA), 2024. *UAE energy diversification*. Available at: https://www.uae-embassy.org/discover-uae/climate-and-energy/uae-energy-diversification (Accessed 5 October 2025).

UAE Government, 2024. *UAE energy strategy 2050 (updated)*. Abu Dhabi: Ministry of Energy & Infrastructure (Official Portal). Available at: https://u.ae/en/about-the-uae/strategies-initiatives-and-awards/strategies-plans-and-visions/environment-and-energy/uae-energy-strategy-2050 (Accessed 5 October 2025).

UNEP/Cool Coalition, 2023. *Development of National Cooling Action Plans (NCAPs) to deliver sustainable cooling* (presentation), 31 October. Available at: https://www.ccacoalition.org/sites/default/files/resources/files/Manjeet-NCAPs.pdf

UNFCCC (United Nations Framework Convention on Climate Change), 2023. *Decision 1/CMA.5: Outcome of the first global stocktake under the Paris agreement*. Bonn: UNFCCC Secretariat. Available at: https://unfccc.int/sites/default/files/resource/1_CMA.5.pdf

United Nations Development Programme (UNDP), 2023. *Summary of national cooling action plans*. New York: UNDP. Available at: https://www.undp.org/sites/g/files/zskgke326/files/2023-11/undp-summary-of-national-cooling-action-plans.pdf

United Nations Development Programme (UNDP), 2025. *Egypt | climate promise*. Available at: https://climatepromise.undp.org/what-we-do/where-we-work/egypt

U.S. Department of Commerce, 2025. *United Arab Emirates—renewable energy and clean energy*. International Trade Administration 2025. Available at: https://www.trade.gov/country-commercial-guides/united-arab-emirates-renewable-energy-and-clean-energy-0

Van Hespen, R., Hu, Z., Borsje, B., De Dominicis, M., Friess, D.A., Jevrejeva, S., Kleinhans, M.G., Maza, M., Van Bijsterveldt, C.E., Van der Stocken, T. and Van Wesenbeeck, B., 2023. Mangrove forests as a nature-based solution for coastal flood protection: Biophysical and ecological considerations. *Water Science and Engineering*, *16*(1), pp.1–13.

Wei, J., Feng, L., Tong, Y., Xu, Y. and Shi, K., 2023. Long-term observation of global nuclear power plants thermal plumes using Landsat images and deep learning. *Remote Sensing of Environment*, *295*, p.113707. https://doi.org/10.1016/j.rse.2023.113707

World Nuclear Association, 2024. *Nuclear power in Saudi Arabia*, 28 March. Available at: https://world-nuclear.org/information-library/country-profiles/countries-o-s/saudi-arabia

Yang, L., Barhemmati, N., Zhao, W., Ji, D., Rajoo, S. and Romagnoli, A., 2024. *Thermal energy storage and cooling load response*. ADBI Working Paper, No. 1431. Tokyo: Asian Development Bank Institute (ADBI). https://doi.org/10.56506/ZFJY2792

Zavadski, K., 2023. *Climate change adaptation in Abu Dhabi*. https://www.degruyterbrill.com/document/doi/10.18574/nyu/9781479805716.003.0021/html

5 The Global Economic Tapestry

Connectedness of ESG, Climate, Geopolitical Risk, and Equity Indices

Yedhu Harikumar, S. Aiswarya, and M. Muthumeenakshi

5.1 INTRODUCTION

The global economy is a complex and interconnected system, where various factors intertwine to shape the economic landscape. The economy today is created from a tapestry of interdependent forces, among the most influential are environmental, social, and governance (ESG) considerations, climate change, geopolitical risk (GPR), and the dynamics of equity indices. These elements have become deeply interconnected, shaping investment flows, market stability, and policy priorities across countries. The current global economy is a comprehensive system with interconnected elements, where all modern economic systems are to a certain extent involved in international economic activities and contribute to its development (Popkova et al., 2016). Hence, whenever there is an event in one country or region, it affects the other region's financial assets or market as a whole. The events can be economic policy uncertainties (Popkova et al., 2016), geopolitical tensions (Babcock-Lumish & Clark, 2007), or climate-related shocks (Choucri, 1993). For instance, the recent global financial crisis showed a strong connection between the economies of various countries, where the economic systems cannot seek only their own interests but must correspond to global priorities of economic development (Popkova et al., 2016).

Globalization has increased the level of interconnectedness among national economies, leading to a multidimensional network of economic, social, and political connections (Popkova et al., 2016). Companies now feel the need to invest in projects that are environmentally friendly, socially responsible, and adhere to the government norms. This has led to a greater focus on ESG factors, which have become an integral part of the investment decision-making process.

The COVID-19 pandemic has further highlighted the importance of ESG practices, as it has caused a physical shock to the ecological system, leading to a social and financial crisis in the form of a large number of deaths, social distancing policies, lockdowns, and business closures. In the aftermath of the pandemic, there has been a significant increase in the number of firms paying attention to sustainable investment as a part of their mission, vision, and compliance with Sustainable Development Goals (Zahid et al., 2023).

Scholars have explored the relationship between ESG practices and corporate financial performance, finding that the consideration of ESG factors can have a material impact on a company's results, through enhanced customer satisfaction, employee engagement, or better risk management (Kim & Li, 2021). Moreover, ESG integration has been shown to enhance financial returns, as companies aligned with long-term sustainability issues not only reduce investment risk but also create growth opportunities (Hughes et al., 2021).

At the same time, the global energy landscape is undergoing a significant transformation, with the shift toward renewable energy sources and the growing emphasis on reducing carbon emissions.

DOI: 10.1201/9781003600152-5

This transition has far-reaching implications for the global economy, with geopolitical rivalries, resource competition, and the impact of climate change on energy supply and demand.

In this context, it is crucial to understand the interconnections between ESG factors, climate change, geopolitical tensions, and their collective impact on equity indices. By analyzing these complex relationships, this research paper aims to provide insights into the global economic tapestry, highlighting the interdependence of various factors and the implications for policymakers, investors, and corporate decision-makers.

5.2 LITERATURE REVIEW

The existing literature on the topic of the global economic tapestry and the interconnectedness of ESG, climate, geopolitical factors, and equity indices provides valuable insights.

One study found that ESG integration as an investment strategy seems to perform better than screening or divestment, and that ESG investing can provide asymmetric benefits, especially during social or economic crises (Atz et al., 2022). This suggests that the consideration of ESG factors can have a tangible impact on financial performance, underscoring the significance of ESG in the investment decision-making process.

An increase in climate change risk-related issues has led to more discussions around ESG integration and the impact on corporate financial performance.

ESG factors have moved to the forefront of financial decision-making, reflecting investor demand for sustainable, ethical, and well-governed corporate practices. At the same time, the increasing impacts of climate change-manifested in extreme weather events, resource scarcity, and regulatory shifts have introduced new dimensions of risk and opportunity for global markets. Overlaying these trends are GRPs, including conflicts. Trade disputes and shifting alliances have proven to be major drivers of volatility in equity markets and broader economic performance. High GRP periods are often associated with lower equity returns and heightened market volatility in energy markets and disturb global supply chains, with plunging effects on equity indices and investment patterns. Moreover, the effectiveness of climate action and sustainable finance is contingent on international cooperation, which is subject to decline and regrowth of geopolitical tensions.

For instance, studies have shown that companies with a stronger focus on environmental sustainability tend to have better financial results, as they are better positioned to manage climate-related risks and capitalize on emerging opportunities (Liu et al., 2022). Research on climate risk looks at the several difficulties environmental elements create for financial markets (Ali & Naz, 2024).

Few studies in this domain consider physical climatic factors such as precipitation, wind speed, and temperature (Bonato et al., 2023). The great influence of climate change on world economic activity drives this focus on climate risk since it forces governments to actively implement climate policy. Consequently, the relevance of climate policy in financial markets has attracted a lot of attention since studies in this field show how policies and rules connected to climate change could affect market dynamics and investment decisions (Ali & Naz, 2024). Most recently, Wang et al. (2022) investigate the effect of climate policy on financial system stability and discover that indirect risk contagion greatly magnifies negative effects on investment fund markets.

In addition to this, there is another important factor which needs to be addressed is GRP, which is thought to influence financial markets and economic cycles; central bankers, corporate investors, and the financial press frequently name GPR as one of the factors influencing investment decisions (Caldara & Iacoviello, 2018). At return quantiles below the median, GPRs are found to affect stock market volatility measures rather than returns, therefore indicating their function as a driver of negative volatility in these markets (Balcilar et al., 2018). Geopolitical risk has a short-term effect on various sectors (Choi, 2024).

A study conducted by Antonakakis et al. (2017) investigates the connection between the price of oil and categorical GPRs and stock markets. It is discovered by the authors that there is a statistically significant inverse link between GPRs and the return and volatility of the market. In addition to that, Bouras et al. (2018) demonstrate that different types of GPRs exert a statistically relevant influence on market volatility, although they do not affect returns. Furthermore, the stock markets of emerging economies are more affected by global geopolitical shocks than by domestic factors. A new study utilizes the autoregressive Markov-switching GARCH-MIDAS model to predict stock market volatility based on geopolitical concerns (Segnon et al., 2023). The investigation was carried out with the assistance of a substantial dataset that spans 22 years and includes daily DJIA returns, monthly GPRs, and macroeconomic factors. Based on the findings, it can be concluded that the GPR index has not had a substantial impact on the stock market in the United States. Another examination into the impact of different categories of the GPR on stock market volatility finds that risks deriving from war-like scenarios are specifically important for predicting the realized volatility of the US equity market (Niu et al., 2023).

The influence of GPR on stock market volatility is multifaceted. An increase in the GPR heightens the investment risk in financial markets. It impedes the decision-making process of investors, hence slowing trading activity in stock markets (Salisu et al., 2024). The GPR is favorably correlated with stock market volatility (Zhang et al., 2022), and the global GPR considerably influences stock market volatility (Christou et al., 2017). Furthermore, the GPR facilitates the transfer of capital from emerging nations to established markets.

This literature review has examined the complex interrelations among ESG, climate, geopolitical variables, and equity indices. The analysis demonstrates a complex interaction among these aspects, underscoring the increasing significance of ESG factors in investing choices (Krappel et al., 2021) and portfolio development (Giese et al., 2019). Analyzing the approaches for ESG scoring (Liu et al., 2023) and the effects of ESG debates (Jucá et al., 2024) highlights the necessity for rigorous and dependable evaluation frameworks. Comprehending these interrelated phenomena is essential for maneuvering through the intricacies of the contemporary financial environment and developing robust, sustainable investment strategies.

Here, in this study, we consider the US Equity Market Index (S&P 500), S&P ESG Index, Climate Policy Uncertainty Index, and Geopolitical Risk Index to explore the interconnections among these variables, which could assist in comprehending their behavior during a crisis period so that investors could make better investment decisions.

The S&P 500 is a market capitalization-weighted index comprising the 500 largest publicly listed firms in the United States. It is broadly regarded as a standard for overall US market performance. Companies are chosen according on market size, liquidity, and industry representation.

We quantify the uncertainty in climate-related policies using the CPU index created by Gavriilidis (2021). The CPU data is sourced from policy uncertainty website. Twice examining eight well-known American newspapers, *The Wall Street Journal*, *New York Times*, *Los Angeles Times*, *Chicago Tribune*, *Boston Globe*, *Miami Herald*, *Tampa Bay Times*, and *USA Today*, helps one to develop the CPU index. Selecting stories from various sources and searching for key terms like "carbon dioxide, "climate," "climate risk, "green energy," "greenhouse gas emissions," among others, helps one construct an index, which gauges the degree of uncertainty about climate policy. Different source data go through standardizing and normalizing procedures. The series then passes more normalizing and averaging in preparation for study.

The study has resorted to monthly GPR data available to measure the impact of geopolitical events. The GPR index construction involves definition, measurement, and validation. Geopolitical risk is the possibility, manifestation, and intensification of negative circumstances related to conflicts, acts of terrorism, and other disputes between nations and political entities that impact the peaceful global affairs (Caldara & Iacoviello, 2018). They have constructed the index using an algorithm that counts the number of articles related to geopolitical events published in the 11 leading newspapers of the United States, UK, and Canada.

We utilized transfer entropy to investigate the link between the variables and determine the information flow from one variable to another. The TVP-VAR method is used to estimate time-varying coefficients, representing the changing nature of causal linkages between the S&P ESG index, Climate policy uncertainty index, Geopolitical Risk Index, and US Equity Market Index over time. The structure of the chapter is arranged in the following order: Section 5.2 deals with the review of available literature, followed by the methodology adopted in Section 5.3. Section 5.4 deals with results and discussion, and Section 5.5 concludes.

5.3 MATERIALS AND METHODS

5.3.1 Transfer Entropy

We used transfer entropy to investigate the link between the variables and determine the information flow from one variable to the next. Operating independently of models, this measurement technique is a method that can detect statistical correlations outside of linear dynamics (Caserini & Pagnottoni, 2021; Assaf et al., 2021). Using transfer entropy eliminates the assumptions included in conventional models, such as cointegration in the current context. It can be used more easily in several academic disciplines. Efficient transfer entropy also helps us reduce the influence of prejudices caused by limited sample sizes, which are present in traditional transfer entropy systems. Furthermore, it lets us draw conclusions about the projected information exchanges between financial time series running concurrently.

$$TE(X \rightarrow Y) = x \in X \, \Sigma \, y \in Y \, \Sigma \, P(y|x) \log \left(P(y) \, P(y|x)\right)$$

where $TE(X \rightarrow Y)$ denotes transfer entropy from variable X and Y. This analysis helps quantify the direction of information flow between the selected variables.

5.3.2 TVP-VAR Connectedness

A strong tool for investigating relationships between certain variables is TVP-VAR or vector autoregression. This approach allows us to monitor the evolution of interactions over time, thereby enhancing our understanding of how shocks propagate and the likelihood of systematic risk materializing. It does this by considering time-varying elements. Traditional VAR models believe that factors remain constant over time, which cannot correctly reflect how interdependencies evolve in fast and frequently changing marketplaces. This approach allows us to determine time-varying coefficients. These coefficients indicate how the direct connections between the S&P ESG Index, the Climate Policy Uncertainty, the Geopolitical Risk Index, and the US Equity Market Index evolve with time. Understanding how market links alter in reaction to various shocks, such as when the economy declines or when the government intervenes, requires this capacity to evolve.

Let's denote the multivariate time series as Yt, where t represents time, and it has N variables. The TVP-VAR model can be expressed as follows:

$$Y_t = \mu_t + A_t Y_{t-1} + \varepsilon_t$$

where

- Y_t is an $N \times 1$ vector of variables at time t.
- μ_t is an $N \times 1$ vector representing time-varying means.
- A_t is an $N \times N$ matrix of time-varying coefficients.
- ε_t is an $N \times 1$ vector of white noise residuals.

5.3.3 Data

The study investigates the links among the US Equity Market Index, S&P ESG Index, Climate Policy Uncertainty, and Geopolitical Risk Index by means of return analysis. Covering the observation period from January 2015 to December 2024, the dataset was sourced from investing.com, a trusted financial data provider. We had to undertake a lot of preprocessing, including outlier and missing number checks, to ensure the data was complete before the study.

5.4 RESULTS AND DISCUSSION

Table 5.1 displays the descriptive statistics of datasets. All of the return series are stationary, according to the Augmented Dickey–Fuller (ADF) test, which is essential for the connectivity analysis's validity.

Figure 5.1 is the graphical representation of stationarity for the selected datasets. Stationary series can be forecasted and generalizable.

Table 5.2 shows the results of the transfer entropy. The findings indicate that there is noticeable transmission of information between the GPR and S&P indices (S&P ESG and S&P 500). Because of a mix of investor behavior, market sensitivity, and the interdependence of world financial systems,

TABLE 5.1
Descriptive Statistics

Sum. Statistics	Mean	Median	Minimum	Maximum	Std. Dev.	ADF
SPESG	0.0095254	0.014936	−0.12369	0.11946	0.044526	0.00
SP500	0.0090857	0.015771	−0.13367	0.11942	0.044373	0.00
CPU	0.008725	−0.0085344	−0.94387	1.2327	0.35032	0.00
GPR	0.0019283	−0.015268	−0.60015	0.69627	0.21137	0.00

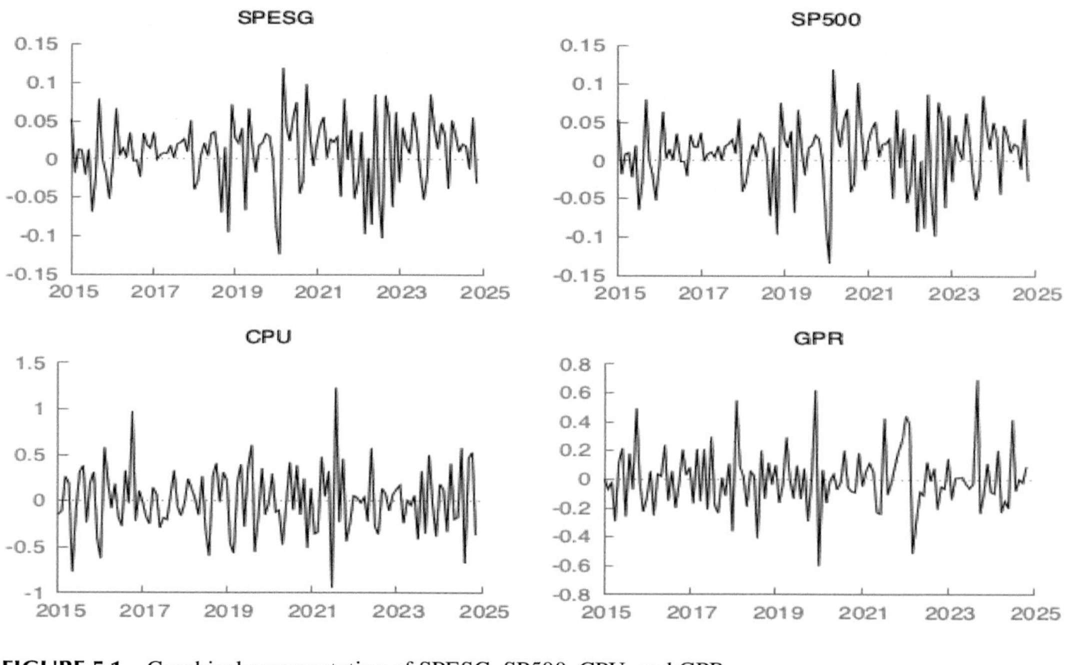

FIGURE 5.1 Graphical representation of SPESG, SP500, CPU, and GPR.

the results on transfer entropy show that GPR greatly affects the S&P indices, notably the S&P 500 and the S&P ESG. Geopolitical events include conflicts, terrorism, and other political difficulties create uncertainty in the market that influences investor mood directly and hence influences asset valuations (Mahjoubi & Henchiri, 2023). Investors' risk-averse behavior resulting from this increased uncertainty causes them to review their portfolios and sometimes withdraw into safer assets or sectors thought to be less impacted by geopolitical turmoil. Consequently, the S&P indices, which cover a large portion of the market, are prone to swings connected to the events, clearly shown through big drawdowns. Furthermore, media coverage and news cycles that impact public opinion help to explain the information flow between GPR and the S&P indices; investors react to changing geopolitics and their expected economic consequences. In addition, globally diversified corporations included in these indices generally have activities in politically sensitive areas, hence increasing their vulnerability to geopolitical events. Changes in geopolitics are so closely connected with equity market volatility and returns since expected effects on supply networks, legal systems, and corporate governance policies in impacted areas directly influence GRP. Emphasizing the need for systematic risk variables in global finance, the dependency of markets on geopolitical information aggravates the reactivity of indices to changing GPR. Consequently, the visible information flow between GPR and S&P indices shows a complicated interaction of market psychology with worldwide economic dynamics.

TVP-VAR connectivity results are presented in Table 5.3. The results of the net pairwise connection are depicted in Figure 5.2. The dynamic connections and spillover effects among several variables in a system throughout time are investigated in this study. Here, the variables are S&P 500 ESG index (S&PESG), S&P 500 index (S&P500), CPU index (CPU), and GPR index (GPR). The diagonal elements stand for the own connectedness, that is, the effect a variable has on itself over time. High values mean that the future movements of a variable can be mostly described by its own historical behavior. CPU, for example, has the highest own connectivity (77.16), implying that its past values are rather good forecasters of its future values. The "FROM" column lists the overall impact a variable has on every other variable in the system. S&PESG, for instance, has a "FROM" connectivity of 55.19, meaning that, combined, the variance in the forecast errors of the other three variables (S&P500, CPU, and GPR) explains 55.19%. The "TO" row displays the overall effect of a variable experience on all other factors. With a "TO" connectivity of 59.19, for instance, S&PESG explains 59.19% of the variance in its forecast inaccuracy by the other three variables. Although not

TABLE 5.2
Results of Transfer Entropy

S. No	Hypothesis	*P* Value
1	No directed information transfer between S&P ESG and S&P 500	0.1069
2	No directed information transfer between S&P 500 and S&P ESG	0.1173
3	No directed information transfer between S&P ESG and CPU	0.4567
4	No directed information transfer between CPU and S&P ESG	0.3633
4	No directed information transfer between S&P ESG and GPR	0.2867
5	No directed information transfer between GPR and S&P ESG	0.0233*
6	No directed information transfer between S&P 500 and CPU	0.5267
7	No directed information transfer between CPU and S&P 500	0.8233
8	No directed information transfer between S&P 500 and GPR	0.2600
9	No directed information transfer between GPR and S&P 500	0.0133*
10	No directed information transfer between CPU and GPR	0.5533
11	No directed information transfer between GPR and CPU	0.2300

* As a statistical rule, a value less than or equal to 0.05 is assumed to be significant.

TABLE 5.3
TVP-VAR Connectedness

Connectedness	S&PESG	S&P500	CPU	GPR	FROM
S&PESG	44.81	44.88	4.32	6	55.19
S&P500	44.57	45.18	3.92	6.33	54.82
CPU	10.67	9.26	77.16	2.91	22.84
GPR	3.95	4.36	1.46	90.24	9.76
TO	59.19	58.5	9.69	15.23	
Inc.Own	104	103.68	86.85	105.47	cTCI/TCI
NET	4	3.68	−13.15	5.47	47.54/35.65
NPT	1	2	0	3	

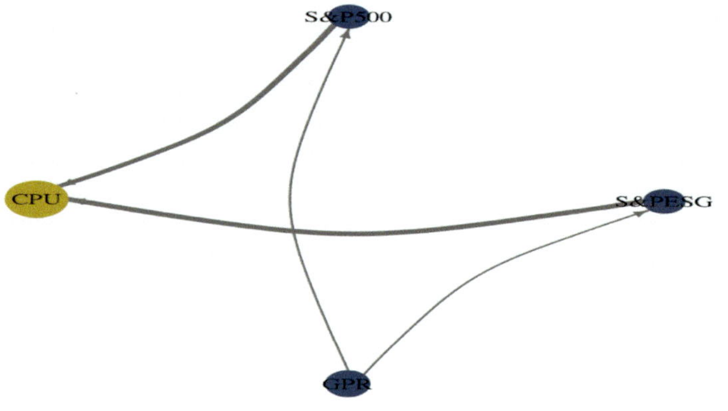

FIGURE 5.2 Net pairwise connectedness.

stated specifically in this table, net connectedness can be computed by subtracting "FROM" from "TO" connectivity. It tells whether a variable is a net transmitter or a net receiver of connectivity. A positive net connectivity suggests that the variable is a net transmitter, that is, influences more than it is influenced; a negative value shows it is a net receiver. S&PESG and S&P500 exhibit a clear two-way correlation according to the table. S&P500 gives S&PESG 44.57; S&P500 is then sent with 44.88. This points to a close relationship between the ESG index and the general market index, therefore implying mutual influence between both. CPU has quite low connectivity with other variables (77.16), although a high own connectedness. Suggesting it is quite isolated in terms of connection relative to the other variables, it receives the least from others (TO=9.69) and transmits the least to others (FROM=22.84). The GPR index exhibits a 'TO' connectivity of 15.23, contributing to the overall system-wide total connectedness (cTCI) of 142.61. Its "FROM" connectedness, which is fairly low (9.76), indicates, nevertheless, that it is more of a receiver of influence than a driver. Examining the personal, directional, and net connectivity helps one to understand the dynamic interactions and spillover effects among these variables.

5.5 CONCLUSION

Using transfer entropy and TVP-VAR connectedness, the interrelations among the US Equity Market Index, S&P ESG Index, Climate Policy Uncertainty, and Geopolitical Risk Index are analyzed to provide important new perspectives on the dynamics controlling these financial indices. The results show that information is getting through clearly, especially when it comes to global issues and the S&P indices. This complicated connection shows how global events like political unrest, terrorism, and wars have a direct effect on how investors feel and market prices. When there is more GRP, the S&P indexes drop a lot. This shows that investors seek out safer assets when they are unsure, which affects their portfolios and market trends in general. In addition, the results show how complicated relationships between public opinion and media stories can have a big effect on how investors react to changes in geopolitics.

As for policy, the results make it clear that lawmakers and investors need to think about GRP when making financial decisions. Policymakers should think about how to make regulatory systems better at dealing with the weaknesses in the market that make it more vulnerable to global shocks. Setting rules for financial diversification could help to keep geopolitical worries from affecting portfolios too much, especially for businesses that are located in places that are sensitive to geopolitics. During geopolitical situations, better ways to communicate and share information could also help control how investors feel and act, which would help to better control market reactions and lower volatility during uncertain times.

Furthermore, this effort opens up fresh paths for study. The time-varying character of the links discovered by TVP-VAR analysis indicates that future studies should emphasize the long-term dynamics of these connections rather than only the short-term ones. Investigating how the link evolves under various market conditions, bull, bear, and quiet periods, could help one better grasp how the market responds to particular geopolitical events. Future studies would find it fascinating to explore more the interactions between domestic market conditions and internal political elements with these worldwide indices, particularly in view of the growing populism and polarization seen in many countries. The outcomes seem to be somewhat interdependent; thus, it would be helpful to connect the sites of companies across various indexes to particular geopolitical risks. Finally, broadening the emphasis to include other important elements like environmental risks or economic policy uncertainty could provide a more complete picture of the interdependence of global financial data, therefore helping to develop strategic investment frameworks that give resilience top priority among geopolitical issues.

All things considered, our results underline the important influence geopolitical threats have on market dynamics and contribute to our knowledge of the complex interactions between several financial indices. Including these ideas into policy frameworks and following research actions will enable stakeholders to improve their plans for negotiating the complex interrelated worldwide financial scenario.

5.5.1 Policy Implications

A combination of GRP evaluations into regulatory systems should be given top priority by lawmakers. For industries with significant worldwide exposure, especially the energy and materials sectors, which show clear sensitivity to geopolitical disturbances, this is especially important. Improved stress-testing techniques connected to dynamic geopolitical events such as wars and trade conflicts may help to reduce systemic weaknesses in equity and ESG markets. The results highlight, second, the need for diversification plans to protect portfolios from sudden geopolitical disturbances. While requiring openness in ESG reporting to match investor decisions with climate resilience goals, regulatory authorities can provide financial incentives for investments in industries like healthcare and utilities, which are less sensitive to GRP. The changing interaction between GRP and market indices emphasizes the importance of having policy instruments able to react to changing circumstances. During crises, including energy price shocks or changes in climate policy, central banks and financial agencies should use TVP-VAR models and real-time monitoring systems to forecast structural changes in market mood. Fourthly, to reduce market volatility caused by panic, governments and investors must enhance their communication channels. Public groups' sharing of thorough risk assessments of world events could help to offset false information and harmonize media narratives with factual data. Eventually, international cooperation is very necessary since emerging markets are disproportionately affected by cross-border spillovers resulting from uncertainties about climate policies and geopolitical concerns. Multinational agreements on uniform ESG standards, along with coordinated monetary responses during geopolitical crises, could help to stabilize linked financial systems. The frameworks should be strengthened in the future to include economic policy uncertainties and sector-specific climate risks. This will allow the development of thorough policies balancing the needs of sustainability with financial stability.

ACKNOWLEDGEMENTS

In the course of preparing this work, the author(s) employed Quillbot to improve the fluency and coherence of the text. Subsequent to utilizing this tool, the author(s) evaluated and revised the text as necessary and accept full responsibility for the publication's content.

REFERENCES

Ali, H., & Naz, S. (2024). Forecasting equity premium in the face of climate policy uncertainty. *Journal of Forecasting*. https://doi.org/10.1002/for.3206

Antonakakis, N., Chatziantoniou, I., & Filis, G. (2017). Oil shocks and stock markets: Dynamic connectedness under the prism of recent geopolitical and economic unrest. *International Review of Financial Analysis, 50*, 1–26. https://doi.org/10.1016/j.irfa.2017.01.004

Assaf, A., Bilgin, M. H., & Demir, E. (2021). Using transfer entropy to measure information flows between cryptocurrencies. *Physica a Statistical Mechanics and Its Applications, 586,* 126484. https://doi.org/10.1016/j.physa.2021.126484

Atz, U., Van Holt, T., Liu, Z. Z., & Bruno, C. C. (2022). Does sustainability generate better financial performance? Review, meta-analysis, and propositions. *Journal of Sustainable Finance & Investment, 13*(1), 802–825. https://doi.org/10.1080/20430795.2022.2106934

Babcock-Lumish, T. L., & Clark, G. L. (2007). Pricing the economic landscape: Global financial markets and the communities and institutions of risk management. In *The Geojournal library* (pp. 145–166). https://doi.org/10.1007/978-1-4020-5849-3_8

Balcilar, M., Bonato, M., Demirer, R., & Gupta, R. (2018). Geopolitical risks and stock market dynamics of the BRICS. *Economic Systems, 42*(2), 295–306. https://doi.org/10.1016/j.ecosys.2017.05.008

Bonato, M., Cepni, O., Gupta, R., & Pierdzioch, C. (2023). Climate risks and state-level stock market realized volatility. *Journal of Financial Markets, 66*, 100854. https://doi.org/10.1016/j.finmar.2023.100854

Bouras, C., Christou, C., Gupta, R., & Suleman, T. (2018). Geopolitical risks, returns, and volatility in emerging stock markets: Evidence from a Panel GARCH Model. *Emerging Markets Finance and Trade, 55*(8), 1841–1856. https://doi.org/10.1080/1540496x.2018.1507906

Caldara, D., & Iacoviello, M. (2018). Measuring geopolitical risk. International Finance Discussion Paper, 2018.0(1222), 1–66. https://doi.org/10.17016/ifdp.2018.1222

Caserini, N. A., & Pagnottoni, P. (2021). Effective transfer entropy to measure information flows in credit markets. *Statistical Methods & Applications, 31*(4), 729–757. https://doi.org/10.1007/s10260-021-00614-1

Choi, S. (2024). Sectoral responses to economic policy uncertainty and geopolitical risk in the US stock market. *Journal of Multinational Financial Management, 76*, 100874. https://doi.org/10.1016/j.mulfin.2024.100874

Choucri, N. (1993). Political economy of the global environment. *International Political Science Review, 14*(1), 103–116. https://doi.org/10.1177/019251219301400107

Christou, C., Cunado, J., Gupta, R., & Hassapis, C. (2017). Economic policy uncertainty and stock market returns in PacificRim countries: Evidence based on a Bayesian panel VAR model. *Journal of Multinational Financial Management, 40*, 92–102. https://doi.org/10.1016/j.mulfin.2017.03.001

Gavriilidis, K. (2021). Measuring climate policy uncertainty. *SSRN Electronic Journal*. https://doi.org/10.2139/ssrn.3847388

Giese, G., Lee, L. E., Melas, D., Nagy, Z., & Nishikawa, L. (2019). Performance and risk analysis of index-based ESG portfolios. *The Journal of Index Investing, 9*(4), 46–57.

Hughes, A., Urban, M. A., & Wójcik, D. (2021). Alternative ESG ratings: How technological innovation is reshaping sustainable investment. *Sustainability*, *13*(6), 3551. https://doi.org/10.3390/su13063551

Jucá, M. N., Muren, P. D., Valentinčič, A., & Ichev, R. (2024). The impact of ESG controversies on the financial performance of firms: An analysis of industry and country clusters. *Borsa Istanbul Review*, *24*(6), 1305–1315. https://doi.org/10.1016/j.bir.2024.08.001

Kim, S., & Li, Z. (2021). Understanding the impact of ESG practices in corporate finance. *Sustainability*, *13*(7), 3746. https://doi.org/10.3390/su13073746

Krappel, T., Bogun, A., & Borth, D. (2021). Heterogeneous ensemble for ESG ratings prediction. arXiv. *arXiv preprint arXiv:2109.10085*.

Liu, W., Li, X., Liu, C., Wang, M., & Liu, L. (2022). Resilience assessment of the cobalt supply chain in China under the impact of electric vehicles and geopolitical supply risks. *Resources Policy*, *80*, 103183. https://doi.org/10.1016/j.resourpol.2022.103183

Liu, Y., Osterrieder, J., Misheva, B. H., Koenigstein, N., & Baals, L. (2023). Navigating the Environmental, Social, and Governance (ESG) landscape: Constructing a robust and reliable scoring engine—insights into data source selection, indicator determination, weighting and aggregation techniques, and validation processes for comprehensive ESG scoring systems. *Open Research Europe*, *3*, 119. https://doi.org/10.12688/openreseurope.16278.1

Mahjoubi, M., & Henchiri, J. E. (2023). Policy uncertainty and the volatility of the S&P 500: Before and after the launch of the S&P 500 ESG. *International Journal of Economics and Finance*, *15*(11), 28. https://doi.org/10.5539/ijef.v15n11p28

Niu, Z., Wang, C., & Zhang, H. (2023). Forecasting stock market volatility with various geopolitical risks categories: New evidence from machine learning models. *International Review of Financial Analysis*, *89*, 102738. https://doi.org/10.1016/j.irfa.2023.102738

Popkova, E., Chechina, O., & Sultanova, A. (2016). Structural and logical model of contemporary global economic system. *European Research Studies Journal*, *19*(2), 218–227. https://doi.org/10.35808/ersj/536

Salisu, A. A., Isah, K. O., & Cepni, O. (2024). Conventional and unconventional shadow rates and the US state-level stock returns: Evidence from non-stationary heterogeneous panels. *The Quarterly Review of Economics and Finance*, *97*, 101890. https://doi.org/10.1016/j.qref.2024.101890

Segnon, M., Gupta, R., & Wilfling, B. (2023). Forecasting stock market volatility with regime-switching GARCH-MIDAS: The role of geopolitical risks. *International Journal of Forecasting*, *40*(1), 29–43. https://doi.org/10.1016/j.ijforecast.2022.11.007

Wang, H., Li, S., & Ma, Y. (2022). Climate policy and financial system stability: Evidence from Chinese fund markets. *Climate Policy*, *23*(4), 395–408. https://doi.org/10.1080/14693062.2022.2104790

Zahid, M., Naqvi, S. U., Jan, A., Rahman, H. U., & Wali, S. (2023). The nexus of environmental, social, and governance practices with the financial performance of banks: A comparative analysis for the pre and COVID-19 periods. *Cogent Economics & Finance*, *11*(1). https://doi.org/10.1080/23322039.2023.2183654

Zhang, Y., He, J., He, M., & Li, S. (2022). Geopolitical risk and stock market volatility: A global perspective. *Finance Research Letters*, *53*, 103620. https://doi.org/10.1016/j.frl.2022.103620

6 Exploring the Impact of Climate Change on Life Expectancy and Youth Death in BRICS Countries

Empirical Evidence Through the Pooled Mean Group Approach

Karnati Kiran Kumar, Naveen Kumar and P. Raja Lingam Goud

6.1 INTRODUCTION

According to Intergovernmental Panel on Climate Change (IPCC), climate change refers to a long-term shift in the global weather pattern, whether by natural variability, that is, change in solar radiation, change in Earth's circumgyration, change in the ocean and atmospheric circulation, changes in thermohaline circulation and pyroclastic flow and by anthropogenic activities, that is, deforestation, exploiting hydrocarbons, the emanation of GHGs, metropolitanism, intensification in the industrial sector and change in land use pattern (McNutt, 2013; Hegerl et al., 2019; Fakana, 2020). These natural and anthropogenic factors transmute the energy flow of Earth, which eventually leads to climate change. All the countries are experiencing some impacts of climate variation. Its consequences are far-ranging and significant in the context of ecosystems and humans. Climate change is observed through rising temperatures, variations in precipitation and an increase in the frequency of extreme climatic events. In past and present due to these events several incidents such as the extreme temperature in Russia Federation in 2010; heat wave in Australia in 2014; wildfire in Australia in 2019; the extreme temperature in Brazil in 1975 (World Meteorological Department, 2022); floods in China in 1996, 1998, 2010 and 2016; floods in India in 2014; a tropical storm in India 1971, 1977 and 1999; floods in South Africa in 1987 and drought in South Africa in 1990 and 2017; floods in Brazil in 1984 and 2011; drought in Brazil in 1978 and 2014; and drought in Russian Federation in 2010 and 2012 (World Meteorological Department, 2022) have occurred. By 2100, it is expected that the global mean temperature will rise by 1.0–3.5°C (IPCC, 2022), and sea levels will rise by 1.3–1.6 m (IPCC, 2022). The frequency and intensity of flooding, droughts and extreme climatic events will increase significantly (IPCC, 2022). All these events will endanger the ecosystem and human life. As the global mean temperature increases by 5°C, 3–48% of species in the biodiversity will be highly vulnerable to extinction (IPCC, 2022). Also, due to variation in precipitation patterns, there is a high probability of water availability and water-related risk by 2100 (IPCC, 2022), extreme climatic events and a rise in global mean temperature by 2°C or higher will lead to a higher risk of food security especially, in South Asia, South and Central Asia and sub-Saharan Africa (IPCC, 2022). In addition to the health impact predicted by IPCC due to extreme climatic scenarios, a valid

DOI: 10.1201/9781003600152-6

rise in infirm health and premature deaths will be there in the mid to long term, that is, 2040–2100 (IPCC, 2022), also it is predicted by WHO that 2,50,000 additional deaths per year will be caused due to heat waves, malnutrition and water-borne diseases and US$2–UD$4 billion per year by 2030 is estimated as the direct damage cost to health due to climate change (WHO, 2019). CRED EM-DAT recorded 22,326 extreme climatic events from 1970 to 2019, which disclosed 4,607,671 deaths and an economic loss of US$4.92 trillion (World Meteorological Department, 2022). It is perspicuous that climate change by itself is a substantial threat to human life.

6.2 REVIEW OF LITERATURE

6.2.1 Climate Change and Health

Climate inconstancy is a considerable jeopardy to the well-being of the anthropoid. Climate change's impact on health depends on geographical location, stage of life, access to healthcare facilities and healthcare configuration (Patz and Kovats, 2002). It influences human health directly, that is, through extreme climatic events like floods, droughts, extreme temperatures and storms and indirectly, that is, by influencing the network on which human health relies (McMichael and Haines, 1997; Jay and Marmot, 2009; Smith, 2014; Sen et al., 2017). From 1970 to 2019, EM-DAT (Emergency Events Database) estimated that 50% of the calamities from 1970 to 2019 comprised meteorological, climatological and hydrological hazards and accounted for 45%, that is, 2.06 million deaths of all recorded deaths and 74%, that is, US$3.6 trillion of all recorded economic losses (World Meteorological Department, 2022). From 1970 to 2019, in India, 551 extreme climatic events were recorded which attributed to 134,037 deaths and an economic loss of US$16.9 billion, in China 907 extreme climatic events were recorded which attributed to 90,624 deaths and an economic loss of US $596 billion, in Brazil, 193 extreme climatic events were recorded which attributed an economic loss of US $42 billion (World Meteorological Department, 2022). Health impacts of climate variation are observed globally and are anticipated to intensify in the future until significant adaptation strategies toward climate change are implemented.

6.2.2 Extreme Events and Health Vulnerability

Human health is vulnerable to all extreme climatic events, and its impact is disproportionate depending on the geography, social, political and economic conditions and resilience (Ebi et al., 2006; Furgal and Seguin, 2006; Mills, 2009; Spickett et al., 2011; Smith, 2014). Low-income countries and their subset are susceptible to unpropitious health effects of climate change; this discrepancy is expected to escalate in the future because of the intensity of climate change, economic conditions and dependency on local natural resources (McMichael et al., 2008). Extreme climatic events like a flood, hurricanes, heat waves and storms make humans infirmly ill by dissuading the previously non-endemic areas at risk, which leads the way for major vector-borne diseases and water-borne diseases, climate change also affects the mental state of humans which leads to anxiety, stress and depression, extreme event, unpredicted precipitation and temperature affects the crop yield leading to food and water security and malnutrition (McMichael and Haines, 1997; McMichael et al., 2008; Jay and Marmot, 2009; Mills, 2009; Bush et al., 2011; Smith, 2014). Deaths attributed to diarrhoeal diseases were 828,651, malaria was 354,924, dengue was 38,315, malnutrition was 28,194 and drowning was 233,890 in 2016 (WHO, 2019).

6.2.3 Pollution of Air, Climate Change and Health

Genesis, shift, dissemination and accumulation of air pollutants are affected by the scope of climate variables like temperature and precipitation. Superfluity emission of carbon dioxide, methane, short-lived climate pollutants and other greenhouse gases because of anthropogenic activities accumulates in the atmosphere and escalates the radioactive force and harms the Earth's surface and air quality, extreme events like heat waves and wildfires affect the air quality, that is, deteriorates it (McMichael and Haines, 1997; Spickett et al., 2011; Kuylenstierna et al., 2020). The association between pollution of the air and climate variation is such that one amplifies the other. Its impact depends on the subjection of area, reactivity and magnitude of the flexibility of humans, so its impacts are disseminated unevenly (Mills, 2009; Spickett et al., 2011). On anthropoids, its impact is dire as it escalates the rate of mortality, escalates the probability of acute myocardial infarction and respiratory conditions like asthma, and escalates the incidence of lung cancer and skin diseases. Exposure to it has a significantly high health cost (McMichael and Haines, 1997; Spickett et al., 2011; Kuylenstierna and Malley, 2020). Global CO_2 emission is estimated to be 640 Mt in 2021, attributed to global coal use, and CO_2 emissions attributed to natural gas combustion is expected to escalate by 215 Mt in 2021 (IEA, 2021). Deaths attributed to acute respiratory infections were 370,370 in 2016 (WHO, 2019).

6.2.4 Climate Change and Child Health

Children are a particularly vulnerable group that is pretentious by climate change. Children in underdeveloped economies are immensely vulnerable to climate change. It expedites children gravely infirm by engendering various health complications, that is, nettle rash, swoon, enervation and heat stroke due to extreme temperature events like a heat wave, hyperosmolar, hyponatraemia, gastric and intestinal diseases, malaria, dengue and Lyme disease due to hydrological events like floods and droughts, respirational problems due to air pollution (Bunyavanich et al., 2003; Shea & Committee on Environmental Health, 2007; Ahdoot & Pacheco, 2015; Kistin et al., 2010; Seal and Vasudevan, 2011; Helldén et al., 2021), because of their diminutive physique, lack of rationality and reliance on custodian and idiosyncrasy (Shea, 2007; Ahdoot and Pacheco, 2015). The World Health Organization (WHO) estimated 88% of a load of ailments as a result of climate variation in children below 5 years. About 160 million children are living in an area which is extremely vulnerable to drought, and 530 million children are living in an area where the frequency of occurrence of floods is high (UNICEF, 2015). Malaria is responsible for over two-thirds of global deaths in a cohort of children under 5 years, and over 800 children under 5 per day, dengue fever affects 50 million lives are affected by dengue and causes 15,000 deaths per year. Alfresco air pollution recorded 3.7 million deaths per year globally, out of which 3% of children were under 5 years, and indoor air pollution accounted for 4.3 million deaths per year, out of which 13% of children were under 5 years (UNICEF, 2015).

6.2.5 Public Perception of Climate Change and Health

Perception of climate change and its impact on health depends on the level of education—a higher level of education results in a higher level of perception and vice versa (Buzelli et al., 2022). The public is aware of climate change, that is, temperature rise, variation in precipitation and occurrence of extreme climatic events and their impact on health (Akerlof et al., 2010). People connect that effluvium of squandering, deforestation, an emanation of GHGs, enlightenment and slighter subjection to research-based verity are the root cause for climate change and its impact on health (Kabir et al., 2016). People perceive that climate change will impact flora and fauna, the upcoming generation, agricultural yield and third-world countries drastically and is also associated with numerous health problems like nettle rashes, cancer, heat stroke, cardiovascular disease and respiratory problems, water-borne diseases and vector-borne diseases (Akerlof et al., 2010; Kabir et al., 2016; Anderko et al., 2020; Buzelli et al., 2022).

6.3 MATERIALS AND METHODS

6.3.1 Data Used

The current study examines the impact of climate change on the health aspect of humans by using a panel data sample from 2000 to 2022 for Brazil, Russia, India, China and South Africa (BRICS). Data regarding health variables, that is, life expectancy at birth, number of deaths (ages 10–14), number of deaths (ages 15–19) and number of deaths (ages 20–24), were obtained from the World Bank. Data for climate variables, that is, precipitation, maximum temperature and minimum temperature, were extracted from the climate change knowledge portal. Moreover, other variables that affect the health of a human, like health expenditure, nutrition, pollution and renewable energy, were considered and data regarding these variables were obtained from the World Bank (see Table 6.1). For the analysis of the panel dataset, initially, simple descriptive statistics and correlation were used. For further analysis Fisher-type unit root test was applied to check for stationarity, Padroni cointegration test was executed to find out whether there is a long-term relation or not. The panel dataset was also checked for serial autocorrelational, heteroskedasticity and multicollinearity. Afterwards the Pooled Mean Group (PMG) model was used to check for the long-run and short interdependency between the climate variables and health.

6.3.2 Model Specification

In the current study,

Model 1,

$$LE = f\left(pr, tmax. tmin, he, co, re, un\right)$$

TABLE 6.1
Variable Description

Category of the Variables		Variable	Sign	Source
Dependent variables	Health	Life expectancy at birth (in years) Number of deaths ages 10–14 years Number of deaths ages 15–19 years Number of deaths ages 20–24 years		World Development Indicators (World Bank)
Independent variables	Climate	Precipitation (in mm)	+/–	Climate Change Knowledge Portal
		Maximum temperature (in degrees Celsius)	+/–	
		Minimum temperature (in degrees Celsius)	+/–	
	Health	Current health expenditure per capita PPP (in international $)	+	World Development Indicators (World Bank)
	Pollution	Carbon intensity of GDP (kg CO_2 per 2021 PPP $ of GDP)	–	
	Renewable energy	Renewable energy consumption (% of total final energy consumption)	+	
	Health and nutrition	Prevalence of undernourishment (% of population)	–	

Model 2,

$$d(10-14) = f(pr, tmax. tmin, he, co, re, un)$$

Model 3,

$$d(15-19) = f(pr, tmax. tmin, he, co, re, un)$$

Model 4,

$$d(20-24) = f(pr, tmax. tmin, he, co, re, un)$$

The linear regression equation is as follows,

$$Y_i = \alpha + \beta_1 X_1 + \beta_2 X_2 + \ldots + \beta_n X_n + \varepsilon_t$$

Linear regression equation for Model 1,

$$LE_{it} = \alpha + \beta_1 pr_{it} + \beta_2 tmax_{it} + \beta_3 tmin_{it} + \beta_4 he_{it} + \beta_5 co_{it} + \beta_6 re_{it} + \beta_7 un_{it} + \varepsilon_{it}$$

Linear regression equation for Model 2,

$$d(10-14)_{it} = \alpha + \beta_1 pr_{it} + \beta_2 tmax_{it} + \beta_3 tmin_{it} + \beta_4 he_{it} + \beta_5 co_{it} + \beta_6 re_{it} + \beta_7 un_{it} + \varepsilon_{it}$$

Linear regression equation for Model 3,

$$d(15-19)_{it} = \alpha + \beta_1 pr_{it} + \beta_2 tmax_{it} + \beta_3 tmin_{it} + \beta_4 he_{it} + \beta_5 co_{it} + \beta_6 re_{it} + \beta_7 un_{it} + \varepsilon_{it}$$

Linear regression equation for Model 4,

$$d(20-24)_{it} = \alpha + \beta_1 pr_{it} + \beta_2 tmax_{it} + \beta_3 tmin_{it} + \beta_4 he_{it} + \beta_5 co_{it} + \beta_6 re_{it} + \beta_7 un_{it} + \varepsilon_{it}$$

In the current study results are long-run and short-run results are computed using Pesaran et al. (1999) PMG model and is expressed as,

For Model 1,

$$\begin{aligned}
\Delta LE_{it} = \; & \alpha_{it} + \delta_1 LE_{it-1} + \delta_2 pr_{it-1} + \delta_3 tmax_{it-1} + \delta_4 tmin_{it-1} + \delta_5 he_{it-1} \\
& + \delta_6 co_{it-1} + \delta_7 re_{it-1} + \delta_8 un_{it-1} + \sum_{n=1}^{p} \alpha_1 \Delta LE_{it-n-1} \\
& + \sum_{n=1}^{q} \beta_1 pr_{it-n} + \sum_{n=1}^{r} \beta_2 tmax_{it-n} + \sum_{n=1}^{s} \beta_3 tmin_{it-n} + \sum_{n=1}^{t} \beta_4 he_{it-n} \\
& + \sum_{n=1}^{u} \beta_5 co_{it-n} + \sum_{n=1}^{v} \beta_6 re_{it-n} + \sum_{n=1}^{v} \beta_7 un_{it-n} + \varepsilon_{it}
\end{aligned}$$

For Model 2,

$$\begin{aligned}\Delta d(10-14)_{it} &= \alpha_{it} + \delta_1 d(10-14)_{it-1} + \delta_2 pr_{it-1} + \delta_3 tmax_{it-1} + \delta_4 tmin_{it-1} \\ &+ \delta_5 he_{it-1} + \delta_6 co_{it-1} + \delta_7 re_{it-1} + \delta_8 un_{it-1} + \sum_{n=1}^{p} \alpha_1 \Delta d(10-14)_{it-n-1} \\ &+ \sum_{n=1}^{q} \beta_1 pr_{it-n} + \sum_{n=1}^{r} \beta_2 tmax_{it-n} + \sum_{n=1}^{s} \beta_3 tmin_{it-n} + \sum_{n=1}^{t} \beta_4 he_{it-n} \\ &+ \sum_{n=1}^{u} \beta_5 co_{it-n} + \sum_{n=1}^{v} \beta_6 re_{it-n} + \sum_{n=1}^{v} \beta_7 un_{it-n} + \varepsilon_{it}\end{aligned}$$

For Model 3,

$$\begin{aligned}\Delta d(15-19)_{it} &= \alpha_{it} + \delta_1 d(15-19)_{it-1} + \delta_2 pr_{it-1} + \delta_3 tmax_{it-1} + \delta_4 tmin_{it-1} \\ &+ \delta_5 he_{it-1} + \delta_6 co_{it-1} + \delta_7 re_{it-1} + \delta_8 un_{it-1} + \sum_{n=1}^{p} \alpha_1 \Delta d(15-19)_{it-n-1} \\ &+ \sum_{n=1}^{q} \beta_1 pr_{it-n} + \sum_{n=1}^{r} \beta_2 tmax_{it-n} + \sum_{n=1}^{s} \beta_3 tmin_{it-n} + \sum_{n=1}^{t} \beta_4 he_{it-n} \\ &+ \sum_{n=1}^{u} \beta_5 co_{it-n} + \sum_{n=1}^{v} \beta_6 re_{it-n} + \sum_{n=1}^{v} \beta_7 un_{it-n} + \varepsilon_{it}\end{aligned}$$

For Model 4,

$$\begin{aligned}\Delta d(20-24)_{it} &= \alpha_{it} + \delta_1 d(20-24)_{it-1} + \delta_2 pr_{it-1} + \delta_3 tmax_{it-1} + \delta_4 tmin_{it-1} \\ &+ \delta_5 he_{it-1} + \delta_6 co_{it-1} + \delta_7 re_{it-1} + \delta_8 un_{it-1} + \sum_{n=1}^{p} \alpha_1 \Delta d(20-24)_{it-n-1} \\ &+ \sum_{n=1}^{q} \beta_1 pr_{it-n} + \sum_{n=1}^{r} \beta_2 tmax_{it-n} + \sum_{n=1}^{s} \beta_3 tmin_{it-n} + \sum_{n=1}^{t} \beta_4 he_{it-n} \\ &+ \sum_{n=1}^{u} \beta_5 co_{it-n} + \sum_{n=1}^{v} \beta_6 re_{it-n} + \sum_{n=1}^{v} \beta_7 un_{it-n} + \varepsilon_{it}\end{aligned}$$

Here, LE = Life expectancy at birth in years, d(10–14) = Number of deaths ages 10–14 years, d(15–19) = Number of deaths ages 15–19 years, d(20–24) = Number of deaths ages 20–24 years, pr = Precipitation, tmax = Maximum temperature, tmin = Minimum temperature, he = health expenditure, co = Carbon intensity, re = Renewable energy consumption, un = Prevalence of undernourishment.

6.4 EMPIRICAL RESULTS AND DISCUSSION

6.4.1 Descriptive Statistics

Table 6.2 presents the descriptive statistics, revealing that the mean life expectancy in BRICS countries is 68.93 years, with a range of 53.96 to 78.58 years. The average number of deaths (10–14 years)

TABLE 6.2
Descriptive Statistics

Variables	Obs.	Mean	Standard Deviation	Minimum	Maximum
Life expectancy	115	68.93	6.077503	53.96	78.587
Number of deaths (10–14 years)	115	26,653.7	37,777.44	1695	132,901
Number of deaths (15–19 years)	115	44,657.83	54,344.03	4006	186,725
Number of deaths (20–24 years)	115	62,073.56	65,476.53	6614	222,541
Precipitation	115	893.551	511.3971	374.29	1,956.06
Maximum temperature	115	20.59217	11.58989	0.27	31.63
Minimum temperature	115	8.724174	10.85949	–9.38	20.58
Health expenditure	110	765.15	504.7017	86	2530
Carbon intensity	115	0.4002971	0.2017788	0.1176496	0.7786349
Renewable energy	110	22.68364	16.87277	3.2	50
Undernourishment	105	6.637143	5.46498	2.5	22

Source: Calculated by the author.

during the study period is 26,653.7, with a minimum value of 1,695 and a maximum value of 132,901. The average number of deaths (15–19 years) is 44,657.83 with a minimum value of 4,006 and a maximum value of 186,725. Moreover, the average number of deaths (20–24 years) is 62,073.56, with the minimum and maximum values being 6,614 and 222,541, respectively. Average precipitation in BRICS countries from 2000 to 2022 is 893.551 mm, ranging from 374.29 to 1,956.06 mm. The average maximum and minimum temperatures are 20.59 and 8.72 degrees Celsius, respectively. The average health expenditure per capita in BRICS countries is US$765.15, with a minimum value of US$86 and a maximum of US$2,530. The average prevalence of undernourishment (percentage of population) is 6.63% in the BRICS countries, ranging from 2.5% to 22%.

At the individual country level, China has the highest average life expectancy (75.70 years), followed by Brazil (72.89 years), Russia (69.03 years), India (67.11 years) and South Africa (59.95 years) (Figure 6.1). The number of deaths in all age groups, that is, age group 10–14 years, 15–19 years and 20–24 years, is highest in India, followed by China, Brazil, South Africa and Russia (see Figure 6.2).

The average precipitation in BRICS countries from 2000 to 2022 was 893.55mm. Brazil and India received the highest average precipitation among the BRICS countries (1,779.18 and 1,128.50 mm, respectively). However, South Africa received the lowest precipitation, that is, 466.79 mm (see Figure 6.3). The average maximum temperature in India and Brazil was 31.06 and 30.92 degrees Celsius, respectively. For Russia, the average maximum temperature was 1.30 degrees Celsius. However, the average minimum temperature was –8.35 degrees Celsius, and in China average minimum temperature was 1.85 degrees Celsius (see Figure 6.4).

Among the BRICS countries, Brazil (US$1,140.39) and Russia (US$1,127.86) have the highest average current health expenditure per capita. India (US$152.50) has the lowest current health expenditure per capita (see Figure 6.5). It suggests that the two major economies with the highest population, that is, India and China, are spending the least on health care among the BRICS countries. As for the carbon intensity, it is highest in China (0.62) and lowest in Brazil (0.13) (see Figure 6.6).

The average renewable consumption in Brazil (45.36) and India (38.50) is the highest within the BRICS countries. However, Russia (3.41) has the lowest average renewable energy consumption

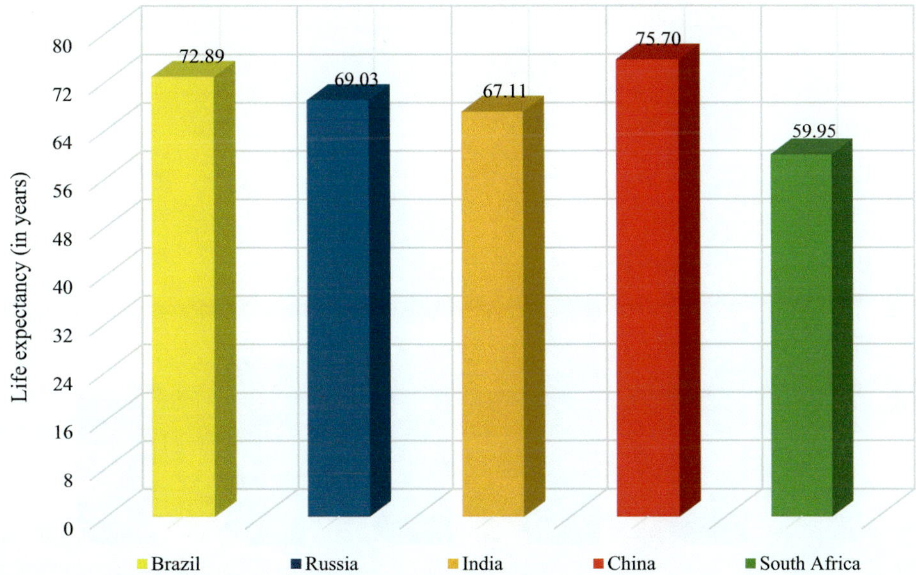

FIGURE 6.1 Average life expectancy in BRICS countries from 2000 to 2022.

Source: World Development Indicators (WDI).

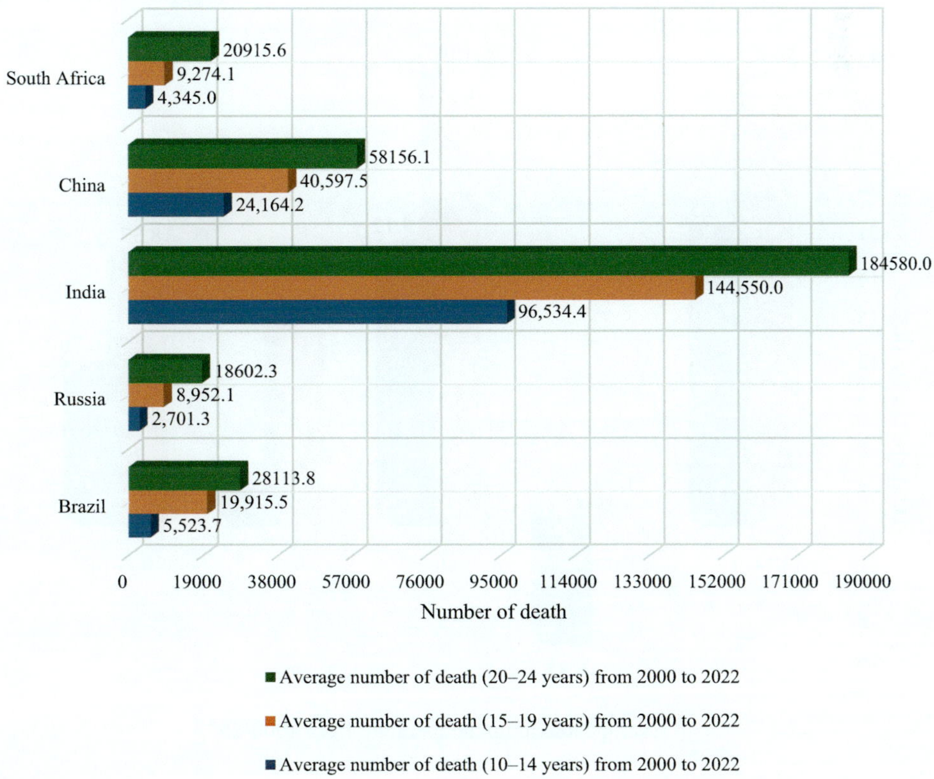

FIGURE 6.2 Average number of deaths in age groups (10–14 years, 15–19 years and 20–24 years) in BRICS countries from 2000 to 2022.

Source: World Development Indicators (WDI).

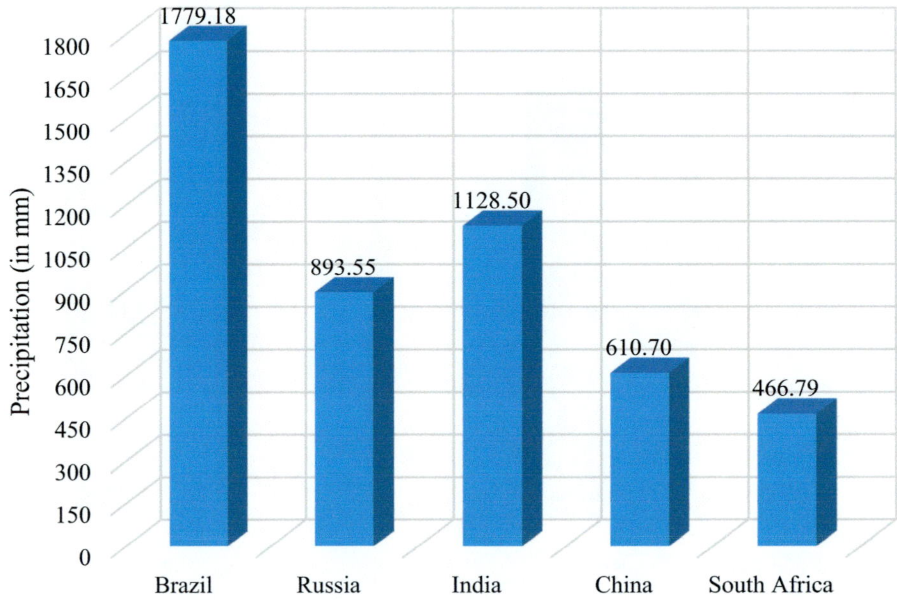

FIGURE 6.3 Average precipitation in BRICS countries from 2000 to 2022.

Source: Climate Change Knowledge Portal.

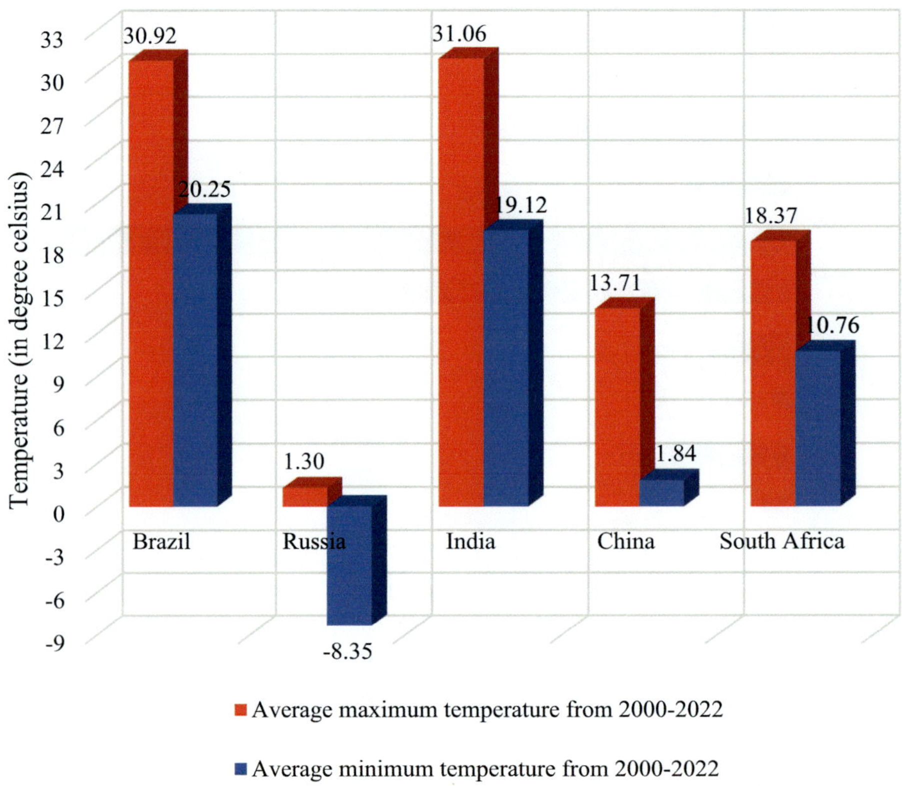

FIGURE 6.4 Average maximum and minimum temperatures in BRICS countries from 2000 to 2022.

Source: Climate Change Knowledge Portal.

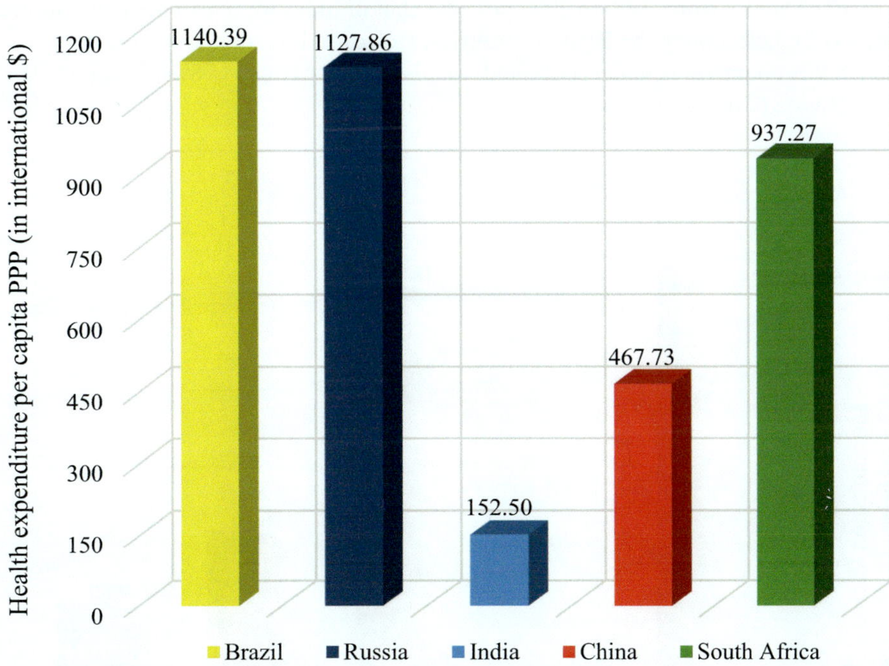

FIGURE 6.5 Average health expenditure in BRICS countries from 2000 to 2022.

Source: World Development Indicators (WDI).

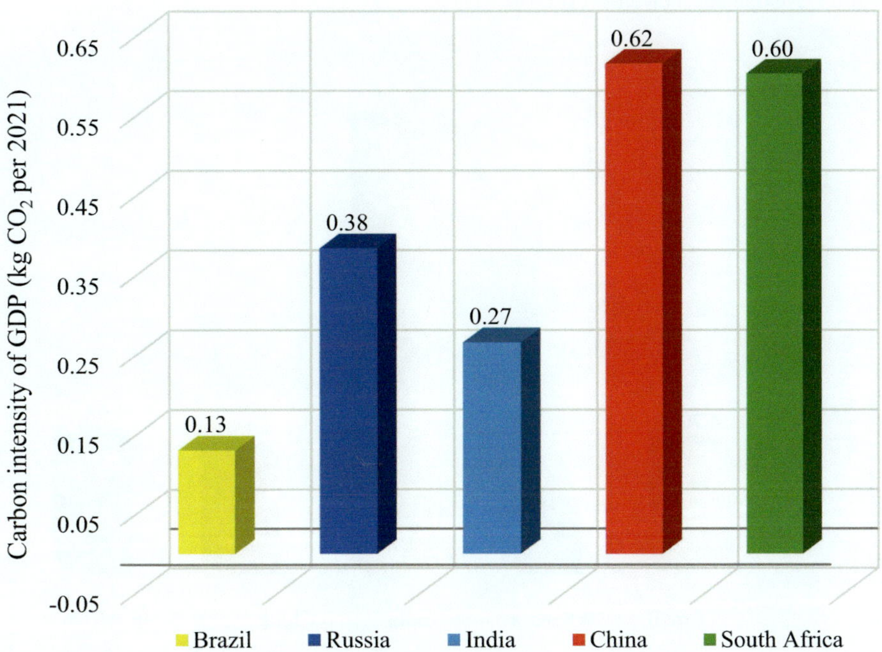

FIGURE 6.6 Average carbon intensity in BRICS countries from 2000 to 2022.

Source: World Development Indicators (WDI).

(see Figure 6.7). Dataset from 2000 to 2022 portrays that in India 16.53% of the population is undernourished, it is highest among the BRICS countries. However, less than 5% of the population of the rest of the BRICS countries is undernourished, that is, South Africa (4.98%), Brazil (4.70%), China (4.36%) and Russia (2.62%) (see Figure 6.8).

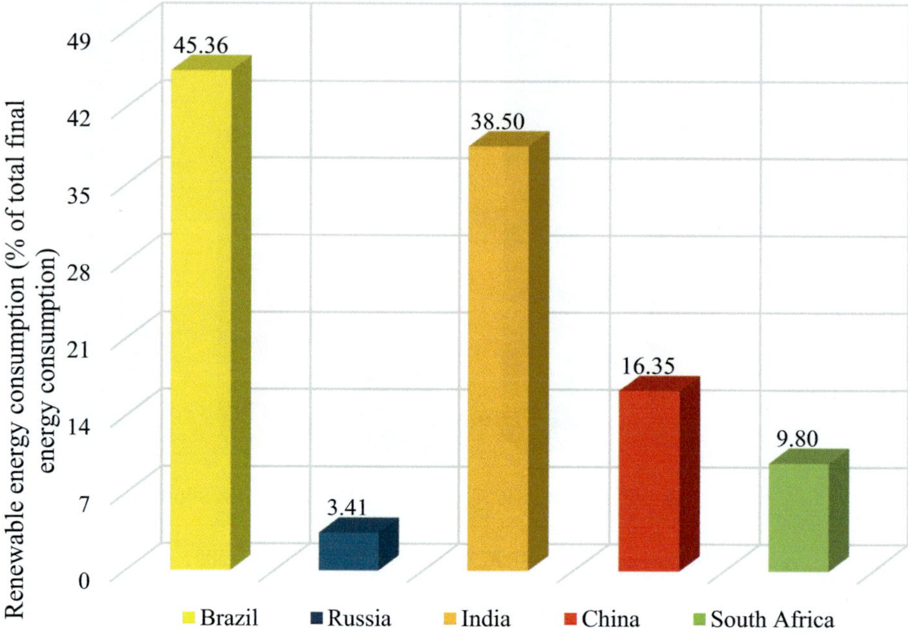

FIGURE 6.7 Average renewable energy consumption in BRICS countries from 2000 to 2022.

Source: World Development Indicators (WDI).

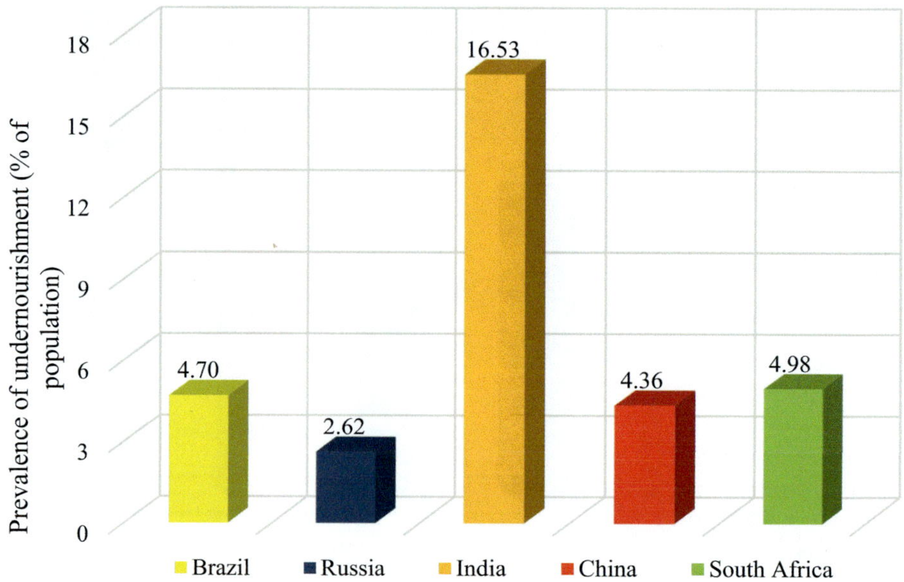

FIGURE 6.8 Average prevalence of undernourishment in BRICS countries from 2000 to 2022.

Source: World Development Indicators (WDI).

6.4.2 Correlation Analysis

Table 6.3 displays the correlation matrix of Model 1, Model 2, Model 3 and Model 4. The correlation matrix for Model 1 reveals the correlation between life expectancy and other independent variables, suggesting that life expectancy has a weak negative correlation with precipitation and carbon intensity and a moderate negative correlation with health expenditure and undernourishment. Furthermore, life expectancy has a weak positive correlation with the maximum temperature, minimum temperature and renewable energy consumption. The correlation matrix for Model 2 portrays a weak negative correlation between the number of deaths (10–14 years), maximum temperature and minimum temperature. The number of deaths (10–14 years) has a moderate negative correlation with precipitation and carbon intensity. Additionally, there is a moderate positive correlation between the number of deaths (10–14 years), health expenditure, renewable energy consumption and undernourishment. The correlation matrix for Model 3 puts forward that the number of deaths (15–19 years) has a weak negative correlation with maximum and minimum temperature and a moderate negative correlation with health expenditure, renewable energy consumption and undernourishment. Moreover, there is a moderate positive correlation between the number of deaths (15–17 years), precipitation and carbon intensity. The correlation matrix for Model 4 suggests that there is a weak negative correlation between the number of deaths (20–24 years) and precipitation. There is a weak positive correlation between the number of deaths (20–24 years), health expenditure and carbon intensity. Moreover, there is a moderate positive correlation between the number of deaths (20–24 years), maximum temperature, minimum temperature, renewable energy consumption and undernourishment.

6.4.3 Unit Root Test Results

The Fisher-type unit root test proposed by Choi (2001) was used to test for stationarity with the null hypothesis (H_0): Panel is not stationary and alternative hypothesis (H_a): Panel is stationary. Test statistics and P-values were computed using/considering the following equations:

$$Y_{it} = a_{it} + x_{it}$$

Y_{it} is composed of a nonstochastic process a_{it} and a nonstochastic process x_{it}, where

$$a_{it} = \beta_{i0} + \beta_{i1}t + \beta_{i2}t + \ldots + \beta_{in}t^{n_i}$$

$$x_{it} = p_i x_{i(t-1)} + u_{it}$$

The P-value for the proposed Fisher-type test is obtained using the given formula:

$$P = -2\sum_{i=1}^{N} \ln p_i$$

Table 6.4 displays the results of the Fisher-type unit root test. Life expectancy, number of deaths (10–14 years), health expenditure, carbon intensity, renewable energy and undernourishment were found to be stationary at first difference. The number of deaths (15–19 years) and precipitation were stationary at the level. The number of deaths (20–24 years) was stationary at the second difference. Maximum and minimum temperatures were found to be stationary at the level and first difference. The problem of multicollinearity was found in maximum and minimum temperature at level; hence, the variable series stationary at first difference was considered for further analysis.

TABLE 6.3
Correlation Matrix

Model 1 (Life Expectancy)

Variables	le	pr	tmax	tmin	he	co	re	un
le	1.0000							
pr	−0.0483	1.0000						
tmax	0.0642	−0.0604	1.0000					
tmin	0.1247	−0.0365	0.8874	1.0000				
he	−0.2220	−0.1441	−0.0314	−0.0372	1.0000			
co	−0.0310	0.2278	0.0442	0.0233	−0.0618	1.0000		
re	0.0173	0.1342	−0.0123	0.0315	0.1123	−0.3139	1.0000	
un	−0.2165	−0.1542	−0.0530	−0.0561	0.0716	−0.1196	0.2311	1.0000

Model 2 (Number of death, 10–14 years)

Variables	d(10–14)	pr	tmax	tmin	he	co	re	un
d(10–14)	1.0000							
pr	−0.1249	1.0000						
tmax	−0.0146	−0.0604	1.0000					
tmin	−0.0247	−0.0365	0.8874	1.0000				
he	0.2663	−0.1441	−0.0314	−0.0372	1.0000			
co	−0.1796	0.2278	0.0442	0.0233	−0.0618	1.0000		
re	0.3117	0.1342	−0.0123	0.0315	0.1123	−0.3139	1.0000	
un	0.1444	−0.1542	−0.0530	−0.0561	0.0716	−0.1196	0.2311	1.0000

Model 3 (Number of death, 15–19 years)

Variables	d(15–19)	pr	tmax	tmin	he	co	re	un
d(15–19)	1.0000							
pr	0.2188	1.0000						
tmax	−0.0224	−0.0604	1.0000					
tmin	−0.0067	−0.0365	0.8874	1.0000				
he	−0.3018	−0.1441	−0.0314	−0.0372	1.0000			
co	0.1214	0.2278	0.0442	0.0233	−0.0618	1.0000		
re	−0.2705	0.1342	−0.0123	0.0315	0.1123	−0.3139	1.0000	
un	−0.1129	−0.1542	−0.0530	−0.0561	0.0716	−0.1196	0.2311	1.0000

Model 4 (Number of death, 20–24 years)

Variables	d(20–24)	pr	tmax	tmin	he	co	re	un
d(20–24)	1.0000							
pr	−0.0248	1.0000						
tmax	0.1144	−0.0604	1.0000					
tmin	0.1346	−0.0365	0.8874	1.0000				
he	0.0564	−0.1441	−0.0314	−0.0372	1.0000			
co	0.0030	0.2278	0.0442	0.0233	−0.0618	v1.0000		
re	0.1803	0.1342	−0.0123	0.0315	0.1123	−0.3139	1.0000	
un	0.1985	−0.1542	−0.0530	−0.0561	0.0716	−0.1196	0.2311	1.0000

Source: Calculated by the author.

TABLE 6.4
Fisher-Type Unit Root Test Result

	Fisher-Type Unit Root Test			
Variables	**Level**	**First Difference**	**Second Difference**	**Result**
Life expectancy	9.3851	48.5545*	—	Stationary at first difference
Number of deaths (10–14 years)	12.2394	31.2186*	—	Stationary at first difference
Number of deaths (15–19 years)	46.4323*	—	—	Stationary at level
Number of deaths (20–24 years)	39.5849*	1.5490	19.0107*	Stationary at second difference
Precipitation	99.6161*	—	—	Stationary at level
Maximum temperature	57.7769*	165.7195*	—	Stationary at first difference
Minimum temperature	46.0327*	115.7615*	—	Stationary at first difference
Health expenditure	2.7164	39.3132*	—	Stationary at first difference
Carbon intensity	4.0667	43.6243*	—	Stationary at first difference
Renewable energy	11.4250	64.3938*	—	Stationary at first difference
Undernourishment	72.6058*	20.7693*	—	Stationary at first difference

Source: Calculated by the author.
Note: * Significant at 1% at level; ** significant at 5% at level; *** significant at 10% level.

6.4.4 Cointegration Test Results

To test the existence of long-term interdependency among the variables Pedroni (1999) cointegration test was executed with the null hypothesis (H_0): No cointegration and alternative hypothesis (H_a): All panels are cointegrated. The following equations were considered for the test and to obtain the Panel ADF (Augmented Dickey–Fuller t) value,

$$y_{it} = \alpha_i + \beta_1 x_{1it} + \beta_2 x_{2it} + \beta_3 x_{3it} + \ldots + \beta_{ni} x_{nit} + e_{it}$$

$$\text{Panel ADF}\left(\text{Augmented Dickey} - \text{Fuller t}\right) = \left(\widetilde{S}_{nt}^{*2} \sum_{i=1}^{n} \sum_{i=1}^{t} \widehat{L}_{11i}^{-2} \widehat{e}_{i,t-1}^{*2}\right)^{-\frac{1}{2}} \sum_{i=1}^{n} \sum_{i=1}^{t} \widehat{L}_{11i}^{-2} \widehat{e}_{i,t-1}^{*2} \Delta \widehat{e}_{i,t}^{*2}$$

Cointegration test results are displayed in Table 6.5. P-values for Augmented Dickey–Fuller t and modified Phillips-Perron t in the case of Model 1, Model 2, Model 3 and Model 4 are less than 0.05, rejecting the null hypothesis (No cointegration) and concluding that the panel dataset is cointegrated; that is, there exists a long-term association between the dependent and independent variables.

6.4.5 Results of Heteroskedasticity, Autocorrelation and Multicollinearity Test

The diagnostic results from the simple regression suggest that in all four models, there is no potential threat of multicollinearity. Model 1, Model 2 and Model 3 suffer from the problem of heteroskedasticity. In Model 4, there is no problem of heteroskedasticity. In Model 1, there is no serial autocorrelation problem, whereas in Model 2, Model 3 and Model 4, there is a problem of serial autocorrelation (see Appendix Tables 6.7, 6.8 and 6.9).

TABLE 6.5
Results of Pedroni Test for Cointegration

Pedroni Test for Cointegration			
Model 1 (Life Expectancy)			
Test	Statistics	P-Value	Result
Augmented Dickey–Fuller t	2.0168	0.0219	Cointegration
Model 2 (Number of Deaths (10–14 years))			
Augmented Dickey–Fuller t	2.4906	0.0064	Cointegration
Model 3 (Number of Deaths (15–19 years))			
Augmented Dickey–Fuller t	5.3511	0.0000	Cointegration
Model 4 (Number of Deaths (20–24 years))			
Augmented Dickey–Fuller t	1.6756	0.0469	Cointegration

Source: Calculated by the author.

6.4.6 Results of PMG Model

The panel dataset cointegration is present for Model 1, Model 2, Model 3 and Model 4, suggesting that there is an existence of a long-run relationship between life expectancy, number of deaths (10–14 years), number of deaths (15–19 years) and number of deaths (20–24 years) with their respective variables, that is, climate variables and other variables. Hence, the PMG model is suitable to investigate the long-run and short-run relationship between life expectancy, number of deaths (10–14 years), number of deaths (15–19 years) and number of deaths (20–24 years) with their respective climate variables and other variables. Moreover, the PMG model takes into consideration the problem of heteroskedasticity and imposes homogeneity on the long-run coefficients and the averaging across groups is used to obtain the mean of the estimated error-correction coefficient and the other short-run parameters of the model (Pesaran et al., 1999).

Table 6.6 reports the results of the PMG for Model 1, Model 2, Model 3 and Model 4. In Model 1, long-run and short-run estimates for life expectancy, climate variables and other variables have been computed. The long-run projections report that precipitation and maximum temperature have a negative impact on life expectancy with coefficient values of −0.0016132 and −3.912491, respectively, and both are statistically significant at the 1% level. The results reveal that precipitation and maximum temperature enhances the risk to life expectancy which is consistent with other studies such as Osei et al. (2023) and Ingole et al. (2012). The minimum temperature has a positive impact on life expectancy with a coefficient value of 4.883653 and is statistically significant at a 1% level. Furthermore, long-run estimates suggest a favourable impact of health expenditure on life expectancy; the result aligns with Linden and Ray (2017); Radmehr and Adebayo (2022). A 1% rise in health expenditure will increase life expectancy by 0.01 years. Consuming more of renewable energy has a positive influence on life expectancy, result of the current analysis are aligns with Liu and Zhong (2022); Karimi Alavijeh et al. (2024). The long-run coefficient value suggests that a 1% increase in renewable energy consumption can increase life expectancy by 0.16 years. The prevalence of undernourishment has a positive coefficient value. The coefficient value for the carbon intensity is statistically insignificant. In the short-run none of the coefficient values was statistically significant.

Model 2 explains the long-run and short-run interdependency between the number of deaths (10–14 years), climate variables and other independent variables. Long-run estimates suggest a negative and statistically insignificant impact of precipitation and health expenditure on the number of deaths (10–14 years). Carbon intensity has a positive but statistically insignificant impact on the number of deaths (10–14 years). Maximum temperature has an unfavourable impact on the number of deaths (10–14 years) with a coefficient value of −657.5037. However, the minimum temperature has a favourable impact on the number of deaths (10–14 years) with a coefficient value of 707.6165. The coefficient values of maximum and minimum temperature are statistically significant at 5% level. The renewable energy consumption is having a positive impact on the number of deaths (10–14 years) with a coefficient value of 527.4925 and is statistically significant at 1% level. Prevalence of undernourishment has a positive impact on the number of deaths (10–14 years) with a coefficient value of 954.3028 suggesting that 1% rise in undernourishment will lead to a 954% rise in the number of deaths (10–14 years). The prevalence of undernourishment is statistically significant at 1% level. In the short-run, all the independent variables are statistically insignificant except for renewable energy consumption which has a negative impact on the number of deaths (10–14 years).

Maximum has a negative impact on the number of deaths (15–19 years), and minimum temperature has a positive impact on the number of deaths (15–19 years). Both maximum and minimum temperatures are statistically insignificant. Precipitation and health expenditure have a positive impact on the number of deaths (15–19 years) with coefficient values of 209.6161 and 19.56464, respectively. Their coefficient value is also statistically significant at 1% level. Carbon intensity is also having a positive impact on the number of deaths (15–19 years) with a coefficient value of 201,059.9, suggesting that a 1% rise in carbon intensity will increase the number of deaths (15–19 years) by 201,059. Renewable energy consumption has a negative impact on the number of deaths (15–19 years) with a coefficient value of −9,303.95, indicating that a 1% rise in renewable energy consumption can reduce the number of deaths (15–19 years) by 9,303%. Both carbon intensity and renewable energy consumption are statistically significant at 1% level. The prevalence of undernourishment has a negative impact on the number of deaths (15–19 years) and is statistically insignificant. In the short run, all the variables are statistically insignificant in the BRICS countries.

In the case of the number of deaths (20–24 years), precipitation and maximum temperature have a positive impact and their coefficient values suggest that a 1% rise in the precipitation will increase the number of deaths (20–24 years) by 0.47% and a 1-degree Celsius rise in the maximum temperature will increase the number of deaths (20–24 years) by 608.56%. The minimum temperature has a negative impact on the number of deaths (20–24 years) with a coefficient value of −1,012.402. The coefficient value of precipitation is statistically significant 10% level, whereas maximum and minimum temperatures are statistically significant at 1% level. The coefficient value of health expenditure is statistically insignificant. Carbon intensity and prevalence of undernourishment are statistically significant at 1% level and have a positive impact on the number of deaths (20–24 years) with the coefficient values of 19,061.86 and 95.28916, indicating that 1% rise in the carbon intensity will lead to an increase in the number of deaths (20–24 years) by 19,061; moreover, 1% rise in undernourishment is expected to rise the number of deaths (20–24 years) by 95.28%. The coefficient value of renewable energy is statistically significant at the 10% level and asserts that 1% rise in renewable energy consumption is expected to reduce the number of deaths (20–24 years) by 41.59%. In the short run, precipitation, maximum temperature and minimum temperature are statistically significant at 1% level. Precipitation and maximum temperature have a negative impact on the number of deaths (20–24 years) with the coefficient values of −0.3459203 and −105.5698, respectively. However, the minimum temperature has a positive impact on the coefficient value of 320.5698. health expenditure, carbon intensity, renewable energy consumption, and prevalence of undernourishment are statistically insignificant in the BRICS nations.

TABLE 6.6
Results of Pooled Mean Group Model

	Model 1 (Life Expectancy)	Model 2 (Number of Deaths, 10–14 Years)	Model 3 (Number of Deaths, 15–19 Years)	Model 4 (Number of Deaths, 20–24 Years)
Long-Run Estimates				
Variables	Coefficient (Std. Err.)	Coefficient (Std. Err.)	Coefficient (Std. Err.)	Coefficient (Std. Err.)
Pr	−0.0016132*	−2.045819	209.6161*	0.4755867***
	(0.0006205)	(2.62226)	(72.7568)	(0.2956754)
Tmax	−3.912491*	−657.5037**	−2,557.359	608.5667*
	(1.877622)	(386.0971)	(7,530.279)	(235.4143)
Tmin	4.883653*	707.6165**	4,313.261	−1,012.402*
	(2.340321)	(351.8367)	(7,352.583)	(257.4965)
He	0.0192424*	−0.2778195	19.56464*	0.0750683
	(0.0065388)	(0.2081416)	(8.631216)	(0.2868735)
Co	5.574578	1,163.417	201,059.9*	19,061.86*
	(8.864438)	(3,131.518)	(89,247.46)	(3,047.675)
Re	0.1625024*	527.4925*	−9,303.95*	−41.59135***
	(0.065946)	(139.1719)	(1,467.419)	(24.60205)
Un	0.0040115*	954.3028*	−3,976.57	95.28916*
	(0.0459548)	(262.1825)	(3,058.644)	(38.84212)
Short-run estimates				
__ec	0.3241366	−0.3248751	−0.0415414	−0.3397424**
	(0.3055374)	(0.233888)	(0.0334002)	(0.1679958)
Pr	0.0008678	0.6761945	−4.282889	−0.3459203*
	(0.0011887)	(0.7664087)	(3.753883)	(0.1240407)
Tmax	−0.7865115	173.5429	229.197	−105.0307*
	(0.649913)	(334.184)	(550.9224)	(50.47795)
Tmin	1.137994	−593.5188	−487.2993	320.5698*
	(0.8077189)	(728.41)	(627.4505)	(140.7419)
He	0.001489	5.781069	−0.1828971	−7.573461
	(0.008913)	(4.756636)	(3.229999)	(7.847064)
Co	−2.57582	−4,886.849	−2,355.279	−2,503.193
	(6.549781)	(5,150.762)	(10,544.71)	(6,347.913)
Re	0.0903034	−98.77058**	121.6557	131.6587
	(0.0682322)	(59.23601)	(170.7715)	(186.0039)
Un	−0.0361829	−33.97983	−196.9237	−156.0219
	(0.3820361)	(76.86568)	(1,213.158)	(489.9623)
_cons	−0.7880039	201.446	−6,454.987*	−95.27194
	(0.6800751)	(156.5887)	(2,949.916)	(124.3156)

Source: Calculated by the author.
Note: * Significant at 1% at level; ** significant at 5% at level; *** significant at 10% level.

6.5 CONCLUSION

The current study explored the impact of climate change on life expectancy and the number of deaths (in the age groups 10–14 years, 15–19 years and 20–24 years) in the BRICS countries. The results of the present study suggest that human health in BRICS countries is vulnerable to climatic events. Coefficient values of the PMG approach reveal a negative impact of precipitation and maximum temperature negatively impacts the life expectancy, which might be due to the infrastructural damage caused by high precipitation, increasing temperature and disease outbreaks. However, minimum temperature positively impacts life expectancy. Furthermore, the increase in health expenditure and the consumption of more renewable energy can enhance the life expectancy of humans in BRICS countries. In the case of the number of deaths in the age group 15–19 years and 20–24 years, precipitation has a positive and significant impact. Maximum temperature has a negative impact on the number of deaths (ages 10–14 years) and a positive impact on the number of deaths (ages 20–24 years). Minimum temperature impacts number of deaths (ages 10–14 years) positively; conversely, it has a negative impact on the number of deaths (ages 20–24 years). Additionally, undernourishment makes the health of the youth more vulnerable to climate change especially in the age group 10–14 years and age group 20–24.

In accordance with the results, the current study suggests that it is important to invest in the health sector, especially in emerging economies like BRICS, to increase the health of their human capital. Furthermore, to reduce the impact of climate change on human health, adaptation and mitigation strategies are required. Among the adaptation and mitigation strategies primary focus should be on reducing carbon emissions, as it is the leading cause of rising temperatures. It can be done by consuming more renewable energy, increasing the investment in renewable energy and conducting research in sustainable energy. High carbon emission and climate change affects the physiology of crop plants, reducing the quality and quantity of the crops, leading to the problem of food insecurity that ultimately results in undernourishment. This affects the health of youth, especially in the age group 10–14 years and 20–24 years. Hence, it is important to execute the adaptive and mitigative strategies to provide humans a secure and healthy lifestyle so that the life expectancy increases and the number of deaths reduces.

REFERENCES

Ahdoot, S., Pacheco, S. E., Council on Environmental Health, Paulson, J. A., Ahdoot, S., Baum, C. R., . . . & Trasande, L. (2015). Global climate change and children's health. *Pediatrics*, *136*(5), e1468–e1484.

Akerlof, K., DeBono, R., Berry, P., Leiserowitz, A., Roser-Renouf, C., Clarke, K. L., . . . & Maibach, E. W. (2010). Public perceptions of climate change as a human health risk: Surveys of the United States, Canada and Malta. *International Journal of Environmental Research and Public Health*, *7*(6), 2559–2606. https://doi.org/10.3390/ijerph706255p9

Anderko, L., Chalupka, S., Du, M., & Hauptman, M. (2020). Climate changes reproductive and children's health: A review of risks, exposures, and impacts. *Pediatric Research*, *87*(2), 414–419. https://doi.org/10.1038/s41390-019-0654-7

Bunyavanich, S., Landrigan, C. P., McMichael, A. J., & Epstein, P. R. (2003). The impact of climate change on child health. *Ambulatory Pediatrics*, *3*(1), 44–52. https://doi.org/10.1367/1539-4409(2003)003<0044:TIOCCO>2.0.CO;2

Bush, K. F., Luber, G., Kotha, S. R., Dhaliwal, R. S., Kapil, V., Pascual, M., . . . & Hu, H. (2011). Impacts of climate change on public health in India: Future research directions. *Environmental Health Perspectives*, *119*(6), 765–770. https://doi.org/10.1289/ehp.1003000

Buzelli, L., Cameron, G., Duxbury, K., Gardner, T., Rutherford, S., Williamson, S., & Alderwick, H. (2022). Public perceptions of health and social care. *The Health Foundation*. https://www.health.org.uk/sites/default/files/upload/publications/2022/What%20the%20new%20government%20should%20know_RGB_WEB.pdf

Choi, I. (2001). Unit root tests for panel data. *Journal of international money and Finance*, *20*(2), 249–272. https://dx.doi.org/10.1016/S0261-5606(00)00048-6

Ebi, K. L., Lewis, N. D., & Corvalan, C. (2006). Climate variability and change and their potential health effects in small island states: Information for adaptation planning in the health sector. *Environmental Health Perspectives*, *114*(12), 1957–1963. https://doi.org/10.1289/ehp.8429

Fakana, S. T. (2020). Causes of climate change. *Global Journal of Science Frontier Research: H. Environmental & Earth Science*, *20*(2), 7–12.

Furgal, C., & Seguin, J. (2006). Climate change, health, and vulnerability in Canadian Northern Aboriginal communities. *Environmental Health Perspectives*, *114*(12), 1964–1970. https://doi.org/10.1289/ehp.8433

Hegerl, G. C., Brönnimann, S., Cowan, T., Friedman, A. R., Hawkins, E., Iles, C., . . . & Undorf, S. (2019). Causes of climate change over the historical record. *Environmental Research Letters*, *14*(12), 123006. https://doi.org/10.1088/1748-9326/ab4557

Helldén, D., Andersson, C., Nilsson, M., Ebi, K. L., Friberg, P., & Alfvén, T. (2021). Climate change and child health: A scoping review and an expanded conceptual framework. *The Lancet Planetary Health*, *5*(3), e164–e175. https://doi.org/10.1016/S2542-5196(20)30274-6

IEA (2021), *Global Energy Review 2021*, IEA, Paris https://www.iea.org/reports/global-energy-review-2021.

Ingole, V., Juvekar, S., Muralidharan, V., Sambhudas, S., & Rocklöv, J. (2012). The short-term association of temperature and rainfall with mortality in Vadu Health and Demographic Surveillance System: A population level time series analysis. *Global Health Action*, *5*(1), 19118. https://doi.org/10.3402/gha.v5i0.19118@zgha20.2012.5.issue-s1

International Energy Agency. (2021). *Global energy review 2021—assessing the effects of economic recoveries on global energy demand and CO2 emission in 2021*, International Energy Agency.

IPCC, (1996). *Climate Change 1995: The Science of Climate Change*. Contribution of Working Group I to the Second Assessment Report of the Intergovernmental Panel on Climate Change. Cambridge University Press. http://www.ipcc.ch/site/assets/uploads/2018/02/ipcc_sar_wg_I_full_report.pdf

IPCC. (2022). *Climate change 2022. Impacts, adaptation and vulnerability*. Sixth assessment report, Intergovernmental Panel on Climate Change. https://www.ipcc.ch/report/sixth-assessment-report-working-group-ii/

Ipsos MORI. (2021). *Public perceptions of health and social care*, The Health Foundation.

Jay, M., & Marmot, M. G. (2009). Health and climate change. *The Lancet*, *374*(9694), 961–962. 10.1016/S0140-6736(09)61603-2

Kabir, M. I., Rahman, M. B., Smith, W., Lusha, M. A. F., Azim, S., & Milton, A. H. (2016). Knowledge and perception about climate change and human health: Findings from a baseline survey among vulnerable communities in Bangladesh. *BMC Public Health*, *16*, 1–10. https://doi.org/10.1186/s12889-016-2930-3

Karimi Alavijeh, N., Ahmadi Shadmehri, M. T., Dehdar, F., Zangoei, S., & Nazeer, N. (2024). The role of renewable energy on life expectancy: Evidence from method of moments quantile regression based on G-7 countries data. *International Journal of Energy Sector Management*, *18*(4), 767–788. https://doi.org/10.1108/IJESM-11-2022-0001

Kistin, E. J., Fogarty, J., Pokrasso, R. S., McCally, M., & McCornick, P. G. (2010). Climate change, water resources and child health. *Archives of Disease in Childhood*, *95*(7), 545–549. https://doi.org/10.1136/adc.2009.175307

Kuylenstierna, J., Malley, C., Büker, P., & Marmon, T. (2020). *Air pollution and its impact on human health: An important driver for achieving the 1.5°C goal of the Intergovernmental Panel on Climate Change*. Stockholm Environment Institute. http://www.jstor.org/stable/resrep28402

Linden, M., & Ray, D. (2017). Life expectancy effects of public and private health expenditures in OECD countries 1970–2012: Panel time series approach. *Economic Analysis and Policy*, *56*, 101–113. https://doi.org/10.1016/j.eap.2017.06.005

Liu, H., & Zhong, K. (2022). Relationship between health spending, life expectancy and renewable energy in China: A new evidence from the VECM approach. *Frontiers in Public Health*, *10*, 993546. doi: 10.3389/fpubh.2022.993546

McMichael, A. J., & Haines, A. (1997). Global climate change: the potential effects on health. *Bmj*, *315*(7111), 805–809. https://doi.org/10.1136/bmj.315.7111.805

McMichael, A. J., Friel, S., Nyong, A., & Corvalan, C. (2008). Global environmental change and health: impacts, inequalities, and the health sector. *Bmj*, *336*(7637), 191–194. doi:10.1136/bmj.39392.473727.AD

McNutt, M. (2013). Climate change impacts. *Science*, *341*(6145), 435–435. https://doi.org/10.1126/science.1243256

Mills, D. M. (2009). Climate change, extreme weather events, and us health impacts: What can we say? *Journal of Occupational and Environmental Medicine*, 26–32. https://doi.org/10.1097/JOM.0b013e31817d32da

Osei, B., Kunawotor, M. E., & Appiah-Konadu, P. (2023). Mortality rate and life expectancy in Africa: The role of flood occurrence. *International Journal of Social Economics*, *50*(7), 910–924. https://doi.org/10.1108/IJSE-07-2022-0508

Patz, J. A., & Kovats, R. S. (2002). Hotspots in climate change and human health. *BMJ*, *325*(7372), 1094–1098. https://doi.org/10.1136/bmj.325.7372.1094

Pedroni, P. (1999). Critical values for cointegration tests in heterogeneous panels with multiple regressors. *Oxford Bulletin of Economics and statistics*, *61*(S1), 653–670. https://doi.org/10.1111/1468-0084.0610s1653

Pesaran, M. H., Shin, Y., & Smith, R. P. (1999). Pooled mean group estimation of dynamic heterogeneous panels. *Journal of the American statistical Association*, *94*(446), 621–634. https://doi.org/10.2307/2670182

Radmehr, M., & Adebayo, T. S. (2022). Does health expenditure matter for life expectancy in Mediterranean countries? *Environmental Science and Pollution Research*, *29*(40), 60314–60326. https://doi.org/10.1007/s11356-022-19992-4

Seal, A., & Vasudevan, C. (2011). Climate change and child health. *Archives of disease in childhood*, *96*(12), 1162–1166. https://doi.org/10.1136/adc.2010.186213

Sen, B., Dhimal, M., Latheef, A. T., & Ghosh, U. (2017). Climate change: health effects and response in South Asia. *bmj*, *359*. doi:10.1136/bmj.j5117

Shea, K. M., & Committee on Environmental Health. (2007). Global climate change and children's health. *Pediatrics*, *120*(5), e1359–e1367. https://doi.org/10.1542/peds.2007-2646

Smith, J. (2014). The health impacts of climate change. *BMJ*, *348*. doi:10.1136/sbmj.g2874

Spickett, J. T., Brown, H. L., & Rumchev, K. (2011). Climate change and air quality: The potential impact on health. *Asia Pacific Journal of Public Health*, *23*(2_suppl), 37S–45S. doi:10.1177/1010539511398114

UNICEF. (2015). *Unless We Act Now*. UNICEF, New York. https://sustainabledevelopment.un.org/index.php?page=view&type=400&nr=2161&menu=1515

World Health Organization. (2019). *Water, sanitation, hygiene and health: A primer for health professionals* (No. WHO/CED/PHE/WSH/19.149), World Health Organization. https://www.who.int/publications/i/item/WHO-CED-PHE-WSH-19.149

World Meteorological Department. (2022). *State of the global climate 2022*. https://library.wmo.int/records/item/56300-state-of-the-global-climate-2021

World Meteorological Organization (WMO). (2021). *State of the global climate 2020*. WMO, Geneva, WMO-No. 1264. https://library.wmo.int/idurl/4/56247.

Appendix

TABLE 6.7
Results of Heteroskedasticity Test

Heteroskedasticity Test		
Model 1 (Life Expectancy)		
Chi-square	P-Value	Result
22.08	0.0000	Heteroskedasticity
Model 2 (Number of Deaths, 10–14 Years)		
Chi-square	P-Value	Result
4.90	0.0269	Heteroskedasticity
Model 3 (Number of Deaths, 15–19 Years)		
Chi-square	P-Value	Result
3.59	0.0581	Heteroskedasticity
Model 4 (Number of Deaths, 20–24 Years)		
Chi-square	P-Value	Result
2.55	0.1106	No heteroskedasticity

Source: Calculated by the author.

TABLE 6.8
Results of Autocorrelation Test

Autocorrelation Test		
Model 1 (Life Expectancy)		
F-Value	P-Value	Result
5.536	0.0783	No autocorrelation
Model 2 (Number of Deaths, 10–14 Years)		
F-Value	P-Value	Result
31.408	0.0050	Autocorrelation
Model 3 (Number of Deaths, 15–19 Years)		
F-Value	P-Value	Result
412.398	0.0000	Autocorrelation
Model 4 (Number of Deaths, 20–24 Years)		
F-Value	P-Value	Result
16.909	0.0147	Autocorrelation

Source: Calculated by the author.

TABLE 6.9
Results of Multicollinearity Test

Multicollinearity Test		
Model 1 (Life Expectancy)		
Mean VIF	2.19	No potential threat of multicollinearity
Model 2 (Number of Deaths, 10–14 Years)		
Mean VIF	2.19	No potential threat of multicollinearity
Model 3 (Number of Deaths, 15–19 Years)		
Mean VIF	2.19	No potential threat of multicollinearity
Model 4 (Number of Deaths, 20–24 Years)		
Mean VIF	2.19	No potential threat of multicollinearity

Source: Calculated by the author.

7 The Corporate Sector's Contribution to Achieving SDGs and Climate Action Goals

Pankaj Singh Rawat

7.1 INTRODUCTION

In 2015, the world saw two landmark multilateral agreements—adopting sustainable development goals, which must be achieved by 2030, and a legally binding climate treaty, which aims to limit the global temperature rise by 1.5 degrees above the pre-industrial level (United Nations 2015; UNFCCC 2018). For each of these agreements, countries had set targets for themselves and aligned their development agenda to achieve these goals. Achieving these targets is difficult to accomplish by government initiatives alone; it will require the participation of the private sector (Palau-Pinyana, Llach, and Bagur-Femenías 2024). The public sector has to sign up for most of the financing challenges (Cingolani 2024); however, the private sector can contribute immensely to the implementation of sustainable goals (Mohieldin 2017).

The corporate sector's contribution is key to achieving the United Nations' Sustainable Development Goals (SDGs). It can do so by adopting sustainable business practices by integrating sustainability into their operations; firms can minimise environmental harm (Schneider 2014). Actions such as reducing carbon emissions, conserving resources, and adopting circular economy principles directly support the SDG goals (Baumgartner and Ebner 2010). Further, the corporate sector can also foster economic development by generating employment, supporting small businesses, and investing in sustainable technologies like renewable energy and green innovations (Bocken 2021; Zakari et al. 2022).

Through Corporate Social Responsibility (CSR) initiatives, firms contribute to social and environmental well-being, progressively aligning their efforts with the SDGs (Scheyvens, Banks, and Hughes 2016). Transparency in sustainability incentives reporting enables businesses to measure and communicate their impact effectively (Tsang, Frost, and Cao 2023). Multiple standards exist for climate-related disclosure and measuring environmental, social, and governance (ESG) activities. These standards improve verification rates in corporate ESG-related disclosure (Darnall et al. 2022). However, due to the voluntary nature of these disclosures, compliance remains limited to large corporations (KPMG 2022).

Despite many advantages, the corporate sector's SDG engagement faces multiple limitations. The major one is access to climate finance in developing countries, where the effects of climate change are most severe (Qi and Qian 2023). There are also concerns with reporting corporate environmental disclosures objectively, including selective reporting for favourable information (Lashitew 2021). Other challenges include the lack of innovative leadership in firms (Garrido-Ruso, Aibar-Guzmán, and Suárez-Fernández 2023) and a lack of an effective monitoring and evaluation framework (Yan et al. 2022).

This chapter discusses the importance of the corporate sector for implementing SDGs and climate action targets. It links the corporate sector's contribution to specific sustainable goals and provides

DOI: 10.1201/9781003600152-7

real-world examples of its implementation. It also discusses the area where the private sector can contribute and the challenges associated with measuring its contribution. The remainder of the chapter is structured as follows. Section 7.2 discusses the importance of the corporate sector in adopting the SDGs. Section 7.3 discusses why the corporate sector should care about climate change and its impact on them. Section 7.4 discusses the potential areas where corporate contributions can help achieve SDGs. Section 7.5 touches on regulations and the measurement of corporate contributions toward a sustainable environment. Section 7.6 discusses the limitations and challenges in attaining significant contributions from the corporate sector. Finally, Section 7.7 concludes the study.

7.2 IMPORTANCE OF THE PRIVATE SECTOR FOR SDGS

As significant players in the global economy, corporations are responsible and capable of contributing to these goals, particularly around climate action (Park 2018). The corporate sector not only drives innovation and economic growth but also shapes societal behaviour, consumption patterns, and policy frameworks, making it essential to realise the SDGs, especially in mitigating climate change (Rosati and Faria 2019). In addition to reducing their environmental impact, businesses can use their influence, innovation, and financial power to combat climate change more broadly (Tsang, Frost, and Cao 2023). A survey of the world's top 250 firms found that 96 per cent of firms report their business impact on SDGs; however, only 10 per cent report on all goals (KPMG 2022). This statistic is even bleaker when it comes to small and medium-scale businesses.

Integrating the corporate sector into the SDGs is necessary due to its extensive economic influence, innovation capacity, and global operational scale. The corporate sector accounts for a significant share of global GDP and employment, making it crucial in mobilising financial resources and driving sustainable economic growth (Chen, Karabarbounis, and Neiman 2017; Park 2018). Private sector investments play a critical role in bridging financial shortfalls, particularly in developing economies, where government resources are often inadequate (Kalinowski 2024).

Private corporations are key in advancing technological innovation, promoting sustainable energy solutions, circular economies, and eco-friendly production methods. Given their substantial environmental impact, it is imperative for businesses to align with SDG targets to reduce carbon emissions, minimise resource depletion, and contribute to climate change mitigation. The collaboration between the private sector and governments is one of the proposed tools by the United Nations to strengthen the implementation of the SDGs (Wang and Ma 2021). It does so by leveraging corporate expertise, financial contributions, and operational efficiencies to scale up sustainability efforts (Castelblanco and Guevara 2022). Aligning business strategies with SDG principles also enhances long-term corporate resilience, helping companies adapt to climate risks, regulatory changes, and evolving consumer expectations (Jomo et al. 2016). Thus, integrating sustainability into the private sector's core operations can contribute to global development objectives.

The public–private partnerships (PPPs) are also key in delivering the SDGs goals. PPP works best in areas where private finance lacks incentives for investment. Cingolani (2024) stated that private finance lacks incentives to invest in SDGs due to the involvement of a substantial public good. In such cases, the PPP route can be followed, where public funding can provide support for the private sector's SDG implementation. Wang and Ma (2021) show that there is significant interest in sustainability-oriented PPP, which indicates a stakeholders' growing interest in sustainability.

7.3 IMPACT OF CLIMATE CHANGE ON THE CORPORATE SECTOR

Climate change also significantly impacts how businesses operate (Sautner et al. 2023). While researchers have created advanced models to predict the effects of greenhouse gas (GHG) emissions on the global climate, there is limited knowledge on how it will affect jobs, innovation, and the day-to-day operability of firms. The task becomes even more difficult when it comes to measuring the impact on individual firms, as the effects are complex and arise from multiple sources (Giglio,

Kelly, and Stroebel 2021). Climate change can potentially disturb the supply chain in one sector, like agriculture, which will then have ripple effects, directly or indirectly, in other sectors of the economy.

Regulatory measures to combat climate change can increase costs for certain businesses, while they can also create opportunities for others (Cox et al. 2022). Interestingly, these opportunities and threats can sometimes be visible within a sector. For example, in the energy sector, regulations aimed at reducing GHG emissions increase scrutiny on businesses working in areas related to fossil fuels. However, businesses that use renewable and clean energy can benefit from these regulations. Climate change poses an aggregate risk not only to the broader economy but also to individual firms. And those firms that do not respond proactively to such climate change transition risks observe a drop in their business value (Li et al. 2024). Climate risk can reduce earnings by increasing the cost of major global companies due to the adoption of carbon pricing (S&P Global 2024).

There are several factors that contribute to firms' limited climate risk adaptability. First, climate change is an ongoing and unpredictable process that makes it difficult for firms to anticipate how future changes will impact businesses. It adds a considerable amount of uncertainty to corporate decision-making (Barnett, Brock, and Hansen 2020). Another factor that limits firms' ability to respond is their hesitation to disclose climate risk exposure in the first place (Shapiro 2021). Even after regulators adopted the stricter disclosure norms, compliance remains low. Hummel and Schlick (2016) suggest that a reason for such behaviour by some firms is to disguise their true performance and to simultaneously protect their legitimacy.

Although historical sustainability disclosure data are necessary to assess a company's past performance, forward-looking corporate policy is more valuable for a firm's climate exposure risk mitigation plan (Weinhofer and Busch 2013). An increasing number of today's investors take non-financial data on ESG disclosures just as seriously as financial data. It is believed that those companies that measure and report their ESG risks are also more likely to manage these risks better (Duan et al. 2025). So, it is becoming increasingly important for firms to focus on the impact of climate change on their businesses.

EXHIBIT 7.1 EXAMPLES OF CORPORATE SECTOR CONTRIBUTION FOR SUSTAINABLE CONSUMPTION

RECYCLING E-WASTE

Manufacturing of digital devices requires a host of rare earth minerals. Mining these minerals is usually environmentally detrimental, but there is a sustainable way, that is, to recycle minerals from end-of-life electronics products. Corporations can step up and support the recycling of e-waste. A few firms which are engaged in recycling are ERI[1] in the United States and Ecovera[2] in India. Another way to reduce e-waste is by extending the life of electronics by refurbishing and reusing them. Cashify[3] adopted such a business model in India.

UPCYCLING

Many firms use recycled material in their production process, which not only reduces waste but also extends the lifecycle of the materials. For example, Patagonia,[4] a fashion brand, uses recycled materials in its clothing. Opposite to fast fashion, it encourages consumers to use the product longer.

Aurbis[5] is a German metal company. Apart from metal production for metal ores, it has established metal recycling processes. It collects scrap metal and electronics and extracts copper from them. This process helps in creating a circular economy.

7.4 FOCUS GOALS FOR THE CORPORATE SECTOR

The most direct way corporations can help address climate change is by reducing their emission. Companies can adopt various strategies to reduce GHG emissions and energy consumption. They can use energy-efficient technologies to reduce their energy footprint. Energy efficiency adoption by corporations is essential for achieving long-term emissions reduction targets (Mahi 2025). Further, corporations can transition toward using renewable energy and decarbonise their supply chains (Lee and Wen 2025). Corporations can adopt circular economy principles to minimise waste and maximise the reuse, repair, and recycling of products and materials to reduce environmental harm. There is a strong relationship between circular economy practices and the targets of SDGs for Clean Water and Sanitation, Affordable and Clean Energy, Decent Work and Economic Growth, Responsible Consumption and Production, and protecting Life on Land and Sea (Schroeder, Anggraeni, and Weber 2019). Simple changes like designing products for durability and long-term use can reduce waste and lower products' overall long-term environmental impact (Dantas et al. 2021).

Although all 17 SDG goals are important, there are three goals where corporations' contribution can have the most impact. These three SDG goals are Goal 8—Decent Work and Economic Growth, Goal 12—Responsible Consumption and Production, and Goal 13—Climate Action. These goals are the ones where most firms report their disclosures (KPMG 2022). Table 7.1 summarises the role of corporations in achieving these three SDGs.

EXHIBIT 7.2 CORPORATE SECTOR AND CLIMATE ACTION

CARBON OFFSET

Corporations which can compensate their unavoidable emissions by funding clean and renewable energy, supporting forestry, or clean water projects. These projects can reduce the overall climate impact of business operations. Corporations like Google[7] and Microsoft[8] actively purchase carbon offsets for their carbon emissions. These carbon offsets are invested in developing new green technologies and reducing emissions through projects unrelated to their core business.

REDUCE EMISSIONS

Corporations can reduce their scope 1, 2 and 3 emissions by adopting net-zero targets. For example, Coca-Cola HBC set a target to reduce scope 1 and 2 emissions in its manufacturing process and scope 3 from its suppliers in packaging and ingredients. The firm has shown that by adopting target-based reduction, corporations can achieve sustainability.[9] Similarly, Unilever has also adopted a climate transition action plan,[10] which aims to reduce its operations and value chain. They have adopted a holistic approach, from raw material suppliers to the final consumption, by improving processes and reducing emissions.

CLIMATE ADVOCACY

Corporations are a strong advocacy group that supports climate-related intervention in markets. Firms with stronger CSR records offer policymakers justification for their policy stances and provide evidence to support policy choices (Favotto and Kollman 2025). Environment Bank[11] works toward restoring biodiversity and creating new habitats in the UK. It influences the environmental policy and collaborates with other corporations to combine climate action.

TABLE 7.1
Business Contribution to SDGs

SDG Goals	Corporate Contribution to Support the SDG[6]	Corporate Contribution Description
Goal 8—Promote sustained, inclusive, and sustainable economic growth, full and productive employment, and decent work for all	Creating Employment Opportunities and Guaranteeing a Living Wage	Corporations can drive economic growth by generating jobs and ensuring a living wage, particularly for underrepresented and marginalised groups (McBride and Mustchin 2013).
	Investing in Education and Skill Development	Investing in education is crucial for developing a skilled workforce capable of contributing to economic growth (Finegold, Benson, and Mohrman 2002).
	Promoting Diversity and Inclusion	A diverse and inclusive workforce fosters innovation and reflects a commitment to social equity (Norbash and Kadom 2020; Fitzsimmons et al. 2023).
	Investing in Sustainable Infrastructure	Sustainable infrastructure investments are vital for long-term economic growth. Constrained by debt and deficits, governments do not have the means to meet this global infrastructure gap. Public–private partnerships can help bridge that gap (Hebb 2019).
	Supporting Community Welfare	Corporations can enhance economic growth by investing in community development and welfare initiatives. A local economy organised around smaller-scale and diversified enterprises has more favourable societal outcomes (Tsang, Frost, and Cao 2023).
Goal 12—Ensure sustainable consumption and production patterns	Adoption of circular economy Practices	Adopting circular economy principles enables organisations to obtain more sustainable results while reducing impacts (Schroeder, Anggraeni, and Weber 2019; Dey et al. 2022).
	Adoption of Green and Sustainable Supply Chain Management	Businesses adopting sustainable procurement policies ensure a decrease in waste and a shorter lead time from manufacturing locations to markets (Rohit Agrawal et al. 2022).
	Optimisation of Resource Efficiency	Energy efficiency adoption by corporations is essential for achieving long-term emissions reduction targets (Mahi 2025). Adoption of energy efficiency has a positive relationship with sustainable development (Zakari et al. 2022).
	Stakeholder Education and Engagement	An important and often neglected part is consumer awareness and responsible marketing by firms (Grewal, Kopalle, and Hulland 2024). Businesses should continuously improve engagement by evaluating and learning from past efforts to reach and interact with stakeholders, strengthening sustainable development efforts (Monteiro et al. 2024).

TABLE 7.1 (*Continued*)
Business Contribution to SDGs

SDG Goals	Corporate Contribution to Support the SDG[6]	Corporate Contribution Description
Goal 13—Take urgent action to combat climate change and its impacts	Transition to Renewable Energy	Adoption of renewable energy, such as solar and wind energy, by businesses can reduce their reliance on fossil fuels. Corporate renewable energy procurement could accelerate the transition to clean energy (O'Shaughnessy et al. 2021).
	Carbon Footprint Reduction	Firms' direct action to reduce carbon footprint leads to ecological sustainability (Penz and Polsa 2018). Implementing carbon capture, setting net-zero targets, and reducing emissions, among other actions, can decarbonise firms' operations (Lee and Wen 2025).
	Green Innovation	Corporations can lead the development of low-carbon technologies such as EVs, eco-friendly materials, and biodegradable products. Green innovation has also been found to be a significant positive predictor of corporate sustainable development (Wang et al. 2022).
	Climate Advocacy	Climate governance involves many stakeholders with competing interests. Corporations can influence policies to promote sustainability awareness and public–private partnerships (Böhler, Hanegraaff, and Schulze 2022).

7.4.1 Role of Smaller Corporations in Achieving SDGs

Studies often focus on the importance of large corporations in providing financing and technologies to support SDGs. It is understandable because they generate a large portion of the world's output by value (ILOSTAT 2019). However, when it comes to employment, they just contribute a small piece of the entire pie (OECD 2024). In terms of employment, the contribution of micro, small, and medium-scale enterprises (MSMEs) is much higher. And that contribution increases further in developing economies (UNDESA 2020). MSMEs play an important role in creating employment and income, and are usually the service providers at the tail end (ILO 2019). Their significant numbers make them unmissable participants in achieving the SDG targets. The contribution of MSMEs is linked to all seventeen SDGs (UNDESA 2020). Moreover, as the tail end of the service providers, their role is even more important in implementing the SDGs.

Elaborating on some of the contributions of small, and medium-scale enterprises, they can help eradicate poverty by generating employment and a decent work environment (Beck, Demirguc-Kunt, and Levine 2005). Small businesses can reach groups in which larger organisations are not interested (Beck, Demirgüç-Kunt, and Singer 2013). Small firms can provide food security by operating a sustainable supply chain (Reardon, Liverpool-Tasie, and Minten 2021). SMEs can help optimise existing healthcare infrastructure and reduce overall delivery costs (Horgan, van Kranen, and Morré 2018; Akinola and Obokoh 2024). One way to reduce delivery costs is by integrating ICT into health service delivery (Pfister and Lehmann 2023).

Similarly, for water and sanitation, which are generally serviced by the public sector, engaging SMEs can improve outcomes (Gero et al. 2014). The private sector engagement supported by public funding can deliver these services more efficiently by providing innovative solutions. Although individual SME firms have a lower environmental footprint, their overall number still adds to a significant total. They have a significant potential to achieve higher energy efficiency by adopting green processes and technologies (Agrawal et al. 2023).

EXHIBIT 7.3 EXAMPLES OF SMES' CONTRIBUTION TO SDGS

RECYCLE AND REUSE

GPS Renewables[12] produces biofuel from waste in India. It collects municipal and agricultural waste and generates biofuel out of it. It also builds and manages biofuel plants for other firms. It started as a startup and now operates the entire waste management cycle, covering collection, treatment and disposal. Similarly, Deluxe recycling[13] converts waste plastics and packaging materials into composite board, industrial pallets and corrugated sheets in India.

SKILL DEVELOPMENT AND RURAL EMPLOYMENT

Rural Shores[14] is engaged in creating employment in rural India. It trains youth in rural areas and provides them with employment in the service delivery operations of corporations. It provides youth employment in their villages, promoting economic development in such areas.

SUSTAINABLE SUPPLY CHAIN

Cotton Connect[15] works to create a sustainable supply chain for the textile industry. It trains and educates local farmers to grow cotton sustainably, which not only helps increase their income but also reduces the impact of climate change on agriculture. It also helps global brands connect with local farmers for sourcing, creating markets for sustainably grown products. It also provides supply chain visibility to corporations and helps them make informed decisions.

SME enterprises usually lack the resources to transition into green operations. In collaboration with the government, corporate financial institutions can support SMEs through green financing for mitigation and adaptation projects. An example of such a mechanism is the Green Finance Scheme[16] operated by SIDBI[17] in India. It raises funds through the markets to support green financing. This mechanism ensures that capital risk remains distributed for private investors, and no single investor has unusually large exposure to individual SMEs (Aliano, Cestari, and Madonna 2024).

7.5 MEASURING CORPORATE SECTOR CONTRIBUTION

There are multiple standards for climate-related disclosure and measuring ESG activities. The most common reporting standards are the Global Reporting Initiative, Task Force on Climate-related Financial Disclosures, Sustainability Accounting Standards Board and UN SDGs reporting. These reporting standards are optional and are usually adopted by multinationals. However, regulators in many countries have adopted reporting regulations that make climate-related disclosures mandatory for listed firms. Many of these regional reporting standards are designed to capture corporations' responsibility toward sustainable development. India's BRSR framework is one such example of mandatory regulation (Khalid and Rawat 2025). Table 7.2 in the appendix summarises India's BRSR principal linkages to the SDGs.

Materiality should be the foundation of sustainability-related disclosures, general financial reporting, and a practical starting point for companies of all sizes. It helps determine whether the omission or misstatement of certain information in a financial report would impact a reasonable user's decision-making. If any ESG-related information is significant, it should be material, and, if not, then it is said to be immaterial (Khan, Serafeim, and Yoon 2016). The materiality process assesses an ESG-related topic's impact within a specific context. Several external rating agencies evaluate and rate firms' ESG practices. Bloomberg,[18] MSCI,[19] and S&P Global,[20] and so on are a few examples of external rating agencies that evaluate firms based on their sustainability performance and provide visibility to investors and other stakeholders.

Another important aspect related to the measurement of corporate contribution is its assurance, which comes through corporate disclosure assessment by an independent external agency. It adds assurance that sustainability reporting information enhances the credibility of the reported information (Simnett, Vanstraelen, and Chua 2009; Maroun 2020). These independent audits enhance the firm's capabilities to navigate climate policy uncertainties (Mohy-ud-Din, Shahbaz, and Du 2025). A proactive sustainability disclosure also communicates their legitimacy to operate to the stakeholders (Michelon 2011; Sehgal, Garg, and Singh 2023). Further audit committee, in addition to the sustainability committee, adds credibility and improves sustainability reporting (Al-Shaer and Zaman 2018).

Corporate-level sustainability committees also play an important role in a firm's CSR investments and effective climate change mitigation strategies (Mohy-ud-Din, Shahbaz, and Du 2025). Although the audit committee remit encompasses monitoring of sustainability issues, there are potential complementary and substitution issues between governance mechanisms. Additional spending on assurance can be an additional financial burden on smaller firms; however, it adds credibility to the sustainability reporting through their expertise and oversight (Pozzoli, Pagani, and Paolone 2022).

7.6 CHALLENGES AND LIMITATIONS

However, there are limitations to which the corporate sector can help. In a study of the world's 70 largest firms, Gneiting and Mhlanga (2021) observed that firms reap positive reputational benefits by marketing their SDGs contribution; however, on the ground, it is business as usual. One of the problems they pointed out is the voluntary nature of reporting corporate contributions, which allows companies to be selective in their reporting. By selective reporting, they can set a positive agenda, shield core business areas from the SDGs and veil the existence of conflicting interests.

Another challenge with corporate sustainability reporting is the failure to adhere to the materiality principle in sustainability disclosures (Jebe 2019). Regulators in a few countries have issued guidelines for corporate sustainability-related disclosures. However, compliance remains limited to a few large farms due to the optional nature of these regulations. For SMEs, the challenges are further extreme, as many countries lack regulation on corporate sustainability disclosure, and, wherever it exists, supervision is inadequate (Setyaningsih, Widjojo, and Kelle 2024).

A challenge firms face, especially those operating in the most climate-vulnerable regions, is the financing challenge. Corporations in developing countries face financing challenges for both physical risk, a risk due to the direct impact of climate change, and transition risk, a risk that arises due to a possible transition to a low-carbon economy (Giglio, Kelly, and Stroebel 2021). Due to climate change, firms in developing countries face uneven labour demand, productivity decline, and increased variability in earnings, leading to more uncertainty in the business environment (Feriga, Gracia, and Serneels 2025). This uncertainty will make climate financing even more difficult for such corporations, as they already work with thinner profit margins. The SMEs that are critical for service delivery in many SDGs are also vulnerable to financing risk (OECD 2024).

Over the years, it has been observed that GHG emissions have moved from developed to emerging economies; however, climate financing has not followed a similar path (Allen et al. 2025).

Despite contributing the least, developing countries suffer the most and lack resources for mitigation. A potential solution to this financial problem is through financial instruments like green bonds (Banga 2019). Financial markets in these countries are usually also not fully developed, making mobilising funds through such innovative financial instruments challenging.

Lack of climate leadership is another issue affecting firms' quick adoption of sustainable practices (Rosati and Faria 2019). The lack of formality regarding the sustainability commitment at the highest decision-making level is related to a low level of SDG integration in the reporting and management systems (Ferrero-Ferrero et al. 2023). Innovative leadership improves the business outcome in environmental innovations (Chassagnon and Haned 2015).

7.7 CONCLUSION

The corporate sector's contribution to achieving the SDGs is pivotal, especially in those regions where governments lack resources. As an important sector of the economy, the corporate sector's contribution is essential to achieving targets for the SDGs. Achieving the SDGs requires strong collaboration between governments, businesses, and civil society. Firms that prioritise sustainability not only contribute to climate action but also safeguard their long-term success. Corporate sectors can help in achieving all seventeen SDG goals. Achieving these requires contributions from large corporations and engagement from SMEs, which usually act as end-point service delivery.

The private sector can significantly contribute to responsible consumption and production. Financing institutions can provide incentives to firms based on their ESG performance. They can incentivise firms that incorporate long-term sustainability into their growth strategy. Integration of ESG with financial market incentives can help bridge the funding gap. Further, developing a robust and transparent reporting framework for ESG performance gives firms and external stakeholders visibility. Push for creating an ecosystem with industry peers that supports long-term sustainable investments. The private sector is important in public policy advocacy for climate action; it can provide real-world evidence supporting policy changes.

Despite huge potential, the corporate sector faces multiple challenges in orienting its operations to implement SDG targets. The major one is access to climate finance in developing countries, where the effects of climate change are most severe. A possible solution to this financing challenge is usually an external multilateral funding. However, large corporations from developed nations can also provide solutions, serve as carriers of public policies, and harmonise global climate regulations across regions. They can implement climate solutions more efficiently than governments by coordinating actions, diffusing best practices, and funding renewable energy projects.

The other limitation is associated with reporting corporate environmental disclosures objectively. Many countries have adopted mandatory disclosure policies; however, they are limited to a selected list of large firms and do not cover SMEs. Despite these mandatory disclosure policies, the assurance and audit requirements for these disclosures required further strengthening. An effective monitoring and evaluation framework adopting principles of materiality will help firms of all sizes. Finally, firms also face challenges in adopting sustainability practices due to a lack of innovative leadership; studies have shown that top-level management commitments are often key to driving corporate sustainability efforts. Finally, despite many limitations, contributions from corporations, small to large, are key to reducing their environmental impact; they also hold the power to drive systemic change through leadership, innovation, and collaboration.

To conclude, the corporate sector contribution toward SDGs can be summarised in the following fundamental points, which are as follows:

- Adoption of clear target-based sustainable business goals, such as the Science Based Targets initiative (SBTi) for reduction in emissions.
- Work toward sustainability in the entire value chain by partnering with upstream and downstream firms. It can partner with them to decarbonise their supply chain.

- Adopt robust and transparent corporate sustainability reporting.
- Mobilise resources to create an ecosystem to support SDGs. They can support the development and adoption of clean technologies.

NOTES

1 https://eridirect.com
2 https://ecoverva.com
3 https://www.cashify.in/about-us
4 https://www.patagonia.com/our-footprint/
5 https://www.aurubis.com/en/about-us/group/business-model
6 This is not an exhaustive list of activities that corporations can perform to achieve sustainable development targets.
7 https://sustainability.google/reports/google-carbon-offsets/
8 https://www.microsoft.com/en-us/corporate-responsibility/sustainability/carbon-removal-program
9 https://www.coca-colahellenic.com/en/a-more-sustainable-future/performance
10 https://www.unilever.com/sustainability/climate/our-climate-transition-action-plan/
11 https://www.environmentbank.com
12 https://gpsrenewables.com
13 https://ecolinkindia.com
14 https://www.ruralshores.com/copy-of-home
15 https://www.cottonconnect.org/our-impact
16 https://www.sidbi.in/green-finance
17 Small Industries Development Bank of India (SIDBI) is a financial institution that caters to the financing and development needs of the MSME sector.
18 Bloomberg ESG Data. https://www.bloomberg.com/professional/products/data/enterprise-catalog/esg/
19 MSCI ESG rating. https://www.msci.com/data-and-analytics/sustainability-solutions/esg-ratings
20 S&P Global Climate Transition Assessment (CTA). https://www.spglobal.com/ratings/en/products-benefits/products/climate-transition-assessment

REFERENCES

Agrawal, Rohit, L. De Tommasi, P. Lyons, S. Zanoni, G. K. Papagiannis, C. Karakosta, A. Papapostolou, et al. 2023. "Challenges and Opportunities for Improving Energy Efficiency in SMEs: Learnings from Seven European Projects." *Energy Efficiency* 16 (3): 17. https://doi.org/10.1007/s12053-023-10090-z.

Agrawal, Rohit, Abhijit Majumdar, Kirty Majumdar, Rakesh D. Raut, and Balkrishna E. Narkhede. 2022. "Attaining Sustainable Development Goals (SDGs) through Supply Chain Practices and Business Strategies: A Systematic Review with Bibliometric and Network Analyses." *Business Strategy and the Environment* 31 (7): 3669–3687. https://doi.org/10.1002/bse.3057.

Akinola, Segun, and Lawrence Obokoh. 2024. "Opportunities and Challenges of Empowering SMEs for Sustainable Healthcare Delivery." *Journal of Infrastructure, Policy and Development* 8 (2). https://doi.org/10.24294/jipd.v8i2.2579.

Aliano, Mauro, Greta Cestari, and Salvatore Madonna. 2024. "Financial Instruments for Green Finance Tailored to SMEs." In *Sustainable Finance for SMEs*, edited by Mauro Aliano, Greta Cestari, and Salvatore Madonna, 73–91. Sustainable Finance. Cham: Springer Nature Switzerland. https://doi.org/10.1007/978-3-031-74193-7_5.

Allen, Franklin, Adelina Barbalau, Erik Chavez, and Federica Zeni. 2025. "Leveraging the Capabilities of Multinational Firms to Address Climate Change: A Finance Perspective." *Journal of International Business Studies* (January). https://doi.org/10.1057/s41267-024-00748-w.

Al-Shaer, Habiba, and Mahbub Zaman. 2018. "Credibility of Sustainability Reports: The Contribution of Audit Committees." *Business Strategy and the Environment* 27 (7): 973–986. https://doi.org/10.1002/bse.2046.

Banga, Josué. 2019. "The Green Bond Market: A Potential Source of Climate Finance for Developing Countries." *Journal of Sustainable Finance & Investment* 9 (1): 17–32. https://doi.org/10.1080/20430795.2018.1498617.

Barnett, Michael, William Brock, and Lars Peter Hansen. 2020. "Pricing Uncertainty Induced by Climate Change." Edited by Harrison Hong. *The Review of Financial Studies* 33 (3): 1024–1066. https://doi.org/10.1093/rfs/hhz144.

Baumgartner, Rupert J., and Daniela Ebner. 2010. "Corporate Sustainability Strategies: Sustainability Profiles and Maturity Levels." *Sustainable Development* 18 (2): 76–89. https://doi.org/10.1002/sd.447.

Beck, Thorsten, Asli Demirguc-Kunt, and Ross Levine. 2005. "SMEs, Growth, and Poverty: Cross-Country Evidence." *Journal of Economic Growth* 10 (3): 199–229. https://doi.org/10.1007/s10887-005-3533-5.

Beck, Thorsten, Asli Demirgüç-Kunt, and Dorothe Singer. 2013. "Is Small Beautiful? Financial Structure, Size and Access to Finance." *World Development* 52 (December): 19–33. https://doi.org/10.1016/j.worlddev.2013.05.014.

Bocken, Nancy. 2021. "Sustainable Business Models." In *Decent Work and Economic Growth*, edited by Walter Leal Filho, Anabela Marisa Azul, Luciana Brandli, Amanda Lange Salvia, and Tony Wall, 963–975. Encyclopedia of the UN Sustainable Development Goals. Cham: Springer International Publishing. https://doi.org/10.1007/978-3-319-95867-5_48.

Böhler, Heike, Marcel Hanegraaff, and Kai Schulze. 2022. "Does Climate Advocacy Matter? The Importance of Competing Interest Groups for National Climate Policies." *Climate Policy* 22 (8): 961–975. https://doi.org/10.1080/14693062.2022.2036089.

Castelblanco, Gabriel, and Jose Guevara. 2022. "Building Bridges: Unraveling the Missing Links between Public-Private Partnerships and Sustainable Development." *Project Leadership and Society* 3 (December): 100059. https://doi.org/10.1016/j.plas.2022.100059.

Chassagnon, Virgile, and Naciba Haned. 2015. "The Relevance of Innovation Leadership for Environmental Benefits: A Firm-Level Empirical Analysis on French Firms." *Technological Forecasting and Social Change* 91 (February): 194–207. https://doi.org/10.1016/j.techfore.2014.02.012.

Chen, Peter, Loukas Karabarbounis, and Brent Neiman. 2017. "The Global Rise of Corporate Saving." *Journal of Monetary Economics* 89 (August): 1–19. https://doi.org/10.1016/j.jmoneco.2017.03.004.

Cingolani, Massimo. 2024. "Public and Private Financing of Sustainable Development Goals (SDGs)." *Review of Political Economy* 36 (2): 792–826. https://doi.org/10.1080/09538259.2022.2063513.

Cox, Emma, Colm Kelly, Barry Murphy, and Nicole Röttmer. 2022. "Time to Get Serious about the Realities of Climate Risk." *Strategy+Business, PwC*, May 16. https://www.pwc.com/gx/en/services/sustainability/publications/risks-and-opportunities-of-climate-change-on-business.html.

Dantas, T. E. T., E. D. de-Souza, I. R. Destro, G. Hammes, C. M. T. Rodriguez, and S. R. Soares. 2021. "How the Combination of Circular Economy and Industry 4.0 Can Contribute towards Achieving the Sustainable Development Goals." *Sustainable Production and Consumption* 26 (April): 213–227. https://doi.org/10.1016/j.spc.2020.10.005.

Darnall, Nicole, Hyunjung Ji, Kazuyuki Iwata, and Toshi H. Arimura. 2022. "Do ESG Reporting Guidelines and Verifications Enhance Firms' Information Disclosure?" *Corporate Social Responsibility and Environmental Management* 29 (5): 1214–1230. https://doi.org/10.1002/csr.2265.

Dey, Prasanta Kumar, Chrysovalantis Malesios, Soumyadeb Chowdhury, Krishnendu Saha, Pawan Budhwar, and Debashree De. 2022. "Adoption of Circular Economy Practices in Small and Medium-Sized Enterprises: Evidence from Europe." *International Journal of Production Economics* 248 (June): 108496. https://doi.org/10.1016/j.ijpe.2022.108496.

Duan, Dingkang, Ran Wei, Chao Wang, and Bingyu Xia. 2025. "Opportunity or Obstacle? Climate Risk Disclosure and Corporate ESG Performance." *International Review of Economics & Finance* 100 (June): 104101. https://doi.org/10.1016/j.iref.2025.104101.

Favotto, Alvise, and Kelly Kollman. 2025. "Corporate Responsibility and the State: Business Lobbying on Climate Change in the UK." *Interest Groups & Advocacy* (May). https://doi.org/10.1057/s41309-025-00230-4.

Feriga, Moustafa, Nancy Lozano Gracia, and Pieter Serneels. 2025. "The Impact of Climate Change on Work: Lessons for Developing Countries." *The World Bank Research Observer* 40 (1): 104–146. https://doi.org/10.1093/wbro/lkae002.

Ferrero-Ferrero, Idoya, María Jesús Muñoz-Torres, Juana María Rivera-Lirio, Elena Escrig-Olmedo, and María Ángeles Fernández-Izquierdo. 2023. "SDG Reporting: An Analysis of Corporate Sustainability Leaders." *Marketing Intelligence & Planning* 41 (4): 457–472. https://doi.org/10.1108/MIP-07-2022-0332.

Finegold, David, George S. Benson, and Susan A. Mohrman. 2002. "Harvesting What They Grow: Can Firms Get a Return on Investments in General Skills?" *Organizational Dynamics* 31 (2) [New York, American Management Association]: 151–164.

Fitzsimmons, Stacey, Mustafa F. Özbilgin, David C. Thomas, and Stella Nkomo. 2023. "Equality, Diversity, and Inclusion in International Business: A Review and Research Agenda." *Journal of International Business Studies* 54 (8): 1402–1422. https://doi.org/10.1057/s41267-023-00642-x.

Garrido-Ruso, María, Beatriz Aibar-Guzmán, and Óscar Suárez-Fernández. 2023. "What Kind of Leaders Can Promote the Disclosure of Information on the Sustainable Development Goals?" *Sustainable Development* 31 (4): 2694–2710. https://doi.org/10.1002/sd.2539.

Gero, Anna, Naomi Carrard, Janina Murta, and Juliet Willetts. 2014. "Private and Social Enterprise Roles in Water, Sanitation and Hygiene for the Poor: A Systematic Review." *Journal of Water, Sanitation and Hygiene for Development* 4 (3): 331–345. https://doi.org/10.2166/washdev.2014.003.

Giglio, Stefano, Bryan Kelly, and Johannes Stroebel. 2021. "Climate Finance." *Annual Review of Financial Economics* 13 (1): 15–36. https://doi.org/10.1146/annurev-financial-102620-103311.

Gneiting, Uwe, and Ruth Mhlanga. 2021. "The Partner Myth: Analysing the Limitations of Private Sector Contributions to the Sustainable Development Goals." *Development in Practice* 31 (7): 920–926. https://doi.org/10.1080/09614524.2021.1938512.

Grewal, Dhruv, Praveen K. Kopalle, and John Hulland. 2024. "Addressing the Greatest Global Challenges (UN SDGs) with a Marketing Lens." *Journal of the Academy of Marketing Science* 52 (5): 1263–1272. https://doi.org/10.1007/s11747-024-01049-2.

Hebb, Tessa. 2019. "Investing in Sustainable Infrastructure." In *Challenges in Managing Sustainable Business*, edited by Susanne Arvidsson, 251–273. Cham: Springer International Publishing. https://doi.org/10.1007/978-3-319-93266-8_11.

Horgan, Denis, Henk J. van Kranen, and Servaas A. Morré. 2018. "Optimising SME Potential in Modern Healthcare Systems: Challenges, Opportunities and Policy Recommendations." *Public Health Genomics* 21 (1–2): 1–17. https://doi.org/10.1159/000492809.

Hummel, Katrin, and Christian Schlick. 2016. "The Relationship between Sustainability Performance and Sustainability Disclosure—Reconciling Voluntary Disclosure Theory and Legitimacy Theory." *Journal of Accounting and Public Policy* 35 (5): 455–476. https://doi.org/10.1016/j.jaccpubpol.2016.06.001.

ILO. 2019. "The Power of Small: Unlocking the Potential of SMEs." October. https://webapps.ilo.org/infostories/en-GB/Stories/Employment/SMEs.

ILOSTAT. 2019. "Employment by Economic Activity." *International Labor Organization*. https://ilostat.ilo.org/.

Jebe, Ruth. 2019. "The Convergence of Financial and ESG Materiality: Taking Sustainability Mainstream." *American Business Law Journal* 56 (3): 645–702. https://doi.org/10.1111/ablj.12148.

Jomo, Kwame Sundaram, Anis Chowdhury, Krishnan Sharma, and Daniel Platz. 2016. "Public-Private Partnerships and the 2030 Agenda for Sustainable Development: Fit for Purpose?" *DESA Working Paper No. 148*. United Nations. https://www.un.org/esa/desa/papers/2016/wp148_2016.pdf.

Kalinowski, Thomas. 2024. "The Green Climate Fund and Private Sector Climate Finance in the Global South." *Climate Policy* 24 (3): 281–296. https://doi.org/10.1080/14693062.2023.2276857.

Khalid, Ahmad Mohd, and Pankaj S. Rawat. 2025. "Corporate Environmental Disclosures and Role of Top Management: Evidence Based on the Business Responsibility and Sustainability Reporting in India." *Corporate Social Responsibility and Environmental Management* (April). https://doi.org/10.1002/csr.3205.

Khan, Mozaffar, George Serafeim, and Aaron Yoon. 2016. "Corporate Sustainability: First Evidence on Materiality." *The Accounting Review* 91 (6): 1697–1724. https://doi.org/10.2308/accr-51383.

KPMG. 2022. "Big Shifts, Small Steps." *Survey of Sustainability Reporting*. KPMG International. https://assets.kpmg.com/content/dam/kpmg/se/pdf/komm/2022/Global-Survey-of-Sustainability-Reporting-2022.pdf.

Lashitew, Addisu A. 2021. "Corporate Uptake of the Sustainable Development Goals: Mere Greenwashing or an Advent of Institutional Change?" *Journal of International Business Policy* 4 (1): 184–200. https://doi.org/10.1057/s42214-020-00092-4.

Lee, Chien-Chiang, and Huwei Wen. 2025. "Global Value Chain Embedding and Enterprise Energy Efficiency: A Worldwide Firm-Level Analysis." *Renewable and Sustainable Energy Reviews* 207 (January): 114955. https://doi.org/10.1016/j.rser.2024.114955.

Li, Qing, Hongyu Shan, Yuehua Tang, and Vincent Yao. 2024. "Corporate Climate Risk: Measurements and Responses." Edited by Itay Goldstein. *The Review of Financial Studies* 37 (6): 1778–1830. https://doi.org/10.1093/rfs/hhad094.

Mahi, Masnun Al. 2025. "Gaining More from Energy Efficiency: The Role of Firms' Environmental Concerns." *Journal of Environmental Management* 373 (January): 123918. https://doi.org/10.1016/j.jenvman.2024.123918.

Maroun, Warren. 2020. "A Conceptual Model for Understanding Corporate Social Responsibility Assurance Practice." *Journal of Business Ethics* 161 (1): 187–209. https://doi.org/10.1007/s10551-018-3909-z.

McBride, A., and S. Mustchin. 2013. "Creating Sustainable Employment Opportunities for the Unemployed." *Policy Studies* 34 (3): 342–359. https://doi.org/10.1080/01442872.2013.804302.

Michelon, Giovanna. 2011. "Sustainability Disclosure and Reputation: A Comparative Study." *Corporate Reputation Review* 14 (2): 79–96. https://doi.org/10.1057/crr.2011.10.

Mohieldin, Mahmoud. 2017. "The Sustainable Development Goals and Private Sector Opportunities." *United Nations*. http://pubdocs.worldbank.org/en/394231501877501769/The-Sustainable-Development-Goals-and-Private-Sector-Opportunities.pdf.

Mohy-ud-Din, Kamran, Muhammad Shahbaz, and Anna Min Du. 2025. "Corporate Social Responsibility and Climate Change Mitigation: Discovering the Interaction Role of Green Audit and Sustainability Committee." *Corporate Social Responsibility and Environmental Management* 32 (1): 1198–1212. https://doi.org/10.1002/csr.3011.

Monteiro, Sónia, Beatriz Aibar-Guzmán, and Alice Loureiro. 2024. "Embedding Stakeholder Engagement into SDG Reporting Processes." In *The Routledge Handbook of Accounting for the Sustainable Development Goals*, 58–73. Routledge. https://doi.org/10.4324/9781003404118-5.

Norbash, Alexander, and Nadja Kadom. 2020. "The Business Case for Diversity and Inclusion." *Journal of the American College of Radiology* 17 (5): 676–680. https://doi.org/10.1016/j.jacr.2019.11.018.

OECD. 2024. "Financing SMEs and Entrepreneurs 2024: An OECD Scoreboard." *Financing SMEs and Entrepreneurs*. OECD. https://doi.org/10.1787/fa521246-en.

O'Shaughnessy, Eric, Jenny Heeter, Chandra Shah, and Sam Koebrich. 2021. "Corporate Acceleration of the Renewable Energy Transition and Implications for Electric Grids." *Renewable and Sustainable Energy Reviews* 146 (August): 111160. https://doi.org/10.1016/j.rser.2021.111160.

Palau-Pinyana, Erola, Josep Llach, and Llorenç Bagur-Femenías. 2024. "Mapping Enablers for SDG Implementation in the Private Sector: A Systematic Literature Review and Research Agenda." *Management Review Quarterly* 74 (3): 1559–1588. https://doi.org/10.1007/s11301-023-00341-9.

Park, Sang-Bum. 2018. "Multinationals and Sustainable Development: Does Internationalization Develop Corporate Sustainability of Emerging Market Multinationals?" *Business Strategy and the Environment* 27 (8): 1514–1524. https://doi.org/10.1002/bse.2209.

Penz, Elfriede, and Pia Polsa. 2018. "How Do Companies Reduce Their Carbon Footprint and How Do They Communicate These Measures to Stakeholders?" *Journal of Cleaner Production* 195 (September): 1125–1138. https://doi.org/10.1016/j.jclepro.2018.05.263.

Pfister, Paul, and Claudia Lehmann. 2023. "Returns on Digitisation in SMEs—a Systematic Literature Review." *Journal of Small Business & Entrepreneurship* 35 (4): 574–598. https://doi.org/10.1080/08276331.2021.1980680.

Pozzoli, Matteo, Alessandra Pagani, and Francesco Paolone. 2022. "The Impact of Audit Committee Characteristics on ESG Performance in the European Union Member States: Empirical Evidence before and during the COVID-19 Pandemic." *Journal of Cleaner Production* 371 (October): 133411. https://doi.org/10.1016/j.jclepro.2022.133411.

Qi, Ji, and Haoqi Qian. 2023. "Climate Finance at a Crossroads: It Is High Time to Use the Global Solution for Global Problems." *Carbon Neutrality* 2 (1): 31. https://doi.org/10.1007/s43979-023-00071-7.

Reardon, Thomas, Lenis Saweda O. Liverpool-Tasie, and Bart Minten. 2021. "Quiet Revolution by SMEs in the Midstream of Value Chains in Developing Regions: Wholesale Markets, Wholesalers, Logistics, and Processing." *Food Security* 13 (6): 1577–1594. https://doi.org/10.1007/s12571-021-01224-1.

Rosati, Francesco, and Lourenço Galvão Diniz Faria. 2019. "Business Contribution to the Sustainable Development Agenda: Organizational Factors Related to Early Adoption of SDG Reporting." *Corporate Social Responsibility and Environmental Management* 26 (3): 588–597. https://doi.org/10.1002/csr.1705.

Sautner, Zacharias, Laurence Van Lent, Grigory Vilkov, and Ruishen Zhang. 2023. "Firm-Level Climate Change Exposure." *The Journal of Finance* 78 (3): 1449–1498. https://doi.org/10.1111/jofi.13219.

Scheyvens, Regina, Glenn Banks, and Emma Hughes. 2016. "The Private Sector and the SDGs: The Need to Move Beyond 'Business as Usual'." *Sustainable Development* 24 (6): 371–382. https://doi.org/10.1002/sd.1623.

Schneider, Tina. 2014. "Responsibility for Private Sector Adaptation to Climate Change." *Ecology and Society* 19 (2) [Resilience Alliance Inc.]. http://www.jstor.org/stable/26269584.

Schroeder, Patrick, Kartika Anggraeni, and Uwe Weber. 2019. "The Relevance of Circular Economy Practices to the Sustainable Development Goals." *Journal of Industrial Ecology* 23 (1): 77–95. https://doi.org/10.1111/jiec.12732.

SEBI. 2020. "Consultation Paper on the Format for Business Responsibility and Sustainability Reporting." *Reports for Public Comments*. Securities and Exchange Board of India. https://www.sebi.gov.in/reports-and-statistics/reports/aug-2020/consultation-paper-on-the-format-for-business-responsibility-and-sustainability-reporting_47345.html.

Sehgal, Varsha, Naval Garg, and Jagvinder Singh. 2023. "Impact of Sustainability Performance & Reporting on a Firm's Reputation." *International Journal of System Assurance Engineering and Management* 14 (1): 228–240. https://doi.org/10.1007/s13198-022-01782-3.

Setyaningsih, Santi, Rosita Widjojo, and Peter Kelle. 2024. "Challenges and Opportunities in Sustainability Reporting: A Focus on Small and Medium Enterprises (SMEs)." *Cogent Business & Management* 11 (1): 2298215. https://doi.org/10.1080/23311975.2023.2298215.

Shapiro, Joseph S. 2021. "The Environmental Bias of Trade Policy." *The Quarterly Journal of Economics* 136 (2): 831–886. https://doi.org/10.1093/qje/qjaa042.

Simnett, Roger, Ann Vanstraelen, and Wai Fong Chua. 2009. "Assurance on Sustainability Reports: An International Comparison." *The Accounting Review* 84 (3): 937–967. https://doi.org/10.2308/accr.2009.84.3.937.

S&P Global. 2024. "S&P Truecost." https://www.marketplace.spglobal.com/en/datasets/trucost-environmental-(46).

Tsang, Albert, Tracie Frost, and Huijuan Cao. 2023. "Environmental, Social, and Governance (ESG) Disclosure: A Literature Review." *The British Accounting Review* 55 (1): 101149. https://doi.org/10.1016/j.bar.2022.101149.

UNDESA. 2020. "Micro-, Small and Medium-Sized Enterprises (MSMEs) and Their Role in Achieving the Sustainable Development Goals." *Report on MSMEs and the Sustainable Development Goals*. United Nations. https://sdgs.un.org/sites/default/files/2020-07/MSMEs_and_SDGs.pdf.

UNFCCC. 2018. "Paris Agreement." *United Nations Framework Convention on Climate Change*. https://unfccc.int/documents/184656.

United Nations. 2015. "Transforming Our World: The 2030 Agenda for Sustainable Development." *United Nations*. https://docs.un.org/en/A/RES/70/1.

Wang, Nannan, and Minxun Ma. 2021. "Public–Private Partnership as a Tool for Sustainable Development—What Literatures Say?" *Sustainable Development* 29 (1): 243–258. https://doi.org/10.1002/sd.2127.

Wang, Shouwen, Jawad Abbas, Muhammad Safdar Sial, Susana Álvarez-Otero, and Lucian-Ionel Cioca. 2022. "Achieving Green Innovation and Sustainable Development Goals through Green Knowledge Management: Moderating Role of Organizational Green Culture." *Journal of Innovation & Knowledge* 7 (4): 100272. https://doi.org/10.1016/j.jik.2022.100272.

Weinhofer, Georg, and Timo Busch. 2013. "Corporate Strategies for Managing Climate Risks." *Business Strategy and the Environment* 22 (2): 121–144. https://doi.org/10.1002/bse.1744.

Yan, Min, Fu Jia, Lujie Chen, and Fangxu Yan. 2022. "Assurance Process for Sustainability Reporting: Towards a Conceptual Framework." *Journal of Cleaner Production* 377 (December): 134156. https://doi.org/10.1016/j.jclepro.2022.134156.

Zakari, Abdulrasheed, Irfan Khan, Duojiao Tan, Rafael Alvarado, and Vishal Dagar. 2022. "Energy Efficiency and Sustainable Development Goals (SDGs)." *Energy* 239 (January): 122365. https://doi.org/10.1016/j.energy.2021.122365.

Appendix

Business Responsibility and Sustainability Reporting (BRSR) ensure that corporations must report the social and environmental impact of their business actions (SEBI 2020). BRSR principles are closely related to the Sustainable Development Goals, which are important in ensuring corporations adopt sustainable development. Table 7.2 shows the link between BRSR principles and SDGs.

TABLE 7.2
Linking India's BRSR Principles with the SDGs Goals

BRSR Principles (SEBI 2020)	Linked SDGs
1. Ethical, Transparent, and Accountable Governance Businesses should conduct and govern themselves with integrity and in an ethical, transparent, and accountable manner.	SDG 16
2. Sustainable and Safe Goods & Services Businesses should provide goods and services in a sustainable and safe manner.	SDG 12, SDG 13
3. Employee Well-being and Fair Treatment Businesses should respect and promote the well-being of all employees, including those in their value chains.	SDG 8, SDG 10
4. Stakeholder Engagement & Responsiveness Businesses should respect the interests of and be responsive to all its stakeholders.	SDG 17
5. Respect for Human Rights Businesses should respect and promote human rights.	SDG 10, SDG 16
6. Environmental Protection & Restoration Businesses should respect and try to protect and restore the environment.	SDG 13, SDG 15
7. Responsible Public Policy Advocacy Businesses, when engaging in influencing public and regulatory policy, should do so in a manner that is responsible and transparent.	SDG 16
8. Inclusive Growth & Equitable Development Businesses should promote inclusive growth and equitable development.	SDG 1, SDG 5, SDG 8
9. Responsible Consumer Engagement Businesses should engage with and provide value to their consumers in a responsible manner.	SDG 12

8 Status of Data Reporting and Required Reforms for Effective Monitoring and Verification

Shailendra Singh

8.1 INTRODUCTION

Climate change is one of the most critical global issues, requiring immediate and focused interventions to solve this issue. Timely and accurate data is required for effective decision-making, policy formulation, and progress tracking (Li, 2025). The Climate Action Goal (SDG 13) under the United Nations Sustainable Development Goals (SDGs) highlights the demand for prompt actions to fight climate change and its impacts (Arora & Mishra, 2023). Achieving this goal requires systematic data reporting, monitoring, and verification to ensure transparency, accountability, and targeted policy-making.

Despite considerable international and national efforts to improve climate data collection and reporting, significant gaps persist in data availability, quality, standardization, and accessibility. Many countries face issues related to incomplete datasets, inconsistency in data availability, inadequate statistical capacity, and lack of synergy with global reporting frameworks. These issues become obstacles in effective assessment of progress and evidence-based policy-making (IPCC, 2022; Sachs et al., 2024).

This chapter highlights the current state of data reporting and required reforms for effective monitoring and verification of climate action goals. It examines the data reporting status of countries based on different classifications like geographical, income level, and development status. It also discusses the classifications of reported data based on their nature, periodicity, and age. It examines the challenges associated with data reporting, particularly in those countries which are not reporting data. On the other hand, this chapter highlights the practices followed by those countries which are reporting data consistently. Additionally, this chapter examines the role of the national statistical plan, legal framework, technology, and international cooperation in enhancing data collection and analysis. Furthermore, this chapter reviews the statistical performance indicators of various countries. By identifying critical gaps and proposing necessary reforms, this chapter aims to contribute to the development of robust data systems that can support evidence-based climate policy-making to achieve the climate action goal.

8.2 DATA REPORTING STATUS

Climate change is a dynamic and rapidly escalating threat. Its interconnected nature with other critical issues, such as land degradation, water crisis, biodiversity loss, and food insecurity, requires immediate and comprehensive action. To effectively combat climate change, we need a holistic and sustainable approach (Arora & Mishra, 2023). To provide solutions to climate change, we need data to quantify climate change. For SDG 13, we have eight indicators to measure the performance of climate change. Out of these eight indicators, three indicators (13.2.1, 13.a.1, and 13.b.1) do not have data reporting for any country, revealing an absolute gap in the implementation and tracking of countries' nationally determined contributions (NDCs), financial commitments towards finance,

DOI: 10.1201/9781003600152-8

and support for least developed countries and small island states. Data reporting for the rest of the indicators is different across regions and income levels. This section provides global data reporting status for different SDG13 indicators at different geographical and income levels.

8.2.1 Data Reporting Status at Different Geographical Levels

UN publishes SDG reports not only at the country level but also at the regional level. These regions are Eastern Europe and Central Asia (27 countries), East and South Asia (21), Latin America and the Caribbean (LAC) (29 countries), Middle East and North Africa (MENA) (17 countries), Oceania (12 countries), Organisation for Economic Co-operation and Development (OECD) (38 countries), and sub-Saharan Africa (49 countries)(Sachs et al., 2024). Based on data availability in the UN database,[1] data reporting status at different geographical levels is explained in the following sections.

8.2.1.1 At Global Level

Table 8.1 shows the data reporting status as well as the number of countries for which data is reported. Data reporting is not consistent across all indicators. Some have much higher reporting rates than others. Indicators 13.2.1, 13.b.1, and 13.a.1 show no data reporting for a single country, indicating absolute data and action gaps pertaining to countries' NDCs, financial commitments towards finance, and support for least developed countries and small island states. Indicators 13.1.1 and 13.1.2 show high data reporting, reflecting substantial global engagement in tracking disaster impact and national disaster risk strategies reporting. Moderate data reporting is observed for indicators 13.2.2 and 13.1.3, indicating the need for better greenhouse gas emission reporting and more local-level disaster risk strategy implementation. Lastly, indicator 13.3.1 shows a low data reporting rate, reflecting gaps in mainstreaming global citizenship and sustainable development education.

Overall, out of eight climate action-related indicators, data is being reported for five indicators, covering between 33% and 85% of countries. The highest data reporting (85%) by countries is about disaster-related deaths and affected persons, which is indicator 13.1.1, while the lowest reporting (33%) is about the inclusion of sustainable development education in academic curricula, which is indicator 13.3.1. We do not have data about countries' national climate strategies and financial mobilization.

8.2.1.2 At Regional Level

Figure 8.1 shows the data reporting status for different indicators for different geographical regions. Data reporting is inconsistent for different indicators in different regions. Following is the indicator-wise data reporting breakdown:

Indicator 13.1.1 shows high data reporting across all geographical regions, with 100% reporting by OECD and Oceania, indicating a strong global commitment to tracking disaster-related impacts such as the number of deaths, missing, and directly affected persons.

Indicator 13.1.2 shows strong data reporting, especially by Oceania, East and South Asia, and OECD countries, reflecting substantial engagement in tracking national disaster risk reduction strategies. While other regions such as Eastern Europe and Central Asia, Latin America, the Middle East and North Africa, and sub-Saharan Africa show slightly less data reporting.

Indicator 13.1.3 shows 100% data reporting for Oceania, while moderate data reporting for OECD, Eastern Europe and Central Asia, East and South Asia, and the Middle East and North Africa. Latin America and the Caribbean and sub-Saharan Africa show weak data reporting. This reflects the level of local-level disaster risk reduction strategies implementation and tracking.

Indicator 13.2.2 shows high data reporting by OECD (100%) and Eastern Europe and Central Asia (89%), moderate data reporting by East and South Asia, Latin America and the Caribbean, and the Middle East and North Africa, while sub-Saharan Africa and Oceania show weak data reporting. This indicates the level of tracking of greenhouse gas emissions.

TABLE 8.1
Data Reporting for Different Indicators at the Global Level

Indicator	Data Reporting	No. of Countries for Which Data Is Reported	Percentage of Countries for Which Data Is Reported
13.1.1 Number of deaths, missing persons, and directly affected persons attributed to disasters per 100,000 population	Yes	164	85
13.1.2 Number of countries that adopt and implement national disaster risk reduction strategies in line with the Sendai Framework for Disaster Risk Reduction 2015–2030	Yes	136	70.5
13.1.3 Proportion of local governments that adopt and implement local disaster risk reduction strategies in line with national disaster risk reduction strategies	Yes	119	61.7
13.2.1 Number of countries with nationally determined contributions, long-term strategies, national adaptation plans, and adaptation communications, as reported to the secretariat of the United Nations Framework Convention on Climate Change	No	0	0
13.2.2 Total greenhouse gas emissions per year	Yes	130	67.4
13.3.1 Extent to which (i) global citizenship education and (ii) education for sustainable development are mainstreamed in (a) national education policies; (b) curricula; (c) teacher education; and (d) student assessment	Yes	64	33
13.a.1 Amounts provided and mobilized in United States dollars per year in relation to the continued existing collective mobilization goal of the $100 billion commitment through to 2025	No	0	0
13.b.1 Number of least developed countries and small island developing States with nationally determined contributions, long-term strategies, national adaptation plans and adaptation communications, as reported to the secretariat of the United Nations Framework Convention on Climate Change	No	0	0

Indicator 13.3.1 shows the lowest data reporting across all the regions. Among these also sub-Saharan Africa and Oceania have minimal data reporting. This reflects a significant global gap in tracking the mainstreaming of global citizenship and sustainable development education.

Overall data reporting is highest for disaster-related indicators (13.1.1 and 13.1.2). Significant gaps exist in tracking mainstreaming global citizenship and sustainability education (13.3.1) and implementation of local-level disaster risk reduction strategies (13.1.3). OECD countries perform best across all indicators, while sub-Saharan African countries have room for improvement.

8.2.2 Data Reporting Status at Different Income Levels

For analytical purposes, the World Bank categorizes the countries into four income groups—low, lower-middle, upper-middle, and high income. This categorization relies on the country's per capita gross national income in U.S. dollars. This income-level categorization is used by the United

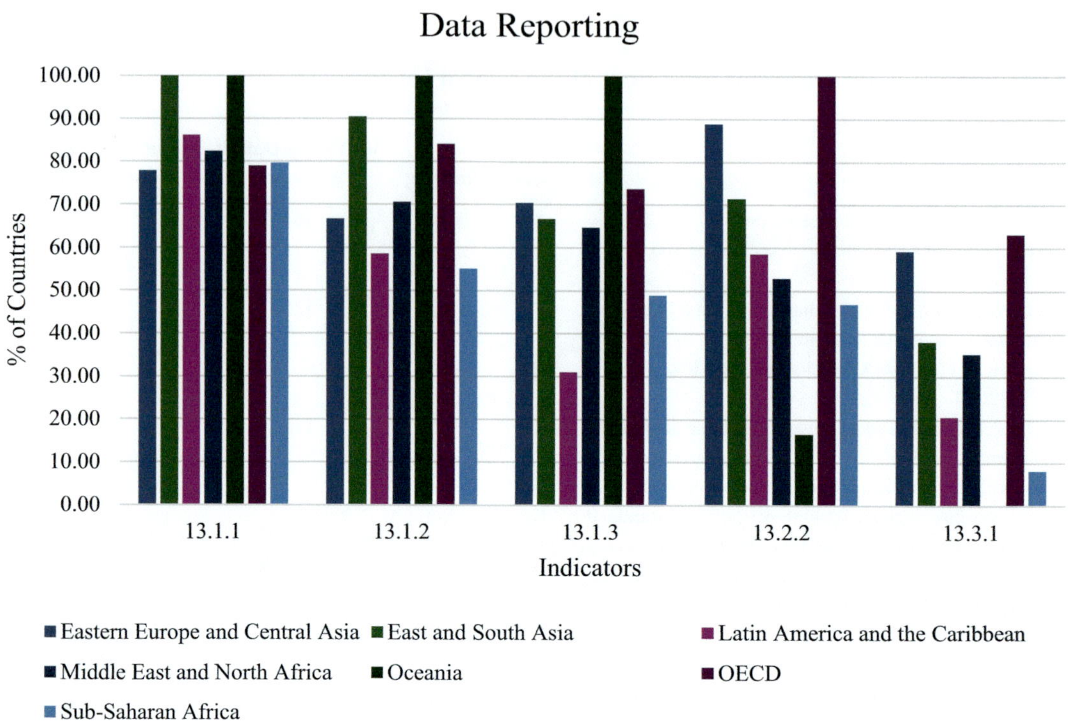

FIGURE 8.1 Data reporting status for different regions.

Nations for SDGs' performance analysis. In this chapter we also used this categorization to analyze the data reporting status of SDG 13 (Climate Action)-related indicators.

Figure 8.2 shows the data reporting status for different indicators for countries at different income levels. Data reporting is inconsistent for different indicators for different income-level countries. The following is the indicator-wise data reporting breakdown:

For indicator 13.1.1 data reporting is high, more than 75%, for all the income group countries. Here low-income countries slightly outperform other income group countries, indicating strong disaster impact data collection.

Indicator 13.1.2 shows a decrease in data reporting with a decrease in income level. More than 70% of high-income, upper-middle-income, and lower-middle-income countries are reporting data, whereas only 50% of the low-income countries are reporting data, reflecting their level of tracking of national disaster risk reduction strategies.

Indicator 13.1.3 shows moderate data reporting across all the income group countries, ranging from 65% (high-income) to 54% (low-income). This indicates the implementation and tracking of local-level disaster risk reduction strategies.

More than 70% of the high-income and upper-middle-income countries are reporting data, whereas only 54% of lower-middle-income and approximately 42% of low-income countries are reporting data for indicator 13.2.2. This reflects that high-income countries have strong tracking while low-income countries have weak tracking of greenhouse gas emissions.

Indicator 13.3.1 shows moderate data reporting for high-income countries while low data reporting for other income group countries, indicating global neglect in tracking of mainstreaming of global citizenship and sustainable development education.

Overall, the disaster impact-related indicator (13.1.3) is reported very well for all the income group countries. The indicator related to mainstreaming of global citizenship and sustainability education (13.3.1) had the lowest data reporting, highlighting the global issues in integrating and

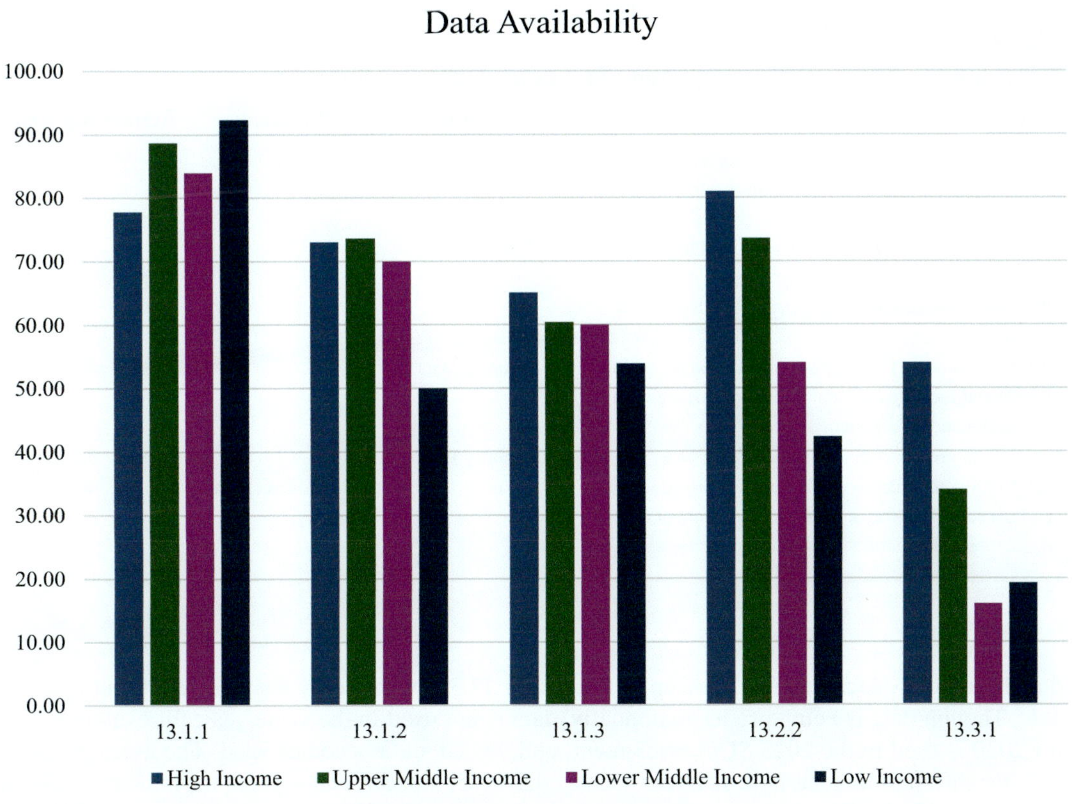

FIGURE 8.2 Data reporting status at different income levels.

tracking the same. For all the indicators with data, except for indicator 13.1.1, high-income countries show better data reporting, and low-income countries show poor data reporting.

8.2.3 Data Reporting Status Based on Nature, Periodicity, and Age

Based on the nature of the data, the UN classified data into five categories, which are defined here:[2]

- Country data (C): Generated and shared by the country (including data modified by the country according to international standards).
- Country data adjusted (CA): Gathered and shared by the country but modified by the international agency according to international standards, definitions, and classifications.
- Estimated (E): Calculated based on the country data (surveys or administrative records, etc.) or based on other sources by the international organizations when national data is not consistently available, when more than one sources exist, or when there is data quality concern.
- Modelled (M): Modelled using other related variables when no direct data exist for the variable.
- Global monitoring data (G): Generated consistently by the global agency for global monitoring. It is based on country data but is not available at the country level.

Table 8.2 shows that for all the indicators, country data, which is compiled and shared by the country, that is, country data (C), is being reported. Table 8.2 also shows that for all the indicators,

TABLE 8.2
Nature, Periodicity, and Average Age of Data for Different Indicators

Indicator	Nature of the Data	Periodicity	Average Age (year)
13.1.1 Number of deaths, missing persons and directly affected persons attributed to disasters per 100,000 population	Country Data	1	4.5
13.1.2 Number of countries that adopt and implement national disaster risk reduction strategies in line with the Sendai Framework for Disaster Risk Reduction 2015–2030	Country Data	1	3.65
13.1.3 Proportion of local governments that adopt and implement local disaster risk reduction strategies in line with national disaster risk reduction strategies	Country Data	1	3.6
13.2.2 Total greenhouse gas emissions per year	Country Data	1	5.7
13.3.1 Extent to which (i) global citizenship education and (ii) education for sustainable development are mainstreamed in national education policies	Country Data	Single data point	4

data is being produced annually except for indicator 13.3.1, where only a single data point is available. Though data is being produced annually, data is not used in the same year; for example, data for 2020 is used in the 2023 SDG assessment, and, hence, data becomes aged. The average age of data for different indicators is different such as 4.5, 3.65, 3.6, 5.7, and 4 years for indicators 13.1.1, 13.1.2, 13.1.3, 13.2.2, and 13.3.1, respectively, as shown in Table 8.2.

Though the periodicity of the data collection for indicators is annual, it creates a deceptive sense of timeliness of data availability, as data for most of the indicators are aged between 3.6 and 5.7 years. The aged data can be potentially less relevant in decision-making for current scenarios. This considerable time lag points out the inefficiency of the data system, from collection to analysis and its application. This reflects the lack of agility to provide near-real-time data required for effective climate action and policy adjustments. The average data age indicates a considerable delay in understanding the present status and impact of interventions.

8.3 NATIONAL STATISTICAL LEGISLATION AND PLAN ADOPTION

National Statistical Legislation provides a legal framework and mandate for the national statistical system. It establishes the roles and responsibilities of statistical agencies, defines data collection standards, ensures data quality, and protects the confidentiality of statistical information (UN, 2024). A robust legal framework is needed for creating a strong and independent statistical system. It provides the authority and resources needed to collect, process, and disseminate reliable and timely statistics.

The National Statistical Plan outlines the strategic objectives and priorities for the national statistical system. It identifies key statistical areas, sets targets for data collection and dissemination, and allocates resources accordingly. The plans align with the principles and objectives of the national statistical legislation. It translates the legal framework into concrete actions and programs.

Strong national statistical legislation and well-designed national statistical plans provide the foundation for good statistical performance. They ensure that the statistical system has the necessary resources, authority, and guidance to produce high-quality statistics. To measure the countries' national statistical legislation and plan adoption, SDG 17 (Partnerships for the Goals) provides two indicators, 17.18.2 (countries with national statistical legislation exist that comply with the

Fundamental Principles of Official Statistics) and 17.18.3 (countries with a national statistical plan that is fully funded and under implementation), listed in Table 8.3. Indicator 17.18.3 is broken down into two sub-indicators, 17.18.3(a) and 17.18.3(b), due to separate data availability for these sub-indicators. Indicator 17.18.2 ensures the legal foundation for sound statistical systems, while indicators 17.18.3(a) and 17.18.3(b) ensure the financial commitment for statistical development and the progress in the statistical plan implementation, respectively.

8.3.1 National Statistical Legislation and Plan Adoption at Global Level

Table 8.3 shows national statistical legislation and plan adoption status at the global level. The following is the indicator-wise explanation of the status of the national statistical legislation and plan adoption.

Indicator 17.18.2 highlights that 74% of countries have national statistical legislation that adheres to the fundamental principles of official statistics, indicating a widespread acknowledgement of the need for legal frameworks to ensure credible, transparent, and standardized national statistics.

Indicator 17.18.3(a) shows that only 53.4% of countries have adopted national statistical plans that are fully funded and under implementation. This indicates that while plans may exist, more than 46% of countries do not have sufficient funding, which hinders the implementation of the national statistical plans.

Indicator 17.18.3(b) shows that 79% of countries have implemented national statistical plans, reflecting their intention to produce high-quality statistics. However, not all of these plans are fully funded, highlighting issues about the long-term viability of the plans.

Strong global progress is evident in the adoption and implementation of national statistical legislation and plans. However, a critical funding gap may create bottlenecks between planning and effective execution. The difference between fully funded and under-implementation statistical plans reveals a disconnect between the intention and resource allocation.

8.3.2 National Statistical Legislation and Plan Adoption at Different Geographical Levels

Figure 8.3 provides information about the status of national statistical legislation and plan adoption across different geographical regions. The following is the region-wise overview of the status of the national statistical legislation and plan adoption based on the indicators given in Table 8.3.

Eastern Europe and Central Asia (EECA) show strong performance across all indicators of national statistical legislation and plan adoption. Eighty-five percent of EECA countries in this region have national statistical legislation that complies with the UN's fundamental principles of official statistics. Further, 81.5% of EECA countries have fully funded national statistical plans, and 89% of EECA countries have national statistical plans under

TABLE 8.3
Percentage of Countries with National Statistical Legislation and Plan Adoption at Global Level

Indicator	No. of Countries	Percentage of Countries
17.18.2 Countries with national statistical legislation exist that comply with the Fundamental Principles of Official Statistics	143	74
17.18.3(a) Countries with a national statistical plan that is fully funded	103	53.4
17.18.3(b) Countries with a national statistical plan that is under implementation	152	79

implementation. This indicates a strong legal framework, sufficient funding for statistical plans, and effective implementation of the plans in the region.

East and South Asia (ESA) display a relatively high percentage of ESA countries (76%) that have national statistical legal frameworks, and 100% of ESA countries in this region have implemented national statistical plans. However, only 47.6% of the ESA countries have their national statistical plan fully funded, indicating a potential obstacle in turning these plans into actions due to a lack of sufficient funds.

Latin America and the Caribbean (LAC) show lower performance across all indicators. Only 45% of LAC countries have a national statistical legal framework that adheres to the UN's fundamental principles of official statistics. Further, only 41.4% of LAC countries have fully funded national statistical plans, and 58.6% have national statistical plans under implementation. This reflects the shortcomings in the statistical legal framework, financial commitment to statistical plans, and the extent of implementation of statistical plans.

The Middle East and North Africa (MENA) display moderate performance across all the indicators. Performance of MENA on these indicators ranges between 59% and 65%, suggesting a need for improvement to strengthen the statistical capacity in this region.

Oceania exhibits the lowest performance among all the regions in terms of fully funded statistical plans. Only 16.7% of the Oceania countries have fully funded statistical plans. It shows moderate performance, with 50% of the countries having a statistical legal framework. Further, only 41.7% of the Oceania countries have statistical plans under implementation. This reflects a major financial barrier hindering statistical development in the region.

OECD shows the highest performance across all the indicators and among all the regions. Performance of the OECD on these indicators ranges between 89.5% and 94.7%, reflecting a strong commitment to adopt and establish a robust national statistical system in this region.

Sub-Saharan Africa (SSA) exhibits mixed performance for these indicators. Although a good percentage (77.6%) of SSA countries have both national statistical legal frameworks and national statistical plans under implementation, a very low percentage (26.5%) of SSA countries have national statistical plans fully funded. This indicates a significant challenge in securing the necessary funding to support national statistical plans.

Overall, the data reflects considerable differences across regions in terms of national statistical capabilities. Generally, OECD countries are leading, with Eastern Europe and Central Asia following closely behind. Sub-Saharan Africa and Oceania face significant challenges in securing sufficient funding for their national statistical plan, despite showing good performance in terms of having a statistical legal framework and statistical plans under implementation. East and South Asia show high statistical plan implementation but low statistical plan funding. Latin America and the Caribbean, along with the Middle East and North Africa, need to strengthen all three areas: national statistical legislation, national statistical plans funding, and their implementation.

8.3.3 National Statistical Legislation and Plan at Different Income Levels

Figure 8.4 presents data about the status of national statistical legislation and plan adoption across different income group countries. Details of the status of the national statistical legislation and plan adoption, based on the indicators given in Table 8.3, for different income groups, are given in the following text.

High-income countries show the highest level of performance across all three indicators. 81% of High-income countries have a national statistical legal framework, 78% have fully funded national statistical plans, and 82.5% have national statistical plans under

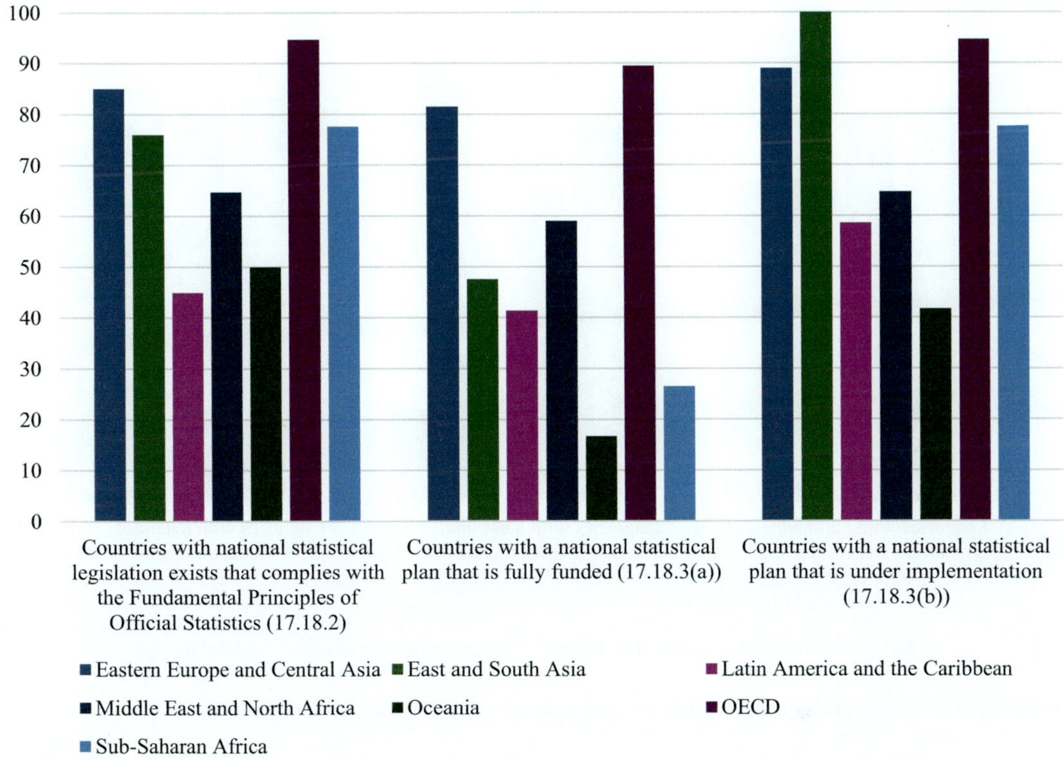

FIGURE 8.3 National statistical legislation and plan adoption at different geographical levels.

implementation. This indicates that High-income countries have robust national statistical legal frameworks, sufficient funding for statistical plans, and effective implementation of statistical plans.

Upper-middle-income countries perform slightly lower than high-income countries but still show relatively high performance. 79% and 81% of Upper-middle-income countries have a national statistical legal framework and statistical plan under implementation, respectively, while only 64% of these countries have their statistical plan fully funded. This highlights the significant gap in sufficient funding to implement the statistical plans.

Lower-middle-income countries have mixed performance on these three indicators. While 66% of the Lower-middle-income countries have national statistical legal frameworks and 78% have plans under implementation, showing moderate to high performance, only 30% of these countries with fully funded national statistical plans, reflecting low performance. This highlights the significant challenges in securing sufficient funds to implement national statistical plans.

Low-income countries also show mixed performance for these three indicators. Although 65.4% of Low-income countries have national statistical legislation and 75% have their national statistical plans under implementation, showing moderate to little high performance, only 19% of these countries with fully funded national statistical plans, indicating very low performance. This reflects that a lack of sufficient funds can pose a severe constraint on the effectiveness of the statistical legislation and implementation of national statistical plans.

It is clear from the previous discussion that there is a clear positive correlation between a country's income group and its national statistical capacity across all three indicators. More high-income

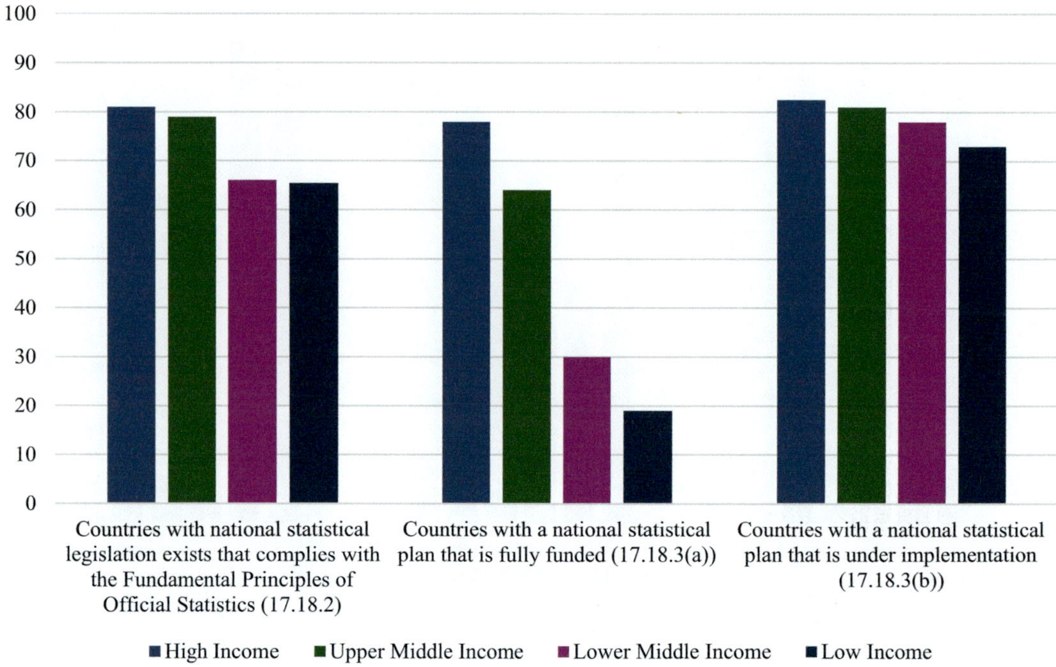

FIGURE 8.4 National statistical legislation and plan adoption at different income levels.

countries have national statistical legal frameworks and national statistical plans that are fully funded and under implementation than other income group countries. As income level decreases, performance across these indicators also decreases, with Low-income countries facing the most significant challenges, particularly in securing sufficient funding for the implementation of their statistical plans.

8.4 RELATIONSHIP BETWEEN NATIONAL STATISTICAL LEGISLATION AND PLANS AND DATA REPORTING

8.4.1 At Global Level

At the global level, although the majority of countries have their national statistical legislation and many have implemented statistical plans, the lack of sufficient funding for these statistical plans is a major challenge to comprehensive data reporting for SDG 13 indicators, especially for indicator 13.3.1.

8.4.2 At Different Geographical Levels

National statistical legislation and plan adoption and data reporting for SDG 13 indicators vary across different regions. Based on Figures 8.5 and 8.6, relationships between national statistical legislation and plan adoption and data reporting for SDG 13 indicators for different regions are explained here:

Eastern Europe and Central Asia show relatively good data reporting across most SDG 13 indicators and strong performance in terms of national statistical legislation and plan adoption. This indicates a positive relationship where a strong statistical legal framework, statistical plans, and implementation contribute to better data reporting for SDG 13 indicators.

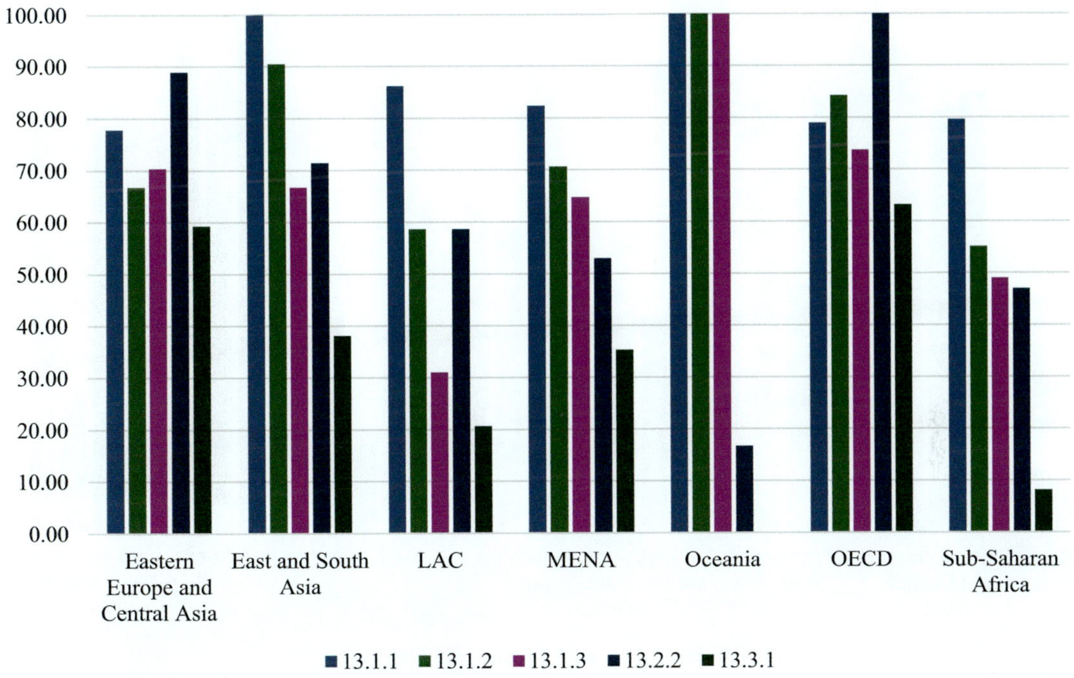

FIGURE 8.5 Data reporting for SDG 13 indicators in different regions.

East and South Asia exhibit mixed data reporting for SDG 13 indicators. In terms of national statistical legislation and plan adoption, this region shows mixed performance, high for national statistical legislation and statistical plan implementation and low in terms of funding the statistical plan. This reflects a partially positive relationship, where national statistical legislation and statistical plan implementation potentially contribute to good data reporting, but funding limitations might hinder the comprehensiveness and consistency of the data reporting.

Latin America and the Caribbean show lower data reporting, mostly across all the SDG 13 indicators. The region also shows a lower performance in terms of national statistical legislation and plan adoption. This indicates a weaker positive relationship, where low adoption of statistical legislation, statistical plan implementation, and funding contribute to poor data reporting.

The Middle East and North Africa display moderate data reporting for all the SDG 13 indicators and moderate national statistical legislation and plan adoption. This suggests a moderately positive relationship, where the existing statistical legislation and plan support a certain level of data reporting, but improvements in statistical legislation and funding and implementation of statistical plans could lead to better data reporting

Oceania shows very low data reporting for most of the SDG 13 indicators. This region shows moderate adoption of national statistical legislation and statistical plans implementation, but low funding for these plans. This clearly reflects the negative impact of low funding on effective data reporting in the region. This also highlights how, even with statistical legislation and plan implementation, low funding can severely impact data reporting.

OECD displays the best data reporting across all SDG 13 indicators. This region also shows a very high level of performance in terms of national statistical legislation and plan adoption. This exhibits a strong positive relationship where robust statistical legislation and fully funded and implemented statistical plans support excellent data reporting for climate action.

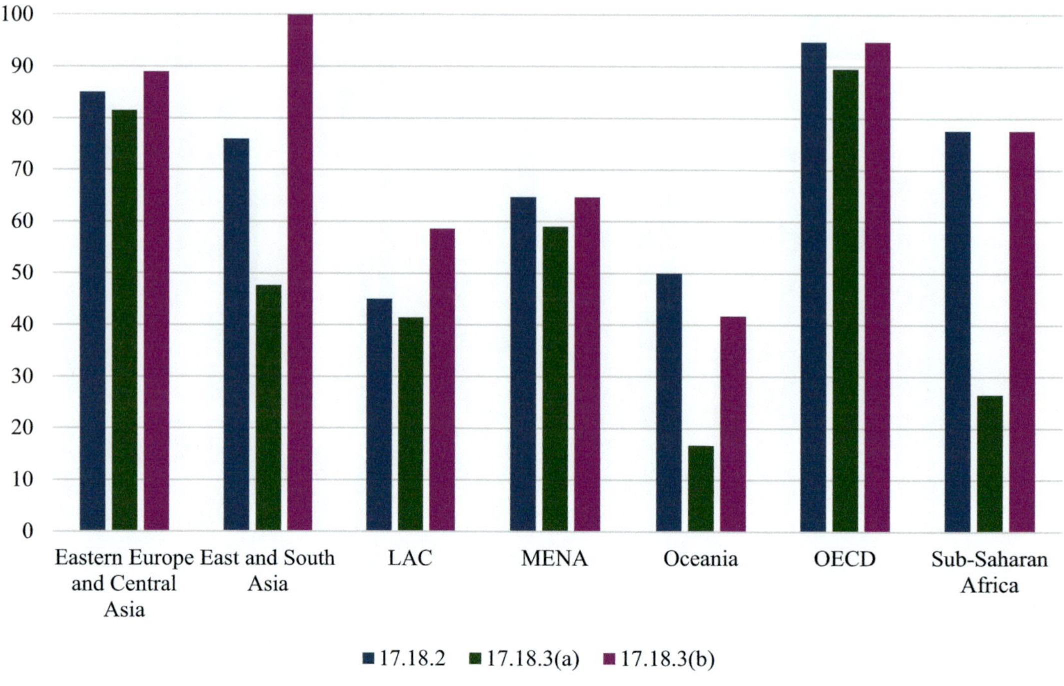

FIGURE 8.6 National statistical legislation and plan adoption in different regions.

Sub-Saharan Africa shows a mixed picture for SDG 13 data reporting, with relatively better reporting for 13.1.1 and 13.3.1 but lower for others. The region also shows reasonable performance in the adoption of statistical legislation and statistical plan implementation but very low funding for these plans. This again reflects a limited positive relationship where efforts are made through the adoption of statistical legislation and plan implementation; the lack of financial resources limits comprehensive and high-quality data reporting.

Overall analysis shows a positive correlation between national statistical legislation and plan adoption (particularly funding and implementation) and the status of data reporting for SDG 13 indicators. Regions like OECD and Eastern Europe and Central Asia show strong statistical legislation and plan adoption and that helps to have better data reporting. Oceania and sub-Saharan Africa are facing challenges related to funding and implementation of statistical plans and show poorer data reporting.

8.4.3 At Different Income Levels

The adoption of national statistical legislation and plan, and data reporting for SDG 13 indicators differs for different income groups. Based on Figures 8.7 and 8.8, relationships between these elements for different income groups are explained here:

High-income countries show good data reporting across most SDG 13 indicators. These income group countries are also shown to have national statistical legislation and fully funded and implemented statistical plans. This reflects a strong positive relationship where strong statistical legislation and a fully funded statistical plan support good data reporting.

Upper-middle-income countries show slightly lower data reporting for most of the SDG 13 indicators, compared to high-income countries. This income group also shows slightly

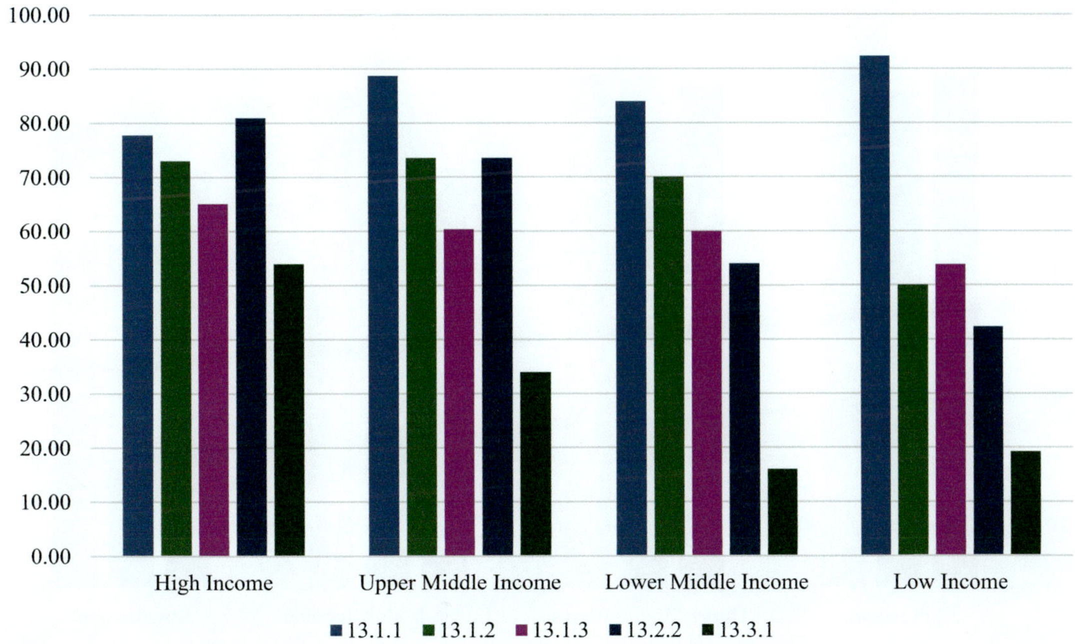

FIGURE 8.7 Data reporting for SDG 13 indicators in different income groups.

lower performance than the high-income group in terms of statistical legislation and plan adoption, particularly in terms of fully funded statistical plans. This indicates a positive relationship but with some limitations due to insufficient funding.

Lower-middle-income group shows lower data reporting compared to the high-income and upper-middle-income groups for SDG 13 indicators. This income group also shows low performance in terms of having fully funded statistical plans and moderate performance in terms of having statistical legislation and implemented plans. This reflects that insufficient funding is becoming a major constraint to this income group of countries for effective data reporting, despite having statistical legislation and implemented statistical plans.

Low-income group displays the lowest data reporting across all SDG 13 indicators. This income group also shows the lowest performance for national statistical legislation and plan adoption, particularly for fully funded statistical plans. This strongly highlights the positive correlation between weak statistical legislation and plan adoption, especially insufficient funding, and poor data reporting.

The data clearly demonstrates a positive correlation between a country's income level and the level of statistical legislation and plan adoption and data reporting for SDG 13 indicators. Higher-income group countries tend to have stronger statistical legislation and plan adoption, especially fully funded statistical plans, and better data reporting. Lower-income group countries have weaker statistical legislation and plan adoption, especially insufficient funding for statistical plans, and poorer data reporting.

8.5 PROPOSED REFORMS FOR EFFECTIVE MONITORING AND VERIFICATION

Effective monitoring and verification are important for monitoring progress, ensuring data accuracy, and targeted policy implementation. For effective monitoring and verification, a country

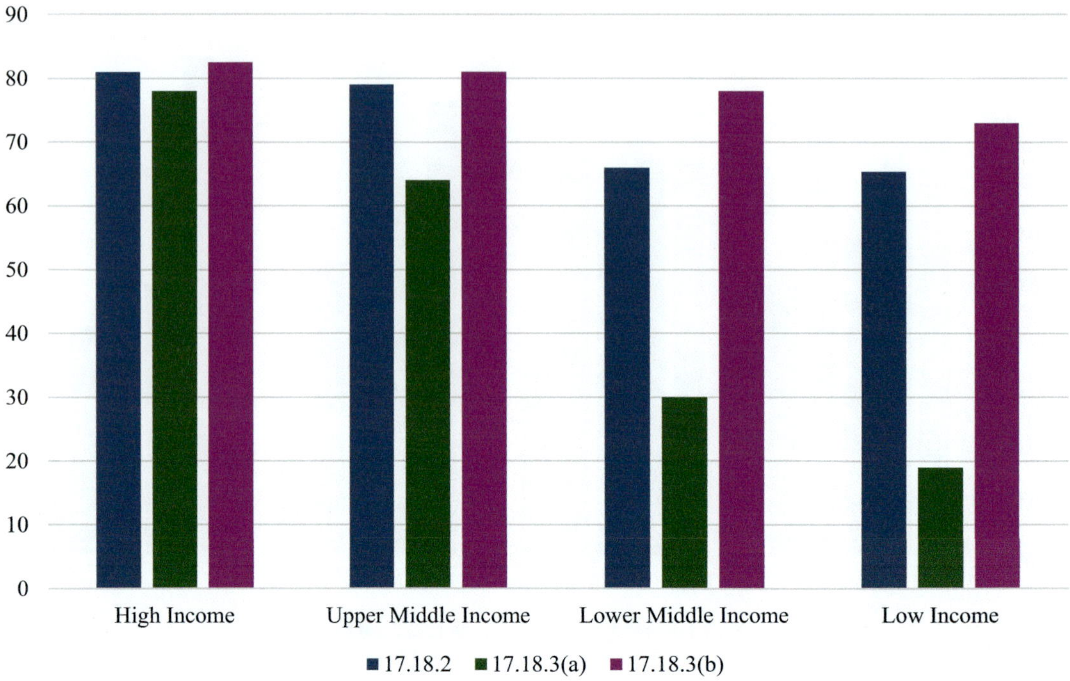

FIGURE 8.8 National statistical legislation and plan adoption in different regions.

should have a strong legal framework, flawless plans, and sufficient funding for national statistics. Section 8.2 shows the data reporting status for countries of different regions and income groups. Section 8.3 shows data about countries with national statistical legislation and national statistical plans for different regions and income groups. Based on the data available in Sections 8.2 and 8.3, the following reforms are proposed for effective monitoring and verification.

8.5.1 Create and Strengthen Statistical Legal Framework

The legal framework provides the mandate for the national statistical system and establishes the roles and responsibilities of statistical agencies, defines data collection standards, ensures data quality, and protects the confidentiality of statistical information (UN, 2024). The United Nations should promote the universal adoption and implementation of national statistical legislation that adheres to the Fundamental Principles of Official Statistics. Countries with sound statistical systems should provide technical support to countries needing to update or develop their statistical legal frameworks. Countries, especially Latin American and Caribbean, Oceania, lower-middle-income and low-income countries, with no data reporting and no national statistical legislation, should create new or strengthen existing statistical legal frameworks for robust statistical systems.

8.5.2 Implement and Strengthen Statistical Plans

Figure 8.3 and Figure 8.4 show the percentage of countries with a national statistical plan under implementation for different regions and income groups. Overall, 21% of the countries do not have national statistical plans. A larger number of Latin American and Caribbean, Oceania, and lower-middle-income and low-income countries do not have a national statistical plan and do not report data. Hence, countries, especially Latin American and Caribbean, Oceania, and lower-middle-income

9.4.2 Closing the Infrastructure Divide for AI-Driven Climate Solutions

Without reliable access to high-speed internet, IoT sensors, and consistent electricity, these communities remain highly vulnerable to climate risks and alerts. However, efforts to close this gap can be informed by existing initiatives such as Colombia's "Vive Digital" program, which has successfully expanded broadband access in rural areas, improving digital inclusion (Góngora 2014). Similarly, Rwanda's deployment of solar-powered IoT sensors has demonstrated how off-grid regions can strengthen climate resilience using renewable energy (Alotaibi and Nassif 2024). On a larger scale, China's use of 5G connected drones and IoT networks for environmental monitoring illustrates the transformative potential of next-generation connectivity, while Mexico's collaboration with Huawei to expand 5G infrastructure marks a promising regional step (Barrios-Ulloa et al. 2021).

Hence, to accelerate progress, this chapter recommends that LAC governments explore expanding broadband access using LEO satellites (e.g., SpaceX Starlink) and subsidize solar-powered AI sensors for real-time climate monitoring. Policymakers should strengthen public–private partnerships (PPPs) as mentioned before, to accelerate 5G adoption and introduce AI-focused green bonds to drive investments in climate-tech infrastructure (Global Center on Adaptation 2019).

9.4.3 Enhancing AI Governance to Drive Climate Adaptation

Another major barrier to AI adoption for climate resilience in LAC is the lack of cohesive, regionally aligned AI policies, thus leading to fragmented efforts, duplicated initiatives, and a slower uptake of AI-based climate strategies.

Drawing on successful international examples, the European Green Deal, for instance, incorporates AI to support carbon neutrality targets, while Singapore's National AI Strategy aligns AI development with its sustainability goals (Global Center on Adaptation 2019). Japan also demonstrates how AI can be embedded into national disaster response systems, with AI-driven prediction models now supporting emergency preparedness. The Caribbean Disaster Emergency Management Agency has begun piloting similar AI-based forecasting models. Brazil's National AI Strategy was developed through widespread public consultations to help promote transparency and fairness in its governance framework (Filgueiras and Junquilho 2023).

To move forward, LAC should develop a Regional AI-Climate Strategy, modeled after Singapore, to align AI with national climate goals. Governments should train emergency responders in AI-driven disaster modeling, while policymakers establish AI ethics councils that include indigenous and environmental experts. Additionally, LAC should launch a climate AI-network or innovation hub to foster cross-border research collaboration.

9.4.4 Expanding Computing Resources to Power AI Innovations for Climate Resilience

Many LAC nations lack high-performance computing (HPC) facilities and cloud access, limiting their ability to build AI solutions tailored for climate resilience. While other regions offer strong examples of how to bridge this divide. India's National Supercomputing Mission (NSM) has demonstrated the value of building regional HPC infrastructure to support AI research and innovation. Meanwhile, open-source initiatives such as TensorFlow Lite have made it possible to deploy AI models on low-power devices, making AI more accessible in low-resource settings (Alotaibi and Nassif 2024). Hardware-focused efforts like the Open Compute Project have further driven down costs by promoting open-source AI hardware development. Additionally, programs like Microsoft's AI for Earth have expanded access to cloud computing by offering grants for environmental research.

Hence, LAC should invest in shared AI supercomputing hubs, modeled after NSM, to support climate AI projects. Universities should develop open-source AI libraries, while governments

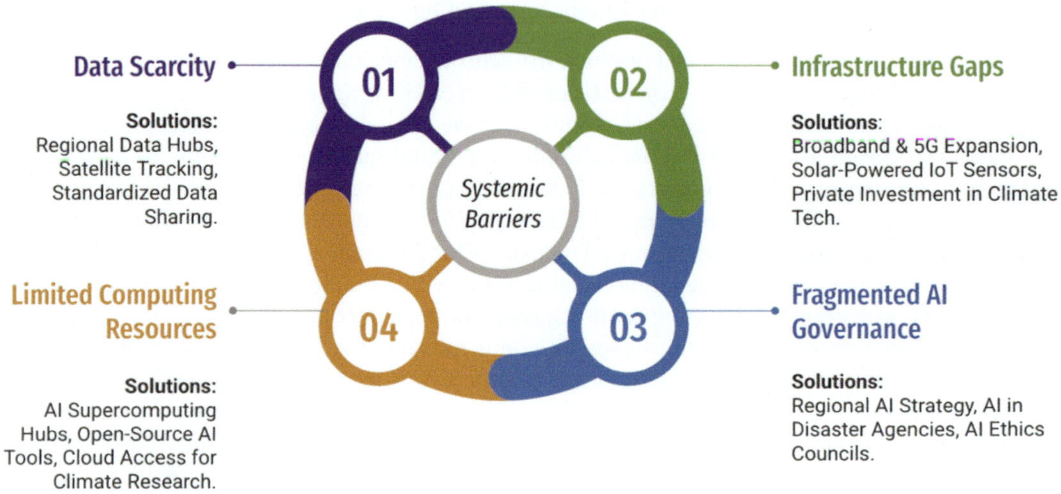

FIGURE 9.4 Systemic barriers and solutions for AI-driven climate resilience.

Source: Figure by authors.

subsidize low-cost AI hardware and partner with cloud providers (e.g., Google Cloud, AWS) to ensure affordable AI cloud access for climate researchers.

Figure 9.4 shows the systemic barriers hindering AI-driven climate resilience, including data scarcity, infrastructure gaps, fragmented AI governance, and limited computing resources. It also presents strategic solutions such as regional data hubs, broadband expansion, AI ethics councils, and cloud access to overcome these challenges, emphasizing the proposed solution for innovative AI solutions to build a sustainable future.

9.5 STRATEGIES FOR EFFECTIVE IMPLEMENTATION

When looking at strategies for effective implementation of AI policies within vulnerable regions such as LAC and small island developing states, a strong, ethical, and adaptable policy framework is essential to overcoming systemic barriers while ensuring frameworks remain scalable and inclusive. This section outlines strategies that align with international SDGs initiatives by leveraging AI technologies to foster resilience in climate-sensitive areas and their alignment with SDGs 9, 11, and 13.

9.5.1 Governance and Policy Development

To coordinate a well-planned strategy for climate resilience, a well-designed and unified governance framework is essential. Looking at successful use cases, some of the challenges that persist are data gaps and infrastructure limitations. However, in Africa, there are AI applications that are already being used for early warning systems, satellite monitoring of deforestation, and sustainable water management. Initiatives like the Smart Africa Alliance and AUDA-NEPAD are working to integrate AI into governance while ensuring transparency and inclusivity. These efforts not only advance SDG 13 by enhancing adaptive capacity but also support SDG 9 through technological development in vulnerable regions. Hence, it is crucial that the research, design, and implementation team that makes up the AI task forces mentioned in Section 9.3 comprises technologists, policymakers, environmentalists, and community leaders that holistically drive effective governance to implement climate policies and sustainable development aligned with SDG 11 and ensure that governance models reflect the needs of all.

9.5.2 Infrastructure Investment

Investing in digital infrastructure, such as IT networks and communication systems for real-time monitoring in remote areas, is essential for success for AI-driven climate resilience, directly supporting SDGs 9 and 13. Continuous connectivity and operational efficiency can be achieved through renewable energy solutions such as solar-powered IoT devices that can overcome energy deficits in rural regions (Kano et al. 2021). Projects like Google Loon and Microsoft's Airband Initiative demonstrated the need for governmental support and PPPs that play a critical role in facilitating SDG 9 and 13 initiatives, as well as ensuring the firms and stakeholders receive a return on investment for long-term continuity and partnership.

9.5.3 Capacity Building

While investments are a massive aid to climate resilience and AI projects, to maximize the benefits of the investment model, capacity building must target all levels of stakeholders. Hence, education and training for community leaders, local governments, and private organizations will aid in understanding AI tools and shape the regional culture to accept new innovations that will bridge any knowledge gaps. Specialization and emphasis should be placed on technologies like digital twins, LLMs, agentic AI, and hands-on experience that foster a culture of innovation and self-resilience (European Commission 2022). Collaborative efforts between academia and industry can provide the technical expertise and research needed to design and implement sustainable solutions directly supporting SDG9 through skill development and technological advancement.

9.5.4 Ethical AI Implementation

Anything that influences society must be ethically sound to ensure fairness and inclusiveness; this includes AI applications. This research suggests leveraging explainable AI models to address concerns about bias and decision-making, as they provide a layer of transparency and interpretability, as highlighted previously through XAI. According to the General Data Protection Regulation, compliance guidelines for data privacy and security must strictly enforce the public trust ("General Data Protection Regulation (GDPR) Compliance Guidelines" 2018), further aligning with SDG 13 by fostering trusted and community-centered climate solutions. Additionally, data equity principles should ensure that underserved communities benefit equally from AI-driven solutions (World Bank, 2021).

9.5.5 Adaptability and Scalability

It is imperative that the long-term success of these AI-driven climate resilience initiatives requires adaptable and scalable solutions to survive any curveballs, unpredictable circumstances, and changes in the predicted parameters. Simple proof of concept and low-cost modular pilot projects allow teams to test the feasibility of strategies across varying simulator circumstances and diverse contexts that will give valuable insights for broader implementation. Lessons from international case studies, as mentioned before, would inform the development of context-specific solutions. Following this approach, AI technologies can strengthen climate resilience, directly contributing to SDG 13 and SDG 9 by fostering technological experimentation and infrastructure readiness, while also aligning with SDG 11 through localized, inclusive applications tailored to community needs.

9.5.6 AI-Driven Climate Resilience Implementation Framework

The flowchart shown in Figure 9.5 illustrates a structured, phased approach to implementing AI-driven strategies for climate resilience (Table 9.2). The framework is divided into three key

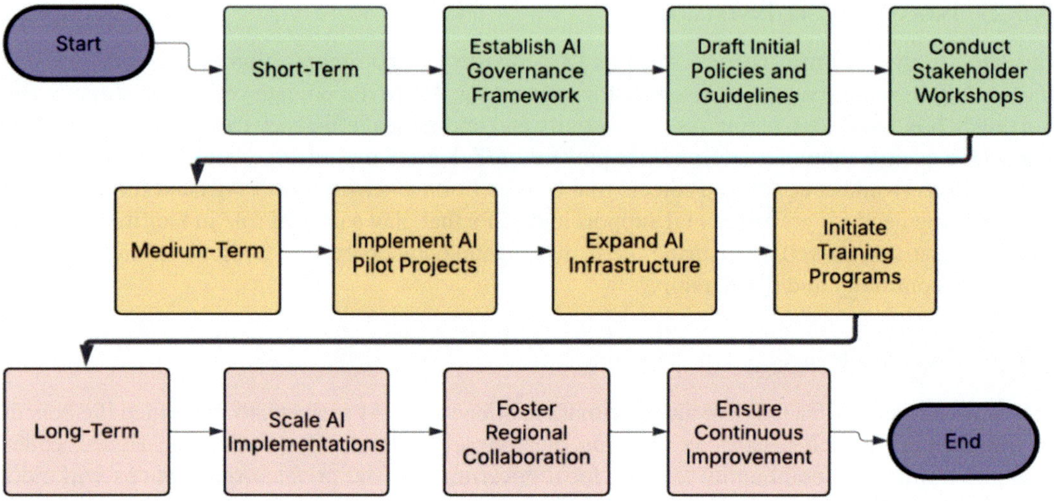

FIGURE 9.5 Flow chart of AI-driven climate resilience implementation framework.

Source: Figure by authors.

TABLE 9.2
AI-Driven Climate Resilience Implementation Framework

Phase	Initiatives
Short-Term (0–1 Year)	Establish AI Governance Framework: Engaging key stakeholders to define foundational policies for AI-driven climate initiatives.
	Draft Initial Policies and Guidelines: Developing a regulatory and strategic foundation to guide AI applications in climate resilience.
	Conduct Stakeholder Workshops: Gathering feedback from various sectors to refine the governance approach.
Medium-Term (2–3 Years)	Expand AI Infrastructure: Enhancing technological capabilities, such as IoT networks and digital platforms, to support climate monitoring.
	Train Workforce on AI Tools: Conducting targeted training programs to equip professionals with AI skills relevant to climate resilience.
	Initiate Training Programs: Implementing structured educational initiatives to build capacity among stakeholders.
	Develop Pilot Infrastructure: Deploying initial AI-driven projects to test and refine real-world applications.
Long-Term (5+ Years)	Scale AI Implementations: Expanding successful pilot projects into large-scale deployments for widespread climate adaptation benefits.
	Foster Regional Collaboration: Strengthening inter-regional partnerships to enhance knowledge-sharing and policy alignment.
	Ensure Continuous Improvement: Adapting AI policies and technologies to address emerging climate challenges effectively.

Source: Table by authors.

timeframes: short-term (0–1 year), medium-term (2–3 years), and long-term (5+ years) to ensure gradual and sustainable progress.

Each milestone represents a critical step in integrating AI into climate resilience efforts, ensuring that solutions are scalable, inclusive, and continuously evolving. The ultimate goal is to create a sustainable, AI-enabled climate adaptation framework that can be replicated across different regions.

9.6 HOW TO MEASURE ADAPTATION AND SUCCESS

Understanding the overall evaluation of the project adaptation and success allows stakeholders to maximize the impact, results, and output by altering and refining their approaches (EcoAdapt 2008). This section is a brief overview of the metrics and methodologies that can be used to assess how effective AI solutions are in supporting sustainable development as well as mitigating climate risks in practical implementation.

9.6.1 Quantitative and Qualitative Metrics for Measuring Success

Quantitative indicators evaluate the success of AI interventions by utilizing measurable numerical benchmarks. For example, when comparing baseline data to outcomes, AI-driven early warning systems allow overseers to review the impact on community safety and preparedness (Dhanani et al. 2024). Hence, some key metrics would include reductions in disaster-related fatalities, financial and economic losses, and overall social recovery times.

In addition, qualitative metrics allow for observation from a broader social and institutional perspective of the impacts of AI adoption. Feedback such as sentiment readings, along with general community engagement and their satisfaction, is crucial and a vital indicator of whether our framework addresses local requirements and is inclusive to marginalized groups. This feedback can help with improvements in governance by facilitating enhanced collaborations between government and private organizations, where the result is a stronger, systematic resilience.

9.6.2 Methodologies for Assessment

9.6.2.1 Financial Justifications

Positive financial metrics often provide a substantial rationale for increasing the scale of AI-driven solutions (Dhanani et al. 2024). Return on investments can be quantified by a cost–benefit analysis that shows the upfront costs of the deployment of AI technologies, along with long-term savings and financial gain achieved through disaster prevention and climate mitigation (Dhanani et al. 2024). For example, AI-powered early warning systems have already been proven to significantly reduce costs associated with evacuation and minimize infrastructure damage (Reuters 2025).

9.6.2.2 Performance Metrics

Precision, accuracy, and timeliness are one of the most prominent performance metrics used to evaluate predictive AI systems (Jennifer 2025). The larger scope of these metrics would include the precision of disaster forecasts and its speed at which emergency responses are made as an intervention of those in harm's way, while simultaneously providing insights into the effectiveness of AI technologies. These metrics and their associated data can be collected from real-time analytics that perform a crucial role in enabling stakeholders to assess whether performance aligns with the expected results by monitoring the outcomes (Jennifer 2025).

9.6.2.3 Dynamic Decision-Making

AI technologies, especially those in the agentic-AI systems, often thrive on dynamic feedback loops that allow continuous improvement, adaptability, and scalability (LSE Grantham Institute 2023), enabling dynamic decision-making by stakeholders and policymakers to refine the strategies based on outcomes. For example, if discrepancies are observed with predictive models from what was originally predicted to the actual disaster impacts, the machine learning algorithms can be tweaked and retrained to enhance their precision and accuracy for long-term reliability (Dhanani et al. 2024).

9.6.2.4 Case Studies as Proof of Concept

Successful deployments demonstrated the power of a metric-driven approach. For example, as mentioned before, farmers are using AI to make better planting decisions that reduce debts and increase incomes in rural India (Reuters 2025). Likewise, during Hurricane Maria, a flood prediction system was utilized in the Dominican Republic that reduced evacuation times, saved lives, and gave a clear image of how to measure success (Dhanani et al. 2024). While the use case may not be exactly for small island developing states, the similarities and differences and lessons learned from such projects give us and teams across the world actionable insights for replicating similar success in our local context.

9.7 ACTIONABLE RECOMMENDATIONS

To fully leverage AI in climate resilience, a multi-stakeholder approach is needed, one that integrates governance, infrastructure, and capacity building into a strategic roadmap for success. To do this, our framework puts an emphasis on the collaboration between policymakers, technologists, and local communities to enable effective and sustainable climate solutions.

9.7.1 Establishing AI Governance Hubs

A governing committee is recommended for AI governance in the region, whose purpose will be to ensure ethical and responsible utilization and deployment of AI in climate resistance. This holistic community of talent should include policymakers, AI researchers, stakeholders, and industry leaders, but most importantly, community representatives who will work together to develop ethically sound guidelines for AI in disaster prediction and environmental monitoring. It is imperative that they establish regulatory frameworks that are compliant with industry-standard standard AI policies that promote transparency and trust in the AI-driven decision-making processes with the public.

9.7.2 Deploying Low-Cost IoT Climate Monitoring and AI-Powered Early Warning Systems

Data is vital; hence this chapter recommends real-time environmental data collection that will aid in detection and response to climate threats. It is recommended that IoT sensors be deployed in vulnerable regions that will provide continuous monitoring. Necessary climate indicators such as temperature, humidity, and water levels will also provide open-source AI models with fresh datastreams to analyze hidden trends but also detect any potential negative anomalies that would indicate potential climate disasters. Lastly, sharing real-time data can enhance collaboration among researchers, scientists, policymakers, and local communities, thereby fostering innovation among all parties involved.

9.7.3 Capacity Building Through AI Training Programs

Regardless of how advanced an AI-based climate solution may be, its effectiveness depends on a skilled workforce capable of deploying and advancing it. Therefore, building AI literacy is critical for the successful integration of AI into climate resilience efforts. To enable this, stakeholders should facilitate both short-term and long-term learning events like AI workshops that target policymakers, community leaders, and emergency responders while partnering with research departments from universities and tech companies to ensure the highest quality of training to all. Most importantly, training should be accessible and hands-on to give equitable knowledge distribution and sustainable implementation.

9.7.4 Public–Private Collaboration for AI Infrastructure

Robust infrastructure investments should also be prioritized to ensure equitable access to these AI-driven climate solutions. Therefore, PPPs play a critical role in funding AI infrastructure in

TABLE 9.3
AI-Driven Climate Resilience Pilot Implementation

Initiative	Timeframe	Actions
Community-Based Climate Monitoring	0–1 year	Deploy IoT sensors in flood-prone or drought-affected regions. Train local teams to maintain, analyze, and act on collected data. Develop real-time dashboards to facilitate community engagement and proactive response planning.
AI-Driven Agriculture Advisory System	2–3 years	Implement AI models that provide small-scale farmers with real-time climate insights. Collaborate with agricultural cooperatives to drive adoption and maximize impact. Develop mobile applications that deliver AI-generated recommendations to enhance crop yields and resilience.
Scalable AI-Powered Disaster Response	5+ years	Expand AI-driven early warning systems to multiple regions with varying climate risks. Integrate AI-based risk mapping into national disaster management strategies. Develop autonomous AI tools for rapid post-disaster assessment and response coordination.

Source: Table by authors.

the areas that are most vulnerable to climate change by leveraging existing projects to facilitate AI adoption and connectivity. Finally, centers for research development and the deployment of the solutions that come from them should be built, such as AI-Innovation hubs.

9.7.5 Pilot Implementations

Table 9.3 presents a phased approach to AI-driven climate resilience in LAC, detailing short-, mid-, and long-term solutions to manage climate risks and empower local communities.

The multi-stakeholder approach gives confidence in enabling AI to become a cemented pillar in climate resilience. By giving importance to governance, real-time data collection, and public–private collaboration, this framework gives power to communities to make a positive stride in climate adaptation.

9.8 CONCLUSION

As explored in this chapter, AI technologies can support climate resilience efforts across LAC through real-time environmental monitoring, predictive analytics, and proactive disaster management to reduce vulnerability to climate hazards, but, despite these opportunities, the integration of AI in the LAC region continues to face significant challenges, such as limited data availability, inadequate infrastructure, and fragmented policy frameworks.

This chapter demonstrates that aligning AI-driven climate resilience strategies with the SDGs can help address these issues in a meaningful and forward-looking way. These recommendations aim to contribute directly to these global objectives such as improving early warning systems and adaptive responses supporting SDG 13 on climate action. Similarly, using digital twins and smart planning tools to strengthen urban resilience aligns with SDG 11, while investments in digital infrastructure and emerging technologies further the aims of SDG 9.

Moving forward, AI in climate resilience will require proactive ethical decision-making, inclusive stakeholder involvement, and scalable solutions, and with a coordinated and thoughtful approach, technologies such as digital twins, LLM, 5G, 6G, and agentic AI can become key enablers of long-term sustainability and regional resilience in LAC.

REFERENCES

Alotaibi, Emran, and Nadia Nassif. 2024. "Artificial Intelligence in Environmental Monitoring: In-Depth Analysis." *Discover Artificial Intelligence* 4 (1). https://doi.org/10.1007/s44163-024-00198-1.

Bakker, Bas. 2024. "What Can Artificial Intelligence Do for Stagnant Productivity in Latin America and the Caribbean?" *IMF Working Papers* 2024 (219): 1. https://doi.org/10.5089/9798400290770.001.

Barrios-Ulloa, Alexis, Dora Cama-Pinto, Johan Mardini-Bovea, Jorge Díaz-Martínez, and Alejandro Cama-Pinto. 2021. "Projections of IoT Applications in Colombia Using 5G Wireless Networks." *Sensors* 21 (21): 7167. https://doi.org/10.3390/s21217167.

Bauer, Peter, Bjorn Stevens, and Wilco Hazeleger. 2021. "A Digital Twin of Earth for the Green Transition." *Nature Climate Change* 11 (2): 80–83. https://doi.org/10.1038/s41558-021-00986-y.

Brocca, Luca, Silvia Barbetta, Stefania Camici, Luca Ciabatta, Jacopo Dari, Paolo Filippucci, Christian Massari, et al. 2024. "A Digital Twin of the Terrestrial Water Cycle: A Glimpse into the Future through High-Resolution Earth Observations." *Frontiers in Science* 1 (March). https://doi.org/10.3389/fsci.2023.1190191.

Cavazos, Tereza, Maria Laura Bettolli, Donovan Campbell, Roberto A. Sánchez Rodríguez, Michelle Mycoo, Paola A. Arias, Juan Rivera, et al. 2024. "Challenges for Climate Change Adaptation in Latin America and the Caribbean Region." *Frontiers in Climate* 6 (April). https://doi.org/10.3389/fclim.2024.1392033.

Colverd, Grace, Paul Darm, Leonard Silverberg, and Noah Kasmanoff. 2023. "FloodBrain: Flood Disaster Reporting by Web-Based Retrieval Augmented Generation with an LLM." *arXiv* [cs.AI]. https://doi.org/10.48550/ARXIV.2311.02597.

CRED. 2025. *EM-DAT: The International Disaster Database*. Accessed January 28, 2025. https://www.emdat.be/.

Darley, James. 2024. "How Google Is Using AI to Forecast Floods." *Bizclik Media Ltd.*, November 12. https://sustainabilitymag.com/articles/google-expands-ai-powered-flood-forecasting.

Dhanani, Qahir, Stefano Niavas, Greg Fischer, Charlotte Saint-Bonnet, and Emily Weeks. 2024. "New Measures of Success for Climate Adaptation and Resilience." *BCG Global*, July 23. https://www.bcg.com/publications/2024/new-framework-for-climate-adaptation-and-resilience.

Di Noia, Jlenia. 2022. "Agent-Based Models for Climate Change Adaptation in Coastal Zones: A Review." *SSRN Electronic Journal*. https://doi.org/10.2139/ssrn.4180554.

Ding, Chao, Jing Ke, Mark Levine, and Nan Zhou. 2024. "Potential of Artificial Intelligence in Reducing Energy and Carbon Emissions of Commercial Buildings at Scale." *Nature Communications* 15 (1): 5916. https://doi.org/10.1038/s41467-024-50088-4.

EcoAdapt. 2008. *Meeting the Challenges of Climate Change*. Accessed February 19, 2025. https://ecoadapt.org/.

European Commission. 2022. *Legal Framework of EU Data Protection*. Accessed January 28, 2025. https://commission.europa.eu/law/law-topic/data-protection/legal-framework-eu-data-protection_en.

Filgueiras, Fernando, and Tainá Aguiar Junquilho. 2023. "The Brazilian (Non)perspective on National Strategy for Artificial Intelligence." *Discover Artificial Intelligence* 3 (1). https://doi.org/10.1007/s44163-023-00052-w.

GDPR.eu. 2018. *General Data Protection Regulation (GDPR) Compliance Guidelines,* June 18. https://gdpr.eu/.

Ghaith, Maysara, Ahmed Yosri, and Wael El-Dakhakhni. 2022. "Synchronization-Enhanced Deep Learning Early Flood Risk Predictions: The Core of Data-Driven City Digital Twins for Climate Resilience Planning." *Water* 14 (22): 3619. https://doi.org/10.3390/w14223619.

Global Center on Adaptation. 2019. *Adapt Now: A Global Call for Leadership on Climate Resilience*. Accessed February 9, 2025. https://gca.org/reports/adapt-now-a-global-call-for-leadership-on-climate-resilience/.

Góngora, Gina Paola Maestre. 2014. "Public Policy on Information Technology in Colombia: Bet on the Future for the Use and Ownership of IT in Society." In *2014 XL Latin American Computing Conference (CLEI)*, 1–8. IEEE. https://doi.org/10.1109/clei.2014.6965109.

House Committee on Science, Space, and Technology. 2024. *House Bipartisan Task Force on Artificial Intelligence Delivers Report*. Accessed February 10, 2025. https://science.house.gov/2024/12/house-bipartisan-task-force-on-artificial-intelligence-delivers-report.

International Trade Administration. 2024. *Finland—Digital Economy*. Accessed January 21, 2025. https://www.trade.gov/country-commercial-guides/finland-digital-economy.

IPCC—Intergovernmental Panel on Climate Change. 2023. Accessed February 19, 2025. https://www.ipcc.ch/.

Jennifer. 2025. "Climate Finance Evidence Series: Use of AI in Climate Change Evaluations." *Adaptation Fund*, January 21. https://www.adaptation-fund.org/use-of-ai-in-climate-change-evaluations/.

Jobin, Anna, Marcello Ienca, and Effy Vayena. 2019. "The Global Landscape of AI Ethics Guidelines." *Nature Machine Intelligence* 1 (9): 389–399. https://doi.org/10.1038/s42256-019-0088-2.

Kano, Masaki, Hiroyuki Morioka, and Shigeru Yamaguchi. 2021. "HumanAI Collaboration in Graphic Design: A Survey on Creative Practice and Challenges." *Digital Creativity* 33 (4): (December): 322339. https://doi.org/10.1016/j.digcre.2021.1001169.

Loghin, Vasile. 2017. "Copernicus Program of the European Union for Environmental Monitoring and Civil Security." *ResearchGate*. Accessed January 28, 2025. https://www.researchgate.net/publication/316456497_Copernicus_Program_of_the_European_Union_for_Environmental_Monitoring_and_Civil_Security.

LSE Grantham Institute. 2023. *What Opportunities and Risks Does AI Present for Climate Action?* Accessed July 4, 2023. https://www.lse.ac.uk/granthaminstitute/explainers/what-opportunities-and-risks-does-ai-present-for-climate-action/.

Lu, Wenjing, Qiang Ma, Tao Sun, Xiaolei Zhang, Qiuling Yao, Ronghua Liu, Bingshun He, and Philippe Gourbesville. 2024. "Digital Twin Technologies for Flood Management in Large Catchment: Challenges and Operational Solution." In *Springer Water*, 57–66. Springer Nature. https://doi.org/10.1007/978-981-97-4072-7_4.

Luo, Peihang, Erika Parn, and Ioannis Brilakis. 2024. "ChatTwin: Enabling Natural Language Interactions with Infrastructure Digital Twins." In *Computing in Construction*, 5:0–0. European Council for Computing in Construction. https://doi.org/10.35490/ec3.2024.259.

Maghsoudi, Mehrdad, Navid Mohammadi, and Mohammadreza Bakhtiari. 2025. "Artificial Intelligence and Sustainable Development: Public Concerns and Governance in Developed and Developing Nations." *Cleaner Environmental Systems* 19 (December): 100340. https://doi.org/10.1016/j.cesys.2025.100340

Martelo, Rafaela, Kimia Ahmadiyehyazdi, and Ruo-Qian Wang. 2024. "Towards Democratized Flood Risk Management: An Advanced AI Assistant Enabled by GPT-4 for Enhanced Interpretability and Public Engagement." *arXiv* [cs.AI]. https://doi.org/10.48550/ARXIV.2403.03188.

Mittelstadt, Brent Daniel, Patrick Allo, Mariarosaria Taddeo, Sandra Wachter, and Luciano Floridi. 2016. "The Ethics of Algorithms: Mapping the Debate." *Big Data & Society* 3 (2): 1–21. https://doi.org/10.1177/2053951716679679.

Ogunsesan, Abimbola, Ademola Akanmu, and Umar Obafemi Salisu. 2024. "Sustainable Flood Disaster Mitigation Measures: The Nexus Between Policy and Regulations for Spatial Development in Ibadan, Nigeria." *ResearchGate*. Accessed January 28, 2025. https://www.researchgate.net/publication/385175744.

Patil, Dimple. 2024. "Explainable Artificial Intelligence (XAI): Enhancing Transparency and Trust in Machine Learning Models." *ResearchGate*. Accessed January 28, 2025. https://www.researchgate.net/publication/385629166_Explainable_Artificial_Intelligence_XAI_Enhancing_transparency_and_trust_in_machine_learning_models.

Peters, Linda. 2022. "Climate Change Prompts Grenada to Create the First National Digital Twin." *Esri*, June 16. https://www.esri.com/about/newsroom/blog/grenada-digital-twin-climate-change/.

Putsoane, Tebello, Johannes Itai Bhanye, and Abraham Matamanda. 2024. "Extreme Weather Events and Health Inequalities: Exploring Vulnerability and Resilience in Marginalized Communities." In *Developments in Environmental Science*, edited by Lakshmi Sivaramakrishnan, Bharat Dahiya, Madhuri Sharma, Saswati Mookherjee, and Ranita Karmakar, 15:225–248. Elsevier. https://doi.org/10.1016/b978-0-443-21948-1.00011-x.

Rajendran, Sruthie. 2024. "Development of Digital Twin Framework for Decision Support in Flood Risk Management." *University of Twente*. https://essay.utwente.nl/101928/.

Rashid, Abdur, Parag Biswas, AKM Ahasan Habib, AFM Zainul Abadin, Arnob Ghosh, and S.M.A. Motakabber. 2025. "Review on 6G Communication Network Technology for Sustainable Renewable Energy System: Implementations, Constraints, and Recommendations." *Results in Engineering* 28 (December): 107933. https://doi.org/10.1016/j.rineng.2025.107933.

Reuters. 2025. "Comment: How Empowering Smallholder Farmers with AI Tools Can Bolster Global Food Security." *Reuters*, January 10. https://www.reuters.com/sustainability/land-use-biodiversity/comment-how-empowering-smallholder-farmers-with-ai-tools-can-bolster-global-food-2025-01-10/.

Reyer, Christopher P. O., Sophie Adams, Torsten Albrecht, Florent Baarsch, Alice Boit, Nella Canales Trujillo, Matti Cartsburg, et al. 2017. "Climate Change Impacts in Latin America and the Caribbean and Their Implications for Development." *Regional Environmental Change* 17 (6): 1601–1621. https://doi.org/10.1007/s10113-015-0854-6.

Ricciardi, Guglielmo, and Guido Callegari. 2022. "Digital Twins for Climate-Neutral and Resilient Cities: State of the Art and Future Development as Tools to Support Urban Decision-Making." In *Technological Imagination in the Green and Digital Transition*, 617–626. Cham: Springer International Publishing. https://doi.org/10.1007/978-3-031-29515-7_55.

Roberts, H., E. Hine, M. Taddeo, and L. Floridi. 2024. "Global AI Governance: Barriers and Pathways Forward." *International Affairs* 100 (3): 1275–1286. https://academic.oup.com/ia/article-abstract/100/3/1275/7641064.

Rolnick, David, Priya L. Donti, Lynn H. Kaack, Tatjana Choucri, et al. 2019. "Tackling Climate Change with Machine Learning." *arXiv*. https://doi.org/10.48550/arXiv.1906.05433.

Shinde, Yogesh. 2025. "Agentic AI in Smart Cities Market to Hit USD 58 Bn By 2034." *Market Scoop*, February 7. https://scoop.market.us/agentic-ai-in-smart-cities-market-news/.

Tuazon, Veronica, Xinying Gao, and Natalie Yimei Gahre. 2024. "Developing a Framework for the Integration and Use of Large Language Models in Disaster Risk Management." *CIMA Foundation*. Accessed February 19, 2025. https://www.cimafoundation.org/wp-content/uploads/2024/07/CIMA-Capstone-Report-MID-2023-2024.pdf.

United Nations Office for Disaster Risk Reduction (UNDRR). 2023. *Overview of Disasters in Latin America and the Caribbean (2000–2022)*. Accessed February 19, 2025. https://www.undrr.org/publication/overview-disasters-latin-america-and-caribbean-2000-2022.

United Nations Office for Disaster Risk Reduction (UNDRR). 2024. *Artificial Intelligence for Climate Action in Developing Countries: Opportunities, Challenges and Risks*. Accessed February 19, 2025. https://www.preventionweb.net/publication/artificial-intelligence-climate-action-developing-countries-opportunities-challenges.

Uzoka, A., Cadet, E., & Ojukwu, P. U. 2024. "Applying Artificial Intelligence in Cybersecurity to Enhance Threat Detection, Response, and Risk Management." *Computer Science & IT Research Journal* 5, no. 10 (2024): 2511–2538. https://doi.org/10.51594/csitrj.v5i10.1677.

Wu, Y., H. Zhang, M. Li, and Q. Chen. 2024. "Optimizing AI-Driven Disaster Management through LLMs." *Preprints*. https://www.preprints.org/manuscript/202407.1446/v1.

Zhu, Jun, Pei Dang, Yungang Cao, Jianbo Lai, Yukun Guo, Ping Wang, and Weilian Li. 2024. "A Flood Knowledge-Constrained Large Language Model Interactable with GIS: Enhancing Public Risk Perception of Floods." *Geographical Information Systems*, February, 1–23. https://doi.org/10.1080/13658816.2024.2306167.

10 Building Economically Robust, Technologically Advanced, and Environmentally Sustainable Higher Education Institutions

A Glimpse into the Case of a Turkish Private University

Mahfoozur Rahman, Mehmet Akın Bulut, and Adem Yurdunkulu

10.1 INTRODUCTION

The number of students attending higher education institutions (HEIs) has doubled from around 100 million in 2000 to close to 250 million in 2020 (UNESCO, 2022), which also indicates the rapidly increasing demand for such institutions globally. Simultaneously, HEIs are facing economic issues ranging from global recession, public budget cuts, and rising cost of operations to obsolete funding models (Chinyoka & Mutambara, 2020; Mossman, 2018; Salazar-Clemeña, 2011); technology-related cost-burden resulting from rapid adoption of online education, artificial intelligence (AI), and other technology-enhanced learning platforms (Comi et al., 2023; Mahajan et al., 2025); and environmental issues like biodiversity loss, climate change, and increasing pollution (Dyke & Monbiot, 2024; Matesanz et al., 2023). So, to efficiently tackle the ever-rising demand for tertiary education, more economically healthy, technologically agile, and environmentally responsible HEIs are required.

Turkey has had its own share of challenges and threats on the way to sustainability in higher education. Due to a recent surge in their numbers, Turkish universities, particularly the public ones, are experiencing a shortage of qualified faculty, reduced funding, and a decline in the quality of education (Acer & Güçlü, 2017; Akar, 2010; Özoğlu et al., 2016). Many Turkish HEIs lack adequate technical infrastructure and face a digital divide (including internet access) across various regions (Dindar et al., 2022; Özoğlu et al., 2016). They are also faced with the technological integration challenge due to scarce software and hardware laboratories (Kilinc et al., 2018), and, to make the situation worse, non-compliance of the teaching staff in adopting technological tools in teaching and learning remains a problem (Kilinc et al., 2018; Kurt, 2014). Moreover, although environmental awareness-related courses have been added to curricula in some Turkish universities (Saribas et al., 2016; Talay et al., 2004), there is a gap between the theoretical knowledge and its practical application regarding environmental sustainability among the students (Önder, 2006). Another environmental sustainability issue is the lack of funding for environmentally friendly initiatives, particularly in less-developed regions (Özoğlu et al., 2016).

Despite the aforementioned challenges, HEIs, at both the global and local (Turkish) level, are striving to continue in a more sustainable style with respect to finance, environment, and

DOI: 10.1201/9781003600152-10

technology. A range of HEIs worldwide are taking staunch steps toward enhancing the financial sustainability of their institutions, in both the short and long term, as reflected in the cases in some studies (Allevato & Ribeiro, 2017; Cooper & Gorman, 2018; Francis & Moore, 2019; Zou et al., 2015). In addition, studies like Coronel and Trigos (2024), Harini et al. (2024), and Skiba et al. (2024) suggest technology-backed improvements in HEI administration, leading to better efficiency and service quality, improved academic quality because of easier access to personalized training, technical resources, and interactive teaching methods, and the use of information and communication technology (ICT) for education has boosted the quality of education services and overall operational efficiency. Similarly, some universities are adopting environmentally sustainable endeavors such as utilizing efficient energy systems, waste management, and sustainable transportation (MAAME et al., 2025); pursuing participatory and collaborative approaches to foster environmental awareness and responsibility (Durão et al., 2025); generating transformative change by integrating sustainability in leadership roles, teaching practices and policy decision of HEIs (Liu & Curtin, 2024); and promoting environmental sustainability in tertiary education institutions by opting for Human Resource Management (HRM) practices like green recruitment and eco-training (Imtiaz & Iqbal, 2024).

Lately, the influence and importance of private universities have increased, both in Turkiye and globally, as they tend to offer higher quality education, considering their smaller class size, more advanced teaching methods, and more dynamic faculty (Eser & Birkan, 2005). According to Collins et al. (2021), international students feel more satisfied at private universities as they perceive those institutions as having better teaching standards, extracurricular activities, and support services. There are 75 private universities in Turkiye, which is above one-third of the total universities in the country (CoHE, n.d.).

This study focuses on a case study of Ibn Haldun University, a Turkish foundation (private) university established in 2017 in Istanbul, Turkiye. This case has been selected with the crucial nature of private universities, both locally and internationally, in view. Additionally, despite being a relatively new institution, the case of Ibn Haldun University presents a picture of an emerging HEI that is promising innovative ways to tackle economic, technological, and environmental issues in a smart and responsible way, mostly aligning with the SDGs set by the United Nations.

This study aims to describe how a new private university in a metropolitan city strives to provide solutions for economic, technological, and environmental challenges. It also focuses on contributing to the existing literature on sustainability and sustainable practices in private universities. Furthermore, empirically, the research will help the Turkish Council of Higher Education and other higher education policy-making agencies to analyze and adopt reliable, realistic, and sustainable ways to turn HEIs into self-sufficient organizations, technologically, economically, and environmentally.

10.2 LITERATURE REVIEW

10.2.1 Economic Sustainability in Higher Education

10.2.1.1 Meaning and Background

Sustainability is the ability of an institution to sustain itself in its environment. Similarly, financial sustainability in higher education means the capability of an HEI to attain its strategic aims while maintaining its financial strength (Al-Filali et al., 2024; Omazic & Zunk, 2021).

The key components of financial sustainability in an HEI cover many points, some of which are explained here. According to Makki and Al-Filali (2024), one of the fundamental elements of economic sustainability is strategic planning and governance. Practices backed by good governance, legislation, and regulations are essential to achieving financial sustainability goals, as they ensure transparent and efficient resource management. Some of the prevalent strategic planning tools include the business model canvas, strategy map, and balanced scorecard (BSC). Utilizing these tools leads to an enhanced identification of resources, improved financial capacity indication, and easy access to the economic sustainability roadmap (Al-Filali et al., 2024). Next, revenue diversification and cost management are crucial for gaining and maintaining financial prowess. These practices are advised to reduce dependence on tuition fees, increase the number of educational (non-regular tuition) and non-educational income streams, and avoid waste in resource utilization through an efficient financial management system (Mahamad et al., 2016; Nik Ahmad et al., 2019). Also, without an active and transparent financial reporting and accountability system, no financial strategy can render the desired result. Dragija Kostić et al. (2017) purport that it is crucial to accommodate a reporting mechanism in the institutional system of governance that can furnish a detailed and comprehensive set of institutional financial activities, which in turn will also facilitate full cost tracking in a transparent and accountable manner. Almagtome et al. (2019) add that financially accountable institutions enjoy legitimacy in the eyes of the law and can win investors' confidence easily. In addition, the economically flexible and robust nature of HEIs will allow them to navigate through internal and external financial threats maturely. Some of the financial flexibility tools suggested by experts include adopting reforms like shifting to accrual accounting and adopting financial ratios and frameworks (Di Carlo et al., 2019).

10.2.1.2 Importance of Economic Sustainability for Universities

Universities must be economically sustainable to perform their educational, research, and social duties flawlessly. In other words, Rogers (2019) says only a financially sound HEI can survive and undertake its prime activities in an ever-changing environment. He adds that a lack of financial stability may make it difficult for a university to retain quality students and employ staff with desired qualities, adversely impacting the overall financial figures and institutional image. Caballero et al. (2004) suggest that universities' quest for economic sustainability should encourage them to pay urgent attention to the efficient allocation and utilization of critical resources by employing tools like multiple criteria decision-making (MCDM) and data envelopment analysis (DEA). At the same time, Elliott and Lewis (2016), Maragakis et al. (2017), and Marriott (2007) mention that students have to endure a heavy debt burden for higher education that negatively affects their graduate and post-graduation lives. In these conditions, economically sustainable universities can financially support deserving students by granting scholarships and fee waivers. Moreover, HEIs that have considerable financial resources actively engage in activities that support SDGs set by the United Nations, as these require ample finances (Blasco et al., 2020; Rammel et al., 2015). Many HEIs experience the challenge of balancing their financial growth with their social and environmental responsibilities. Economically robust institutions find it comparatively easier to cope with such challenges and generally consider the quality factor in their not-for-profit ventures, too (Bui et al., 2023). Lastly, according to Tiron-Tudor et al. (2020) and Von Hauff and Nguyen (2014), financially healthier HEIs naturally perform with high confidence in terms of both academic and non-academic institutional activities, which include investments in contemporary research and education programs to gain institutional and social benefits. Ruben et al. (2023) also say that HEIs with sustainable finances lead to developing policies and systems that can bear economic fluctuations and compete with their contemporaries. These HEIs achieve a leadership role in the market by tackling internal and external financial challenges.

In a nutshell, an economically sustainable university can effectively handle its financial performance, ensure its graduates' bright future, and reinforce its SDGs, leading to overall institutional efficiency. Moreover, such institutions can succeed at efficient resource management, reduce student debt, and be exemplary leaders in the sector.

10.2.1.3 Some of the Sustainable Funding Models for Universities

This section discusses popular and effective modes of earning for HEIs. It enlists income generated through revenue diversification, public financing, private partnerships, and entrepreneurial ventures.

10.2.1.3.1 Diversification of Revenue Streams

HEIs worldwide tend to adopt varied income diversification strategies to bolster their short-term and long-term economic performance. Zilwa (2005) expresses that one of the primary sources of income for many universities around the globe, especially in Australia, is earnings from international students. However, over-reliance on such a source of income may be risky in the long run. Research and consultancy engagements are also potential monetary revenue sources, though they remain underutilized in some HEIs (Ngcobo et al., 2024; Zilwa, 2005). Irvine and Ryan (2019) and Zilwa (2005) suggest that setting up university endowments and their effective management as a strategic financial management strategy may lead to a reliable long-term income stream. Short courses and professional certifications are another source of academic earnings that can benefit vast populations of working professionals (Ngcobo et al., 2024). Directing alumni's attention toward university donations and fundraising campaigns may prove financially lucrative, as observed in the case of monetarily successful American universities that have turned alumni-based funds into strong economic support for education (Chenghua et al., 2023). Other diversified income sources are collaborations with international industrial organizations to boost research and lift financial returns. Introducing and developing innovation-based courses curated for the market's demands will open doors to improved tuition revenue (Ngcobo et al., 2024). According to Yurchyshena et al. (2024), maintaining a viable mix of varied funding sources, like government grants, donations from individuals and private organizations, and commercial operations, can make HEIs more economically sustainable. After obtaining adequate funding, managing and utilizing old and new resources efficiently is crucial. These resources range from infrastructure and facilities to human resources (Nik Ahmad et al., 2019). While managing resource utilization and other university operations, it is necessary to carefully screen and prepare for the legal and financial risks, specifically in monetary consultancy and transfer of technologies (Rahman, 2025; Van Hoorebeek & Marson, 2005).

10.2.1.3.2 Public Funding and Private Sector Partnerships

Government grants and university funding have significantly dropped in recent years in different regions, adding to the already financially challenging higher education sector, struggling with the higher cost of education services (Broadbent & Middlehurst, 2013; Drezner & Villarreal, 2023; Jaafar et al., 2023). Thus, research shows that more universities are focusing on establishing public–private partnerships (PPPs) to enhance the infrastructural and innovative research needs of HEIs in different regions, like Russia (Takmazyan et al., 2018), where such initiatives resulted in raising revenues and monetary stability in universities. PPPs also aid HEIs in Europe and other regions in many ways, such as through technology transfer to industry with the help of public funding (Anderson & Odei, 2018; Thune & Gulbrandsen, 2011). Private sector partnerships in the form of university-industry alliances also assist educational institutions in striding through increasing competition and financial threats, as they are research and innovation-rich relationships with high monetary potential (Jongbloed, 2015; Prigge, 2005; Yarkent et al., 2023). Klein (2021) says that public–public partnerships (PuPs) are partnerships similar to PPPs, except they involve partners from only the public sector, that is, partnerships between public universities and municipal corporations. Arizona State University's Downtown Phoenix Campus project has been a popular example of a successful PuP. Some business houses pay attention to corporate social responsibility projects and actively collaborate with HEIs to participate in pro bono activities, extend financial support to HEIs (Clevenger & MacGregor, 2021), and provide employment opportunities to university students and graduates (Yarkent et al., 2023).

10.2.1.3.3 Entrepreneurial Universities: Fostering Innovation for Economic Growth

To address the financial issues of HEIs, initiatives like locally developed startups and incubation centers, an increased number of entrepreneurial courses, student-driven innovation, and monetization of university assets may generate a good mix of diversified income. Redondo and Camarero (2019), Scagnelli et al. (2019), and Villalba et al. (2024) suggest that HEIs host incubators that help with the generation of innovative business ideas and transforming them into lucrative ventures, which will eventually lead them to an enhanced university image and advantageous investments. According to Maxwell and Levesque (2011), these startups will benefit the local community by creating value for the local people in many ways, creating massive employment opportunities. These university-based startups can also capture financial support from the regional economic development agencies by taking these steps.

10.2.1.4 Efficient Resource Management and Cost Optimization

A healthy financial management system is incomplete without an effective cost-control mechanism because it ensures the optimal allocation of resources for profit maximization and long-term monetary sustainability (Chen, 2023; He, 2015; Ma, 2011). An effective resource management model may choose to adopt intelligent measures for institutional resource allocation, such as DEA, MCDM, and other optimization algorithms, which support the rational allocation of financial resources while protecting transparency and efficiency in institutional functioning (Caballero et al., 2004; Di Leo et al., 2024; Guo, 2022). Some scholars like López (2006) claim universities can establish a thriving autonomous financial management culture by following normative models for performance-based budgeting and efficient resource allocation. Implementing budget evaluation systems powered by AI can also reduce wastage and effective utilization of funds in universities (Wang & Song, 2021). Finally, by following the previously mentioned measures, HEIs may succeed in reducing the waste of resources, leading their institutions to be leaders in resource efficiency and thus strengthening their support to core educational and research activities by thwarting any economic threats. Such universities will also reinforce their support for SDGs like SDG 12 (Responsible consumption and production) and SDG 4 (Quality education) (United Nations, n.d.).

10.2.1.5 Challenges and Risks in Economic Sustainability

This section covers some key economic challenges and risks facing HEIs worldwide. The primary challenge for the higher education system is regular budgetary reductions and declining government funding in many countries, threatening quality instruction and infrastructure (Abankina et al., 2018; Chinyoka & Mutambara, 2020; Giroux, 2015; Wangenge-Ouma, 2011). The economic crises have further worsened the HEIs' worries, as they broaden their revenue needs for survival and risk their financial stability (Bruininks et al., 2010; Chen & He, 2018). The second looming challenge is the constantly rising cost of operations resulting from the high costs of providing pensions, healthcare, and campus facilities, forcing HEIs to continue with reduced resources and employees (Chinyoka & Mutambara, 2020; Freer, 2019).

According to Chen and He (2018), many universities urgently require an adequate financial management system as they are under immense debt pressure, mainly because of their infrastructure and campus expansion projects. HEIs in regions witnessing negative migration and population decline face the threat of under-enrollment, leading them to contemplate their financial survival strategies deeply (Bruininks et al., 2010; Tsymbalenko et al., 2019). The recent trend of privatization of higher education also threatens the role of education as a public service, exacerbating the financial woes of universities as economic pressure forces them to yield to market demands (Giroux, 2015). The rapid technological advancements in HEIs have created an extra financial burden, as they need a regular outflow of revenue to facilitate the need for hardware and software and their continuous maintenance and updates (Arbabisarjou & Allameh, 2012). The technological requirements of HEIs may also force them to rely on new external funding sources, putting heavier financial pressure on them (Di Carlo et al., 2019; Ngcobo et al., 2024). Thus, a robust risk management system is imperative to successfully navigate higher education's internal and external financial challenges (Pavlova & Shaposhnikov, 2019; Aud et al., 2013).

10.2.1.6 Brief Case Studies and Best Practices From Resilient Universities

Many HEIs have efficiently integrated sustainability models into their operations to lead their institutions to sustainable development and improve their financial performance. Here, we will briefly discuss some of the initiatives of various universities (see Table 10.1).

According to Francis and Moore (2019), Charles Sturt University, with continuous sustainability efforts like energy saving and sustainable energy production, has demonstrated savings of $500,000 per year. Monash University announced its planned spending of AUD$135 million till 2030 to shift to 100% renewable energy and achieve Passive House Certification for its buildings. The planned on-site energy project will cover 20% of the total requirement, and the remaining will be outsourced via agreements with external renewable energy suppliers. The study also mentions the case of the Royal Melbourne Institute of Technology (RMIT), which designated AUS$128 million in 2010 to reduce water consumption, energy use, and greenhouse gas emissions. RMIT's efforts under the "Sustainable Urban Precincts Program" (SUPP) include water-saving measures, solar renewable energy generation, energy-efficient heating and cooling assets, and consolidated building management systems. SUPP is supposed to lead RMIT to a 263 million kWh of energy requirement reduction (over 8 years), reduced greenhouse gas emissions (a 32,000 ton cut from about 78,000 tons in 2015), and 53 million L/year of water use savings. RMIT's various buildings across campuses have also been included in the Energy Performance Contract. The fourth case is of the University of New South Wales, which has integrated a real-time resource consumption system into its buildings with the help of 650 smart meters tracking water, energy, and gas usage, allowing its users to save 5–10% of energy while consuming building services. Lastly, the One Planet Ecological Project (EF) by Macquarie University aims to achieve the EF standard set by the Bioregional organization by 2030. To achieve the target, the university has devised strategies to meet 5-star Green Star performance and 4.5-star NABERS energy and water ratings for its buildings. Most of these initiatives at the Australian universities seem to support SDGs like SDG 7 (Affordable and clean energy), SDG 12 (Responsible consumption and production), and SDG 13 (Climate action) (United Nations, n.d.).

In Brazil, the Federal University of Rio de Janeiro created the Green Fund to finance sustainable projects at the university and enhance environmental sustainability and management. The Environmental Plan of the Federal University of Lavras consists of many tools and strategies to help plan and screen sustainability projects to strengthen the university's overall sustainability. University of the Sinos River Valley (Unisinos)'s Environmental Management System ensures long-term environmental and economic gains by coupling sustainability practices with university operations (Allevato & Ribeiro, 2017). These initiatives appear to underpin SDG 7, SDG 12, and SDG 13 (United Nations, n.d.).

Cooper and Gorman's study of the University of Edinburgh in the United Kingdom (2018) sheds light on the university's "Living Lab" approach, which aims to connect its teaching, research, and other operations to sustainability. The university's social responsibility and sustainability department works on seven crucial sustainability areas covering climate change and energy; supply chains and fair trade; resource efficiency and circular economy; learning, teaching, and research; responsible investment; sustainability reporting; and community and public engagement. The research also suggests that the University of Edinburgh's case reflects its dedication to many SGDs (see Table 10.1).

Some other best practices have been identified at Indiana University, Bloomington (IUB), in the United States of America (Zou et al., 2015), and a study on 19 Italian universities by Di Carlo et al. (2019). While Indiana University's flexible institutional dynamic encourages broad participation in its detailed sustainability projects containing economic, environmental, and social strategies, adopting accrual accounting systems in Italian universities has rendered positive financial results (support SDG 12).

Finally, according to George and Wooden (2023), integration of technological tools like AI and quantum technologies can lead to enhanced academic and administrative performance. Moreover, the adoption of optimization technology can help mitigate waste and improve economic returns in

TABLE 10.1
Initiatives Impacting the Financial Sustainability of Universities

S. No.	Country	Universities	Best Practices	Aligning SDGs
1.	Australia	Charles Sturt University; Monash University; RMIT University; University of New South Wales; Macquarie University	Carbon Neutral Offsets, Net Zero Carbon Project, Energy Performance Contracting & "Sustainable Urban Precincts Program" (SUPP), Live Consumption Feedback, and One Planet Ecological Footprint (EF) (Francis & Moore, 2019).	SDG 7: Affordable and clean energy SDG 12: Responsible consumption and production SDG 13: Climate action
2.	Brazil	Federal University of Rio de Janeiro; Federal University of Lavras; University of the Sinos River Valley (Unisinos)	The Green Fund, the Environmental Plan, and the Environmental Management System (Allevato & Ribeiro, 2017).	SDG 7: Affordable and clean energy SDG 12: Responsible consumption and production SDG 13: Climate action
3.	Italy	19 Italian Universities	Accrual accounting makes universities financially robust (Di Carlo et al., 2019).	SDG 12: Responsible consumption and production
4.	UK	University of Edinburgh	The "Living Lab" approach helps integrate sustainability into its operations, research, and teaching. It works in seven key areas: circular economy, supply chain, responsible investment, sustainability reporting, and others (Cooper & Gorman, 2018).	SDG 9: Industry, innovation and infrastructure SDG 12: Responsible consumption and production SDG 17: Partnerships for the goals
5.	USA	Indiana University, Bloomington (IUB)	A sustainability scheme that equally values environmental, economic, and social aspects is needed (Zou et al., 2015).	SDG 12: Responsible consumption and production

Source: Author's illustration.

HEIs (Mukerjee, 2014). Since maintaining economic sustainability in the modern university systems isn't possible without paying attention to technological advancements, the forthcoming section is dedicated to the role of technology in HEIs.

10.3 THE ROLE OF TECHNOLOGY IN ORGANIZATIONAL SUSTAINABILITY AND RESILIENCE

10.3.1 Organizational Resilience and Technology in Higher Education

Organizational resilience is the ability of an institution to be prepared, react to, and recover from disruptions. In higher education, this resilience is critical for adapting to challenges like pandemics, economic crises, and rapid or disruptive technological changes. Universities face these challenges regularly, and technology plays a pivotal role in reinforcing their ability to adapt and recover. As tertiary-level education institutions increasingly integrate digital tools into their workflows, they become more able to continue operations, even during tough times.

10.3.2 Technological Advancements and Organizational Resilience

Technology has become a signpost of resilience in HEIs. Tertiary-level schools that actively adopt digital transformation are ready to tackle unexpected disruptions and enhance operational efficiency. The integration of digital tools and innovative systems, such as AI, deep learning, and machine learning, enables institutions to create systems that are more robust and adaptable to disruptions. To exemplify, digital transformation has paved the way to the development of adaptive cultures within universities, improving their resilience by preparing them to handle novel situations swiftly (Gull et al., 2023). Thus, the role of novel and emerging technologies is multivariate. ICT equips universities with the strategies to design new learning models, personalizing the learning experience, and promoting student engagement. Likewise, AI, deep learning, and machine learning have redesigned educational management by automating simple tasks like grading and attendance, giving faculty and administrative staff more time to focus on more strategic activities (Ajuwon et al., 2024). By embracing these tools, universities not only scale up their resilience but also improve the quality.

10.3.3 Strategic Digital Transformation

For HEIs to build resilience through technology, they need to adopt structured techniques for digital transformation. Aligning digital innovations with the strategic goals is pivotal to success. For instance, numerous German media companies' strategies shed light on how matching digital transformation with business objectives aided them in managing challenges effectively (Hess et al., 2016). HEIs can adopt a similar approach by incorporating technology into academic and administrative workflows, guaranteeing that it aligns with the institution's strategic vision.

A case study at The Open University of China suggested that the adoption of digital tools not only enhanced student–teacher interactions but also assisted in tailoring curricula to more effectively align with sectoral needs (Wang & Li, 2024). The University of Porto's digital transformation centered on setting up a Shared Services Center that redesigned university operations to be more efficient, offering a valuable framework for universities seeking to optimize their internal systems (Faria & Nóvoa, 2017).

10.3.4 Influence on Institutional Excellence

The impact of digital transformation on institutional excellence is significant. Digital adopter organizations tend to stand a greater chance of meeting their strategic goals, maintaining a high level of academic quality, and becoming more competitive on the international stage. For instance, at Zarqa University in Jordan, digital transformation assisted in improving leadership, knowledge, and human resources (Masadeh et al., 2024). Likewise, in Nigeria, universities that emphasized enhancing ICT infrastructure and staff education managed to transcend the most pressing operational challenges (Usman & Bolaji, 2023). These instances provide evidence of the inexorable correlation between technology adoption and institutional excellence, with digital tools enhancing both administrative effectiveness and academic quality.

10.3.5 Fostering Resilience Through Innovation and Collaboration

Besides adopting digital technologies, universities need to promote innovation and collaborations to ensure enhanced resilience. The collaborations should be both internal (for instance, between disciplines or departments) and external (for instance, with business partners). Digital tools such as the digital suite offered by Google have allowed universities to scale up distance learning and encourage collaborations among students and lecturers even during times of difficulty (Shkatula et al., 2023). This is an aspect that shows the significance of not just incorporating technology but establishing collaborations that improve the learning environment.

There are challenges, too, particularly in developing nations, where universities should ensure they break down the barriers to adopting technologies. Developing nations have specific challenges, such as limited infrastructure and resources, that make it even more challenging to implement AI and other technologies. According to Jácome et al. (2017), universities in such environments should embrace best practices, establish collaborations, and implement staff capacity building in an effort to counter the challenges.

The convergence of SDGs and the digitization of university higher education is a complex process of ensuring that the practice of education is aligned with the goals of sustainability. Universities are key players in this convergence by using digital technologies to advance learning outcomes and facilitate sustainable development. This convergence discusses how digitization is incorporating SDGs in the context of higher education as an opportunity and a challenge.

Using digital technologies, AI is capable of playing a significant role in the realization of multiple SDGs by enabling innovation, equity, and digital literacy. The case of the University of Ljubljana shows how AI makes 11 SDGs and 28 targets possible, with the potential risks of protection of data and accessibility equally (Savec & Jedrinović, 2024). The digital learning transition in response to the COVID-19 pandemic in Saudi Arabia emphasized the role of e-learning in sustainable development. The transition reinforced its strengths in improving capacity for improvisation and organizational preparedness toward SDG realization (Alotaibi, 2022).

SDG incorporation into the syllabus is a key area as well, with universities incorporating the principles of sustainability into all disciplines. This interdisciplinary approach is significant as it enables the learning of sustainability issues from diverse angles (Rajabifard et al., 2021). A transdisciplinary pedagogical approach for digital sustainability further enables critical thinking and prepares learners to be co-leaders of sustainable solutions (Lampoltshammer et al., 2021).

Digitalization also inspires social commitment and the participation of learners in sustainability. Literature indicates that digital tools make the moral aspirations of learners and their professional activities more productive toward SDG realization (Lan, 2025). Moreover, the digitalization of universities is nurturing digital citizenship, as well as encouraging learners' active roles in political and social issues toward sustainable development (Lozano-Díaz & Prados, 2020).

Challenges persist. The transfer of SDGs in the higher education context is impeded by resource limitations, commercialization of education, and the challenge of measuring outcomes (McCowan, 2023). Implementation of SDGs is not balanced across disciplines and regions and requires a more comprehensive approach toward incorporating sustainability into curricula (Alcántara-Rubio et al., 2022). Digital transformation in higher education provides magnificent potential but needs to be managed meticulously to ensure inclusivity, equity, and quality education. Universities need to internalize the social responsibility culture and digital practice ethics if they want to make a meaningful contribution toward sustainable development and bring solutions to problems on a global scale (Adiguzel et al., 2023; Bulut et al., 2026; Karanfiloğlu & Bulut, 2025; Yurdunkulu et al., 2025).

10.4 ENVIRONMENTAL SUSTAINABILITY IN HEIS

Environmental sustainability in HEIs is a sophisticated venture involving the integration of sustainable uses into campus proceedings, curriculum, and community engagement. HEIs are unambiguously positioned to influence tomorrow's leaders and contribute to worldwide sustainability efforts through education, study, and policy development. That union fails various dimensions of sustainability in HEIs, highlighting successful initiatives, challenges, and future directions.

10.4.1 Key Areas of Sustainability in HEIs

Sustainable campus actions are basic for reducing the environmental impact of HEIs. Initials such as energy efficiency, waste management, and sustainable transportation have been successfully applied at institutions like the University of Copenhagen and the University of Nottingham, leading to important reductions in carbon emissions and energy use. However, gaps remain, as seen at AAMUSTED, where formalized waste management and energy conservation policies are lacking (MAAME et al., 2025). Environmental education plays a pivotal role in transforming students' green behavior. Studies show that Green Campus initiatives, institutional ecosystems, and support systems importantly influence students' green intentions and behaviors (Cai et al., 2025). Programs like the Healthy Campus initiative at IPBeja in Portugal show the influences of participant and collaborative ways in raising environmental cognizance and liability (Durão et al., 2025). Effective leaders and policy frames are necessary for embedding sustainability in HEIs. The addition of sustainability to leadership roles, policy decisions, and teaching uses can drive transformative change (Liu & Curtin, 2024). Institutional commitment and governance structures are basic in getting over challenges such as resource restrictions and interdisciplinary strength (Shih et al., 2025). HRM uses, including green enlisting and eco-training, can encourage environmental sustainability within HEIs. In Karachi, Pakistan, HRM use has been shown to have a positive impact on eco-friendly behaviors and sustainability initiatives, especially when supported by a strong organizational culture (Imtiaz & Iqbal, 2024). Similarly, in Jordan, Green HRM aligns with public sustainability goals, raising a culture of environmental responsibility (Almetramia et al., 2024).

10.4.2 Challenges and Future Directions

In spite of advancement, challenges such as resource constrictions and the need for long-term planning persist. Tailoring sustainability plans to particular organizational and cultural buildings is a basic step in overcoming these barriers (Sütünç, 2025). Comparative studies and amended data collection methods are necessary to raise decision-making and policy progress (Oliveira & Proença, 2025). Raising partnerships against fields and taking technological advancements are vital for disseminating sustainability in HEIs. Successful initiatives frequently include stakeholder partnerships and innovative solutions that align with the SDGs (Shih et al., 2025). While HEIs have made significant strides in promoting environmental sustainability, challenges such as institutional resistance and limited resources continue to hinder advancement. Addressing these challenges needs a holistic strategy that puts sustainability into all perspectives of HEI operations, from leadership and policy development to student conflict and HRM uses. After that, HEIs can become pivotal drivers of sustainable development, forming a future that aligns with global sustainability goals. HEIs play a basic role in forward environmental sustainability, acting as catalysts for sustainable development through education, study, and operational uses. That result synthesizes insights from 20 study papers to provide a comprehensive overview of environmental sustainability in HEIs, focusing on curriculum integration, operational uses, student engagement, policy frames, digital transformation, community partnerships, and assessment of sustainability initiatives.

10.4.3 Curriculum Integrating and Education for Sustainable Development

The integration of sustainability into university curricula is a basic step toward raising a sustainable future. At the University of Vermont, a successful implementation of a university-wide sustainability requirement provided a model for other institutions. This first involved a multi-stakeholder way, emphasizing organizational dynamics and values to guide curricular change (Hill & Wang, 2018). Education for Sustainable Development (ESD) increases academic success, motivating, and problem-solving abilities. Almost 60% of students in ESD programs show greater motivation compared to 50% in customary programs. However, bars such as being without resources and teacher training hinder implementation (Bonilla-Jurado et al., 2024). A study of 13 HEIs determined that authorities vary in adopting sustainability competencies. The D-RAPID frame and Disciplinary Multi-dimensional Sustainability Influence Change for Academia (D-MuSICA) frame were proposed to increase understanding and collaboration across disciplines (Lozano et al., 2021). Additionally,

K-means clustering showed the highest integration of green and low-carbon principles in Cluster 3, indicating strong uses in green education (Meng et al., 2024).

10.4.4 Operational Uses and Green Campus Initials

HEIs are increasingly adopting sustainable uses in their operations, including energy efficiency, waste management, and green infrastructure. Data mining techniques have been used to analyze variables such as waste generation, energy, and water use in Chinese HEIs, improving resource efficiency and sustainability (Liu, 2025). A frame for implementing Green IT in HEIs was proposed to reduce the environmental impact of IT operations, highlighting sustainable IT uses in higher education (Segara et al., 2023). Grand Valley State University has successfully positioned sustainability principles into its services and operations, receiving national recognition for sustainability innovation (Akbulut-Bailey et al., 2011). At Universiti Teknologi MARA, a pilot study discovered that as students accepted the significance of sustainability, they were unwilling to lose ease for energy-saving behaviors (Zakaria et al., 2024).

10.4.5 Student Engagement and Empowerment

Students play a serious part in forward sustainability practices in HEIs through curricular and extracurricular activities. At California State University, Fresno, student activism has contributed to setting sustainability initiatives, though challenges such as organizational and student load remain (Acosta et al., 2022). In accordance with Krasny and Delia (2014), the case study showed that the impact of adaptive common management involved students in nature field management, where students showed social learning, social capital, and shared activity, leading to policy contributions. On the other hand, Blythe (2017) expresses that programs such as "Make a Difference" and university–community partnerships have been shown to lead to youth sustainability leadership and make sustainable schools. A questionnaire administered among Brazilian students showed that environmental, social, and ethical elements positively affect their perceptions of organizational sustainability (da Silva, 2024).

10.4.6 Policy Frames and Institutional Support

Effective policy frames and organizational assists are a necessity for forward sustainability in HEIs. A literature review and case study at Universidad de La Frontera highlighted the importance of policies to reduce gender gaps in STEM education and promote sustainable development (Bustamante-Mora et al., 2024). A qualitative study across HEIs found that institutional support and policies are pivotal for embedding sustainability into academic and operational uses. Bright principles and top-level leadership are essential for raising a cultural shift toward sustainability (Park, 2024). Portuguese HEIs have amended their sustainability uses and reporting, positioning themselves with UN SDGs and enhancing their positioning in sustainability rankings (Caria et al., 2022).

10.4.7 Digital Transformation and Innovation

The digital transformation in higher education is reframing sustainability efforts, and emerging technologies are playing an important role in this regard. The integration of digital technologies has led to the development of sustainable curricula and increased innovation in higher education. Shenkoya and Kim (2023) predict that the future of higher education will include developments in gamification, datafication, and AI-focused reforms. Besides this, according to Popowska (2024), European HEIs, such as the University of Groningen and Lund University, have made important progress in sustainable development, with a focus on corporate social responsibility and Green Campus initiatives.

10.4.8 Community Partnerships and Collaborative Initiatives

Collaborative partnerships between HEIs, communities, and stakeholders are vital for forward sustainability. The Federal Institute of Santa Catarina in Brazil has established outreach programs that empower students as agents of sustainability. These programs have proven effective in promoting local sustainability and raising student leadership (Fuchs et al., 2023). The project of RUCAS, including 12 universities, has run curricular alterations and structural improvements to embed sustainability in higher education across Europe and the Middle East (Makrakis, 2014). In an Indian university, SDG instruction programs led to affirmative behavioral changes in students, enhancing socio-environmental awareness and advocacy for sustainability (Prakasha et al., 2021).

10.5 METHODOLOGY

The study presents the case of Ibn Haldun University, a foundation (private) university in the city of Istanbul, Turkey. As per the study objectives, the university website and some university reports have been screened to depict a vivid picture of the university's financial, technological, and environmental initiatives underpinning its sustainable existence in the long run. A similar methodology has been used before in works like those of Hasim et al. (2018) and Arias-Valle and Marimon (2024). While doing the website content analysis, reports relevant to this research (attached to the website in document form) have also been analyzed. Due to time constraints, triangulation methods like surveys, interviews, print document analysis, and others couldn't be performed.

CASE STUDY AND BEST PRACTICES AT IBN HALDUN UNIVERSITY

ECONOMIC SUSTAINABILITY

This section lists the activities of Ibn Haldun University that underpin its commitment to financial sustainability. The operations mentioned, directly or indirectly, contribute to the institution's sustainable economic development and numerous SDGs (see Table 10.2).

TABLE 10.2
Initiatives Impacting the Financial Sustainability of Ibn Haldun University

Institution	Best Practices	Aligning SDGs
Ibn Haldun University	Active Donation & Fundraising Policy	SDG 3: Good health and well-being
	Strong Foundation Support	SDG 4: Quality education
	Solar Energy Project	SDG 7: Affordable and clean energy
	Private and Public Sector Collaborations	SDG 12: Responsible consumption and production
	Financially Strong Board of Directors	SDG 13: Climate action
	High Enrollment Rate	SDG 17: Partnerships for the goals (United Nations, n.d.)
	Real Estate and Campus Services	
	Accreditation	
	Project Office Activities	

Source: Author's illustration.

ACTIVE DONATION AND FUNDRAISING POLICY (SUPPORTS SDG 4)

Ibn Haldun University became one of the top private universities in Turkey to receive the highest amount of donations (Ibn Haldun University, 2021). The university actively engages in fundraising activities, and on their Global Education Fund webpage, they explain that they accept corporate donations, regular donations (periodic), and donations in the form of 70 types of cryptocurrencies (Ibn Haldun University, n.d.-g). On another web page, they ask for donations by demonstrating how successful their new university has been in terms of quality education with multiculturalism, facilities to students and employees, international student ratio, students from over 80 countries, its modern and green infrastructure, and the scholarship rate (Ibn Haldun University, n.d.-a).

STRONG FOUNDATION SUPPORT (SUPPORTS SDG 4)

Ibn Haldun University was established in 2015 by the Youth and Education Service Foundation of Türkiye (TÜRGEV), founded in 1996. TÜRGEV caters to a large section of people, and to date, it has provided dormitory service to 80,871 students, scholarships to around 15,000 students, and educational courses to over 30,000 learners. It has graduated over 40,000 beneficiaries over the years (TÜRGEV, n.d.). In Türkiye, as the primary source of economic support for private universities (foundation universities) is their parent foundations (Soyşekerci & Erturgut, 2010), TÜRGEV plays a vital role in Ibn Haldun University's economic sustainability.

SOLAR ENERGY PROJECT (SUPPORTS SDGS 7, 12, 13)

To generate energy from renewable and zero-emission resources under the Green Campus Mission, Ibn Haldun University began installing solar panels on its buildings at the beginning of 2023, and it started to operate in the same year. According to the initial calculations, the university will produce 20% more energy using solar panel installations than its total requirement. The energy generated will be used by campus heating, cooling, and lighting systems (Ibn Haldun University, 2023a, 2023b).

PRIVATE AND PUBLIC SECTOR COLLABORATIONS (SUPPORTS SDG 17)

The university has established relationships with many private, public, and civil society institutions to ensure quality education, better student career opportunities, improved economic benefits from collaborative projects, and enhanced university image. One of the renowned collaborations of the university is with Turkish Airlines and Airbus to run a master's degree program specializing in Air Transport Management, with the prime aim of training more workforce to be employment-ready by leading airline companies (Ibn Haldun University, 2018). Some other collaborations are with the Turkish Council of State (Danıştay), prime business conglomerates, industrial associations, notable NGOs, schools, education foundations, and marketing firms (Ibn Haldun University, n.d.-d). Moreover, IHU is making strides toward becoming a world University by expanding its networks and cooperation agreements with global universities and think tanks in countries like Japan, the USA, Malaysia, Jordan, Qatar, the United Kingdom, and Morocco (Ibn Haldun University, n.d.-b).

STRONG BOARD OF TRUSTEES (SUPPORTS SDGS 4, 17)

The board of trustees plays a significant role in an HEI's governance and overall success, as they are responsible for making critical decisions affecting the institution's future operational direction, policies, and overall mission (Barringer & Riffe, 2018). IHU's board of trustees has a broad spectrum, ranging from an ex-member of parliament, giant business conglomerate owners, and a chief financial officer from a giant airliner to the president of the district municipality (Ibn Haldun University, n.d.-i). The presence of such a varied group of people on the board can undoubtedly enhance the university's decisions affecting its financial sustainability.

HIGH ENROLLMENT RATE (SUPPORTS SDG 4)

According to an internal audit report of IHU (Ibn Haldun University, 2021), the university has the fifth-highest enrollment rate and ranks first in course selection by students from different programs (elective/non-major courses). Along with the domestic student enrollments, the university has successfully attracted students from over 50 countries, amounting to more than 30% of the total student base (Ibn Haldun University, n.d.-b).

REAL ESTATE AND CAMPUS SERVICES (SUPPORTS SDGS 3, 4)

The university has a fully functional state-of-the-art sports center founded on October 12, 2020, to serve university students, academic staff, administrative staff, and the local people in the region. It covers an area of 9,560 m² that accommodates a 215-seat stadium, an indoor basketball court, an indoor AstroTurf field, an indoor swimming pool, an indoor tennis court, a wrestling hall, gyms, a squash playroom, an archery area, a sauna, a salt room, and a steam room. The sports center also has a cafe, a restaurant, a prayer room, and a first aid room (Ibn Haldun University, n.d.-e).

IHU, in collaboration with a real estate company, is building a guest house for visiting scholars, researchers, and other university members as needed. The guest house has more than 50 rooms with different specifications for people with varying needs (Ibn Haldun University, 2024a).

The university is also home to a Turkish language instruction Center (İHÜ TÖMER), which started functioning in 2018 and helps students learn Turkish, with more than 35% of its total students as international candidates. Another popular language program of the university is its summer language school, open to both local and international students, which provides courses in Modern Turkish, Ottoman Turkish, Arabic, Persian, and Chinese during the university summer breaks (Ibn Haldun University, n.d.-f, n.d.-h).

ACCREDITATION (SUPPORTS SDG 4)

Ibn Haldun University initiated the FIBAA (Foundation for International Business Administration Accreditation) process in September 2022, which resulted in the accreditation of a mix of 18 undergraduate and graduate programs until March 2029 (Ibn Haldun University, 2024b). FIBAA is an experienced education and science accreditor working in line with the "European Standards and Guidelines for Quality Assurance," which are applicable throughout Europe (FIBAA, 2025).

PROJECT OFFICE ACTIVITIES (SUPPORTS SDGS 4, 17)

The Ibn Haldun University Project Office has successfully partnered domestically and internationally with public, private, and civil society organizations like the Scientific and Technological Research Council of Turkey (TÜBİTAK) and the European Union. These projects bring dual benefits of academic and industrial knowledge and funding to the researchers and the university.

The project office has carried out projects amounting to approximately 91 million Turkish Liras (approx. 2.5 million USD). The projects mainly fall into the following categories (Ibn Haldun University, n.d.-c):

- Infrastructure Service Projects
- Academic Consultancy Projects
- Research and Development Projects
- Capacity Building Projects
- Youth & Cooperation Projects
- Youth Support Projects
- Thesis Support Projects
- Scientific Research Projects

10.6 TECHNOLOGICAL AGILITY AND SUSTAINABILITY

This section logs the activities at the Ibn Haldun University that emphasize adherence and dedication to technological sustainability. The operations given in Table 10.3, primarily or secondarily, add to the university's digital sustainable development and relevant SDGs.

As the table explicitly illustrates, Ibn Haldun University is proactively driving its technological sustainability by embedding digital transformation in the different aspects of its work, with an emphasis on resilience, innovation, and equity. The actions of the university dovetail into several of the SDGs, with a specific emphasis on support for quality education (SDG 4), economic growth (SDG 8), and driving innovation in the industries (SDG 9). By using technologies such as AI, the institution is not

TABLE 10.3
Initiatives Impacting the Technological Sustainability at Ibn Haldun University

Institution	Best Practices	Aligning SDGs
Ibn Haldun University	Strengthening Resilience through Digital Transformation Leveraging Emerging Technologies for Resilience Aligning Digital Strategies with Institutional Goals Path to Excellence Through Digital Tools Innovating University Operations with Technology Ensuring Equity in Technology Integration AI for Enhanced Student Engagement AI Ethics in Education Digital Integration for Administrative Efficiency AI in Academic Research and Innovation Bridging the Digital Divide in Higher Education Building Resilience Through Digital Transformation Ethical AI for Transparent Education Systems Innovations in Administrative Efficiency	SDG 4—Quality Education SDG 8—Decent Work and Economic Growth SDG 9—Industry, Innovation, and Infrastructure SDG 10—Reduced Inequalities SDG 16—Peace, Justice, and Strong Institutions SDG 17—Partnerships for the Goals

Source: Author's illustration.

merely enhancing administrative efficiency and research but is also ensuring that its digital practices are ethical, open, and equitable, and thereby contributes toward both SDG 16 and SDG 17.

10.6.1 Ibn Haldun University as a Success Story in Digitization

Ibn Haldun University is a leading social sciences university with a full scholarship and tuition-free model that welcomes top-ranked national and international students. The institution has invested heavily in digitization. Table 10.4 shows a number of the university's initiatives on digital education, education technology, AI in education, and innovation themes (Table 10.5).

The events and activities at Ibn Haldun University demonstrate a dynamic and constructive approach to incorporating AI and digital tools into higher education, with a particular emphasis on pedagogical innovation and interdisciplinary collaboration. These initiatives display the university's commitment to improving the learning experience through AI, shedding light on skill development, interdisciplinary cooperation, and the successful adoption of advanced digital methodologies in education.

10.6.2 Center for Innovative Learning and Teaching-AI

The university founded one of the first centers excelling, researching, practicing, and publishing on AI in education. Called "Center for Innovative Learning and Teaching-Artificial Intelligence (CILT-AI)," this center gives training on AI and digital education not only to faculty members but also to students. Offering needs-based training programs for university-, faculty-, and department-wide participation and individual consultation, CILT-AI is a cornerstone of the IHU digital initiative. So far, over 1,000 events, activities, written-oral-multimedia-based content, and numerous publications have been conducted. CILT-AI also sets an example for other universities on a national and international scale.

10.6.3 Advancing AI in Education: Dissemination and Awareness

The bulletins by the Center for Innovative Learning and Teaching—AI keep faculty and students up to date on the most recent advancements in AI and digital education. This bulletin serves as a

TABLE 10.4
Ibn Haldun University Operations on Digitalization and Technology-Based Innovation

Type of Event	Count
Publication	346
Seminar	42
Training	37
Bulletin	28
Workshop	29
Digital Education	15
Project	13
Paper Presentation	13
Meeting	26
Competition	5
Symposium	4
Announcement	4
Certificate Program	3
Conference	2

TABLE 10.5
Ibn Haldun University Operations and Definitions

Section	Name of Event	What Is It?
Bulletin	Artificial Intelligence—Innovative Learning and Teaching Coordination Bulletin Issue	Announcement of the 6th issue of the AI-Innovative Learning and Teaching Coordination Bulletin for readers.
Workshop	"Mind Mapping, Timeline, and Data Visualization Applications" Workshop	A workshop focusing on mind mapping, timeline management, and data visualization skills for students.
Training	Business and Psychology Departments Join AI-Supported Innovative Education Process	Integration of Business and Psychology departments into the AI-driven, innovative transformation in education.
Meeting	Innovative Education Approaches with AI and Digital Tools in Collaboration with the New Media and Communication Department	A meeting to explore innovative educational approaches using AI and digital tools in collaboration with the New Media and Communication Department.
Workshop	Step-by-Step Time and Workspace Management Workshop for Productive Study	A workshop aimed at teaching effective strategies for time and workspace management to enhance productivity.
Meeting	Productive Meeting on the Use of AI in Education with the Psychological Counseling and Guidance Department	A meeting will be held to discuss productive strategies and applications of AI in education in collaboration with the Psychological Counseling and Guidance department.
Training	Creative and Effective Material Design with Canva	A training session on creating innovative and effective materials using Canva as a tool for digital education.
Workshop	Participation in the 1st National Workshop on Learning and Teaching Science in Higher Education	CILT-AI's participation in the first national workshop dedicated to the science of learning and teaching in higher education.
Seminar	AI and Digital Tools in Education Seminar	A seminar aimed at integrating artificial intelligence and digital tools effectively into the educational process.
Meeting	Innovative Steps in Digital Education: Collaboration with the Faculty of Social Sciences	A meeting to discuss innovative steps in digital education through tools like Canvas and active learning methodologies in partnership with the Faculty of Social Sciences.

valuable tool for assisting academic institutions in navigating the digital transformation by providing updates on research, best practices, and innovative AI-powered teaching approaches. What's more, the AI and Digital Tools *Training Seminar* offers an exchange forum for in-depth dialogues and conversations regarding AI applications. It motivates faculty members to utilize AI in lesson plans by touching upon real-world issues, moral paradoxes, ethical dilemmas, and the integration of digital tools into learning and teaching.

10.6.4 Skill Development Through Workshops and Training

To equip faculty members and students with the tools needed to negotiate a technology-enriched learning environment, numerous seminars, and training sessions have been organized. Promoting cognitive tools that help students efficiently organize and present information—skills vital for academic research and project management—is addressed in the *Mind Mapping, Timeline, and Data Visualization Applications* session. In an increasingly digitized academic setting, the *Step-by-Step Time and Workspace Management Workshop for Productive Study* aids students in optimizing their study habits and teaches them coping skills with efficiency issues. The *Creative and Effective Material Design with Canva* workshop brings faculty members face-to-face with digital content creation tools, so that they can enrich creative and attention-grabbing teaching resources. These events support faculty and students in improving their digital competencies, thus serving the university's strategic objectives of digital transformation (Adıgüzel et al., 2023; Birgili et al., 2025; Bulut et al., 2025).

10.6.5 Interdisciplinary Collaboration and Institutional Integration

The university also stresses the importance of interdisciplinary collaboration through a series of workshop events and training sessions to incorporate AI-driven educational strategies across all academic departments. For instance, the workshop *Business and Psychology Departments Join AI-Supported Innovative Research Atelier* initiative shows the use of AI in non-technical fields, highlighting AI's growing impact across diverse academic fields. The *Innovative Education Approaches with AI and Digital Tools in Collaboration with the New Media and Communication Department* emphasizes AI's place in media studies, where digital transformation is swiftly changing the way content is formed and consumed. The *Productive Meeting on the Use of AI in Education with the Psychological Counseling and Guidance Department* showcases AI's potential to improve personalized learning and mental health guidance for students. Likewise, the *Innovative Steps in Digital Education: Collaboration with the Faculty of Islamic Sciences* explores how AI and digital tools could be incorporated into religious sciences, displaying the potential of AI-enriched learning across diverse disciplines.

10.6.6 National and Global Engagement

By taking an active role in the *1st National Workshop on Learning and Teaching Science in Higher Education*, Ibn Haldun University keeps itself aligned with international dialogue on the future of tertiary education. This involvement contributes to forming scientific research networks and helps to create a shared vision for the future of digital transformation. All in all, these events and activities are reflections of the university's well-studied strategy to incorporate AI and digital tools into education, bringing together theoretical ground with applied practice. This approach not only equips students and faculty for the revolutionary landscape of education but also positions the university as an active producer of knowledge in the global dialogue on AI-led pedagogical innovation.

10.7 ENVIRONMENTAL SUSTAINABILITY

The Ibn Haldun University Complex, designed and constructed by the principles of a "Green Campus," features an orchard with 124 trees, dedicated bicycle paths, and parking areas, as well as interior and exterior courtyards that reflect the architectural heritage of Turkish-Islamic tradition. High-quality landscaping arrangements further enhance the campus environment. Currently, the complex accommodates 100 bicycle parking spaces. In the first phase of development, the campus includes a total of 65,000 square meters of green space, providing 65 square meters of greenery per student. This achievement positions Ibn Haldun University among the leading institutions in Türkiye in terms of environmentally conscious campus planning.

The Ibn Haldun University Complex effectively implements a comprehensive "Zero Waste" management system. In this context, 150 recycling waste stations have been installed across all buildings, enabling the separate collection of waste batteries, glass, plastic, metal, paper, and household waste. Additionally, waste management procedures are meticulously carried out in 20 designated waste collection depots and a dedicated composting facility, ensuring sustainable and efficient waste processing. On the campus, the irrigation needs of green spaces are met through rainwater harvesting and underground irrigation reservoirs within the complex, promoting sustainable water management. Additionally, all on-campus transportation relies exclusively on bicycles, scooters, and electric vehicles, reflecting the university's commitment to reducing its carbon footprint.

The Ibn Haldun University Complex serves as an academic hub for students from around the world while also hosting a diverse collection of valuable plants and trees within its campus. Among these are rare and remarkable species such as Pirnal Oak, Magnolia, and Frankincense, contributing to the campus's rich ecological landscape. Exploring the variety and unique characteristics of these trees offers an opportunity to appreciate the natural beauty that surrounds us—trees under which we pass daily, seek shade, and place birdhouses on their branches. As part of the Europe-based EuroVelo project, the bicycle path passing through Ibn Haldun University

Chenghua, L., Weikang, L., & Yanjie, X. (2023). How capital campaigns support the development of universities: A multiple-case study from top American universities. *International Journal of Educational Development, 100*, 102789. https://doi.org/10.1016/j.ijedudev.2023.102789.

Chinyoka, A., & Mutambara, E. (2020). The challenges of revenue generation in state universities: The case of Zimbabwe. *Cogent Social Sciences, 6*(1). https://doi.org/10.1080/23311886.2020.1748477.

Clevenger, M. R., & MacGregor, C. J. (2021). Corporate philanthropy and beyond. *Philanthropy & Education, 5*(1), 31–71. https://dx.doi.org/10.2979/phileduc.5.1.02

Colás-Bravo, P., Conde-Jiménez, J., & Reyes-de-Cózar, S. (2021). Sustainability and digital teaching competence in higher education. *Sustainability, 13*(22), 12354. https://doi.org/10.3390/SU132212354

Collins, A., Sanal, Z. G., & Takır, A. (2021). The qualitative study of factors influencing to international students' satisfaction: The case of a private university in Turkey. In *Global Perspectives on Recruiting International Students: Challenges and Opportunities* (pp. 155–172). Emerald Publishing Limited. https://doi.org/10.1108/978-1-83982-518-720211010

Comi, A., Gruenwald, N., Danchuk, V., Kunytska, O., Vakulenko, K., & Zakrzewska, M. (2023). How cities and universities approach the sustainable development goals. In *Handbook of Sustainability Science in the Future: Policies, Technologies and Education by 2050* (pp. 1–21). Springer International Publishing.

Cooper, L., & Gorman, D. (2017). A holistic approach to embedding social responsibility and sustainability in a university—Fostering collaboration between researchers, students and operations. In *Handbook of sustainability science and research* (pp. 177–192). Cham: Springer International Publishing. https://doi.org/10.1007/978-3-319-63007-6_11

Coronel, R., & Trigos, F. (2024). Reframing Higher Education Management: A Transdisciplinary Digital Framework for University Administration. *Engineering For Social Change*, 947–956.

da Silva, R. G. L. (2024). The advancement of artificial intelligence in biomedical research and health innovation: challenges and opportunities in emerging economies. *Globalization and health, 20*(1), 44. https://doi.org/10.1186/s12992-024-01049-5

Di Carlo, F., Modugno, G., Agasisti, T., & Catalano, G. (2019). Changing the accounting system to foster universities' financial sustainability: First evidence from Italy. *Sustainability, 11*(21), 6151. https://doi.org/10.3390/su11216151

Di Leo, S., Avenali, A., Daraio, C., & Wolszczak-Derlacz, J. (2024). Climbing university rankings under resources constraints: A combined approach integrating DEA and directed Louvain community detection. *Annals of Operations Research*, 1–26. https://doi.org/10.1007/s10479-024-06219-7

Dindar, M., Çelik, I., & Muukkonen, H. (2022). # WedontWantDistanceEducation: A thematic analysis of higher education students' social media posts about online education during Covid-19 pandemic. *Technology, Knowledge and Learning, 27*(4), 1337–1355. https://doi.org/10.1007/s10758-022-09621-x

Dragija Kostić, M., Jovanović, T., & Letica, M. (2017). Perspectives on and obstacles to the internal reporting reform at higher education institutions–Case of Bosnia and Herzegovina, Croatia and Slovenia. *Management: Journal of Contemporary Management Issues, 22*(Special Issue), 129–143.

Drezner, N. D., & Villarreal, R. C. (2023). Engaging the Latino community: Enhancing Hispanic-Serving Institutions' Latino donor base. In *Hispanic-Serving Institutions in American Higher Education* (pp. 178–193). Routledge.

Durão, A., Borralho, T., & Raposo, A. (2025). Healthy Campus: A Contribution to the Environment, Sustainability, and Social Responsibility Practices at the IPBeja in Portugal. *Sustainability, 17*(3), 862. https://doi.org/10.3390/su17030862

Dyke, J. G., & Monbiot, G. (2024). What is the role of universities at a time of climate and ecological crisis? *Geo: Geography and Environment, 11*(2), e00146.

Elliott, W., & Lewis, M. (2016). Student debt effects on financial well-being: Research and policy implications. In *A Collection of Reviews on Savings and Wealth Accumulation* (pp. 33–57). Wiley Blackwell. https://doi.org/10.1002/9781119158424.ch3

Eser, Z., & Birkan, I. (2005). Marketing education in Turkey: A comparative study between state and private universities. *Journal of Teaching in International Business, 16*(2), 75–101. https://doi.org/10.1300/J066v16n02_05

Eskinat, A., & Teker, S. (2024). Financial sustainability of higher education system. *PressAcademia Procedia, 18*(1), 71–72. http://doi.org/10.17261/Pressacademia.2023.1854

FIBAA. (2025). FIBAA: Your partner for excellence in higher education. *Welcome to FIBAA*. https://www.fibaa.org/en/

Ferk Savec, V., & Jedrinović, S. (2025). The Role of AI Implementation in Higher Education in Achieving the Sustainable Development Goals: A Case Study from Slovenia. *Sustainability, 17*(1), 183. https://doi.org/10.3390/su17010183

Francis, M., & Moore, T. (2018). University buildings: the push and pull for sustainability. In *Energy Performance in the Australian Built Environment* (pp. 131–147). Singapore: Springer Singapore. https://doi.org/10.1007/978-981-10-7880-4_9

Francis, N. J., Jones, S., & Smith, D. P. (2025). Generative AI in higher education: Balancing innovation and integrity. *British Journal of Biomedical Science*. https://doi.org/10.3389/bjbs.2024.14048

Freer, L. (2019, October). Strategies for career resilience in the face of economic uncertainty. In *Proceedings of the 2019 ACM SIGUCCS Annual Conference* (pp. 20–22). https://dl.acm.org/doi/pdf/10.1145/3347709.3347811

Fuchs, P.G., de Aguiar Dutra, A.R., Scalia, M.D., de Andrade Guerra, J.B.S.O. (2023). The Role of Higher Education Institutions in the Promotion of Collaborative Community Partnerships and Engagement: The Case of Federal Institute of Santa Catarina, Brazil. In: Leal Filho, W., Lange Salvia, A., Pallant, E., Choate, B., Pearce, K. (eds) *Educating the Sustainability Leaders of the Future. World Sustainability Series*. Springer, Cham. https://doi.org/10.1007/978-3-031-22856-8_19

George, B., & Wooden, O. (2023). Managing the strategic transformation of higher education through artificial intelligence. *Administrative Sciences*, *13*(9), 196. https://doi.org/10.3390/admsci13090196

Giroux, H. A. (2015). Higher education and the promise of democracy. In *Future of Higher Education* (pp. 53–62). Routledge.

Gull, M., Parveen, S., & Sridadi, A. R. (2024). Resilient higher educational institutions in a world of digital transformation. *Foresight*, *26*(5), 755–774. https://doi.org/10.1108/FS-12-2022-0186

Guo, S. (2022). Research on multi-objective intelligent optimization of financial resource management and allocation methods in colleges and universities. *Mathematical Problems in Engineering*, *2022*(1), 1906031. https://doi.org/10.1155/2022/1906031

Harini, H., Ripki, A. J. H., Sulistianingsih, S., Herlina, H., & Putri, A. R. (2024). Digital transformation: The utilization of information and communication technology to enhance educational management efficiency in the Modern Era. *Jurnal Minfo Polgan*. https://doi.org/10.33395/jmp.v13i2.14195

Hasim, M. S., Hashim, A. E., Ariff, N. R. M., Sapeciay, Z., & Abdullah, A. S. (2018). Commitment to sustainability: A content analysis of website for university organisations. In *IOP conference series: Earth and environmental science* (Vol. 117, No. 1, p. 012046). IOP Publishing. https://doi.org/10.1088/1755-1315/117/1/012046

He, Y. (2015). The study of the reflection of cost control on financial management. *Open Cybernetics & Systemics Journal*, *9*, 1846–1848.

Hess, T., Matt, C., Benlian, A., & Wiesböck, F. (2020). Options for formulating a digital transformation strategy. In *Strategic information management* (pp. 151–173). Routledge. https://doi.org/10.4324/9780429286797-7

Hill, L. M., & Wang, D. (2018). Integrating sustainability learning outcomes into a university curriculum: A case study of institutional dynamics. *International Journal of Sustainability in Higher Education*. https://doi.org/10.1108/IJSHE-06-2017-0087

Ibn Haldun University. (2018). *Air Transport Management Master Program Is Launched by Ibn Haldun University*. https://www.ihu.edu.tr/en/hava-tasimaciligi-yonetimi-yuksek-lisans-programi-nin-lansmani-yapildi

Ibn Haldun University. (2021). *Kurum iç değerlendirme raporu [Internal Evaluation Report of the Institution]*. https://mis.yokak.gov.tr/images/report/kidr-1093-2021.pdf

Ibn Haldun University. (2023a). *Ibn Haldun University SES Started Operations*. https://yesilkampus.ihu.edu.tr/en/ibn-haldun-university-ses-started-operations

Ibn Haldun University. (2023b). Ibn Haldun University will produce more than it consumes with the solar energy panels—green campus—Ibn Haldun University. *Green Campus*. https://yesilkampus.ihu.edu.tr/en/ibn-haldun-university-will-produce-more-than-it-consumes-with-the-solar-energy-panels

Ibn Haldun University. (2024a). *Groundbreaking Ceremony of Ramazan Bülbül Guest House Has Been Held*. https://www.ihu.edu.tr/en/groundbreaking-ceremony-of-ramazan-bulbul-guest-house-has-been-held

Ibn Haldun University. (2024b). *18 Programs Awarded International Accreditation by FIBAA*. https://www.ihu.edu.tr/en/18-programs-awarded-international-accreditation-by-fibaa

Ibn Haldun University. (n.d.-a). *Bağış Yapın [Donate]*. https://www.ihu.edu.tr/en/bagis-yapin

Ibn Haldun University. (n.d.-b). *Global Competition*. https://www.ihu.edu.tr/en/kuresel-rekabet

Ibn Haldun University. (n.d.-c). Homepage. *Project Support Office*. https://pdo.ihu.edu.tr/en

Ibn Haldun University. (n.d.-d). Institutions we collaborate with—Ibn Haldun University. *Career Center*. https://kariyer.ihu.edu.tr/en/institutions-we-have-contracted-with

Ibn Haldun University. (n.d.-e). *Sports Center*. https://www.ihu.edu.tr/en/sports-center

Ibn Haldun University. (n.d.-f). *Summer School*. https://summer.ihu.edu.tr/en

Ibn Haldun University. (n.d.-g). Support Ibn Haldun's global education fund. *Department of International Relations*. https://international.ihu.edu.tr/en/landing

Ibn Haldun University. (n.d.-h). *Tomer—Center for Turkish Language Instruction*. https://tomer.ihu.edu.tr/en

Ibn Haldun University. (n.d.-i). *University Board of Trustees*. https://www.ihu.edu.tr/en/mutevelli-heyeti

Imtiaz, Q., & Iqbal, S. (2024). The role of HRM practices in promoting environmental sustainability in higher education institutions: A study in Karachi, Pakistan. *Global Management Sciences Review*. https://doi.org/10.31703/gmsr.2024(ix-iv).04

Irvine, H., & Ryan, C. (2019). The financial health of Australian universities: Policy implications in a changing environment. *Accounting, Auditing & Accountability Journal*, *32*(5), 1500–1531. https://doi.org/10.1108/AAAJ-03-2018-3391

Jaafar, J. A., Latiff, A. R. A., Daud, Z. M., & Osman, M. N. H. (2023). Does revenue diversification strategy affect the financial sustainability of Malaysian public universities? A panel data analysis. *Higher Education Policy*, *36*(1), 116–143. https://doi.org/10.1057/s41307-021-00247-9

Jácome, E. P., Enciso, E. M. V., Arias, A. V., & Hernández, C. N. J. (2017). Analysis of university management of emerging technologies and recommendations for developing countries. *Turkish Online Journal of Educational Technology-TOJET*, *16*(2), 1–10.

Jongbloed, B. (2015). Universities as hybrid organizations: Trends, drivers, and challenges for the European university. *International Studies of Management & Organization*, *45*(3), 207–225. https://doi.org/10.1080/00208825.2015.1006027

Jor, J. (2025). Impact of artificial intelligence on engagement, motivation, work efficiency, and academic outcomes. In *Connecting Ideas, Cultures, and Communities*. Routledge. https://doi.org/10.1201/9781003591511-79

Karanfiloğlu, M., & Bulut, M. A. (2025). Techno-pedagogical communication, ed-tech and media professionals: Crossroads for enhancing instructional quality. *Interactive Learning Environments*, *33*(7), 4193–4217. https://doi.org/10.1080/10494820.2025.2459180

Kilinc, E., Tarman, B., & Aydin, H. (2018). Examining turkish social studies teachers' beliefs about barriers toTechnology integration. *TechTrends*, *62*, 221–223. https://doi.org/10.1007/s11528-018-0280-y

Klein, M. W. (2021). Public-public partnerships: Expanding higher education capacity through economic redevelopment. In *Re-Envisioning Higher Education's Public Mission: Global Perspectives* (pp. 35–57). Springer AG Switzerland. https://doi.org/10.1007/978-3-030-55716-4_3

Krasny, M. E., and J. Delia. 2014. Campus sustainability and natural area stewardship: student involvement in adaptive comanagement. *Ecology and Society*, *19*(3), 27. http://dx.doi.org/10.5751/ES-06787-190327

Kurt, S. (2014). Creating technology-enriched classrooms: implementational challenges in Turkish education. *Learning, Media and Technology*, *39*(1), 90–106. https://doi.org/10.1080/17439884.2013.776077

Kurt, T., & Gümüş, S. (2015). Trends of financing higher education in the world and recommendations for turkey. *Journal of Higher Education and Science*, *5*(1), 14. https://doi.org/10.5961/JHES.2015.105

Lampoltshammer, T. J., Albrecht, V., & Raith, C. (2021). Teaching Digital Sustainability in Higher Education from a Transdisciplinary Perspective. *Sustainability*, *13*(21), 12039. https://doi.org/10.3390/su132112039

Lan, H. (2025). Sustainable development of social responsibility in universities in the digital era: Perceptions and advances. *Information Development*. https://doi.org/10.1177/02666669241303124

Liu, Q. (2025). Enhancing green smart campus development with data mining technology. *Journal of Machine and Computing*. https://doi.org/10.53759/7669/jmc202505041

Liu, Y. & Curtin, J. (2025). Sustainability in Higher Education: Integrating Leadership, Policy, and Teaching Practices for a Brighter Future. In S. Saeed, O. Rana, & R. Dhanaraj (Eds.), *Higher Education and Quality Assurance Practices* (pp. 123–158). IGI Global Scientific Publishing. https://doi.org/10.4018/979-8-3693-6765-0.ch005

López, M. J. G. (2006). Towards decentralized and goal-oriented models of institutional resource allocation: The Spanish case. *Higher Education*, *51*(4), 589–617. https://doi.org/10.1007/s10734-004-1905-y

Lozano, R., Barreiro-Gen, M., Pietikäinen, J., Gago-Cortes, C., Favi, C., Jimenez Munguia, M. T., Monus, F., Simão, J., Benayas, J., Desha, C., Bostanci, S., Djekic, I., Moneva, J. M., Sáenz, O., Awuzie, B., & Gladysz, B. (2022). Adopting sustainability competence-based education in academic disciplines: Insights from 13 higher education institutions. *Sustainable Development*, *30*(4), 620–635. https://doi.org/10.1002/sd.2253

Lozano-Díaz, A., & Fernández-Prados, J. S. (2020). Educating Digital Citizens: An Opportunity to Critical and Activist Perspective of Sustainable Development Goals. *Sustainability*, *12*(18), 7260. https://doi.org/10.3390/su12187260

Ma, N. B. (2011). Analysis of the influence of costs on profit and corporate efficiency of the enterprise: A case study. *Recebimento dos originais*, 77–89.

Maame, a. D. D. A. M., Acheampong, m., & Wepea, w. A. M. (2025). Sustainability practices at higher education institutions, current efforts and future directions: the case study of aamusted. *International journal of research*, *8*, 1093–1102. Https://doi.Org/10.47772/Ijriss.2024.8120092

Mahajan, S., Pandey, P. K., Pandey, P. K., & Guleria, N. (2025). Leveraging Artificial Intelligence in Indian Higher Education Institutions to Foster Sustainable Practices: A Comprehensive Analysis. In *Adaptive*

Strategies for Green Economy and Sustainability Policies (pp. 301–320). IGI Global Scientific Publishing. https://doi.org 10.4018/979-8-3693-7570-9.ch018

Mahamad, N. A., Mohamad Shahimin, M., Romle, A. R., Mohamad Zabri, M. A. H., Mohd Rodzi, S., & Ideris, M. S. K. (2016). Financial management analysis in the higher education institutions: Experience from Universiti Malaysia Perlis (UniMAP)(2003-2012). *World Applied Sciences Journal*, *34*(5), 571–580. https://doi.org/10.5829/idosi.wasj.2016.34.5.15695

Makki, A. A., & Al-Filali, I. Y. (2024). Modeling the Strategic Enablers of Financial Sustainability in Saudi Higher Education Institutions Using an Integrated Decision-Making Trial and Evaluation Laboratory–Interpretive Structural Modeling Approach. *Sustainability*, *16*(2), 685. https://doi.org/10.3390/su16020685

Makrakis, V. (2014). Transforming university curricula towards sustainability: A Euro-Mediterranean initiative. In *Handbook of research on pedagogical innovations for sustainable development* (pp. 619–640). IGI Global Scientific Publishing. doi.org/10.4018/978-1-4666-5856-1.ch032

Maphalala, M. C., & Ajani, O. A. (2025). Leveraging artificial intelligence as a learning tool in higher education. *Interdisciplinary Journal of Education Research*. https://doi.org/10.38140/ijer-2025.vol7.1.01

Maragakis, A., van den Dobbelsteen, A., & Maragakis, A. (2016). Validating the Need to Include the Economic Returns of Graduates as a Metric of a Higher Education Institutions Level of Sustainability. *International Journal of Higher Education*, 5(4), 237–246. https://doi.org/10.7480/abe.2017.3.3661

Marriott, P. (2007). An analysis of first experience students' financial awareness and attitude to debt in a post-1992 UK University. *Higher Education Quarterly*, *61*(4), 498–519. https://doi.org/10.1111/j.1468-2273.2007.00369.x

Masadeh, M. A., Alzeaideen, K., Haimour, F. O., Masadeh, F. A., Alsabatin, H., & Hakami, H. (2024). *The Impact of Digital Transformation on Institutional Excellence in Private Jordanian Universities—Case Study: Zarqa University*. https://doi.org/10.1007/978-3-031-36895-0_19

Matesanz, M. M., da Silva Caeiro, S. S. F., & Bacelar Nicolau, P. (2023). Anticipating Future Needs in Key Competences for Sustainability in Two Distance Learning Universities of Spain and Portugal. *Sustainability*, *15*(5), 4444. https://doi.org/10.3390/su15054444

Maxwell, A., & Levesque, M. (2011). Technology incubators: Facilitating technology transfer or creating regional wealth? *International Journal of Entrepreneurship and Innovation Management*, *13*(2), 122–143. https://doi.org/10.1504/IJEIM.2011.038855

McCowan, T. (2023). The crosscutting impact of higher education on the Sustainable Development Goals. *International Journal of Educational Development*, *103*, 102945. https://doi.org/10.1016/j.ijedudev.2023.102945

Meng, J., Abed, A. M., Elsehrawy, M. G., Al Agha, A., Abdullah, N., Elattar, S., ... & Assilzadeh, H. (2024). Nano-integrating green and low-carbon concepts into ideological and political education in higher education institutions through K-means clustering. *Heliyon*, *10*(10). https://doi.org/10.1016/j.ijedudev.2023.102945

Mossman, A. P. (2018). Retrofitting the ivory tower: Engaging global sustainability challenges through interdisciplinary problem-oriented education, research, and partnerships in US higher education. *Journal of Higher Education Outreach and Engagement*, *22*(1), 35–60.

Mukerjee, S. (2014). Organizational agility in universities: Tensions and challenges. In *Advancing Knowledge in Higher Education: Universities in Turbulent Times* (pp. 15–25). IGI Global. https://doi.org/10.4018/978-1-4666-6202-5.ch002

Ngcobo, X. M., Marimuthu, F., & Stainbank, L. J. (2024). Revenue sourcing for the financial sustainability of a university of technology: an exploratory study. *Cogent Education*, *11*(1). https://doi.org/10.1080/2331186X.2023.2295173

Nik Ahmad, N. N., Ismail, S., & Siraj, S. A. (2019). Financial sustainability of Malaysian public universities: officers' perceptions. *International Journal of Educational Management*, *33*(2), 317–334. https://doi.org/10.1108/IJEM-06-2017-0140

Oliveira, M. C., & Proença, J. (2025). Sustainable Campus Operations in Higher Education Institutions: A Systematic Literature Review. *Sustainability*, *17*(2), 607. https://doi.org/10.3390/su17020607

Omazic, A., & Zunk, B. M. (2021). Semi-Systematic Literature Review on Sustainability and Sustainable Development in Higher Education Institutions. *Sustainability*, *13*(14), 7683. https://doi.org/10.3390/su13147683

Önder, S. (2006). A survey of awareness and behaviour in regard to environmental issues among selcuk university students in Konya, Turkey. *Journal of Applied Sciences*, *6*(2), 347–352. https://doi.org/10.3923/jas.2006.347.352

Özoğlu, M., Gür, B. S., & Gümüs, S. (2016). Rapid expansion of higher education in Turkey: The challenges of recently established public universities (2006–2013). *Higher Education Policy*, *29*, 21–39. https://doi.org/10.1057/hep.2015.7

Park, Y. (2024). *Sustainable Education Practices: Voices from Higher Education Institutions*. https://doi.org/10.21203/rs.3.rs-3923563/v1

Pavlova, X. L., & Shaposhnikov, S. O. (2019, January). Risk management for university competitiveness assurance. In *2019 IEEE Conference of Russian Young Researchers in Electrical and Electronic Engineering (EIConRus)* (pp. 1440–1443). IEEE. https://doi.org/10.1109/EIConRus.2019.8657275

Popowska, M. (2024). *Greening Higher Education in Europe: Institutional Transitions to Sustainable Development*. Routledge. https://doi.org/10.4324/9781003433033

Prakasha, G. S., Chakraborty, R., & Kenneth, A. (2021, September). Energy saving, waste management, and pollution free steps for university campuses. In *AIP Conference Proceedings* (Vol. 2396, No. 1, p. 020021). AIP Publishing LLC. https://doi.org/10.1063/5.0066325

Prigge, G. W. (2005). University—industry partnerships: What do they mean to universities? A review of the literature. *Industry and Higher Education, 19*(3), 221–229. https://doi.org/10.5367/0000000054300486

Rahman, M. (2025). Revenue Diversification in Higher Education Institutions: A Systematic Literature Review. *BABUR Research*, 4(1), 157–181.

Rajabifard, A., Kahalimoghadam, M., Lumantarna, E., Herath, N., Hui, F. K. P., & Assarkhaniki, Z. (2021). Applying SDGs as a systematic approach for incorporating sustainability in higher education. *International Journal of Sustainability in Higher Education, 22*(6), 1266–1284. https://doi.org/10.1108/IJSHE-10-2020-0418

Rammel, C., Velazquez, L., & Mader, C. (2015). Sustainability assessment in higher education institutions: What and how? In *Routledge Handbook of Higher Education for Sustainable Development* (pp. 331–346). Routledge.

Redondo, M., & Camarero, C. (2019). University business incubators: Mechanisms to transform ideas into businesses. In *The Role and Impact of Entrepreneurship Education* (pp. 275–299). Edward Elgar Publishing. https://doi.org/10.4337/9781786438232.00024

Rogers, N. M. (2019). Why change? A practitioner's perspective on why and how universities tackle organisational change (or don't). *Perspectives: Policy and Practice in Higher Education, 23*(4), 152–157. https://doi.org/10.1080/13603108.2019.1581850

Ruben, B. D., De Lisi, R., & Gigliotti, R. A. (2023). *A Guide for Leaders in Higher Education: Concepts, Competencies, and Tools*. Taylor & Francis.

Salazar-Clemeña, R. M. (2011). Higher education counseling: Keeping pace with rapidly changing learning environments. In *The Emergent Knowledge Society and the Future of Higher Education* (pp. 161–176). Routledge.

Saribas, D., Doganca Kucuk, Z., & Ertepinar, H. (2016). Evaluating effects of an exhibition visit on pre-service elementary teachers' understandings of climate change. *Journal of Turkish Science Education, 13*(1), 19–30. https://doi.org/10.12973/tused.10154a

Scagnelli, S., Vasile, L., & Apostolov, M. (2019). Survival drivers of post-incubated start-ups: The effect of academic governance. *International Journal of Innovation Management, 23*(07), 1950062. https://doi.org/10.1142/S1363919619500622.

Segara, P., Hikmawati, E., & Surendro, K. (2023, October). Towards sustainable higher education: A framework for implementing green IT strategies. In *2023 International Conference on Electrical Engineering and Informatics (ICEEI)* (pp. 1–6). IEEE. doi.org/https://doi.org/10.1109/ICEEI59426.2023.10346921

Shenkoya, T., & Kim, E. (2023). Sustainability in Higher Education: Digital Transformation of the Fourth Industrial Revolution and Its Impact on Open Knowledge. *Sustainability, 15*(3), 2473. https://doi.org/10.3390/su15032473

Shih, Y. H., Hsu, M. C., & Chang, C. L. (2025). Sustainability transformations in higher education: global perspectives on the challenges and solutions. *International Journal of Education and Humanities, 5*(1), 126–139. https://doi.org/10.58557/(ijeh).v5i1.290

Shkatula, O., Zbanatska, O., Doroshenko, Y., Varenko, V., & Grinberg, L. (2023). Google Digital Tools for Higher Education: Exploitation, Results, Prospects. *Journal of Higher Education Theory & Practice, 23*(5). https://doi.org/10.33423/jhetp.v23i5.5924

Singh, P. (2025). Artificial Intelligence and Student Engagement: Drivers and Consequences. In *Cases on Enhancing P-16 Student Engagement With Digital Technologies* (pp. 201–232). IGI Global Scientific Publishing. doi.org/10.4018/979-8-3693-5633-3.ch008

Skiba, M., Turganbaeva, A. R., Skaburskiene, N., & Pavalkis, D. (2024). To the question of digitalization of universities in the context of striving for academic excellence. Bulletin of Abai KazNPU. *Series of Physical and Mathematical sciences, 88*(4), 270–277. https://doi.org/10.51889/2959-5894.2024.88.4.027

Soyşekerci, S., & Erturgut, R. (2010). Anatomy of a period: Structural transformation in the Turkish higher education system between 1981 and 2009. *Procedia-Social and Behavioral Sciences, 2*(2), 3477–3482. https://doi.org/10.1016/j.sbspro.2010.03.538

Sütünç, H. S., Tsai, P. L., & Kuo, W. C. (2025). Green campus initiatives: Assessing the impact and necessity of sustainable practises in higher education institutions. *Journal of Anatolian Environmental and Animal Sciences, 10*(1), 15–24.

Takmazyan, A. S., Rukina, S. N., Samoylova, K. N., & Gerasimova, K. A. (2018). Public-private partnership in the Russian educational system. *European Research Studies Journal, 21*(Special issue 1), 479–485.

Talay, I., Gunduz, S., & Akpinar, N. (2004). On the status of environmental education and awareness of undergraduate students at Ankara University, Turkey. *International Journal of Environment and Pollution, 21*(3), 293–308. https://doi.org/10.1504/IJEP.2004.004196

Thune, T., & Gulbrandsen, M. (2011). Institutionalization of university-industry interaction: An empirical study of the impact of formal structures on collaboration patterns. *Science and Public Policy, 38*(2), 99–107. https://doi.org/10.3152/030234211X12924093660110

Tiron-Tudor, A., Zanellato, G., & Moise, E. M. (2020). Higher education institutions' sustainable development: Performance and reporting. *Journal of Environmental Protection and Ecology, 21*(1), 211–221.

Treve, M. (2024). Integrating artificial intelligence in education: Impacts on student learning and innovation. *International Journal of Vocational Education and Training Research, 10*(2), 61–69. https://doi.org/10.11648/j.ijvetr.20241002.14

Tsymbalenko, N., Tarasenko, I. O., & Bielialov, T. (2019). The impact of demographic processes on forming student body in Ukraine. *Revista Espacios. 40*(12).

TÜRGEV. (n.d.). *Anasayfa [Homepage]*. Türgev. https://turgev.org/tr

UNESCO. (2022). *Higher Education Figures at a Glance*. https://uis.unesco.org/sites/default/files/documents/f_unesco1015_brochure_web_en.pdf

United Nations. (n.d.). The 17 goals | sustainable development. *United Nations*. https://sdgs.un.org/goals

Usman, H. A., & Bolaji, H. O. (2023). Optimizing Administrative Efficiency: The Impact of ICT Adoption and Change Management in Educational Institutions. *JUMPA: Jurnal Manajemen Pendidikan, 4*(2), 97–108. https://doi.org/10.33650/jumpa.v4i2.9388

Van Hoorebeek, M., & Marson, J. (2005). Teaching and technology transfer as alternative revenue streams: A primer on the potential legal implications for UK universities. *International Journal of Educational Management, 19*(1), 36–47. https://doi.org/10.1108/09513540510574939

Villalba, C. I. C., Sánchez, M. J. H., Bravo, I. F. B., & Mosquera, Á. E. B. (2024). Transformación de universidades incubadoras a creadoras directas de empresas Spin-Off. *Revista De Ciencias Sociales, 30*(2), 305–319.

Von Hauff, M., & Nguyen, T. (2014). Universities as Potential Actors for Sustainable Development. *Sustainability, 6*(5), 3043–3063. https://doi.org/10.3390/su6053043.

Wang, D., Song, G. (2021). Construction of Budget Evaluation Index System for Application-Oriented Undergraduate Universities Based on Artificial Intelligence. In: Atiquzzaman, M., Yen, N., Xu, Z. (eds) *Big Data Analytics for Cyber-Physical System in Smart City. BDCPS 2020. Advances in Intelligent Systems and Computing*, vol 1303. Springer, Singapore. https://doi.org/10.1007/978-981-33-4572-0_80.

Wang, X., & Li, J. (2024). Digital transformation in higher education: An archi mate-based business architecture case study. *IEEE Access*. https://doi.org/10.1109/access.2024.3520829

Wangenge-Ouma, G. (2011). Managing resource dependence difficulties in African higher education: The case of multiple exchange relationships. *Higher Education Policy, 24*, 167–184. https://doi.org/10.1057/hep.2010.30

Yarkent, Ç., Mutaf, T., Temel, S., Vardar Sukan, F., Oncel, S.S. (2023). University-Industry Collaboration: A Way to New Technologies. In: Oncel, S.S. (eds) *A Sustainable Green Future*. Springer, Cham. https://doi.org/10.1007/978-3-031-24942-6_3.

Yurchyshena, L., Dluhopolskyi, O., Vechirko, I., Kozlovskyi, S., & Lavrov, R. (2024). Prerequisites of hybridization of university financing as a tool for ensuring sustainability and strategic development. *Natsional'nyi Hirnychyi Universytet. Naukovyi Visnyk* (5), 184–191. https://doi.org/10.33271/nvngu/2024-5/184

Yurdunkulu, A., Bulut, M. A., & Göçen, A. (2025). From academics to Aidemics: Unpacking the human–AI symbiosis in higher education. *Acta Psychologica, 261*, 105796. https://doi.org/10.1016/j.actpsy.2025.105796

Zakaria, Z., Isa, B., Ismail, M. H., Wahid, N. T. A., Ismail, A. M., Yahaya, M. A., ... & Alias, N. D. (2024). Minds and Megawatts: University Students' Engagement with Energy Efficiency and Sustainability Practices. *International Journal of Academic Research in Business & Social Sciences, 14*, 771–787. http://dx.doi.org/10.6007/IJARBSS/v14-i6/21739

Zilwa, D. D. (2005). Using entrepreneurial activities as a means of survival: Investigating the processes used by Australian universities to diversify their revenue streams. *Higher Education, 50*, 387–411. https://doi.org/10.1007/s10734-004-6359-8

Zou, Y., Zhao, W., Mason, R., & Li, M. (2015). Comparing Sustainable Universities between the United States and China: Cases of Indiana University and Tsinghua University. *Sustainability, 7*(9), 11799–11817. https://doi.org/10.3390/su70911799

11 Harmonising Goals of Sustainable Development and Climate Action

The Function of Legal and Policy Frameworks

Siti Hafsyah Idris

11.1 INTRODUCTION

The climate issue is no longer a far-off concern; its effects are felt globally and have terrible effects on human civilisations and the environment. Concurrent with this ambitious agenda meant to eradicate poverty, advance social justice, and guarantee environmental sustainability, the Sustainable Development Goals (SDGs), by 2030, have been pledged by the worldwide community (United Nations, 2025). These two agendas, the SDGs and climate action, are intimately connected. While attempts to reach the SDGs might either aggravate or help to slow down climate change, it is a significant danger to their achievement (Anteneh G. Dagnachew et al., 2021).

The defining issues of the twenty-first century are the intersection of the growing climate change crisis and the ambitious framework of the SDGs. The Intergovernmental Panel on Climate Change has repeatedly underlined that human activities have warmed the land, sea, and atmosphere, causing extensive and fast alterations in the systems of the Earth. Unchecked, these developments run the danger of permanent and irreversible effects, especially if global temperature rises above 1.5 degrees Celsius (United Nations, 2025). Concurrent with this, the United Nations started the 2030 Agenda for Sustainable Development, a thorough strategy to reach sustainability and resilience for the earth and humans. Comprising 17 linked objectives, this agenda is inherently related to climate since it acknowledges that climate action is not only a goal (SDG 13) but also a basic requirement for reaching many other SDGs (United Nations, 2025).

Climate action and the more general sustainable development agenda are not only coincidental but also very tightly entwined and mutually reinforcing. As the United Nations Environment Program (UNEP) rightly notes, attaining the lofty goals set out in the SDGs depends on climate action, underlining and guiding principles. Improving synergies between climate action and sustainable development policies will help achieve the Paris Agreement, the main international legislative instrument for tackling climate change, as well as the 2030 Agenda (UNEP, 2024). The United Nations Secretary-General has indeed underlined that the main guiding ideas for all climate change policies and projects must be the goals of the 2030 Agenda and their SDGs (Josephina et al., 2024).

Legal and policy frameworks give the basis for guiding and controlling human activity, therefore influencing the decisions taken by governments, companies, and people. Including climate issues in legal and policy systems will help us build a road towards a resilient, fair, and sustainable future. In this regard, law and policy become essential tools for coordinating the SDGs with climate action (Yang, Hongbo, et al., 2022). These instruments set the necessary structures for responsibility, create the frameworks for action, and help to integrate these two related goals. Investigating the

DOI: 10.1201/9781003600152-11

causes of these links emphasises the intricate interactions among the SDGs that can directly guide governance and policies (Advocates for International Development (A4ID), 2021). This knowledge enables the creation of rules and policies addressing climate change and so promoting advancement in all spheres of sustainable development objectives.

Reducing and adjusting to climate change depends mostly on the legal sphere, which includes national laws and international accords like the Paris Agreement, so directly supporting the accomplishment of SDG 13 and its linked objectives. Therefore, the function of law and policy goes beyond simple implementation; it is vital to create the links and ensure the coherence required to effectively tackle the linked challenges of climate change and sustainable development.

This chapter explores the crucial part legal and policy frameworks play in closing the distance between the urgent need to reach the SDGs and ambitious targets for climate action. It contends that in view of the growing climate problem, harmonising these linked agendas through strong legal and policy structures is not just a need but also a must. The chapter critically examines the strengths and shortcomings of current systems, including national laws and the Paris Agreement, going beyond a summary of them. It highlights important areas for reform and innovation by pointing out key difficulties and possibilities within these frameworks.

11.2 THE INTERTWINED NATURE OF CLIMATE ACTION AND SDGS

SDG 13 highlights the close relationship between climate action and the other SDGs. Efforts to control global temperature increases could greatly help achieve targets on poverty (SDG 1), hunger (SDG 2), water availability (SDG 6), and the health of terrestrial (SDG 15) and marine (SDG 14) life. Reducing poverty depends on climate action and fundamental environmental services and livelihood protections. Initiatives aimed at improving resilience and adaptation usually help vulnerable communities, the most impacted by climate change, most of all. Still, climate change is a threat multiplier, perhaps aggravating necessary sustainable development measures like health, poverty, hunger, inequality, and ecological protection (United Nations, 2025). With its 17 interrelated and indivisible SDGs, the 2030 Agenda acknowledges the need for a comprehensive approach to these linked problems. Clear synergies between climate action and the SDGs can be observed across all 17 goals, especially in sectors like energy, water, and agriculture, where significant opportunities for mutual benefits exist. The connections are robust (UNEP, 2024).

Achieving climate targets will also further other SDGs; conversely, sustainable development can facilitate climate initiatives. Investing in clean technology and renewable energy reduces greenhouse gas emissions, fosters innovation and economic diversification, and slows climate change. The latter goal directly relates to SDG 3 (Good Health and Well-being) and SDG 8 (Decent Work and Economic Growth) since it generates new employment in green sectors and reduces air pollution, improving public health. Furthermore, aiming for universal power access in sub-Saharan Africa by 2030 (SDG 7) proves more affordable when it aligns with climate plans (UN Climate Press Release, 2023). This positive feedback loop shows how well-combined legal and policy systems can create virtuous cycles of sustainable development in which developments in one area inspire developments in another.

Achieving harmony will prove difficult when specific development goals conflict with climate projects, resulting in several possible trade-offs. Minimising negative trade-offs and exploiting synergies will help us to meet the Paris Agreement and the 2030 Agenda (Josephina Wouda Kuipers and Upalat Korwatanasakul, 2024). Policy coherence problems arise when conflicts between climate goals and other national interests, such as boosting gas supplies alongside renewable energy in Australia, reconciling climate actions with fossil fuel-based economic growth in Kenya, or balancing mining and agricultural needs with Colombia's climate initiatives, are present. Climate-related crises could undo efforts toward poverty reduction and seriously hinder the achievement of the SDGs (Zoha Shawoo, 2020). In urban planning, tensions can also arise when land usage for development takes centre stage over transportation, which can lead to lower carbon emissions. Some climate technologies, such as carbon capture and use (CCU), may unintentionally produce trade-offs by distorting responsibilities to sectors like public health and resource depletion.

Moreover, massive bioenergy generation and other climate-friendly policies might result in changes in land use that compromise biodiversity, water availability, and food security. Even in the search for clean energy, the development of bioenergy and hydropower raises possible hazards to managing water and land resources and biodiversity (Vaidya et al., 2021; Phemelo Tamasiga et al., 2025). These potential conflicts emphasise the importance of thorough legal and policy planning to ensure climate initiatives do not inadvertently compromise other vital aspects of sustainable development. Therefore, integrated assessments and mitigation strategies are crucial in navigating these complex interrelations.

11.3 LAW AND POLICY'S ROLE IN HARMONISING SDGS WITH CLIMATE ACTION

11.3.1 Legal Frameworks

11.3.1.1 International Law and Agreements' Function

Establishing a coordinated effort for climate change and the SDGs heavily depends on the current state of international law. The Paris Agreement assumes a central role as it serves as a crucial legal and political platform for aligning these two goals. This agreement underlines the need to use climate and SDG synergies to meet the 2030 Agenda and the related targets. The objectives of the Paris Agreement and the SDGs are integrally intertwined; advancement in one area significantly affects the other. Particularly concerning lowering emissions and improving climate resilience, the legally enforceable Paris Agreement, signed by 196 parties, offers a framework for the climate activities needed to meet the SDGs. Nations may reach their targets more quickly and efficiently by including sustainable development under the Paris Agreement, the 2030 Agenda, and climate action (Josephina Wouda Kuipers and Upalat Korwatanasakul, 2024). Globally aligning climate projects with the SDGs can result in millions of green jobs, significant financial gains, and many annual deaths prevented. Based on the UN Framework Convention on Climate Change (UNFCCC), the Paris Agreement brings nations together to quickly lower greenhouse gas emissions and increase national resilience and climate adaptation capacity. Realising pledges taken under the Paris Agreement into significant home results depends on national legal systems for climate change. The Global Stocktake of the Paris Agreement points out the urgent need for a comprehensive and coordinated strategy to tackle biodiversity loss and climate change while implementing the SDGs (Nilsson et al., 2024).

Effective alignment of the SDGs and climate action depends on the UNFCCC. Designed with the UN Department of Economic and Social Affairs (UN DESA), the Climate Action and SDG Synergies Platform is a significant initiative to help use these linked systems. Understanding the inherent link between the SDGs, the UNFCCC concentrates on group initiatives associated with them. One ongoing UNFCCC project examines the adaptation and mitigation strategies outlined in Parties' Nationally Determined Contributions (NDCs) and their connections to specific SDGs. The UNFCCC organises the "Climate Actions and Impact Assessment: A Unified Approach to Sustainable Development Goals" event to support integrated planning and policymaking. The UNFCCC enhances the development and dissemination of knowledge regarding these essential interrelations by publishing global reports and co-convening the Expert Group on Climate and SDG Synergy (United Nations, 2025). The UNFCCC supports regional climate weeks, allowing attendees to investigate local synergies between SDGs and climate action. Under the Paris Agreement, the UNFCCC also maintains the register for NDCs, a crucial instrument for nations to match their development strategy with their climate objective.

Apart from the Paris Agreement and the UNFCCC, many international legal mechanisms help the SDGs align with environmental action. Laying the groundwork for sustainable, low-carbon, resilient development in a changing climate depends critically on the Paris Agreement, the 2030 Agenda, and the Sendai Framework for Disaster Risk Reduction. Earlier agreements, such as the Kyoto Protocol (1997), helped open the path for the more all-encompassing policy shown in the Paris Agreement by starting worldwide attempts to lower greenhouse gas emissions. Almost 3,500 multilateral and bilateral environmental treaties involving UN member states emphasise the vast legal terrain addressing environmental concerns, many of which are directly linked to climate change and the pursuit of the SDGs.

Regional human rights systems such as the African and Inter-American systems have acknowledged, within sustainable development, the right to a healthy environment, highlighting the link between environmental protection and human rights (Astrid Puentes Riaño, 2025). Furthermore, it underlined the relevance of procedural rights in environmental governance, which is necessary to guarantee environmental justice within the limits of climate action and the SDGs, instruments like the Escazú Agreement (2018) in Latin America and the Caribbean, and the Aarhus Convention (1998), which focuses on access to information, public participation in decision-making, and justice in environmental affairs. The dedication of nations like the United States to the UN SDGs shows even more national support for the more general sustainability agenda, which naturally incorporates climate action.

Transforming the SDGs and climate action pledges into real-world outcomes depends on strong legal systems. Achieving the SDGs depends critically on legal frameworks since they create enforcement systems to guarantee compliance and encourage responsibility through judicial and non-judicial agencies. With courts progressively acknowledging a legal need for governments and companies to increase their efforts to address climate change, stakeholders enforcing international law against corporations are following a growing trend. Suggested as a foundation for achieving climate targets and motivating governments to carry out preventative actions to save the environment is the international law concept of "due diligence." Moreover, as underlined in SDG 16, public information access enables people and stakeholders to hold governments and businesses responsible for their environmental and sustainable promises (Figure 11.1).

11.3.1.2 National Legal and Policy Models for Coordinated Use

Incorporating the SDGs into their climate policies becomes more and more important for national governments if they are to improve synergistic implementation. Countries under the Paris Agreement and the SDGs concentrate on the interactions between their NDCs to negotiate internal political environments and prevent policy fragmentation. Many entities see the critical need for adaptation in reaching the SDGs and the assistance sustainable development offers for adaptation efforts (Nilsson et al., 2024). They also perceive the link between their adaptation, mitigating strategies, and the SDG frameworks inside their NDCs. With institutional support from sites like the Climate Action and SDGs Synergies Platform, the general idea is that the objectives of the 2030 Agenda should be the fundamental direction for all national climate change policies and projects. Viewed as a means of implementing NDC adaptation priorities, directly helping to achieve the SDGs, National Adaptation Plans (NAPs) are being developed (Anne Hammill and Hayley Price-Kelly, 2016). For example, Mexico and Kenya have started projects matching their NDCs with the SDGs, therefore illustrating pragmatic methods of this integration. Mexico wants more influence and lower costs by means of integration; Morocco's policies are presented as "win-win solutions for climate and sustainable development," matching its renewable energy pledges with SDG 7. Kenya's National Planning Framework successfully combines NDC and SDG goals; Colombia has created a road map for SDG application. Integrated planning tactics in Morocco, Colombia, and Nigeria expose the co-benefits of climate action that complement national development goals, therefore encouraging larger support for these projects (Chris Malley and Jason Veysey, SEI, 2023). With the SDG approach providing a

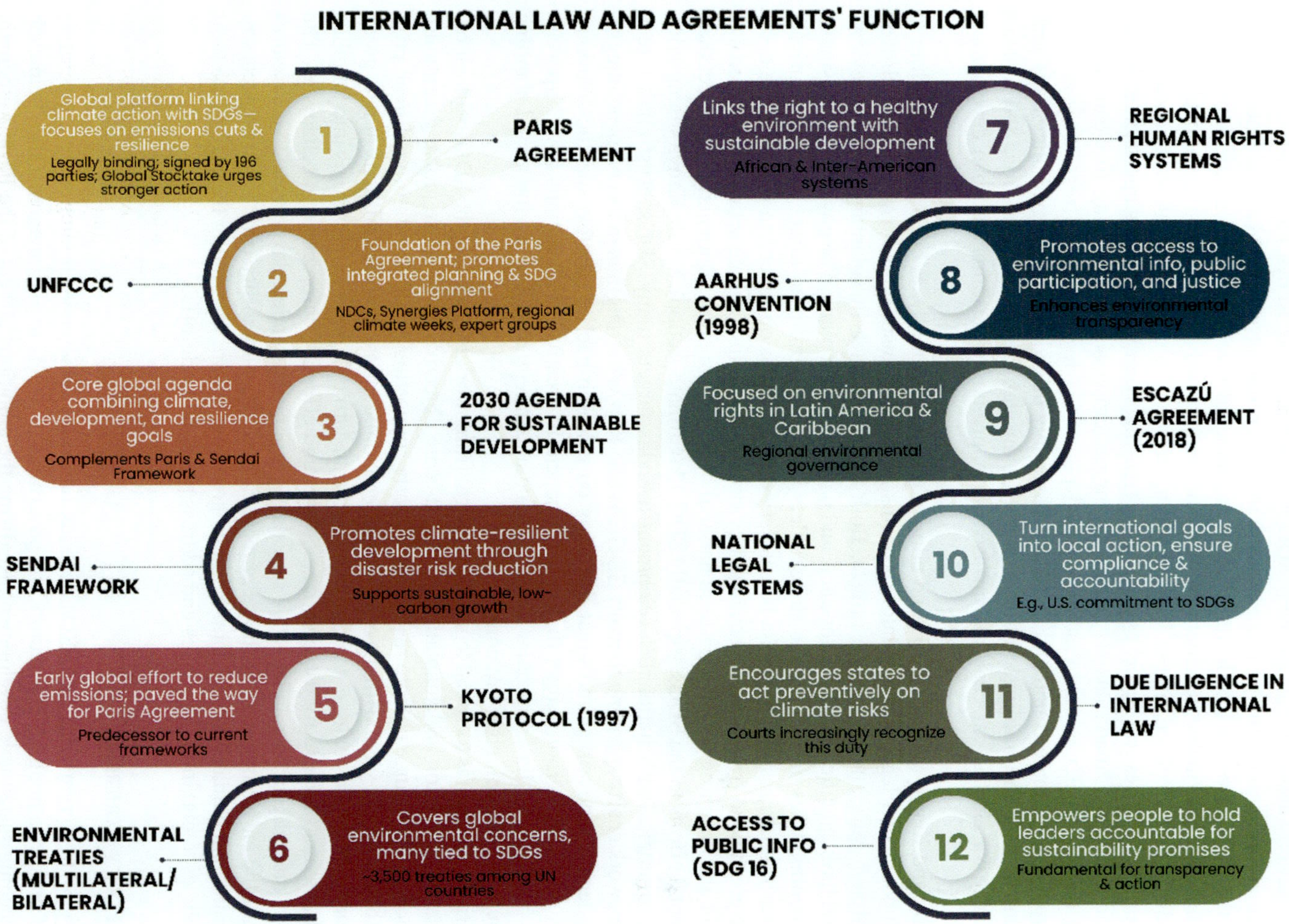

FIGURE 11.1 International law and adjustments' function.

useful framework for decision-making, the development and finalisation of NDCs are progressively recognised as opportunities to support sustainable development.

National laws highlighting a dedication to climate action and the SDGs abound in nations including Australia and Kenya. Though their stated link to other SDGs may vary, their current climate change legislation shows interaction with SDG 13. Australia and Kenya have used studies and impact evaluations to match their climate plans to the larger SDG structure (Mathilde Bouyé, 2018). Although the SDGs are not legally required at the national level, governments are supposed to carry out national sustainable development plans and monitor advancement using national indicators, including climate-related targets. The expanding presence of climate initiatives inside national agendas connected to several SDGs exposes an increasing synergy of various approaches. Under SDG 7, for example, Canada has set particular targets emphasising switching to clean energy, improving energy efficiency, and eradicating coal power via legislative and regulatory actions, thereby supporting climate action and access to reasonably priced, clean energy. Policy tools seek to lower emissions from buildings and transportation through fiscal incentives, road toll rebates, and assistance for electric car charging infrastructure, promoting sustainable urban development (Figure 11.2).

Australia and Kenya address SDGs and climate change differently. While Kenya is leading in building fundamental legal and regulatory systems in Africa, Australia is formalising its climate pledges through laws and aggressive targets. Both nations engage in voluntary national reviews (VNRs) and have specific climate change acts and national action plans. The government, business, and local communities are among the stakeholders involved. Aiming to direct public and private money towards green results, Australia boasts very defined emission reduction targets and a comprehensive green bond framework. Kenyan policy stresses including climate action in all spheres of government and matching it with long-term national development goals. Both nations connect their environmental efforts to the SDGs, especially SDG 13 (Climate Action). While Kenya's integration of climate strategies with national development agendas targets many SDGs, Australia's Green Bond Framework charts spending to many SDGs. While Kenya's strength is in its complete integration, early legal frameworks, and great emphasis on research-backed, context-specific solutions, Australia's approach is more target-driven and financially creative. Both nations have important lessons to teach; Australia's particular climate financing strategy and Kenya's integrated planning and research-to-policy pipeline are noteworthy.

The Paris Agreement is a crucial legal and political framework for aligning the SDGs with global climate change mitigation and adaptation efforts. Its core objective is to reduce global greenhouse gas emissions and enhance climate resilience, directly supporting SDG 13. Aligning climate projects with the SDGs can yield trillions of dollars in economic gains, generate millions of new green jobs, and improve public health outcomes. However, translating these ambitious goals into tangible, real-world outcomes require robust legal frameworks and effective enforcement mechanisms. The principle of "due diligence" mandates that states and corporations undertake preventative measures to safeguard the environment and achieve climate targets. Accountability calls for open access to environmental and climate-related data. National governments are integrating the SDGs into their climate policies to maximise synergies and address potential trade-offs.

11.3.2 Policy Coherence

Matching climate action with the SGDs depends on policy coherence. It entails the cooperative creation of policies among several sectors and governmental levels to guarantee that they help each other in reaching sustainable development. Realising the ambitious targets of the Paris Agreement and the 2030 Agenda depends on implementing coherent policies that improve synergies and lower negative trade-offs (Anteneh G. Dagnachew et al., 2021). Managing possible conflicts and trade-offs between climate policies and other development goals depends on this coherence. Conflicts can develop, for instance, between supporting renewable energy and maintaining the fossil fuel

FIGURE 11.2 Australia vs. Kenya: climate change and SDG implementation comparison.

industries or between balancing climate projects with the necessity of fast economic development. Under a cooperative policymaking approach, one ensures a strong integration of domestic aims with international commitments and takes into account the broader consequences of climate and sustainable development policies, particularly their spill-overs. This method guarantees that remedies in one domain do not have detrimental effects in another by requiring a comprehensive perspective to solve the core causes of entwined concerns such as climate change, biodiversity loss, pollution, poverty, and inequality.

Dealing with such conflicts and trade-offs requires a detailed awareness of the complex links among several objectives. For example, although climate action depends on renewable energy, as Australia shows, its deployment may contradict goals related to gas supply and the economic benefits of the coal sector. Kenya's attempts to address climate change can similarly contradict its reliance on fossil fuels for poverty reduction and economic development. In urban environments, conflicts can arise between protecting green areas for liveability and biodiversity and developing compact city development to help mitigate climate change. Trade-offs also exist in land use since allocating land for bioenergy generation targeted at climate action may conflict with water resources and food security. Technologies like CCU can unintentionally compromise other facets of sustainability, even within the field of climate change. These events emphasise the need for thorough planning and combined evaluations to minimise possible negative results.

Policymakers can help to increase coherence between climate activities and the SDGs through several instruments and approaches. Forecasting the effects of policies and making sure domestic agendas and global responsibilities are connected calls for an integrated strategy. Policy combinations, a thorough economic perspective, matching adaptation and mitigating goals, and deftly combining climate initiatives, SDGs, and domestic concerns can all be part of this. A range of tools, such as textual analyses, technical guidelines, integrated models, and visualisation software, can help to expose interrelations and improve policy formation. While technical guidelines offer methodical directions for assessing synergy, textual analysis techniques expose conceptual commonalities between agendas. Integrated models enable thorough quantitative studies on linkages, and visualisation tools assist in showing the found connections. These instruments, taken together, can help governments steer toward more coherent and successful plans for matching climate action with the SDGs.

11.3.3 Multi-Level Government: Promoting Group Action

Harmonising climate action with the SDGs calls for coordinated efforts across all spheres of government, from international to local. For this merging, national governments have to build a strong legal and policy framework. With help from international venues like those supported by UN DESA, they may use the synergies between climate projects and the SDGs to create a sustainable future. National governments may lead the green transition by changing energy and industrial systems, raising energy efficiency, and safeguarding natural resources using several policy tools and financial measures. They are also crucial in collaborating with significant UN agencies to leverage frameworks and resources supporting sustainable development and climate change (Catherine Friday et al., 2022). Identifying, assessing, and putting integrated climate and sustainable development plans and policies into action falls to senior government officials. Making significant advancements in the broader SDG agenda depends on national-level enhancement of climate action.

Local and subnational authorities are becoming more and more important in helping to achieve SDGs and in contributing to climate action. Especially in poorer nations, their capacity for applying climate solutions has to be strengthened immediately. Furthermore, as they account for a significant share of public expenditure targeted at environmental goals, regional and municipal governments in wealthy countries also carry great responsibility. Executing national climate policy depends on these subnational bodies, which also use important local climate data to address community-specific contributions to both adaptation and mitigation actions (UN-Habitat, 2024). Given their direct relevance in these domains, SDGs 9 (infrastructure) and 11 (sustainable cities and communities) highlight the need to empower various governance levels for climate resilience. Achieving the 17 SDGs, under the control of domains including education, social welfare, healthcare, access to water, housing, transportation, and climate action, depends mostly on local and regional authorities. Realising global goals such as the SDGs and the Paris Agreement depends on their participation in multilateral decision-making processes and their capacity to supplement central government projects. Accelerating the localisation of SDGs calls for appreciation that many goals directly connect to the operations of local and regional authorities, which is essential for improving policy coherence and integration (Thibaud Voïta, 2020). Particularly, cities are essential in preparing for and facing climate challenges through local climate strategies and resilient infrastructure projects. Their capacity to translate the broad 2030 Agenda into specific localised measures sets them at the core of success in sustainable development.

Global cooperation and knowledge sharing are vital to improving group efforts in climate change and reaching the SDGs. Developing nations need financial support and technical transfers from wealthy countries to achieve sustainable industrialisation and the transition to greener energy sources. The Paris Agreement emphasises the need for global collaboration in addressing climate change and calls on all governments to act considering their situation. Working with other international organisations, the UNFCCC stresses the need to give developing countries enough help to increase their resilience and enable their adaptation to the effects of climate change. Global cooperation must be strengthened to drive investments in infrastructure, technology, and clean energy innovation if we are to meet specific goals, such as Zero Hunger (SDG 2) and Affordable and Clean Energy (SDG 7).

11.3.4 Emphasising Environmental Justice and Community Involvement

Aligning environmental activity with the SDGs calls for the application of inclusive and fair climate policy. Dealing with climate change mostly depends on equality and inclusiveness since it affects the issues of economic development for present and future generations, and the reduction of poverty. Inequalities in equity in climate projects could isolate underprivileged groups. Therefore, climate action should combine fair and sensible solutions, prioritising social responsibility so that the simultaneous progress of social integration, environmental sustainability, and economic growth is ensured (Avesahemad et al., 2024). An ethical concept aiming at solving the unequal consequences of climate change on underprivileged groups and advancing fair remedies is climate justice. Environmental justice holds that, independent of background, everyone and every community has the right to protect against negative environmental situations. Emphasising pollution reduction, guaranteeing access to reasonably priced energy, and building quality jobs inside a pollution-free economy can help a national climate agenda prioritise the needs of vulnerable populations. A shift to low-carbon energy should meet the needs of every country, regardless of income, fostering equitable and sustainable growth (Equitable & Just, 2024). Reducing vulnerability and improving climate resilience for all people depends on applying sensible and fair adaptation and mitigation techniques. Dealing with climate change is essentially related to the search for a fairer society and calls for a "just" transition, guaranteeing inclusivity for every person.

Promoting harmony depends mostly on public access to knowledge and participation in decision-making. Using collective intelligence, involving the public in climate policy promotes positive communication and highlights shared goals for action. Informed participation depends on including citizens in the SDGs and addressing false information. Within the UNFCCC structure, public participation is a fundamental component of Action for Climate Empowerment, therefore appreciating the value of every person's contribution to climate-related decision-making. The 2030 Agenda emphasises the need for inclusive, participative, and representative decision-making (SDG 16.7), therefore orienting public involvement as vital to sustainable development. Particularly addressing procedural environmental rights, information access, involvement in decision-making, and access to justice, SDG 16 Encouragement of public information access (SDG 16.10) helps stakeholders and citizens hold businesses and governments responsible, hence boosting development toward other SDGs. Rio Declaration's Principle 10 underlines the need for public access to environmental information and participation in decision-making procedures. Achieving the SDGs depends on information access, which also helps people to monitor government actions and make wise decisions.

11.3.5 Using Environmental Impact Assessments to Pursue Combined Goals

Achieving integrated outcomes in climate action and the SDGs depends on environmental impact (EI) assessments. Emphasising the significant influence of climate change on all 17 SDGs, we ensure that policies are not just environmentally sustainable but also socially fair by integrating climate action with impact assessments. Stakeholders can use the SDG Impact Assessment Tool as a useful tool to assess their influence on every SDG, therefore promoting a holistic perspective that goes beyond mere environmental issues (Martin Eriksson et al., 2020). Informed decision-making depends critically on EIs, who use methodical approaches of identifying, forecasting, and analysing possible environmental effects. Their main goals are to maintain biodiversity, lower pollution, and mitigate climate change, enabling the realisation of SDGs (Velosi Author, 2023). This will help to minimise environmental footprints. Generally incorporated in the larger Environmental and Social Impact Assessment framework, climate change impact assessments are essential for determining how a project would affect climate change and its vulnerability to expected variations. They also evaluate indirect effects and national greenhouse gas emissions compliance. Examining climate projects for their wide-ranging social, financial, and environmental effects helps EIAs direct decision-makers toward laws that would help people as well as the planet. Moreover, including SDG ideas into legal systems, especially in EIAs, helps some industries, notably food systems, become more resilient against climate change.

11.3.6 Focusing on Community Involvement and Environmental Justice

We must follow inclusive and equitable climate policies to align environmental action with the SDGs. Dealing with climate change largely focuses on equity and inclusiveness since it affects the issues of economic development for present and future generations and the reduction of poverty. Inequalities in equity in climate projects could separate impoverished populations. Thus, climate action should combine reasonable and fair solutions, giving social responsibility top priority so that the simultaneous progress of social integration, environmental sustainability, and economic growth is guaranteed (Victoria et al., 2024). Climate justice is a moral idea meant to solve the unfair effects of climate change on vulnerable people and to forward equitable solutions. Environmental justice maintains that everyone, wherever they are, has the right to defend against unfavourable environmental conditions, regardless of background. Emphasising pollution reduction, assuring access to moderately priced energy, and establishing quality jobs within a pollution-free economy can enable a national climate agenda to prioritise the needs of vulnerable populations. Regardless of income, a switch to low-carbon energy should satisfy the demands of every nation, thereby promoting fair and sustainable development. Applying reasonable and fair adaptation and mitigating strategies can help everyone reduce vulnerability and increase climate resilience. Addressing climate change necessitates a "just" transition that ensures inclusivity for all individuals and fundamentally aligns with the pursuit of a more equitable society.

Encouragement of harmony largely depends on public understanding and involvement in decision-making. Involving the public in climate policy through collective intelligence helps foster good communication and emphasises common objectives for action. Including people in the SDGs and correcting misleading information will help define informed participation.

Notwithstanding these advances, national-level integration still faces significant challenges. Different government agencies handle each agenda; thus, institutional divisions sometimes result in a limited link between climate efforts and SDG aspirations. Different national priorities and the effect of vested interests cause conflicts in policy coherence. Dealing with the conflicts inherent in complex policy procedures and deftly negotiating the political terrain remain essential tasks for government officials aiming for coordinated results.

11.4 DIFFICULTIES AND POSSIBILITIES IN APPROACHING PERFECT HARMONY

Harmonising environmental activity with the SDGs presents several difficulties. Maintaining policy coherence across many sectors and government levels is difficult; there are fundamental trade-offs among several goals. Often, integrated techniques are hampered by institutional inertia and compartmentalised government activities.

Policy coherence definition and measuring can be challenging since reaching coherence does not guarantee fair results. There is still much to learn about fully appreciating the intricate trade-offs and synergies between the SDGs and climate control. Notwithstanding good intentions, certain programmes could have policy drift and execution challenges. The extent of the difficulty is highlighted by the limited worldwide progress toward reaching the SDGs, with only a small proportion of targets now on track (Josephina Wouda Kuipers and Upalat Korwatanasakul, 2024). Often, disjointed government structures and policymaking procedures cause policies linked to SDGs to be disconnected from climate efforts. Competing authorities hamper the integration of strategies, vague directives, department-specific agendas, and transient political cycles. Destroying these institutional silos and encouraging efficient cooperation and coordination amongst several sectors is essential if we are to get synergistic benefits. Moreover, guaranteeing development depends on clearly assigning responsibility for reaching targets related to climate and SDGs. Therefore, establishing platforms for consultation and dialogue, enhancing transparency and access to information, and incorporating participatory decision-making processes are vital to addressing the problem of low stakeholder engagement and public participation. These actions can enhance the legitimacy and effectiveness of policies by considering the diverse perspectives and needs of stakeholders in the decision-making process.

Political and economic ambitions could impede efforts toward a more coherent and sustainable development route. The trade-offs and conflicts that could develop between the quest for particular SDGs and climate action programmes provide a significant obstacle (Yang et al., 2022). While vital in the fight against climate change, some well-meaning mitigating strategies could, if poorly planned or executed, impede progress in areas including poverty reduction, food security, and general well-being, especially in vulnerable regions like sub-Saharan Africa, South Asia, and Southeast Asia. Negative impacts could show up as higher food and energy costs or damage to livelihoods. Similarly, ambitious climate projects requiring land-use changes for bioenergy or extensive afforestation may undermine food and water security, generate land rights disputes, and cause loss of biodiversity. Furthermore, the economic change away from fossil fuels could provide challenges for nations depending on their production and exports, therefore aggravating inequality. For low-income households, there are also particular technological trade-offs, including the high upfront expenses connected with increasing energy efficiency in buildings or the big hydroelectric dams that might cause displacement of populations and lower biodiversity. These possible conflicts highlight the need for a careful approach to policy formation. Examining the wider sustainability effects of climate policies is important since every decision can have far-reaching effects.

Other obstacles include financing, technological transfer, and capacity-building issues, especially in underdeveloped nations (Amanda Scheichet, 2024). Significant challenges include financial limitations, especially in underdeveloped countries and at the subnational level. Attaining a low-carbon and sustainable future requires significant financial investment, yet many developing nations lack the necessary resources to address climate change while striving for integrated development goals. Moreover, the scarcity of appropriate climate-friendly technologies and the inability to implement effective solutions exacerbate these challenges. Addressing these obstacles demands robust international cooperation and support. Increased climate financing, efficient technology transfer, and targeted capacity-building initiatives are essential for empowering developing countries in their integration efforts. The urgency for these global initiatives cannot be overstated.

Apart from that, external shocks, including crises in the climate and conflicts, can aggravate problems in these spheres (Mathilde Bouyé, 2018). The diverse national contexts and global development priorities add complexity to the harmonisation challenge. There is no one-size-fits-all solution, as countries exhibit varying levels of economic development, resource availability, social priorities, and vulnerability to climate change impacts. Legal and policy frameworks should be tailored to these specific contexts to ensure their effectiveness and equity. Balancing the urgent need for climate action with immediate development goals, such as poverty alleviation and energy access, requires careful consideration and the development of solutions that are context-specific and nationally driven.

Based on the previously mentioned issues, it is evident that justice issues are not given enough integration. Thus, addressing the inadequate integration of justice issues depends on adopting just transition strategies, offering focused support for underprivileged populations, and guaranteeing fair access to climate solutions. These steps can ensure that underprivileged areas are involved in the change toward a sustainable future, therefore fostering social justice and equality (Figure 11.3).

Notwithstanding challenges, the close links between climate action and the SDGs offer enormous opportunities. Dealing with climate change could drive sustainable development by inspiring invention and creating green job opportunities. Combining the SDGs with climate action will have significant financial benefits and help promote job creation. Visible synergies among all 17 SDGs open many avenues for coordinated action. More nations see the need for searching for solutions that simultaneously solve the problems of climate change and sustainable development. Emerging countries can aim to be leaders in environmentally friendly economies through technological

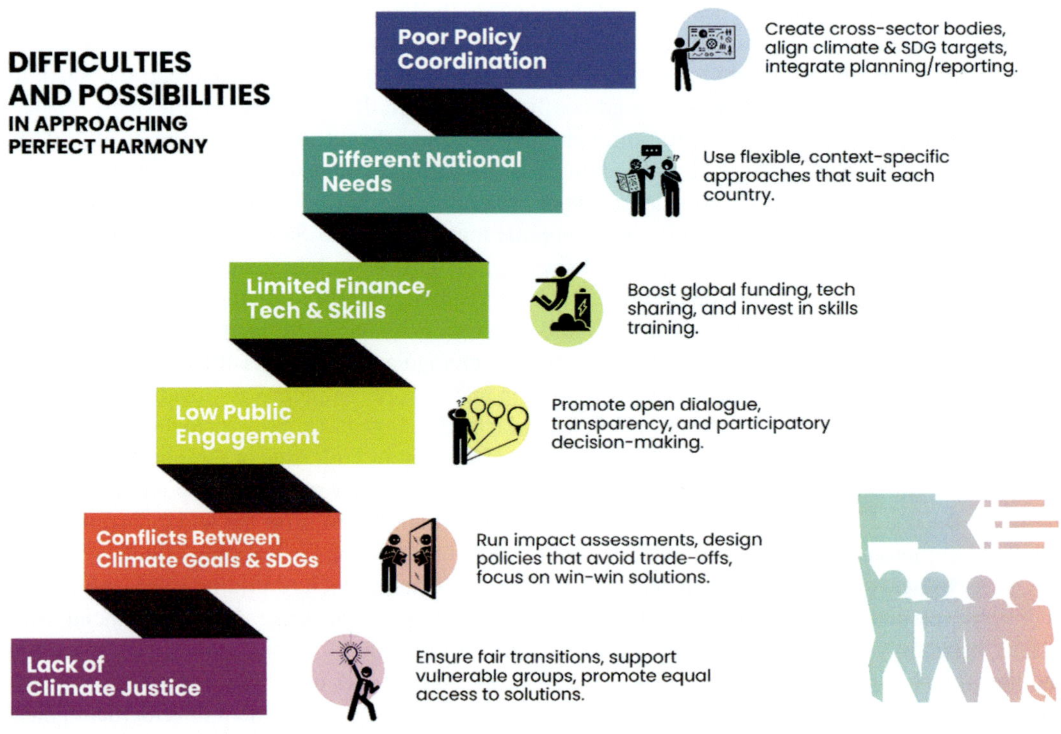

FIGURE 11.3 Difficulties and possibilities in approaching perfect harmony.

transfer and international cooperation. Well-designed adaptation and mitigation strategies can significantly increase resilience. Furthermore, using science, technology, and creativity can magnify synergies through new energy sources that both meet environmental goals and support sustainable economic development.

While concurrently achieving several SDGs, a range of policy instruments can efficiently drive climate action (Huang et al., 2024). Increasing the share of renewable energy in power generation, for instance, generates significant synergies with many SDGs, including improving air quality and public health (SDG 3), expanding access to reasonably priced and clean energy (SDG 7), and generating favourable jobs and economic growth prospects (SDG 8). Strict energy efficiency rules across many sectors, especially in buildings and industry, also help to significantly lower greenhouse gas emissions while improving energy affordability (SDG 7), industrial competitiveness (SDG 9), and supporting more sustainable cities and communities (SDG 11). By improving carbon storage and supporting biodiversity (SDG 15), water quality (SDG 6), and possibly food security and livelihoods in rural areas (SDG 2), encouraging sustainable land management practices, such as reducing deforestation and increasing reforestation, helps combat climate change. Emphasising these synergistic policy measures can maximise co-benefits, therefore enabling a more integrated and efficient method of reaching both climate and SDGs (Gerd Müller, 2024).

Integrated planning, evaluation, and the execution of climate and SDG projects all depend on legal frameworks to a great extent (Anteneh et al., 2021). Under the Paris Agreement and VNRs for the SDGs, and by including synergistic targets or co-benefits into national reporting systems like NDCs, policymaking becomes more holistically approached. Many nations are already making progress in this regard; some are even including the SDGs in their NDCs, which would lead to ever more socially conscious changes. Laws in nations such as Finland and Sweden compel government reports to demonstrate how their policies complement the SDGs and their NDCs, thereby institutionalising these relationships. Moreover, encouraging vertical coordination among several government levels, as Colombia's attempts to include the Paris Agreement and the SDGs into municipal planning and budgeting procedures show, helps to guarantee policy coherence and improve the efficiency of implementation. Legal requirements for integrated planning and evaluation can help to produce more cohesive and robust policies.

Significant synergistic results are likely from market-based initiatives, financial incentives, and legislative changes. Using carbon pricing policies, such as emissions trading programmes or carbon taxes, helps lower emissions and generates money for social projects supporting the SDGs, green technologies, or other uses. Fiscal incentives, such as tax breaks for energy-efficient devices or subsidies for renewable energy projects, can help to further encourage the acceptance of climate-friendly innovations with more benefits for sustainability. Stricter car emissions rules or required building energy code compliance help to reduce both pollution and energy consumption, therefore supporting climate initiatives and improving public health and air quality through these regulations (Figure 11.4).

11.5 LESSONS FROM LEGAL AND POLICY INTERVENTIONS GLOBALLY

Effective interventions show how possible it is to reach goals related to sustainable development and climate. Regulations meant to lower emissions in North America have helped lower coal use, improving air quality and public health (SDG 3). Europe's tougher regulations promoting energy efficiency gains in energy-intensive sectors have helped support both decent work and economic growth (SDG 8) and climate targets. Driven by lowering costs, which supports reasonably priced and clean energy (SDG 7) and promotes economic diversification, some Middle Eastern and North African nations have set ambitious targets for the growth of renewable energy. With robust enforcement and public publication of air quality data, China's 2013 Air Pollution Prevention and Control Action Plan significantly improved air quality (SDG 3 and 11) while lowering coal use. Laws passed in Finland and Sweden demonstrate a dedication to integrated governance and call for government reporting on policy alignment with the SDGs and NDCs. Strict automobile emission rules in

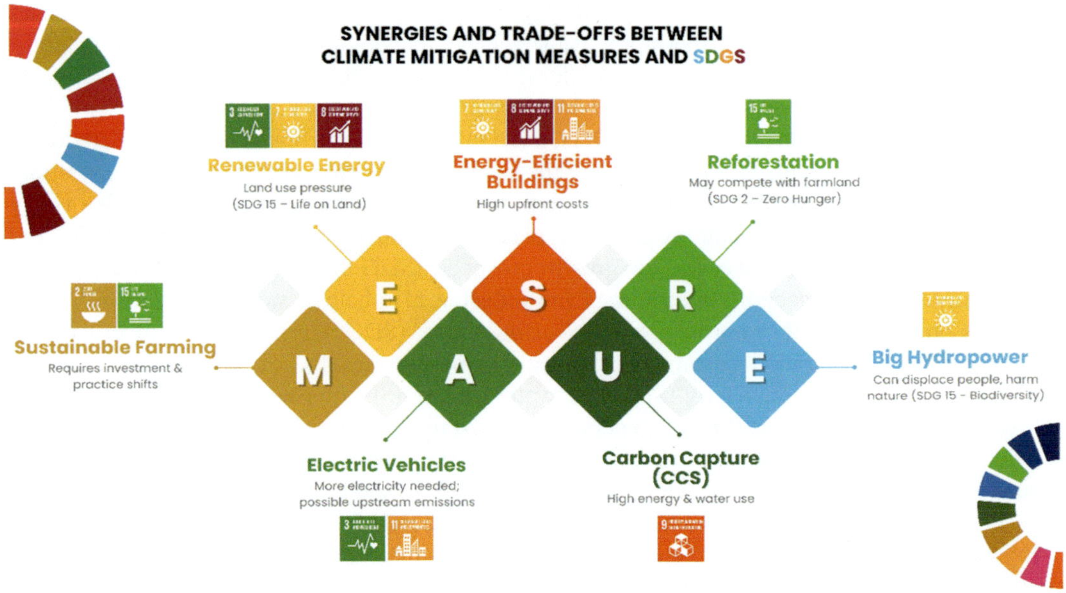

FIGURE 11.4 Synergies and trade-offs between climate mitigation measures and SDGs.

London's Ultra Low Emission Zone have resulted in significant declines in air pollution (SDGs 3 and 11) and public transport reinvestment. These several cases underline the spectrum of legal and policy instruments available for the efficient application to produce synergistic results.

Many times, the outcome of these treatments depends on various important elements and supporting conditions. As China's forceful response to air pollution shows, strong political will and dedication are fundamental. Crucially, efficient governance systems that support policy consistency and institutional coordination, including initiatives at vertical coordination in Colombia. A strong basis for harmonisation is provided by solid legal systems requiring coordinated planning and reporting, as in Finland and Sweden. As stressed in many situations, active participation of many stakeholders, including government agencies, industry, civil society, and local communities, ensures that a variety of points of view shape policies and have better public support. Moreover, effectively executing these interventions depends on having enough technical and financial capability and long-term sustainability.

On the other hand, less effective intervention cases highlight the difficulties and the need for thorough preparation and evaluation of possible effects. For instance, although with an eye towards climate reduction and adaptation, Bangladesh's National Adaptation Program for Action apparently had unanticipated detrimental social implications in one case. This emphasises the requirement of extensive social effect analyses and the possibility of ignoring such trade-offs. Emphasises the need for context-specific solutions and the need for meaningful involvement with local communities in policy design and implementation, since it is realised that synergies are not always guaranteed and top-down, one-size-fits-all approaches may not be effective at the local level. Refining methods and guaranteeing more fair and efficient harmonising efforts depend on learning from these less successful examples.

11.6 RECOMMENDATIONS ON ENHANCING LEGAL AND POLICY FRAMEWORKS

Harmonising climate action with the SDGs calls for a multifaceted strategy, including numerous important initiatives. Using analytical tools to create policies that acknowledge and maximise the linkages between climate measures and the SDGs would help governments give integrated policy coherence top priority, therefore minimising trade-offs and maximising synergies. Complementing aims and targets depends on the SDGs being included in national climate policies, including NDCs and NAPs.

Effective local-level implementation depends on subnational and local authorities being empowered by national frameworks that offer resources, capacity-building, and authority. Equity and fair transitions should be the main priorities of climate policies, thereby addressing the effects on workers and underprivileged areas during the change to a low-carbon economy. Transparency and responsibility in policymaking depend on improving legal systems to guarantee public engagement and information availability, therefore fostering accountability. Comprehensive evaluations of suggested solutions will be facilitated by strengthening EI assessments to include climate change issues and match with all SDGs.

While supporting sustainable development, the deliberate use of economic tools, including carbon pricing, renewable energy requirements, and incentives, can propel climate action. Establishing strong legal systems will help to guarantee responsibility and enforcement connected to sustainability and climate pledges. By means of higher financial support and sharing of superior technologies with poor nations, developed nations should improve international collaboration and technology transfer. Finding answers that meet demands for sustainable growth, as well as climate change, calls for constant research and investment in invention.

11.7 CONCLUSION—TOWARDS A COHERENT AND SUSTAINABLE FUTURE

Approaching the linked problems of climate change and sustainable development requires a basic change in direction towards a coherent and sustainable future. This study has underlined how important law and policy are in enabling the necessary harmonisation between SDGs and climate action. Although the first legal and policy frameworks have been developed and much progress has been made in realising the interdependence of various agendas, considerable difficulties still exist in putting this knowledge into practical use on the ground. Policy coherence and institutional coordination problems, financing and capacity-building constraints, possible trade-offs between climate measures and particular SDGs, and the complexity resulting from different national circumstances all demand careful thought and creative solutions.

Not only is it necessary to harmonise climate action with the SDGs through law and policy, but it also guarantees a sustainable and resilient future for all by means of this process. The study shows the great interdependence between these two objectives and the many synergies that might be used to get several advantages in social, environmental, and economic spheres. The chances for coordinated action are great even if major obstacles still exist, including possible trade-offs, policy incoherence, and institutional constraints. The world community can approach a more integrated and effective response to the linked crises of climate change and sustainable development by strengthening legal and policy frameworks at all levels of governance, prioritising equity and public participation, strategically using economic instruments, and supporting international cooperation. The road ahead calls for a dedication to policy coherence, an awareness of the several roles that various actors play, and an unflinching attention to creating a future in which the earth may flourish as well as humans may.

Still, there are great chances for releasing strong synergies via combined legal and policy strategies. Promising directions forward are policy tools that simultaneously advance climate objectives and multiple SDGs, use legal frameworks to support integrated planning and assessment, and deliberately apply market-based processes and fiscal incentives. The increasing volume of effective case studies from all over the world offers insightful teachings. It shows that harmonisation is feasible and has major co-benefits in several spheres and situations.

The fundamental foundations of stakeholder involvement, public participation, and justice will ultimately determine the degree of success of our endeavours. Shining, fair and efficient laws and policies reflecting the interests and viewpoints of all stakeholders depend on inclusive and participatory procedures. Including justice issues guarantees a fair shift to a sustainable future that leaves no one behind.

REFERENCES

Agbakwuru, Victoria, Obidi, Peter Ofuje, Salihu, Ojonimi, and Ogwu, Chinelo (2024). The Role of Renewable Energy in Achieving Sustainable Development Goals. International Journal of Engineering Research Updates 7(2):13–27. https://doi.org/10.53430/ijeru.2024.7.2.0046

Bouyé, Mathilde (2018). INSIDER: Bridging Implementation of SDGs and NDCs: Examples and Early Lessons from Country Experiences. World Resources Institute at https://www.wri.org/technical-perspectives/insider-bridging-implementation-sdgs-and-ndcs-examples-and-early-lessons-country

Dagnachew, Anteneh G., Hof, Andries F., van Soest, Heleen, van Vuuren, Detlef P. (2021). Climate Change Measures and Sustainable Development Goals. PBL Netherlands Environmental Assessment Agency. PBL Publishers at https://sdgs.un.org/sites/default/files/2023-01/pbl-2021-climate-change-measures-and-sustainable-development-goals_4639.pdf

Equitable & Just (2024). A Vision for An Equitable and Just Climate Future. National Climate Platform at https://ajustclimate.org/

Eriksson, Martin, Ahlbäck, Anders, Silow, Nina, and Svane, Maria (2020). SDG Impact Assessment Tool at https://files.unsdsn.org/SDG-Impact-Assessment-Tool-Guide-1-0_final_ver02_mini%20(1).pdf

Friday, Catherine, Mills, Meghan, and McQueen, Julie (2022). Six Ways That Governments Can Drive the Green Transition. EY Publications. EY refers to the global organization, and may refer to one or more, of the member firms of Ernst & Young Global Limited, each of which is a separate legal entity. At https://www.ey.com/en_gl/insights/government-public-sector/six-ways-that-governments-can-drive-the-green-transition

Hammill, Anne, and Price-Kelly, Hayley (2016). Using NDCs, NAPs and the SDGs to Advance Climate-Resilient Development. The National Adaptation Plan (NAP) Global Network at https://ndcpartnership.org/using-ndcs-naps-and-sdgs-advance-climate-resilient-development

Huang, L., Swain, R. B., Jeppesen, E., Cheng, H., Zhai, P., Gu, B., Barceló, D., Lu, J., Wei, K., Luo, L., Wang, F., Wang, H., Zeng, J., and Guo, H. (2024). Harnessing Science, Technology, and Innovation to Drive Synergy Between Climate Goals and the SDGs. Innovation (Cambridge (Mass.)) 5(6):100693. https://doi.org/10.1016/j.xinn.2024.100693

Husainy, Avesahemad S. N., Sumit S. Mangave, Chetan C. Patil, and Sudesh D. Mane (2024). Harmonizing Responsibilities: Challenges and Opportunities in the Integration of Climate Change and Sustainable Development Goals (SDGs). The Asian Review of Civil Engineering 13(1):30–43. https://doi.org/10.70112/tarce-2024.13.1.4232

Kuipers, Josephina Wouda, and Korwatanasakul, Upalat (2024). Interlinkages Between Climate Action and the Sustainable Development Goals: Trends from the Recent Decade. SAGE Open. July–September:1–15. https://doi.org/10.1177/21582440241267354. https://www.researchgate.net/publication/383032252_Interlinkages_Between_Climate_Action_and_the_Sustainable_Development_Goals_Trends_from_the_Recent_Decade/citations

Legal Guide to the Sustainable Development Goals (SDGs) Was First Published By Advocates for International Development (A4ID) (2021). London, United Kingdom at https://www.a4id.org/wp-content/uploads/2021/11/SDG-Legal-Guide_Chapter-13_Final.pdf

Malley, Chris, and Veysey, Jason, SEI (2023). Integrating Climate and Development Planning Through Two-in-one Approach. IISD. SDG Knowledge Hub. https://sdg.iisd.org/commentary/guest-articles/integrating-climate-and-development-planning-through-two-in-one-approach/

Marta, Stefano, Chatry, Isabelle, Gross, Lorenz, Lafortune, Guillaume, and Drumm, Eamon (2024). Financing Our Future: How Subnational Governments Can Mobilise the Resources to Deliver on the SDGs. The OECD Centre for Entrepreneurship, SMEs, Regions and Cities (CFE) at https://oecdcogito.blog/2024/05/22/financing-our-future-how-subnational-governments-can-mobilise-the-resources-to-deliver-on-the-sdgs/

Müller, Gerd (2024). Climate Action That Protects the Right to Development. SGD Action at https://sdg-action.org/climate-action-that-protects-the-right-to-development/

Ochs, Alexander, Indriunaite, Ieva, and Engstroem, Sara, Muschner, Christoph, Gioutsos, Dean, and Roman, Kathia (2019). Towards Policy Coherence An Assessment of Tools Linking the Climate, Environment and Sustainable Development Agendas. Deutsche Gesellschaft für Internationale Zusammenarbeit (GIZ) GmbH. https://www.giz.de/de/downloads/1911_giz_Towards%20policy%20coherence_tool%20analysis.pdf

Ozdamirova, Laura, Dzhamaldinova, Marina, and Samatova, Chulpan (2024). The Role of Legal Mechanisms in The Implementation of The Sustainable Development Goals (SDGs). RT&A Special Issue № 6(81), Part-3, Volume 19, December.

Potočnik, Janez, and Khosla, IRP Ashok (2015). Policy Coherence of the Sustainable Development Goals—A Natural Resource Perspective. An International Resource Panel Report. United Nations Environment Programme (UNEP). At https://www.resourcepanel.org/sites/default/files/documents/document/media/policy_coherenceofthe_sustainable_developmentgoals.pdf

Riaño, Astrid Puentes (2025). Healthy Environment: A Human Right and Customary International Law at https://sdg.iisd.org/commentary/guest-articles/healthy-environment-a-human-right-and-customary-international-law/

Scheichet, Amanda (2024). Is Sustainable Finance the Solution to Funding Sustainable Development Goal 13: Climate Action? Kogod Sustainability Review at https://kogod.american.edu/news/is-sustainable-finance-the-solution-to-funding-sustainable-development-goal-13-climate-action

Shawoo, Zoha, Associate Scientist, Stockholm Environment Institute (2020). Addressing Climate Change Equitably Requires a Better Understanding of Policy Coherence. Experiences from Kenya, Colombia and Australia show why. United Nations Development Programme at https://sdgintegration.undp.org/addressing-climate-change-equitably-requires-better-understanding-policy-coherence

Tamasiga, Phemelo, Dzingai, Valentine Munyaradzi, Onyeaka, Helen, Tchonkouang, Rose Daphnee, Siyanbola, Kehinde Favour, Genesis, Ulakom, and Mudimu, George T. (2025). Energy transition effects on food security amidst climate change and progress toward sustainable development goals. Energy and Climate Change 6, 100222. https://doi.org/10.1016/j.egycc.2025.100222

Taylor, Russell (2024). Sustainable Development Goals and the Impact of Global Conflict, Extreme Poverty and Climate-Related Emergencies. House of Lords Library. UK Parliament. At https://lordslibrary.parliament.uk/sustainable-development-goals-and-the-impact-of-global-conflict-extreme-poverty-and-climate-related-emergencies/

UN Climate Press Release (2023). Governments Must Seek Win-Win Synergies By Addressing the Crises of Climate Change and Sustainable Development Together, Urges the Expert Group Report at https://unfccc.int/news/governments-must-seek-win-win-synergies-by-tackling-climate-and-sustainable-development-crises

UNEP (2024). SDG 13 and Interlinkages with Other SDGs—Climate Action at https://www.unep.org/news-and-stories/statements/sdg-13-and-interlinkages-other-sdgs-climate-action.

Vaidya, Ramesh Ananda, Molden, David James, Shrestha, Arun Bhakta, Wagle, Nisha, and Tortajada, Cecilia (2021). The role of hydropower in South Asia's energy future. International Journal of Water Resources Development 37(3):367–391. https://doi.org/10.1080/07900627.2021.1875809

Velosi Author (2023). Sustainability Goals in Relation to Environmental Impact Assessment. VELOSI at https://velosiaims.com/sustainability-goals-in-relation-to-environmental-impact-assessment/

Voïta, Thibaud (2020). Insight Brief: Engaging Subnational Governments in Climate Action Lessons Learned from the NDC Partnership Support Unit. NDC Partnership at https://ndcpartnership.org/sites/default/files/2023-09/engagling-subnational-governments-climate-actioninsight-briefmarch-2020.pdf

Yang, Hongbo, Dietz, Thomas, Li, Yingjie, Dou, Yue, Wang, Yujun, Huang, Qiongyu, Zhang, Jindong, Songer, Melissa, and Liu, Jianguo (2022). Unraveling Human Drivers Behind Complex Interrelationships Among Sustainable Development Goals: A Demonstration in a Flagship Protected Area. Ecology and Society. The Resilience Alliance. 27-3-7-15 at https://www.ecologyandsociety.org/vol27/iss3/art15/ DOI-10.5751/ES-13275-270315.

12 Aligning Multilateral Coal Transition Initiatives

A Case Study of Indonesia and Vietnam

E.M. Gui, C. Ayas and M.R. Kresnawan

12.1 INTRODUCTION: BACKGROUND AND MULTILATERAL COAL TRANSITION INITIATIVES

All coal-fired power stations in industrialising countries must be closed by 2040 for the world to achieve the Paris Agreement (IPCC, 2018; IEA, 2021). However, against a backdrop of significant support from governments, multilateral banks and philanthropies, there has been progress and setbacks in phasing down coal-fired power stations (CFPPs) in Southeast Asia and globally.

The approach taken by governments for coal transition in Southeast Asian countries varies considerably. In 2024, Indonesia's president announced plans to retire all coal-fired power plants within the next 15 years and build over 75 GW of renewable energy capacity by 2040 (Satriastanti, 2025). There has been political momentum in Indonesia to achieve these ambitious goals through international collaboration and assistance, for example, the launch of the $20 billion Just Energy Transition Partnership (JETP) agreement. The Indonesian JETP aims to help mobilise critical financial resources to achieve the country's net-zero power sector and phase-out of unabated coal by 2050, making the power sector emissions peak in 2030 (from the initial target in 2037) with a GHG emissions cap of 290 Mt CO_2 (from the initial baseline of 363 Mt CO_2).

Concurrently, Vietnam has committed to phase out unabated coal power by the 2040s, or as early as possible thereafter, to align with its target of carbon neutrality by 2050 (Do and Burke, 2023). This requires Vietnam to cease the issuance of permits for new coal-fired power plants, as committed in the JETP agreement between Vietnam and the International Partners Group (IPG) countries. Although there are no plans yet to retire existing coal-fired power plants, the cancellation of some new projects has reduced the total capacity of ongoing coal-fired power generation projects listed in the Power Development Plan 8 to around 5,000 MW. Table 12.1 lists key parameters for coal transition in Indonesia and Vietnam, including coal-fired power station capacity and current and projected renewable energy capacity.

In this context, multilateral coal transition initiatives have emerged as influential drivers to catalyse critical financial resources and technical assistance to address the complex and multifaceted challenges in transitioning away from coal. Three key initiatives, aiming to facilitate regional and global collaboration, mobilise public and private funding and strengthen the regulatory-enabling environment, are emphasised and explored in this chapter. They include:

Energy Transition Mechanism (ETM)—a market-based programme developed by the Asian Development Bank (ADB), to provide concessional loans and commercial capital to country-specific funds, which are used to buy out coal-fired power plants from asset owners to ensure early retirement and develop renewable energy projects (ADB, 2021).

DOI: 10.1201/9781003600152-12

TABLE 12.1
A Summary of Key Parameters for Coal Transition in Indonesia and Vietnam

	Indonesia	Vietnam
Total installed electricity capacity (all sources, on-grid) by 2024 (GW)	101.1	82.4
Operating CFPPs (on-grid) as of January 2025 (GW)	54.7	27.2
Cancelled CFPP capacity (since 2010) as of Jan 2025(GW)	51.5	60.9
RE added (since 2010, on-grid) by 2024 (GW)	8	40
Estimated added RE capacity by 2040	75 GW committed	120 GW (Estimated for BAU scenario); 200 GW (Estimated for net-zero scenario)
Coal transition commitments	Committed to phase-out all CFPPs by 2040	Retire inefficient CFPPs by 2030; stop new CFPPs
Planned plant retirement	Cirebon-1 coal-fired power plant in West Java (660 MW) Pelabuhan Ratu coal-fired power plant in West Java (1,050 MW)	Prospective plants for closure: Pha Lai (440 MW) Ninh Binh plants (100 MW)
The year committed to achieving net zero	2050	2050
JETP finance committed (in billion US dollars)	20	15
Finance needed to retire all CFPPs in billion US dollars (estimated)	165	70

JETP—launched at the 26th Convention of Parties (COP26) in Glasgow, aiming to promote a partner country-led model in phasing out/down[1] unabated coal-fired power generation and ceasing issuance of new permits and construction of new unabated coal-fired power plants in industrialising countries reliant on fossil fuels, such as South Africa (2021), Indonesia (2022), Vietnam (2022) and Senegal (2023).[2]

Glasgow Finance Alliance on Net Zero (GFANZ)—a global coalition of leading financial institutions committed to accelerating the decarbonisation of the economy, mobilising a coalition of blended finance sources, leveraging catalytic and concessionary capital sources and selecting the right mix of debt and equity finance.

The momentum derived from these multilateral coal transition initiatives has brought to the fore a number of important questions:

1. What are the key elements of multilateral coal transition initiatives to help accelerate the transition away from coal?
2. How can multilateral coal transition initiatives be effectively aligned to enhance system-wide impacts beyond individual project implementation?

12.2 A THEORY OF CHANGE FOR MULTILATERAL COAL TRANSITION INITIATIVES

In seeking answers to these questions, we apply Qualitative Comparative Analysis (QCA) (Ragin, 1987, 2000) to help identify factors and patterns across multiple coal transition initiatives and cases

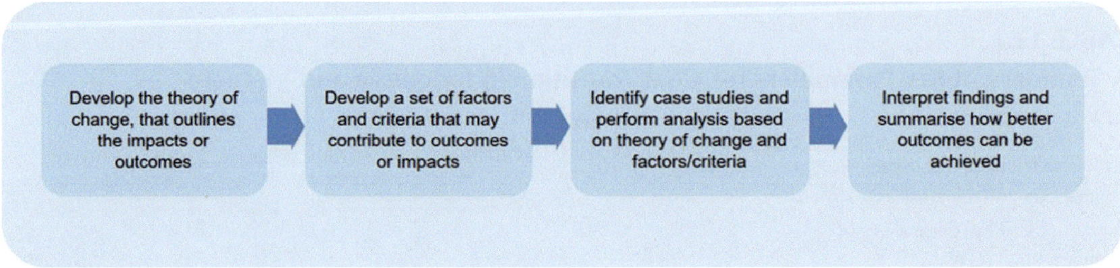

FIGURE 12.1 An overview of steps taken in Chapter 12 as suggested by qualitative comparative analysis.

Source: Author.

to better understand their potential to facilitate more effective planning and implementation, to realise greater performance and impacts. We follow the steps suggested by Simister and Scholz (2017), as illustrated in Figure 12.1.

A theory of change, as illustrated in Figure 12.2, is first envisaged to help identify priority outcomes and impacts from these multilateral coal transition initiatives in Southeast Asia, such as ETM, JETP and GFANZ.

Outcomes from coal transition initiatives have been identified in various contexts (IEA, 2021; World Bank, 2021; ILO, 2022; Ayas *et al.*, 2023) and can be summarised in four categories:

- The systemic dimension, focusing on energy security, accessibility and the adoption of low-emission technologies.
- The financial-economic dimension, emphasising the need for sustainable private and public financial flows to bolster renewable energy investments and diversify coal-reliant economies.
- The governance dimension, emphasising institutional capabilities to align energy and climate objectives and the meticulous management of resources at all levels.
- The socio-environmental dimension, addressing the social protection schemes for poverty alleviation, job creation in alternative sectors, environmental benefits and protection.

Longer-term impacts that these multilateral coal transition initiatives aim to achieve include accelerating multiple sustainable development goals (Nerini *et al.*, 2018), such as

- SDG 3, SDG 11, SDG 13 and SDG15: Reducing the coal use by 15% in Asia could save billions of tonnes of greenhouse gas emissions (Chen and Mauzerall, 2021), which is crucial for meeting the Paris Agreement and limiting global warming, mitigating environmental and health impacts associated with air pollution and greenhouse gas emissions from coal mining and coal combustion, enhancing social benefit and harmony in the coal dominant region.
- SDG7 and SDG12: Reducing the dependence on a single energy source, such as coal, while increasing clean energy shares in the generation mix, supported by energy efficiency measures, helps to increase supply diversification, thus enhancing energy security, energy resilience and sustainability of energy systems and infrastructure (Leong *et al.*, 2024; Sahid *et al.*, 2018). Facilitated through cross-border energy trade and regional supply chain development, supply diversification can also support more efficient use of resources and energy management and more responsible production and consumption.
- SDG 8, SDG11 and SDG17: Decarbonisation, digitalisation, decentralisation bring opportunities for innovation, international cooperation and partnerships, facilitating knowledge,

technology and resource sharing for the development of modern, cleaner and more efficient energy systems and infrastructure. While inflexibility, high stranded asset costs and the carbon-intensive nature of coal-fired power plants have made coal assets increasingly risky and economically unfeasible investments for investors and financiers, compared to cleaner technologies, including solar and wind.

Based on the aforementioned theory of change and the QCA methodology, we explore the key features of these multilateral initiatives in two case studies in Indonesia and Vietnam and discuss the success criteria for aligning these initiatives for an orderly transition away from coal that leaves no one behind.

12.3 QCA FOR MULTILATERAL COAL TRANSITION INITIATIVES

Successful coal transition initiatives call for an interdisciplinary and multi-dimensional strategy integrating policy-making, mechanism design and planning, implementation capacities and financial and resource mobilisation. In the following subsections, we explore key characteristics and differences of major multilateral coal transition initiatives, as well as success criteria for aligning these initiatives.

12.3.1 Key Characteristics of Major Multilateral Coal Transition Initiatives

There are clear distinctions among the focus of multilateral coal transition initiatives. ETM has an activity-level focus (transaction or project-based) (Azhgaliyeva *et al.*, 2022), JETP has an entity-level focus (system-wide resource allocation) and GFANZ has a greater focus on financial innovation and investment models (GFANZ, 2023). Table 12.3 in the Appendix provides an overview of the key features of the multilateral initiatives, focusing on objectives, decision-making, coverage of funding and technical assistance, financing mechanisms, governance and monitoring impact measurement.

To facilitate their implementation, designated public institutions can be made responsible for translating the multilateral initiatives' mandate into local actions, as the national governments are the natural owners overseeing the investment and resource mobilisation plans and alignment with the fiscal policies of the multilateral initiatives. These institutions act as central bodies to carry out administration and coordination functions, creating the regulatory-enabling environment and investment planning. Hence, ensuring alignment with national socio-economic objectives facilitates their operationalisation. For example, the Vietnam JETP Secretariat is hosted by the Ministry of Natural Resources and Environment (MONRE) and has set up four working groups, including Coordination (led by MONRE); Governance, Policy and Investment (led by Ministry of Planning and Investment (MPI)); Technology and Energy (led by Ministry of Industry and Trade (MOIT)); and Finance (led by Ministry of Finance (MOF)).

12.3.2 Key Differences Between Multilateral Coal Transition Initiatives

Table 12.2 shows the comparison of the major multilateral coal transition initiatives in relation to their alignment with national policies and plans, country governance structure and financing mechanism platforms, in-country implementation and how progress is measured and monitored.

Focusing on project implementation, ETM faces smaller coordination challenges; however, its project-based approach may have limited influence in the wider energy transition eco-system and policy-making. ETM has also employed a two-tier structure for coordinating initiatives across countries. At a higher level, the Southeast Asia ETM Partnership Platform has been established to support development financing and operationalisation at the regional level with three pilot countries and assigned a country-specific entity through the ETM country platform (Dalal-Clayton *et al.*, 2022). At the country level, the ETM country platform is supervised by the Steering Committee,[3]

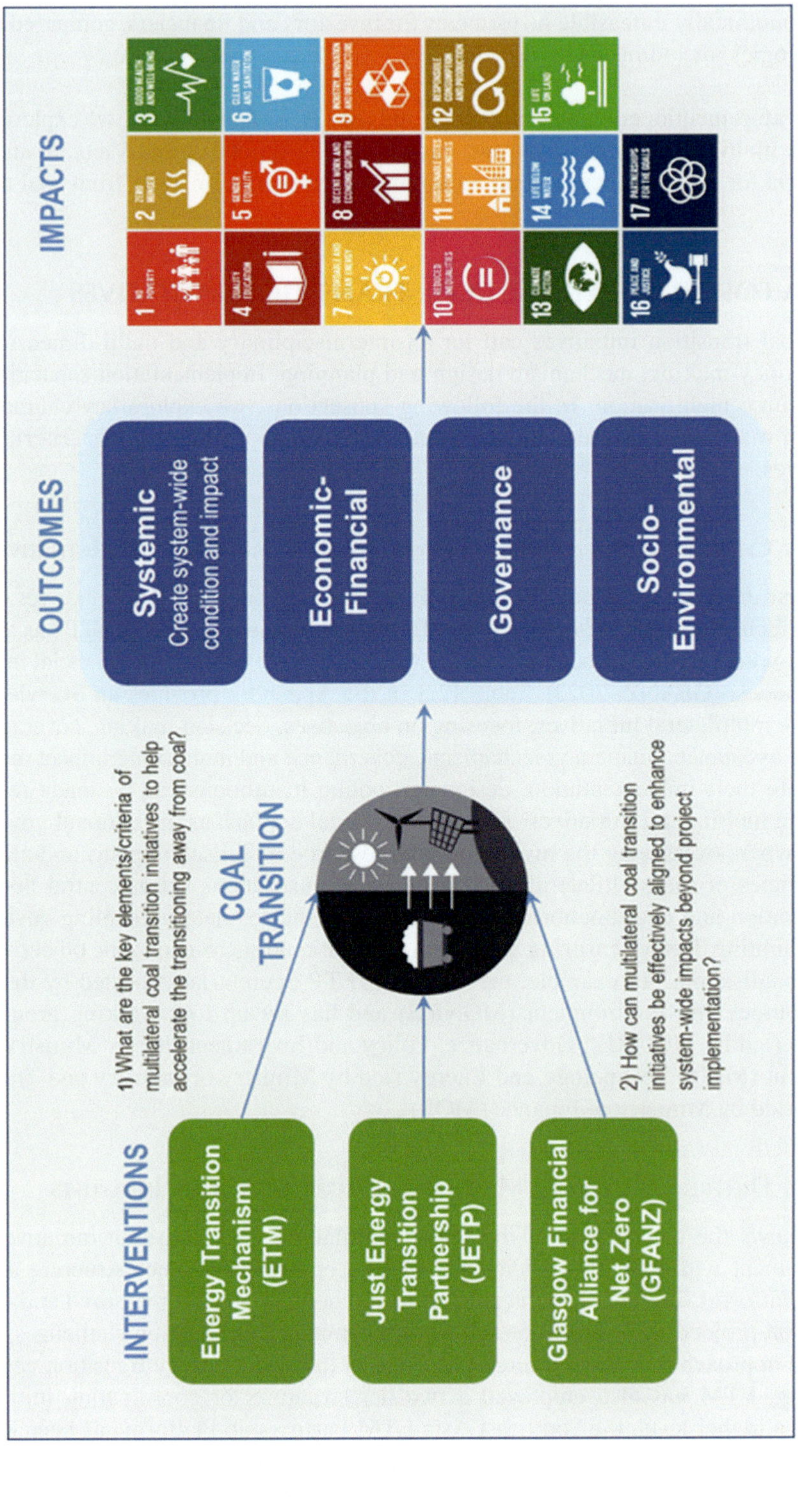

FIGURE 12.2 A visual representation of the theory of change for multilateral coal transition initiatives.

Source: Author.

which is responsible for determining ETM's overall strategic direction and overseeing its implementation. The decision-making is mostly exercised through ADB in consultation with the coordinating ministries' Steering Committee.

JETP has greater potential to align with countries' national development plans, socio-economic policies, NDCs and long-term strategies, net-zero objectives and sectoral roadmaps and to be managed alongside budgetary processes, for wider systemic impacts (Winkler *et al.*, 2023). Cross-ministerial coordination bodies are established in Indonesia and Vietnam to coordinate policies, develop projects, structure their financing and ensure coherence between public entities overseeing the legal, financial, policy-making and regulatory aspects. However, the involvement of international donors, financial institutions, private sector participants, technical assistance partners and multilateral agencies in a country-led approach suggests complex interrelation and interdependency of a multi-stakeholder structure. This complexity and plurality of JETP structure and governance may signal risks in decision-making and operational complexity, leading to potential delay and ineffectiveness in JETP implementation.

GFANZ operates through regional networks and sector-specific initiatives and does not have a collective country approach. GFANZ's governance involves a multi-layered member institutions, including a coordinating committee composed of representatives from member institutions providing overall guidance and direction for GFANZ and working groups of experts from member institutions collaborating on specific areas such as mobilising capital or financing. Most GFANZ members make decisions based on the best alignment with their company strategy when selecting individual projects and targeted countries.

Despite the different strategies across these initiatives, aligning multilateral efforts with each country's needs and plans is critical to increasing involvement from decision-makers and social acceptance, while achieving effective allocation of human and capital resources and improving the effectiveness of the assistance effort (Gui and Inal, 2025).

12.3.3 Synergies, Cross-Learning and Knowledge Intermediary

The various strategies utilised by these initiatives present a diverse array of opportunities for mutual learning. Sharing experiences, strategies and best practices fosters an environment conducive to innovation and advancement, thereby driving the region towards the collective objective of achieving a low-carbon economy. By engaging in collaborative endeavours and collectively applying acquired knowledge, these initiatives have the potential to enhance their influence and facilitate profound transformations in the regional process of accelerating the phase-down of coal-fired power plants to achieve countries' NDCs and net-zero goals.

A number of organisations are building experience across multiple initiatives, such as ADB, IPG and GFANZ. These organisations could be pivotal in bringing in holistic experience and perspectives from different approaches and mechanisms, proposing potential solutions in addressing their challenges and opportunities for more effective implementation of coal transition. Better alignment of objectives and actions from multilateral development banks and local financial institutions can also bring down barriers and fast-track the progress of coal transition in the region.

12.3.4 Success Criteria for Aligning Multilateral Coal Transition Initiatives

A transition away from coal necessitates a careful balance of two parts of the investment needs: channelling finances and incentives in reducing coal power plant capacity while driving investments in clean technologies. The size and scale of renewable energy projects can indicate a readiness and commitment to phase out fossil fuels. Increased capital flow into clean energy projects and technologies, such as energy storage solutions and grid connections, and smart, clean technologies and their supply chains, can help to further reduce the value of coal assets and drive down the costs of coal transition (Gui *et al.*, 2024b).

TABLE 12.2
Key Metrics of ETM, JETP and GFANZ Initiatives

Key Metrics	ETM	JETP	GFANZ
Policy plan and regulatory strategies	X	X	
Alignment with national development plan, socio-economic policies, NDCs, LTS		X	
Alignment with net-zero objectives		X	X
Consideration for the budgetary process		X	
Improve the enabling environment	X	X	X
Initiative-wide coordination platform	X	X	X
In-country platform	X	X	
Financing platform	X	X	X
Development of a sectoral roadmap		X	
Metrics and targets for monitoring progress		X	X
Performance-based payment	X		
Knowledge sharing and dissemination	X	X	X
Social protection	X	X	X
Gender equity		X	X

Based on our theory of change, we assess the success criteria for aligning multilateral coal transition initiatives in phasing down coal-fired power plants from systemic, socio-environmental, economic-financial and governance dimensions. Figure 12.3 lists key dimensions and indicators in evaluating the impact of multilateral coal phase-out initiatives.

12.3.4.1 Systemic Dimensions

Multilateral initiatives are anticipated to create system-wide impact through mobilising finance, enhancing regulatory frameworks, facilitating effective policy and market instruments and building a nurturing eco-system to incentivise private sector participation and stakeholder coordination and consultation.

An integrated planning approach ensures a more coherent and comprehensive coal transition strategy, including the implementation of an orderly transition that minimises economic disruption and safeguards countries' fiscal health. This should be well aligned and considered in national development plans and socio-economic policies, as well as when developing NDCs and long-term strategies.

Clear targets and pathways towards decarbonisation, underpinned by ambitious policies, are a fundamental starting point. Regular reporting on the implementation of these policies ensures transparency and accountability. The introduction of carbon pricing can provide a financial incentive for reducing greenhouse gas emissions, while setting emission standards can ensure that industries limit their greenhouse gas emissions. The removal of coal subsidies can make renewable energy sources more competitive.

Other aspects include strengthening the standards and disseminating best practices for governments across the region for more efficient use and conservation of natural resources, as well as wider capacity development for officials and decision-makers to best design policy, programmes and projects for renewable energy development and just transition (Gui *et al.*, 2024a).

12.3.4.2 Economic-Financial Dimensions

The economic aspects of multilateral coal transition initiatives are critical to ensuring the financial viability and attractiveness of switching from coal to renewable energy sources. Key economic indicators such as return on investment (ROI) and net present value (NPV) are important metrics for

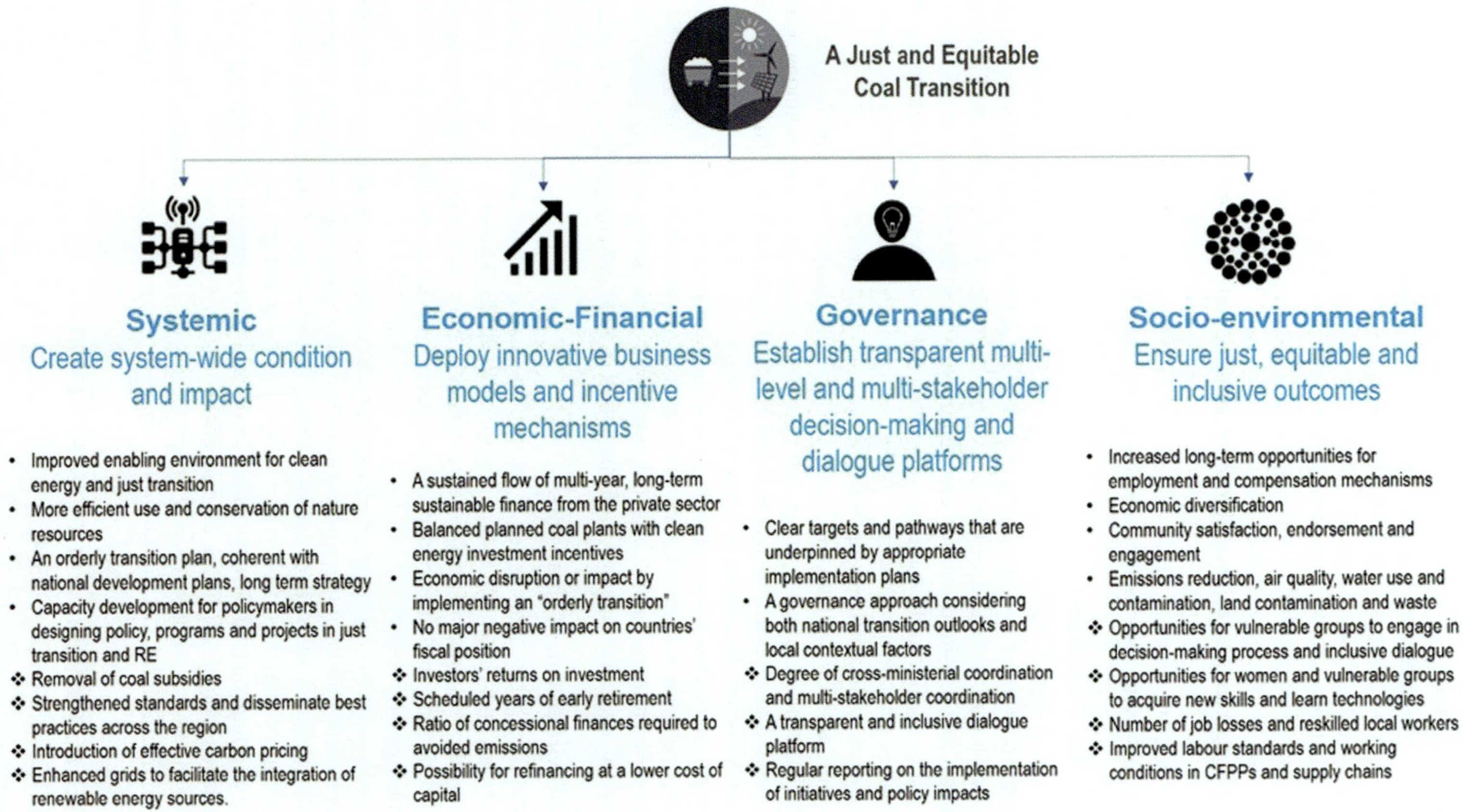

FIGURE 12.3 A summary of key dimensions and indicators for aligning multilateral coal transition initiatives.

Source: Author.

evaluating the economic impact of these investments. These metrics help manage investor expectations by providing a clear picture of financial returns over time, which boosts investor confidence and encourages participation.

Incentivising these investments necessitates a multifaceted approach, including tax breaks and subsidy programmes to encourage supply chain participation, long-term policy frameworks for renewable energy to level the playfield and reduce risk, clear planning guidelines for land use allocation, enabling grid connection and building a skilled workforce (Paredes *et al.*, 2023; Manych and Jakob, 2021). These incentives are intended to speed up the adoption of clean energy alternatives while improving their cost-effectiveness, increasing ROI and potentially improving NPV calculations for these projects. Financing strategies can be strengthened by refinancing opportunities at a lower cost of capital, improving both ROI and NPV metrics and making the coal transition projects more economically appealing. The protracted timescale in phasing down coal-fired power plants also implies that attracting a steady flow of multi-year, long-term sustainable finance from the private sector is essential.

12.3.4.3 Governance Dimensions

The importance of a well-structured governance of the coal transition mechanism cannot be understated. It serves as the backbone for successful implementation at later stages, providing clear guidelines and regulations for all stakeholders involved in the coal transition process. Well-structured governance ensures a fair transition for workers and communities, helping to minimise the environmental impact of the coal transition and providing certainty for investors and companies in the energy sector. It also establishes mechanisms for monitoring and evaluating the progress of the coal transition and promotes public participation in the decision-making process.

The existing coal transition mechanisms have a two-fold governance structure: multi-level and multi-stakeholder. The multi-level governance will promote vertical collaboration among actors at various decision-making levels, such as national and regional governments, representatives from donor and recipient countries, international financial institutions through member banks and asset managers and the private sector. Meanwhile, multi-stakeholder governance (Ratner *et al.*, 2022) brings together experts from businesses, civil society, non-governmental organisations (NGOs), think tanks, research institutions and the affected communities inside the working groups, to participate in dialogues, decision-making and implementation of solutions.

12.3.4.4 Socio-Environmental Dimensions

The success of multilateral mechanisms relies on ensuring just, equitable and inclusive outcomes for consumers, workers and vulnerable communities (ACE, 2025; Diluiso *et al.*, 2021). The communities affected by the retirement of coal-fired power plants must participate actively in decision-making. Their participation further enhances a transparent and inclusive dialogue platform and increases community buy-in.

Economic diversification of coal-reliant regions offers development opportunities in alternative sectors. For example, skill-building programmes in clean technologies empower local communities, preparing them for a future in alternative sectors. Further, unemployment benefits, early retirement schemes and retraining programmes need to be readily available to mitigate the potential disruption in local communities reliant on coal. Strengthening health and safety standards in the supply chains will also improve labour conditions during the transition.

In the next section, we use Indonesia and Vietnam as case studies to demonstrate the paths for greater alignment of multilateral coal transition mechanisms for a more successful coal transition.

12.4 CASE STUDY: TOWARD A SUCCESSFUL COAL TRANSITION IN INDONESIA

As of the end of January 2025, the current coal power capacity in Indonesia is 54.7 GW. As a part of its coal phase-out pledge, the Indonesian government is partnering with ADB and JETP to assess

13 coal-fired power plants with a total capacity of 4.8 GW for potential early retirement (Antara, 2023). On the other hand, new captive coal power plants are being built at a rapid rate, with capacity tripling between 2019 and 2025, reaching 16.6 GW, as Indonesia is prioritising the extraction and production of mineral resources to develop a new energy supply chain (Setiawan, 2025). In this context, multilateral coal transition initiatives such as ETM, JETP and GFANZ are pivotal in helping countries move away from coal-dependent power generation. The more their mandate is coordinated and aligned, the more efficient their in-country implementation and the wider their system impacts.

There are two coal power plants targeted for decommissioning in Indonesia, including

Cirebon-1 coal-fired power plant in West Java (660 MW): ADB has signed a memorandum of understanding (MoU) with Indonesian state-owned power utility Perusahaan Listrik Negara (PT PLN), the Indonesian Investment Authority and Cirebon Electric Power to close down the Cirebon-1 coal-fired power plant (ADB, 2021). This MoU sets out the terms for ending the plant's obligation to supply electricity and shortening the duration of the power purchase agreement. Under the ETM mechanism, the plant is planned to cease operations in December 2035, nearly seven years ahead of its original decommissioning timeline in July 2042. The ADB initially aimed to finalise the financial closure by the end of 2024, but the complexity of managing this pioneering project has resulted in a postponement (Karyza, 2024). The early retirement plan is currently being reconsidered.

Pelabuhan Ratu coal-fired power plant in West Java (1,050 MW): PT PLN is also exploring the early retirement of the Pelabuhan Ratu coal-fired power plant, using buy-out models (Silviana, 2025). This initiative was formalised through a Principal Framework Agreement signed in October 2022. Under the proposal, the plant's operational period would be reduced from 24 years to 15 years, bringing forward its retirement to 2037. This reduction is expected to result in a significant decrease in carbon dioxide emissions, estimated at approximately 51 million tonnes. Given the plant's role as a key electricity supplier in southern Java, its early retirement requires careful planning to maintain energy security and ensure economic benefits. At present, the Pelabuhan Ratu power plant remains operational and is co-firing biomass as part of its transition efforts. Meanwhile, PT Sarana Multi Infrastruktur (PT SMI) is leading efforts to finalise the purchase of the plant.

12.4.1 Effective Country Platform for Project Implementation

The national governments play a critical role in establishing an effective coal transition country platform for project implementation. Inter-ministerial committees and task forces can be formed, comprising representatives from diverse sectors and levels of government, along with stakeholders from civil society, the private sector and international partners. These entities should be responsible for coordinating the implementation of policies, monitoring the progress and ensuring that they align with the initiatives' overall objectives to retire coal-fired power plants. In addition, governments can establish specialised divisions within relevant ministries to oversee the technical and financial aspects of energy transition projects. These divisions should collaborate closely with international financial institutions and multilateral development banks involved in multilateral coal transition initiatives, such as ETM, JETP and GFANZ. Building upon existing work from ETM and JETP initiatives in Indonesia (Fiscal Policy Agency, 2022), Figure 12.4 illustrates the proposed governance of the coal transition country platform and its utilisation in energy transition projects in Indonesia.

A number of key elements to support the coal transition implementation are essential, as listed here.

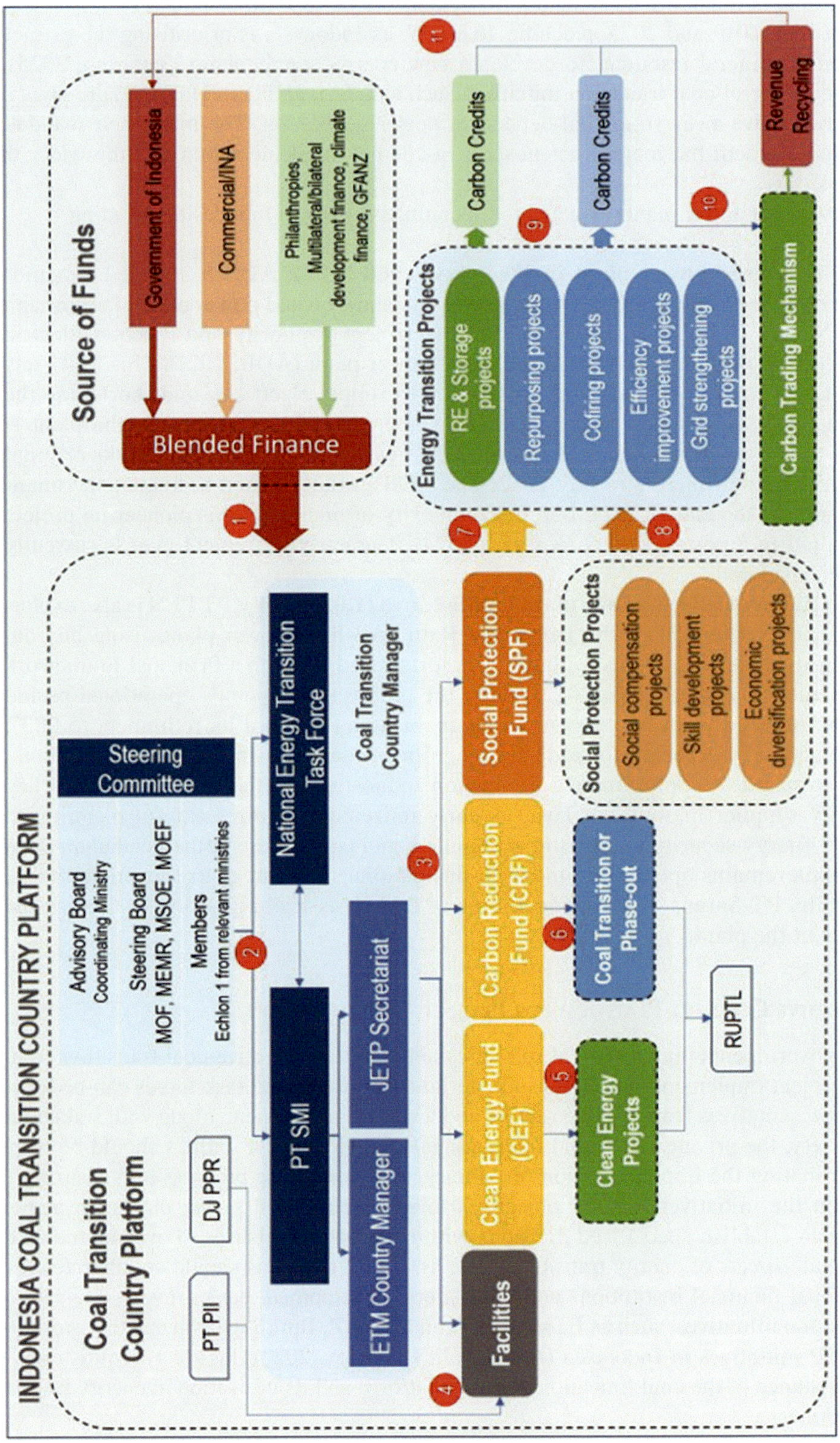

FIGURE 12.4 An illustration of Indonesia's coal transition country platform.

Source: Adapted from Fiscal Policy Agency (2022).

Flows of Information

1. Investment flows from blended finance into country platform manager PT SMI.
2. Program priority direction from SC to country platform manager.
3. Country platform conduct early retirement process based on roadmap.
4. Finance support model from MOF to the country platform.
5. CEF mobilisation from coal power plant to RE according to RUPTL.
6. CRF mobilisation to retrofit or repurpose to improve emission standards or efficiency according to RUPTL
7. RE projects or energy transition projects produce carbon credits; potential to be sold in the carbon market.
8. CEF or CRF mobilisation to reduce GHG through repurposing, co-firing, emission reduction from coal power plants.
9. Workers (supported through SPF) mobilisation to RE projects or energy transition projects.
10. Carbon credits sold to the carbon market.
11. Revenue recycle through Non-Tax State Revenue (PNBP).

MoF: Ministry of Finance

MEMR: Ministry of Energy and Mineral Resources

MSOE: Ministry of State-owned Enterprise

MOEF: Ministry of Environment and Forestry

DJPPR: Director General of Budget Financing and Risk Management

PT RII: PT Indonesia Infrastructure Guarantee Fund

PT SMI: PT Sarana Multi Infrastruktur

RUPTL: Indonesia National Power Supply Plan

INA: Indonesia Investment Authority governing Indonesia Sovereign Wealth Fund

FIGURE 12.4 (Continued)

The Coal Transition Country Platform—provides the central function for coal transition implementation and is responsible for carrying out the processes necessary for early retirement in accordance with the roadmap, and the Ministry of Finance provides financial support. The Platform also manages investment flows from blended finance and goes through the coal transition country manager to support coal transition project implementation.

The Steering Committee—provides the coal transition country platform with priority direction, as well as policy support.

The Coal Transition Country Manager—PT Sarana Multi Infrastruktur (PT. SMI) (PT Sarana Multi Infrastruktur, 2022) serves as Indonesia's coal transition country manager. PT. SMI implements Indonesia's coal transition in-country financing platform and acts as fund manager and managing agency for ETM and the JETP. It also hosts the ETM country manager and the JETP Secretariat that serve as the coordinator for internal and external stakeholders while serving the project development function. Together with the National Energy Transition Task Force, they are ensuring cohesion between public entities overseeing the legal, financial, policy-making and regulatory aspects of renewables development, coal phase-out and fiscal allocation.

The Clean Energy Fund (CEF)—mobilises funding to develop renewable energy and energy storage projects to replace coal-fired power plants.

The Carbon Reduction Fund (CRF)—mobilises funding for retrofitting or repurposing coal-fired power plants, plant efficiency improvements and strengthening emission standards or grids.

The Social Protection Fund (SPF)—mobilises funding for worker compensations, skill development, job replacement or economic diversification for the local economy.

The Emissions Trading Mechanism—produces carbon credits that can be sold on the carbon market.

Revenue Recycling—Non-tax state revenue is responsible for managing the revenue cycle associated with coal transition projects.

This coordinated and targeted approach demonstrates how institutions like PT. SMI can play a pivotal role in translating and implementing regional energy transition initiatives at the local level, ensuring alignment with national and global coal transition commitments as well as national infrastructure development plans.

In particular, we would highlight the importance of a Social Protection Fund to provide financial support to workers and communities affected, as well as strengthen green and clean energy skill development to support the long-term effort of coal transition. This is further elaborated in Section 12.5.

12.4.2 Designing Greater Alignment for Coal Transition Mechanisms

To have a significant influence on the entire system, governments need to create comprehensive frameworks that leverage the benefits of various multilateral coal transition initiatives that can both align with national and regional energy and climate planning strategies and guide the practical project implementation. Figure 12.5 illustrates the interactions between national planning and multilateral coal transition planning and project implementation.

For successful planning and implementation of multilateral coal transition strategies, a Comprehensive Investment and Policy Plan (CIPP) established through JETP can serve as a technical roadmap that helps to formulate decarbonisation strategies, identify strategic areas and priority projects and financing mechanisms (JETP Secretariat, 2023). A number of essential components include the Power Development Scenario, Emissions Projection, Investment Areas and the "just" transition framework. These will, in turn, help to guide the project implementation for the timely and orderly decommissioning of targeted coal-fired power plants.

A significant step involves a Strategic Environmental and Social Assessment (SESA) process, participated in by key stakeholders, including government agencies, NGOs and local communities.

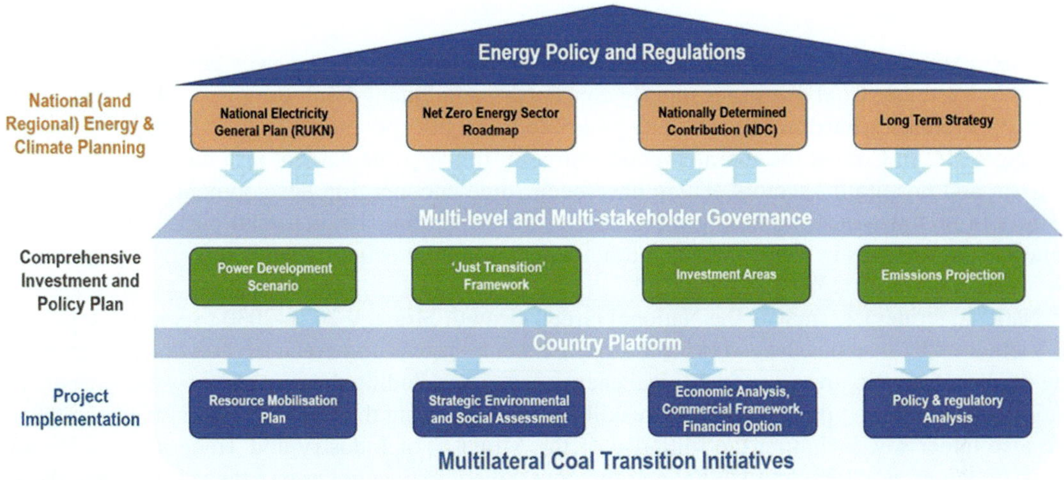

FIGURE 12.5 An illustration of the multilateral coal transition initiatives planning and implementation framework.

Source: Author.

SESA includes policy reform, feasibility studies, pilot, asset identification, asset negotiations, asset transactions, research and impact assessments, vocational education for upskilling and reskilling existing coal plant workers, public meetings, focus group discussions and information disclosure (Dalal-Clayton *et al.*, 2022).

An effective planning and implementation framework should incorporate mechanisms for stakeholder engagement, guaranteeing that the perspectives of impacted communities, industry representatives and environmental organisations are acknowledged and taken into account during decision-making procedures. Through the promotion of inclusive and transparent governance, governments can cultivate trust and garner support for the transition away from coal.

We will dive into the social protection mechanisms in Vietnam and Indonesia as a case study to demonstrate the social component of these coal transition initiatives.

12.5 SOCIAL PROTECTION MECHANISMS

A number of international collaborations have been formed to support coal regions in Southeast Asia, such as Indonesia, the Philippines, Thailand and Vietnam, in planning and implementing their just energy transition pathways, while exchanging knowledge and best practices around social protection mechanisms in coal-dependent regions. These include the Coal Regions in Transition Platform, the Powering Past Coal Alliance and the Interregional Platform for the Just Energy Transition in Coal Regions, Community of Practice on Just Energy Transition in Asia and a knowledge hub (Energy Transition Partnership, 2025; Powering Past Coal Alliance, 2025; GIZ, 2025; ILO, 2025). The JETP agreements in Indonesia and Vietnam have also developed comprehensive strategies in promoting gender equality and inclusiveness as part of diversifying regional economies and creating jobs in coal-dependent regions, outlined in its Resource Mobilisation Plan (RMP) and CIPP.

Despite the high-level commitments, these initiatives have shown positive though limited progress to date in diversifying regional economies and creating jobs in coal-dependent regions. The challenges hinge on acquiring adequate financial resources for social protection, as well as discovering and realising viable economic transformation opportunities for these regions.

12.5.1 Vietnam

Approximately 150,000 jobs are currently tied to coal mining and power generation (CEED, 2023) in Vietnam. Coal mining is mainly concentrated in Quang Ninh province, which produces 99% of Vietnam's hard coal and employs around 100,000 people in the coal sector (Nguyen and Ngo, 2022). Coal is the main revenue source in this coal-dependent province, supported by tourism and hospitality sectors. There have been some encouraging developments in alternative sectors in this province, including the potential for developing 16 industrial parks, which could generate 20,000 jobs in electronics manufacturing, automotive and logistics sectors (Nguyen and Ngo, 2022; Minh, 2023).

National strategies, such as the National Green Growth Strategy 2021–2030 (Socialist Republic of Vietnam, 2021) emphasises workforce development by promoting green job creation and upskilling workers transitioning from traditional energy sectors, including the coal industry. Several international collaboration programmes for reskilling and upskilling the workforce and green economy are also underway, including the initiative of the Ministry of Industry and Trade and the World Bank (Doan *et al.*, 2023) and UNDP and the government of Japan. Despite these initiatives, there is limited evidence of widespread participation in these reskilling programmes in coal-dependent regions, since the government does not yet have a firm plan to retire a large capacity of coal-fired power plants. Other potential challenges include capacity gaps in supplying green skills, institutional challenges in implementing green job initiatives and financial constraints and limited digital infrastructure, which might collectively hinder participation.

In line with the National Green Growth Strategy and other workforce and green economy programmes, the RMP under the JETP Vietnam focuses on enhancing the regulatory framework to facilitate the energy transition, with a significant part of the plan dedicated to accelerating the deployment of renewable energy sources (Ayas *et al.*, 2025). The plan specifies the design and implementation of development programmes in regions that are negatively impacted by the energy transition, including encouraging the participation of youth, women, ethnic minorities and micro, small and medium enterprises (MSMEs) as entrepreneurs or labourers in JETP initiatives. RMP includes provisions to address the transition's impacts on workers, communities and businesses by outlining investment priorities, such as social security mechanisms, training for green jobs and support for Vietnamese businesses to participate in renewable energy value chains. It also proposes regulations to enable multi-purpose land use to mitigate land use impacts and targeted development programmes for MSMEs (CEED, 2023; PAGE, 2023).

12.5.2 Indonesia

Over 400,000 jobs in Indonesia are dependent on coal mining and power generation, of which 250,000 are directly employed in the coal mining sector (IESR, 2022). Unlike Vietnam, in which coal mining is predominantly for domestic consumption, Indonesia also has a vibrant coal export sector, contributing to around 6% of its GDP. This presents a significant barrier in decarbonising its economy and transition away from coal and fossil fuels. In an effort to mitigate the negative effects of the coal phase-out on employment and regional economies, the government is conducting impact assessments (Klaikaew, 2023). The IPG committed US$200 million in grants for just transition assessment intervention, of which US$18 million will be allocated for capacity building, scoping studies and project piloting (Larasati and Mafira, 2023).

To support economic diversification, the government has introduced the coal revenue-sharing scheme along with corporate social responsibility programmes. The coal revenue-sharing scheme (*Dana Bagi Hasil, DBH*) allocates funds derived from taxes and royalties on coal mining activities to regional governments to support local development and economic diversification efforts. DBH scheme has been applied in Paser Regency and Muara Enim Regency regions and contributed 20%

of the total revenue budget of the Muara Enim Regency Government in 2023 and 27% of the total revenue of the Paser Regency Government in 2013–2020 (IESR, 2023). Although the funds do not contribute significantly to the income of coal industry workers at present, the benefits from local economic development opportunities it generates can indirectly flow to the local community.

Indonesia's JETP has established the Standard 9 programme to support economic diversification and promote new job opportunities in coal-dependent regions, which includes training and support programmes for reskilling and upskilling of the workforce, targeting women and marginalised communities, including youth and vulnerable groups (Larasati and Mafira, 2023). There is potential to create 96,000 high-skilled jobs in coal-producing regions if JETP initiatives are implemented, particularly by replacing existing coal plants with solar energy facilities (Setyawati and Setiawan, 2024). The agreement also recognises the participation of women employed in the coal sector in decision-making processes and requires consultation with them. However, the delivery partners formed under the JETP governance structure to date do not have a specific gender mandate and are selected primarily on their subject matter expertise on decarbonisation.

Learnings from Indonesia and Vietnam experiences suggest that social protection for coal workers and economic diversification must be addressed in concerted efforts through government programmes and multilateral initiatives to accelerate regional socio-economic transformation away from coal. A clear timeline committed by the government to retire coal-fired power plants is critical to help communities prepare for the transition mentally and physically.

12.6 CONCLUSION AND STRATEGIES FOR ACCELERATING COAL TRANSITIONS

Multilateral coal transition initiatives are catalytic in bringing in necessary capital and human resources, knowledge and innovation to accelerate the phase-down of coal-fired power generation at scale and in line with the timeline required to achieve the Paris Agreement and Net Zero by mid-century.

To amplify the impact and local endorsement of these initiatives, multilateral coal transition initiatives will need to achieve the following outcomes.

12.6.1 Closely Align Their Systemic Outcomes and Project- and Entity-Level Strategies

The project-level strategies are important to ensure better socio-economic and environmental outcomes and financial viability to help bring in private sector finance and satisfy local community needs. Meanwhile, policy, regulatory, legal and financing eco-systems can be better improved through the entity-level strategies that aim for more efficient system-wide resource allocation and wider system impact. These strategies enable coal transition plans and activities to be better integrated into and align with the country's transition pathways, national development plans, socio-economic policies, NDCs and long-term strategies, net-zero objectives, sectoral roadmaps, and so on.

12.6.2 Establish Central Coordination Bodies and Co-Management Frameworks for Institutionalised Forms of Collaboration

A central platform for collaboration among coal transition initiatives can help to achieve more efficient governance, implementation and deployment of finances across platforms in accelerating the transition of coal-fired power plants. Multi-ministerial involvement and governance can help strengthen the collaboration among different line ministries and improve a more coherent policy, regulatory, legal and investment eco-system for transitioning from coal to renewable energy. This will also enable resource sharing, cross-platform learning and knowledge dissemination and the development of best practices.

12.6.3 Actively Address Social Protection for Local Communities and Pursue New Economic Development Opportunities for Coal-Dependent Regions

Effective social protection and compensation mechanisms and plans must be put in place, beyond techno-economic considerations, when decommissioning the coal-fired power plants. A blueprint for local economic prosperity for all, including women, youth and vulnerable groups, will help to bring down social resistance and gain long-term support from communities in transition.

12.6.4 Enhance Regional Cross-Learning Opportunities

Opportunities for cross-learning and knowledge exchange can be gained through exploring different approaches and mechanisms in accelerating the retirement of coal assets, such as market-based strategies, partner-country-led integrated financing platforms and constituency building for political momentum. All aim to better achieve coal transition objectives, as well as ensure more efficient financial resource allocation, better mobilisation of private finance and better integration with national policies and pathways and improve the enabling environment for a low-carbon transition. Their strengths and weaknesses need to be better understood for a more collaborative approach and more substantial collective impact on the coal transition in the region.

"All existing coal plants must be retired by 2030 in the world's richest countries and by 2040 everywhere" (Global Energy Monitor, 2025). Financial institutions and multilateral coal transition initiatives play a critical role in facilitating governments and coal businesses to transition away from coal into clean energy to meet the Paris Agreement Goals. Collaboration and cross-learning are key in gaining knowledge and experiences in accelerating the phase-down of coal-fired power plants. It is our hope that this chapter will provide some insights and food for thought for more coordinated action among multilateral coal transition initiatives towards a cleaner future for the generations to come.

ACKNOWLEDGEMENT

The authors thank the editors and two anonymous reviewers for their valuable comments to improve the quality of this chapter. We also thank Mr. Michael Williamson from the United Nations Economic and Social Commission for Asia and the Pacific, and Mr. Beni Suryadi and Dr. Zulfikar Yurnaidi from the ASEAN Centre for Energy to provide valuable external reviews. The conceptualisation of this work and an early draft of this paper was supported through Climateworks Centre. However, the findings do not necessarily reflect the views of any of the institutions involved and are solely the responsibility of the authors.

NOTES

1 *Phase-out* refers to the systematic reduction and complete elimination of coal consumption over time, aligning with the calls from the United Nations Secretary-General for a full phase-out of coal and other fossil fuels. This is the ultimate goal for achieving global climate objectives. *Phase-down* involves a gradual reduction of coal usage to lower levels than currently consumed, which can serve as an immediate and actionable step for countries facing significant energy security and economic challenges (Gui et al., 2024b).

2 Despite the withdrawal from JETP by the new U.S. administration in March 2025, which had co-led the JETP agreements alongside Japan, several countries in the European Union, particularly Germany and France, reaffirmed their commitments and launched the Indonesia Energy Transition Facility.

3 The committee in Indonesia is made up of an Advisory Board from Coordinating Ministries, a Steering Board from the Ministries of Finance (MOF), Ministry of Energy and Mineral Resources (MEMR), Ministry of State-owned Enterprises (MSOE), and Ministry of Environment and Forestry (MOEF), as well as members from relevant ministries (Fiscal Policy Agency, 2022).

4 The market readiness, potential impact and scalability of the main tools, being explored in APAC—asset purchases at discounts to book in order to increase IRR due to low purchase price, improving debt/equity ratio and cost of debt to increase IRR by reducing cost of capital, blending concessionary capital to increase IRR and reduce IRR hurdles if de-risking tools used, and generating carbon credits from avoided emissions to increase IRR by creating alternative revenue from—are heterogeneous.

REFERENCES

ACE. 2025. A Guide to a Just and Inclusive Energy Transition in ASEAN. ASEAN Centre for Energy (ACE). https://aseanenergy.org/publications/a-guide-to-a-just-and-inclusive-energy-transition-in-asean (accessed 1 Jan, 2026).

Antara. 2023. Indonesia, ADB Partner to Speed Up Coal Plant Retirement. Available at: https://en.antaranews.com/news/300360/indonesia-adb-partner-to-speed-up-coal-plant-retirement. (accessed 1 Jan, 2026).

Asian Development Bank (ADB). 2021. What Is ETM? How Asia and the Pacific Can Make a Clean Energy Transition? Asian Development Bank (ADB). Available at: https://www.adb.org/news/videos/what-etm-how-asia-and-pacific-can-make-clean-energy-transition-adb-insight-full-episode. (accessed 1 Jan, 2026).

Ayas, C., Dolan, M., Nguyen, T. and Giwangkara, J. 2023. Energy Transitions in Vietnam and Indonesia: Building Blocks for Successful Just Energy Transition Partnerships. Melbourne: Climateworks Centre. Available at: https://www.climateworkscentre.org/resource/energy-transitions-in-vietnam-and-indonesia-building-blocks-for-successful-just-energy-transition-partnerships/. (accessed 1 Jan, 2026).

Ayas, C., Khuat, T. A., Giwangkara, J. and Nguyen, T. 2025. Progress on Just Energy Transitions in Vietnam and Indonesia. Melbourne: Climateworks Centre. Available at: https://www.climateworkscentre.org/resource/progress-on-just-energy-transitions-in-vietnam-and-indonesia/. (accessed 1 Jan, 2026).

Azhgaliyeva, D., Hollis, A., Pogge, T., Rahut, D. and Yao, Y. 2022. Financing a Green Future: The Energy Transition Mechanism (ETM) and the Green Impact Fund for Technology (GIFT). Global Solutions Initiative. Available at: https://www.global-solutions-initiative.org/publication/financing-a-green-future-the-energy-transition-mechanism-etm-and-the-green-impact-fund-for-technology-gift. (accessed 1 Jan, 2026).

Centre for Energy, Ecology and Development (CEED). 2023. Op-Ed: Analysis of the Vietnam JETP Resource Mobilisation Plan. Centre for Energy, Ecology and Development (CEED). Available at: https://energytracker.asia/vietnam-jetp-resource-mobilisation-plan/. (accessed 1 Jan, 2026).

Chen, X. and Mauzerall, D. L. 2021. The expanding coal power fleet in Southeast Asia: Implications for future CO_2 emissions and electricity generation. Earth's Future, 9, e2021EF002257. https://doi.org/10.1029/2021EF002257

Dalal-Clayton, B., Brown, S. et al. 2022. Accelerating the Clean Energy Transition in Southeast Asia: Regional Scoping Report for Strategic Environmental and Social Assessment Applied to the Energy Transition Mechanism in Southeast Asia. Project Number: 55124-001. Asian Development Bank. Available at: https://www.adb.org/projects/documents/reg-55124-001-tacr. (accessed 1 Jan, 2026).

Deutsche Gesellschaft für Internationale Zusammenarbeit (GIZ). 2025. Transforming Coal Regions for a Just Energy Transition—Interregional Platform. Deutsche Gesellschaft für Internationale Zusammenarbeit (GIZ). Available at: https://www.giz.de/en/worldwide/124208.html. (accessed 1 Jan, 2026).

Diluiso, F., Walk, P., Manych, N., Cerutti, N., Chipiga, V., Workman, A., Ayas, C., Cui, R. Y., Cui, D., Song, K., Banisch, L. A., Moretti, N., Callaghan, M. W., Clarke, L., Creutzig, F., Hilaire, J., Jotzo, F., Kalkuhl, M., Lamb, W. F., . . . Minx, J. C. 2021. Coal transitions – Part 1: A systematic map and review of case study learnings from regional, national, and local coal phase-out experiences. *Environmental Research Letters*, *16*(11), article 113003. https://doi.org/10.1088/1748-9326/ac1b58

Do, T. N. and Burke, P. J. 2023. Phasing out coal power in a developing country context: Insights from Vietnam. Energy Policy, 176, 113512. https://doi.org/10.1016/j.enpol.2023.113512.

Doan, D., Trang, L., Nguyen, N. and Safir, A. 2023. Green jobs: Upskilling and reskilling Vietnam's workforce for a greener economy. World Bank. Available at: https://documents1.worldbank.org/curated/en/099091823063517461/pdf/P1781121e19c8506719ff81042e6be4e00b.pdf. (accessed 1 Jan, 2026).

Energy Transition Partnership (ETP). 2025. Just Coal Transition Platform (JCTP)—Southeast Asia Energy Transition Partnership. Available at: https://www.energytransitionpartnership.org/jctp/. (accessed 1 Jan, 2026).

Fiscal Policy Agency. 2022. CIF Accelerating Coal Transition (ACT): Indonesia Country Investment Plan (IP). Jakarta: Ministry of Finance (MOF) Indonesia. Available at: https://fiskal.kemenkeu.go.id/docs/CIF-INDONESIA_ACT_IP-Proposal.pdf. (accessed 1 Jan, 2026).

Glasgow Financial Alliance for Net Zero (GFANZ). 2023. Financing the Managed Phaseout of Coal-Fired Power Plants in the Asia Pacific. Glasgow Financial Alliance for Net Zero (GFANZ). Available at: https://assets.bbhub.io/company/sites/63/2023/05/gfanz_consultation_managed-phaseout-of-coal-in-Asia-Pacific.pdf. (accessed 1 Jan, 2026).

Global Energy Monitor. 2025. Global Coal Phase Out. Available at: https://globalenergymonitor.org/projects/global-coal-plant-tracker/coal-phaseout-tracking-retirements-and-paris-aligned-goals/. (accessed 1 Jan, 2026).

Gui, E. M., Overland, I., Suryadi, B. and Yurnaidi, Z. 2024a. Bridging the implementation gap for climate mitigation in ASEAN: A comprehensive capacity-building framework. Fulbright Review of Economics and Policy, 4(2), 154–179. https://doi.org/10.1108/FREP-08-2024-0049.

Gui, E. M., Sandhu, J., Paillere, H., Calamala, B. and Gupta, R. 2024b. Navigating the Coal Transition in Asia: Challenges, Lessons, Pathways—Sustainable Development Solutions Network. New York. Available at: https://www.unsdsn.org/resources/navigating-the-coal-transition-in-asia-challenges-lessons-pathways/. (accessed 1 Jan, 2026).

Gui, E. M. and Inal, S. 2025. *Low Carbon Tipping Points. SDG Action.* https://sdg-action.org/low-carbon-tipping-points/ (accessed 1 Jan, 2026).

IEA. 2021. Net Zero by 2050: A Roadmap for the Global Energy Sector. OECD Publications, Paris. Available at: https://doi.org/10.1787/c8328405-en

Institute for Essential Services Reform (IESR). 2022. Redefining Future Jobs: Implication of Coal Phase-Out to the Employment Sector and Economic Transformation in Indonesia's Coal Region. Jakarta: Institute for Essential Services Reform (IESR). Available at: https://www.climate-transparency.org/wp-content/uploads/2022/07/Redefining-Future-Jobs-IESR-CT.pdf. (accessed 1 Jan, 2026).

Institute for Essential Services Reform (IESR). 2023. Just Transition in Indonesia's Coal Producing Regions, Case Studies: Paser and Muara Enim Regency (Summary for Policymakers). Jakarta: Institute for Essential Services Reform (IESR). Available at: https://www.sipet.org/images/document/14092023109671.pdf. (accessed 1 Jan, 2026).

International Labour Organization (ILO). 2022. Just Transition Finance Tool for Banking and Investing Activities. International Labour Organization (ILO). ISBN 9789220384916. Available at: https://www.ilo.org/publications/just-transition-finance-tool-banking-and-investing-activities. (accessed 1 Jan, 2026).

International Labour Organization (ILO). 2023. Green Jobs and Just Transition Policy Readiness Assessment in the Energy Sector in Indonesia. Partnership for Action on Green Economy. Available at: https://www.ilo.org/publications/green-jobs-and-just-transition-readiness-assessment-energy-sector-indonesia. (accessed 1 Jan, 2026).

International Labour Organization (ILO). 2025. Asia-Pacific Advances Just Energy Transition: Community of Practice Drives Coal Phase-Out and Promotes Labour and Social Solutions. International Labour Organization (ILO). Available at: https://www.ilo.org/resource/news/asia-pacific-advances-just-energy-transition-community-practice-drives-coal. (accessed 1 Jan, 2026).

IPCC. 2018. Summary for Policymakers. In: Global Warming of 1.5°C. An IPCC Special Report on the impacts of global warming of 1.5°C above pre-industrial levels and related global greenhouse gas emission pathways, in the context of strengthening the global response to the threat of climate change, sustainable development, and efforts to eradicate poverty, Masson-Delmotte, V., P. Zhai, H.-O. Pörtner, D. Roberts, J. Skea, P.R. Shukla, A. Pirani, W. Moufouma-Okia, C. Péan, R. Pidcock, S. Connors, J.B.R. Matthews, Y. Chen, X. Zhou, M.I. Gomis, E. Lonnoy, T. Maycock, M. Tignor, and T. Waterfield (eds.). Cambridge University Press, Cambridge and New York, pp. 3–24. https://doi.org/10.1017/9781009157940.001

The Jakarta Post. 2024. Government To Shut Down 13 Coal-Fired Power Plants. Available at: https://www.thejakartapost.com/business/2024/08/22/government-to-shut-down-13-coal-fired-power-plants.html. (accessed 1 Jan, 2026).

JETP Secretariat. 2023. Introduction to JETP. Available at: https://caseforsea.org/wp-content/uploads/2024/10/Session-2_EE_JETP-Intro-for-EE-training.pdf. (accessed 1 Jan, 2026).

Karyza, D. 2024. ADB Moves Back Funding Timeline for First Coal Plant Shutdown. The Jakarta Post. Available at: https://www.thejakartapost.com/business/2024/12/13/adb-moves-back-funding-timeline-for-first-coal-plant-shutdown.html. (accessed 12 May, 2025).

Klaikaew, S. C. 2023. The Just Energy Transition in Coal Regions – Interregional Platform (JET-CR Platform): Regional Workshop for Asia Pacific, Deutsche Gesellschaft für Internationale Zusammenarbeit (GIZ). United Nations Conference Centre, Bangkok, 13 September 2023. Available at: https://unfccc.int/sites/default/files/resource/5_2_GIZ.pdf

Larasati, L. K. and Fajrian, S. 2024. JETP Resource Mobilization Plan: How Viet Nam Can Turn Ambition into Action. Available at: https://www.climatepolicyinitiative.org/jetp-resource-mobilization-plan-how-viet-nam-can-turn-ambition-into-action/. (accessed 1 Jan, 2026).

Larasati, L. K. and Mafira, T. 2023. Highlights from Indonesia's JETP Comprehensive Investment and Policy Plan. Climate Policy Initiative (CPI). Available at: https://www.climatepolicyinitiative.org/highlights-from-indonesias-jetp-comprehensive-investment-and-policy-plan/. (accessed 1 Jan, 2026).

Leong, W. Y., Leong, Y. Z. and Leong, W. S. 2024. Energy management policy and strategies in ASEAN. Advances in science, Technology and Engineering Systems Journal, 9(4), 102–109.

Manych, N. and Jakob, M. 2021. Why coal?—the political economy of the electricity sector in the Philippines. Energy for Sustainable Development, 62, 113–125. https://doi.org/10.1016/j.esd.2021.03.012

Minh, D. 2023. Đẩy mạnh kết nối, tạo cơ hội việc làm cho người lao động. Quang Ninh. Available at: https://baoquangninh.vn/dam-bao-muc-tieu-tao-ra-it-nhat-20-000-lao-dong-tang-them-nam-2023-3236046.html. (accessed 1 Jan, 2026).

Nerini, F., Tomei, J., To, L. S., Bisaga, I., Parikh, P., Black, M., Borrion, A., Spataru, C., Broto, V. C., Anandarajah, G., Milligan, B. and Mulugetta, Y. 2018. Mapping synergies and trade-offs between energy and the sustainable development goals. Nature Energy, 3, 10–15. https://doi.org/10.1038/s41560-017-0036-5.

Nguyen, T. M. A. and Ngo, T. T. N. 2022. Just Transitions and Coal Power Phase Out in Vietnam. Climate Strategies. Available at: https://climatestrategies.org/wp-content/uploads/2023/03/Just-Transitions-and-Coal-Power-Phase-Out-in-Vietnam_v2.pdf. (accessed 1 Jan, 2026).

Paredes, J. M., Lee, W.-J., Gui, E. M., Barros, L., Zerouali, A. and La Marca, C. 2023. Best Practices for Variable Renewable Energy (VRE) Integration. Sustainable Development Solutions Network (SDSN), New York, NY, USA, 20 p. Available at: https://www.unsdsn.org/resources/best-practices-for-variable-renewable-energy-vre-integration. (accessed 1 Jan, 2026).

Powering Past Coal Alliance. 2025. The End of Coal Is in Sight—PPCA. Available at: https://poweringpastcoal.org/. (accessed 1 Jan, 2026).

PT Sarana Multi Infrastruktur (SMI). 2022. Indonesia Launches ETM Country Platform to Accelerate Just and Affordable Energy Transition. Available at: https://ptsmi.co.id/indonesia-launches-etm-country-platform-to-accelerate-just-and-affordable-energy-transition. (accessed 1 Jan, 2026).

Ragin, C. C. 2000. Fuzzy-Set Social Science. University of Chicago Press.

Ragin, C. C. 1987. The Comparative Method: Moving Beyond Qualitative and Quantitative Strategies. University of California Press. Available at: http://www.jstor.org/stable/10.1525/j.ctt1pnx57

Ratner, B. D., Larson, A. M., Sarmiento Barletti, J., ElDidi, H., Catacutan, D., Flintan, F., Suhardiman, D., Falk, T. and Meinzen Dick, R. 2022. Multistakeholder platforms for natural resource governance: Lessons from eight landscape-level cases. Ecology and Society, 27 (2), 2. https://doi.org/10.5751/ES-13168-270202.

Sahid, E. J. M., Sin, T. C. and Hock, G. C. 2018. Energy security in ASEAN region: Its challenges. IOP Conference Series: Earth and Environmental Science, 268. https://doi.org/10.1088/1755-1315/268/1/012168

Satriastanti, F. E. 2025. Is Indonesia's Pledge to End Coal Use by 2040 Realistic? Dialogue Earth. Available at: https://dialogue.earth/en/energy/is-indonesias-pledge-to-end-coal-use-by-2040-achievable/. (accessed 1 Jan, 2026).

Setiawan, D. 2025. Captive Coal Expansion Plan Could Undermine Indonesia's Climate Goals. EMBER. Available at: https://ember-energy.org/app/uploads/2025/02/EN-Indonesia-RUKN-2025_14022025.pdf. (accessed 1 Jan, 2026).

Setyawati, D. and Setiawan, D. 2024. Indonesia's Expansion of Clean Power Can Spur Growth and Equality. Ember. Available at: https://ember-climate.org/insights/research/indonesias-expansion-of-clean-power-can-spur-growth-and-equality/#supporting-material. (accessed 1 Jan, 2026).

Silviana, S., Trihastuti, N., Darsanto, S. Q. G. 2025. Early Retirement Policy of Power Plant In Pelabuhan Ratu: A Multi-Level Governance Analysis To Support The Energy Transition In Indonesia. *Asian Journal of Social and Humanities*, 3 (5). https://doi.org/10.59888/ajosh.v3i5.503.

Simister, N. and Scholz, V. 2017. Qualitative Comparative Analysis (QCA). Intrac. Available at: https://www.intrac.org/app/uploads/2024/12/Qualitative-comparative-analysis.pdf. (accessed 1 Jan, 2026).

Socialist Republic of Vietnam. 2021. National Green Growth Strategy for 2021–2030, Vision towards 2050. Available at: https://en.baochinhphu.vn/national-green-growth-strategy-for-2021-2030-vision-towards-2050-11142515.htm. (accessed 1 Jan, 2026).

Winkler, H., Tyler, E., Keen, S. and Marquard, A. 2023. Just transition transaction in South Africa: An innovative way to finance accelerated phase out of coal and fund social justice. Journal of Sustainable Finance & Investment, 13(3), 1228–1251. https://doi.org/10.1080/20430795.2021.1972678.

World Bank. 2021. Supporting Transition in Coal Regions: A Compendium of the World Bank's Experience and Guidance for Preparing and Managing Future Transitions. © World Bank. Available at: https://doi.org/10.1596/35323

Appendix

TABLE 12.3
An overview of ETM, JETP and GFANZ initiatives (ADB, 2021; Larasati and Fajrian, 2024; GFANZ, 2023)

Key Aspects	ETM	JETP	GFANZ
Approach	A market-based programme tailored to the unique needs and circumstances of each nation	A partner country-led model	Adoption of a set of principles, aligning with individual company strategy
Key objectives	• Expedite the retirement or repurposing of CFPPs and • Replace them with clean energy alternatives	• Early retirement of on-grid CFPPs (Indonesia) and • Cease the issuance of new CFPPs (Vietnam)	Committed to accelerating the decarbonisation of the economy and embedding net-zero objectives in the core evaluation and decision-making framework
Decision-making mechanism	Primary decision-making rights by ADB and other financial partners, with participation from in-country stakeholders	Negotiation between partner countries and IPG members in JETP in planning and implementation	Syndicate or individual company-based decision-making, aligning with GFANZ principles
Coverage of activities	• Early retirement and repurpose CFPPs • GHG reduction from CFPPs by early retirement them early • GHG reduction from CFPPs by repurposing them for clean energy production • Enhanced grid capacity • Policy and regulatory strategies for coal to clean transition • Just transition	• Early retirement and repurpose CFPPs • Increase investment in renewable energy and energy storage • Improve the power generation technology • Expansion and modernisation of the transmission and distribution grids and storage infrastructure • Increase energy efficiency • Enabling environment, including policy framework, innovation, technology transfer • Just transition	• Phase-out coal from its portfolio • Prohibit asset-level or general corporate financing for new thermal coal mines and CFPPs • Set out the expectation for electric utilities to diversify and reduce the share of coal in the power mix
Financing	• Provides concessional loan and commercial capital to country-specific funds • A public–private finance vehicle to be set up comprising (1) the Carbon Reduction Facility (CRF) and decommission CFPPs on an accelerated schedule; (2) the Clean Energy Facility (CEF) to catalyse and channel investment in new renewable energy generation, energy storage infrastructure and requisite grid and storage upgrades	Country-specific fund management, catalysing private capital using grants and loans with below-market interest rates, with 50% of the capital expected to come from private finance	GFANZ's approach has three levers[4]: - Asset valuation and pricing: factoring material climate risks such as stranded assets into valuation - Cost of capital: lowering capital costs to maintain an acceptable risk-adjusted return to investors through blended project finance of concessional sources and commercial capital, selecting the right mix of debt and equity - Exploring additional revenue streams: carbon credits for emissions avoided to make up for loss revenue from the sale of coal-generated electricity

(Continued)

TABLE 12.3 (*Continued*)
An overview of ETM, JETP and GFANZ initiatives (ADB, 2021; Larasati and Fajrian, 2024; GFANZ, 2023)

Key Aspects	ETM	JETP	GFANZ
Governance	• Southeast Asia ETM Partnership Platform to incorporate the considerations of government authorities, multilateral banks, investors, asset owners and utilities and to support the development financing and operation across the three pilot countries in Southeast Asia • A country-specific entity through the ETM country platform, such as the special mission vehicle of the Ministry of Finance (MOF) of Indonesia for managing various energy transition activities	A cross-ministerial coordination body in-country is in place to ensure coherence between public entities overseeing the legal, financial, policy-making and regulatory aspects of renewables development, coal phase-out and fiscal allocation	• Work with local government authorities to facilitate a just transition process and provide new economic opportunities for affected communities • The governance approach may differ for each project and is organised by each leading financial institution
Monitoring	Monitored through a set of indicators by third parties, including • the number of coal-fired power plants retired or repurposed • the amount of new renewable energy capacity installed • the reduction in greenhouse gas emissions	• Each partner country identifies milestones in the resource mobilisation plan and investment and policy plan, which will be monitored and evaluated annually and biannually • Scrutinise how the funds are channelled and how the JETP Secretariat is governed to monitor the progress	The consistent framework to monitor the metrics and targets is established to drive the execution of the net-zero transition plan that focuses on: • aligning financial activity that supports the net-zero transition at a whole-of-economy level • executing the transition plans • monitoring the changes within the client and portfolio GHG emissions
Social protection and gender equity	Safeguards and just transition principles to mitigate environmental and socio-economic impacts and ensure an inclusive process for sharing the benefits • conduct environmental and social impact assessments • engage with local communities and other stakeholders to identify and mitigate potential impacts, • invest in projects that create jobs and opportunities for workers in the fossil fuel sector	Minimise social and economic impacts while ensuring just, equitable and inclusive outcomes for consumers, workers and community groups with a specific focus on vulnerable communities in coal-dependent regions	• Proactively engage with excluded and marginalised populations, including low-income communities and women • Some organisations are working to develop guidelines on how climate action can address just transition concerns and elevate the voices of impacted populations, for example, how gender-related considerations in climate planning can be better integrated

13 Governance in Climate and Sustainable Development Goals (SDGs) in Gulf Council Countries (GCC)—An Overview

Gowhar Meraj

13.1 INTRODUCTION

Governance is a key driver in climate action and Sustainable Development Goals (SDGs). This is more so in the Gulf Cooperation Council (GCC) countries, namely, Bahrain, Kuwait, Oman, Qatar, Saudi Arabia, and the United Arab Emirates (UAE), which are unique in terms of their sustainable development challenges and opportunities. These hydrocarbon-rich countries have traditionally relied on oil to sustain their economies, resulting in high per capita carbon emissions and ecological footprints.[1] Nevertheless, they are also at high risk of climate change, including extreme heat, water scarcity, and coastal risks, which further underscores the importance of climate action for sustainable development (Caro et al., 2017; Varela et al., 2020; Robaa and El Kenawy, 2024). Over the past few years, GCC governments have come to appreciate the fact that good governance is the key to balancing economic growth and environmental management. The six countries are signatories to international regimes such as the Paris Agreement on climate change and the UN 2030 Agenda for Sustainable Development. Following a phase of doubt about international climate initiatives, GCC states are now embracing the sustainability agenda on a global scale by endorsing the SDGs and integrating the goals within their national development plans (Burt et al., 2017; Abulibdeh et al., 2019; Al-Saidi and Saliba, 2019). They have developed new institutions, policies, and partnerships to ensure that climate action and sustainable development are incorporated into their governance systems. The chapter discusses the roles of governance arrangements, policy frameworks, and coordination mechanisms in the GCC in supporting climate action and SDG implementation, including both regional and standout national initiatives. The analysis encompasses institutional structures, policy coordination, regulatory tools, and mechanisms for public and stakeholder engagement across the GCC to define best practices and areas for improvement. Finally, improving governance is key to enabling GCC countries to meet their climate goals and deliver on the SDGs, leveraging their resources and capacities to shift toward a more sustainable, low-carbon economy.

13.2 SUSTAINABLE DEVELOPMENT AND CLIMATE ISSUES IN THE GCC SETTING

The GCC countries share a common background of rapid growth driven by oil and gas, which makes their challenges in sustainable development unique. On the one hand, they have high human development indicators and income levels; on the other hand, these benefits have been accompanied

DOI: 10.1201/9781003600152-13

by environmental costs and resource-intensive lifestyles. GCC states have some of the highest per-capita CO_2 emissions and ecological footprints; Qatar and the UAE are among the highest in per-capita emissions and materials consumption (Dargin, 2021). These countries also exceed multiple planetary boundaries, meaning that they place pressures beyond key biophysical limits, while still falling short of several social-development thresholds, including equality and civic participation[2] (Schuetze, 2024). In other words, GCC countries have established high living standards by compromising their ecological sustainability, as is the case with many advanced global economies. This unsustainable trend poses a two-fold challenge: reducing the environmental footprint, particularly greenhouse gas emissions, while maintaining or even increasing the industry's social and economic benefits.

The direct threat to GCC development aspirations is climate change. The region is already experiencing extreme heat events, chronic water stress, increased drought, and flash floods, which are expected to worsen under current global warming trends (Shapland, 2023). Scientific evaluations caution that some areas of the Gulf would be near the margins of human tolerance to heat and humidity by the end of the century in business-as-usual conditions (Pal and Eltahir, 2016). Water is in short supply, and the arid climate of the Arabian Peninsula means that groundwater can be depleted and is hard to replenish, posing a threat to the water security of both people and industries (Odhiambo, 2017). Rising sea levels and biodiversity loss, including coral reef mortality in the Gulf, pose additional threats to coastal urban centers and ocean life (Chefaoui et al., 2021). Such climate effects pose a risk to a range of SDGs, including food and water security (SDGs 2 and 6), health (SDG 3), and sustainable cities (SDG 11), reiterating that climate action (SDG 13) is part and parcel of the SDG Agenda.[3] Thus, good governance must consider climate risks alongside economic diversification and societal growth strategies.

Nevertheless, GCC countries are in a more favorable position than most developing countries to address and respond to climate change due to their wealth and institutional strength. Their adaptive capabilities, including infrastructure and disaster response, are much better than those of poorer regions, including sub-Saharan Africa (Saidin and O'Neill, 2022). For example, the GCC states can afford to invest in desalination, air conditioning, and modern infrastructure to help guard against some of the climate stresses. But wealth is not a silver bullet; the governance decisions determine how well resources are deployed to resilience and sustainability. More importantly, the whole state's heritage in the Gulf poses a governance quandary: how to preserve lavish welfare and high economic growth rates with the shift to a low-carbon, sustainable state? The high rates of population growth and urbanization in the GCC have increased domestic energy and water demand, putting pressure on the conventional subsidy system and fossil-based systems (Al-Saidi, 2022a).

It has become necessary for GCC governments to diversify their economies and make them sustainable. The instability of oil prices and the exhaustibility of fossil fuel resources make the existing economic models unsustainable in the long term. In fact, the GCC states have been dreaming of diversifying beyond oil for a long time, but this is only now happening (El-Kharouf et al., 2010; Hvidt, 2013; Albassam, 2015; Al-Saidi and Elagib, 2018). Faced with financial constraints and the global trend toward clean energy, they are increasingly looking to invest in green sectors as a win–win strategy for economic renewal. The shift to sustainable energy and green technologies will assist in dismantling the large share of GCC economies that are bound to oil and gas and establish new industries. This understanding, in combination with international pressure and technological change, is one of the factors that have led the Gulf to accept climate action and the SDGs. The GCC has faced increasing international criticism of its environmental record over the past few decades, prompting it to enhance its global image by pledging to address climate change. At the same time, clean technology has become cheaper and more widely available. Solar energy, especially, is now very cost-effective in the Gulf, with world-record-low costs, for example, as low as \$0.01 per kWh for solar PV, due to favorable solar resources and economies of scale (Bellini, 2021). These opportunities allow GCC policymakers to pursue environmental projects without compromising development; many talk about ecological modernization and the use of eco-innovations, such as green buildings and electric transport, to maintain a modern image

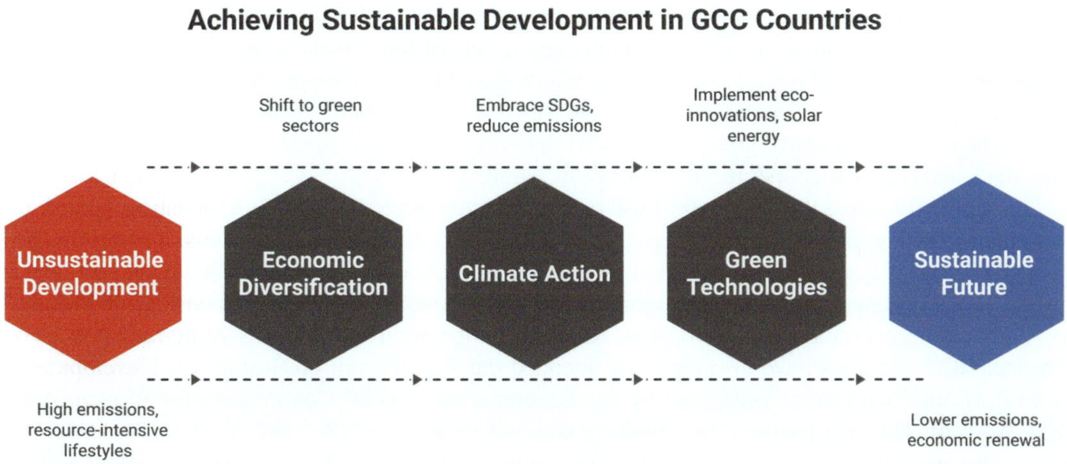

FIGURE 13.1 Governance pathway for GCC sustainability: moving from resource-intensive growth to a low-carbon "sustainable future" via economic diversification, climate action, and green technologies.

Source: Author's synthesis, adapted from literature sources discussed in this chapter.

and attract investment (Al-Saidi and Elagib, 2018). Overall, GCC states are navigating a staged transition, from resource-intensive growth to low-carbon development, through diversification, climate action, and green technologies (Figure 13.1).

13.3 GOVERNANCE FRAMEWORKS OF SUSTAINABLE DEVELOPMENT IN GCC

In GCC countries, institutional frameworks to guide sustainable development and climate action have been put in place, with long-term national visions often serving as a guiding force. These vision documents, which include Saudi Arabia Vision 2030,[4] the UAE Centennial 2071,[5] Oman Vision 2040,[6] New Kuwait 2035,[7] Qatar National Vision 2030,[8] and the Bahrain Economic Vision 2030,[9] have very broad development priorities that closely match the SDGs. In many instances, the SDGs were directly aligned with national visions and development plans, meaning the global goals were translated into domestic action plans. For example, Kuwait has a Vision 2035 that is fully aligned with and structurally embedded in all 17 SDGs and 169 targets, anchoring the SDGs framework within its seven development pillars.[10] A comparable level of alignment can be seen in Bahrain, where the national development program was evaluated to cover about 78 percent of SDG targets.[11] This integration is an indicator of political will among the leaders of the GCC: they see the 2030 Agenda as a convenient framework to support their national reforms and international collaboration. GCC states have already adopted the SDG agenda, including all its reporting and monitoring tools, and have attempted to integrate this agenda with their national visions, effectively mainstreaming sustainable development into their policy planning. There have been institutional mechanisms established or restructured to harmonize the implementation of the SDGs within the various wings of government. The majority of GCC governments have established senior committees and commissions to monitor the 2030 Agenda. For example, the UAE established a National Committee on the SDGs in 2017, tasked with ensuring their implementation in the country. The committee is chaired by the Minister of State for International Cooperation and serves as a multistakeholder platform comprising 15 federal entities. The Federal Competitiveness and Statistics Center of the UAE is the Secretariat of the Committee, responsible for tracking progress and reporting on SDG indicators annually. This central node provides integration of data collection and policy work and coordination of individual efforts of ministries with the objectives of the national adaptation plan (NAP).[12] The

Committee has embraced a whole-of-society approach where it has been able to facilitate cross-sector collaborations and consultations to integrate inputs of the private sector, youth, non-governmental organizations (NGOs), academia, and international partners into the policy-making process. This wide participation is intended to make the SDGs owned by more than just the government, as part of the ideals of inclusive governance. SDG delivery in the GCC hinges on mandates, coordination, data, and regional cooperation, as shown in Figure 13.2.

One similarity across the GCC is the push to connect international objectives with national development plans and to establish special governance units to monitor progress. The success of such arrangements is mixed, yet some common good practices can be identified: senior political support, clear institutional mandates, inter-ministry coordination, integration of data and monitoring, and efforts to involve non-governmental stakeholders. Regional cooperation mechanisms further support governance: GCC countries have also regularly attended the Arab Forum on Sustainable Development, as well as other workshops organized by the Economic and Social Commission for Western Asia (ESCWA), to share experiences on voluntary national reviews (VNRs) and SDG implementation. There have even been suggestions of a voluntary regional review of the Arab region, similar to the EU's joint SDG report, to take stock of regional progress.[13] This is indicative of an increasing realization that GCC and Arab states can borrow governance innovations and coordinate on transboundary issues. In fact, researchers point out that regional environmental cooperation plays a significant supporting role in the GCC in increasing commitments and addressing transnational issues (such as climate change,

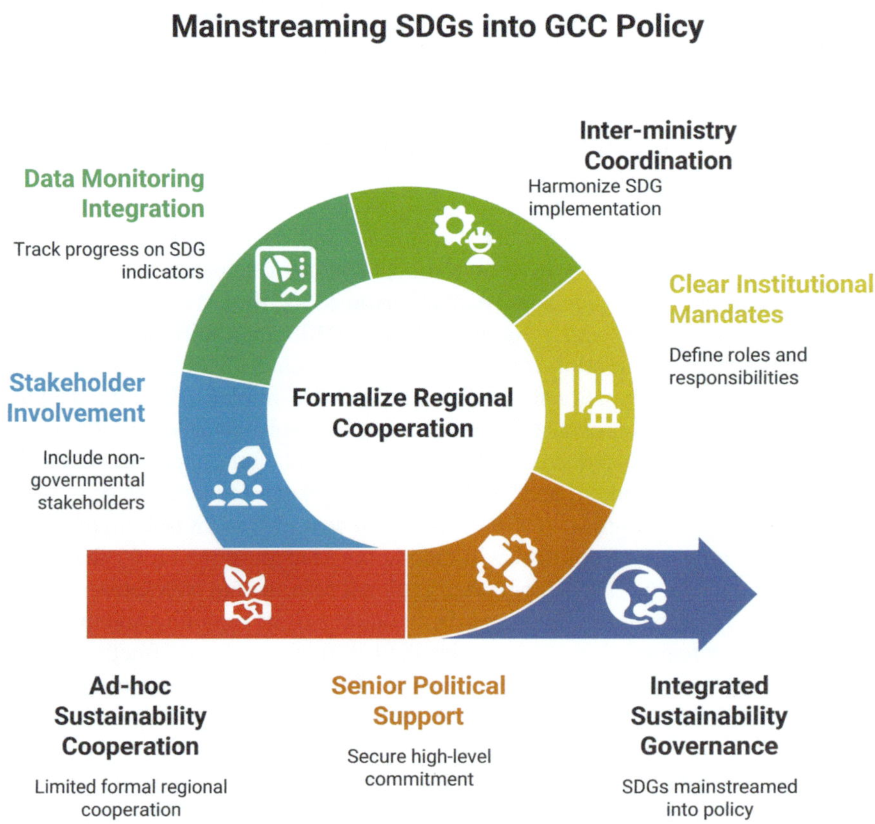

FIGURE 13.2 Mainstreaming SDGs in the GCC: from ad hoc cooperation to integrated sustainability governance via senior political support, clear institutional mandates, inter-ministerial coordination, stakeholder involvement, data-monitoring integration, and formalized regional cooperation.

Source: Author's synthesis, adapted from literature sources discussed in this chapter.

marine pollution, and sandstorms), which affect all countries. So far, much of the GCC cooperation on sustainability has been ad hoc (exchange of knowledge, common declarations) or project-oriented (such as the Gulf electricity grid). Still, the infrastructure is in place to formalize cooperation (Al-Saidi, 2022a). All in all, the GCC is rapidly changing in terms of the governance frameworks, as presented by the twin pressures of international agendas and local reforms. The following sections explore these frameworks in greater detail, as applied to climate action and policy coordination, and how both the public and the private are being integrated into the process.

13.4 NATIONAL POLICIES AND STRATEGIES TO ADDRESS CLIMATE

Climate governance in the GCC has been gaining momentum over the last five years, with the adoption of national climate strategies and legal reforms, as well as international pledges that would have been unheard of a decade ago. All six GCC countries have ratified the Paris Agreement and submitted Nationally Determined Contributions (NDCs) as part of their climate commitments. The ambition of these NDCs has, in the beginning, been lower than the global averages, but the more recent updates are more ambitious. The most recent trend is the trend of net-zero emissions commitments by GCC states, which provide long-term directional objectives and inform policy frameworks. The UAE was the first Gulf country to announce a net-zero target (by 2050), followed by Oman (net zero by 2050 as well), Saudi Arabia and Bahrain (net zero by 2060).[14] In 2022, Kuwait followed this trend with a net-zero 2050 commitment in terms of CO_2 and 2060 in terms of other greenhouse gases, which is indicative of a convergence of long-term climate ambition across the region (Koul, 2024). Such pledges can be substantial signals of governance, they create a uniting mission within government agencies and across economic sectors, leading to the development of specific plans and accountability frameworks to achieve the objective. As an example, following the announcement of the UAE to achieve net zero by 2050, the government established the UAE Net Zero by 2050 Strategic Initiative and is formulating a federal climate law to incorporate climate goals into the law (Ahmed, 2023). Likewise, Saudi Arabia has its 2060 commitment, which is being driven by the Circular Carbon Economy strategy as part of the Saudi Green Initiative (SGI), with new policies and institutions to implement carbon-reducing and carbon-removing projects.[15] The net-zero commitments also drive inter-ministerial collaboration, for example, within energy, environment, and finance, as energy ministries that have traditionally been focused on oil and gas are now mandated with expanding renewables and hydrogen, and the environment ministries are scaled up on conservation and carbon sink programs, as well as finance ministries are exploring green budgeting and investment frameworks. Net-zero commitments across the GCC now set mid-century direction for policy and investment (Figure 13.3).

Climate strategies in the GCC region center on renewable energy expansion, and governance has been modified to support the transition. Quantitative targets of clean energy in the national plans have been established in all GCC states. Saudi Arabia seeks to generate half of its power supply with renewable sources by 2030 (Zabelin, 2024), a big jump considering the country had a near-zero level of renewable energy generation in the recent past. Oman also aims at 30 percent electricity generation through renewables by 2030 (Leal-Arcas et al., 2025). The UAE Energy Strategy 2050 aims to have half of the energy mix clean by 2050, and the country already has three of the largest solar plants in the world, with record-low solar tariffs being reached by conducting competitive auctions (Hummieda et al., 2023). Kuwait, in its turn, has targeted 15 percent of renewables in the power mix by 2030 and has initiated the so-called Shagaya complex of renewable energy (Zumbraegel, 2022). To achieve such targets, a set of strong governance measures is necessary: regulatory changes that allow the entry of independent power producers, subsidies, or incentives to renewables, grid infrastructure improvements, and procurement plans. GCC governments have established special units or companies to promote renewable projects, for example, Masdar in the UAE, ACWA Power in Saudi Arabia, and a Renewable Energy Development Office under the Saudi Ministry of Energy. Competitive tenders have been successful policy instruments, the auction

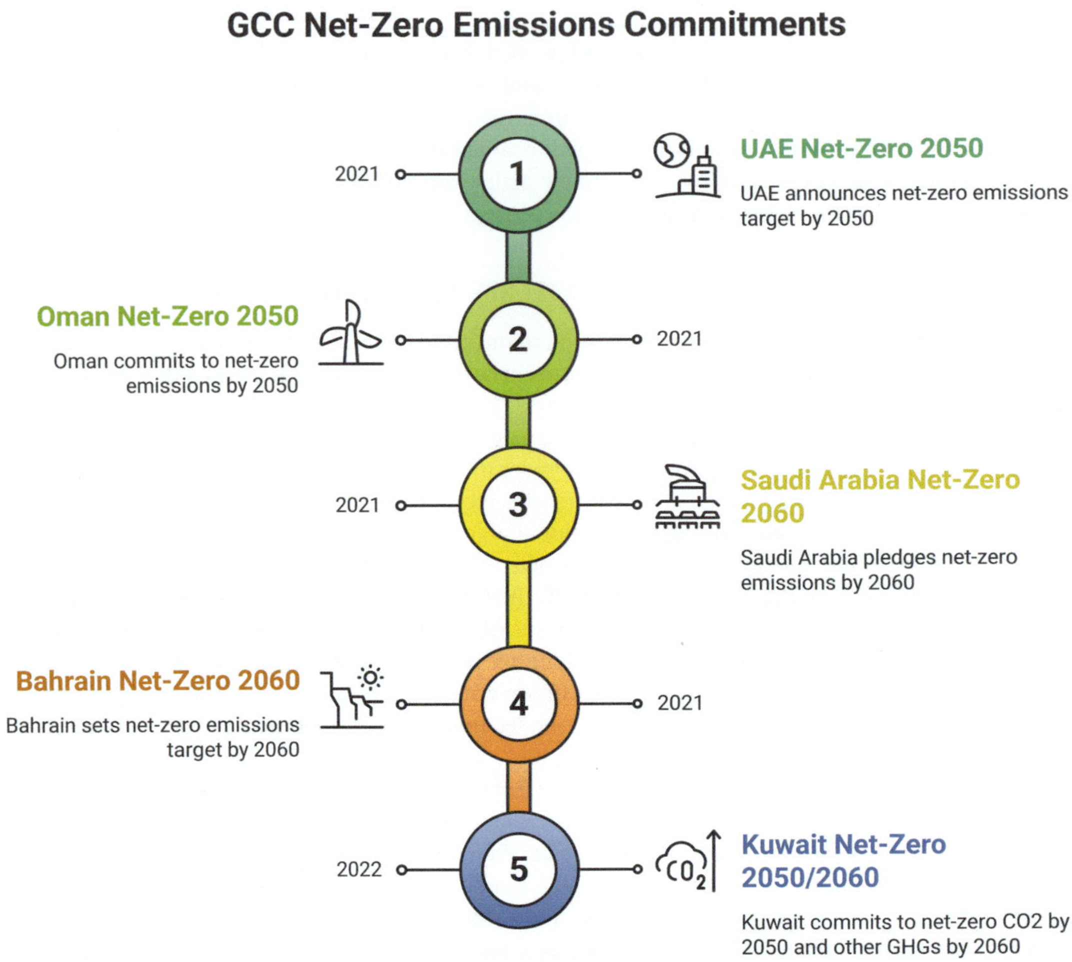

FIGURE 13.3 Timeline of GCC net-zero pledges: UAE (2050), Oman (2050), Saudi Arabia (2060), Bahrain (2060), and Kuwait (CO_2 by 2050; all GHGs by 2060). Qatar has not announced a national net-zero year.

Source: Author's synthesis, adapted from literature sources discussed in this chapter.

format has brought international developers into the region and helped reduce the costs of solar and wind power. Also, sovereign wealth funds are investing in clean energy both locally and internationally to integrate climate goals into economic governance. The example is the Mubadala and the Saudi Public Investment Fund, which are spending significant capital to invest in renewable energy and green technologies as part of their diversification plans (McPherson-Smith, 2021).

In addition to power generation, GCC climate governance encompasses energy efficiency, emissions control, and adaptation initiatives. A key area of concern has been the reform of fossil fuel subsidies, which has long been a politically difficult issue but vital to climate and fiscal sustainability. Most GCC countries have, in the last decade, attempted to roll back or rationalize subsidies on gasoline, electricity, and water to minimize wasteful consumption and cut budgetary costs. The UAE and Oman have already introduced changes to fuel prices based on market prices; Saudi Arabia has gradually increased domestic energy prices and introduced efficiency standards. Also, in 2016, Kuwait passed a law to raise electricity and water prices to promote the idea of conservation and relieve the pressure on state budgets.[16] The reforms were major governance changes because they entail a need to balance out the acceptance of the population with the long-term policy objectives. The gradual phase-out, communication campaigns to the population, and compensations of the households with low income have been used as a part of the governance toolkit to ensure that the reform of the subsidies does not have a serious social backlash. The outcome has been a deceleration of energy demand growth in the country and the reallocation of funds that can be used in sustainable investments (Al-Saidi, 2022b). In fact, research has observed that the saved revenues on energy subsidy cuts can be diverted to renewables development or social welfare schemes, which create a positive cycle and help in achieving SDGs (Abdel Gelil et al., 2017).

GCC countries are also venturing into new international climate regimes, which is also a shift in their external governance stance. It is interesting that many have signed the Global Methane Pledge to reduce the level of methane emissions by 30 percent by 2030, considering the large oil and gas activity in the region (Dargin, 2021). This will involve new laws and technologies to minimize flaring and leakages in the hydrocarbon sector. Also, carbon markets and offsets are gaining more attention: Qatar is home to the Global Carbon Council, a voluntary carbon credit scheme that issues credits against emissions reduction initiatives,[17] and Saudi Arabia has developed a regional voluntary carbon market, which held its first carbon credit auction in 2022 (Hanieh, 2024). Market-based governance tools such as these are intended to attract private investment in emissions-reduction projects and to show that GCC economies are gearing up to live in a carbon-constrained world. To complement these initiatives, institutional capacity is being strengthened, for example, new climate change departments in ministries, technical committees to work on greenhouse gas inventories and measurement systems, and partnerships with international organizations to transfer knowledge.

The other pillar of climate action strategies is adaptation and resilience-building, given the inevitability of certain climate impacts. In this case, some of the governance measures include the formulation of NAPs, the incorporation of climate risks into urban planning, infrastructure design, and disaster management. For example, GCC governments are investing in more efficient desalination technologies and water reuse programs in response to the recognition that water scarcity is a major problem. Bahrain and Oman have been working on coastal zone management to address the risks of sea level rise and storm surges, reinforcing mangrove planting and coastal defenses (Al-Afifi, 2018; Abbas, 2021). GCC states are also increasing green areas in urban environments, in part to address the urban heat island effect and dust, such as through large-scale tree-planting initiatives. The SGI of Saudi Arabia has a target of planting 10 billion trees in the country in the coming decades and 40–50 billion in the wider Middle East as part of the Middle East Green Initiative (Abdelraouf and Nagasawa, 2023). The initiative, in addition to mitigating carbon sequestration and desertification (contributing to SDG 15 on land degradation), also necessitates a governance structure that cuts across ministries such as environment, agriculture, and water, and that engages countries of the region to enable afforestation across regional boundaries. In the UAE, the government has already required risk assessments of climate risks to key sectors and is considering insurance and early warning systems to support resilience; for example, Abu Dhabi has a Climate Risk Management framework that guides local entities (Abidin et al., 2020). Another example is Qatar, which has invested in research and development of climate-smart agriculture, such as hydroponics and salt-tolerant crops, to ensure food security under climate stress (Karanisa et al., 2021).

An interesting feature of GCC climate governance is the incorporation of climate goals into the fundamental economic and industrial planning, frequently in the form of green growth or low-carbon development. According to the World Bank, green growth is rapidly becoming a national policy platform in the GCC across key sectors such as energy, water, food, urban planning, and transportation (World Bank, 2021). Instead of viewing climate action as an environmental problem, the governments of the Gulf countries treat it as a chance to diversify their economies and modernize. The example is a promotion of clean industries such as renewable energy, hydrogen fuel, carbon capture, utilization, and storage (CCUS), and electric mobility, which is perceived as the path to new skilled jobs and economic productivity gains. The International Finance Corporation also estimates that by 2030, trillions of dollars in investment opportunities and millions of jobs will be created as low-carbon investments are embraced across emerging markets.[18] The GCC leaders are interested in capturing a portion of this, and, therefore, governance has changed to allow the participation of the private sector in green sectors. Even state-owned oil companies are shifting gears: the Abu Dhabi National Oil Company aims to sequester 10 million tons of CO_2 per year through CCUS and is pouring significant money into renewables through its subsidiaries.[19] Saudi Aramco is also investing in carbon capture and clean fuels development, and even green hydrogen, generated by renewable energy, is being considered. The planned NEOM city will host one of the largest green hydrogen plants in the world (Schimmel et al., 2022). These actions are supported by new governance mechanisms, such as hydrogen roadmaps, standards governing hydrogen production, and regional collaborations, such as the UAE and Oman, which have recently announced national hydrogen strategies and are working on regulatory regimes to certify green hydrogen for trade.

In the meantime, the inclusion of climate in economic ambitions, such as through the Circular Carbon Economy (CCE), which Saudi Arabia has promoted to highlight the reduction, reuse, recycle, and remove carbon in a comprehensive manner, offers a narrative that fits the ambitions of the region to retain energy leadership as it switches. The concept of governance here is more about trade-offs: minimizing emissions without compromising energy security or economic growth. The initial indications are that as long as there is effective leadership and coordination, GCC countries can, in fact, embark on the process of decoupling growth and emissions. A lot will, however, be dependent on the implementation capacity and cross-sector collaboration, which we explore next in the context of policy coordination and stakeholder engagement.

13.5 POLICY COORDINATION AND MULTI-LEVEL GOVERNANCE

To achieve climate and SDG objectives, it is necessary to get out of old silos and coordinate not only horizontally, such as between sectors and ministries, but also vertically, such as between levels of the government. In the centralized systems of governance found in the GCC, horizontal coordination is the key; most of the GCC states have unitary governments with limited local powers, except for the federal system of the UAE and municipal functions in other countries. Therefore, governance reforms are centered on national-level policy integration across sectors.

Inter-ministerial councils or task forces on climate change and sustainable development have been one of the mechanisms of horizontal coordination. For example, the UAE government established a Council on Climate Change and Environment to bring together representatives from relevant ministries, such as energy, infrastructure, and economy, as well as the seven emirates, to discuss and synchronize climate policies (Alkhaldi et al., 2023). This council is chaired by the Minister of Climate Change and Environment and acts as a coordination center for the pursuance of the UAE National Climate Change Plan and green growth initiatives. In Saudi Arabia, strategic committees under SGI are able to effectively coordinate since the various ministries sit at the table under an agenda. In addition, the Vision 2030 is executed in the form of Vision Realization Programs (VRPs), some of which are cross-ministerial, as is the case with the National Transformation Program and the Environmental Sustainability Program, which have shared objectives with the ministries (Mohiuddin et al., 2023). These VRPs are managed by committees that comprise all concerned ministers and therefore institutionalize inter-sector policy coherence. The unified platform created by Oman through the SDG committee, which represents the public, the private, and the local, also helps overcome policy overlaps or gaps. The benefit of such high-level coordination mechanisms is that they can address conflicts, such as those between a national utility's plans and environmental policy, and that climate/SDG considerations are integrated into sectoral plans, such as energy, transport, industry, and finance.

The other important element is the incorporation of climate and SDG targets in core economic decision-making processes, including the national budget and approval of development projects. In this case, there are some GCC countries that are on track to implement sustainability characteristics in their public investment planning. The UAE and Saudi Arabia have already begun experimenting with green budgeting, whereby climate-related spending in the national budget is tagged and tracked to make sure it is consistent with climate objectives (Almheiri, 2015; Abdelraouf and

Nagasawa, 2023). Also, mega infrastructure projects, a hallmark of GCC development, are increasingly being screened in terms of environmental impact assessments and sustainability. In another example, Saudi Public Investment Fund initiatives, such as NEOM and Red Sea Development, have sustainability requirements embedded into their design (Touri, 2024), as they are required to be approved by cross-ministerial teams, such as tourism, energy, and environment ministries. In Dubai, the government has Green Building Regulations and the 2040 Urban Master Plan to coordinate the efforts of the municipality, electricity and water authority, and environmental department to meet the sustainable infrastructure goals of the city (Saradara et al., 2023). These cases demonstrate that coordination involves not only committees but also the institutionalization of sustainability in the major operating procedures of governance, whether in planning commissions or financial regulators.

Vertical coordination, that is, ensuring that national goals are attained through action at subnational or local levels, is less evident in the GCC due to the centralized context, but it is starting to become a point of interest. In the UAE, the central government collaborates with individual emirates, in particular, Abu Dhabi and Dubai, which enjoy a large amount of independence and resources. The emirates also have local sustainability plans that integrate with the federal ones. For example, Dubai has the Clean Energy Strategy to achieve 75 percent clean energy by 2050, and this aligns with the overall UAE plan (Alnaqbi and Alami, 2023). Federal bodies give guidelines and frameworks, though many projects, such as public transport and waste management, are implemented locally and are often innovative; for example, Dubai Roads & Transport Authority or Abu Dhabi Department of Energy coordinate with the federal level related to EV policies or renewable targets. To coordinate the activities, national SDG committees usually have representatives of local authorities; the UAE had representatives of the entities of Abu Dhabi and Dubai to ensure the two-way alignment.

In Oman, where governance is restricted locally, vertical coordination is more about inclusion of governorates or provinces in national plans, as in the case of Oman in the push for data at the governorate level (Al-Sarihi and Mason, 2020). The National Information Committee (NIC) of Bahrain employed a multi-sectoral task force that may consist of local stakeholders to deliver SDGs under the Government Action Plan.[20] Nevertheless, there are still obstacles; one of the criticisms of the initial SDG work in Kuwait was the absence of locals, such as governorate participation, meaning that municipal institutions or regional authorities did not participate enough. Such a governance gap may impede localized solutions such as city-specific climate adaptation, for example, flood management in Kuwait City, as locals are not included in the planning.[21] In recent developments, ESCWA is encouraging GCC states to enhance vertical integration, perhaps through the creation of SDG focal points at a municipal level or the drafting of local sustainable development plans that are fed into the national strategy.

The coordination of GCC countries on climate and SDGs is a developing sphere. The GCC has in the past been relatively monolithic in its approach to international climate negotiations, often presenting a common front based on shared interests as oil exporters, for example, insisting on equity, seeking to include carbon capture as a viable option. This is followed up in the Arab Group in UN climate negotiations, where GCC members tend to speak with one voice. But regional collaboration is also moving toward practical undertakings. The Middle East Green Initiative (MGI), initiated by Saudi Arabia, is a good example of collective action: it brings together leaders of states of the GCC and neighboring countries to coordinate regional climate initiatives, including a project to plant billions of trees in the Middle East and establish a regional carbon credit market (Abdelraouf and Nagasawa, 2023). At the 2023 MGI Summit, Gulf leaders talked collectively about the efforts to decrease regional emissions, such as collaborating on an interconnected grid that can transport clean electricity across borders, as well as share research on climate resilience.[22] Also, the GCC Secretariat has reaffirmed interest in environmental collaboration; environment ministers of GCC nations hold regular meetings within the GCC framework to discuss concerns such as mitigating dust storms, protecting the Gulf marine environment, and responding to disasters in a coordinated manner. The concrete product could be the initiation of a GCC-wide early warning network of extreme weather events, which is under consideration as climate risks increase. There is also progress on aligning standards; for example, GCC standardization authorities have discussed unified energy efficiency standards on appliances and buildings (Papadopoulou et al., 2013), which would increase the baseline sustainability across all member states.

Among regional projects in the field of sustainable development, one can mention the GCC Electricity Grid Interconnection, which, despite being primarily driven by energy security, can facilitate the sharing of renewable energy across nations (Al Shaikh et al., 2022). When one country has surplus solar energy during the day, it can sell to the neighbors and buy when it needs solar energy, this type of trading needs mutual regulations and trust. The grid is already functional and has been utilized in emergency support; in the future, governance agreements may extend to the use of the grid to facilitate planned energy trades to facilitate renewable integration in the region. Similarly, water security in the form of desalinated water grids or shared solar desalination research is an area where GCC coordination can provide value, and there are ongoing knowledge exchanges through GCC Water Cooperation committees. Transboundary problems such as dust storms and marine pollution are also being discussed in the region since they cannot be addressed by a single country. In response to this, GCC countries have endorsed a Regional Environmental Information Network to exchange information on air quality and the movement of sandstorms, and they liaise through the Regional Organization for the Protection of the Marine Environment to address oil spills and marine waste in the Gulf (Al-Saidi, 2023).

Despite the advancements, political divisions have hindered full coordination in the region. Nonetheless, the opportunity to act together on climate and SDGs in the GCC is growing with reconciliation. Every nation knows that numerous SDGs, including climate action, clean energy, and biodiversity, are transboundary in nature and require collaboration. Even the global community is promoting this type of collaboration, as seen in the World Bank's Arab Coordination Group, which announced funding specifically for green projects in the region; these green transformation commitments by the GCC countries. This implies that governance of climate and the SDGs in the GCC is no longer only a matter of national ministries but also of cooperation with regional financial institutions, multilateral agencies, and cross-border collaborations between the public and private sectors. To sum this section, here I conclude that regional climate–SDG progress in the GCC depends on reconciliation, cross-border cooperation, and external support (Figure 13.4).

13.6 INVOLVING THE STAKEHOLDERS: PRIVATE AND PUBLIC INVOLVEMENT

A feature of good governance in sustainable development is the involvement of stakeholders other than the government. The GCC, in which the government assumes a dominant role in economic and social affairs, has traditionally been reluctant to engage the private sector, academia, civil society, and the populace in the climate and SDG initiatives. However, it is becoming acknowledged that this kind of engagement is required to utilize the full extent of the available resources and expertise and to achieve the support of policies by the population. This is especially important to the GCC with regard to diversification, in that the contribution of the private sector in climate and sustainability is essential. Governments cannot achieve the funding required to make the green transition alone, and most of the funding will have to be delivered by businesses and investors. The World Bank added that the private sector plays a pivotal role in GCC countries in advancing the low-carbon transition and green investment attraction. As another example, the UAE revised its NDC to include an investment requirement of approximately $36 billion in total by 2030, of which a large portion is likely to be financed through private capital mobilization.[23]

Governments under GCC have therefore made some efforts to establish an enabling environment for the green business by way of reforms and incentives. These are public–private schemes on clean energy, green financing, and the encouragement of start-ups in sustainability businesses. The

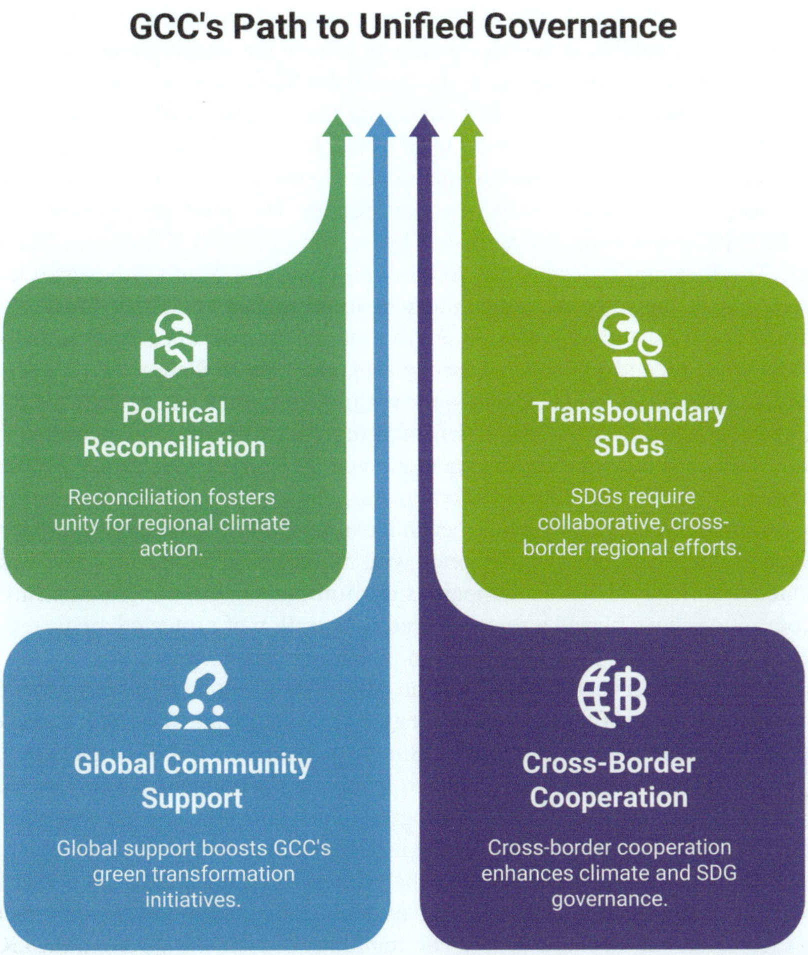

FIGURE 13.4 Pathways to more unified GCC governance: political reconciliation, cross-border cooperation, transboundary SDG action, and international support reinforce regional climate–SDG delivery.

Source: Author's synthesis, adapted from literature sources discussed in this chapter.

regulatory environment is also changing to promote corporate sustainability: examples include the introduction of guidelines on Environmental, Social, and Governance (ESG) disclosure by listed companies in stock exchanges in the GCC (Al Hosani et al., 2025). This compels corporations to disclose their carbon footprints, energy consumption, and social contributions, thereby indirectly promoting the SDG goals at the corporate level. The largest firms in the region have established their own sustainability objectives, including Saudi Aramco (Jrad, 2024) and SABIC (Ahmad and Bulbul, 2024), which have established carbon intensity targets and are investing in renewables to supply their operations; Emirates NBD has embraced the principles of sustainable finance to finance green projects (Salem and Elkhattat, 2023). Governments have increasingly engaged the private sector in policy making through business councils or consultation rounds on new policies. For example, the Economic Diversification committees in Oman incorporate members of the business community who also look at the sustainable development aspects of the policy (Al Balushi and Matriano, 2024).

Policy formulation in GCC countries is not widely open to public participation as in liberal democracies, but there are increasingly many spaces for public participation in policy-making, especially in the development arena. There are more consultations and publicity campaigns on climate and environmental matters. The UAE, as an example, launched national campaigns ahead of Expo 2020, with sustainability as a major theme, to inform the people about the SDGs and invite their ideas, the Global Goals Week at the Expo made the SDGs real and personal to the masses, framing them as the Worlds To-Do List and collecting tens of thousands of promises of action on the part of visitors.[24] Education is also being used to teach sustainability and climate change, with these topics included in school curricula and universities in most GCC states, in an attempt to instill a culture of environmental responsibility in young people. The youth engagement is a particular emphasis, as the GCC population is dominated by people under 30 (Elessawy, 2023). The UAE has designated Youth Ambassadors to the SDGs and created a Youth Advisory Council to the SDG committee to have the voice of young people heard in implementation.[25] Similarly, SGI Saudi Arabia organized youth-related activities on environmental stewardship, for example, "SGI Day," which involved volunteers and students in tree-planting and idea challenges.[26]

Civil society and NGOs in the GCC can work within limitations, but in the area of sustainable development, they have made little progress in making contributions. The community efforts of environmental NGOs, usually semi-governmental, such as Emirates Nature-WWF in the UAE[27] or the environmental unit of the Bahrain Women Association,[28] complement governmental work by organizing community activities such as beach clean-ups, wildlife conservation, and awareness workshops. Such groups occasionally collaborate with ministries in campaigns, as Qatar authorities have collaborated with local NGOs in campaigns to minimize single-use plastics. In Oman, local community committees have played a role in the administration of protected areas, reflecting SDG 15 and providing opportunities for people with an interest in conservation.

Major policies have not yet been subject to broad consultation, but there are examples: when Saudi Arabia was framing its National Environment Strategy, it consulted academics, NGOs, and the business community at workshops to advise on priorities. Similarly, the second VNR in 2023 by Bahrain was developed in a participatory manner with a view to involving government, the private sector, NGOs, and academia in the process of assessing SDG progress. Such consultations not only make the policy content rich but also create greater support for the implementation. However, there are still considerable problems with the participation of people. One criticism in a regional study is that GCC states, and Libya, are yet to reach some social limits, such as democratic quality and equality; that is, they are economically successful, but they lack democratic traditions of inclusive governance. Lack of public involvement and excessive centralization can be a barrier to climate and SDG initiatives, since such policies may fail to consider local requirements or creative solutions in society. To take an example, a water conservation policy may not be effective without community buy-in; an adaptation strategy may not be effective without consulting farmers and businesses. The Carnegie Climate Governance report observes that in MENA (including GCC), poor coordination with local governments and a lack of citizen participation are impediments to the implementation of climate-resilient infrastructure and other initiatives.[29] In recognition of this, a few GCC governments are expanding civic participation within manageable limits. One example of increasing participation without questioning political structures is increased use of technology, such as online portals, where citizens can give feedback on municipal services or report environmental problems. As an example, waste or water leak reporting smart city apps in Dubai indirectly help to achieve SDG targets by crowdsourcing governance. Notably, the idea of the engagement of the whole of society is repeated in GCC sustainability rhetoric. This means that not only the government agencies but also the businesses, academia, youth, and residents are to play their roles. We can find good indicators: universities in the GCC are opening sustainability research centers and involving students in SDG research. The voice of the private sector is getting louder: large companies publish sustainability reports in line with SDGs and join international initiatives, such as the UN Global Compact local networks in the UAE and Saudi Arabia. Moreover, religious and cultural authorities have also started to preach environmental sustainability in the light of Islamic ideologies of conservation that can affect the grassroots level of the population.

One of the biggest events facing the general population was the COP28 climate conference held in the UAE in 2023, and it put an unprecedented spotlight on climate concerns in the Gulf. The

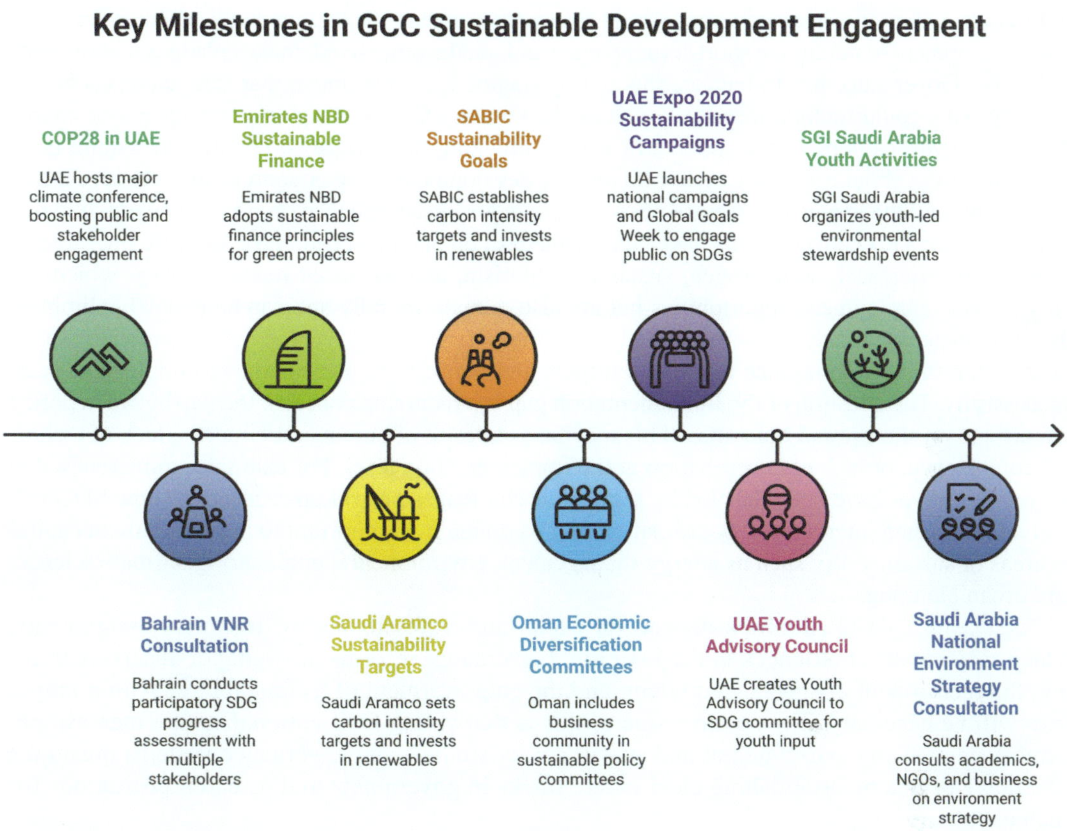

FIGURE 13.5 Milestones in GCC sustainable development engagement: high-profile events (e.g., COP28 UAE), ESG finance steps, national consultations/VNRs, youth councils and campaigns, and diversification committees that broaden public–private participation.

Source: Author's synthesis, adapted from literature sources discussed in this chapter.

government of the UAE organized discussions among policymakers, business executives, youth activists, and global experts in the run-up to COP28. This provided a platform where non-state actors could interact with officials at COP28, UAE youth delegates had a platform to present their views. The COP also stimulated local civil society practices (such as environmental groups and communities holding small-scale teach-ins and climate marches). The legacy of such events includes increased participation and awareness among the people. Saudi Arabia has also created space by hosting the annual Middle East Green Initiative Summit, to which not only government representatives but also NGOs and international civil society attend side meetings. Public–private and youth engagement in the GCC has expanded through flagship events, ESG finance, and participatory platforms (Figure 13.5).

13.7 CHALLENGES AND OPPORTUNITIES IN THE FUTURE

Although the path to the full realization of the climate and SDG goals in the GCC is evident, it is not easy. There are still significant governance issues that must be resolved to maintain momentum, as detailed in the following text.

The main issue is the incompatibility between the GCC's economic model, which is based on oil and gas revenues, and the climate action requirements. Although diversification is in progress, oil and gas continue to play a significant role in the GDP and government budgets in these nations.

A threat to policy incoherence arises when, say, a government, on the one hand, encourages fossil fuel development to maximize short-term revenue and, on the other hand, makes pledges of emissions reduction. Governance has to balance this out by staging a gradual move that safeguards economic stability. This could include measures such as the Circular Carbon Economy, which would enable the use of hydrocarbons in the short term while mitigating their impacts. It will also require open communication about economic trade-offs and the development of other sources of revenue, such as tax reforms and the expansion of non-oil sectors, to reduce fiscal reliance on oil. The potential in this case is that the GCC governments have significant cash on hand and can make strategic investments in new industries such as hydrogen, sustainable tourism, and advanced manufacturing, which not only provide employment opportunities but are also climate-friendly and can help slowly eliminate the oil-climate dilemma.

It is one thing to announce ambitious targets and plans, but quite another to implement them successfully. The question of the implementation gap is a recurring concern: the gap between policy intents and on-the-ground outcomes. This may be as a result of bureaucratic inertia, lack of technical competence, or lack of coordination at the implementation level. For example, some renewable energy initiatives have been delayed by regulatory challenges or grid un-readiness; some SDG initiatives stall when inter-agency coordination breaks down. It is important to develop human capital in areas of sustainability such as energy management, environmental engineering, climate science, and urban planning.

Today, most GCC countries depend on expatriate professionals for technical assignments, which may create a bottleneck in the long run. Governance reforms involving civil service training, the recruitment of specialized talent, and the empowerment of technical units within ministries will be necessary. The positive side of this is that young Gulf nationalists are increasingly earning degrees in environmental and sustainability studies, and governance systems must take advantage of this by establishing clear career tracks in government and research institutions for such experts.

Although there has been improvement in data collection on SDGs and emissions, weaknesses remain in data quality, consistency, and transparency. In other instances, the absence of baseline data makes it difficult to establish realistic targets or tracking of progress, for example, data on biodiversity or certain social indicators. Also, not all information is published or readily available, which reduces accountability. For example, the possibility of independent verification of emissions or evaluation of policy performance by academia or civil society is constrained by a lack of data. Enhancing national statistical systems and open data initiatives will build trust and enable third-party contributions to solving problems. The possibility is that digital transformation is a priority in every GCC state, and if the SDG agenda is connected with it, investments in big data, AI, and Internet of Things can significantly improve monitoring, reporting and verification and environmental monitoring (smart grids, air quality sensors, water use tracking) and, therefore, guide more effective policies.

Culturally, one of the most daunting tasks will be to effect behavior change in cultures accustomed to living energy- and resource-intensive lives, big air-conditioned houses, personal vehicles, and high water consumption, often subsidized. Governance can also implement pricing reforms and regulations, but to maintain them, there must be buy-in and understanding from the people. When individuals are inclined to interpret climate actions as risks to their comfort or economic stability (such as subsidy removal leading to sharp increases in the cost of living), then they can become resistant, even where political dissent is not explicitly channeled in these regimes. Thus, a cuddly aspect of governance, communication, education, and inclusion becomes decisive. Constant education of the population about the necessity of particular measures, such as conserving water or taking public transport, to serve the common good can slowly change norms. The high level of citizen benefits in GCC states can be leveraged, for example, by linking new housing grants or social benefits to adherence to green building standards. The relatively small number of citizens may be a plus: they can be reached out to individually through local majlis meetings, youth forums, or social media in Arabic and encouraged to participate in national sustainability efforts. This is a challenge requiring long-term effort, as well as an opportunity, as a small population is easier to engage and manage.

Although centralization offers benefits such as faster decision-making and a unified direction, it may stifle local initiative and innovation. Top-down governance can miss the specifics of local context, such as a centrally decided energy policy that may not be appropriate for the needs of a remote rural community in Oman or a densely populated district in Dubai. It could be more effective to leave municipalities or local branches of ministries to be more innovative and create solutions that fit their needs. Local leadership has already been demonstrated in some GCC cities, for example, Dubai with its own carbon-reduction strategy, and Riyadh with its public transport and greening project under the Royal Commission. It would be beneficial to establish a healthy interplay between national and subnational initiatives, in which local initiatives are scaled up, and national frameworks enable local action.

Even stability and cooperation in the region itself are factors. The 2017–2021 Qatar crisis showed how political divisions can interfere with trade, cooperation, and trust, all of which are necessary to take joint climate action, for example, a shared electricity market or climate research (Bani Salameh and Darabseh, 2025). Ensuring good diplomatic ties in the GCC is therefore a side-product of maintaining SDG and climate gains. On the one hand, a stronger GCC can exercise greater influence in international climate negotiations, promote sustainable development financing, and, perhaps, negotiate more favorable terms for green technology transfer or climate finance as a group. The revival of GCC unity and future opportunities, such as the possibility of Saudi Arabia hosting Expo 2030 or other global events, can serve as platforms for advancing collective sustainability agendas.

Nevertheless, the GCC has great opportunities that lie ahead. They possess considerable budgetary assets, and when those are geared toward sustainability, as shown by initiatives such as the $24 billion pledge by the Arab Coordination Group toward green projects, the financing bottlenecks that bedevil other developing worlds become less prevalent. The GCC can also serve as a global testbed of climate solutions in arid regions, innovating in solar power generation, where they already lead in cost, green hydrogen production, carbon capture in oil and gas, and cooling technologies to address extreme heat. Effective practices invented in the Gulf, such as large-scale solar-powered desalination or carbon-neutral petrochemical operations, can be sold to other nations with similar climates, transforming the GCC into an exporter of climate know-how rather than just oil. This corresponds with SDG 17 on global partnerships, the notion that GCC states can share their expertise and contribute to the sustainable development of other countries, which many have already begun to do in the form of aid and investment.

Also, the next decade is critical in the global context, as 2030 is the deadline for the SDGs and an important milestone for emissions reductions. The actions of the GCC during this time will be under the international spotlight, yet under the international support. The world is ready to support and collaborate with the Gulf on sustainability through initiatives such as the Global Methane Pledge and new carbon markets in which the GCC is active. Should improvements in governance persist, GCC countries may also be in a better position to access climate finance mechanisms, such as the Green Climate Fund or adaptation funds, to which they have not historically contributed much, though again, specific projects, such as adaptation to vulnerable communities or technology pilot projects, may be eligible, relieving budget pressures at home.

Lastly, demographics may present an opportunity: a young, well-educated population in the GCC may be a catalyst for innovation and social change. Numerous young Gulf citizens are well-connected and technologically savvy; with proper empowerment, they will be able to start green businesses, hold governments accountable in positive ways through social media to promote sustainability, and adopt more sustainable lifestyles faster than older generations. This energy can be harnessed through government programs that incubate youth-led sustainability initiatives or through youth advisory councils. To sum up, progress is constrained by delivery gaps, incoherent signals, weak data, social resistance, centralization, and regional politics (Figure 13.6).

FIGURE 13.6 Core governance barriers to GCC climate–SDG delivery: implementation gaps, policy incoherence, data weaknesses, cultural resistance, centralization risks, and political divisions, each tied to a hydrocarbon-dependent model.

Source: Author's synthesis, adapted from literature sources discussed in this chapter.

13.8 CONCLUSION

Governance will ultimately determine whether GCC countries make only big promises or deliver real climate action and sustainable development outcomes. As the analysis in this chapter demonstrates, the GCC has traveled a long way in a short period of time: climate and sustainability are no longer regarded as marginal issues but as central to the national vision and regional projects. Political commitment is also evident through new institutions, strategies, and laws that resonate with the SDGs and the Paris climate goals. In the six countries, governance is being redesigned, ministries are aligning on cross-cutting issues, data systems are being used to monitor progress, and stakeholders are gradually being brought on board. These initiatives have already borne fruit: the fast-growing renewable energy output, increased incorporation of sustainability into education and planning, and more active participation by GCC states as positive actors in climate diplomacy. To effectively achieve climate action and the SDGs in the GCC, governance needs to further develop toward coherence, inclusiveness, and forward-looking flexibility. It demands maintaining the pace of reforms, not just in individual projects or international events. The key recommendations are as follows: improving institutional capacities to implement, through training and clear mandates, improving transparency and data disclosure to allow oversight and establish trust, incorporating climate and SDG criteria into all government decision-making, such that every policy or project is assessed on its sustainability effects, and institutionalizing stakeholder engagement to allow non-government actors to contribute ideas and help execute solutions. Significantly, governance in the GCC must leverage the region's strengths, including financial muscle, long-term planning capacity, and relative stability, to expand the scope of what is possible in sustainable development. This may imply exceedingly ambitious goals, testing radical ideas such as smart zero-carbon cities, and investing massively in human capital to spearhead the green transition. The regional approach as a whole is promising as well. By acting collectively, the GCC countries could produce regional public goods, such as a shared renewable energy market or coordinated climate adaptation strategies for the Gulf ecosystem, that individual countries may struggle to deliver on their own. Research indicates that regional environmental cooperation is significant for increasing commitments, assisting in addressing transboundary problems, and raising overall ambition. A harmonized GCC sustainability vision would support national efforts and prevent duplication. For example, harmonizing regulations on emissions and efficiency across the GCC would make implementation easier for companies operating in more than one country and would motivate laggards to close the gap with leaders. Overall, governance is and can be catalytic and enabling of GCC climate and SDG action. It establishes direction through visions and targets, mobilizes resources through policies and institutions, and secures accountability through monitoring and review. The GCC experience teaches us that the top-down commitment should be accompanied by bottom-up participation to attain long-term change. By 2030, the world will be watching as the Gulf states balance their oil-driven wealth with the need for sustainability. There is hope that things will improve: these nations have demonstrated that they can change quickly when an objective is made a priority and sustainability is now high on their agenda. Continued governance gains will allow the GCC not only to achieve most of its SDG and climate goals but also to set an example for other regions by showing how a petro-state can pursue a green growth model. It will not be an easy trip, but it has begun, and effective, adaptive governance will lead the way to its success.

DECLARATION STATEMENT

The author declares that this research work is original and has duly acknowledged all the sources of information used in the chapter. There are no conflicts of interest.

ACKNOWLEDGMENTS

The author gratefully acknowledge the organisations that produced and maintain the data reports and supporting datasets used in this chapter. All such sources are cited in the chapter text and reference list. The author also acknowledge the American University of Sharjah for providing full institutional support that enabled the completion of this book contribution. Appreciation is extended to the Japan Society for the Promotion of Science (JSPS) for JSPS fellowship support and for providing a formative opportunity to broaden academic and international perspectives. The author further acknowledge the support and professional exposure provided through The University of Tokyo (UTokyo), UNU-IAS, Tokyo University of Agriculture and Technology (TUAT), Hokkaido University, the Department of Science and Technology (Government of India), and the World Bank, which collectively contributed to the author's academic development and professional trajectory. The views expressed in this chapter are those of the author and do not necessarily reflect the views of the above organisations.

ETHICAL APPROVAL

Not applicable.

CONSENT TO PARTICIPATE

Agree to participate.

CONSENT TO PUBLISH

Agree to publish.

AUTHOR'S CONTRIBUTIONS

Conceptualization, Methodology, Software, Formal analysis, Investigation, Resources, Writing—original draft preparation, Writing—review and editing: Gowhar Meraj.

FUNDING

This work received no external funding.

AVAILABILITY OF DATA AND MATERIALS

All the data is presented in the chapter.

DECLARATION OF GENERATIVE AI AND AI-ASSISTED TECHNOLOGIES IN THE WRITING PROCESS

During the preparation of this work, the author used an AI service only (paperpal/grammarly) to correct grammar and expression errors. After using this service, the author reviewed and edited the content as needed and takes full responsibility for the content of the publication.

NOTES

1 https://carnegieendowment.org/research/2025/02/climate-governance-mena-africa?lang=en
2 https://aps.aucegypt.edu/en/articles/118/cairos-hidden-urban-water-scarcity
3 https://unfccc.int/topics/cooperative-activities-and-sdgs/action-on-climate-and-sdgs
4 https://www.vision2030.gov.sa/en
5 https://u.ae/en/about-the-uae/strategies-initiatives-and-awards/strategies-plans-and-visions/innovation-and-future-shaping/uae-centennial-2071
6 https://www.oman2040.om/?lang=en
7 https://www.mofa.gov.kw/en/pages/kuwait-vision-2035
8 https://www.gco.gov.qa/en/state-of-qatar/qatar-national-vision-2030/our-story/
9 https://www.mofne.gov.bh/en/project-initiatives/bahrain-economic-vision-2030/
10 https://www.idea.int/un-and-democracy/library/kuwait-voluntary-national-review-2023
11 https://www.undp.org/sites/g/files/zskgke326/files/2022-09/Bahrain_VNR_EN.pdf
12 https://sustainabledevelopment.un.org/content/documents/29836Compilation_of_Main_Messages_2022_VNRs.pdf
13 https://www.unescwa.org/sites/default/files/event/materials/9th%20VNR%20Report-English.pdf
14 https://blogs.worldbank.org/en/arabvoices/embracing-climate-opportunities-greener-gcc
15 https://www.vision2030.gov.sa/en/explore/projects/saudi-green-initiative
16 https://www.arabianbusiness.com/politics-economics/kuwait-reveal-increased-electricity-water-fees-within-weeks-652076
17 https://mecouncil.org/publication/gulf-transition-to-low-carbon-economies-the-role-of-carbon-markets
18 https://ppp.worldbank.org/library/climate-change-ifc
19 https://www.adnoc.ae/en/towards-net-zero/decarbonizing-our-operations
20 https://www.undp.org/sites/g/files/zskgke326/files/2022-09/Bahrain_VNR_EN.pdf
21 https://zcralliance.org/blogs/multi-hazards-kuwait
22 https://www.sgi.gov.sa/about-mgi/mgi-targets/reducing-emissions/
23 https://blogs.worldbank.org/en/arabvoices/embracing-climate-opportunities-greener-gcc
24 https://virtualexpodubai.com/listen-watch/events/sdgs-in-action

25 https://economymiddleeast.com/news/uae-holds-global-youth-dialogue-to-shape-future-of-sustainable-development-beyond-2030
26 https://economymiddleeast.com/news/saudi-green-initiative-day-what-to-expect-on-march-27
27 https://www.leadersofchange.ae/member-badge/
28 https://iucn.org/fr/node/34507
29 https://carnegieendowment.org/research/2025/02/climate-governance-mena-africa?lang=en

REFERENCES

Abbas, M., 2021. *Resilience Based Restoration of Mangrove Ecology in the Kingdom of Bahrain* (Master's thesis, Eesti Maaülikool).

Abdel Gelil, I., Howarth, N. and Lanza, A., 2017. *Growth, Investment and the Low-Carbon Transition: A View from Saudi Arabia*. Riyadh: King Abdullah Petroleum Studies and Research Center. https://www.kapsarc.org/research/publications/growth-investment-and-the-low-carbon-transition-a-view-from-saudi-arabia/ (Accessed September 22, 2025).

Abdelraouf, M. and Nagasawa, A., 2023. Saudi green initiatives and their role in achieving environmental goals in the Middle East. *Gulf Research Center*. https://grc.net/public/documents/6530de99c93a7SaudiGreenInitiativesandtheirRole.pdf

Abidin, R.Z.U., Qammar, R. and Alkhawaldah, A., 2020. Evaluation of risk management strategies in government institutions: Case study on Abu Dhabi Municipalities. *Evaluation*, *56*.

Abulibdeh, A., Zaidan, E. and Al-Saidi, M., 2019. Development drivers of the water-energy-food nexus in the Gulf Cooperation Council region. *Development in Practice*, *29*, 582–593. https://doi.org/10.1080/09614524.2019.1602109

Ahmad, K. and Bulbul, A., 2024. Case study of sabic: Saudi company producing petrochemicals for global market needs. *Journal of Integrated Sciences*, *5*(1).

Ahmed, A., 2023, March. Developing strategy for integrating sustainability in UAE higher education institutions towards UAE Strategic Initiative Net Zero 2050. In *2023 International Conference on Computational Intelligence and Knowledge Economy (ICCIKE)* (pp. 370–375). IEEE. https://doi.org/10.1109/ICCIKE58312.2023.10131825

Al Balushi, A.K. and Matriano, M.T., 2024. *Renewable Energy Investments Impact on Economic Diversification and Strategic Growth: A Review of Literature in the Context of Oman Vision 2040*. https://pdfs.semanticscholar.org/c987/f6cc3095eedba8956d48a27abf5f05a2c387.pdf

Al Hosani, N.H., Nobanee, H. and Ellili, N.O.D., 2025. The impact of board gender diversity on financial performance of non-financial companies of the UAE: The moderating role of environmental, social, and governance (ESG) disclosure. *Corporate Governance: The International Journal of Business in Society*, *25*(8), 176–194.

Al Shaikh, T., Bonita, O. and Khan, F.J., 2022, December. The role of GCC interconnection for renewable energy penetration. In *2022 Saudi Arabia Smart Grid (SASG)* (pp. 1–6). IEEE.

Al-Afifi, Z.M.M., 2018. *The Conservation, Sustainable Use and Management of Mangrove Habitats in Oman* (Doctoral dissertation, University of York).

Albassam, B.A., 2015. Economic diversification in Saudi Arabia: Myth or reality? *Resources Policy*, *44*, 112–117. https://doi.org/10.1016/j.resourpol.2015.02.005

Alkhaldi, M., Moonesar, I.A., Issa, S.T., Ghach, W., Okasha, A., Albada, M., Chelli, S. and Takshe, A.A., 2023. Analysis of the United Arab Emirates' contribution to the sustainable development goals with a focus on global health and climate change. *International Journal of Health Governance*, *28*(4), 357–367.

Almheiri, A.K., 2015. *Is the United Arab Emirates Capable to Create a Green and Sustainable Future?* (Master's thesis, The British University in Dubai).

Alnaqbi, S.A. and Alami, A.H., 2023. Sustainability and renewable energy in the UAE: A case study of Sharjah. *Energies*, *16*(20), 7034.

Al-Saidi, M., 2022a. Disentangling the SDGs agenda in the GCC region: Priority targets and core areas for environmental action. *Frontiers in Environmental Science*, *10*, 1025337.

Al-Saidi, M., 2022b. Energy transition in Saudi Arabia: Giant leap or necessary adjustment for a large carbon economy? *Energy Reports*, *8*, 312–318. https://doi.org/10.1016/j.egyr.2022.01.015

Al-Saidi, M., 2023. *Collective Environmental Action to Preserve the Gulf's Marine Ecosystem* (pp. 15–24). London: Bourse & Bazaar Foundation.

Al-Saidi, M. and Elagib, N.A., 2018. Ecological modernization and responses for a low-carbon future in the Gulf Cooperation Council countries. *Wiley Interdisciplinary Reviews: Climate Change*, *9*, e528. https://doi.org/10.1002/wcc.528

Al-Saidi, M. and Saliba, S., 2019. Water, energy and food supply security in the gulf cooperation Council (GCC) countries—a risk perspective. *Water*, *11*, 455. https://doi.org/10.3390/w11030455

Al-Sarihi, A. and Mason, M., 2020. Challenges and opportunities for climate policy integration in oil-producing countries: The case of the UAE and Oman. *Climate Policy*, *20*(10), 1226–1241.

Bani Salameh, M. and Darabseh, M., 2025. Strategic shifts: The role of the Gulf Crisis in transforming Qatari-Turkish relations (2017–2021). *Review of Economics and Political Science*, 1–18.

Bellini, E., 2021. *Saudi Arabia's Second PV Tender Draws World Record Low Bid of $0.0104/kWh*. Berlin: PV Magazine. https://pv-magazine-usa.com/2021/04/12/saudi-arabias-second-pv-tender-draws-world-record-low-bid-of-0-0104-kwh/ (Accessed September 25, 2025).

Burt, J.A., Ben-Hamadou, R., Abdel-Moati, M.A., Fanning, L., Kaitibie, S., Al-Jamali, F., et al., 2017. Improving management of future coastal development in Qatar through ecosystem-based management approaches. *Ocean & Coastal Management*, *148*, 171–181. https://doi.org/10.1016/j.ocecoaman.2017.08.006

Caro, D., Pulselli, F.M., Borghesi, S. and Bastianoni, S., 2017. Mapping the international flows of GHG emissions within a more feasible consumption-based framework. *Journal of Cleaner Production*, *147*, 142–151. https://doi.org/10.1016/j.jclepro.2017.01.106

Chefaoui, R.M., Duarte, C.M., Tavares, A.I., Frade, D.G., Cheikh, M.S., Ba, M.A. and Serrao, E.A., 2021. Predicted regime shift in the seagrass ecosystem of the Gulf of Arguin driven by climate change. *Global Ecology and Conservation*, *32*, e01890. https://doi.org/10.1016/j.gecco.2021.e01890

Dargin, J., 2021. The pathway to a green Gulf: A review and analysis of the evolution of Saudi Arabia, Qatar, and the United Arab Emirates' climate change positions. *Carbon & Climate Law Review*, *15*, 313–341. https://doi.org/10.21552/cclr/2021/4/7

Elessawy, F.M., 2023. Development and fertility decline in the arabian gulf cooperation council countries: The case of the United Arab Emirates. *The Arab World Geographer*, *26*(3–4), 316–331.

El-Kharouf, F., Al-Qudsi, S. and Obeid, S., 2010. The gulf corporation council sovereign wealth funds: Are they instruments for economic diversification or political tools? *Asian Economic Policy Review*, *9*, 124–151. https://doi.org/10.1162/asep.2010.9.1.124

Hanieh, A., 2024. Laundering carbon—The Gulf's' new scramble for Africa. *Middle East Report*.

Hummieda, A., Bouabid, A., Moawad, K. and Mayyas, A., 2023. The UAE's energy system and GHG emissions: Pathways to achieving national goals by 2050. *Clean Energy*, *7*(5), 962–980.

Hvidt, M., 2013. *Economic Diversification in GCC Countries: Past Record and Future Trends*. London, UK: London School of Economics and Political Science.

Jrad, M., 2024. The role of technology in promoting sustainability and business performance in industries and business models: The case of Saudi Aramco. In *Navigating the Intersection of Business, Sustainability and Technology* (pp. 201–225). Singapore: Springer Nature.

Karanisa, T., Amato, A., Richer, R., Majid, S.A., Skelhorn, C. and Sayadi, S., 2021. Agricultural production in Qatar's hot arid climate. *Sustainability*, *13*(7), 4059. https://doi.org/10.3390/su13074059

Koul, S., Energy, S. and Net Zero Emission 2050, 2024, October 13. Available at SSRN. http://dx.doi.org/10.2139/ssrn.5165738

Leal-Arcas, R., Almashloum, S., Jazzar, R., Saleh, N.B. and Almunifi, R., 2025. Energy policy of Oman: Pursuing decarbonization. *Energies*, *18*(5), 1270. https://doi.org/10.3390/en18051270

McPherson-Smith, O., 2021. Diversification, Khashoggi, and Saudi Arabia's public investment fund. *Global Policy*, *12*(2), 190–203. https://doi.org/10.1111/1758-5899.12917

Mohiuddin, K., Nasr, O.A., Miladi, M.N., Fatima, H., Shahwar, S. and Naveed, Q.N., 2023. Potentialities and priorities for higher educational development in Saudi Arabia for the next decade: Critical reflections of the vision 2030 framework. *Heliyon*, *9*(5).

Odhiambo, G.O., 2017. Water scarcity in the Arabian Peninsula and socio-economic implications. *Applied Water Science*, *7*(5), 2479–2492. https://doi.org/10.1007/s13201-016-0440-1

Pal, J.S. and Eltahir, E.A., 2016. Future temperature in southwest Asia projected to exceed a threshold for human adaptability. *Nature Climate Change*, *6*(2), 197–200. https://doi.org/10.1038/nclimate2833

Papadopoulou, A.G., Al Hosany, N., Karakosta, C. and Psarras, J., 2013. Building synergies between EU and GCC on energy efficiency. *International Journal of Energy Sector Management*, *7*(1), 6–28.

Robaa, S.M. and El Kenawy, A.M., 2024. Recent study of urbanization growth effects on the local climatic changes over Greater Cairo Region (Egypt) using satellite images. In *Hydroclimatic Extremes in the Middle East and North Africa* (pp. 249–261). Elsevier. https://doi.org/10.1016/B978-0-12-824130-1.00008-4

Saidin, M.I.S. and O'Neill, J., 2022. Climate change and the diversification of green social capital in the international political economy of the Middle East and North Africa: A review article. *Sustainability*, *14*(7), 3756. https://doi.org/10.3390/su14073756

Salem, S.A.S. and Elkhattat, D., 2023, February. Environmental responsibility of the banking sector in UAE: Practices to confront climate change risks. In *Scientia Moralitas Conference Proceedings* (No. 01256). Research Association for Interdisciplinary Studies.

Saradara, S.M., Khalfan, M.M.A., Rauf, A. and Qureshi, R., 2023. On the path towards sustainable construction—the case of the United Arab Emirates: A review. *Sustainability*, *15*(19), 14652.

Schimmel, M., Bietenholz, D., Steinbacher, K., Braun, J.F., Shabaneh, R., Roychoudhury, J. and Saxena, S., 2022. *Hydrogen Cooperation Potential between Saudi Arabia and Germany*. Berlin: Federal Ministry for Economic Affairs and Climate Action (BMWK). https://www.bundeswirtschaftsministerium.de/Redaktion/EN/Downloads/J/joint-study-saudi-german-energy-dialogue (Accessed September 26, 2025).

Schuetze, B., 2024. The uneven politics of decarbonization in the Middle East and North Africa. *Middle East Report*. https://www.arnold-bergstraesser.de/en/publications/uneven-politics-decarbonization-middle-east-and-north-africa

Shapland, G., 2023. Climate change and the GCC: Economic and environmental impact. In *GCC Hydrocarbon Economies and COVID: Old Trends, New Realities* (pp. 173–200). Singapore: Springer Nature. https://doi.org/10.1007/978-981-19-5462-7_8

Touri, M., 2024. *Crafting the Future: Pioneering Urban Branding and Development in Neom's Visionary Line City Project, Saudi Arabia* (Doctoral dissertation, Politecnico di Torino).

Varela, R., Rodríguez-Díaz, L. and deCastro, M., 2020. Persistent heat waves projected for Middle East and North Africa by the end of the 21st century. *PLoS One*, *15*(11), e0242477. https://doi.org/10.1371/journal.pone.0242477

World Bank, 2021. Middle East & North Africa climate roadmap (2021–2025)—driving transformational climate action and green recovery in MENA. *World Bank Group*. https://www.worldbank.org/en/region/mena/publication/middle-east-north-africa-climate-roadmap

Zabelin, D., 2024. Green Saudi Arabia: The rise of a renewable energy superpower. *Journal of Iranian Studies*, *8*, 31.

Zumbraegel, T., 2022. Inside the green state: Institutional organization, agency and governance. In *Political Power and Environmental Sustainability in Gulf Monarchies* (pp. 83–136). Singapore: Springer Nature. https://doi.org/10.1007/978-981-19-4431-4_4

14 Empowering Communities for Climate Action

Lessons from South Africa's Participatory Sustainability Models

L. Gaegane and V. Molose

14.1 INTRODUCTION

South Africa, with its complex socio-political history and rapid urbanisation, faces unique socio-economic and environmental pressures that require innovative strategies to enhance sustainability and climate resilience (Kamana, Radoine, and Nyasulu 2024; Kajiita and Kang'ethe 2024). Global change, rising inequalities, and environmental degradation intensify these challenges, necessitating comprehensive pragmatic strategies that address socio-ecological vulnerabilities (Birkmann et al. 2022). Urban resilience is increasingly recognised as a key mechanism for addressing environmental degradation and confronting systemic socio-economic disparities. Islam (2024) states that theories of environmental and climate justice emphasise the need for inclusive governance structures to protect marginalised communities from climate-related impacts.

Usman (2024) argues that environmental injustices disproportionately affect disadvantaged groups, advocating for participatory decision-making frameworks that empower affected communities. Similarly Fatti et al. (2023) explore how power relations, justice, and scale influence urban environmental outcomes, emphasising the political dimensions of governance in shaping equitable urban spaces.

Governance plays a pivotal role in shaping resilience and adaptation across both urban and rural settings, which has been seen in case studies of the National Siltation Management Strategy for Dams in South Africa, where community-based adaptation (CBA) projects have been implemented (Tshabalala et al. 2024). Walker et al. (2024) state that effective governance frameworks must incorporate climate justice principles to ensure equitable access to the benefits of climate adaptation initiatives. However, South Africa faces specific challenges that hinder the scalability and integration of adaptation measures, including limited financial resources, fragmented institutional coordination, and complex socio-political dynamics (Donaldson 2015; Singh et al. 2022).

While CBA offers a valuable grassroots pathway to building climate resilience, its broader implementation is constrained by scalability and weak integration into formal governance systems (Donaldson 2015; Singh et al. 2022). Key barriers include limited financial and human resources, poor coordination among government levels, and complex socio-political dynamics that hinder policy alignment and continuity (Selje, Schmid, and Heinz 2024). Successful CBA initiatives, therefore, need to be documented and shared to facilitate replication and scale-up. According to Keziah (2022) and Selje, Strikker, and Heinz (2024), the strength of CBA lies in its use of local knowledge and participatory approaches, which align with the socio-ecological systems theory that recognises the interconnectedness of human and ecological systems. Yet, embedding these community-driven models within formal policy frameworks remains difficult due to institutional fragmentation and entrenched governance challenges (Leal Filho et al. 2022). Participatory sustainability models have

 DOI: 10.1201/9781003600152-14

emerged as transformative tools in South Africa's approach to climate action and as catalysts for successful and meaningful CBA. These models enable urban and rural communities to lead adaptation initiatives by prioritising inclusive decision-making that promotes community ownership from the outset. This ensures that environmental sustainability projects are locally relevant and widely supported (Palmer et al. 2022). By integrating local knowledge with scientific expertise, these models enhance adaptive capacity and support the effective implementation of sustainable development initiatives (Hedelin et al. 2017; Yanou et al. 2023).

Inclusive governance is essential for ensuring equitable participation of marginalised and underrepresented groups in decision-making processes (Sánchez-Soriano et al. 2024). It is based on transparency, accountability, and broad participation, ensuring that all voices are heard. This approach fosters the development of policies that are both fair and resilient.

An example of this can be seen in Clarens, Free State, where community members collaborated with the Water Research Commission (WRC) to co-develop siltation management interventions aimed at improving the ecosystem's resilience to climate change (Tshabalala et al. 2024). Workshops and stakeholder engagement allowed different perspectives to be heard. Similarly, Ileka (2018) suggests that municipal development plans are more successful when communities are engaged throughout the planning and implementation process (Mamokhere and Meyer 2023; Gaegane 2024; Fourie and Van der Waldt 2021).

Active participation enhances democratic legitimacy, reduces inequalities, and builds public trust; these are essential for effective climate action (Fung 2015). The Organisation for Economic Co-operation and Development and the United Nations emphasise the importance of inclusive governance in achieving the SDGs (United Nations 2015), while (World Bank 2024) highlights its role in effective public sector management and sustainable development outcomes.

Clarke (2022) defines *climate resilience* as the ability of socio-ecological and economic systems to resist, adapt to, and recover from climate-induced disturbances (Dodman et al. 2022). In urban contexts, resilience entails the ability of cities to maintain critical functions, adapt to change, and transform systems that hinder adaptive capacity (Meerow, Newell, and Stults 2016). Strategies to foster resilience include strengthening ecosystems, investing in resilient infrastructure, and integrating nature-based solutions into urban planning processes. South Africa's vulnerability to climate change underscores the importance of enhancing adaptive capacity by integrating climate-proofing measures into urban and rural development. This approach could have benefited the KwaZulu-Natal region during the floods in 2022 and 2023. Institutionalising environmental sustainability within governance frameworks ensures that resilience-building practices become part of routine development planning, contributing to sustainable economic growth and climate adaptation (Clarke 2022).

South Africa's climate resilience efforts are highlighted in the chapter through two case studies: urban governance in the City of Tshwane (CoT) and rural adaptation in Limpopo. These examples highlight how integrating CBA strategies into governance promotes inclusive, local resilience approaches that support the Sustainable Development Goals (SDGs). In the CoT, participatory sustainability models are used to implement development projects through collaborative governance. The city implemented green infrastructure projects to enhance energy and water efficiency, improve climate resilience, and support economic growth (CoT 2023; Sachs 2015). Various stakeholders participated from the government, the private sector, civil society, and academia, working to ensure alignment of local development priorities with national sustainability objectives. In rural Limpopo, the communities' capacity to adapt to climate variability was strengthened through community-driven initiatives that blended Indigenous knowledge with modern technologies to improve water resource management and agricultural systems (du Toit, Tshogang, and WRC 2024).

The chapter highlights four urban projects implemented by the CoT and two rural projects in Limpopo to highlight how climate resilience was addressed from a participatory approach. This chapter thus demonstrates how the integration of climate resilience and inclusive governance through participatory sustainability models fosters sustainable development while driving economic growth and enhancing climate adaptation. The case studies from Tshwane and Limpopo illustrate that collaborative approaches can effectively climate-proof both urban and rural communities, offering scalable and replicable strategies for global implementation that combine technical expertise with grassroots participation (Taylor 2019; Gaegane 2024).

This chapter further shows how South Africa's approaches contribute to the global dialogue on the synergies and trade-offs between climate action and the SDGs. By embedding inclusive governance, encouraging community participation, and investing in environmental sustainability, South Africa offers valuable lessons for global climate resilience initiatives. These strategies enhance community adaptive capacity and create inclusive, sustainable pathways for economic growth and environmental stewardship. This ensures communities can better navigate climate risks while advancing global sustainability objectives.

14.2 CONCEPTUAL FRAMEWORK: CLIMATE ACTION, SDGS, AND COMMUNITY EMPOWERMENT

14.2.1 Linking Climate Action and the SDGs

Michael (2019) defines *climate change* as changes in temperatures and weather patterns taking place over prolonged periods. While natural processes contribute to these changes, human activities, especially the burning of fossil fuels since the 1800s, have been the main contributors. Many nations are now tackling climate action through two main strategies: *mitigation*, which focuses on reducing greenhouse gas emissions, and *adaptation*, which prepares for the impacts of climate change (Yumagulova 2011). Taking action against climate change is vital for achieving the SDGs. This requires changing harmful behaviours and emphasises sustainable practices to protect resources for current and future generations.

14.2.1.1 SDGs Roadmap

Community participation is central to advancing meaningful climate action and ensuring that the SDGs translate into concrete, locally led interventions. For each SDG, communities must be empowered to contribute not only to implementation but also to defining what these goals mean in their respective contexts. Since the SDGs are interconnected, actions taken toward one target often influence outcomes across several others. For example, SDGs 1–3, 6, and 7 require sustainable management of water, land, and energy resources. Initiatives such as smart agriculture, efficient irrigation, and renewable energy use can help reduce poverty (SDG 1), address hunger (SDG 2), and promote health (SDG 3) while protecting terrestrial ecosystems (SDG 15). Ensuring access to clean water and sanitation (SDG 6) also depends on community involvement in maintaining infrastructure and preventing pollution. The adoption of solar energy and other renewable technologies (SDG 7) further supports long-term resilience. With improved local food and water security, communities become more self-reliant, which may in turn slow urban migration and ease pressure on rapidly growing cities (SDG 11). This holistic, community-driven model of development reinforces the idea that climate adaptation must be locally embedded to be sustainable and transformative. This linkage is shown in Figure 14.1.

14.2.2 The Role of Community Empowerment

Building local resilience is essential for poor and vulnerable communities, as they are most affected by climate change. Their valuable Indigenous knowledge can be enhanced with technology to help them cope with extreme weather patterns such as floods and droughts. Investing in communities' skills development and building their capacities empowers communities to engage in climate action, thus reducing inequalities and strengthening adaptability (Ghimire, Khatri-Chhetri, and Chhetri 2022). Rural communities, for instance, have retained Indigenous knowledge on coping with

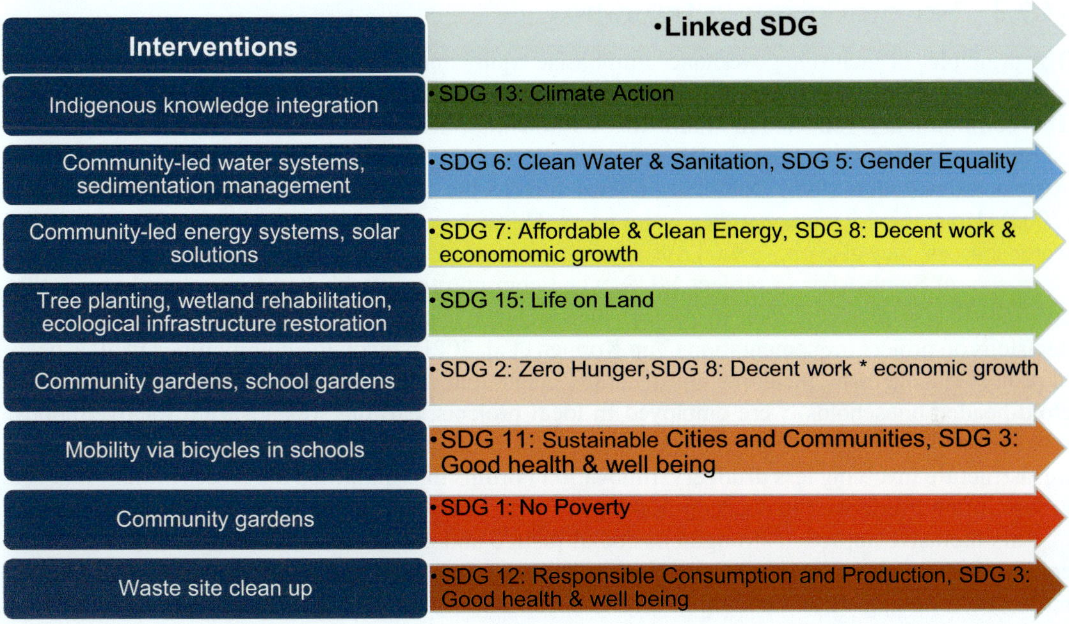

FIGURE 14.1 Alignment of interventions with Sustainable Development Goals (SDGs).

extreme weather by practising effective crop rotation, water conservation methods, and resource management (Jovanovic et al. 2022). When combined with technology, these systems can significantly boost resilience. Thus, community inclusion and participation are key to developing locally based solutions enabling communities to create local adaptive action plans to address climate change challenges.

For example, South Africa's Giyani Local-Scale Climate Resilience Programme (GLSCRP) implemented in Limpopo, involved communities being actively involved in the planning, decision-making and even construction of the projects. They brought to the programme useful insights on Indigenous crops which endure extreme heat and can survive with less irrigation and demonstrated innovative water harvesting techniques during dry seasons (Ghimire, Khatri-Chhetri, and Chhetri 2022). Involving communities in decision-making ensures that their voices are heard throughout project stages; they can contribute their ideas regarding types and locations of infrastructure based on affordability and accessibility, thus promoting sustainability, and ownership of projects. As seen in the "Operationalising Community-Led Multiple-Use Services Project," commonly known as "MUS," community members emphasise active involvement with the phrase "Nothing about us, without us" (Van Koppen et al. 2021). In the community-driven model of project implementation, communities play a greater role in the planning, budgeting, and even construction of projects. Communities (with technical advice from Tsogang Water & Sanitation, a local non-governmental organisation (NGO)) are thoroughly involved in the project management, receiving and inspecting the quality of the materials delivered, reporting on contractor progress in terms of construction schedule, and so on. This is unlike in contractor-based models, where contractors and suppliers report to government agencies, not project beneficiaries.

Training communities in groundwater monitoring and resource management builds active citizenship (Jovanovic and WRC 2024); trusting communities to manage their resources also acknowledges their expertise and experience (Graham and Taylor 2018). If communities are empowered with the skills and tools for monitoring, it enhances their climate action knowledge and involvement.

14.2.3 Inclusive Governance Frameworks

The principles of inclusive governance emphasise equity and fairness, ensuring that all individuals, regardless of their background, have a voice in decisions affecting their development and policies. This commitment is reflected in South Africa's Constitution and Bill of Rights (1996), guiding local government policies to engage communities in determining their service delivery priorities (South African Constitution 1996). National policies such as the Strategic Framework of Water Services (2003) also recognise women as key water users. This is based on their extensive local knowledge of water sources; they also play a vital role in water management in their households and communities.

In a rural South African project MUS, when communities established Innovation Forums that ensured diverse representation in project management, women were elected and participated in decisions on project implementation (Van Koppen et al. 2021). Involving communities in all aspects of project management, including construction, meant that vulnerable households (often female and child-headed households) were employed in local projects, earning stipends vital for livelihoods. Aligned to SDGs 5 and 8 (Figure 14.1), women influenced spending on safer and more user-friendly infrastructure involving less drudgery. Innovation Forums thus amplified the voices of marginalised groups, often overlooked in decision-making.

This inclusive approach strengthens social justice and cohesion by allowing insights from all community members to inform policies. It helps restore individuals' sense of worth and promotes trust among different community groups, reduces conflict, and encourages connection. Furthermore, inclusion upholds human rights and empowerment, as everyone desires to have their rights respected and contribute to shaping their society. Engaging communities builds a sense of ownership, motivating them to sustain beneficial systems. To achieve greater project ownership Innovation Forums were established at the local level. These enable the beneficiaries and stakeholders to work together for the community's benefit. MUS also established four learning platforms at local, district, provincial, and national levels, where stakeholders, including politicians, officials, community members, and traditional authorities, could collaborate on project progress and implementation.

14.2.4 Participatory Sustainable Model

Figure 14.2 illustrates the participatory model used in both the urban and rural case studies and reflects how community members participated in each step from inception. The model highlights how integrating climate resilience and inclusive governance enhances sustainable development, drives economic growth, and improves climate adaptation. It shows how community engagement and participation help define the problem through implementation and skills transfer for long-term sustainability. It draws from urban planning experiences in the CoT alongside practical insights from the two Limpopo initiatives, showing that participatory governance can be an effective pathway to achieving climate resilience and aligning with the SDGs. The model shows that in each process; the community members are actively involved in implementation.

14.3 CASE STUDY 1: URBAN SUSTAINABILITY AND CLIMATE RESILIENCE

Leck et al. (2025) state that urban areas often must balance economic growth and service delivery with effective climate adaptation and mitigation strategies. However, South African cities often struggle with ageing infrastructure, inadequate stormwater drainage (Ruiters and Amadi-Echendu 2022), and water and energy insecurity issues (Canton 2021; World Bank 2020). Urban areas face environmental problems from the early industrialisation of wealthy countries, while dealing with extreme climate conditions exacerbated by social inequalities, governance gaps, and heat island effects (Sachs 2015; Zhai et al. 2018). The combined challenges highlight the critical need to incorporate climate adaptation into urban planning to support resilient infrastructure, secure water and energy access,

FIGURE 14.2 Six-step participatory sustainability model.

Source: Adapted from Adank, van Koppen, and Smits (2012).

and promote inclusive economic growth (United Nations 2024). South Africa's socio-economic disparities make these challenges worse, as many informal settlements do not have basic services like clean water and energy, which increases poverty and unemployment (Statistics South Africa 2022; World Bank 2020; Oskam et al. 2021). The inequalities are worsened by the deteriorating infrastructure and increasing climate-related events, such as floods and droughts (UNEP 2019). To drive meaningful climate action, inclusive and participatory governance models must be adopted, as they empower communities and encourage cross-sector collaboration (Marzo et al. 2023). Thus, African cities must align their local strategies with global sustainability frameworks when addressing climate change and socio-economic inequalities, enabling them to have resilient and inclusive development.

The CoT has demonstrated how sustainable urban transformation can be supported and achieved through participatory governance to enhance climate resilience. This improvement reflects how the city has learned from previous projects. The CoT 2022–2026 Integrated Development Plan (IDP) demonstrates increased stakeholder engagement, improved climate resilience initiatives, and more diverse community engagement platforms. The city is being more collaborative by using WhatsApp messaging to invite communities to its IDP engagement sessions and more platforms to access residents. Rather than complaint-driven engagements during 2017–2021, this has enabled improved engagement, communication, and addressing of challenges.

During the 2022–2026 IDP development phase, the CoT conducted engagement sessions in all its regions; however, the residents of Ward 4 (Centurion) were not pleased with the ward councillor's limited engagement, which led to poor community participation (CoT 2023); they wanted a more inclusive and participatory process. This highlights how critical transparent and inclusive governance is to strengthening public participation. Owing to socio-economic disparities in digital access and institutional capacity, the community engagements in the CoT differ between suburban areas and townships; the townships face logistical barriers (including mobility) and relatively limited digital access, which results in low participation (CoT 2023). The CoT thus included hybrid engagement platforms, door-to-door campaigns, and radio discussions to build trust and enhance inclusive participation. This was particularly effective in Hammanskraal, a township where inadequate service delivery and water issues had led to community unrest (Dube and Telukdarie 2025).

Between the 2017 and 2021 and 2022 and 2026 IDP cycles, the CoT moved from foundational climate planning to a more structured implementation model. The adoption of a project-stage gate mechanism has improved the environmental screening of development proposals and aligned them with sustainability priorities (CoT 2022). Community engagement has strengthened, promoting collaboration on urban resilience projects (CoT 2023). From 2020 to 2024, the city launched climate initiatives where residents, experts, and businesses jointly identified local risks and designed adaptation strategies (CoT 2023). In 2022, the Tshwane Climate Action Plan was introduced, outlining a roadmap to build climate resilience by 2050, particularly in vulnerable communities (CoT 2022; Democratic Alliance 2022). This highlights the value of inclusive governance and offers scalable models for other rapidly urbanising cities (Canton 2021; Baker, Crump, and Harris 2019).

The Pretoria East Urban Biosphere Reserve (PEUBR) initiative, located in the Magalies Biosphere Reserve within the Magaliesberg Mountains (Magaliesberg Biosphere 2025), spans between Rustenburg in the west and Tshwane in the east. The CoT rehabilitated 2,261 hectares of land that had experienced accelerating biodiversity loss and ecosystem degradation from urban expansion and climate variability, thus setting a yardstick for climate resilience. The project thus integrated ecological restoration with citizen science to implement sustainable infrastructure development. It aimed to restore ecological infrastructure and improve urban quality of life through community-led, sustainability-focused interventions.

The participatory approach included extensive citizen science that yielded over 68,000 environmental observations, forming a baseline for future BioFin programming (AIPH 2025). Thorough socio-ecological assessment via multi-stakeholder platforms included government departments, academic institutions, the private sector, civil society, and local communities. It ensured the restoration strategy reflected diverse perspectives and strengthened collective stewardship (Schoeman and Oberholster 2023). The water–energy–food nexus analysis provided an understanding of interdependencies between water, climate, food systems, land, and rewilding priorities. It guided targeted interventions such as afforestation, wetland rehabilitation, and invasive alien plant removal. This resulted in measurable improvements in water quality and ecosystem health (AIPH 2025), while the community was empowered with competencies in ecosystem services mapping, environmental monitoring, and sustainable agriculture through the targeted capacity building. The PEUBR demonstrates that integrated, participatory approaches can align local restoration efforts with global climate action and the SDGs, offering a replicable model for sustainable urban development and enhanced climate resilience across Africa (AIPH 2025).

In 2023, the CoT signed the international net-zero carbon buildings declaration, aiming for net-zero carbon emissions in new buildings by 2030 and all buildings by 2050. This commitment involved training building control officers and implementing the green building development and net-zero carbon building by-law and policy to promote energy-efficient and sustainable building practices. This involves a collaboration between the South African–German Energy Programme, German Ministry for Economic Development and Cooperation, the Swiss State Secretariat for Economic Affairs, South African Local Government Association, the South African National Energy Development Institute, and Sustainable Energy Africa (Berlinton 2023). This is important, as the building and construction sector remains a critical area for achieving global net-zero carbon and resilience goals, accounting for approximately 34% cent of global energy-related CO_2 emissions and over 32% of global energy demand in 2023 (UNEP 2025).

The CoT piloted a project with eight NGOs to implement localised climate action interventions, empowering communities and schools across its seven regions (AIPH 2022). It focused on practical, community-based responses to climate change, rooted in the approved Tshwane Climate Action Plan objectives. It sought to strengthen resilience, protect environmental quality, and encourage more sustainable daily practices. This included planting trees and shrubs, waste minimisation, water harvesting, school gardens, and active mobility education. Trees were planted at forty schools, recycling and cleanup initiatives were set up, water tanks and bicycles were given to the schools, and home gardens were planted in twenty households (AIPH 2022). This collaboration enhanced local climate awareness and provided tangible, cost-effective environmental solutions.

In another initiative, the CoT embarked in 2023 on a participatory project to transform urban blue-green spaces and enhance water security through nature-based solutions. Collaborating with the Universities of Pretoria and Aarhus, it aligned with the city's water security strategy and its partnership with a Danish water company (Aarhus Vand) and focused on restoring ecosystems, addressing sedimentation, and promoting sustainable urban infrastructure. Stakeholder workshops in the diverse settings of Mabopane and Magnolia Park involved participatory community engagement, which allowed residents and students to express their perceptions of water and climate risks. Stakeholders co-designed adaptive measures, including river renaturation, wetlands, and bioretention (thus mitigating flooding and pollution while enhancing urban liveability; Ramadie 2024). This integrated approach not only addresses environmental pressures but also strengthens social inclusion by empowering communities to take ownership of sustainable land use practices. Through active stakeholder engagement with government, industry, academia, and civil society, the project established strong collaborative networks that support the long-term sustainability of its interventions.

Other regions of South Africa have also implemented climate adaptation projects in 2023. In the uMkhomazi catchment in Kwa-Zulu Natal and the Welbedacht catchment area in the Free State, GIS mapping and citizen science initiatives identified sediment hotspots, informing interventions that have contributed to reduced soil erosion and improved water resource management (Tshabalala et al. 2024). Produce from local community gardens was sold to local restaurants, and home gardens were established for the community members.

The CoT approach highlights how participatory sustainability models can serve as a foundation for building climate-resilient urban systems. Through structured intentional collaboration and context-specific adaptation efforts, Tshwane has improved its ability to respond to environmental risks while advancing long-term development priorities.

Case studies from Tshwane and other South African settings demonstrate how integrating Indigenous knowledge, collaborative governance, and technical expertise can produce effective, scalable responses to climate change. These initiatives underscore the value of participatory planning and community-led decision-making in enhancing urban resilience and promoting more equitable development. When aligned with global sustainability frameworks, such locally grounded efforts can drive meaningful transformation. In Tshwane, sustained stakeholder engagement and climate adaptation planning have improved water governance and reduced climate-related risks, reinforcing the importance of inclusive, multi-stakeholder approaches in achieving the Sustainable Development Goals (Ansell and Gash 2008; Healey 2020).

14.4 CASE STUDY 2: RURAL CLIMATE RESILIENCE IN LIMPOPO

The MUS project was a partnership project led by the WRC, with International Water Institute (IWMI) as a knowledge partner, and Tsogang Water and Sanitation (a local NGO) as a socio-technical facilitator. The project was fully funded between 2016 and 2022 by the African Water Facility of the African Development Bank. The implementation sites (six in total) were in Sekhukhune District Municipality (Ga-Mokgotho, Ga-Moela, and Phiring) and Vhembe District Municipality (Tshakuma, Khalavha, and Ha-Gumbu) (Figure 14.3).

The project aimed to operationalise MUS services in South Africa through demonstration investments, awareness, research, and finance mobilisation. From the start, participatory methods shown in Figure 14.2 ensured that communities actively shaped and implemented the project. Involving them in planning, budgeting, construction, and monitoring fostered ownership and affirmed their right to influence decisions. Capacity building and training equipped communities to manage infrastructure, sustain systems, and hold authorities accountable.

14.4.1 Giyani Local-Scale Climate Resilience Programme

The lessons learned from MUS showed it needed to be upscaled, which led to the conceptualisation of the GLSCRP in Giyani, Limpopo. The programme aimed to build local resilience to climate change and provided far-reaching support for marginalised and vulnerable populations, particularly poor and rural communities. The Giyani area faces extreme weather patterns, with rainfall averaging 200–400 mm per annum (Mmbadi 2019). Poor access to water by communities

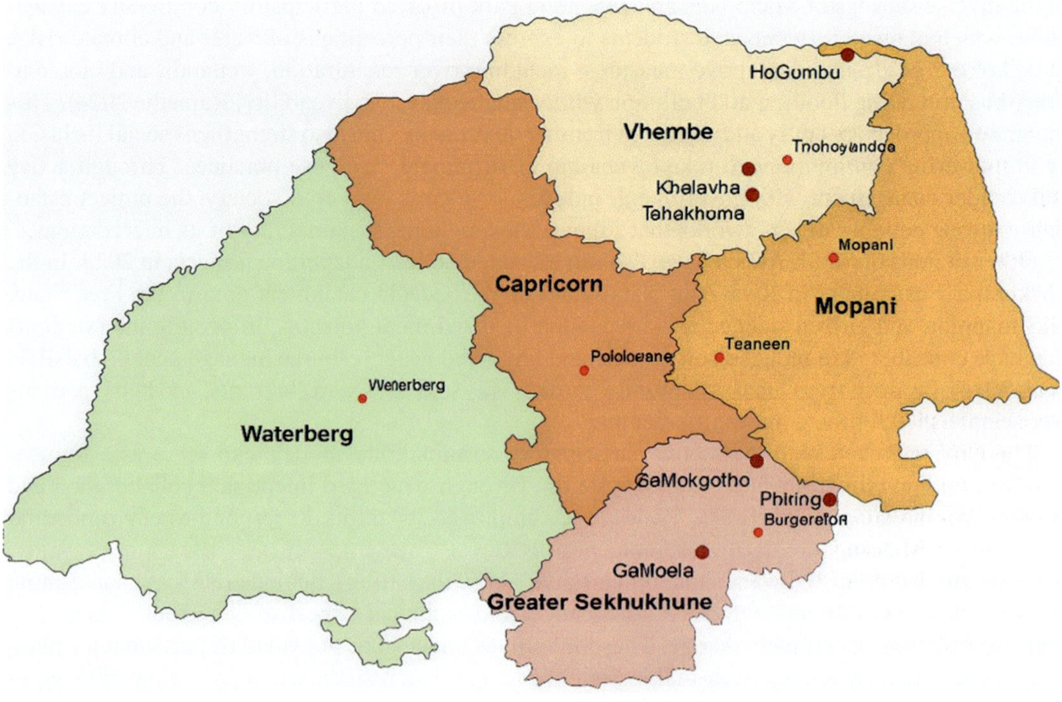

FIGURE 14.3 Operationalising community-led multiple-use services project demonstration sites (Van Koppen et al. 2021).

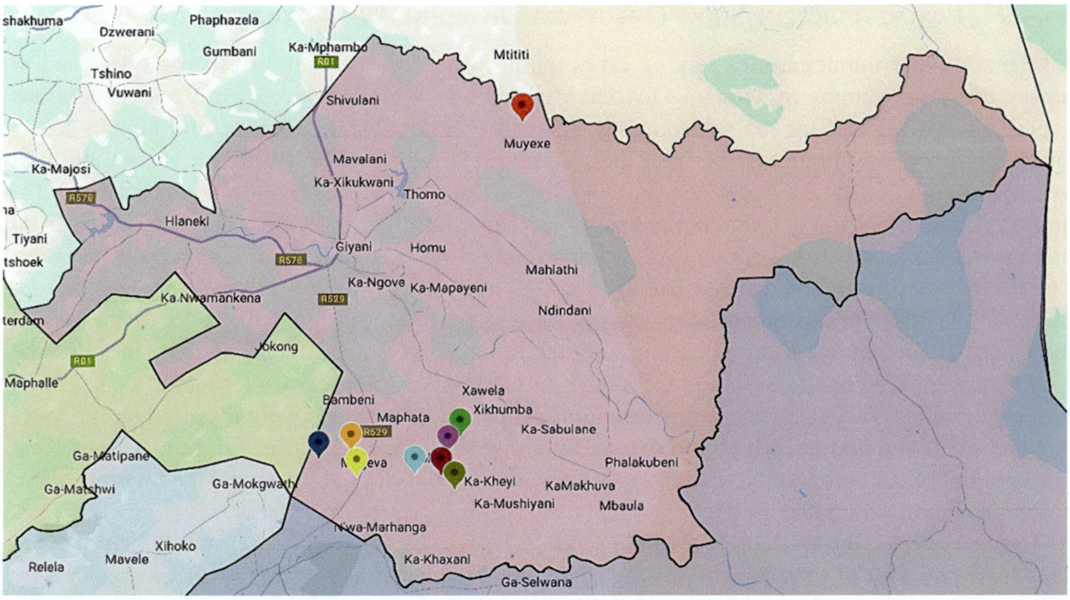

FIGURE 14.4 Map of the GLSCRP sites (du Toit, Tshogang, and WRC 2024).

can be attributed to both reduced rainfall and inadequate municipal service delivery, leading to significant backlogs and distrust in the government's ability to meet communities' needs (du Toit, Tshogang, and WRC 2024) (Figure 14.4). South Africa has also experienced electricity supply failures in recent years, further disrupting water supply, including in rural areas. The Government of Flanders allocated funding to the GLSCRP to demonstrate climate-adaptive solutions and enhance water utilisation practices in Giyani. The programme was led by WRC in collaboration with Tsogang, Association for Water and Rural Development, and the University of the Western Cape (UWC).

As in the MUS project, the six-step participatory framework in the GLSCRP ensured inclusion and participation. A total of nine sites were selected, comprising four community water supply systems and five cooperatives or small-scale farms. The community water supply sites (and respective populations) are Mbhedle Water Supply (1,230); Mayephu Water Supply (1,940); Mzilela Water Supply (1,150); and Matsotsosela Water Supply (2,300). The agriculture or cooperative sites are Muyexe Community Project, and farms at Daniel Ravalela, Dzumeri, Loloka, and Dzumeri (Farm 2).

The programme recognised Mopani District Municipality as the Water Services Authority responsible for water provision in the Giyani area; it also recognised the Department of Water and Sanitation's plans to support the Water Services Authority with the implementation of a large bulk water supply project to serve greater Giyani. Bulk water supply projects are inherently complex with lengthy procurement, so communities can wait years for water from a project. Ongoing municipal service supply disruptions have violated communities' right to water and sanitation, as enshrined in the South African Constitution. This has negative consequences on communities' health, livelihoods, food security, and even social cohesion, as demonstrated by the increasingly destructive service delivery protests.

Setting out to improve access to water for unserved communities, farmers, and cooperatives does not replace the municipal Water Services Authority or the bulk water supply programme of the Department of Water and Sanitation; rather, the GLSCRP aimed to augment these two government initiatives. Jovanovic and WRC (2024) refer to the nine projects as a "no regret" backup action if municipal services fail.

14.4.2 Implementation of Solar Energy and Diversified Water Sources

Like many other municipalities relying on borehole water, diesel, fuel, or electricity is used to power the water pumps in Mopani District Municipality. This dependence on fossil fuels has negative consequences for the environment, air quality, groundwater sources, and human health and well-being. The high cost of fuel, diesel, and electricity makes it difficult for municipalities to provide consistent water services, leading to prolonged water supply interruptions for households that rely solely on municipal supplies. These families often must purchase water from private sources or revert to using unsafe water, which places a financial burden on their incomes. The use of fossil fuels also harms groundwater and air quality. Moreover, power outages disrupt electricity supplies to those boreholes relying on electricity for pumping. To tackle these issues, the GLSCRP, in collaboration with communities, implemented clean solar energy solutions. The NGO Tsogang Water and Sanitation led a transparent process to procure solar materials based on the communities' energy needs. With community support, solar panels were installed on the roofs of borehole pump houses, which are owned by either the Mopani District Municipality or local farmers. In typical community water supply sites, the old diesel pump was replaced with a newer borehole (Figures 14.5 and 14.6).

The boreholes were located close to households, both to bring water nearer and to help prevent vandalism and theft of solar panels. Testing of the borehole water quality and quantity was done by the appointed contractor, with UWC (the programme's knowledge partner) overseeing implementation and taking samples to their laboratory for further testing (Jovanovic and WRC 2024). To ensure safe drinking water, small reverse osmosis plants were provided (Figure 14.6).

To ensure sustainability, it was also essential for the pump operator to be trained on the use of a manual EC/pH/temperature meter to measure water quality, a voltmeter to measure power supply voltage, a dip meter to measure groundwater levels, and a manual plastic rain gauge to record rainfall. Additional monitoring equipment was bought by the programme for monitoring and training was provided to beneficiaries. The operators were encouraged to work cooperatively, sharing lessons and monitoring equipment because there were insufficient funds for all nine sites to have their own equipment.

Jovanovic and WRC (2024) reported that the borehole at that time was pumping for 3–4 hours per day at high yields and with very high discharge rates in the morning that decreased towards the afternoon. The Innovation Forum water committee manages the maintenance of the system; all households benefitting from the scheme had organised themselves into a savings club where

FIGURE 14.5 Final installation scheme of solar-powered borehole pump house (Jovanovic and WRC 2024).

FIGURE 14.6 Installations at Mbhedle Village. Clockwise from top left: Solar panels installed on the roof of borehole H14-1925 pump house; 15 JoJo tanks for bulk water storage; water treatment; water treatment ion exchange vessels; street taps for collection of drinking water; livestock watering (Jovanovic and WRC 2024).

TABLE 14.1
Farm Production at GLSCRP Demonstration Sites (Jovanovic and WRC 2024)

Farm	Period	Total income per area cultivated (ZAR/ha)	Total water used per area cultivated (m^3/ha)	Total income per water used (ZAR/m^3)
Macena	January 2023–November 2023	5,824	132	44.2
Macena	December 2023–August 2024	6,947	141	49.3
A hi Tirheni Farm	June 2023–November 2023	127,106	223	570.8
A hi Triheni Farm	January 2024–August 2024	185,687	1,018	182.4
Duvadzi Farm	11 December 2023–30 August 2024	26,820	1,008	26.6
Matsambbo Ngamba	03 February 2024–30 May 2024	16,475	1,068	15.4

they contribute money monthly to pay for small repairs, including the purchase of resin for maintenance of the small treatment plants. The operator is supposed to be paid by the Mopani District Municipality; however, this is not consistent, thus posing a risk that the operator might abandon the job in search of a more reliable income.

Solar-powered energy systems have proven financially viable; households have saved money they previously used to purchase diesel and fuel. Solar energy has also assisted the Water Service Authority to save millions of rands previously spent on fuelling borehole pumps. Consistent water supply has also enabled increased productivity at cooperatives and farms (see Table 14.1).

14.4.3 Community Engagement and Co-Creation of Low-Cost Water Systems

Participatory approaches to development have been in use for decades and continue to play a vital role in fostering inclusion, ownership, and cohesion in programmes and projects, leading to their sustainability. These approaches recognise that everyone's contributions to a project are valuable, regardless of their background, religion, class, gender, race, or age. The goal is to ensure that everyone can be heard and contribute. These are the principles that the six-step participatory framework is embedded in, as shown in Figure 14.2.

For the GLSCRP, the six-step participatory framework (Figure 14.2) began with stakeholder collaboration to establish project goals and timelines. Initial consultation involved key figures such as the Traditional Authority and the Councillor, followed by discussions with various water user groups to define roles, responsibilities, and execution plans. Local structures, including an Innovation Forum, were created for project oversight (du Toit, Tshogang, and WRC 2024). In the diagnosis stage, stakeholders assessed water challenges, explored source options, and identified issues with existing boreholes. If the quality of a borehole was inadequate, alternative solutions like drilling new boreholes were considered. The socio-technical facilitator, Tsogang Water and Sanitation, promoted transparency in budgeting, helping communities set realistic expectations and prioritise infrastructure improvements over (Van Koppen et al. 2021). During implementation, community members contributed labour, often exceeding the value of their stipends. Stakeholders provided technical expertise, ensuring communities chose affordable and accessible infrastructure options. If a contractor was involved, the Innovation Forum functioned as the project managers to ensure quality, adherence to timelines, and cost-effectiveness.

The last step, "use of infrastructure," occurs when the project is completed and handed to the community. Inspection by all parties ensures it functions properly, and communities receive training on its use and maintenance. Community involvement throughout the project cycle fosters buy-in and builds the necessary skills for managing infrastructure. Community-led MUS water systems have proven to be more time- and cost-effective than contractor-led models, as communities are more invested in achieving durable infrastructure that benefits them for generations.

14.4.4 Skills Development and Long-Term Sustainability

The foundation of resilience is sustainability; long-lasting services can help cushion communities against the negative effects of climate change. Proper operation and maintenance are essential for infrastructure to function effectively for its intended duration. Therefore, it is crucial to invest in the skills of community and municipal officials to manage these systems, allowing communities to withstand extreme weather and other climatic challenges. The GLSCRP prioritised building social cohesion through training workshops on conflict management and the effective operation of local structures like Innovation Forums.

Participants learned project management skills and the roles of committee members. Specialised capacity-building workshops focused on Institutional Social Development, covering topics including conflict resolution, committee constitutions, meeting facilitation, and diversity. Technical training sessions for communities addressed water quality monitoring, basic construction, and the operation of systems like pumps and solar panels. These efforts were aimed at enhancing community capacity to manage systems and themselves during and after project implementation. Additionally, it was essential to challenge entrenched gender roles and encourage reflection on past practices versus current needs.

Municipal officials received essential training on operating and maintaining water systems and supporting Innovation Forums for complex technical needs. Water quality monitoring is also a critical skill owing to the risk of contamination from various sources. Officials learned that while communities can manage basic operations and maintenance, municipal staff must tackle more complex issues. This co-management approach allows communities to manage daily operations, while officials address technical challenges (Van Koppen et al. 2021).

Training and capacity building enhance beneficiaries' skills and foster ownership, which helps mitigate vandalism, a widespread problem in infrastructure projects. The GLSCRP and MUS experiences emphasise the importance of community buy-in instead of over-reliance solely on contractors. Communities received training on securing solar panels, daily management, sourcing parts, and reporting faults, therefore ensuring reliable water access for both domestic and productive use. Such training also created employment opportunities; for instance, Mr Aaron Masolo from Ga-Moela was hired by a local mine after gaining skills in pipe laying in the MUS project. In the GLSCRP, operators improved their understanding of monitoring and reporting systems.

Training in solar-powered systems helps operators determine the best times to pump, ensuring proper maintenance and minimising breakdowns. Operators must know how the systems work and how to troubleshoot issues and source replacement parts. While cooperatives find it easier to assume responsibilities owing to their investment, community water schemes involve shared ownership, which complicates management. Strengthening local structures and Innovation Forums is essential for ensuring a sustainable water supply for communities. The skills and knowledge investment made in communities enables communities to use and operate the systems and to conduct monitoring services for both water (water abstraction quantities and quality) and energy supply. Communities are thus responsible for the daily management of the services, which empowers and affirms their capabilities. The programme's engagement using the six-step participatory framework ensures full participation throughout the stages. This has built the communities' sense of ownership/involvement, being listened to and trusted to make decisions based on their circumstances, knowledge, and context.

14.5 COMPARATIVE ANALYSIS: URBAN VERSUS RURAL APPROACHES TO SUSTAINABILITY AND CLIMATE RESILIENCE

Urban and rural climate resilience projects share core participatory principles yet diverge in approach (Figure 14.7). In urban areas, digital platforms enable efficient stakeholder engagement and data-driven planning, while rural settings often require interpreters and tailored capacity building to overcome lower formal skill levels (Dube and Telukdarie 2025). Both contexts use citizen science and community workshops to co-create adaptive solutions; however, urban projects tend to yield measurable improvements in infrastructure resilience, whereas rural initiatives focus on restoring ecosystems and enhancing livelihoods.

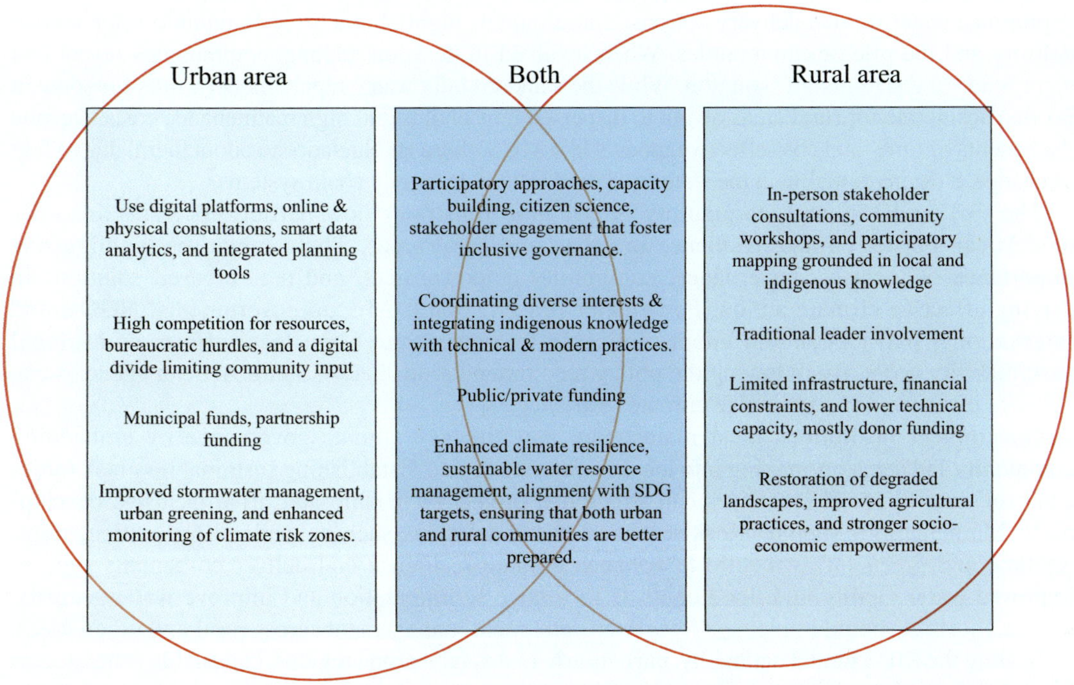

FIGURE 14.7 Comparison, contrasts, and similarities in the urban and rural case studies.

Despite these differences, inclusive governance and multi-stakeholder collaborations remain central, ensuring that locally driven climate adaptation strategies align with global sustainability frameworks (Adeleke et al. 2020). This integrated approach balances technological innovation with community-driven insights, demonstrating that inclusive governance can be effectively tailored to diverse contexts, providing replicable models for sustainable climate action globally (Adeleke et al. 2020).

Based on the CoT's and Limpopo participatory sustainability models and the broader challenges in scaling CBA initiatives in South Africa, several recommendations emerge for both urban and rural contexts. First, CBA initiatives should be integrated within formal governance frameworks with dedicated coordination units that align community efforts with municipal development plans, while extending tailored support to rural communities facing digital access and capacity constraints (Singh 2023; Donaldson 2015). Investment in innovative digital engagement platforms and community-led monitoring systems can drive effective data collection and adaptive management in urban settings, whereas rural areas require on-ground capacity building and mobilisation efforts.

Additionally, targeted training programmes are essential to empower both urban and rural communities with the necessary skills for sustainable climate adaptation (Dube, Maphosa, and Scott-Goldman 2014; Konečný et al. 2024). Finally, robust monitoring, evaluation, and reporting frameworks that combine quantitative and qualitative data will ensure transparency, accountability, and scalability of successful initiatives (Leal Filho et al. 2022). By adopting these integrated strategies, policymakers can replicate Tshwane's successes across diverse settings, fostering sustainable development and climate resilience in line with global SDG targets.

14.6 POLICY IMPLICATIONS AND RECOMMENDATIONS

South African legislation and policies emphasise inclusion and participation, which are vital for effective climate action. Communities have a crucial role in shaping the services they receive, based on their needs and capabilities. Community-led MUS projects have been piloted in over twenty rural communities in Limpopo province, demonstrating the potential for scalability. However, there has been an inadequate political will to adopt these models, even though district municipalities can choose appropriate water service delivery options. These models highlight the need to rethink water service delivery and the role of communities. When involved in decision-making, communities reveal that there is no "one-size-fits-all" solution. While the standard bulk water supply model works for some, it is often unsuitable for rural areas owing to dispersed households and high maintenance costs. Despite the benefits of time- and cost-effective models like MUS, there is reluctance to adopt them, due to fear of change, a desire to maintain the status quo, and inflexible supply chain systems.

The CoT and Limpopo community programme illustrates how participatory sustainability models can bolster climate resilience in urban and rural areas. These initiatives highlight the importance of inclusive governance, community empowerment, and nature-based solutions in driving effective climate action. For greater impact, South African governments, NGOs, and international partners should adopt replicable frameworks to align local efforts with global sustainability goals. By following the policy recommendations below, South African agencies can enhance inclusive and sustainable climate resilience.

Government institutions must mainstream participatory climate governance by embedding community-led decision-making into local climate strategies. Establishing sustainability task forces and providing financial incentives can strengthen project ownership and drive inclusive development. Municipalities should prioritise nature-based solutions such as wetland restoration, reforestation, and green infrastructure to enhance climate resilience and deliver co-benefits such as improved water quality and flood control. To reduce sedimentation and improve water security, local authorities should implement integrated catchment management across rural and urban areas.

Scaling the MUS model nationally, particularly in drought-prone regions, can enable water access for domestic and productive uses, supported by solar-powered pumping and purification systems. Real-time citizen-led water quality monitoring and environmental data collection should be institutionalised through partnerships with academia and the private sector. Additionally, innovation hubs should be created to pilot sustainable infrastructure solutions tailored to local needs. Evidence from community-driven projects, such as the GLSCRP, must be translated into scalable national policy frameworks. These actions will ensure that inclusive, adaptive climate responses are not only locally rooted but nationally supported, contributing meaningfully to both resilience and the achievement of SDG targets.

Ensuring the sustainability of community-led climate adaptation efforts requires both inclusive participation and context-appropriate financing. In underserved and rural areas, limited access to formal funding calls for blended approaches that integrate community, government, and donor resources. A notable model from the GLSCRP involves local savings clubs. These traditional mutual aid systems pool monthly contributions from community members to fund maintenance of water infrastructure, such as solar-powered boreholes, and pay trained local operators responsible for water quality monitoring and system upkeep. This approach enhances ownership, ensures reliability, and supports livelihoods. In scaling such initiatives, small grants and microfinance secured through partnerships with donors, NGOs, and development institutions provide start-up capital, while communities sustain operations through local funds and in-kind support.

Municipalities are also experimenting with co-financing, matching local funds with public contributions to deepen community investment. Financial literacy training and cooperative governance frameworks are embedded in programme design to promote transparency and accountability. Service fees, agreed upon by communities, form part of the cost-recovery model, keeping systems affordable and functional. As capacities grow, oversight shifts from NGOs to local governance bodies, ensuring long-term leadership. This model demonstrates that, with adaptive financing and shared responsibility, community-managed infrastructure can be both scalable and resilient. Examples of financing options that have worked are shown in Table 14.2.

Recommendations for other African and global countries are as follows:

Countries need to emphasise the need to develop national climate resilience roadmaps that incorporate community-driven governance, like South Africa's PEUBR and Giyani initiatives. It is crucial to invest in region-specific participatory water management models, integrating MUS in water-scarce areas across Africa. Strengthening transboundary climate cooperation is also essential, drawing lessons from the joint efforts of South Africa and Lesotho in siltation management governance. Cities should embed nature-based solutions into urban resilience planning, aligning with the UN's New Urban Agenda (Evans et al. 2016). It is important to expand carbon-neutral urban policies, ensuring that new buildings and renovations adhere to net-zero carbon building standards, as adopted in Tshwane. Strengthening urban–rural resilience linkages and integrating catchment-based water governance into metropolitan sustainability plans are critical steps.

In addition, strengthening collaboration in the Global South for climate action is vital. This can be facilitated through knowledge exchange programmes between African, Latin American, and Asian

TABLE 14.2
Financing Options for Scaling Up Community-Based Projects

Funding Source	Description
Village water and investment schemes	Local savings clubs are organised to contribute fairly small amounts of money based on their affordability towards basic repairs and maintenance.
Government grants	Dedicated government streams fund infrastructure and more complex technical repairs and maintenance. Some of the funds can be devolved to communities, based on the capability to manage and account for funding.
Donor/funder support	Working through implementing agents, funders/donors can support infrastructure improvement and/or maintenance.
Private sector support	Private–public partnership lending institutions can work with the government or its entities to invest money in infrastructure improvements or maintenance.

cities to scale participatory climate action frameworks. As developing countries, this collaboration will be beneficial. During the G20 Finance Ministers and Central Bank Governors Meeting in February 2025, held in South Africa, key discussions centred on enhancing financial support for emerging economies and developing countries (DIRCO 2025). The agenda emphasised climate finance, sustainable finance, and establishing a social protection floor. South African President Cyril Ramaphosa highlighted the urgency of mobilising resources for a just energy transition, underscoring the necessity for increased funding to meet the goals of the Paris Agreement (DIRCO 2025). He also addressed the importance of debt sustainability for developing economies, noting that rising debt levels are crowding out essential social services and infrastructure investments (Miriri 2025). Thus, African countries should establish regional climate resilience funds to support community-led adaptation projects, ensuring sustainable financing mechanisms for long-term implementation. Furthermore, it is important to align national climate policies with global sustainability frameworks, leveraging South Africa's experience in embedding local action within the SDGs and the Paris Agreement.

In both the Tshwane urban programmes and the rural Giyani initiatives, women have been positioned not only as beneficiaries but as decision-makers and implementers. In Giyani, women were elected to Innovation Forums tasked with managing infrastructure rollouts and ensuring that design and maintenance protocols reflected the specific needs of women and children. These leadership roles enabled women to influence decisions regarding the safety and usability of infrastructure, while also ensuring their participation in construction and maintenance activities that provided stipends and livelihood support. Similarly, training modules in water quality monitoring and solar energy maintenance were tailored to include unemployed young people, especially women, equipping them with technical skills that enhanced their employability.

Youth participants in Tshwane's climate action schools programme gained firsthand experience in environmental stewardship through initiatives such as school gardens, tree planting, and water harvesting systems. Mentorship was a key component, pairing young people with local leaders, municipal officials, and technical experts, thereby fostering intergenerational learning. These strategies contributed directly to improved resilience outcomes by diversifying the local skill base, increasing system maintenance capacity, and embedding social inclusion into climate governance. The deliberate inclusion of women and youth not only advanced the equity objectives of SDG 5 and SDG 8 but also reinforced community ownership, innovation, and the sustainability of adaptation outcomes.

14.7 CONCLUSION

There is a need to understand how nature-based adaptation approaches can be tailored to meet the specific needs and conditions of South Africa, especially when international partners are involved. Most research studies and design protocols on nature-based solutions originate from developed economies and may not be suitable for countries like South Africa. Therefore, it is essential to contextualise interventions to fit South African environments, given that nature-based adaptation initiatives can function differently in urban areas versus rural settings. Proper contextualisation involves intensive engagement with local communities, using their perspectives and expertise in every aspect from the design of nature-based solutions and restoration activities to the communication of results within and beyond the local context. Additionally, the inclusion of Indigenous knowledge is crucial for success.

Empowering communities and transferring skills are crucial for long-term sustainability. This approach ensures that initiatives can continue even after the project's timeline has ended.

Experience from similar projects indicates that once funding concludes, rehabilitation efforts often falter owing to inadequate community involvement and insufficient capacity building from the outset. It is also essential to recognise communities as valuable resources in the management of water resources and other development initiatives. Communities often have numerous ideas about potential solutions to their development challenges, but they require support through training, capacity building to enhance their skill sets, and technical assistance in areas where they lack expertise.

An ideal co-management arrangement promotes sustainability by assigning communities responsibility for the daily management of simple tasks, while government or municipal officials

FIGURE 14.8 *Five* practical steps for scaling community-led resilience projects.

can manage more complex repairs and maintenance as needed. This creates a win–win situation: municipal or government officials do not have to travel long distances to address simple maintenance issues that the community could manage, and, in turn, communities can rely on government officials for repairs and maintenance that they cannot perform themselves. This collaboration restores trust in state capabilities and fosters a sense of stewardship.

Rural communities in many developing and underdeveloped countries maintain a keen sense of community and social cohesion, which can be eroded by urbanisation. These communities possess a sense of self-reliance and prefer to act for themselves rather than rely solely on government support. This notion, referred to as "letsema" in many southern African countries like Namibia, Botswana, and South Africa, emphasises community mutual support and initiative. In South Africa, efforts are underway to revive this spirit of self-reliance through volunteerism, encapsulated in the concept of "ubuntu," which gained prominence through the country's first democratically elected president, Dr Rolihlahla Nelson Mandela. Similarly, in Tanzania, the late President Julius Nyerere introduced "ujamaa," a system that emphasises self-reliance and communal ownership and labour. Figure 14.8 illustrates an adaptation playbook of practical steps for donors and policymakers for scaling up community-led resilience.

REFERENCES

Adank, Marieke, Barbara van Koppen, and Stef Smits. 2012. "Guidelines for planning and providing multiple-use water services." *IRC International Water and Sanitation Centre and International Water Management Institute: The Hague, The Netherlands*.

Adeleke, Babatunde, Deborah Robertson-Andersson, Gan Moodley, and Simon Taylor. 2020. "Aquaculture in Africa: A comparative review of Egypt, Nigeria, and Uganda vis-a-vis South Africa." *Reviews in Fisheries Science & Aquaculture* 29 (2):167–197.

Ansell, Chris, and Alison Gash. 2008. "Collaborative governance in theory and practice." *Journal of Public Administration Research and Theory* 18 (4): 543–571. https://doi.org/10.1093/jopart/mum032.

Baker, Elaine, John Crump, and Peter Harris. 2019. *Global Environment Outlook (GEO-6): Healthy Planet, Healthy People*. United Nations Environment Programme (UNEP).

Berlinton, Tebatso. 2023. *Net Zero Carbon Buildings Are Essential for Sustainable Service Delivery*. City of Tshwane.

Birkmann, Joern, Emma Liwenga, Rajiv Pandey, Emily Boyd, Riyanti Djalante, François Gemenne, Walter Leal Filho, Patricia Pinho, Lindsay Stringer, and David Wrathall. 2022. "Poverty, livelihoods and sustainable development." In Climate Change 2022: Impacts, Adaptation and Vulnerability. Contribution of Working Group II to the Sixth Assessment Report of the Intergovernmental Panel on Climate Change, eds. H.-O. Pörtner, D. C. Roberts, M. Tignor, E. S. Poloczanska, K. Mintenbeck, et al. Cambridge University Press.

Canton, Helen. 2021. "United Nations environment programme—UNEP." In *The Europa Directory of International Organizations 2021*, 188–214. Routledge.

City of Tshwane (CoT). 2022. *City of Tshwane 2022–2026 Integrated Development Plan*. City of Tshwane.

City of Tshwane (CoT). 2023. *2023–2024 Review of the 2022–2026 Integrated Development Plan*. City of Tshwane.

Clarke, Rebecca. 2022. "Climate resilience: What it means and why it matters: Cities and organizations everywhere need effective climate resilience plans." *Treehugger Sustainability for All*. Last Modified 13 January 2022, accessed 23 May. https://www.treehugger.com/what-is-climate-resilience-5184338.

Democratic Alliance, Democratic Alliance. 2022. "City of Tshwane launches climate action plan to ensure a climate-resilient city by 2050." Last Modified June 2022. https://www.da.org.za/government/where-we-govern/2022/06/city-of-tshwane-launches-climate-action-plan-to-ensure-a-climate-resilient-city-by-2050.

Department of International Relations and Cooperation (DIRCO). 2025. "Chair's summary 1st finance ministers and central bank governors meeting Cape Town, South Africa 26–27 February 2025." *G20 Finance Ministers and Central Bank Governors Meeting*, Cape Town, South Africa, 27 February.

Dodman, D., B. Hayward, M. Pelling, V. Castan Broto, W. Chow, E. Chu, R. Dawson, L. Khirfan, T. McPhearson, and A. Prakash. 2022. "Cities, settlements and key infrastructure." In *Climate Change 2022: Impacts, Adaptation, and Vulnerability*. Contribution of Working Group II to the Sixth Assessment Report of the Intergovernmental Panel on Climate Change.

Donaldson, Christopher. 2015. *Community-Based Adaptation to Climate Change: Scaling It Up*. Taylor & Francis.

du Toit, Derick, Tshogang, and Water Research Commission WRC. 2024. *In-House Capacity Development and Support to Local Government under Giyani Local Scale Climate Resilience Programme*. https://award.org.za/projects/giyani/.

Dube, Renias, Beatrice Maphosa, and Judy Scott-Goldman. 2014. "The role of local community institutions in the adaptation of rural and urban communities to the impacts of climate change on water access and use." *Water Research Commission, WRC Report* (1963/1):14.

Dube, Thabile, and Arnesh Telukdarie. 2025. "Integrated systems approach to enhance rural development: Word2vec analysis." *Cogent Social Sciences* 11 (1):2447904.

Evans, Brian, Pietro Elisei, Orna Rosenfeld, Gulnara Roll, Amie Figueiredo, and Marco Keiner. 2016. "HABITAT III–toward a new urban agenda." *Disp-the Planning Review* 52 (1):86–91.

Fatti, Christina Culwick, Fiona Anciano, Charlotte Lemanski, and Margot Rubin. 2023. "Governance of the just urban transition." *Research Gate*:78.

Fourie, David Johannes, and Gerrit Van der Waldt. 2021. "Participative integrated development planning praxis in local government: The case of selected South African municipalities." *Journal of Local Government Research and Innovation* 2:a43. https://doi.org/10.4102/jolgri.v2i0.43.

Fung, Archon. 2015. "Putting the public back into governance: The challenges of citizen participation and its future." *Public Administration Review* 75 (4):513–522.

Gaegane, Lesego. 2024. "Incorporating sustainability into development plans in selected African cities." *Sustainability* 16 (23):10493.

Ghimire, Rajiv, Arun Khatri-Chhetri, and Netra Chhetri. 2022. "Institutional innovations for climate smart agriculture: Assessment of climate-smart village approach in Nepal." *Frontiers in Sustainable Food Systems* 6:734319.

Graham, Mark, and Jim Taylor. 2018. *Development of Citizen Science Water Resource Monitoring Tools and Communities of Practice for South Africa, Africa and the World*. Water Research Commission.

Healey, Patsy. 2020. Collaborative Planning: Shaping Places in Fragmented Societies, 2nd edn. Bloomsbury Publishing (Imprint: Red Globe Press). https://www.bloomsbury.com/in/collaborative-planning-9781137086006/.

Hedelin, Beatrice, Mariele Evers, Johanna Alkan-Olsson, and Anna Jonsson. 2017. "Participatory modelling for sustainable development: Key issues derived from five cases of natural resource and disaster risk management." *Environmental Science & Policy* 76:185–196.

Ileka, Andreas. 2018. *Assessing Community Participation and Mobilization in Development Planning and Service Delivery in the City of Windhoek, Namibia*. Management College of Southern Africa Durban, Southern Africa.

The International Association of Horticultural Producers (AIPH). 2022. "Tshwane non-profit organizations partnership on climate change and environmental management." https://aiph.org/green-city-case-studies/tshwane-south-africa-partnership/.

The International Association of Horticultural Producers (AIPH). 2025. "Tshwane, South Africa: Pretoria East urban biosphere reserve." *AIPH, The International Association of Horticultural Producers*. https://aiph.org/green-city-case-studies/tshwane-sa-biosphere-reserve/.

Islam, Md Saidul. 2024. "Rethinking climate justice: Insights from environmental sociology." *Climate* 12 (12):203.

Jovanovic, N., and Water Research Commission WRC. 2024. *Adaptive Response and Local Scale Adaptation for Improving Water Security and Increasing Resilience to Climate Change in Selected Communities in Giyani, Limpopo—Improved Energy, Ground and Surface Water Solutions Developed with Communities to Optimise and Diversify Water Sources*. https://award.org.za/projects/giyani/.

Jovanovic, N., A. Willoughby, M. Mpambo, K. Phasha, Z. Rikhotso, and B. Petja. 2022. "Feasibility assessment and sustainability evaluation of water supply options for implementation in drought affected areas of Giyani: Enabling resilience to climate change." *Water Research Commission Report* (3039/1):22.

Kajiita, Robert Mutemi, and Simon Murote Kang'ethe. 2024. "Socio-economic dynamics inhibiting inclusive urban economic development: Implications for sustainable urban development in South African cities." *Sustainability* 16 (7):2803.

Kamana, Alanda A., Hassan Radoine, and Chimango Nyasulu. 2024. "Urban challenges and strategies in African cities—a systematic literature review." *City and Environment Interactions* 21:100132.

Keziah, Mayer. 2022. *Exploring Community & Ecosystem-Based Adaptation through Resilience Theory: Referencing a Lesotho Case*. Stellenbosch University.

Konečný, Ondřej, Ondřej Šerý, Tomáš Zavadil, Barbora Duží, Alice Kozumplíková, Jakub Trojan, Stanislav Martinát, Radek Novák, Ondřej Kotek, and Jiří Lehejček. 2024. "Adapting rural communities to climate change: The undervalued potential of agricultural land." *Journal of Rural Studies* 111:103391.

Leal Filho, Walter, Vanessa Levesque, Subarna Sivapalan, Amanda Lange Salvia, Barbara Fritzen, Ronald Deckert, Valerija Kozlova, Todd Jared LeVasseur, Kay Emblen-Perry, and Ulisses M. Azeiteiro. 2022. "Social values and sustainable development: Community experiences." *Environmental Sciences Europe* 34 (1):67.

Leck, Hayley, Zakiyya Atkins, Luka Dreyer, Yakhuluntu Dubazana, Clara Marais, Lorena Pasquini, Tashi Piprek, Meggan Spires, and Kate Strachan. 2025. "Climate change: Crosscutting report." *Available at SSRN 5200218*.

Magaliesberg Biosphere. 2025. "Magaliesberg biosphere." *Magaliesberg Biosphere*. https://magaliesbergbiosphere.org.za/.

Mamokhere, John, and Daniel Francois Meyer. 2023. "Towards an exploration of the significance of community participation in the integrated development planning process in South Africa." *Social Sciences* 12 (5):256.

Marzo, Roy Rillera, Hana W. Jun Chen, Haryati Anuar, Mohamad Khairuddin Abdul Wahab, Mohd Hairulnizam Ibrahim, Indang Ariati Ariffin, Azral Ismawy Ahmad, Joseph Kawuki, and Mohammed Aljuaid. 2023. "Effect of community participation on sustainable development: An assessment of sustainability domains in Malaysia." *Frontiers in Environmental Science* 11:1268036.

Meerow, Sara, Joshua P. Newell, and Melissa Stults. 2016. "Defining urban resilience: A review." *Landscape and Urban Planning* 147:38–49.

Michael, Patrick S. 2019. "Current evidence and future projections: A comparative analysis of the impacts of climate change on critical climate-sensitive areas of Papua New Guinea." *SAINS TANAH-Journal of Soil Science and Agroclimatology* 16 (2):229–253.

Miriri, Duncan. 2025. "Poor, indebted countries need collective relief, African leaders say." *Reuters*, 27 February.

Mmbadi, Elelwani. 2019. "Water security in rural Limpopo in a changing climate: A study of the Greater-Giyani local municipality, South Africa." MSc Thesis. https://univendspace.univen.ac.za/server/api/core/bitstreams/2ba9213c-4d7d-433b-a3a1-f1a81f9719b1/content.

Oskam, Marieke J., Milena Pavlova, Charles Hongoro, and Wim Groot. 2021. "Socio-economic inequalities in access to drinking water among inhabitants of informal settlements in South Africa." *International Journal of Environmental Research and Public Health* 18 (19):10528.

Palmer, Carolyn G., Anthony Fry, Notiswa Libala, Mateboho Ralekhetla, Nosiseko Mtati, Matthew Weaver, Zintle Mtintsilana, and Patricia-Ann Scherman. 2022. "Engaging society and building participatory governance in a rural landscape restoration context." *Anthropocene* 37:100320.

Ramadie, Oratile. 2024. "Wonders of water." *City of Tshwane*. https://www.tshwane.gov.za/?p=80317.

Ruiters, Cornelius, and Joe Amadi-Echendu. 2022. "Economic costs and investment challenges of water infrastructure in South Africa." *Infrastructure Asset Management* 9 (4):194–206.

Sachs, Jeffrey D. 2015. *The Age of Sustainable Development*. Columbia University Press.

Sánchez-Soriano, Marbella, Paola Miriam Arango-Ramírez, Elvira Iveth Pérez-López, and Iván Antonio García-Montalvo. 2024. "Inclusive governance: Empowering communities and promoting social justice." *Frontiers in Political Science* 6:1478126.

Schoeman, Jolandi, and Paul Johan Oberholster. 2023. "A proposed urban biosphere reserve model concept for the City of Tshwane as a bio-intelligent approach to mitigate climate change." *Inaugural City of Tshwane Climate Change and Research Conference*, Pretoria, March.

Selje, Tom, Lena Anna Schmid, and Boris Heinz. 2024. "Community-based adaptation to climate change: Core issues and implications for practical implementations." *Climate* 12 (10):155.

Selje, Tom, Paul Strikker, and Boris Heinz. 2024. "Community-based adaptation as a transdisciplinary approach and instrument for a social-ecological transformation." *Preprints*:2024041188.

Singh, Abujam Manglem. 2023. "Integrating indigenous knowledge in community-based climate change adaptation." *Indian Journal of Natural Sciences* 14 (81). https://www.researchgate.net/profile/Abujam-Manglem/publication/378497088_Integrating_Indigenous_Knowledge_in_Community-based_Climate_Change_Adaptation/links/65dd9a1fe7670d36abe2a8bb/Integrating-Indigenous-Knowledge-in-Community-based-Climate-Change-Adaptation.pdf.

Singh, Chandni, Soundarya Iyer, Mark G. New, Roger Few, Bhavana Kuchimanchi, Alcade C. Segnon, and Daniel Morchain. 2022. "Interrogating 'effectiveness' in climate change adaptation: 11 guiding principles for adaptation research and practice." *Climate and Development* 14 (7):650–664.

South African Constitution. 1996. "Bill of rights." *Retrieved November* 12:2013.

Statistics South Africa. 2022. "General household survey 2021: Statistics South Africa." *Department of Statistics South Africa*. Last Modified 23 June 2023. https://www.statssa.gov.za/?p=15482.

Taylor, Anna. 2019. *Climate Information for Building Resilient South African Cities*. African Centre for Cities.

Tshabalala, N.N., N. Nhlabathi, B. Van Der Waal, and M. Ntuli. 2024. *Piloting the Siltation Management Model in the Hazelmere Dam and Welbedacht Dam Catchments*. https://natsilt.wrc.org.za/projects/piloting-the-siltation-management-model-in-the-hazelmere-and-welbedacht-dam-catchment-areas/.

United Nations (UN). 2015. "Transforming our world: The 2030 agenda for sustainable development." *New York: United Nations, Department of Economic and Social Affairs* 1:41.

United Nations (UN). 2024. *The Sustainable Development Goals Report 2024*. https://unstats.un.org/sdgs/report/2024/The-Sustainable-Development-Goals-Report-2024.pdf.

United Nations Environment Programme (UNEP). 2019. *Global Environment Outlook Geo-6 Healthy Planet, Healthy People*. https://wedocs.unep.org/20.500.11822/27671.

United Nations Environment Programme (UNEP). 2025. "Centre for Climate Change." *Global Status Report for Buildings and Construction 2024/2025*. https://www.unep.org/resources/report/global-status-report-buildings-and-construction-20242025.

Usman, Khalid. 2024. "Environmental justice and its implications for marginalized communities." *Journal of Psychology, Health and Social Challenges* 1 (3):16–23.

Van Koppen, Barbara, Kenny Phasha, Thando Bophela, Itumeleng Modiba, Malcolm White, Manuel Magombeyi, and I Jacobs-Mata. 2021. "Operationalizing community-led water services for multiple uses in South Africa." *Report to the Water Research Commission by International Water Management Institute and Tsogang Water & Sanitation WRC Report No. TT* 840:20.

Walker, S.E., E.A. Smith, N. Bennett, E. Bannister, A. Narayana, T. Nuckols, K. Pineda Velez, J. Wrigley, and K.M. Bailey. 2024. "Defining and conceptualizing equity and justice in climate adaptation." *Global Environmental Change* 87:102885.

World Bank. 2020. *World Bank Data*. World Bank.

World Bank. 2024. "The institutions global department supports client countries to build capable, efficient, open, inclusive, and accountable institutions through evidence-based reforms." Last Modified 8 August 2024, accessed May. https://www.worldbank.org/en/topic/governance/overview.

Yanou, Malaika P., Mirjam A.F. Ros-Tonen, James Reed, Kaala Moombe, and Terry Sunderland. 2023. "Integrating local and scientific knowledge: The need for decolonising knowledge for conservation and natural resource management." *Heliyon* 9 (11).

Yumagulova, L. 2011. "Adapting cities to climate change: Understanding resilience at the local level." In *Climate: Global Change and Local Adaptation*, 209–235. Springer.

Zhai, P., H.O. Pörtner, D. Roberts, J. Skea, P.R. Shukla, A. Pirani, W. Moufouma-Okia, C. Péan, R. Pidcock, S. Connors, and J.B.R. Matthews. 2018. "Global Warming of 1.5 C. An IPCC Special Report on the impacts of global warming of 1.5 C above pre-industrial levels and related global greenhouse gas emission pathways, in the context of strengthening the global response to the threat of climate change, sustainable development, and efforts to eradicate poverty." *Sustainable Development, and Efforts to Eradicate Poverty* 32.

15 The Role of Indigenous Knowledge in Achieving SDGs and Climate Action

Lone Rafiya Majeed

15.1 INTRODUCTION

Traditional Indigenous knowledge is an invaluable resource that has been passed down through many generations of people's close relationships with the environment. Chaudhuri (2015) also mention that it is often called traditional knowledge or local knowledge. In response to pressing global issues like climate change, biodiversity loss, resource depletion, and environmental degradation, it offers pragmatic and culturally relevant solutions, thus contributing significantly to the cause of sustainable development (Kumar et al., 2021). Unlike the mainstream methods, Indigenous wisdom is firmly based on the values of sustainability, resilience, and environmental stewardship (Figure 15.1). The emphasis is on ecosystem harmony and interdependence, in addition to intergenerational equity and long-term resource management. Innovative and sustainable solutions to the issue of water scarcity can be found in traditional water management practices like qanats, step wells, and rainwater harvesting (Kom et al., 2024). To maintain soil fertility, increase biodiversity, and decrease chemical inputs, farmers use practices like agroforestry, intercropping, and crop rotation. The importance of Indigenous knowledge in attaining sustainability is highlighted by the integration of traditional wisdom with contemporary challenges (Magni, 2017).

Because it considers the interdependencies between all living things from a comprehensive viewpoint, Indigenous knowledge has a bearing on sustainability. By focusing on coexistence instead of exploitation, this approach offers a viable alternative to extractive and unsustainable practices. Controlled harvesting, community-led conservation, and natural regeneration are common Indigenous forest management strategies (Kohsaka et al., 2021). These methods guarantee that ecosystem services and biodiversity will be preserved. The United Nations' Sustainable Development Goals (SDGs)—especially those pertaining to combating climate change, preserving terrestrial life, and building resilient communities—are fully supported by these methods. Individuals striving to adapt to climate change can also find guidance in the inherent resilience of Indigenous knowledge systems (Negi et al., 2023). For example, according to Ghosh et al. (2020), traditional building methods can reduce energy consumption and carbon footprints by utilising locally available climate-resilient materials. These materials also offer natural protection against extreme weather events (Figure 15.1). On top of that, Indigenous communities have a wealth of knowledge about medicinal plants that can aid in the advancement of contemporary medicine and pharmaceuticals while also helping to keep biodiversity intact (Figure 15.1). There are three main objectives that we will be addressing in this chapter. First and foremost, it plans to look into how Indigenous knowledge systems help with sustainable resource management and climate change resilience. This chapter will show how traditional agricultural, water management, and forest conservation practices fit with the global sustainability goals by looking at real-world examples. In addition, it aims to show how Indigenous knowledge can help with modern problems like climate change, food insecurity, and water shortages, with a focus on how Indigenous knowledge can help with community-driven

DOI: 10.1201/9781003600152-15

FIGURE 15.1 Integrating Indigenous knowledge for sustainability.

and culturally appropriate solution development. Lastly, this chapter aims to find practical ways to incorporate Indigenous knowledge into current policy frameworks, educational systems, and technological innovations. This involves tackling issues like Indigenous knowledge loss due to urbanisation and globalisation, advocating for ways to keep Indigenous knowledge alive, and including Indigenous people in decision-making with respect and consideration. By highlighting the intrinsic worth of Indigenous knowledge, we propose a paradigm change in this chapter that goes beyond the traditional methods of sustainability. Rather than being viewed as a static cultural artefact, traditional knowledge is seen as an active component of sustainable development. It sees this as the future. In this chapter, we hope to persuade researchers, communities, and policymakers to tap into Indigenous knowledge's potential for building more just, resilient, and sustainable societies. While they tackle critical global issues, these communities will value cultural diversity.

15.2 INDIGENOUS KNOWLEDGE: DEFINITION AND CHARACTERISTICS

15.2.1 Definition and Scope

Indigenous knowledge is the knowledge, beliefs, and practices that Indigenous and local communities have built up over many generations of living in and caring for their natural environments (Adeyeye, 2019). It is flexible and adaptable, and it is based on spiritual and cultural knowledge. This knowledge covers a wide range of topics, such as healthcare, natural resource management, environmental protection, and farming. In contrast to scientific methods that work everywhere, Indigenous knowledge is specific to the needs, situations, and experiences of the communities that hold it.

15.2.2 Unique Characteristics of Indigenous Knowledge Systems

Indigenous knowledge systems are characterised by their holistic, experiential, and community-oriented nature. They prioritise environmental harmony, emphasising sustainable resource use and intergenerational equity. These systems are communicated through narratives, rituals, and practical demonstrations, making them highly contextual and deeply integrated into cultural practices (Yap et al., 2019). Indigenous knowledge is dynamic, evolving as communities adapt to changing environmental and social conditions. The effectiveness of this approach lies in its ability to offer practical, economical, and contextually relevant solutions to complex challenges, including biodiversity conservation, climate adaptation, and sustainable development.

15.2.3 Evaluation Against Scientific Knowledge

Scientific knowledge is systematic, standardised, and founded on empirical data and experimentation, whereas Indigenous knowledge is intuitive, experiential, and based on extensive observation of local environments (Ellen, 2004). Scientific knowledge is typically intended for universal application, whereas Indigenous knowledge is context-specific, focusing on distinct ecological and cultural environments. Despite their differences, the two systems work well together (Aikenhead et al., 2007). Integrating Indigenous knowledge with scientific methodologies can improve problem-solving skills, resulting in comprehensive and innovative solutions to contemporary environmental and social issues. Understanding the importance of both systems is critical for promoting sustainability, increasing resilience, and ensuring culturally appropriate development (Agrawal, 1995).

15.3 INDIGENOUS KNOWLEDGE AND THE SDGS

15.3.1 Overview of SDGs

The SDGs, adopted by the United Nations in 2015, outline 17 interconnected global objectives aimed at eradicating poverty, protecting the planet, and ensuring peace and prosperity for all by 2030. These goals emphasise the integration of economic, social, and environmental sustainability (Allen et al., 2018). Indigenous knowledge, with its emphasis on sustainable resource management, biodiversity conservation, and resilience, aligns closely with the SDGs and offers valuable insights for achieving these ambitious targets (Figure 15.2) (Rosati et al., 2019).

15.3.2 Indigenous Contributions to SDG 13 (Climate Action)

Indigenous communities are at the forefront of climate adaptation and mitigation because of their close relationship with the environment. Traditional practices like agroforestry, rainwater harvesting, and natural disaster preparedness improve resilience to climate change (Petzold et al., 2020). Indigenous knowledge of weather patterns, seasonal shifts, and ecological indicators contributes to modern climate science by aiding early warning systems and adaptive strategies (Figure 15.2). Furthermore, their forest conservation efforts play an important role in carbon sequestration, making Indigenous practices critical to meeting SDG 13 (Leal et al., 2022).

15.3.3 Contributions to Other SDGs

15.3.3.1 SDG 2: Zero Hunger (Sustainable Agriculture Practices)

Indigenous farming techniques like intercropping, crop rotation, and the use of organic fertilisers improve soil health, reduce reliance on chemical inputs, and ensure food security (Antons, 2024). These practices are essential for achieving sustainable agriculture and reducing hunger (Figure 15.2).

15.3.3.2 SDG 6: Clean Water and Sanitation

Traditional water management systems, such as step wells, rainwater harvesting, and wetland preservation, improve water availability and quality (Figure 15.2). Indigenous approaches to watershed management are consistent with SDG 6 by ensuring equitable and sustainable access to clean water (Priyadarshini and Abhilash, 2019).

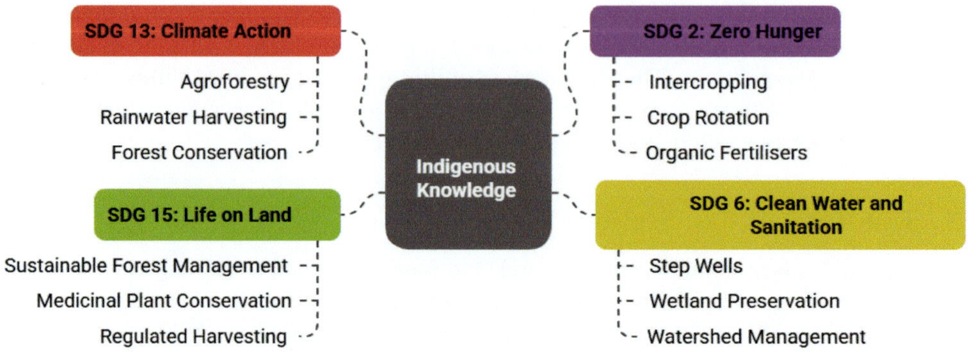

FIGURE 15.2 Indigenous knowledge and Sustainable Development Goals.

15.3.3.3 SDG 15: Life on Land (Biodiversity Conservation)

Indigenous knowledge systems prioritise biodiversity protection via sustainable forest management, conservation of medicinal plants, and regulated harvesting practices. These initiatives play a crucial role in advancing SDG 15, which focusses on the protection of ecosystems and the promotion of sustainable land use (Majeed and Sharma, 2024; Sandoval-Rivera, 2020).

15.4 INDIGENOUS KNOWLEDGE AND CLIMATE ACTION

15.4.1 Indigenous Practices for Climate Resilience

Indigenous knowledge systems are very important for making communities more resistant to climate change, especially those that depend on natural resources a lot. Climate resilience (Nakashima et al., 2018) is the ability of ecosystems and societies to adapt to and recover from adverse weather effects. Because they know so much about the ecosystems and weather patterns in their area, Indigenous communities have long come up with ways to keep the environment in balance and make people less vulnerable to changes in the weather. These habits are based on the ideas of protecting resources, being environmentally friendly, and living in balance with nature (Green et al., 2010). For instance, traditional ways of using land, like shifting cultivation and agroforestry, protect biodiversity and keep the soil healthy at the same time. Native American water conservation methods, such as building reservoirs, qanats, and terraced fields, make it easier to get water during long droughts. Also, Indigenous housing designs, which are usually made with materials that are easy to find and don't break down in bad weather, make people less vulnerable to things like floods and storms (Mafongoya et al., 2017). These tried-and-true methods are used by Indigenous communities to make both natural and human systems more resilient. Indigenous knowledge systems are strong even when there are climate problems because they can keep ecosystem services going while also adapting to changes in the environment.

15.4.2 Traditional Adaptation and Mitigation Practices Examples

Indigenous knowledge offers a variety of adaptation and mitigation strategies aimed at confronting the challenges of climate change. These practices, grounded in observation and experience, have enabled communities to adapt to environmental changes and diminish their ecological footprints over centuries (Filho et al., 2023). Agroforestry systems are utilised in areas including Africa, Southeast Asia, and South America. Agroforestry enhances soil fertility, increases carbon sequestration, and diminishes the need for synthetic fertilisers by integrating trees with crops and livestock. Similarly, conventional wetland conservation methods in areas like the Amazon Basin and South Asia are crucial for regulating water cycles, mitigating floods, and serving as carbon sinks (Moomaw et al., 2018). Indigenous water harvesting systems, such as India's step wells and Morocco's fog-catching systems, provide sustainable solutions to water scarcity in arid and semi-arid areas. The adaptability of Indigenous practices to fluctuating climatic conditions is evidenced by their capacity to secure water availability even in periods of drought. Indigenous forest management techniques, including controlled burning by Aboriginal communities in Australia, mitigate wildfire risks and enhance forest regeneration (Nyong et al., 2007). These practices not only bolster ecosystem resilience but also contribute to the mitigation of greenhouse gas emissions. Similarly, the use of traditional building materials, including bamboo, adobe, and thatch, diminishes the carbon footprint and energy consumption linked to construction activities.

15.4.3 Contribution to Disaster Risk Reduction and Early Warning Systems

Many different approaches to adapting to and reducing the impact of climate change are available in Indigenous knowledge. Adapting to changing environmental conditions and reducing their ecological footprints have been made possible through centuries of these practices, which are based on observation and experience (Hermans et al., 2022; Filho et al., 2023). Many parts of the world, including South America, Southeast Asia, and Africa, use agroforestry techniques. By incorporating trees into agricultural practices alongside crops and livestock, agroforestry improves soil fertility, boosts carbon sequestration, and reduces reliance on synthetic fertilisers. Similarly, traditional approaches to preserving South Asian and Amazonian wetland ecosystems are vital for controlling flooding, regulating water cycles, and acting as carbon sinks (Danladi et al., 2018; Moomaw et al., 2018). Traditional water collection methods, like step wells in India and fog-catching systems in Morocco, offer long-term answers to the problem of water scarcity in dry and semi-arid regions. Indigenous peoples have shown remarkable flexibility in responding to changing weather patterns by ensuring a steady supply of water, even during dry spells. Nyong et al. (2007) found that Indigenous forest management techniques, such as controlled burning, can reduce the likelihood of wildfires and improve the regeneration of forests. One such technique is employed by Aboriginal communities in Australia. In addition to reducing emissions of greenhouse gases, these practices strengthen ecosystem resilience. Traditional building materials, such as thatch, adobe, and bamboo, also reduce energy consumption and the carbon footprint associated with construction.

15.5 CASE STUDIES AND REGIONAL ILLUSTRATIONS

15.5.1 Conventional Customs in the Himalayas

The Himalayan region, home to diverse Indigenous communities, has traditionally relied on a combination of agroforestry and glacier management practices to sustain their livelihoods and ensure water resources (Negi et al., 2023). The traditional agroforestry systems in the region integrate forest management with agricultural practices, thereby improving biodiversity, soil fertility, and water retention (Farooquee et al., 2004; Majeed et al., 2024). The Kullu Valley in Himachal Pradesh, India, employs agroforestry methods that combine the cultivation of crops like wheat, maize, and barley with fruit orchards and medicinal plants (Luharia et al., 2019; Rautela et al., 2024). These systems enhance food security and provide a sustainable income for local farmers by reducing soil erosion and improving water retention during dry spells. Effective glacier management practices in the Himalayas are crucial for sustaining water availability in the context of climate change (Bodh, 2022; Majeed et al., 2024). Indigenous communities possess extensive knowledge of glaciers, their seasonal behaviours, and the timing of snowmelt. Communities in Spiti Valley, Himachal Pradesh, utilise traditional methods to manage glacial meltwater, such as the construction of "kuls," or irrigation channels, which direct meltwater to agricultural fields (Murali et al., 2017). These systems are designed to enhance water efficiency, particularly in areas with limited water resources.

Additionally, communities have developed strategies to manage the accumulation of snow and ice, which is crucial for conserving water resources during the dry season (Singh et al., 2020). Studies indicate that these practices improve adaptive strategies in response to increasing glacial melt due to global warming, thus ensuring sustained water availability for agriculture and daily activities (Maheshwari et al., 2024).

15.5.2 Indigenous Knowledge in the Amazon Basin

Lacerda (2021) emphasises the importance of Indigenous knowledge in the Amazon Basin for resource management and environmental conservation. For centuries, the Tucano and Kayapo tribes have been practising agroforestry, which involves the incorporation of Indigenous plant species into their agricultural practices. The cultivation of crops such as cacao, manioc, and banana beneath the canopy of Indigenous trees, known as "multi-strata" agroforestry systems, contributes to soil conservation, biodiversity enhancement, and deforestation mitigation (Wainaina et al., 2021). These systems improve food security and demonstrate resilience to climatic fluctuations. Additionally, the Indigenous communities of the Amazon have a profound understanding of the waterways of the region. Sustainable food supplies are produced through conventional methods, such as the cultivation of riverine floodplains for agriculture and aquaculture in inundated woodlands (Athayde et al., 2017). These techniques are indispensable for maintaining the equilibrium of river ecosystems, preventing overfishing, and ensuring that local communities can continue to utilise natural resources without depleting them. The adverse effects of climate change, such as erratic rainfall and flooding, are mitigated by Indigenous agricultural systems in the Amazon, according to research (Schmidt et al., 2021).

15.5.3 Indigenous Knowledge in African Communities

The management of natural resources and agricultural systems in sub-Saharan Africa is contingent upon Indigenous knowledge. Historically, communities such as the Maasai in Kenya and Tanzania have relied on livestock mobility to adapt to changing climatic conditions. This practice is now recognised as essential for climate adaptation (Kom et al., 2024). To ensure that their animals have access to sufficient pasture and water, Maasai herders relocate their livestock among a variety of grazing regions in response to seasonal fluctuations and rainfall irregularities. The conventional method of "mobility grazing" is becoming more widely recognised as an effective method for maintaining land productivity, preventing overgrazing, and mitigating droughts (Adeyeye, 2019). Additionally, the agricultural methodologies that have been developed in West Africa by Indigenous knowledge systems are robust and incorporate soil management strategies with crop diversification. To protect the soil from erosion, improve water retention, and maintain fertility, the Fulani and Yoruba communities implement intercropping and mulching (Egeruoh-Adindu, 2022). In regions that are affected by climate variability, the combined use of these methods and conventional drought-resistant crops such as sorghum and millet enhances food security and promotes sustainable agriculture.

15.5.4 Insights From Pacific Island Nations

Particularly by including Indigenous knowledge into modern climate resilience plans, the Pacific Island countries offer important new perspectives on climate adaptation (Lauer, 2017). Because of their sensitivity to sea-level rise and extreme weather events, Indigenous people in places including Kiribati, Tuvalu, and the Solomon Islands have developed unique strategies for disaster risk management and climate change adaptation (Bryant-Tokalau, 2018). The traditional "fale," or sustainable homes built by Pacific Islanders using locally sourced, durable materials including bamboo, palm fronds, and coconut timber, is a major habit. Designed to withstand storms and flooding, they are built to rise above the ground. These homes provide flood protection and help to naturally vent to prevent heat accumulation (Raman et al., 2024). The Polynesian navigators show how early warning systems, modern disaster readiness, and traditional knowledge of ocean currents, celestial patterns, and wind directions might improve each other (Hiruy et al., 2020). These navigational skills are included in community-oriented disaster risk reduction plans following major events of severe storms, so as to improve the timeliness and accuracy of evacuations (Figure 15.3). Moreover, Indigenous farmers in these regions use sustainable fishing methods including "mangrove restoration," to protect coastal ecosystems, guaranteeing food security and acting as barriers against storm surges and tsunamis (Bang, 2024).

These case studies show how well Indigenous knowledge systems around the world provide locally customised solutions for problems, including climate change. Whereas in the Amazon Basin, sustainable land management and water conservation strategies boost biodiversity and food security, in the Himalayan region, agroforestry and glacier management guarantee water accessibility for agriculture. While Pacific Island countries incorporate traditional building techniques and navigation knowledge to reduce disaster risks and increase climate resilience, African communities use livestock mobility and resilient agricultural practices to address drought and soil erosion. Including Indigenous practices in contemporary climate action plans will help communities facing the effects of climate change to be more resilient and sustainable.

15.6 THE CHALLENGES IN INTEGRATING INDIGENOUS KNOWLEDGE

15.6.1 Marginalisation and Loss of Indigenous Knowledge

One of the main challenges to include Indigenous knowledge into modern systems is its marginalisation (Figure 15.3). Regarding land use, natural resource management, and climate policies, Indigenous people have occasionally been excluded from decisions (Shepheard-Walwyn, 2018; Ali et al., 2021). Colonisation, globalisation, and industrialisation have erased traditional knowledge systems as younger generations progressively abandon ancestral practices in favour of urbanisation and modern education (Ocholla, 2007). This marginalisation is exacerbated even more by the relocation of Indigenous people from their territory, resulting from mining, agriculture, or economic development. Many significant customs—knowledge of medicinal plants, water management, sustainable farming—are thus at risk of disappearing. Moreover, the exclusion of Indigenous knowledge causes undervaluation of these systems inside political and administrative structures (Ogunniyi, 2004). Although Indigenous contributions to environmental sustainability are growingly recognised, political will to include these points of view in more general national and international climate action plans still lacks (Jaysawal et al., 2014). This exclusion prevents the full potential of Indigenous knowledge from being realised in addressing modern global concerns, including biodiversity loss and climate change.

15.6.2 Lack of Documentation and Formal Recognition

It is very hard to integrate Indigenous knowledge because these knowledge systems are not officially recognised or written down. People from the same tribe usually pass on their knowledge orally through stories, ceremonies, and hands-on activities (Figure 15.3). Due to the oral tradition, it is hard to record, preserve, and standardise Indigenous knowledge in a way that makes it easy to share or incorporate into academic or policy frameworks (Lwoga et al., 2011). A lot of this information is still out there, but scientists, lawmakers, and international groups can't get to it without formal documentation. Indigenous knowledge is hard to incorporate into everyday environmental and sustainability actions because it is not officially recognised and validated by government and academic institutions (Bow et al., 2019). Most of the time, Western scientific models are used, and Indigenous ways of knowing that focus on full, personal, and situational understandings don't fit well into many of them. This lack of respect not only keeps Indigenous knowledge from being valued but also makes it harder for people in the scientific community to work together with people who hold traditional knowledge.

15.6.3 Intellectual Property Rights and Ethical Concerns

Another significant challenge is the issue of intellectual property rights (IPRs) and the ethical concerns that arise when applying Indigenous knowledge (Brush, 1993). Concerns regarding appropriation and application arise as Indigenous wisdom is acknowledged for its value in areas such as sustainable farming, climate resilience, and biodiversity conservation (Coombe, 1998). Many Indigenous communities face the grave threat of having their traditional knowledge appropriated or commercialised without proper compensation or acknowledgement (Figure 15.3). When companies or academics take advantage of Indigenous peoples' knowledge without respecting their rights or intellectual property, this is called biopiracy (Young, 2006). Additionally, cultural practices may be distorted or stolen because of the commercialisation of Indigenous knowledge. Knowledge ripped from its cultural context and utilised for profit often leads to ethical dilemmas because it loses its original significance and relevance. The commercialisation of Indigenous knowledge is also concerning because it diminishes Indigenous peoples' cultural practices and their ability to decide for themselves (Correa, 2001).

To address ethical concerns, many Indigenous people are fighting for the recognition of their collective IP rights and the establishment of legal frameworks that ensure they benefit from the application of their knowledge. The goal of the Nagoya Protocol on access to genetic resources and the Convention on Biological Diversity is to establish equitable and fair agreements for the sharing of benefits that arise from the use of Indigenous knowledge. However, there are challenges associated with implementing such systems; additional measures are needed to ensure that Indigenous communities retain ownership of their knowledge and cultural practices. Intellectual property concerns, a lack of documentation, and marginalisation are just a few of the challenges that Indigenous knowledge faces when it is integrated into global sustainability initiatives. One way to overcome these obstacles is to recognise and protect Indigenous knowledge systems, ensure their formal inclusion in policy-making, and honour their IPRs. We can create a more equitable and long-term plan to address global issues like climate change and biodiversity loss if we tackle these problems first.

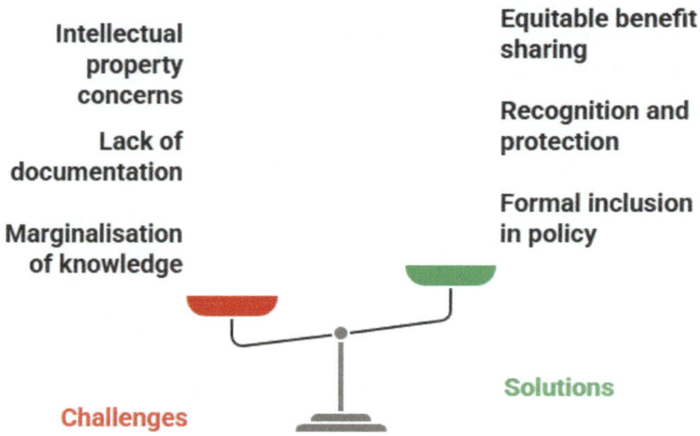

FIGURE 15.3 Balancing challenges and solutions in Indigenous knowledge integration.

15.7 STRATEGIES FOR BRIDGING INDIGENOUS KNOWLEDGE AND MODERN PRACTICES

15.7.1 Participatory Approaches and Community Involvement

Using participatory techniques that actively include Indigenous people in the decision-making process is one of the most successful ways to bridge Indigenous knowledge with modern practices. This method guarantees that Indigenous knowledge is not only acknowledged but also included in projects on sustainable development, climate change action, and conservation (Sarmiento et al., 2020). Indigenous people's direct sharing of their knowledge directly depends on community involvement, which also guarantees that solutions are culturally appropriate and context-specific. Participatory approaches, including community-led research, co-management of natural resources, and dialogue-based seminars, encourage the knowledge exchanging between Indigenous people and outside players, including government agencies, NGOs, and scientists (Alexander et al., 2019). For instance, involving Indigenous people in the management of local wildlife or forest resources results in better results in conservation projects since they have a close knowledge of their ecosystems. This guarantees that development initiatives complement conventional wisdom, promoting mutual respect and trust. Participatory methods also empower Indigenous people by giving them a voice in choices affecting their environment and way of life. Stronger advocacy for the preservation of their knowledge and practices, as well as the sustainability of the ecosystems they depend on, can follow from this empowerment (Orlovic, 2020).

15.7.2 Policy Interventions and Legal Frameworks

Policy interventions and legal frameworks are crucial for the respect, protection, and effective integration of Indigenous knowledge into contemporary governance systems. Governments must formulate policies that recognise the significance of Indigenous knowledge and institutionalise its role within national and international sustainable development frameworks (Figure 15.3) (Chatterjee, 2024). A vital aspect is the formulation of legislation that safeguards IPRs pertaining to Indigenous knowledge. These laws can guarantee that communities receive equitable compensation for their knowledge, especially in sectors such as agriculture, biodiversity conservation, and pharmaceuticals. The Nagoya Protocol, established under the Convention on Biological Diversity, exemplifies a legal framework that facilitates the fair and equitable distribution of benefits arising from the utilisation of genetic resources and traditional knowledge. Moreover, integrating Indigenous governance frameworks into official policies is essential for the sustainable management of natural resources. Indigenous peoples frequently possess distinct systems of land management, resource utilisation, and environmental stewardship that can enhance or reinforce current legal frameworks. Policymakers can improve the efficacy of environmental protection and climate resilience initiatives by acknowledging Indigenous governance frameworks (Weaver, 2023).

15.7.3 Promoting Cross-Disciplinary Collaboration

Bridging Indigenous knowledge with modern practices necessitates cross-disciplinary collaboration among various knowledge systems, such as science, policy, and Indigenous wisdom (Figure 15.3). We can achieve a more comprehensive approach to environmental management and climate adaptation by facilitating dialogue among Indigenous elders, local practitioners, scientists, and policymakers (Sillitoe, 2004). Cross-disciplinary collaboration promotes mutual respect and aids in the identification of areas where traditional knowledge and modern science can complement one another. Indigenous water conservation practices, such as rainwater harvesting and the construction of qanats (underground water channels), can be combined with modern hydrological research to improve water management in drought-prone areas (Singha and Singha, 2024). Similarly, traditional fire management techniques, such as controlled burning, can be combined with modern wildfire risk management strategies to reduce the frequency and severity of forest fires. Building partnerships between academic institutions, Indigenous organisations, and governmental bodies is critical to facilitating this exchange. Collaborative research initiatives, such as participatory mapping and community-based monitoring, enable Indigenous knowledge to inform scientific research and vice versa, ensuring that both systems contribute to SDGs (Görg et al., 2014).

15.7.4 Educational and Training Programmes

Education and training programmes are essential for the transmission of both modern practices and Indigenous knowledge to future generations. Da Silva et al. (2024) propose that these initiatives should endeavour to incorporate Indigenous knowledge into formal education systems and provide instruction on sustainable practices that are derived from Indigenous traditions. The integration of Indigenous knowledge into school curricula has the potential to increase awareness among non-Indigenous communities regarding the importance of traditional practices and to cultivate a greater appreciation for cultural diversity (Muir et al., 2023). Furthermore, the preservation of traditional skills in areas such as agriculture, resource management, and medicinal plant use can be facilitated by training programmes that are specifically designed for younger generations within Indigenous communities. This is particularly critical in light of the potential for younger generations to become increasingly disengaged from traditional practices as a result of urban migration and the impact of modern technology (Bartlett et al., 2012). Additionally, cross-cultural education programmes that unite Indigenous and non-Indigenous students, scientists, and professionals can establish a foundation for mutual learning. These programmes foster collaboration for sustainable development and promote respect for diverse knowledge systems. For instance, universities and research institutions may provide students with the necessary skills to interact with Indigenous communities in a respectful and informed manner by offering certificate programmes in Indigenous environmental management (Gwekwerere, 2014).

A multifaceted approach that encompasses participatory approaches, policy interventions, cross-disciplinary collaboration, and educational programmes is necessary to bridge the gap between Indigenous knowledge and modern practices. By incorporating these strategies, we can guarantee that Indigenous knowledge is actively engaged in the resolution of global challenges, including climate change, biodiversity loss, and sustainable development. This integration not only benefits the environment but also empowers Indigenous communities, ensuring that their cultural practices and wisdom are preserved and respected for future generations.

15.8 THE WAY FORWARD

15.8.1 Recommendations for Policy Makers

Policymakers have to be aggressive in identifying and safeguarding Indigenous knowledge systems if they are to be included in worldwide sustainability projects. First, formal acknowledgement of Indigenous knowledge inside national and international systems is absolutely vital. This covers including SDGs, efforts on biodiversity protection, and traditional practices in climate change plans. Policies should also give land rights and self-determination top priority so that Indigenous people may run their areas and resources according to their customs. Moreover, laws safeguarding IPRs for Indigenous knowledge should be followed to guarantee just pay and stop exploitation.

15.8.2 Role of Technology in Preserving Indigenous Knowledge

The preservation and dissemination of Indigenous knowledge can be significantly influenced by technology. Traditional knowledge can be documented, stored, and shared in a manner that guarantees its accessibility to future generations through the use of digital platforms, databases, and geographic information systems. For instance, Indigenous communities can safeguard their heritage by establishing online archives of sustainable farming methods, ecological practices, and oral traditions. Additionally, mobile applications and digital tools can assist in bridging the gap between conventional knowledge and contemporary scientific methodologies, thereby facilitating the more effective management of natural resources by communities. Nevertheless, it is imperative that Indigenous communities provide their explicit consent to prevent the exploitation of their knowledge.

15.8.3 Fostering Global Partnerships

Incorporating Indigenous knowledge into global sustainability agendas requires international collaborations. Working together, governments, NGOs, scientific institutions, multinational corporations, and Indigenous peoples can establish a forum for knowledge exchange and cooperative problem-solving. It is crucial to make sure that these collaborations are inclusive, equitable, and respectful, and that Indigenous perspectives are given priority when making decisions. These collaborations can support the creation of policies that protect the rights and cultures of Indigenous peoples and encourage the adoption of traditional ecological practices by promoting respect and understanding between the parties.

15.9 CONCLUSION

Indigenous knowledge has been an important part of sustainable practices all over the world for a long time. It gives us important information about how to manage resources, make sure we can handle climate change, and protect biodiversity. This knowledge has been passed down from generation to generation and is deeply connected to the ecosystems in the area. It shows a deep understanding of the environment that stresses balance, sustainability, and respect for nature. Indigenous groups have come up with effective ways to protect biodiversity, use agroforestry, save water, and adapt to climate change. These practices have all made a big difference in how well we take care of the environment. Indigenous knowledge is very important, but it has faced many problems, such as being ignored, not being officially recognised, and concerns about intellectual property. Traditional knowledge systems aren't always included in policy and science discussions, which means they can't help reach global sustainability goals as much as they could. But more and more people are realising that Indigenous knowledge is an important part of fighting climate change, protecting biodiversity, and making sure there is enough food and water for everyone. This means that these practices can be used in new ways in modern environmental strategies. More and more people are realising that we need to connect Indigenous knowledge with modern scientific methods. We can do this through participatory methods, working together across disciplines, and changing policies. Indigenous communities can be given the tools they need to protect their cultural heritage and also help with global efforts to make the world a better place to live. Additionally, new technologies like digital archives and mobile apps are appearing as ways to keep and share Indigenous knowledge, making sure that it remains useful in the modern world.

In the future, Indigenous knowledge will become more and more important in shaping plans for global sustainability. As the world faces bigger problems like climate change, biodiversity loss, and resource depletion, Indigenous practices offer eco-friendly solutions that have been used for a long time and can be used with modern scientific methods. Indigenous knowledge systems may be able to offer useful and creative answers that are long-lasting and fit the needs of the area. These answers could be about things like coping with climate change, farming in a way that doesn't harm the environment, and protecting biodiversity. Indigenous communities will also be able to use their knowledge to help make policies that value both traditional wisdom and modern science as they become more recognised and supported through policy reforms and global partnerships. Indigenous practices will be very important for building resilience and sustainability in the face of climate change and environmental damage. Indigenous knowledge should be included in global sustainability efforts. This will not only make these efforts more effective but also give Indigenous communities more power, protect their rights, and recognise their contributions. This way of working together will make the future more fair, inclusive, and long-lasting for everyone. In the future, Indigenous knowledge will not only help modern practices be more sustainable but also be an important part of how the world's environment is managed.

REFERENCES

Adeyeye, Biliamin Adekunle. "African Indigenous Knowledge and Practices and the 2030 Sustainable Development Goal: Exploring Its Uniqueness for Quality Knowledge Sharing." Journal of Humanities and Education Development (JHED) 1, no. 4 (2019): 147–152.

Agrawal, Arun. "Dismantling the Divide between Indigenous and Scientific Knowledge." Development and Change 26, no. 3 (1995): 413–439.

Aikenhead, Glen S., and Masakata Ogawa. "Indigenous Knowledge and Science Revisited." Cultural Studies of Science Education 2 (2007): 539–620.

Alexander, Steven M., Jennifer F. Provencher, Dominique A. Henri, Jessica J. Taylor, Jed Immanuel Lloren, Lushani Nanayakkara, Jay T. Johnson, and Steven J. Cooke. "Bridging Indigenous and Science-Based Knowledge in Coastal and Marine Research, Monitoring, and Management in Canada." Environmental Evidence 8, no. 1 (2019): 1–24.

Ali, Tahir, Petra Topaz Buergelt, Douglas Paton, James Arnold Smith, Elaine Lawurrpa Maypilama, Dorothy Yuŋgirrŋa, Stephen Dhamarrandji, and Rosemary Gundjarranbuy. "Facilitating Sustainable Disaster Risk Reduction in Indigenous Communities: Reviving Indigenous Worldviews, Knowledge and Practices through Two-Way Partnering." International Journal of Environmental Research and Public Health 18, no. 3 (2021): 855.

Allen, Cameron, Graciela Metternicht, and Thomas Wiedmann. "Initial Progress in Implementing the Sustainable Development Goals (SDGs): A Review of Evidence from Countries." Sustainability Science 13 (2018): 1453–1467.

Antons, Christoph. "The Regulation of Agricultural Knowledge, Seed Policies and the UN Sustainable Development Goal of 'Zero Hunger'." In Implementation of Sustainable Development in the Global South: Strategies, Innovations and Challenges, 173–192. Hart Publishing, 2024.

Athayde, Simone, Jose Silva-Lugo, Marianne Schmink, and Michael Heckenberger. "The Same, But Different: Indigenous Knowledge Retention, Erosion, and Innovation in the Brazilian Amazon." Human Ecology 45, no. 4 (2017): 533–544.

Bang, Henry Ngenyam. "Sustainable Development Goals, Disaster Risk Management, and Indigenous Knowledge: A Critical Assessment of the Interlinkages." Sustainable Earth Reviews 7, no. 1 (2024): 29.

Bartlett, Cheryl, Murdena Marshall, and Albert Marshall. "Two-Eyed Seeing and Other Lessons Learned within a Co-Learning Journey of Bringing Together Indigenous and Mainstream Knowledges and Ways of Knowing." Journal of Environmental Studies and Sciences 2 (2012): 331–340.

Bodh, Vijay K. "Linking Indigenous Traditional Knowledge and Sustainable Development Goals in the North-Western Himalayas." In Economics and Policy of Energy and Environmental Sustainability, 303–313. Springer, 2022.

Bow, Catherine, and Patricia Hepworth. "Observing and Respecting Diverse Knowledge Traditions in a Digital Archive of Indigenous Language Materials." Journal of Copyright in Education and Librarianship 3, no. 1 (2019): 1–36.

Brush, Stephen B. "Indigenous Knowledge of Biological Resources and Intellectual Property Rights: The Role of Anthropology." American Anthropologist 95, no. 3 (1993): 653–671.

Bryant-Tokalau, Jenny. Indigenous Pacific Approaches to Climate Change: Pacific Island Countries. Springer, 2018.

Chatterjee, Ishita. "Bridging Genetic Resources, Traditional Knowledge and Innovation in the Legal Context." Central University of Kashmir Law Review 4 (2024): 93–108.

Chaudhuri, Buddhadeb. "Science in Society: Challenges and Opportunities for Indigenous Knowledge in the Present-Day Context." Global Bioethics 26, no. 2 (2015): 78–85.

Coombe, Rosemary J. "Intellectual Property, Human Rights & (and) Sovereignty: New Dilemmas in International Law Posed by Recognition of Indigenous Knowledge and the Conservation of Biodiversity." Indian Journal of Global Legal Studies 6 (1998): 59.

Correa, Carlos M. "Traditional Knowledge and Intellectual Property." Geneva: The Quaker United Nations Office (QUNO) 17 (2001).

Danladi, Amos, Ho Chin Siong, and Gabriel Ling Hoh Teck. "Importance of Indigenous Knowledge in Flood Risk Reduction: A Review." Journal of Advanced Research in Applied Sciences and Engineering Technology 11, no. 1 (2018): 7–16.

Egeruoh-Adindu, Izuoma. "Leveraging Indigenous Knowledge for Effective Environmental Governance in West Africa." Beijing Law Review 13 (2022): 931.

Ellen, Roy. "From Ethno-Science to Science, or 'What the Indigenous Knowledge Debate Tells Us about How Scientists Define Their Project'." Journal of Cognition and Culture 4, no. 3–4 (2004): 409–450.

Farooquee, Nehal A., B.S. Majila, and C.P. Kala. "Indigenous Knowledge Systems and Sustainable Management of Natural Resources in a High Altitude Society in Kumaun Himalaya, India." Journal of Human Ecology 16, no. 1 (2004): 33–42.

Filho, Walter Leal, Franziska Wolf, Edmond Totin, Luckson Zvobgo, Nicholas Philip Simpson, Kumbirai Musiyiwa, Jokastah W. Kalangu, Maruf Sanni, Ibidun Adelekan, and Jackson Efitre. "Is Indigenous Knowledge Serving Climate Adaptation? Evidence from Various African Regions." Development Policy Review 41, no. 2 (2023): e12664.

Ghosh-Jerath, Suparna, Ridhima Kapoor, Archna Singh, Shauna Downs, Satabdi Barman, and Jessica Fanzo. "Leveraging Traditional Ecological Knowledge and Access to Nutrient-Rich Indigenous Foods to Help Achieve SDG 2: An Analysis of the Indigenous Foods of Sauria Paharias, a Vulnerable Tribal Community in Jharkhand, India." Frontiers in Nutrition 7 (2020): 61.

Görg, Christoph, Joachim H. Spangenberg, Vera Tekken, Benjamin Burkhard, Dao Thanh Truong, Monina Escalada, Kong Luen Heong, Gertrudo Arida, Leonardo V. Marquez, and Jesus Victor Bustamante. "Engaging Local Knowledge in Biodiversity Research: Experiences from Large Inter-and Transdisciplinary Projects." Interdisciplinary Science Reviews 39, no. 4 (2014): 323–341.

Green, Donna, and Gleb Raygorodetsky. "Indigenous Knowledge of a Changing Climate." Climatic Change 100, no. 2 (2010): 239.

Gwekwerere, Yovita. "Bridging Culture, Indigenous and Scientific Ways of Knowing." Cultural and Pedagogical Inquiry 6, no. 2 (2014).

Hermans, Thirze D.G., Robert Šakić Trogrlić, Marc J.C. van den Homberg, Hannah Bailon, Rebecca Sarku, and Anuszka Mosurska. "Exploring the Integration of Local and Scientific Knowledge in Early Warning Systems for Disaster Risk Reduction: A Review." Natural Hazards 114, no. 2 (2022): 1125–1152.

Hiruy, K., and Robyn Eversole. "The Contribution of Research for Development to the Sustainable Development Goals: Lessons from Fisheries Research in Southeast Asia and the Pacific Island Countries." International Journal of Sustainable Development & World Ecology 27, no. 2 (2020): 153–166.

Jaysawal, Dr Neelmani, and Sudeshna Saha. "Marginalisation of Tribal Communities Due to Globalization." Indian Journal of Dalit and Tribal Studies 2, no. 2 (2014): 37–54.

Kohsaka, Ryo, and Marie Rogel. "Traditional and Local Knowledge for Sustainable Development: Empowering the Indigenous and Local Communities of the World." In Partnerships for the Goals, 1261–1273. Springer, 2021.

Kom, Zongho, and Nthaduleni Samuel Nethengwe. "Indigenous and Local Knowledge: Instruments Towards Achieving SDG2: A Review in an African Context." Sustainability 16, no. 20 (2024): 9137.

Kumar, Ajay, Sushil Kumar, Komal, Nirala Ramchiary, and Pardeep Singh. "Role of Traditional Ethnobotanical Knowledge and Indigenous Communities in Achieving Sustainable Development Goals." Sustainability 13, no. 6 (2021): 3062.

Lacerda, Victória Duarte. "The Traditional Ecological Knowledge and Land Use by an Indigenous Community in Brazilian Amazon." Universidade de Aveiro (Portugal) ProQuest Dissertations & Theses (2021): 31478930. https://www.proquest.com/docview/3110426774?pq-origsite=gscholar&fromopenview=true&sourcetype=Dissertations%20&%20Theses

Lauer, Matthew. "Changing Understandings of Local Knowledge in Island Environments." Environmental Conservation 44, no. 4 (2017): 336–347.

Leal Filho, Walter, Jelena Barbir, Juliet Gwenzi, Desalegn Ayal, Nicholas P. Simpson, Lydia Adeleke, Behiwot Tilahun, Innocent Chirisa, Shine Francis Gbedemah, and Daniel M. Nzengya. "The Role of Indigenous Knowledge in Climate Change Adaptation in Africa." Environmental Science & Policy 136 (2022): 250–260.

Luharia, Jaimini, and Haresh Sharma. "Role of Indigenous Women in Securing Sustainable Livelihoods in Western Himalayan Region, India." Scientia Moralitas Research Institute (2019): 77–82. https://www.ceeol.com/search/chapter-detail?id=819743

Lwoga, Edda Tandi, Patrick Ngulube, and Christine Stilwell. "Challenges of Managing Indigenous Knowledge with Other Knowledge Systems for Agricultural Growth in Sub-Saharan Africa." Libri 61 (2011): 226–238. https://doi.org/10.1515/libr.2011.019

Mafongoya, P.L., and O.C. Ajayi. "Indigenous Knowledge and Climate Change: Overview and Basic Propositions." Indigenous Knowledge Systems and Climate Change Management in Africa 17 (2017): 17–28.

Magni, Giorgia. "Indigenous Knowledge and Implications for the Sustainable Development Agenda." European Journal of Education 52, no. 4 (2017): 437–447.

Maheshwari, Alka, and Aruditya Jasrotia. "Development of Ecotourism in Himachal Pradesh: A Case Study of Spiti Valley." In Sustainable Tourism, Part B, 223–232. Emerald Publishing Limited, 2024.

Majeed, Lone Rafiya, Lone Fawad Majeed, Sumaira Rashid, Sartaj Ahmad Bhat, Narendra Kumar, and Vineet Kumar. "Intensification of Contaminants, Hydrology, and Pollution of Hyporheic Zone: The Liver of River Ecology—a Review." Environmental Sustainability 7, no. 2 (2024): 121–133.

Majeed, Lone Rafiya, and Deeplata Sharma. "Unraveling Geomorphometric Anomalies in the Lolab Basin of Kashmir Himalaya and Its Role in Shaping Sustainable Land Use Policies." Discover Water 4, no. 1 (2024): 110.

Moomaw, W. R., G. L. Chmura, G. T. Davies, C. M. Finlayson, B. A. Middleton, S. M. Natali, J. E. Perry, N. Roulet, and A. E. Sutton-Grier. "Wetlands in a Changing Climate: Science, Policy and Management." Wetlands, 38, no. 2 (2018): 183–205. https://doi.org/10.1007/s13157-018-1023-8

Muir, A.M., A.T. Duncan, K. Almack, N. Boucher, E.S. Dunlop, C. Febria, J.T. Ives, R. Lauzon, H. Lickers, and W.P. Mattes. "Sharing across the Space: Introduction to a Special Issue on Bridging Indigenous and Non-Indigenous Knowledge Systems." Journal of Great Lakes Research 49 (2023): S1–11.

Murali, Ranjini, Stephen Redpath, and Charudutt Mishra. "The Value of Ecosystem Services in the High Altitude Spiti Valley, Indian Trans-Himalaya." Ecosystem Services 28 (2017): 115–123.

Nakashima, Douglas, and Igor Krupnik. Indigenous Knowledge for Climate Change Assessment and Adaptation. Vol. 2. Cambridge University Press, 2018.

Negi, Vikram S., Ravi Pathak, Shinny Thakur, Ravindra K. Joshi, Indra D. Bhatt, and Ranbeer S. Rawal. "Scoping the Need of Mainstreaming Indigenous Knowledge for Sustainable Use of Bioresources in the Indian Himalayan Region." Environmental Management 72, no. 1 (2023): 135–146.

Nyong, Anthony, Francis Adesina, and Balgis Osman Elasha. "The Value of Indigenous Knowledge in Climate Change Mitigation and Adaptation Strategies in the African Sahel." Mitigation and Adaptation Strategies for Global Change 12 (2007): 787–797.

Ocholla, Dennis. "Marginalized Knowledge: An Agenda for Indigenous Knowledge Develop-Ment and Integration with Other Forms of Knowledge." The International Review of Information Ethics 7 (2007): 236–245.

Ogunniyi, M.B. "The Challenge of Preparing and Equipping Science Teachers in Higher Education to Integrate Scientific and Indigenous Knowledge Systems for Learners: The Practice of Higher Education." South African Journal of Higher Education 18, no. 3 (2004): 289–304.

Orlovic Lovren, Violeta. "Traditional and Indigenous Knowledge: Bridging Past and Future Sustainable Development." In Life on Land, 1033–1041. Springer, 2020.

Petzold, Jan, Nadine Andrews, James D. Ford, Christopher Hedemann, and Julio C. Postigo. "Indigenous Knowledge on Climate Change Adaptation: A Global Evidence Map of Academic Literature." Environmental Research Letters 15, no. 11 (2020): 113007.

Priyadarshini, Priya, and Purushothaman Chirakkuzhyil Abhilash. "Promoting Tribal Communities and Indigenous Knowledge as Potential Solutions for the Sustainable Development of India." Environmental Development 32 (2019): 100459.

Raman, Raghu, Walter Leal Filho, Hector Martin, Suparna Ray, Dayana Das, and Prema Nedungadi. "Exploring Sustainable Development Goal Research Trajectories in Small Island Developing States." Sustainability 16, no. 17 (2024): 7463.

Rautela, Kuldeep Singh, Vivek Gupta, Juna Probha Devi, Lone Rafiya Majeed, and Jagdish Chandra Kuniyal. "Modeling Stage-Discharge and Sediment-Discharge Relationships in Data-Scarce Himalayan River Basin Dhauliganga, Central Himalaya, Using Neural Networks." Clean–Soil, Air, Water (2024): 2300388.

Rosati, Francesco, and Lourenço G.D. Faria. "Addressing the SDGs in Sustainability Reports: The Relationship with Institutional Factors." Journal of Cleaner Production 215 (2019): 1312–1326.

Sandoval-Rivera, Juan Carlos A. "Environmental Education and Indigenous Knowledge: Towards the Connection of Local Wisdom with International Agendas in the Framework of the Sustainable Development Goals (SDGs)." Diaspora, Indigenous, and Minority Education 14, no. 1 (2020): 14–24.

Sarmiento, Iván, Germán Zuluaga, Sergio Paredes-Solís, Anne Marie Chomat, David Loutfi, Anne Cockcroft, and Neil Andersson. "Bridging Western and Indigenous Knowledge through Intercultural Dialogue: Lessons from Participatory Research in Mexico." BMJ Global Health 5, no. 9 (2020): e002488.

Schmidt, Marcus Vinícius C., Yakuna Ullillo Ikpeng, Tariaiup Kayabi, Rosely Alvim Sanches, Katia Yukari Ono, and Cristina Adams. "Indigenous Knowledge and Forest Succession Management in the Brazilian Amazon: Contributions to Reforestation of Degraded Areas." Frontiers in Forests and Global Change 4 (2021): 605925.

Shepheard-Walwyn, Emma. "Marginalisation of Traditional Groups and the Degradation of Nature." The Sage Handbook of Nature 3 (2018): 1037–1063.

Sillitoe, Paul. "Interdisciplinary Experiences: Working with Indigenous Knowledge in Development." Interdisciplinary Science Reviews 29, no. 1 (2004): 6–23.

Silva, Cláudio da, Fátima Pereira, and José Pedro Amorim. "The Integration of Indigenous Knowledge in School: A Systematic Review." Compare: A Journal of Comparative and International Education 54, no. 7 (2024): 1210–1228.

Singh, Rashmi, Rishi Kumar Sharma, Suresh Babu, and Yash Veer Bhatnagar. "Traditional Ecological Knowledge and Contemporary Changes in the Agro-Pastoral System of Upper Spiti Landscape, Indian Trans-Himalayas." Pastoralism 10, no. 1 (2020): 15.

Singha, Surjit, and Ranjit Singha. "Cross-Disciplinary Collaborations and Partnerships for Sustainability Education: Including Community-Based Learning, Industry Partnerships, and International Collaborations." In Teaching and Learning for a Sustainable Future: Innovative Strategies and Best Practices, 17–37. IGI Global, 2024.

Wainaina, Priscilla, Peter A. Minang, Lalisa Duguma, and Kennedy Muthee. "A Review of the Trade-Offs across Different Cocoa Production Systems in Ghana." Sustainability 13 (2021): 10945. https://doi.org/10.3390/su131910945

Weaver, Kaja. "Bridging Indigenous and Western Knowledge-Systems in Knowledge Co-Production with Amazonian Indigenous Communities: A Systematic Realist Review." Development Studies Research 10, no. 1 (2023): 2203842.

Yap, Mandy Li-Ming, and Krushil Watene. "The Sustainable Development Goals (SDGs) and Indigenous Peoples: Another Missed Opportunity?" Journal of Human Development and Capabilities 20, no. 4 (2019): 451–467.

Young-Ing, Greg. "Intellectual Property Rights, Legislated Protection, Sui Generis Models and Ethical Access in the Transformation of Indigenous Traditional Knowledge" (2006). PhD Thesis, University of British Columbia. http://hdl.handle.net/2429/18517

16 Exploring the Role of Gender and Technology in Shaping Climate Action Strategies

Charlotte Remteng

16.1 INTRODUCTION

Climate action integrated with gender, IM, and technology forms a strong basis towards tackling the social dimension of climate change, thereby ensuring inclusion. This will bring about effective and sustainable solutions that are more inclusive as it will leverage the unique strengths of various groups (vulnerable groups including women and girls, youths, indigenous groups). The International Institute for Sustainable Development article on "Why Gender Matters in Climate Adaptation," by Angie Dazie, views climate change inherently as a justice issue, as those who contribute the least suffer the most consequences, accentuated by the Paris Agreement and other committees. The article further clearly states that effective mitigation and adaptation strategies recognize that women, men, and children experience the impacts of climate change differently depending on where they live, how they sustain their livelihoods, and the roles they play in their families and communities. The World Bank Group Gender Thematic Policy Notes Series 2023 examined the intersection of gender equality and climate change, delving into programmatic experiences related to the gender-climate nexus, and identified promising entry points and solutions, offering recommendations for development practitioners, policymakers, and businesses. Women and disadvantaged groups often face greater impacts from climate change across various dimensions, such as health and livelihoods (CEDAW, 2018). Increasingly, gender gaps are recognized as obstacles to effective mitigation and adaptation strategies, with women being crucial leaders and participants in low-carbon transitions. The UN and other international organizations are promoting a gender perspective, establishing mechanisms for assessing the impact of environmental policies on women and offering policy recommendations (Eva Alonso-Epelde et al., 2025).

IM refers to the understandings, skills, and philosophies developed by societies with long histories of interaction with their natural surroundings. It encompasses language, systems of classification, resource use practices, social interactions, ritual, and spirituality (UNESCO, 2023). This knowledge has helped communities adapt to their agro-ecological and socio-economic environments by systematically observing local conditions, experimenting with solutions, and readapting them to new contexts (Fernandez, 1994; Brouwers, 1993). Its role in environmental conservation is recognized and emphasized under the Convention on Biological Diversity, with growing interest in its documentation for its practicality and potentiality in supporting local climate change adaptation in recent years. A great number of indigenous peoples possess unique knowledge systems and innovations that are intergenerational, rooted in deep understanding and respect for the ecosystems, sustainable resource use, that help create a balance between the natural world and humans, ensuring sustainability, and thus valuable for climate solutions. In more recent times, about 25% land inhabited by indigenous peoples has faced increasing pressure for development, with lands covered with forests, protected by cultures, and the need to safeguard the livelihood of their people, which is essential as carbon sinks. Thus, the twin pressures of climate change and development are adversely affecting the well-being and adaptive capacity of indigenous peoples, presenting risks and opportunities that they can easily identify, which can be quite instrumental in climate actions (UNDP, 2024; UNESCO, 2023).

 DOI: 10.1201/9781003600152-16

Climate technologies help combat climate change by reducing greenhouse gas emissions and also help adaptation to the adverse effects of climate change, such as early warning systems, with the development and transfer of technologies to support national climate action key throughout the UNFCCC process to date. The technology framework under Article 10, paragraph 4, of the Paris Agreement recognizes the need to operationalize it to support the achievement of the Paris Agreement, consistent with the long-term vision for technology development and transfer. It also reiterates support (including financial) for developing country parties in strengthening cooperative actions on technology and transfer.

Multifaceted perspectives that link different but similar aspects of problems or issues have been essential in problem-solving, as they reduce costs and yield interconnected, coherent strategies, actions, policies, and so on that are effective and sustainable. With this method, a wider range of stakeholders is reached, leading to broader ideas and solutions across interlinked sectors. This is the rationale for a transversal view of gender, IM, and technology for innovative solutions, in which IM integrated with climate technologies will make them sustainable, innovative, and culturally accepted. This chapter uses a diagnostic review of gender and IM as social dimensions necessary for climate action. It identifies vulnerabilities, the impacts of climate change on gender and indigenous peoples, and the challenges involved in the uptake of technology by these social groups.

16.2 CLIMATE ACTION: AN OVERVIEW

Climate change is disproportionately affecting all countries, disrupting national economies, people, and communities through its significant impacts, including changing weather patterns, rising sea level, and extreme weather events. Tackling climate change is a sustainable development goal, Goal 13-Climate Action, which is inextricably linked to the other 16 goals. In 2015, countries adopted the Paris Agreement, agreeing to limit global temperature rise well below 2 degrees Celsius, though striving for 1.5 degrees Celsius. The Sustainable Development Goals further highlight the strengthening of resilience and adaptive capacity to climate-related disasters, integration of climate change measures into policies and planning, building knowledge and capacity to meet climate change, implementing the UN Framework Convention on Climate Change, and promoting mechanisms to raise capacity for planning and management as the five targets under climate action (Target 13.1–13.5). With the continuous increase in greenhouse gases over the years, reaching unprecedented levels in recent years, climate action has never been more urgent than now. These actions cut across mitigation, adaptation with climate finance, policy and governance, and education and awareness essential ways to quicker and effective actions. According to the UNFCCC, the goal of mitigative actions is to decrease the concentration of greenhouse gases released to the atmosphere by enhancing sinks, by targeting economic activities and incentivizing cleaner actions. Nationally Appropriate Mitigation Actions are being implemented by countries with communications of quantified economy-wide emissions targets. Adaptation and resilience are clearly mentioned under the target on climate actions, referring to the changes in processes and practices and structures to moderate potential damages, with the need for countries and communities to develop and implement adaptation solutions and actions to respond to current and future climate change impacts, with emphasis on the need for long-term prediction and planning. Climate finance flow highly determines the timely implementation and shortfalls in achieving climate actions. Its availability, timely disbursement, and transparency play a key role in the overall fight against climate change.

Climate change is a technical issue and a major political challenge requiring public action at all levels (Perga et al., 2023). *Policy and governance* play a vital role in political commitment, including the translation of most that most climate policies into actionable outcomes. According to the Intergovernmental Panel on Climate Change (IPCC), effective climate governance enables mitigation and adaptation, with policies picking up the cross-cutting nature of problems and solutions. Additionally, education, training, and awareness empower, equip, and provide the necessary skills for mitigation and adaptation at all levels and therefore a key asset for climate action. It strengthens resilience, shapes behaviour, and spurs innovation (World Bank, 2019, 2024) (Figure 16.1).

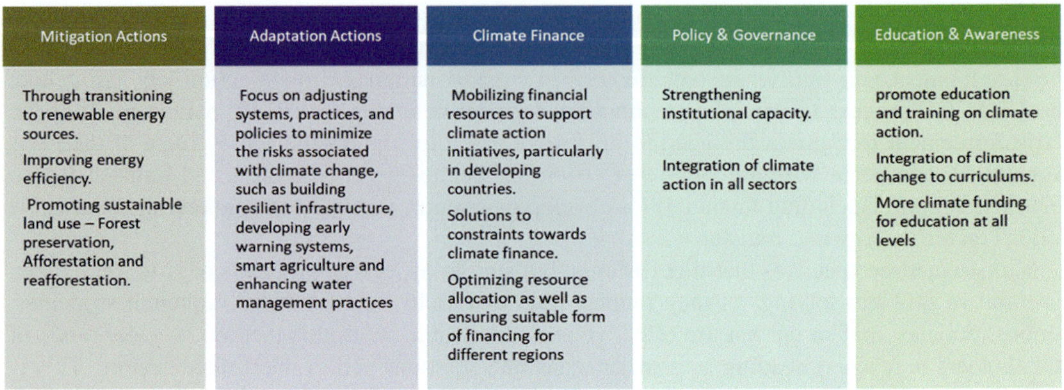

FIGURE 16.1 Key examples of climate actions and components.

Most countries have submitted their Nationally Determined Contributions (NDCs), outlining their plans for reducing greenhouse gas emissions, with contributions reviewed and updated every five years to increase ambition. However, despite these commitments, current NDCs are insufficient to limit global warming to 1.5°C, with a notable gap between current commitments and the levels required to achieve this target (Statista, 2023). A stark report from UNFCCC shows implementation of the combined climate pledges of 193 Parties under the Paris Agreement puts the world on track for a 2.5°C warmer world by the end of the century. The following graph illustrates the change in global surface temperature compared to the long-term average from 1951 to 1980, with the earth's average surface temperature reaching its highest level on record since data collection began in 1880 in 2014, and the last 10 years being the warmest on record (NASA, 2025). The LOWESS (Locally Weighted Scatterplot Smoothing) was projected showing 1.6°C temperature abnormal increase as opposed to about 1.19°C in 2024. The Paris Agreement has also established a robust transparency framework, requiring countries to report their emissions and progress regularly. Many countries, including major emitters like China, the EU, and the US, have set net-zero targets, aiming to balance the amount of greenhouse gases emitted with the amount removed from the atmosphere (UNFCCC, 2023) (Figure 16.2; Table 16.1).

Climate finance is another area where the Paris Agreement has faced challenges, as the $100 billion per year pledge to assist developing nations has not been fully realized, and there are concerns about the transparency and predictability of climate finance (World Economic Forum, 2021, 2024a, 2024b). The withdrawal–re-entry–withdrawal of the US from the agreement and other political not only challenges climate finance but also hinders global cooperation. The European Union has, on the other hand, been a leader in climate action, implementing policies like the European Green Deal, which aims to make Europe climate-neutral by 2050. Asian countries such as China and India are making significant strides in climate action, with China committing to achieve carbon neutrality by 2060 (UNFCCC, 2023). Many African nations are focusing on sustainable development and adaptation strategies, with initiatives like the Africa Renewable Energy Initiative (AREI, 2022) and several institutional reforms, projects, and so on to tackle energy transition and climate resilience.

16.3 GENDER AND CLIMATE ACTION

Climate change impacts are differentiated across genders, with women facing distinct challenges and vulnerabilities (UNFCCC, 2022; Cogswell & Warszawski, 2022). Climate finance remains contentious under the Paris regime: the USD 100 billion per year commitment was missed in 2020 (USD 83.3 billion) and only reached in 2022, and ongoing concerns persist around the transparency, comparability, and predictability of climate finance reporting and delivery (OECD, 2024). The Gender Action Plan under the Paris Agreement, adopted in 2017, is a clear indication that gender-differentiated impacts have been recognized by the global community. This action plan

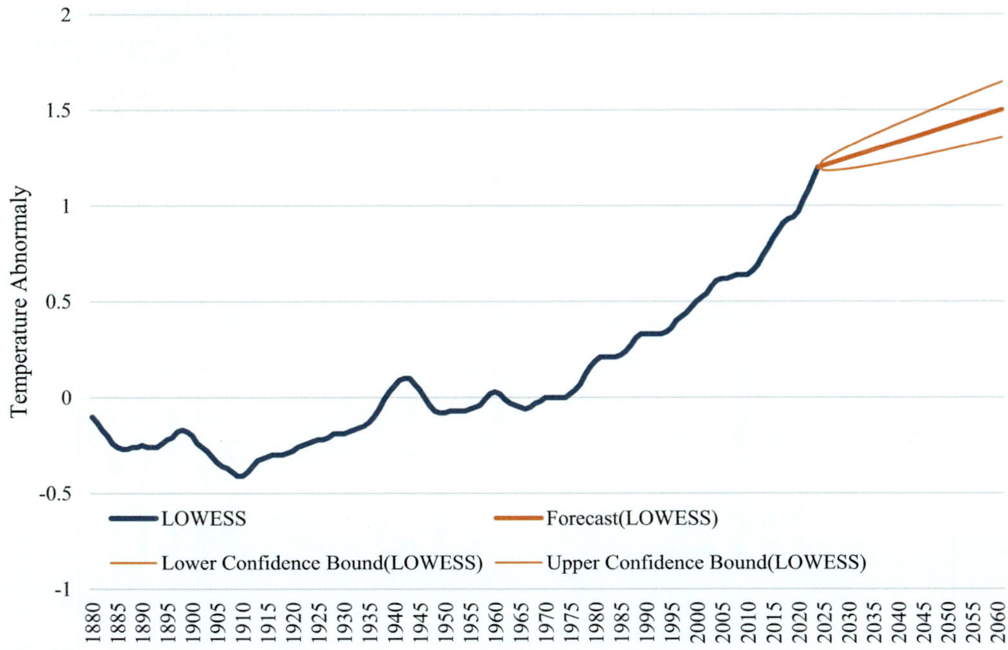

FIGURE 16.2 Global surface temperature changes relative to the long-term average from 1951 to 1980, projected to 2060.

Source: NASA's Goddard Institute for Space Studies (GISS). Credit: NASA/GISS.

further acknowledges country-driven gender-responsive, participatory, and transparent approach in article 7.5. Climate change exacerbates existing gender inequalities, posing unique threats to the livelihoods, health, and safety of women and girls due to their high dependence on natural resources for their daily needs and livelihood (UN Women, 2017). In many regions, women bear the responsibility of securing food, water, and fuel (World Economic Forum, 2024a). With climate change, water and sanitation systems become compromised, putting women and girls at greater health risks. Graham et al. (2016) conducted a study in 24 sub-Saharan African countries, revealing that in households where water collection took over 30 minutes, adult females were the primary collectors, with percentages ranging from 46% in Liberia to 90% in Côte d'Ivoire. Additionally, female children were more often responsible for water collection compared to male children, with a ratio of 62% to 38%, respectively. Across all 24 countries, female children were more likely to be responsible for water collection than male children (62% vs. 38%, respectively) (Graham et al., 2016). Women take on increased domestic and care work as resources disappear, further straining their physical and mental well-being. In conflict-affected settings, climate change escalates social, political, and economic tensions, increasing vulnerabilities to gender-based violence, including conflict-related sexual violence, human trafficking, and child marriage (UN Women, 2020).

Women in developing countries constitute the majority of the world's poor largely due to their reliance on natural resources and agriculture for income, influenced by cultural barriers that limit their access to formal education and employment. Because of their domestic responsibilities, they are usually confined to lower-paying jobs, usually in agriculture, comprising the majority of the workforce in this sector. Women comprise a substantial share of the agricultural labour force, yet women represent only a small share of agricultural landholders and receive a very limited share of agricultural extension services, indicating persistent gender gaps in access to productive resources and support (CFS, 2020). This implies that their agricultural activities might not likely withstand the shock from climate change, as they possess fewer resources for production. The fact that most of their income-generating sources are climate-sensitive makes them highly vulnerable to climate change.

TABLE 16.1
Gender-Related Climate Change Vulnerability Indicators

Economic Vulnerability	Indicator(s)	Social/Political Vulnerability	Indicator	Health Vulnerability	Indicator
-Inadequate access resources of production such as land, finance, technology, and information, which hinders their adaptive capacity impacting their livelihoods, with sectors such as agriculture highly affected	-Measuring access to land, loans/finance and technology by both women and men	-Low participation of women in decision-making has led to their marginalization	-Tracking gender participation in decision-making	-Change in disease patterns due to climate change greatly affects women, especially pregnant women, girls and children.	Monitoring and assessing disease patterns and their effects on various genders
-Income inequality gap for climate-affected income-generating sectors such as agriculture is widened with climate change, leading to less participation of women in the economy.	-Food prices	-Most women have limited access to information and climate technologies which affects their coping capacity	-Tracking access to in information and technology	-Climate change has inevitably led to low crop yields, which have consequently been responsible for malnutrition to mostly among women and children, due to their limited access to resources.	-Monitoring calorie intake and malnutrition rates
		-Displacement and increased conflicts due to scarcity of resources increase vulnerabilities for women	-Assessing movements and displacements due to climate change as	-Migration and conflicts due to climate change have led to poor sanitation/hygiene conditions for women and girls. This has also led to instances of rape.	-Assessing the sanitation/hygiene in migration and displacement camps. -Access to clean water

There is generally an underrepresentation of women in decision-making in climate forums, despite the contributions they are making in addressing climate change at various levels, thus impacting the effectiveness of equity in climate actions. This lack of representation affects the effectiveness of equity in climate actions, leading to initiatives, strategies, and policies that are not inclusive (Eva Alonso-Epelde et al., 2025). The UNFCCC, following the Gender Composition Report, showed that the number of women represented bodies of the UNFCCC was not in line with the efforts to create gender balance in the convention, with only two (02) constituted bodies reported to be near gender balance in the year 2019, as opposed to three in 2018. In 2021, the European Institute for Gender Equality reported that women are woefully underrepresented in decision-making in environment and climate change within the EU member states, following analysis from a time series data set from 2012 to 2021, both at national levels (26.5% women as opposed to 73.2% men), within the national parliaments (29.7% women and 70.3% men). Most countries, however, are making efforts to integrate gender actions and policies in their climate action plans. Analysis of the NDCs (and INDCs) of all African countries shows that over 85% of actions reference gender. At the regional level, West Africa has the highest gender actions in its NDCs (41%), East Africa 25%, Southern Africa 15%, Central Africa 8%, and North Africa 6% (Remteng et al., 2022). Some of the action plans involve considering gender-differentiated impacts of climate change, promoting women's access to land, introducing energy-saving cooking stoves, rainwater harvesting, and groundwater exploration for women. It is hoped that as countries amend their NDCs, upgrading them to NDC 3.0, more ambitious actions that cater for women's needs are included. Asia has also made strides in gendered-climate actions; in India, the Bhungroo project in Gujarat is a gender-just climate solution that supports women in self-help groups to respond to recurring droughts and floods (UNDP, 2019); Bangladesh's Climate Change Strategy and Action Plan includes specific measures to address the vulnerabilities of women and promote their participation in climate adaptation efforts (Government of Bangladesh, 2009; Majumder et al., 2024). In the Americas, Guatemala has developed and implemented a gender strategy to reduce women-specific vulnerabilities to climate change and ensure that women benefit from community-based gender-responsive measures (UN Women, 1995, 2017).

16.4 INDIGENOUS AND LOCAL KNOWLEDGE IN CLIMATE ACTION

For successful climate actions and achieving the goals of the Paris Agreement, there is a need for the full participation of all stakeholders, including communities such as indigenous communities. This should encompass not only inclusion in decision-making but also financial support, capacity building, and integrating IK into climate policies (UNDP, 2024). According to UNESCO, 25% of the earth's land area is traditionally owned, managed, used, or occupied by indigenous people, including 35%of highly conserved ecosystems and 35%of protected areas. Additionally, 15% of global forests are managed as community resources by indigenous communities, with 38 million km^2 managed for fishing, gathering, terrestrial animal harvesting and other uses of wildlife species in 87 countries (UNESCO, 2023). Indigenous knowledge systems originate from indigenous peoples who have accumulated unique knowledge, practices, and innovations over the years passed down through generations (Magni, 2016). These practices/knowledges are usually deeply rooted in understanding of and respect for ecological systems that promote sustainable resources use, such as sustainable agriculture systems, climate resilient water management, safeguarding key ecosystems that act as carbon sinks and protecting biodiversity (UNDP, 2024). These communities have managed to generate food and sustain their livelihoods without degrading the natural environment, which could be crucial input to environmental sustainability. A cross pollination of their knowledge and practices with scientific knowledge can help address climate change challenges. In this context, following the adoption of the Paris Agreement, a Local Communities and Indigenous Peoples Platform was created within the United Nations Framework on Climate Change, with indigenous peoples at the forefront. The objective of this platform is to provide a platform handling issues such as their rights to their role in implementation and knowledge sharing within the framework of the convention. This Platform was further cemented with workplans, the most recent of which contains six (06) collective approaches, including; Gathering of knowledge holders; regional engagement; collaboration with constituted bodies and workstreams of the UNFCCC; enhanced engagement with Parties; and overall strategic planning at COP 29 in Baku spanning 3 years. In addition, youths, women, and person with disabilities are greatly involved in this platform, which helps in ensuring that the needs and vulnerabilities of these groups are taken into consideration.

Africa is particularly affected by climate change due to its exposure to climate hazards, high vulnerability, and low adaptive capacity. Africa possesses Indigenous and Local Knowledge (ILK) that has a long history of informing responses to climatic variability and change. The International World Group for Indigenous Affairs and the African Commission of Human and Peoples' Rights estimated that there are approximately 50 million indigenous people in Africa, with most of them being farmers, pastoralists, agro-pastoralists, and hunter-gatherers, making the continent home to over 30% of the world's indigenous languages, exceptionally rich in ecosystem-specific knowledge on biodiversity, soil systems, and water (Trisos et al., 2022). These people respond and anticipate climate variability by utilizing techniques such as observing animal behaviour, weather patterns, and vegetation changes. For example, natives in Borana (Ethiopia), Lushoto (Tanzania), and Hoima and Rakai (Uganda) use cloud type and cover, direction and strength of wind, very high night temperatures, animal behaviour (birds, insects, and some large animals), flowering of specific trees as indicative of the onset of the rainy season help in seasonal quality forecasting, onset rains, particular storms, and long-term forecasting (Leal Filho et al., 2022). Niger was affected by severe deforestation and desertification in the 1980s with farmers struggling with declining crop yields and environmental degradation, but by applying traditional knowledge to protect and manage naturally regenerating trees on their farmland, rather than cutting down all trees for fuel, selectively pruning certain species to encourage growth, promote soil fertility, and provide shade, over 5 million hectares of degraded land were restored, leading to increased crop yields and improved food security, climate change resilience building by improving soil moisture retention and reducing the risk of drought (Rinaudo, 2021).

Thus, it is clear that IM is quite essential in climate adaptation with traditional practices such as rotational farming, water conservation, and controlled burning that can provide sustainable alternatives to modern agricultural and water practices, reducing vulnerabilities to climate impacts (Galappaththi and Schlingmann, 2023), early warning systems based on natural signs, for example, cloud patterns, animal behaviour, or sea level changes that may enhance community preparedness and minimize risks from extreme weather events like floods, storms, and droughts (Leal Filho et al., 2025). ILK and people have and still experience key challenges that affect their ability to contribute to climate actions, especially adaptation. Land tenure insecurity, policies promoting modernization, and lack of legal recognition have led to loss of traditional grazing routes and reduced resilience to climate change for the Maasai people in Kenya and Tanzania. And, in Mali, the disregard for traditional calendars, external development pressures, and modernization have contributed to reduced crop yields and loss of intergenerational knowledge transfer (Leal Filho et al., 2025). Even though there has been recognition of ILK in some key science and policy processes, such as IPCC, Intergovernmental Science-Policy Platform on Biodiversity and Ecosystem Services, and the Convention on Biological Diversity, there is still a need for improved integration of ILK into scientific knowledge in most processes for effective environmental sustainability.

16.5 TECHNOLOGY AND CLIMATE ACTION

The role of technology in the fight against climate change has evolved over the years, featuring new advancements aimed at reversing carbon emissions and reducing and ultimately preventing climate change. Additionally, innovations such as smart agriculture, early warning systems, flood prediction, and resilient infrastructure have emerged to support adaptation. The COP29 Declaration on Green Digital Action recognizes the importance of digital technologies as key tools for climate monitoring, early warning systems, and overall climate adaptation and mitigation and therefore essential in accelerating the achievement of the 2030 Agenda for Sustainable Development. MethaneSAT, a satellite that tracks emissions of methane, has been launched and is expected to provide data, which will be combined with Google's Artificial intelligence (AI) and infrastructure mapping, thereby creating a better understanding of how to mitigate methane emissions, especially from oil and gas infrastructure. Additionally, Climate TRACE, funded by a Google nonprofit coalition, tracks and analyses carbon dioxide (CO_2) emission from power plants, factories, controlled burns, cargo ships etc., using satellite imagery, infrared imaging, and nitrogen oxide sensors to create a publicly accessible source of real-time emissions data. Tidal stream energy remains at an early commercial stage in the United States, with recent feasibility and pilot initiatives underway in Washington State and proposed scaling efforts in Maine, where grid-connected tidal generation has previously been demonstrated. In the United Kingdom, the Liverpool City Region Combined Authority is progressing the Mersey Tidal Power concept, with project materials indicating that the largest option under consideration could supply electricity for up to one million homes for over 120 years. In parallel, battery innovation is targeting higher energy density and faster charging through silicon-containing anodes and solid-state electrolytes, while ultra-fast-charging lithium-ion variants such as Nyobolt's niobium-tungsten-oxide anode chemistry have demonstrated rapid charge performance in prototype testing (OPALCO, 2024; ORPC, 2012; Toki et al., 2024; Luo et al., 2024; Nyobolt, 2024; Reuters, 2025). Furthermore, emerging technologies such as blockchain, AI, quantum computing, robotics, though not initially developed for sustainability or climate change, are increasingly being leveraged to drive real, measurable impact in global environmental efforts. Blockchain technology, for example, is helping to create transparent and secure carbon markets, where carbon credits can be traded. AI is being used for optimizing energy systems, helping to reduce waste and improve efficiency. With smart grids powered by AI being able to predict energy demand and dynamically adjust energy supply, integrating renewable sources such as solar and wind. Moreso, quantum computing can model the behaviour of molecules and materials to help researchers design more efficient batteries or materials for solar panels, which are crucial for a clean energy future (Bolaji, 2025; Demir et al., 2024).

Climate technology development and innovation are highly concentrated in developed countries; for example, two-thirds of all high-value patented climate adaptation technologies in 2010–2015 were located in China, Germany, Japan, the Republic of Korea, or the United States, with 85% of cross-border trade of adaptation-related patents happening in high-income countries and China (Dechezleprêtre et al., 2022). Developing countries, on the other hand, face several structural bottlenecks, which constrain their ability to tap into these technologies and deploy them at scale. Additionally, these technologies are not deployed quickly enough across countries and regions (UNEP, 2023), resulting in a low diffusion rate of climate technologies, ultimately slowing down climate action. Technology development and transfer as a means of implementation and climate action is clearly highlighted in Article 4.7 of the United Nations Framework Convention on Climate Change (UNFCCC) and Articles 10.5 and 10.6 of the Paris Agreement. Institutions such as the Technology Mechanism have also been set up under the convention and developing countries have undertaken technology needs assessments and developed technology action plans under the UNFCCC. National actions and international cooperation are therefore critical to promote the transfer of technology from developed to developing countries to support effective climate action, which could further "birth" technologies suitable for country-level/regional-level situations.

Technologies provide data that governments and other groups worldwide could use independently to spot illegal polluters, verify compliance with international climate change agreements, and manage carbon cap-and-trade markets (UNEP Climate Technology Progress Report, 2023). Globally, there is a significant potential of 27 PWh yr-1 for rooftop solar photovoltaic systems, achievable at costs ranging from USD 40 to USD 280 per MWh; other renewable sources, wind and geothermal power, will significantly lower their carbon footprint. This electrification will play a significant role in also decarbonizing light-duty vehicles. Several advanced economies have already surpassed a 10% sales share of electric cars in new light-duty vehicle markets (International Energy Agency (IEA), 2022). Buildings are also a major mitigation lever: operational energy use in buildings represents about 30% of global final energy consumption, and in 2022 building operations accounted for around 26% of total energy-system emissions (IEA, 2023). This footprint reflects energy demand for heating and cooling, appliances and lighting, alongside upstream emissions from construction materials, which is why innovation in building design, construction and operations is central to resource efficiency and decarbonisation.

Technological developments and innovations have translated to international collaboration, a key catalyst for effective climate change solutions, also requiring coordination nationally amongst different government agencies or ministries.

16.5.1 The Case of Gender and the Use of Technology in the Rice Farming Fields of Ndop Plain

Rice farming has over the years become an important farming and source of revenue for inhabitants of the Ndop Plain, with evidence of migration from neighbouring communities motivated by the rice farming activity. Ndop is a small Town in the NorthWest Region of Cameroon, specifically in the Ngohketunjia division. Its topography, which is low-lying, its fertile alluvial soils owing to its location along the Cameroon volcanic line, and its humid climate with average rainfall ranging from 1500–2000 mm make this area suitable for rice farming. Economic factors such as demand for rice, both locally and nationally, have also been a key push factor for rice cultivation.

Rice cultivation is mainly rain-fed irrigation, and temperature and rainfall key determinants of productivity. The Upper Noun Valley Development Authority is the development authority overseeing rice production and commercialization in Ndop and estimates that 3,700 hectares of fields have been developed and currently used for rice production, including 2,500 ha of paddy fields, with over 15,000 hectares yet to be exploited. UNDVA is hugely responsible for technology development and transfer to farmers. They produce and distribute tonnes of improved rice seeds (seed multiplication) and train farmers in seed selection techniques. They also practice agricultural extension by assisting farmers in the use of improved farming techniques through the provision of extension services, giving advice to the farmers almost daily.

Rice cultivation is considered very tough due to the use of mainly manual labour in the process, with non-rice cultivators terming it "slavery." It starts with the nursing of good quality seeds for 21days, after which they are transplanted to the rice fields that were already prepared. Weed control is done after 2 weeks, and the application of the first dose of fertilizer (usually the NPK) takes place to boost tillering, after which water level control is done, making sure the water covers one-third of the plant for easy adoption of nutrients and weed prevention. The application of the second dose of fertilizer (urea fertilizer) follows, and then water and weed control is done until the rice is mature and ready for harvesting.

Falling yields over the years, especially yields/hectare, have characterized rice production in Ndop. This has been attributed to climate change (change in seasons, floods and droughts), neglect of the sector by the government, and the ineffective implementation of the agricultural extension approaches, migration of youths and most recently the "anglophone crises"—a conflict between the government of Cameroon and secessionists impacting the North and SouthWest regions of the country. Seed selection has been critical to the yields over the years and is still being used, while the use of rudimentary tools such as hoes, cutlasses, and diggers in the past has seen the addition of more mechanized technologies such as tractors. Rice is still being harvested traditionally with the use of transverse harvest knives, even though there are rice harvesting machines used in other rice-producing areas of the world. Winnowing is usually done manually, where farmers collect the rice grains mixed with weeds and other unwanted dirt in containers and buckets and patiently wait for the winds to winnow, but some very few winnowing machines are available, owned by UNDVA and some farming cooperatives but not easily available to individual farmers. The UNVDA has established itself as a precursor in the process of rice farming, with 30 manual trans-planters, 5 combined harvesters,

7 threshers, 10 winnowers, electronic weighing machines, 2 rice polishers, and 10 motorbikes for planting to harvesting, threshing, winnowing, and polishing. Information sharing and communication have been farmer-to-farmer, individual, and group communication methods. However, modern forms of knowledge sharing and communication, such as the use of smartphones and the creation of farming apps, could be used. As such, while farmers are willing to adopt new technologies and technologies, these are usually not affordable and therefore not readily available to them. The case of low technology uptake in this community is not that of accustoming to traditional practices of farming, but a situation of people who want these technologies and techniques to increase production, thereby increasing their income and reducing the "stress" usually accompanying rice production process but rather that of farmers who lack the necessary finances to acquire modern technologies.

In the past years, women have been greatly involved in rice production, though with limited access to resources such as plots of land for rice farming, access to information, inputs, and finance. In particular, traditional norms peculiar to not only this community but most parts of Africa gave mostly men the privilege to own land. However, recently, with government's recognition of inclusion of female farmers in mainstream commercial agriculture activities and support from UNVDA, international and national organizations, and women rights and human rights group, women have been able to inherit, buy land, and have necessary funds for machineries and inputs (new seeds), as well as marketing opportunities which has seen the increase of women in rice cultivation. This has resulted in women farmers leading the way in commercial rice production, helping them cope with climate stresses and provide for their families.

PICTURE 16.1 Women cultivating rice.

Source: UNDVA Ndop.

PICTURE 16.2 Use of tractors for rice cultivation.

16.6 THE ROLE OF GENDER, IM, AND TECHNOLOGY IN CLIMATE ACTIONS AND STRATEGIES

The multifaceted outlooks of relationships are known to help create links that bring about effective and innovative solutions to problems. Women, due to some cultural barriers and other limitations, have inadequate access to climate technology, while some indigenous communities are often neglected and can't afford certain technologies, or prefer their methods or practices with nature and the environment. This greatly affects the pace and effectiveness of climate actions and strategies. The Sustainable Development Goals call for no one to be left behind as a measure of progress on development goals; thus, social inclusion becomes an important aspect in the equation. Climate technologies could also be enriched and effective if they embed IM, as they will be culturally appropriate, which will ease uptake. A gender consideration in the development of such technologies, on the other hand, is an exploitative angle, as women are often closer to the environment and therefore at the centre of mitigative and adaptive actions and strategies. For this to occur, the participation of all stakeholders in these areas is crucial for developing inclusive climate strategies, ensuring a cross-sectoral solution, as shown in Figure 16.3.

Technological uptake by women and indigenous communities will close the digital and tech divide that exacerbates existing inequalities and hinders the integration of technology with IM and gender perspectives in climate action. The adoption of climate technologies by women and indigenous communities is hindered by certain barriers and challenges, as illustrated in Table 16.2.

Lack of access to financial resources means no resources available for the implementation of climate action initiatives. Limited access to education, training, and capacity building restricts the ability of women and indigenous communities to participate in and benefit from climate action efforts, as well as uptake of climate technologies. Poverty and economic inequality also limit their ability to invest in and adopt new technologies. Additionally, the underrepresentation of indigenous peoples and women in political institutions and climate governance structures limits their ability to influence policy decisions that affect them and, in the grand scheme of things, result in climate policies that do not reflect the diverse perspectives and knowledge systems necessary for effective climate action (Climate Investment Funds, 2021).

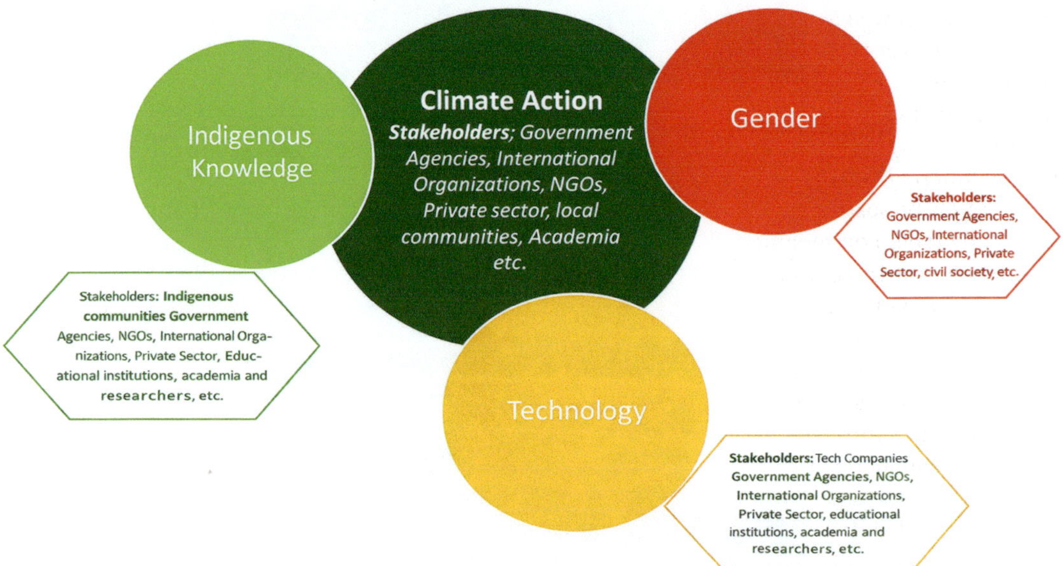

FIGURE 16.3 Stakeholder approach to the intersection of indigenous knowledge, gender, and technology within climate action policies.

TABLE 16.2
Challenges and Barriers for the Uptake of Climate Technology by Women and Indigenous Knowledge Peoples

Key Challenge/Barrier	Gender	Indigenous and Local Knowledge
Discriminatory cultural norms/marginalization	X	X
Limited access to education/capacity Building and training	X	X
Underrepresentation in decision-making in climate action and strategies	X	X
Limited or inadequate financial resources/challenges in accessing credit and finance	X	X
Lack of incentives for the adoption of sustainable and innovative practices	X	X
Side-lining of knowledge/inputs—lack of political will and commitment to recognize and integrate Indigenous Knowledge and gender perspectives in climate policies	X	X
Poverty and economic inequality	X	X

16.7 CONCLUSION

Climate change impacts are not isolated but have differentiated effects on various social groups. The implementation of climate actions and strategies, therefore, will be quite effective if it explores the social dimension, such as diving into gender perspectives, engaging vulnerable and marginalized communities and acknowledging the contribution of indigenous communities who possess key insights into sustainable environmental practices. An integration of such practices with contemporary scientific and technological advancements will lead to strengthened climate resilience and adaptation efforts. Additionally, technology is a key aspect driving climatic actions/strategies and therefore a key angle that can be used to bridge the gaps in social inclusion, translating to gender-sensitive approaches in climate action, leading to the overall improvement in community resilience and adaptive capacity. Thus, it is vital to ensure that there exist key actions at the policy level, community level, as well as individual actions for robustness and transparency.

It is crucial for policymakers to adopt and implement policies that recognize a holistic approach towards climate action to ensure its effectiveness. This includes promoting inclusive governance structures that ensure the active participation of all sectors and communities in decision-making processes, strengthening legal and institutional frameworks. Additionally, there is a need for policies encouraging economic incentives such as subsidies, grants, and also financial support to enable the adoption of sustainable practices and technologies by women, indigenous communities, and other vulnerable communities. Policy makers must also recognize and respect IM, documenting and including it in decision-making, while also ensuring the participation of indigenous peoples and women in political spheres, essential for culturally appropriate, gender-balanced effective, and inclusive climate solutions (Dorji et al., 2024), thereby improving community resilience and adaptive capacity (Climate Investment Funds, 2021). Policy makers need to create enabling environments and encourage investments in research and development of endogenous technologies.

Communities through various initiatives should also advocate for their rights and actively participate in climate decision-making processes, building strong networks and alliances with other communities and organizations, which can amplify their voices and influence policy decisions (Dorji et al., 2024; Climate Investment Funds, 2021). In this light, promoting girl education at the community level will also lead to an empowered generation of women, greatly involved in decision-making in climate actions. Breaking cultural norms and injustices against women should begin at the community level, where it originates, giving women a fair chance to adapt and support the fight towards climate change.

Finally, every individual holds the power to help in the fight against climate change. Actions such as self-education on climate change complexities and solutions, behavioural changes such as contributing to energy efficiency, conscious efforts towards safeguarding and protecting the environment, and encouraging others to join their efforts, amongst others. Individuals can also lend their voices and actively support initiatives that champion inclusivity and sustainability, advocate for policies and programmes that embrace inclusivity, and get involved in community-led or national climate action projects, volunteering time and skills, and donating resources.

REFERENCES

AREI. (2022). Africa Renewable Energy Initiative. https://www.arei.info/eng

Bolaji, G. (2025). Smart Energy Grids with AI for Sustainable Power Distribution: Using AI to Optimize Energy Consumption, Reduce Waste, and Integrate Renewable Sources (Solar, Wind, Hydro) into Smart Grids. https://www.researchgate.net/publication/395020038

Brouwers. (1993). Rural People's Response to Soil Fertility Decline: The Adja Case (Benin), Wageningen Agricultural University Papers 93–4. https://edepot.wur.nl/282865

CEDAW. (2018). General Recommendation No. 37 on the Gender-Related Dimensions of Disaster Risk Reduction in the Context of Climate Change. https://digitallibrary.un.org/record/1626306?v=pdf

CFS. (2020). Gender Equality and Women's Empowerment in the Context of Food Security and Nutrition: Scoping Paper. https://www.fao.org/fileadmin/templates/cfs/Docs1920/Gender/GEWE_Scoping_Paper-FINAL040ct.pdf

Climate Investment Funds. (2021). Empowering Indigenous Women to Integrate Traditional Knowledge and Practices in Climate Action. https://cif.org/sites/cif_enc/files/knowledge-documents/empowering_indigenous_women_summary.pdf

Cogswell, N., & Warszawski, N. (2022). 5 Challenges the UNFCCC Must Overcome to Spur Climate Action. https://www.wri.org/insights/5-challenges-unfccc-must-overcome-climate-action

Dechezleprêtre, A., Gennaioli, C., Martin, R., Muûls, M., & Stoerk, T. (2022). Searching for Carbon Leaks in Multinational Companies. *Journal of Environmental Economics and Management*, Volume 112. https://doi.org/10.1016/j.jeem.2021.102601

Demir, A., Yildiz, E., & Kaya, C. (2024). Application of Quantum Computing in the Design of New Materials for Batteries, Volume 1, No. 6. https://doi.org/10.70177/quantica.v1i6.1700

Dorji, T., Rinchen, K., Morrison-Saunders, A., Blake, D., Banham, V., & Pelden, S. (2024). Understanding How Indigenous Knowledge Contributes to Climate Change Adaptation and Resilience: A Systematic Literature Review. *Environmental Management*, Volume 74, 1101–1123. https://link.springer.com/article/10.1007/s00267-024-02032-x

Eva Alonso-Epelde, E., Rodés, C., & Moyano-Reina, M. (2025). MEDUSA—Modelling Equity and Distributional Impacts for Socioeconomic Analysis. https://doi.org/10.5194/egusphere-egu24-5065

Fernandez, M. E. (1994). Gender and Indigenous Knowledge. *Indigenous Knowledge & Development Monitor*, Volume 2, 6–7.

Galappaththi, K. E., & Schlingmann, A. (2023). The Sustainability Assessment of Indigenous and Local Knowledge-Based Climate Adaptation Responses in Agricultural and Aquatic Food Systems. *Current Opinion in Environmental Sustainability*, Volume 62. https://doi.org/10.1016/j.cosust.2023.101276

Government of Bangladesh. (2009). Climate Change Strategy and Action Plan. https://www.climate.gov.bd/

Graham, J. P., Hirai, M., & Kim, S. S. (2016). An Analysis of Water Collection Labor Among Women and Children in 24 Sub-Saharan African Countries. *PloS One*, Volume 11, No. 6, e0155981.

International Energy Agency (IEA). (2022). Global EV Outlook 2022: Trends in Electric Light-Duty Vehicles. Paris: IEA. https://www.iea.org/reports/global-ev-outlook-2022/trends-in-electric-light-duty-vehicles

International Energy Agency (IEA). (2023). Breakthrough Agenda Report 2023: Buildings. Paris: IEA. https://www.iea.org/reports/breakthrough-agenda-report-2023/buildings

Leal Filho, W., Lütz, M. J., Totin, E., Ayal, D., & Mendy, E. (2025). Obstacles to Implementing Indigenous Knowledge in Climate Change Adaptation in Africa. *Journal of Environmental Management*, Volume 373. https://doi.org/10.1016/j.jenvman.2024.123678

Leal Filho, W., Wolf, F., Totin, E., Zvobgo, L., Philip Simpson, N., Musiyiwa, K., Kalangu, W. K., Sanni, M., Adelekan, I., Efitre, J., Donkor, K. F., Balogun, A., Rui Mucova, A. S., & Yayeh Ayal, D. (2022). Is Indigenous Knowledge Serving Climate Adaptation? *Evidence from Various African Regions*. https://doi.org/10.1111/dpr.12664

Luo, Y., Rao, Z., Yang, X., Wang, C., Sun, X., & Li, X. (2024). Safety Concerns in Solid-State Lithium Batteries: From Materials to Devices. *Energy & Environmental Science*, Volume 17, 7543–7565. https://doi.org/10.1039/D4EE02358G

Magni, G. (2016). *Indigenous Knowledge and Implications for the Sustainable Development Agenda.* UNESCO.

Majumder, Z. H., Alim Shampa, T. A., Islam, A., Deowan, A. S., & Hafiz, F. (2024). Marine Renewable Energy Harnessing for Sustainable Development in Bangladesh: A Technological Review. *Energy Reports*, Volume 11, 1342–1362. ISSN 2352–4847. https://doi.org/10.1016/j.egyr.2024.01.001

NASA. (2025). Global Temperature: Earth Indicator. NASA Science, 25 September. https://science.nasa.gov/earth/explore/earth-indicators/global-temperature/

Nyobolt. (2024). Nyobolt EV Sportscar Prototype Takes Ultra-Fast Charging Out of the Lab and Onto the Road. Nyobolt (Company News Release), 28 June. https://nyobolt.com/resources/news/nyobolt-ev-sportscar-prototype-takes-ultra-fast-charging-out-of-the-lab-and-onto-the-road

OECD. (2024). Climate Finance Provided and Mobilised by Developed Countries in 2013–2022. https://www.oecd.org/content/dam/oecd/en/publications/reports/2024/05/climate-finance-provided-and-mobilised-by-developed-countries-in-2013-2022_8031029a/19150727-en.pdf

OPALCO. (2024). Tidal Energy Pilot Project. Orcas Power and Light Cooperative, 8 October. https://www.opalco.com/pilot-project/2024/10/

ORPC. (2012). Americas First Ocean Energy Delivered to the Grid: ORPC Sells Tidal Power in Maine, 13 September. Ocean Renewable Power Company (Press Release/PDF), ORPC. https://orpc.co/wp-content/uploads/2021/10/americas-first-ocean-energy-delivered-to-the-grid-orpc-sells-tidal-power-in-maine-sept.-13-2012-2012913155.pdf

Perga, M., Sarrasin, O., Steinberger, J., Lane, N. S., & Butera, F. (2023). The Climate Change Research That Makes the Front Page: Is It Fit to Engage Societal Action? *Global Environmental Change*, Volume 80. https://doi.org/10.1016/j.gloenvcha.2023.102675

Remteng, C., Nkem, J., Mofor, L., & Murombedzi, J. (2022). Gender in the Nationally Determined Contributions of African Countries: A Way Forward for Effective Implementation of Adaptation and Mitigation Strategies. *Ecofeminism and Climate Change*, Volume 3, No. 1, 2–22.

Reuters. (2025). Fast-Charging Battery Firm Nyobolt Raises $30 Million for Expansion, 16 April. https://www.reuters.com/technology/fast-charging-battery-firm-nyobolt-raises-30-million-expansion-2025-04-16/

Rinaudo, T. (2021). *Farmer Managed Natural Regeneration in Action: Learnings from the Field. Presentation, GLF Africa 2021*. World Vision. https://www.globallandscapesforum.org/presentation/farmer-managed-natural-regeneration-in-action-learnings-from-the-field/

Statista. (2023). Global GHG Emissions. https://www.statista.com/topics/5770/global-greenhouse-gas-emissions/

Toki, G. F. I., Hossain, M. K., Rehman, W. U., Manj, R. Z. A., Wang, L., & Yang, J. (2024). Recent Progress and Challenges in Silicon-Based Anode Materials for Lithium-Ion Batteries. *Industrial Chemistry & Materials*, Volume 2, No. 2, 226–269.

Trisos, C., Adelekan, I., Totin, E., Ayanlade, A., Efitre, J., Gemeda, A., Kalaba, K., Lennard, C., Masao, C., Mgaya, Y., Ngaruiya, G., Olago, D., Simpson, N., Zakieldeen, R., & Thorn, J. (2022). Africa. In: *Climate Change 2022: Impacts, Adaptation and Vulnerability. Contribution of Working Group II to the Sixth Assessment Report of the Intergovernmental Panel on Climate Change*. Cambridge University Press. https://doi.org/10.1017/9781009325844.011

UN Women. (1995). https://www.un.org/womenwatch/daw/beijing/platform/

UN Women. (2017). Gender Strategy for Climate Change. https://www.unwomen.org/en

UN Women. (2020). *Gender Equality and Women's Empowerment in Climate Action*. UN Women.

UNDP. (2019). Bhungroo Project in Gujarat. https://www.undp.org/india

UNDP. (2024). *Integration of Indigenous Peoples, Local Communities in REDD+ Projects Will Uphold Sustainability*. UNDP Climate and Forests. https://www.climateandforests-undp.org/knowledgeexchange

UNEP. (2023). Goal 13: Climate Action. https://www.unep.org

UNESCO. (2023). Natural Sciences Sector: Local and Indigenous Knowledge Systems (LINKS). https://unesdoc.unesco.org/ark:/48223/pf0000385181

UNFCCC. (2022). *How Technology Can Help Fight Climate Change*. UNFCCC.

United Nations Framework Convention on Climate Change (UNFCCC). (2023). The Paris Agreement. https://unfccc.int/process-and-meetings/the-paris-agreement/the-paris-agreement

World Bank. (2019). *Gender Dimensions of Disaster Risk and Resilience: Existing Evidence*. World Bank.

World Bank Group. (2023). Placing Gender Equality at the Center of Climate Action. Gender Thematic Policy Notes Series: Issues and Practices Note.

World Bank. (2024). Choosing our Future. Education for Climate Action. chrome-extension://efaidnbmnnnibpcajpcglclefindmkaj/https://openknowledge.worldbank.org/server/api/core/bitstreams/9d1c318a-bcd3-49fa-b1c6-cc03e18d4670/content

World Economic Forum. (2021). *How Can Technology Help Combat Climate Change*. World Economic Forum.

World Economic Forum. (2024a). *6 Technologies to Help the World Adapt to Climate Change*. World Economic Forum. https://sustainablereview.com/top-8-green-technology-innovations/

World Economic Forum. (2024b). *Indigenous Communities Face Climate Displacement Challenges*. World Economic Forum. https://www.weforum.org/stories/2024/02/indigenous-challenges-displacement-climate-change/

17 Sustainable Aviation and SDGs in BRICS+ Countries
A Comprehensive Review and Bibliometric Study

Reshma Sucheran

17.1 INTRODUCTION

Sustainable development in the Global South is intrinsically linked to improvements in international tourism and air transport. Tourism is widely acknowledged as a critical driver of economic growth, providing substantial contributions to gross domestic product, creating jobs, and facilitating cross-cultural exchange (Pisa, 2018; Dimitrios & Maria, 2018). Air transport plays a pivotal role in this process by enhancing connectivity, enabling the efficient movement of people and goods, and stimulating trade, investment, and tourism (Hasan et al., 2021; Zhang & Graham, 2020). In addition, the expansion of air transportation enhances employment opportunities in associated industries and supports the local economies with an improved system (Usmani et al., 2021).

The BRICS+ countries have expanded to a geopolitical landscape worldwide. This reflects a determination adopted by the emerging economies to amalgamate with the existing order of a multipolar world. The current expansion of BRICS+ countries has brought together many countries with political, economic, and strategic urgencies. The divergent economic policies of all the BRICS+ countries stipulate mutual collaboration, which may add challenges to aligning aviation and tourism policies. Therefore, intense negotiations and multilateral dialogues are expected among the members of all the states to accommodate their interests. Air transport is important in promoting trade, developing tourism, and facilitating easy cultural exchange to establish cooperation. Aviation supports the recent global trade, acts as the spine of tourism, and promotes cross-border investment. However, the air transport networks cannot be overstated in connection with the BRICS+ countries (Joshan & Maertens, 2020). Air transport needs to be revised and upgraded to build the connectivity between the BRICS+ countries with their initiatives to facilitate the set objectives (Kazelko & Semeghini, 2024).

The aviation industry is a crucial driver for the global transformation of greenhouse gas (GHG) emissions. In 2018, global CO_2 emissions accounted for approximately 2.5%, equating to about 1 billion tonnes of CO_2, and this is seen as a critical challenge to the United Nations Sustainable Development Goals (SDGs) (Katırcıoğlu, 2024). The change in climatic conditions is exacerbated by aviation emissions as CO_2, nitrogen oxides (NO_x), water vapour, and other such matters contribute to radiative forcing responsible for rising temperatures, extreme weather events, and the retreated glaciers (Lee et al., 2021; Terrenoire et al., 2019). Carbon-intensive fossil fuels are the major components of the air transport industry, which poses a significant barrier to the advancement of SDG 13, where immediate action is called upon to combat climate change. Planès et al. (2021) and Liu and Zhang (2023) predict that the aviation sector may contribute up to 25% of worldwide CO_2 emissions by 2050 if current trends persist, thereby jeopardising the Paris Climate Agreement to limit global warming to 1.5°C.

While research on air transportation in BRICS+ countries is evident, a comprehensive analysis of research trends specific to sustainable aviation in these nations remains underexplored. This

DOI: 10.1201/9781003600152-17

chapter will examine this issue. Specifically, the chapter analyses research trends, evaluates sustainable aviation initiatives, identifies challenges and gaps in each BRICS+ country, and proposes strategies for effective sustainable aviation solutions.

17.2 MATERIALS AND METHODS

This study implemented the bibliometric approach to investigate research trends related to sustainable aviation, aviation emissions, and sustainable aviation fuels (SAFs), utilising the Scopus database due to its extensive coverage of peer-reviewed literature. The methodology consisted of a systematic document selection process followed by bibliometric and visualisation analyses, as outlined in Figure 17.1. The document selection process commenced with a search in the Scopus database using the search string "SUSTAINABLE AVIATION" OR "AVIATION

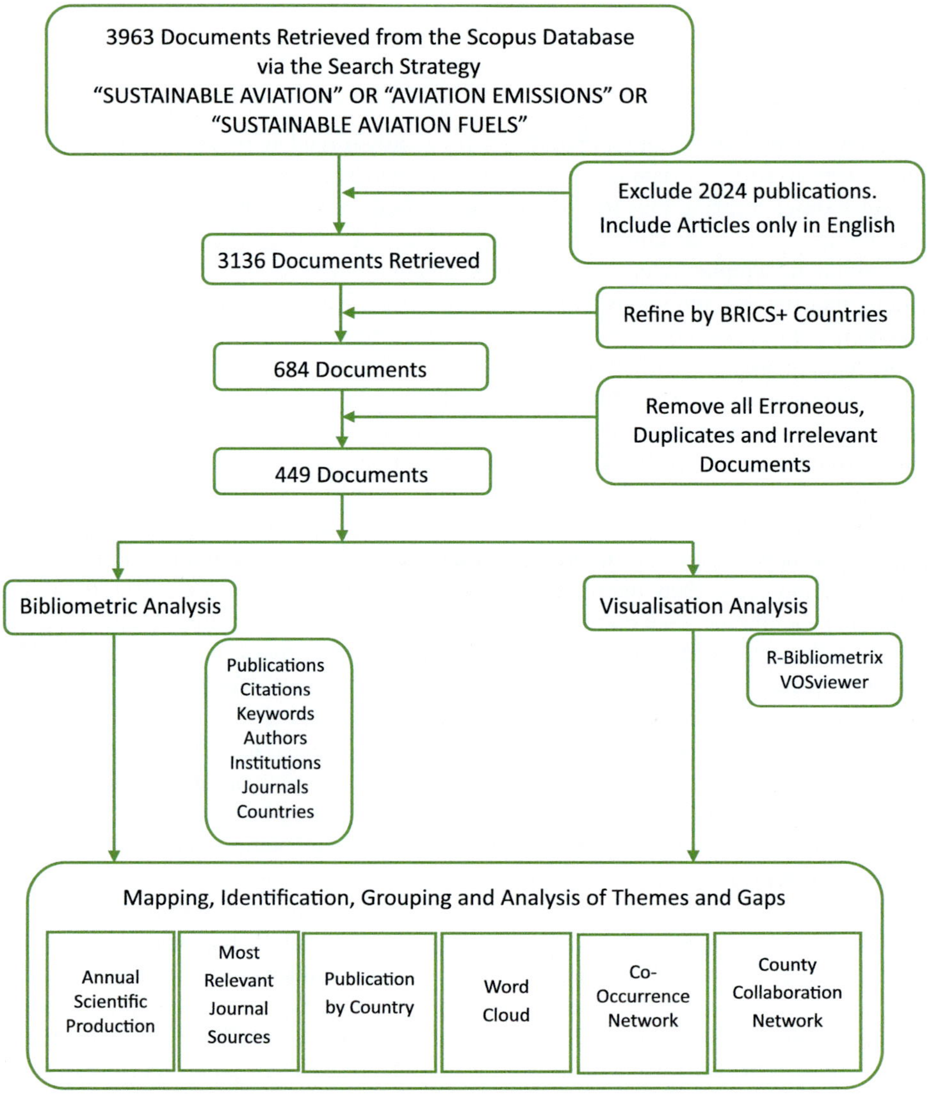

FIGURE 17.1 Bibliometric analysis flow diagram.

Source: Author's compilation.

EMISSIONS" OR "SUSTAINABLE AVIATION FUELS." This initial search retrieved 3,963 documents, including publications spanning various years. The dataset was refined through several filtering steps to ensure relevance and consistency. First, publications from 2024 were excluded, and only articles written in English were retained, which reduced the dataset to 3,136 documents. Thereafter, the scope was narrowed to include only documents originating from BRICS+ countries (Brazil, Russia, India, China, South Africa, Egypt, United Arab Emirates (UAE), Saudi Arabia, Ethiopia, and Iran), resulting in 684 documents. Finally, a manual review was conducted to remove erroneous, duplicate, and irrelevant documents, yielding a final dataset of 449 peer-reviewed articles.

The final dataset of 449 articles served as the foundation for the bibliometric analysis. Concurrently, a visualisation analysis was performed using R-Bibliometrix and VOSviewer software to map and interpret the data visually. The analysis concluded in a thematic mapping and gap identification phase, which involved assessing annual scientific production, identifying the most relevant journal sources, analysing publications by country, generating word clouds to highlight prevalent keywords, constructing co-occurrence networks to explore conceptual relationships, and mapping country collaboration networks to understand international research partnerships. This multi-step methodology ensured a thorough and systematic literature investigation, providing a comprehensive overview of research trends in sustainable aviation.

17.3 RESULTS

17.3.1 Science Mapping

According to Table 17.1, research on sustainable aviation is growing rapidly, with an annual increase of 17.45%. This field of study now includes 449 documents demonstrating global attention to environmentally friendly aviation practices. Each document is cited an average of 13.72 times, and the average age of these documents is 2.78 years, implying that the research reports on current trends. Moreover, the wide range of topics covered by 3,010 Keywords Plus and 1,384 author keywords links sustainable aviation to environmental policy and technology, suggesting a broad, integrated approach. The participation of 1,567 authors denotes substantial interest and activity, with only 9.13% of documents written by a single author. In addition, each document has an average of 4.48 co-authors, further emphasising this collaborative trend. Finally, nearly 30% of the documents involve international co-authorship, which emphasises the global nature of the research and the importance of sharing knowledge across countries.

TABLE 17.1
Document Growth and Output

Timespan	1995–2024
Articles	449
Annual growth rate %	17.45
Document average age	2.78
Average citations per doc	13.72
Keywords plus (ID)	3010
Author's keywords (DE)	1384
Authors	1567
Authors of single-authored docs	37
Single-authored docs	41
Co-authors per doc	4.48
International co-authorships %	29.84

Figure 17.2 illustrates that between 1995 and 2005, research on sustainable aviation in BRICS+ countries was scant, with only a few papers published each year and just one or two articles appearing in 1995, 2001, 2002, 2003, and 2004. This limited output likely reflects the initial lack of emphasis on sustainable aviation. However, starting in 2006, there was a modest increase in publication numbers, which peaked at six papers in 2015. This gradual rise suggests a growing awareness of climate change and an emerging interest in sustainable practices in the aviation sector. Subsequently, from 2016 onward, research experienced a dramatic surge. Publications escalated from just 3 in 2016 to 66 in 2021, further increasing to 103 in 2023 and 106 in 2024. This substantial growth can be attributed to several factors, including heightened global concern about climate change, advancements in green aviation technologies, increased funding from governmental and non-governmental sources, and the interdisciplinary nature of the research. Moreover, the peak in research output in recent years signals a significant shift towards a concentrated focus on sustainable aviation in BRICS+ countries.

Data on the most relevant journal sources for research on sustainable aviation in BRICS-PLUS (Figure 17.3) shows various important contributions from several important journals. The journal

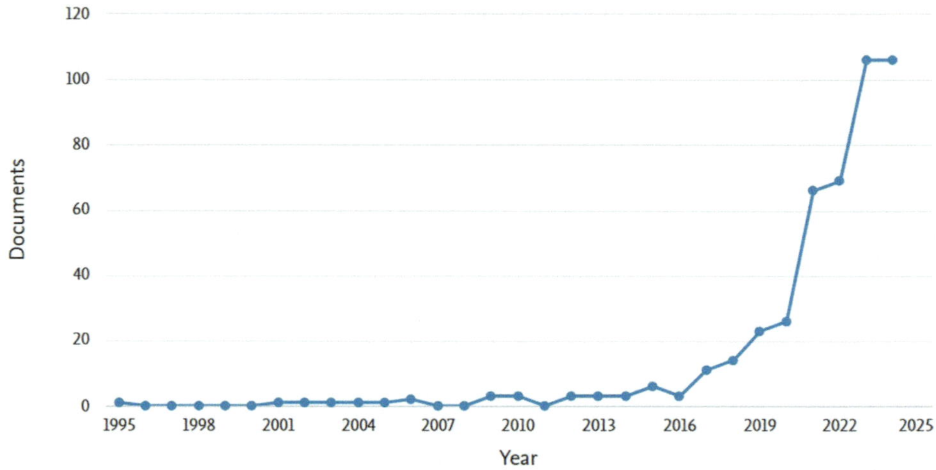

FIGURE 17.2 Annual scientific production.

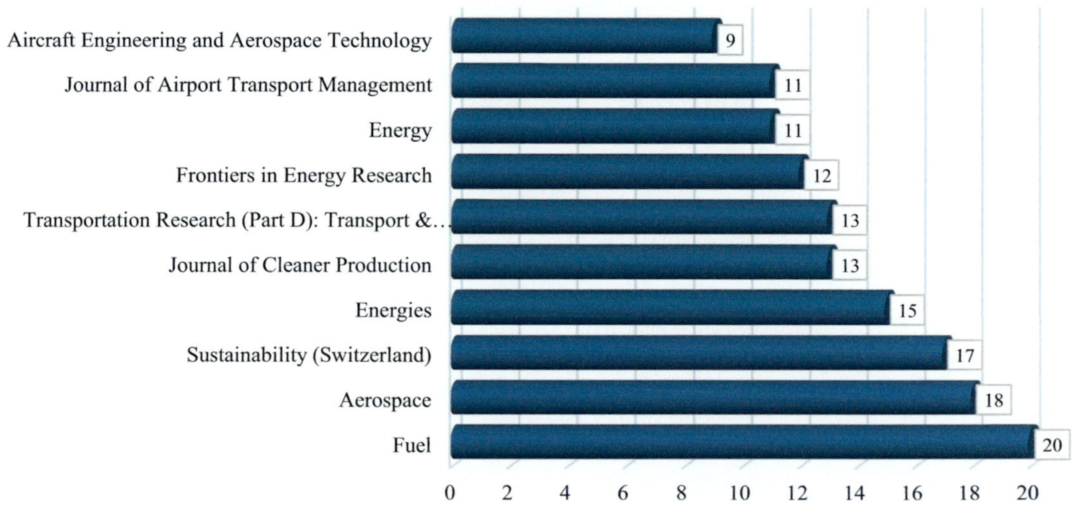

FIGURE 17.3 Most relevant journal sources.

Fuel leads in 20 publications, focusing on alternative fuels and their emission reductions. *Aerospace* follows with 18 publications highlighting the progress of aircraft design and drive systems, which are important for developing sustainable aviation technologies. *Sustainability* (Switzerland) deals with 17 publications on a wide range of sustainability issues, which reflects the interdisciplinary nature of research in this field. Furthermore, with 15 publications, *Energies* contributes significantly through research into energy generation and renewable solutions, which is extremely important for improving the energy efficiency of aviation. The *Journal of Cleaner Production* and *Transportation Research Part D,* with 13 publications each, highlight the importance of clean production technologies and the environmental impact of air traffic. Similarly, *Frontiers in Energy Research* has 12 publications on energy technology, which is extremely important for the further development of sustainable aviation. *Energy* and the *Journal of Air Transport Management* contains 11 articles each, which help understand the energy systems and sustainable practices within air traffic management. With nine publications, *Aircraft Engineering and Aerospace Technology* investigate the progress of aerospace technology that directly supports sustainable aviation goals.

Some countries dominate the field of sustainable aviation research. For example, China is leading with 65 publications, indicating its focus on aviation technology and sustainability. Similarly, the United States follows closely with 63 publications supported by advanced research institutions and important financial resources. With 32 publications, Germany has made a major contribution due to its strong industrial base and commitment to sustainability. Additionally, there are 25 or 20 publications in the UK and Australia for established research institutions, focusing on environmental issues. Other contributors such as Türkiye, Italy, Canada, India and the Netherlands have 13–19 publications. Regarding BRICS+ Countries, China continues to lead in sustainable aviation research. With 13 publications, India has shown considerable progress, reflecting the rapid expansion of the aviation sector. In contrast, Brazil, Russia and South Africa have lower research results than China and India, mainly due to different investment levels and research priorities.

Research on sustainable aviation presents several important themes, as indicated in the frequent mention of specific words in the literature shown in Figure 17.4. This analysis presents words emphasised in the discourse and their links to broader environmental goals. For example, "aviation fuel" is the most frequently mentioned term, highlighting its central role in the discussion of sustainable aviation. Also, SAFs are presented prominently, with an emphasis on the development and introduction of fuels that minimise environmental impacts and reduce aviation CO_2 footprint (Yang & O'Connell, 2020; Said et al., 2022; Manigandan et al., 2023; Rambabu et al., 2023; Zhu et al., 2022; Li et al., 2021; Ershov et al., 2023; Wang et al., 2024a, 2024b; Liu & Yang, 2023; Guo et al., 2023). Additionally, the recurrent mention of "sustainable development" (Ul-Haq et al., 2023; Ryley et al.,

TABLE 17.2
Publications per Country

Top Ten Countries		BRICS+ Countries	
Country	**Publications**	**Country**	**Publications**
China	65	China	65
USA	63	India	13
Germany	32	Brazil	12
United Kingdom	25	Egypt	3
Australia	20	South Africa	3
Turkey	19	Saudi Arabia	2
Italy	17	United Arab Emirates	2
Canada	14	Ethiopia	1
India	13	Iran	1
Netherlands	13		

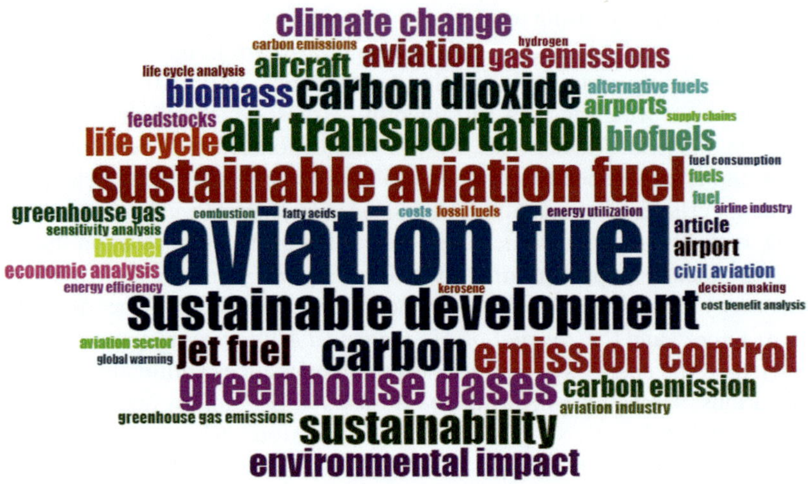

FIGURE 17.4 Word cloud.

2013; Chen et al., 2023) aligns with global sustainability goals to ensure that aviation practices contribute positively to overarching sustainable development objectives. Furthermore, the theme "air transport" occurs repeatedly indicating an emphasis on the entire air travel system that go beyond fuel to reflect on sustainability in efficiency, reductions in emissions, and improved environmental performance of the wider air transport network (Bartle, 2006; Sobieralski, 2023; Chen et al., 2021). Terms such as "carbon," "carbon dioxide," and "GHG" confirm the impact of air transport on climate change and its detrimental gas emissions (Jing et al., 2022; Singh et al., 2018; Han et al., 2022). Moreover, the term "emission control" is frequently mentioned in the literature (Liang et al., 2024; Yang et al., 2023; Cui et al., 2022; She et al., 2023), emphasising efforts to lessen aviation emissions through technological and strategic improvement. Finally, the term "sustainability" refers to the broader goal of ecological aviation, and "jet fuel" stresses the importance of fuel types (Huang et al., 2019; Zhang et al., 2021a, 2021b; Julio et al., 2021).

Figure 17.5 illustrates the co-occurrence network for sustainable aviation, a web of interconnected concepts and key terms. The themes "Aviation" and "SAF" are central to the literature, confirming efforts in using cleaner technologies to reduce the environmental impact of aviation (Rambabu et al., 2023; Zhu et al., 2022; Li et al., 2021; Ershov et al., 2023; Wang et al., 2024a, 2024b; Liu & Yang, 2023; Singh et al., 2024; Guo et al., 2023). The term "sustainability" is also central in the academic discourse, associating climate action with sustainable aviation practices. Other emerging terms such as "hydrogen" (Pinto & Chege, 2024; Suresh et al., 2023; Yusaf et al., 2022; Ardo et al., 2022; Manigandan et al., 2023; Shao et al., 2021; Wei et al., 2019) and "CORSIA" (Thuy, 2021; Prussi et al., 2021; Escobar et al., 2024; Bauen et al., 2022), are also evident. Hydrogen is less central to SAF and is recognised for its potential as an alternative fuel. CORSIA directs policy-related dialogues on aviation emissions. The data also reveal less conspicuous terms such as "electric aircraft" (Ansell, 2022; Eaton et al., 2024; Marciello et al., 2023) and "life cycle assessment" (Marangon et al., 2024; Zhang et al., 2020; Rojas-Michaga et al., 2023; Seber et al., 2022).

Figure 17.6 illustrates the country collaboration network and confirms that different countries show fluctuating levels of influence. For example, China emerges as a major player in sustainable aviation research. In contrast, the USA, Canada, and Spain are grouped in Cluster 1, and while these countries hold important positions, they play a more supportive role than China's dominance. Countries such as China, Australia, India, Malaysia, the UAE, and Egypt, which form part of Cluster 2, are very prominent and active in sustainable aviation research. China's leadership within this group confirms its central role. A collaborative research environment exists in this cluster, with

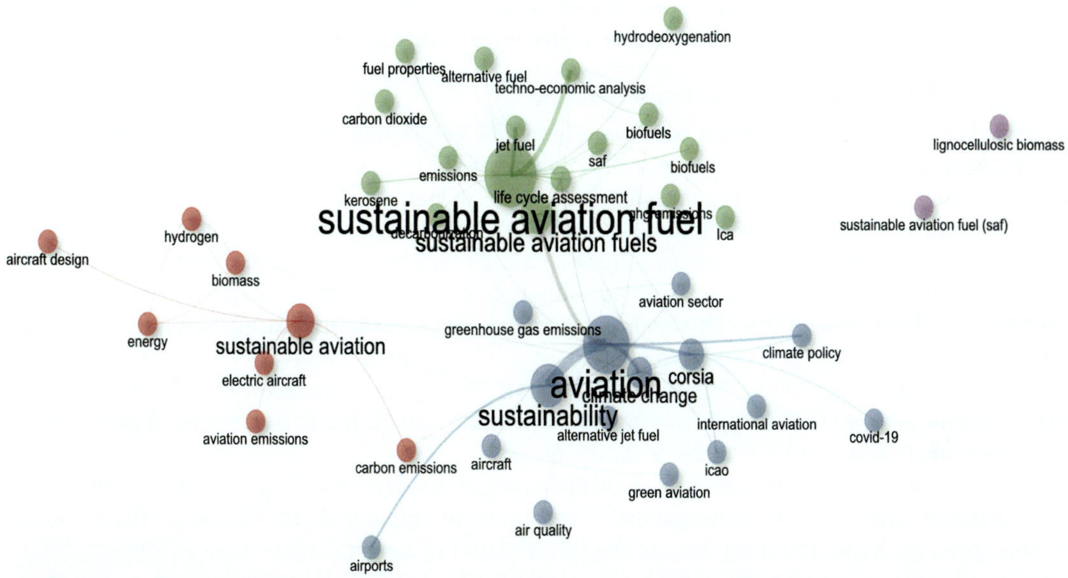

FIGURE 17.5 Co-occurrence network.

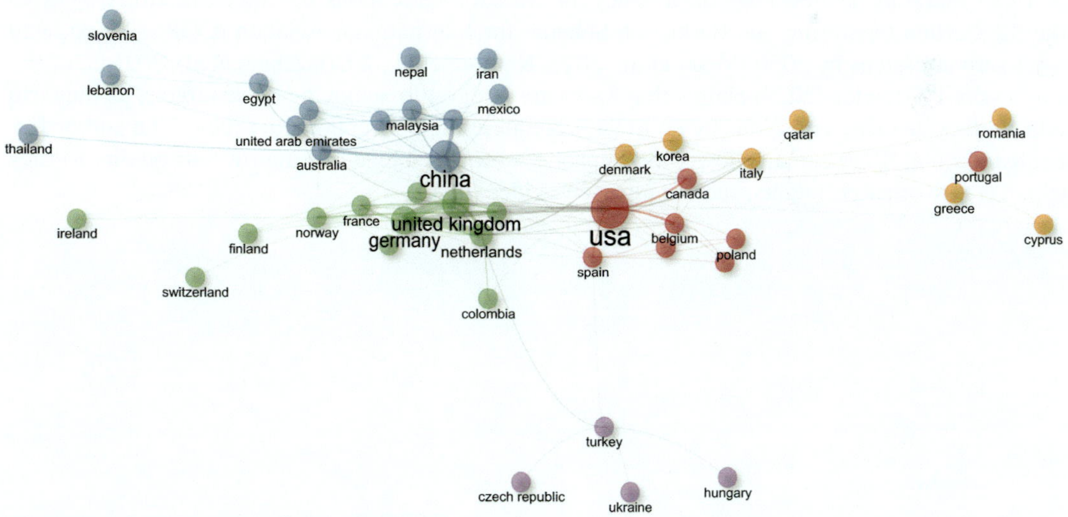

FIGURE 17.6 Country collaboration network

the diversity of countries verifying that these nations work together closely, enhancing their overall impact on sustainable aviation research. The UK, Germany, and Sweden make up Cluster 3, and although they have a significant influence, their impact is secondary to that of other countries.

17.3.2 Thematic Literature Review

The aviation industry's environmental footprint has become a central concern within global sustainability efforts as its contribution to GHG emissions grows. This literature review synthesises key information from academic research to comprehensively understand the current state of sustainable aviation in BRICS+ countries and emerging solutions to reduce aviation's environmental impact.

17.3.2.1 Global CO_2 Emissions From Aviation

Aviation is a sector that contributes substantially to GHG emissions. As reported in Figure 17.7, the International Civil Aviation Organisation (ICAO) (ICAO, 2019) confirms that aviation contributed about 1.04 billion tonnes of CO_2, amounting to 2.5% of global GHG emissions (Li et al., 2021, 2023). The travel sector's heavy reliance on fossil fuels and the long lifecycle of aircraft hinders the transition to alternative energy sources (Grewe et al., 2021; Lee et al., 2021). Since the mid-1980s, CO_2 emissions from aviation have doubled, although the sector's share of global emissions has remained stable at around 2%. This stability conceals the substantial rise in absolute emissions, making aviation one of the fastest-growing sources of emissions (Gößling & Humpe, 2020; Jing et al., 2022; Lyu et al., 2023). The International Air Transport Association (IATA) predicts that international air travel will surpass pre-pandemic levels by 2024, increasing the need for effective emission reduction strategies (Planès et al., 2021).

The aviation sector relies heavily on fossil fuels, which impedes its efforts to reduce GHG emissions. Aircraft have a very long operational lifespan and therefore include older, less efficient models that generate high emissions despite the introduction of newer designs (Liu & Zhang, 2023). Additionally, the move towards sustainable aviation infrastructure (airports and air traffic management systems) is a challenge as these are designed for conventional fuels (Klapmeyer & Marr, 2012). Terrenoire et al. (2019) claim that the demand for air travel is projected to grow exponentially, which in turn will drastically increase the sector's contribution to global emissions. The ICAO and the IATA therefore acknowledge the urgency for critical interventions and are promoting initiatives like the Carbon Offsetting and Reduction Scheme for International Aviation (CORSIA), targeted at net-zero emissions by 2050 (Yusaf et al., 2022; Rathore et al., 2020; Zhang et al., 2021a, 2021b). Kurzawska-Pietrowicz (2024) claims that SAFs are integral to achieving these targets, as they can reduce lifecycle GHG emissions by up to 94% compared to conventional jet fuels. The author further asserts that SAFs can also lower the impact of climate change and improve air quality as they help decrease other pollutants, such as nitrogen oxides (NO_x).

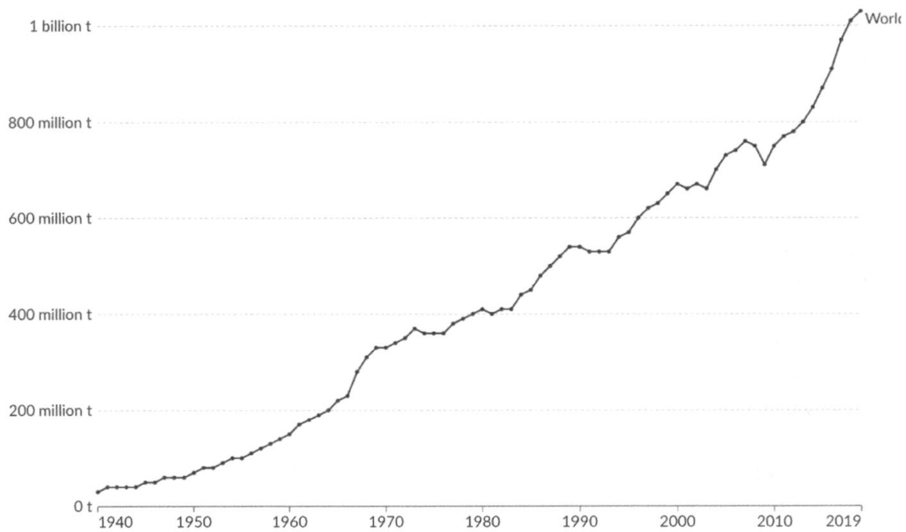

FIGURE 17.7 Global CO_2 emissions from aviation, 1940 to 2019.

Source: Pre-1990 data (Lee et al., 2021); 1990 onwards (Bergero et al., 2023).

17.3.2.2 Sustainable Aviation Fuels

SAFs are made from agricultural waste and used cooking oil (Baxter, 2020; Hasan et al., 2021). SAFs are cleaner than conventional fuels, use renewable resources and can reduce carbon emissions by 50–78%, depending on the raw materials used in their production (Alam et al., 2021). During combustion, SAFs emit far fewer carbon emissions, soot and pollutants than conventional fuels (Khalili et al., 2019; Braun-Unkhoff et al., 2016; Richter, 2024; Riebl et al., 2017). Biofuel is an important type of SAF that can reduce CO_2 emissions by up to 95% (Dyk et al., 2019; Song, 2024) and GHG emissions by nearly 80% over its lifecycle (El-Galad et al., 2018; Johnson et al., 2022).

SAFs also have the advantage of being highly compatible with existing aviation infrastructure. These fuels have a "drop-in" function that blends easily with traditional jet fuels without modifying existing aircraft engines (Zhang et al., 2016; Munirathinam et al., 2023), and many synthetic variants have already obtained regulatory clearance for use in commercial aviation (Silva et al., 2016; Białecki et al., 2021). However, SAFs are not a complete solution as other emissions from aviation, like NOx and contrails, continue to impact the environment adversely (Bergero et al., 2023; Dray et al., 2022). Therefore, the literature confirms a need for broad-based plans to tackle the environmental concerns linked to aviation. Encouragingly, producing SAFs contributes to air quality improvement and generates new economic opportunities (Oluwaseun, 2023). Additionally, many SAFs are derived from residual waste, aligning with the global circular economy model by optimising resource utilisation and reducing landfill dependency (Baxter, 2020). Despite the promise of SAFs, their implementation faces important challenges (Champeecharoensuk, 2023). Production capacity remains limited since the aviation sector relies greatly on fossil fuels. Economic viability is substantially prohibitive as SAFs typically cost more than conventional jet fuels, and are influenced by factors such as feedstock availability, production costs, and market demand (Oluwaseun, 2023). To overcome these economic challenges, Hasan et al. (2021) and Zhang et al. (2020) suggest that supportive policy frameworks be developed to improve the economic feasibility through carbon pricing and blending mandates. Anger (2010) and Champeecharoensuk (2023) add that SAF production infrastructure investments and technological advancements are essential to address supply constraints and cost barriers.

17.3.2.3 Sustainable Aviation Practices in BRICS and Other Nations

17.3.2.3.1 Brazil

As per Brazil's National Civil Aviation Policy (PNAC), sustainable aviation aligns with the ICAO and the Paris Agreement, to reduce emissions and invest in a way to contribute towards green technologies. The observance of Brazil on ICAO's Carbon Offsetting and Reduction Scheme validates its commitment towards sustainable aviation (Staples et al., 2018). Abundant biomass resources, such as sugarcane, position Brazil favourably toward SAF production (Fiorini et al., 2023; Cortez et al., 2015). Additionally, the country's expertise in biofuels, especially for ethanol, provides a solid foundation with which to advance aviation biofuels. According to Carvalho et al. (2020), high reductions in GHG emissions are achievable through SAF production from agricultural residues, such as sugarcane bagasse. Guimarães et al. (2023) and Moreira et al (2014) concur that renewable jet fuel from sugarcane is very sustainable, with lifecycle GHG emissions being much lower than conventional fuel ones. However, commercial viability remains challenging due to high production costs and limited infrastructure in Brazil.

Technological advancements in aircraft efficiency are essential to address climate change challenges. Anholon et al. (2016) and (Rendón et al. (2021) affirm that Embraer, a Brazilian aircraft manufacturer, is pioneering efforts in electric and hybrid aircraft technologies to lower carbon emissions Moreover, Brazil is updating its air traffic management systems to optimise flight paths and enhance fuel efficiency to reduce GHG emissions (Silva et al., 2016). The Brazilian National Agency of Civil Aviation (ANAC) supports biofuel use through a regulatory framework aligned with national and international emission reduction commitments (Cortez et al., 2015). Nonetheless,

Brazil lacks comprehensive policies specifically targeting aviation decarbonisation. Mechanisms like carbon pricing, blending mandates, and incentives for sustainable fuels and technologies are vital for promoting SAF adoption and other green aviation technologies (Hasan et al., 2021). Brazil's biofuel experience and potential for SAF development from waste feedstocks can encourage a circular economy. However, significant investments in research, development, and SAF production infrastructure are necessary to realise this potential on a larger scale.

17.3.2.3.2 Russia

Russia's sustainable aviation strategy encompasses regulatory and technological aspects. The Federal Air Transport Agency oversees aviation policies emphasising environmental sustainability, yet enforcement and integration remain challenging (Ojekunle, 2023). Internationally, Russia participates in forums on aviation sustainability and aims to comply with standards set by the ICAO (Ojekunle, 2023). Despite these efforts, fully adopting these norms is difficult. Technologically, Russia advances in aircraft development, with manufacturers like Sukhoi focusing on fuel-efficient, low-emission aircraft. Moreover, investments in airport infrastructure support green technologies and operational efficiency. These initiatives aim to modernise Russia's aviation sector to meet global emissions standards. However, significant barriers to sustainability persist. The largely state-controlled sector operates many outdated, energy-inefficient aircraft and relies heavily on fossil fuels. Research indicates Russia's aviation infrastructure is not equipped for a large-scale shift to SAFs. Additionally, Russia's vast geography and cold climate complicate the adoption of electric or hydrogen-powered aircraft, requiring specific infrastructure and conditions. SAFs are crucial for sustainable aviation, and Russia is increasingly interested in their development. The Ministry of Science and Higher Education supports biofuel research to reduce aviation-related GHG emissions (Varakin et al., 2020). However, commercial viability is limited, and international certification is necessary for safety and compliance with environmental standards. Government policy is pivotal in promoting sustainable aviation in Russia, and effective policies that balance economic, environmental, and social considerations are essential (Kimeria, 2022). A comprehensive approach involving policymakers, industry leaders, and financial institutions is vital for overcoming the challenges in Russia's aviation sector (Haksevenler et al., 2023).

17.3.2.3.3 India

India's National Civil Aviation Policy (NCAP) focuses on sustainable aviation by reducing carbon footprints and promoting SAF. The focus on producing and adopting SAFs in India involves using domestic feedstocks like agricultural residues and non-food crops (Racha et al., 2024; Sharno & Hiloidhari, 2024). However, Ramakrishnan et al. (2022) and Singh et al. (2022) note that several barriers, such as inadequate infrastructure, limited policy incentives, insufficient research investment, and regulatory difficulties that struggle to align with international standards, impede SAF production in India. Therefore, the country's regulatory framework must be strengthened via increased government intervention to advance sustainability goals in aviation. India's aviation sector's technological innovation includes startups aimed at developing electric and hybrid aircraft technologies to reduce emissions. However, those startups require significant investment along with infrastructure development. India leads in sustainable aviation and green airport infrastructure (Kumar & Gupta, 2020; Ramakrishnan et al., 2022; Sreenath & Sudhakar, 2017), particularly in the integration of renewable energy within aviation operations by way of initiatives like Cochin International Airport, which is the world's first airport fully powered by solar energy (Sukumaran & Sudhakar, 2017). Despite such achievements, the rising air travel demand with slow technological progress in green aviation poses several challenges for the aviation sector to reduce its carbon footprint in India (Korba et al., 2023). According to She (2023) and Zhang et al. (2016), it is predicted that India will likely be the third-largest air passenger market around 2030 as aviation increases greatly, raising concerns about its GHG emissions.

17.3.2.3.4 China

China, the world's largest aviation market, has made significant strides in sustainable technologies through its policies. The country's 14th Five-Year Plan advocates for reducing carbon emissions using SAFs and aims to promote green technologies and electric aircraft (Li et al., 2023; Zhang et al., 2020; Ma et al., 2017; Xiang et al., 2018). In alignment with these targets, China aspires to reach peak carbon emissions by 2030 and achieve carbon neutrality by 2060. According to Wang et al. (2023) the role of government and state-owned organisations in SAF production is vital. China incorporates renewable and sustainable energy systems into aviation infrastructure to promote green airports and enhance energy-efficient designs. Moreover, China is at the forefront of SAF research and development (R&D) and is engaged in several projects investigating biofuels as replacements for conventional aviation fuels. Consequently, the country has developed greener aviation through initiatives like the modern COMAC C919 aircraft, focusing on fuel efficiency and carbon emission reduction. Additionally, China is upgrading its air traffic management systems to assist airlines in reducing their fuel usage and lessening their environmental footprint (Chen et al., 2022). The country also invests in next-generation aircraft powered by varied energy sources to accurately forecast emissions and optimise flight pathways, ensuring complete control over traffic operations (Wan et al., 2022; Shang, 2023). To further encourage sustainable aviation practices, China is offering financial incentives for airlines to adopt environmentally friendly technologies (Afşar, 2023). China hopes to lead the world with its largest aviation market by 2030, with plans for significant changes to improve aircraft designs for better fuel efficiency and sustainable airport infrastructure. Moreover, the country has set targets for emission control to reduce aviation-related CO_2 emissions from 2.9 billion tonnes by 2050 (Wu, 2023).

17.3.2.3.5 South Africa

South Africa's national transport plans concentrate on sustainable aviation to reduce carbon emissions and adopt greener technologies, aligning with the United Nations SDGs (Nkosi et al., 2023). These efforts validate South Africa's commitment to sustainable aviation and environmental conservation. The government supports producing and implementing SAFs. However, high costs and technological barriers hinder the uptake of SAFs in the country. To address this, Mohamed et al. (2014) and Rogerson (2014) argue that financial support through sustainability funds and green financing schemes is integral to expediting the aviation sector's adoption of eco-friendly practices. Additionally, initiatives like the Air Liberalisation Index promote sustainable air travel by enhancing connectivity while prioritising environmental considerations (Surovitskikh & Lubbe, 2015).

Key challenges facing sustainable aviation in South Africa include insufficient investment in SAFs and renewable energy sources and a persistent reliance on fossil fuels. Furthermore, economic constraints that hinder R&D impede the country's ability to transition to SAFs fully (Pinto, 2024). Nevertheless, South Africa has the potential to use its renewable energy resources and establish collaborations to accelerate the transition towards a more sustainable aviation sector (Pinto, 2024). The country possesses diverse renewable energy resources, particularly abundant solar power. The aviation sector can significantly reduce its carbon footprint by integrating solar power into airport operations. Collaborating with countries from the BRICS group and beyond, particularly in sustainable aviation, can expedite the transition to more environmentally friendly practices. Additionally, optimising flight routes and enhancing airport infrastructure are critical strategies that could improve operational efficiency and assist South Africa in achieving its sustainability objectives (Surovitskikh & Lubbe, 2015).

17.3.2.3.6 United Arab Emirates

The UAE has established itself as a global aviation hub through its leading airlines, Emirates and Etihad Airways, and advanced infrastructure at Dubai International and Abu Dhabi International Airports. These hubs facilitate connectivity across Europe, Asia, and Africa, strengthening the

UAE's role in international aviation. The sector significantly contributes to the economy while addressing sustainability challenges to meet global environmental goals (Samour et al., 2022). The country is also increasing investments in SAFs, with Etihad Airways partnering with the Abu Dhabi National Oil Company and Masdar to develop biofuels. Moreover, the "Etihad Greenliner" initiative, utilising Boeing 787 aircraft, tests eco-friendly technologies to support the UAE's net-zero emissions goal by 2050 (Bahman, 2023). Hydrogen fuel adoption further supports sustainable airline practices, balancing economic and environmental priorities (Bahman, 2023). The UAE is also exploring hydrogen-powered and electric aircraft to enhance aviation sustainability. As a leader in clean energy, it integrates solar panels and smart energy systems at Dubai Airports to reduce energy use (Jenčová, 2024). However, limited agricultural land poses challenges for biofuel production, leading to international collaborations for synthetic fuel development and significant investments in energy-efficient solutions (Mahroogi et al., 2022). Corporate social responsibility also strengthens the UAE's aviation strategy, promoting ethical practices and market presence (Mohamed, 2023). Programmes like the Sustainable Bioenergy Consortium and Biojet Abu Dhabi drive SAF adoption, reducing fossil fuel reliance and carbon emissions (Mahroogi et al., 2022).

17.3.2.3.7 Saudi Arabia

Saudi Arabia envisions reshaping the aviation industry through economic diversification and sustainability, aiming to boost tourism with eco-conscious strategies. A key element of this transformation is the ambitious plan to reduce emissions in air travel. The country has initiated trials of SAFs to cut GHG emissions, aligning its aviation sector with environmental objectives (Alshuwaikhat & Mohammed, 2017). Neom Airport exemplifies this vision, with infrastructure designed to minimise ecological impact by integrating advanced technologies for sustainable design. This airport represents a shift in aviation infrastructure towards environmental responsibility by maximising energy conservation (Waheed, 2022). By 2030, the country plans to revolutionise the aviation industry by promoting tourism and expanding its scope through environmentally conscious programmes. The Saudi Green Initiative aims to reduce emissions in the aviation sector significantly. Airlines are experimenting with SAFs to mitigate their climatic impact in line with national ecological goals (Alshuwaikhat & Mohammed, 2017). Neom Airport, an eco-friendly aviation creation, sets a benchmark for sustainable infrastructure by promoting green technologies and optimising renewable energy sources to minimise ecological harm (Waheed, 2022). However, the heavy reliance on oil presents a challenge in balancing economic reform with sustainability. Some foundational systems for SAF remain underdeveloped. Al-Saidi et al. (2021) highlight that circular economy practices, such as efficient resource usage, aircraft lifecycle management, and waste reduction, drive the aviation sector towards sustainable outcomes. Despite challenges in reconciling sustainability with economic diversification, necessary facilities, such as extensive SAF manufacturing and comprehensive implementation, are still in development. Al-Saidi et al. (2021) emphasise that a circular economy focuses on resource recovery, aircraft reuse, and energy optimisation, supporting sustainability targets. Such practices are critical as the aviation sector expands, necessitating more substantial environmental efforts. Integrating green technologies into airport development is a key area of progress. Alhumayn and Chinyio (2017) highlight the role of Building Information Modelling (BIM) in constructing sustainable infrastructure. BIM enhances project management and enables greener airport operations, a vital innovation for improving aviation sustainability. Saudi Arabia increasingly relies on technological advancements to meet its sector-wide sustainability objectives. Government policy is crucial in embedding sustainability within the aviation sector, guided by Vision 2030. Regulatory frameworks are being developed to promote sustainable practices and achieve long-term environmental goals, enhancing air transport efficiency and reducing the sector's ecological footprint (Alshuwaikhat & Mohammed, 2017). Investments in renewable energy projects position Saudi Arabia to increase SAF production, though research on SAFs remains limited. Expanding biofuels and alternative energy sources could reduce reliance on fossil fuels, supporting the country's sustainable aviation objectives and broader green growth ambitions (Waheed, 2022).

17.3.2.3.8 Ethiopia

Ethiopia's aviation sector, led by Ethiopian Airlines, plays a central role in advancing African aviation, focusing on sustainability. The airline has prioritised fuel-efficient aircraft and is pursuing SAFs made from local feedstocks like jatropha and other non-food crops. Bole International Airport in Addis Ababa employs energy-efficient technologies, including solar power, to minimise environmental impact, demonstrating Please correct the citation or add the missing reference. Ethiopia's dedication to green aviation infrastructure. Nevertheless, financial and technological limitations impede progress, confirming the need for enhanced infrastructure to facilitate sustainable aviation practices. Ethiopian Airlines is integrating biofuels into its operations, demonstrating leadership in adopting sustainable technologies. Further incorporation of sustainability into decision-making processes and the adoption of innovative technologies could improve its environmental performance. Orhan (2021) highlights that cohesive government policies are essential for balancing economic and environmental priorities. Through sustainable practices and long-term investments, Ethiopia can establish a resilient, eco-friendly aviation system that reduces emissions across various sectors.

17.3.2.3.9 Iran

Modernising its outdated aircraft fleet and implementing advanced technologies hinders sustainable aviation in Iran. The country's aviation sector is therefore characterised by low energy efficiency and high GHG emissions. This is mainly due to the international restrictions the country faces. Iran is working towards enhancing its fuel efficiency and developing SAFs to align with environmental objectives. However, financial and technological constraints set back the realisation of renewable resources in Iran (Hashemipour, 2023). At Tehran Imam Khomeini International Airport, energy-efficient technologies have been implemented to mitigate environmental impact. The country also uses hydrogen fuel and fuel cells to enhance eco-friendliness, reducing fossil fuel usage and carbon emissions (Baroutaji et al., 2019). The government has established clear and comprehensive guidelines to support environmental protection and enforces regulations to promote green aviation technologies. However, limited resources continue to challenge progress in the air transport sector. Hashemipour (2023) advises that international collaboration is essential in promoting greener aviation through shared efforts. Therefore, by engaging in partnerships and adopting modern technologies, Iran can develop a more sustainable aviation system and improve its environmental impacts from aviation (Hasan et al., 2021).

17.3.2.3.10 Egypt

Egypt's aviation sector has grown steadily, but adopting sustainable practices remains challenging. The country employs solar energy to support its aviation operations. Also, it explores SAFs made from agricultural waste, by-products, or urban waste. Progress in SAFs and solar energy strengthens Egypt's sustainable aviation framework (Ministry of Petroleum and Mineral Resources, 2023), and these efforts reflect the country's commitment to biofuels, SAF expansion, and sustainable infrastructure development. The government collaborates with stakeholders and encourages private sector participation to enhance SAF production and distribution. Partnerships with North African countries further promote green aviation solutions (Ministry of Petroleum and Mineral Resources, 2023). Khalaf and Abdelrady (2023) suggest that digital tools can improve airline efficiency and lower environmental impact. EgyptAir focuses on innovative strategies to reduce fuel consumption, minimise waste, and enhance efficiency to meet the country's environmental objectives. El-Mawardy (2023) asserts that passenger support for sustainable airlines, such as EgyptAir, further strengthens its green branding. However, El Malky et al. (2016) note that noise pollution at Hurghada Airport remains a concern.

17.4 DISCUSSION

This chapter examines sustainability-focused aviation research in BRICS+ countries, providing a framework to evaluate their compliance with global climate initiatives, such as the CORSIA, the goal of net-zero GHG emissions by 2050, and SDG 13 (SDG 13), which prioritises urgent climate action. These initiatives target the aviation industry's environmental impact, which accounts for about 2.5% of global carbon dioxide emissions, with estimates indicating a potential rise to 25% by 2050 if effective alleviation strategies are not adopted (Katırcıoğlu, 2024; Planès et al., 2021).

17.4.1 Sustainable Development Goals

In 2018, the global aviation industry accounted for 2.5% of total CO_2 emissions, the equivalent of approximately 1 billion metric tonnes of carbon dioxide, significantly contributing to human-induced climate change. Moreover, emissions of nitrogen oxides (NO_x), water vapour, and particulate matter are accelerating global temperature rise (Katırcıoğlu, 2024; ICAO, 2019; Lee et al., 2021). These changes lead to glacier melting and extreme weather, threatening the United Nations SDGs, particularly SDG 13 (Climate Action), SDG 12 (Responsible Consumption and Production), SDG 7 (Affordable and Clean Energy), SDG 9 (Industry, Innovation and Infrastructure), and SDG 17 (Partnerships for the Goals). The growing demand for international and domestic air travel complicates efforts to decarbonise the aviation sector. Without intervention, aviation could account for 25% of global CO_2 emissions by 2050, jeopardising global efforts to limit temperature increases to 1.5°C above pre-industrial levels as outlined in the Paris Agreement (Planès et al., 2021; Liu & Zhang, 2023).

Sustainable aviation strategies are essential to address this, particularly among BRICS+ nations (Brazil, Russia, India, China, South Africa) and partners like the United Arab Emirates, Saudi Arabia, Ethiopia, Egypt, and Iran. These countries are developing region-specific solutions aligned with global climate goals. SDG 13 emphasises urgent climate action, including the adoption of SAFs made from renewable and waste sources, such as sugarcane bagasse in Brazil and synthetic fuels in the UAE, which can reduce lifecycle carbon emissions by up to 94% compared to conventional jet fuels (Kurzawska-Pietrowicz, 2024; Carvalho et al., 2020; Mahroogi et al., 2022). China's proactive SAF production aims to cut 2.9 billion metric tonnes of CO_2 by 2050, supporting global sustainability efforts (Wang et al., 2023).

Advancements in aviation technology, including fuel-efficient turbine engines, lightweight composite materials, and electric or hydrogen-powered aircraft, support SDG 9 by promoting sustainable industrialisation and innovation (Klapmeyer & Marr, 2012; Terrenoire et al., 2019). Cochin International Airport in India, the first airport fully powered by solar energy, exemplifies sustainable infrastructure with reduced carbon emissions (Sukumaran & Sudhakar, 2017; Surovitskikh & Lubbe, 2015). The ICAO's CORSIA, launched in 2016, fosters global cooperation under SDG 17 by allowing airlines to offset emissions through carbon reduction projects, aiming to stabilise international aviation emissions at 2020 levels (ICAO, 2016a, 2016b). However, CORSIA's voluntary framework, limited scope, and inconsistent regulations in countries like Russia and India highlight the need for stronger policies to achieve effective climate action (Ojekunle, 2023; Singh et al., 2022).

Sustainable aviation promotes cleaner production and renewable resource use by advancing SDG 12 (Responsible Consumption and Production). Brazil's use of agricultural residues like sugarcane bagasse for SAF production aligns with circular economy principles, reducing waste and enhancing resource efficiency (Guimarães et al., 2023). Similarly, the UAE's integration of solar power into airport operations and Saudi Arabia's Neom Airport, designed with renewable energy systems, exemplify responsible production by minimising environmental degradation (Jenčová, 2024; Waheed, 2022). These initiatives reduce reliance on fossil fuels, supporting SDG 12's goals and SDG 7's emphasis on clean energy. The thematic literature review highlights keywords like "sustainable development" and "emission control," reflecting a research focus on integrating environmental and economic sustainability (Figure 17.3). However, challenges such as high SAF production costs and limited feedstock availability hinder scalability, particularly in financially constrained nations like Ethiopia and Egypt, while Iran's sanctions impede fleet modernisation, limiting access to fuel-efficient technologies (Hashemipour, 2023).

As major contributors to global aviation activity, BRICS+ nations face unique economic, technological, and regulatory challenges in aligning with the SDGs. Economic growth in China and India drives increased air travel demand, amplifying emissions, while technological and financial constraints in Ethiopia and Egypt limit sustainable aviation advancements (Hashemipour, 2023). Despite these disparities, collaborations within BRICS+, such as knowledge-sharing and joint investments in green technologies, align with SDG 17 and can bridge gaps. For instance, China's leadership in SAF research, with 65 of 106 global publications in 2024, demonstrates the potential for regional innovation to drive global progress (Table 17.2). The commitment of airlines and manufacturers to achieve net-zero emissions by 2050, supported by IATA and regional initiatives, further underscores the role of innovation in sustainable aviation (IATA, 2022; Bastinho & de Moura, 2025). To fully align with the SDGs and the Paris Agreement, BRICS+ nations must address these gaps through targeted policies. Incentivising SAF production via subsidies and international partnerships can accelerate progress toward SDG 7 and SDG 13, particularly in resource-constrained countries. As the EU's robust frameworks exemplify, strengthening emissions accounting regulations would support SDG 12 by internalising environmental costs (European Commission, 2021). By integrating SAFs, green infrastructure, advanced technologies, and fostering global cooperation, BRICS+ nations can lead in sustainable aviation, contributing to climate action, responsible consumption, and sustainable development while advancing the broader SDG framework.

There is a rapid growth in the aviation sector in the BRICS+ nations, which is driven by economic expansion and increasing air travel demand. Hence, these countries face several challenges in aligning aviation with the SDGs and climate goals. For instance, limited access to SAF production infrastructure and high costs hamper progress toward SDG 7 and SDG 13, while economic priorities often overshadow environmental regulations, complicating SDG 12 (Ovdiienko et al, 2021). Technological disparities and varying regulatory agendas across BRICS+ countries further challenge the adoption of innovations critical for SDG 9. For example, China and India are investing in SAF research, but Russia's reliance on fossil fuel-based aviation and South Africa's economic constraints limit their contributions to sustainable aviation (Liu & Zhang, 2023). Collaborative efforts within BRICS+, such as knowledge-sharing and joint investments in green technologies, could strengthen SDG 17 and accelerate progress toward zero-carbon aviation.

17.4.2 CORSIA

The CORSIA, introduced by the ICAO in 2016, is a cornerstone of global aviation sustainability. It aims to achieve carbon-neutral growth for international aviation from 2020 onward by requiring airlines to offset emissions through carbon reduction projects. CORSIA promotes SAFs, which can reduce lifecycle GHG emissions by up to 94% compared to conventional jet fuels (Kurzawska-Pietrowicz, 2024). BRICS+ nations, particularly China and Brazil, vigorously support CORSIA through SAF development, with Brazil using its abundant sugarcane biomass for biofuels, while China advances SAF production through several pilot projects (Fiorini et al., 2023). However, CORSIA excludes alignment to domestic flights, posing challenges for countries like India and China with rapidly growing domestic air travel. Moreover, inadequate infrastructure in several BRICS+ nations hinders CORSIA's implementation, while Russia's reliance on outdated aviation systems and fossil fuels limits SAF adoption.

17.4.3 Net-Zero Emissions

The BRICS+ countries attempt to reach net-zero emissions by 2050, aligning with the IATA's global initiative to advance sustainable aviation. Research confirms that the BRICS+ countries are working toward this target by adopting SAFs, electric propulsion systems, and hydrogen-based energy solutions. China incorporates sustainable aviation into its broader goal of attaining carbon neutrality by 2060, through fuel-efficient aircraft like the COMAC C919 and advancing environmentally friendly aviation infrastructure (Wang et al., 2023). Likewise, the UAE's Etihad Airways' Greenliner programme targets SAFs and innovative sustainable aviation technologies to achieve net-zero emissions by 2050 (Bahman, 2023). The co-word analysis of sustainable aviation studies identifies key themes, including SAFs, sustainability, and innovations like hydrogen energy, highlighting low-carbon emission options and decarbonisation methods to meet global net-zero goals.

However, there is varied progress across the BRICS+ countries. Russia's state-controlled aviation sector and Iran's constraints due to sanctions restrict their adoption of advanced sustainable aviation technologies. Similarly, India and South Africa face obstacles due to limited research funding and inadequate infrastructure (Ojekunle, 2023; Hashemipour, 2023). Additionally, the study notes that non-CO_2 emissions, such as contrails and nitrogen oxides, which contribute to radiative forcing, present a constant challenge that SAFs alone cannot fully address (Bergero et al., 2023). Hence, achieving net-zero goals requires a multidimensional approach, including operational efficiencies, alternative propulsion systems, and strong policy frameworks. The country collaboration network on sustainable aviation research reveals China leading Cluster 2 alongside India and the UAE, signifying that international partnerships are fundamental for sharing resources and expertise; however, the study advocates for a BRICS+ Aviation Sustainability Task Force to enhance these efforts further.

17.4.4 The Impact of National Policies and Technological Innovation on Research Output

The study shows significant variability in research output and technological innovation in sustainable aviation across BRICS+ nations, which is likely attributed to differing national policies and economic conditions. Such factors play a fundamental role in shaping the research landscape and technological advancements, with disparities evident among the countries analysed. For instance, China's leadership in publication output (65 documents) and SAF production is likely to be supported by strong governmental support through its 14th Five-Year Plan, which sets ambitious carbon neutrality goals by 2060, and substantial state funding for green technologies. This contrasts greatly with South Africa, where limited investment and economic constraints delay progress, resulting in lower research output and challenges in SAF production despite its renewable energy potential. Similarly, India benefits from supportive policies like the NCAP, yet infrastructure shortfalls and insufficient research funding reduce its drive for innovation, as seen with only 13 publications. As Russia has obsolete infrastructure, Brazil has biofuel with high production cost and thus has inadequate policy incentives. Technology breakthroughs and research productivity rely on the economic stability of the countries; hence, due to fiscal limitations, progress is impeded.

The study found that BRICS+ countries with strong economies, such as the UAE and Saudi Arabia, invest heavily in sustainable aviation projects. The UAE's Etihad Greenliner programme and Saudi Arabia's Neom Airport drive innovation in SAFs, environmentally friendly infrastructure, and advancements in green technologies. Conversely, faced with limited resources, nations like Ethiopia and Iran struggle to adopt SAFs and modern sustainable technologies into their aviation sectors. Although they are introducing efforts within their limited capacity, dependence on outdated aircraft models persists. These variations highlight an uneven global landscape in sustainable aviation progress. International collaboration can help bridge this gap, as approximately 30% of sustainable aviation research stems from international partnerships. Such collaborations promote knowledge-sharing, funding, and infrastructure development, enabling resource-limited countries to advance technologically and contribute more effectively to global sustainability objectives. For example, a partnership between Brazil's biofuel expertise and South Africa's renewable energy potential could accelerate SAF development. At the same time, Iran's inclusion in knowledge-sharing networks could mitigate the impact of sanctions. Globally, this could mean a more coordinated push toward net-zero aviation, aligning with initiatives like CORSIA. However, geopolitical tensions, differing economic priorities, and unequal contributions may hinder effective collaboration. To realise this potential, a structured framework, such as a BRICS+ Aviation Sustainability Task Force, would be essential to develop equitable partnerships and drive collective progress toward a sustainable aviation future.

17.4.5 Towards Sustainable Aviation in BRICS+ Nations

Scholars have identified the following methods for helping the aviation sector achieve the various SDGs and reduce its impact on climate change.

SAFs, from renewable sources, reduce lifecycle GHG emissions and support SDG 7 by promoting clean energy alternatives (Planès et al., 2021). SAFs are a practical and immediate decarbonisation solution since they can be integrated into existing aircraft models. However, increasing production remains challenging because of cost and feedstock limitations (Liu & Zhang, 2023).

Aircraft designers are also innovating using lightweight materials and fuel-efficient engines, which helps realise SDG 9 through advancing technology and promoting sustainable industrialisation (Klapmeyer & Marr, 2012).

Advances in electric and hybrid propulsion systems align with SDG 9 and SDG 13 in reducing GHG emissions, specifically regarding short-haul flights (Terrenoire et al., 2019).

In 2016, the ICAO introduced the CORSIA. This scheme addresses SDG 17 (Partnerships for the Goals) by promoting worldwide collaboration in the anticipation that international aviation emissions remain stable at 2020 levels (ICAO, 2016a, 2016b). Though it gives a structure for joint climate efforts through investment in projects to reduce carbon, the voluntary aspect and constrained reach of CORSIA's offsetting incite concerns about its efficacy.

Commitments from airlines and manufacturers, intended to achieve net-zero emissions at some point by 2050, are additionally backed by both the IATA and the European Commission, and they align with SDG 13 and SDG 9 through technological innovation, operational efficiencies and also investments in SAFs (IATA, 2022; European Commission, 2021; Bastinho & de Moura, 2025).

To address the multifaceted challenges to sustainable aviation in BRICS+ countries identified in the study, policymakers and industry leaders can adopt targeted strategies such as:

- Implement carbon pricing, SAF blending mandates, and subsidies to reduce high SAF costs, while establishing funding mechanisms.
- Invest in cost-effective SAF production, and advocate for incentives to counter fossil fuel dependence in Russia and Saudi Arabia.
- Fund SAF production facilities and research into electric/hydrogen propulsion, particularly in China and the UAE.
- Retrofit outdated fleets in Russia and Iran and address non-CO_2 emissions through optimised flight routing.
- Investment in SAF and airport infrastructure, modelling India's solar-powered Cochin Airport.
- Upgrade refuelling systems and airports, adapting solutions for Russia's climatic constraints.
- Legislate comprehensive decarbonisation policies, align with CORSIA, and strengthen enforcement in India and Russia.
- Manage rising air travel demand with alternatives like rail and enforce noise mitigation.
- Ease sanctions in Iran, fund SAF initiatives in Ethiopia, and integrate sustainability into Saudi Arabia's Vision 2030.
- Form alliances for technology transfer and invest in local projects like the UAE's Biojet Abu Dhabi.
- Establish a BRICS+ Aviation Sustainability Task Force and fund research in low-output nations.

17.5 CONCLUSION, LIMITATIONS, AND FUTURE SCOPE

This bibliometric study illuminates the dynamic landscape of sustainable aviation research in BRICS+ nations, highlighting their critical contributions to global climate objectives and the United Nations SDGs, particularly SDG 12 (Responsible Consumption and Production) and SDG 13 (Climate Action). The analysis, leveraging the Scopus database and Bibliometrix R, reveals a remarkable 17.45% annual growth in research output since 2016, with China leading through 65 publications in 2024, driven by advancements in SAFs, electric and hydrogen propulsion, and green airport infrastructure. Innovations like Brazil's sugarcane-based SAFs, India's solar-powered Cochin International Airport, and the UAE's Etihad Greenliner programme exemplify regional efforts to reduce aviation's 2.5% share of global CO_2 emissions. However, progress is uneven due to economic constraints, inadequate infrastructure, and inconsistent regulatory frameworks, particularly in nations like Ethiopia, Iran, and South Africa. High SAF production costs, limited feedstock availability, and sanctions-related technological barriers impede scalability. By bridging these gaps, BRICS+ nations can lead the transition to a carbon-neutral aviation sector, reinforcing global sustainability and climate resilience.

This bibliometric study highlights trends and collaborative networks in sustainable aviation research across BRICS+ countries but has some limitations. Its reliance on the Scopus database and English-language, peer-reviewed publications may exclude contributions in other languages or non-indexed local journals, particularly from countries like Russia and Iran, where English is not the dominant academic language. This bias risks underrepresenting scholarship from non-English-speaking regions, reinforcing the prominence of high-output nations like China and the USA. The study's focus on quantitative bibliometric indicators, such as publication volumes, citation patterns, co-authorship networks, and keyword prevalence, effectively captures numerical trends but overlooks qualitative aspects of sustainability research. It does not address socio-political factors, public perceptions, or policy implementation barriers influencing sustainable aviation in BRICS+ countries. For example, Brazil's progress in sugarcane-based SAFs and South Africa's economic challenges receive limited focus on practical industry obstacles, essential for evaluating sustainability efforts. Consequently, the quantitative approach may oversimplify complex issues, such as environmental trade-offs in SAF production or the shift to greener aviation systems, hindering a complete understanding of sustainability outcomes. Future research should integrate qualitative and quantitative methods. Case studies, like Ethiopia's airline initiatives or the UAE's Etihad Greenliner programme, could reveal innovation and operational challenges. Comparative policy analyses, such as India's NCAP, could illuminate diverse governmental strategies. Including non-English academic works and grey literature from national repositories would reduce language bias, despite challenges with visibility or quality. These combined approaches would provide a more balanced and comprehensive portrayal of sustainable aviation research, linking empirical data with practical implementation complexities.

ACKNOWLEDGEMENTS

The study acknowledges the National Institute for Humanities and Social Sciences through the DUT BRICS Research Institute Grant Ref: BRI22/1215 for supporting this research.

REFERENCES

Afşar, B. (2023). Analyzing the EU ETS, challenges, and opportunities for reducing greenhouse gas emissions from the aviation industry in Europe. *Sustainability*, *15*(24), 16874. https://doi.org/10.3390/su152416874

Alam, A., Masum, F., & Dwivedi, P. (2021). Break-even price and carbon emissions of Carinata-based sustainable aviation fuel production in the southeastern United States. *GCB Bioenergy*, *13*(11), 1800–1813. https://doi.org/10.1111/gcbb.12888

Alhumayn, S., & Chinyio, E. (2017). The barriers and strategies of implementing BIM in Saudi Arabia. https://doi.org/10.2495/bim170061

Al-Saidi, M., Das, P., & Saadaoui, I. (2021). Circular economy in basic supply: Framing the approach for the water and food sectors of the Gulf Cooperation Council countries. *Sustainable Production and Consumption*, *27*, 1273–1285.

Alshuwaikhat, H., & Mohammed, I. (2017). Sustainability matters in national development visions: Evidence from Saudi Arabia's Vision for 2030. *Sustainability*, *9*(3), 408. https://doi.org/10.3390/su9030408

Anger, A. (2010). Including aviation in the European emissions trading scheme: Impacts on the industry, CO2 emissions and macroeconomic activity in the EU. *Journal of Air Transport Management*, *16*(2), 100–105. https://doi.org/10.1016/j.jairtraman.2009.10.009

Anholon, R., Quelhas, O. L. G., Leal Filho, W., de Souza Pinto, J., & Feher, A. (2016). Assessing corporate social responsibility concepts used by a Brazilian manufacturer of airplanes: A case study at Embraer. *Journal of Cleaner Production*, *135*, 740–749.

Ansell, P. J. (2022). Hydrogen-electric aircraft technologies and integration: Enabling an environmentally sustainable aviation future. *IEEE Electrification Magazine*, *10*(2), 6–16. https://doi.org/10.1109/mele.2022.3165721

Ardo, F. M., Lim, J. W., Ramli, A., Lam, M. K., Kiatkittipong, W., Abdelfattah, E. A., . . ., & Sahrin, N. T. (2022). A review in redressing challenges to produce sustainable hydrogen from microalgae for the aviation industry. *Fuel*, *330*, 125646. https://doi.org/10.1016/j.fuel.2022.125646

Bahman, N. (2023). Evaluating the effectiveness of sustainable aviation in the Middle East: A study of two UAE-based carriers. *Journal of Airline Operations and Aviation Management*, *2*(1), 96–107. https://doi.org/10.56801/jaoam.v2i1.6

Baroutaji, A., Wilberforce, T., Ramadan, M., & Olabi, A. (2019). Comprehensive investigation on hydrogen and fuel cell technology in the aviation and aerospace sectors. *Renewable and Sustainable Energy Reviews*, *106*, 31–40. https://doi.org/10.1016/j.rser.2019.02.022

Bartle, J. R. (2006). The sustainable development of U.S. air transportation. *Public Works Management & Policy*, *10*(3), 214–224. https://doi.org/10.1177/1087724x06287494

Bastinho, S., & de Moura, I. S. (2025). Sustainable aviation fuels: A pathway to decarbonization in aviation—The case study of SWISS International Airlines. *South Florida Journal of Development*, *6*(7), e5581. https://doi.org/10.46932/sfjdv6n7-035

Bauen, A., Harris, A., Sim, C., Gudde, N., Prussi, M., & Scarlat, N. (2022). CORSIA lower carbon aviation fuels: An assessment of the greenhouse gas emission reduction potential. *Applied Sciences*, *12*(22), 11818. https://doi.org/10.3390/app122211818

Baxter, G. (2020). The use of aviation biofuels as an airport environmental sustainability measure: The case of Oslo Gardermoen Airport. *MAD—Magazine of Aviation Development*, *8*(1), 6–17. https://doi.org/10.14311/mad.2020.01.01

Bergero, C., Gosnell, G., Gielen, D., Kang, S., Bazilian, M., & Davis, S. (2023). Net-zero emissions aviation. https://doi.org/10.21203/rs.3.rs-1871023/v1

Białecki, T., Dzięgielewski, W., Kowalski, M., & Kulczycki, A. (2021). Reactivity model as a tool to compare the combustion process in aviation turbine engines powered by synthetic fuels. *Energies*, *14*(19), 6302. https://doi.org/10.3390/en14196302

Braun-Unkhoff, M., Riedel, U., & Wahl, C. (2016). About the emissions of alternative jet fuels. *CEAS Aeronautical Journal*, *8*(1), 167–180. https://doi.org/10.1007/s13272-016-0230-3

Carvalho, A. F. A., de Figueiredo, F. C., Campioni, T. S., Pastore, G. M., & de Oliva Neto, P. (2020). Improvement of some chemical and biological methods for the efficient production of xylanases, xylooligosaccharides, and lignocellulose from sugar cane bagasse. *Biomass and Bioenergy*, *143*, 105851.

Champeecharoensuk, A. (2023). Climate change mitigation in Thailand's domestic aviation: Mitigation options analysis towards 2050. *Energies*, *16*(20), 7199. https://doi.org/10.3390/en16207199

Chen, D., Yin, J., Xu, F., Huang, C., & Li, Z. (2023). A market-based framework for CO2 emissions reduction in China's civil aviation industry. *Transport Policy*, *143*, 150–158. https://doi.org/10.1016/j.tranpol.2023.09.018

Chen, J., Hu, X., Razi, U., & Rexhepi, G. (2021). The sustainable potential of efficient air-transportation industry and green innovation in realizing environmental sustainability in G7 countries. *Economic Research-Ekonomska Istraživanja*, *35*(1), 3814–3835. https://doi.org/10.1080/1331677x.2021.2004190

Chen, X., Zhang, J., & Liu, Y. (2022). Air traffic management innovations in China. *Journal of Aviation Technology and Engineering*, *33*(5). https://ieeexplore.ieee.org/abstract/document/9940157

Cortez, L. A. B., Nigro, F. E. B., Nogueira, L. A. H., Nassar, A. M., Cantarella, H., Moraes, M. A. F. D. d., . . ., & Baldassin, R. (2015). Perspectives for sustainable aviation biofuels in Brazil. *International Journal of Aerospace Engineering*, *2015*, 1–12. https://doi.org/10.1155/2015/264898

Cui, Q., Hu, Y., & Yu, L. (2022). Can the aviation industry achieve carbon emission reduction and revenue growth simultaneously under the CNG2020 strategy? An empirical study with 25 benchmarking airlines. *Energy*, *245*, 123272. https://doi.org/10.1016/j.energy.2022.123272

Dimitrios, D., & Maria, S. (2018). Assessing air transport socio-economic footprint. *International Journal of Transportation Science and Technology*, *7*(4), 283–290.

Dray, L., Schäfer, A., Grobler, C., Falter, C., Allroggen, F., Stettler, M., . . ., & Barrett, S. (2022). Cost and emissions pathways towards net-zero climate impacts in aviation. *Nature Climate Change*, *12*(10), 956–962. https://doi.org/10.1038/s41558-022-01485-4

Dyk, S., Su, J., Ebadian, M., O'Connor, D., Lakeman, M., & Saddler, J. (2019). Potential yields and emission reductions of biojet fuels produced via hydrotreatment of biocrudes produced through direct thermochemical liquefaction. *Biotechnology for Biofuels*, *12*(1). https://doi.org/10.1186/s13068-019-1625-2

Eaton, J., Naraghi, M., & Boyd, J. G. (2024). Regional pathways for all-electric aircraft to reduce aviation sector greenhouse gas emissions. *Applied Energy*, *373*, 123831.

El-Galad, M., El-Khatib, K., Abdelkader, E., El-Araby, R., Diwani, G., & Hawash, S. (2018). Empirical equations and economical study for blending biofuel with petroleum jet fuel. *Journal of Advanced Research*, *9*, 43–50. https://doi.org/10.1016/j.jare.2017.10.005

El Malky, M. G., El Bardisi, M. M., & Khallaaf, E. K. (2016). Noise levels and mitigations at Hurghada airport and it's surrounding. *Journal of Environmental Science*, *35*(2), 23–47. https://jes.journals.ekb.eg/article_28030.html

El-Mawardy, G. (2023). The effect of airlines' green image on the Egyptian air travelers' behavioural intentions to adopt green practices. *Journal of Association of Arab Universities for Tourism and Hospitality*, *24*(2), 732–749. https://doi.org/10.21608/jaauth.2023.219410.1479

Ershov, M. A., Savelenko, V. D., Burov, N. O., Makhova, U. A., Mukhina, D. Y., Aleksanyan, D. R., . . ., & Ali Abdelkareem, M. (2023). Incorporating innovation and new interactive technology into obtaining sustainable aviation fuels. *Energy*, *280*, 128156. https://doi.org/10.1016/j.energy.2023.128156

Escobar, N., Seber, G., Skalsky, R., Wögerer, M., Jung, M., & Malina, R. (2024). Spatially-explicit land use change emissions and carbon payback times of biofuels under the carbon offsetting and reduction scheme for international aviation (CORSIA). *Science of the Total Environment*, *948*, 174635. https://doi.org/10.1016/j.scitotenv.2024.174635

European Commission. (2021). Sustainable aviation fuels. *European Commission*. https://ec.europa.eu

Fiorini, A. C. O., Angelkorte, G. B., Portugal-Pereira, J., Fonte, C., Vicente, C. d. S., Maia, P. L. B., . . ., & Portugal-Pereira, J. (2023). Sustainable aviation fuels must control induced land use change: An integrated assessment modelling exercise for Brazil. *Environmental Research Letters*, *18*(1), 014036. https://doi.org/10.1088/1748-9326/acaee1

Gößling, S., & Humpe, A. (2020). The global scale, distribution and growth of aviation: Implications for climate change. *Global Environmental Change*, *2020*, 102194. https://doi.org/10.1016/j.gloenvcha.2020.102194

Grewe, V., Rao, A., Grönstedt, T., Xisto, C., Linke, F., Melkert, J., . . ., & Dahlmann, K. (2021). Evaluating the climate impact of aviation emission scenarios towards the Paris Agreement including COVID-19 effects. *Nature Communications*, *12*(1). https://doi.org/10.1038/s41467-021-24091-y

Guimarães, H. R., Bressanin, J. M., Motta, I. L., Chagas, M. F., Klein, B. C., Bonomi, A., . . ., & Watanabe, M. D. B. (2023). Decentralization of sustainable aviation fuel production in Brazil through biomass-to-liquids routes: A techno-economic and environmental evaluation. *Energy Conversion and Management*, *276*, 116547. https://doi.org/10.1016/j.enconman.2022.116547

Guo, X., Ning, C., Shen, Y., Yao, C., Chen, D., & Cheng, S. (2023). Projection of the co-reduced emissions of CO2 and air pollutants from civil aviation in China. *Sustainability*, *15*(9), 7082. https://doi.org/10.3390/su15097082

Haksevenler, B. H. G., Temel, N. U., & Topal, Y. E. (2023). Civil aviation policies and practices in Turkey in a global context through sustainable aviation. *Global Challenges*, *7*(12). https://doi.org/10.1002/gch2.202300194

Han, R., Li, L., Zhang, X., Lu, Z., & Zhu, S. (2022). Spatial-temporal evolution characteristics and decoupling analysis of influencing factors of China's aviation carbon emissions. *Chinese Geographical Science*, *32*(2), 218–236.

Hasan, M. A., Mamun, A. A., Rahman, S. M., Malik, K., Al Amran, M. I. U., Khondaker, A. N., . . ., & Alismail, F. S. (2021). Climate change mitigation pathways for the aviation sector. *Sustainability*, *13*(7), 3656. https://doi.org/10.3390/su13073656

Hashemipour, B. (2023). Sustainability in the civil aviation industry supply chain based on attracting international entrepreneurs: A case study of a civil aviation company in Iran. *In Attracting International Entrepreneurs*, 311–329. https://doi.org/10.1108/978-1-80382-233-420231015

Huang, E., Zhang, X., Rodríguez, L. F., Khanna, M., Jong, S. d., Ting, K. C., . . ., & Lin, T. (2019). Multi-objective optimization for sustainable renewable jet fuel production: A case study of corn stover based supply chain system in Midwestern U.S. *Renewable and Sustainable Energy Reviews*, *115*, 109403. https://doi.org/10.1016/j.rser.2019.109403

ICAO. (2016a). CORSIA: A global market-based measure for international aviation. *International Civil Aviation Organization*. https://www.icao.int

International Air Transport Association (IATA). (2022). Air passenger market analysis. *IATA*. https://www.iata.org

International Civil Aviation Organization (ICAO). (2016b). CORSIA: Carbon Offsetting and Reduction Scheme for International Aviation. *International Civil Aviation Organization*. https://www.icao.int

International Civil Aviation Organization (ICAO). (2019). Environmental report 2019. *International Civil Aviation Organization*. https://www.icao.int

Jenčová, E. (2024). Aviation industry in the United Arab Emirates in the post-COVID period. *Acta Avionica Journal*, 83–88. https://doi.org/10.35116/aa.2024.0010

Jing, L., El-Houjeiri, H. M., Monfort, J., Littlefield, J., Al-Qahtani, A., Dixit, Y., . . ., & Bergerson, J. (2022). Understanding variability in petroleum jet fuel life cycle greenhouse gas emissions to inform aviation decarbonization. *Nature Communications*, *13*(1). https://doi.org/10.1038/s41467-022-35392-1

Johnson, J., Carpenter, M., Okeke, N., Nnabuife, S., & Mai, N. (2022). Investigation of water droplet size distribution in conventional and sustainable aviation turbine fuels. *SAE International Journal of Fuels and Lubricants*, *15*(3). https://doi.org/10.4271/04-15-03-0016

Joshan, S., & Maertens, S. (2020). Low cost carriers in the Middle East and North Africa (MENA) region: Emergence and barriers to development. *Journal of Transport Geography*, *87*, 102799.

Julio, A. A. V., Batlle, E. A. O., Trindade, A. B., Nebra, S. A., Reyes, A. M. M., & Palácio, J. C. E. (2021). Energy, exergy, exergoeconomic, and environmental assessment of different technologies in the production of bio-jet fuel by palm oil biorefineries. *Energy Conversion and Management*, *243*, 114393. https://doi.org/10.1016/j.enconman.2021.114393

Katırcıoğlu, S. (2024). Aviation and climate change: Trends, challenges, and opportunities. *Journal of Air Transport Management*, *105*, 102–112. https://doi.org/10.1016/j.jairtraman.2023.102112

Kazelko, A., & Semeghini, U. S. (2024). Expansion of brics: Implications for global energy markets. *BRICS Journal of Economics*, *5*, 53–67.

Khalaf, A., & Abdelrady, H. (2023). Impact of digital transformation technologies on EgyptAir sustainability. *Journal of Association of Arab Universities for Tourism and Hospitality*, *24*(1), 44–61. https://jaauth.journals.ekb.eg/article_299860.html

Khalili, S., Rantanen, E., Bogdanov, D., & Breyer, C. (2019). Global transportation demand development with impacts on the energy demand and greenhouse gas emissions in a climate-constrained world. *Energies*, *12*(20), 3870. https://doi.org/10.3390/en12203870

Kimeria, T. B. (2022). Policy formulation and advocacy towards sustainability of aviation industry in Kenya: A literature based review. *Journal of Strategic Management*, *4*(1), 1–15. https://doi.org/10.53819/81018102t2093

Klapmeyer, M., & Marr, L. (2012). CO2, NOx, and particle emissions from aircraft and support activities at a regional airport. *Environmental Science & Technology*, *46*(20), 10974–10981. https://doi.org/10.1021/es302346x

Korba, P., Sekelová, I., Koščáková, M., & Behúnová, A. (2023). Passengers' knowledge and attitudes toward green initiatives in aviation. *Sustainability*, *15*(7), 6187. https://doi.org/10.3390/su15076187

Kumar, A., A, A., & Gupta, H. (2020). Evaluating green performance of the airports using hybrid BWM and VIKOR methodology. *Tourism Management*, *76*, 103941. https://doi.org/10.1016/j.tourman.2019.06.016

Kurzawska-Pietrowicz, P. (2024). Exhaust emissions from a jet engine powered by sustainable aviation fuel calculated at various cruising altitudes. *Combustion Engines*, *198*(3), 62–67. https://doi.org/10.19206/ce-186211

Lee, D. S., Fahey, D., Skowron, A., Allen, M., Burkhardt, U., Chen, Q., . . ., & Wilcox, L. (2021). The contribution of global aviation to anthropogenic climate forcing for 2000 to 2018. *Atmospheric Environment*, *244*, 117834. https://doi.org/10.1016/j.atmosenv.2020.117834

Li, J., et al. (2021). Aviation emissions and climate change: A review of the current state and future prospects. *Environmental Science & Policy*, *123*, 1–12. https://doi.org/10.1016/j.envsci.2021.12.004

Liang, Z., Liu, H., Fan, Y., Salehi, F., Zhang, Z., & Wang, C. (2024). Applying Fischer Tropsch and its pentanol blends into an aviation compression ignition engine for PM emissions control. *Journal of the Energy Institute*, *116*, 101742.

Liu, X., & Zhang, Y. (2023). What drives the decoupling progress of China's civil aviation transportation growth from carbon emissions? A new decomposition analysis. *PLoS One*, *18*(3), e0282025. https://doi.org/10.1371/journal.pone.0282025

Liu, Z., & Yang, X. (2023). The potential GHGs reduction of co-processing aviation biofuel in life cycle. *Bioresources and Bioprocessing*, *10*(1). https://doi.org/10.1186/s40643-023-00674-z

Lyu, C., Liu, X., Wang, Z., Yang, L., Líu, H., Yang, N., . . ., & Cai, B. (2023). An emissions inventory using flight information reveals the long-term changes of aviation CO2 emissions in China. *Energy*, *262*, 125513. https://doi.org/10.1016/j.energy.2022.125513

Ma, S., Wang, S., Zhang, C., & Zhang, S. (2017). A method to improve the efficiency of an electric aircraft propulsion system. *Energy*, *140*, 436–443. https://doi.org/10.1016/j.energy.2017.08.095

Mahroogi, F., Narayan, S., Kaisan, M., Aliyu, A., Yahuza, I., Grujić, I., . . ., & Stojanović, N. (2022). A critical review of biofuels as an alternative fuel to diesel in Gulf Regional Corporation (GCC) region: Current market trends and future opportunities. *Communications—Scientific Letters of the University of Zilina*, *24*(1), B67–B73. https://doi.org/10.26552/com.c.2022.1.b67-b73

Manigandan, S., Praveenkumar, T. R., Ryu, J. I., Verma, T. N., & Pugazhendhi, A. (2023). Role of hydrogen on aviation sector: A review on hydrogen storage, fuel flexibility, flame stability, and emissions reduction on gas turbines engines. *Fuel*, *352*, 129064. https://doi.org/10.1016/j.fuel.2023.129064

Marangon, B. B., Castro, J. D. S., Assemany, P. P., Machado, N. A., & Calijuri, M. L. (2024). Wastewater-grown microalgae biomass as a source of sustainable aviation fuel: Life cycle assessment comparing hydrothermal routes. *Journal of Environmental Management*, *360*, 121164. https://doi.org/10.1016/j.jenvman.2024.121164

Marciello, V., Stasio, M. D., Ruocco, M., Trifari, V., Nicolosi, F., Meindl, M., . . ., & Caliandro, P. (2023). Design exploration for sustainable regional hybrid-electric aircraft: A study based on technology forecasts. *Aerospace*, *10*(2), 165. https://doi.org/10.3390/aerospace10020165

Ministry of Petroleum and Mineral Resources. (2023). *Honeywell Company Discusses the Sustainable Aviation Fuel (SAF) Project*. https://www.petroleum.gov.eg/en/media-center/news/news-pages/Pages/mop_20102023_02.aspx

Mohamed, M. (2023). Implementing innovative CSR practices to improve the company's perceived value and market share in the context of aviation industry in the UAE. *International Conference on Pioneer and Innovative Studies*, *1*, 173–184. https://doi.org/10.59287/icpis.825

Mohamed, N., Maitho, E., Masvikeni, E., Fourie, R., Tilly, M., & Zondi, N. (2014). The green fund of South Africa: Origins, establishment and first lessons. *Development Southern Africa*, *31*(5), 658–674. https://doi.org/10.1080/0376835x.2014.935295

Moreira, M. M. R., Gurgel, A., & Seabra, J. E. A. (2014). Life cycle greenhouse gas emissions of sugar cane renewable jet fuel. *Environmental Science & Technology*, *48*(24), 14756–14763. https://doi.org/10.1021/es503217g

Munirathinam, B., Schlüter, N., & Schröder, D. (2023). Recent advances in metallic catalyst materials for electrochemical upgrading of furfurals to drop-in biofuels: A mini review. *Energy Technology*, *11*(8). https://doi.org/10.1002/ente.202300307

Nkosi, M., Jablonski, M., & Duze, N. C. (2023). Lessons learned from evaluating options using multi-criteria analysis in two corridors within the city of Johannesburg. In *41st Southern African Transport Conference* (pp. 1–13). Pretoria, South Africa.

Ojekunle, A. O. (2023). The role of ICAO in the resolution of conflicts arising from war-induced aviation disruptions and damage: The Russia–Ukraine imbroglio in view. *Issues Aviation L. & Pol'y*, *23*, 389.

Oluwaseun, A. (2023). Carbon-neutral aviation in Nigeria: Assessing the feasibility and impacts of sustainable aviation fuel adoption. *Proceedings of the Nigerian Academy of Science*, *16*(1s), 72–83. https://doi.org/10.57046/xfva9454

Orhan, G. (2021). The effects of airline strategies on environmental sustainability. *Aircraft Engineering and Aerospace Technology*, *93*(8), 1346–1357. https://doi.org/10.1108/aeat-03-2021-0078

Ovdiienko, O., Hryhorak, M., Marchuk, V., & Bugayko, D. (2021). An assessment of the aviation industry's impact on air pollution from its emissions: Worldwide and Ukraine. *Environmental & Socio-Economic Studies*, *9*(2), 1–10. https://doi.org/10.2478/environ-2021-0006

Pinto, J. (2024). Regulating green and low-carbon hydrogen in Africa: A case study of South Africa. *Advances in Science and Technology*. https://doi.org/10.4028/p-pv7uh9

Pinto, J., & Chege, K. (2024). Regulating green and low-carbon hydrogen in Africa: A case study of South Africa. *Advances in Science and Technology*. https://doi.org/10.4028/p-pv7uh9

Pisa, N. (2018). Causal relationship between air transport, tourism and economic growth: Joinpoint regression and granger causality analysis. *Euro Economica*, *37*(2), 164–176.

Planès, T., Delbecq, S., Pommier-Budinger, V., & Bénard, E. (2021). Simulation and evaluation of sustainable climate trajectories for aviation. *Journal of Environmental Management*, *295*, 113079. https://doi.org/10.1016/j.jenvman.2021.113079

Prussi, M., Lee, U., Wang, M., Malina, R., Valin, H., Taheripour, F., . . ., & Hileman, J. (2021). CORSIA: The first internationally adopted approach to calculate life-cycle GHG emissions for aviation fuels. *Renewable and Sustainable Energy Reviews*, *150*, 111398. https://doi.org/10.1016/j.rser.2021.111398

Racha, A., Kumar, L., Pai, S., Samanta, C., Newalkar, B. L., & Thota, C. (2024). Highly selective hydrodeoxygenation catalyst for sustainable aviation fuel production from used cooking oil. *Catalysis Today*, *442*, 114895. https://doi.org/10.1016/j.cattod.2024.114895

Ramakrishnan, J., Liu, T., Yu, R., Seshadri, K., & Gou, Z. (2022). Towards greener airports: Development of an assessment framework by leveraging sustainability reports and rating tools. *Environmental Impact Assessment Review*, *93*, 106740. https://doi.org/10.1016/j.eiar.2022.106740

Rambabu, K., Bharath, G., Sivarajasekar, N., Velu, S., Sudha, P. N., Wongsakulphasatch, S., . . ., & Banat, F. (2023). Sustainable production of bio-jet fuel and green gasoline from date palm seed oil via hydroprocessing over tantalum phosphate. *Fuel*, *331*, 125688. https://doi.org/10.1016/j.fuel.2022.125688

Rathore, H., Nandi, S., & Jakhar, S. K. (2020). The future of Indian aviation from the perspective of environment-centric regulations and policies. *IIMB Management Review*, *32*(4), 434–447. https://doi.org/10.1016/j.iimb.2020.11.003

Rendón, M. A., Sánchez, R. C. D., Gallo, M. J., & Anzai, A. H. (2021). Aircraft hybrid-electric propulsion: Development trends, challenges, and opportunities. *Journal of Control, Automation and Electrical Systems*, *32*(5), 1244–1268.

Richter, S. (2024). Evaluation of the applicability of synthetic fuels and their life cycle analyses. *Energies*, *17*(5), 981. https://doi.org/10.3390/en17050981

Riebl, S., Braun-Unkhoff, M., & Riedel, U. (2017). A study on the emissions of alternative aviation fuels. *Journal of Engineering for Gas Turbines and Power*, *139*(8). https://doi.org/10.1115/1.4035816

Rogerson, J. (2014). Green commercial property development in urban South Africa: Emerging trends, emerging geographies. *Bulletin of Geography Socio-Economic Series*, *26*(26), 233–246. https://doi.org/10.2478/bog-2014-0056

Rojas-Michaga, M. F., Michailos, S., Cardozo, E., Akram, M., Hughes, K. J., Ingham, D., . . ., & Pourkashanian, M. (2023). Sustainable aviation fuel (SAF) production through power-to-liquid (PtL): A combined techno-economic and life cycle assessment. *Energy Conversion and Management*, *292*, 117427. https://doi.org/10.1016/j.enconman.2023.117427

Ryley, T., Elmirghani, J. M. H., Budd, T., Miyoshi, C., Mason, K., Moxon, R. W., . . ., & Zanni, A. (2013). Sustainable development and airport surface access: The role of technological innovation and behavioral change. *Sustainability*, *5*(4), 1617–1631. https://doi.org/10.3390/su5041617

Said, Z., Nguyen, T. H., Sharma, P., Li, C., Ali, H. M., Nguyen, V. N., . . ., & Truong, T. H. (2022). Multi-attribute optimization of sustainable aviation fuel production-process from microalgae source. *Fuel*, *324*, 124759. https://doi.org/10.1016/j.fuel.2022.124759

Samour, A., Baskaya, M., & Türsoy, T. (2022). The impact of financial development and FDI on renewable energy in the UAE: A path towards sustainable development. *Sustainability*, *14*(3), 1208. https://doi.org/10.3390/su14031208

Seber, G., Escobar, N., Valin, H., & Malina, R. (2022). Uncertainty in life cycle greenhouse gas emissions of sustainable aviation fuels from vegetable oils. *Renewable and Sustainable Energy Reviews*, *170*, 112945.

Shang, M. (2023). Advancements and trends in global low-carbon aviation: A Citespace knowledge map analysis. *Journal of Green Economy and Low-Carbon Development*, *2*(4), 213–231. https://doi.org/10.56578/jgelcd020404

Shao, S., Hu, X., Dong, W., Li, X., Zhang, H., Xiao, R., . . ., & Cai, Y. (2021). Integrated C–C coupling/hydrogenation of ketones derived from biomass pyrolysis for aviation fuel over Ni/Mg–Al–O/AC bifunctional catalysts. *Journal of Cleaner Production*, *282*, 124331. https://doi.org/10.1016/j.jclepro.2020.124331

Sharno, M. A., & Hiloidhari, M. (2024). Social sustainability of biojet fuel for net zero aviation. *Energy for Sustainable Development*, *79*, 101419. https://doi.org/10.1016/j.esd.2024.101419

She, Y. (2023). From takeoff to touchdown: A decade's review of carbon emissions from civil aviation in China's expanding megacities. *Sustainability*, *15*(24), 16558. https://doi.org/10.3390/su152416558

She, Y., Deng, Y., & Chen, M. (2023). From takeoff to touchdown: A decade's review of carbon emissions from civil aviation in China's expanding megacities. *Sustainability*, *15*(24), 16558. https://doi.org/10.3390/su152416558

Silva, L., Silva, V., Barros, M., & Arroyo, P. (2016). Zeolites as potential structures in obtaining jet fuel through the Fischer-Tropsch synthesis. https://doi.org/10.5772/63662

Singh, J., Sharma, S. K., & Srivastava, R. (2018). Managing fuel efficiency in the aviation sector: Challenges, accomplishments, and opportunities. *FIIB Business Review*, *7*(4), 244–251. https://doi.org/10.1177/2319714518814073

Singh, R. K., Panda, D., & Singh, S. (2024). Deoxygenation pathways for sustainable aviation fuel from used cooking oil: A review on catalyst and operating parameters. *Journal of Hazardous, Toxic, and Radioactive Waste*, *28*(4), 04024022.

Sobieralski, J. B. (2023). Sustainable air transportation through the operational use of a social cost index. *Journal of Cleaner Production*, *385*, 135663. https://doi.org/10.1016/j.jclepro.2022.135663

Song, Z. (2024). Comparison of emission properties of sustainable aviation fuels and conventional aviation fuels: A review. *Applied Sciences*, *14*(13), 5484. https://doi.org/10.3390/app14135484

Sreenath, S., & Sudhakar, K. (2017). Fully solar powered airport: A case study of Cochin International Airport. *Journal of Air Transport Management*, *62*, 176–188. https://doi.org/10.1016/j.jairtraman.2017.04.004

Staples, M., Malina, R., Suresh, P., Hileman, J., & Barrett, S. (2018). Aviation CO2 emissions reductions from the use of alternative jet fuels. *Energy Policy*, *114*, 342–354. https://doi.org/10.1016/j.enpol.2017.12.007

Sukumaran, S., & Sudhakar, K. (2017). Fully solar powered airport: A case study of Cochin International Airport. *Journal of Air Transport Management*, *62*, 176–188.

Suresh, R., Rajendran, S., Dutta, K., Khoo, K. S., & Soto-Moscoso, M. (2023). An overview on light-assisted techniques for waste-derived hydrogen fuel towards aviation industry. *Fuel*, *334*, 126645. https://doi.org/10.1016/j.fuel.2022.126645

Surovitskikh, S., & Lubbe, B. (2015). The air liberalisation index as a tool in measuring the impact of South Africa's aviation policy in Africa on air passenger traffic flows. *Journal of Air Transport Management*, *42*, 159–166. https://doi.org/10.1016/j.jairtraman.2014.09.010

Terrenoire, E., Hauglustaine, D., Gasser, T., & Penanhoat, O. (2019). The contribution of carbon dioxide emissions from the aviation sector to future climate change. *Environmental Research Letters*, *14*(8), 084019. https://doi.org/10.1088/1748-9326/ab3086

Thuy, D. T. (2021). Revising the EU ETS and CORSIA in times of the COVID-19 pandemic: Challenges for reducing global aviation emissions. *Climate Policy*, *21*(10), 1357–1367. https://doi.org/10.1080/14693062.2021.1966358

Ul-Haq, J., Imran, K., Oad, S., & Visas, H. (2023). Nexus between air travel and environmental degradation in BRICS countries: A tourism perspective. *Chinese Journal of Population, Resources and Environment*, *21*(3), 189–196. https://doi.org/10.1016/j.cjpre.2023.09.008

Usmani, G., Akram, V., & Praveen, B. (2021). Tourist arrivals, international tourist expenditure, and economic growth in BRIC countries. *Journal of Public Affairs*, *21*(2), e2202.

Varakin, A., Gulyaeva, L. A., Ishutenko, D. I., Kulinich, A., Nikulshin, P., Pimerzin, A. A., . . ., & Васидов, Р. Г. (2020). Development of technologies and prospects for the introduction of aviation biofuels. *Biotekhnologiya*, *36*(5), 13–30. https://doi.org/10.21519/0234-2758-2020-36-5-13-30

Waheed, R. (2022). Energy challenges, green growth, blue indicators, and sustainable economic growth: A study of Saudi Arabia. *Evaluation Review*, *47*(6), 983–1024. https://doi.org/10.1177/0193841x221134653

Wan, J., Zhang, H., Lyu, W., & Zhou, J. (2022). A novel combined model for short-term emission prediction of airspace flights based on machine learning: A case study of China. *Sustainability*, *14*(7), 4017. https://doi.org/10.3390/su14074017

Wang, B., Ting, Z. J., & Zhao, M. (2024a). Sustainable aviation fuels: Key opportunities and challenges in lowering carbon emissions for aviation industry. *Carbon Capture Science & Technology*, *13*, 100263. https://doi.org/10.1016/j.ccst.2024.100263

Wang, K., Zhang, Y., & Wei, Y. (2023). China's aviation passenger transport can reduce CO2 emissions by 2.9 billion tons by 2050 if certain abatement options are implemented. *One Earth*, *6*(8), 1050–1065. https://doi.org/10.1016/j.oneear.2023.07.003

Wang, Y., et al. (2024b). The impact of aviation growth on global CO2 emissions: A comprehensive analysis. *Transportation Research Part D: Transport and Environment*, *120*, 103–115. https://doi.org/10.1016/j.trd.2023.103115

Wei, H., Liu, W., Chen, X., Yang, Q., Li, J., & Chen, H. (2019). Renewable bio-jet fuel production for aviation: A review. *Fuel*, *254*, 115599. https://doi.org/10.1016/j.fuel.2019.06.007

Wu, G. (2023). Low-carbon development of civil aviation industry. *Strategic Study of CAE*, *25*(5), 165. https://doi.org/10.15302/j-sscae-2023.05.015

Xiang, S., Liu, Y., Tang, G., Zhao, W., Sheng-xi, T., & Li, Y. (2018). An improved propeller design method for the electric aircraft. *Aerospace Science and Technology*, *78*, 488–493. https://doi.org/10.1016/j.ast.2018.05.008

Yang, H., & O'Connell, J. F. (2020). Short-term carbon emissions forecast for the aviation industry in Shanghai. *Journal of Cleaner Production*, *275*, 122734. https://doi.org/10.1016/j.jclepro.2020.122734

Yang, L., Hu, Y., Wang, H., Li, C., Tang, B., Wang, B., . . ., & Cui, H. (2023). Uncertainty quantification of CO2 emissions from China's civil aviation industry to 2050. *Journal of Environmental Management*, *336*, 117624. https://doi.org/10.1016/j.jenvman.2023.117624

Yusaf, T., Fernandes, L., Talib, A., Altarazi, Y. S., Alrefae, W., Kadirgama, K., . . ., & Laimon, M. (2022). Sustainable aviation: Hydrogen is the future. *Sustainability*, *14*(1), 548. https://doi.org/10.3390/su14010548

Zhang, C., Hui, X., Lin, Y., & Sung, C. (2016). Recent development in studies of alternative jet fuel combustion: Progress, challenges, and opportunities. *Renewable and Sustainable Energy Reviews*, *54*, 120–138. https://doi.org/10.1016/j.rser.2015.09.056

Zhang, F., & Graham, D. J. (2020). Air transport and economic growth: A review of the impact mechanism and causal relationships. *Transport Reviews*, *40*(4), 506–528.

Zhang, J., Zhang, S., Wu, R., Duan, M., Zhang, D., Wu, Y., . . ., & Hao, J. (2021a). The new CORSIA baseline has limited motivation to promote the green recovery of global aviation. *Environmental Pollution*, *289*, 117833. https://doi.org/10.1016/j.envpol.2021.117833

Zhang, L., Dang, Y., Zhou, X. H., Gao, P., Bavel, A. P. V., Wang, H., . . ., & Sun, Y. (2021b). Direct conversion of CO2 to a jet fuel over CoFe alloy catalysts. *The Innovation*, *2*(4), 100170. https://doi.org/10.1016/j.xinn.2021.100170

Zhang, Q., Wang, M., Li, H., & Cui, T. (2020). Research on control characteristics of three-phase permanent magnet synchronous motor for electric aircraft propulsion. In *2020 39th Chinese Control Conference (CCC)*. https://doi.org/10.23919/ccc50068.2020.9188977

Zhu, X., Xiao, J., Wang, C., Zhu, L., & Wang, S. (2022). Global warming potential analysis of bio-jet fuel based on life cycle assessment. *Carbon Neutrality*, *1*(1). https://doi.org/10.1007/s43979-022-00026-4

Index

D

E

F

G

H

T

U

V

W

Y

Z

For Product Safety Concerns and Information please contact our
EU representative GPSR@taylorandfrancis.com Taylor & Francis
Verlag GmbH, Kaufingerstraße 24, 80331 München, Germany

Stephan Lunau (Ed.)

Christian Staudter
Clemens von Hugo
Philipp Bosselmann
Jens-Peter Mollenhauer
Renata Meran
Olin Roenpage

Design for Six Sigma $^{+Lean}$ Toolset

Mindset for Successful Innovations

Second Edition

Springer

Editor
Stephan Lunau
UMS (Consulting) GmbH
Frankfurt, Germany

Authors
Christian Staudter
Clemens von Hugo
Philipp Bosselmann
Jens-Peter Mollenhauer
Renata Meran
Olin Roenpage

UMS (Consulting) GmbH
Frankfurt, Germany

Additional material to this book can be downloaded from http://extras.springer.com

ISSN 2192-8096 ISSN 2192-810X (electronic)
ISBN 978-3-642-41454-1 ISBN 978-3-642-41455-8 (eBook)
DOI 10.1007/978-3-642-41455-8
Springer Heidelberg New York Dordrecht London

Library of Congress Control Number: 2013953000

Printed on acid-free paper

Springer is part of Springer Science+Business Media (www.springer.com)

Table of Contents

APPENDIX

Foreword to the Second Edition

Innovation has been developed to a key focus of our work. Driven by this focus we have upgraded, refreshed and extended our advice about the mind and tool set for the Design for Six Sigma methodology within the new edition of the DFSS+Lean Toolset.

In the future, the Toolset will help you identify and select good ideas and transform them into commercially successful product and process innovations. We have added some topical new elements to the Toolset and developed a common guideline so that you will find it easier to solve those more complex challenges. Still based on the popular DMADV model we have created a question based approach which will help you select the best tools for your purpose thus avoiding getting lost in the "Tool Jungle".

What we hope will be a very useful addition to this edition are the practical tips drawn from our project work which serve to reinforce the tools contained within each phase. There is also one example that is carried through the duration of the text to provide the continuity that is so important to building understanding.

New to this edition are some reflections on items that need to be taken into account during the process development, chief among these is the "Process Design Workout" which is designed to accelerate process development projects.

Migrating from a tool based approach to a question based approach is a decisive success factor in our opinion enabling firstly, increased efficiency of project work for the Project Leader, his team and the associated Stakeholders, and secondly, significantly increasing the probability of success for the respective innovation projects.

Try it out – we are looking forward to sharing your experiences and incorporating them into future editions of this book.

We, the authors of the UMS Team, hope you will have fun and success when applying the methodology.

Frankfurt am Main, August 2013

Stephan Lunau

Foreword to the First Edition

Every company relies on innovation to compete globally. However, creative ideas are mostly insufficient if you want to translate an innovative spirit into commercial success. The ability to put a new product or a new process on the market as quickly as possible is becoming increasingly important.
Systematic management is necessary for developing cost-effective and successful products based on market realities and customer requirements. Especially open innovation, which is currently intensively discussed and widely implemented, requires consideration. Only a sensible interface and information management is capable of generating overall success from a variety of good ideas.

Design for Six Sigma^{+Lean} is an approach for such a systematic innovation management.
This concept was developed to achieve a target-oriented realization of innovations and is strongly associated with the Six Sigma^{+Lean} methodology, currently applied globally to optimize existing processes. DFSS^{+Lean} synthesizes a number of key factors, including the active integration of employees, customer-oriented development, the reduction of complexity in products and processes, and controlling of innovation in terms of a standardized procedure.

The present toolset represents the proven approach UMS takes when putting Design for Six Sigma^{+Lean} into practice. Its individual tools are assigned to the process model Define, Measure, Analyze, Design, and Verify in a clear and manageable structure. This structure can be considered as a red thread and makes it easier to apply the tools in practice and organize an innovative product and process development that is target-oriented and efficient.

Besides the whole UMS team, I would like to thank the authors, who along with their expertise and experience have shown enormous commitment in putting this book together. My thanks also go to Mariana Winterhager for the graphic layout of the material and Astrid Schmitz for the translation work.

I wish everyone great success in implementing innovations.

Frankfurt am Main, October 2008

Stephan Lunau

DESIGN FOR SIX SIGMA+LEAN TOOLSET

INTRODUCTION

Introduction

"The capacity to innovate determines our fate."
(Federal President Roman Herzog in his speech "Emergence into the 21st Century" on April 26, 1997)

Today, the capacity to innovate is one of the key success factors of a company. According to a benchmarking study carried out by American Productivity and Quality Control (APQC), companies that display strong growth rates generate one third of their turnover from products which are less than three years old. Moreover, over the last 50 years there has been a drastic reduction in product life cycles.[1)] Those who don't improve continuously and quickly enough are shaken off by global competition. This applies particularly to companies from the key industrial nations since they cannot stand up to competitors from Asia in terms of costs let alone the fact that these companies are also working at full speed on their own innovative strength.[2)] Whilst most companies recognize this circumstance, many have not yet instigated action plans which focus on innovation to a sufficient extent. In a study from the year 2008, 87% of the medium to large sized companies surveyed evaluated innovative strength as the key value driver. In their own opinion only 60% of them reacted to this situation appropriately.[3)] Politicians also heeded the pressure to act. The European Commission provided a budget amounting to $ 8 billion to support innovative strength in the year 2011 an increase of 12% compared to 2010.[4)]

Pursuing excellence in innovative capability is the goal of this book. However, the activities connected to this pursuit, must be put in a wider context, since excellent innovative strength is only possible in a successfully functioning corporate structure. Business Excellence, i.e. the maximizing performance capability in all areas of economic affairs, must be the prime goal. Operational Excellence which realizes effective and efficient business processes, must be combined with Commercial Excellence i.e. optimal market development and exploitation of customer potential, before Business Excellence can be achieved. People Excellence ultimately underpins all of this. It ensures competence and motivation of employees and thus enables best performance for a company's success in all areas.

---

1 American Productivity & Quality Center (2003): Improving New Product Development Performance and Practices. Houston (TX): APQC

2 Cf. Der Spiegel, Issue 34/2010: Die Rivalen. China gegen Deutschland – Kampf um die Weltmärkte

3 Christoph Wamser/Klaus Deimel/Karsten Heinrich: Studie über Werttreiber in Unternehmen, MBA Research Institute at Bonn-Rhine-Sieg University of Applied Sciences, 2008

4 Press release by the European Commission: http://ec.europa.eu/research/fp7/index_en.cfm as of 09/13/2010

Illustration: Overview Business Excellence

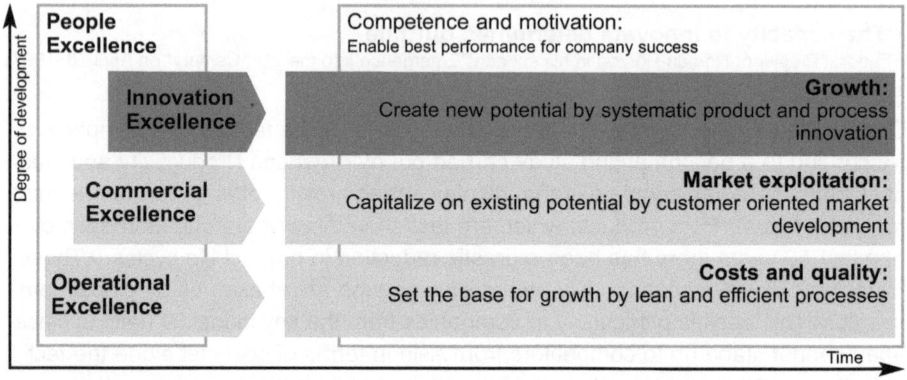

How far away most companies are from the excellence claim is exemplified by the fact that for every 100 research and development projects only about 37 are concluded with a successful market launch and from which only two out of three reach the set targets for time and cost.[5] What is needed to overcome this state of resource waste can be described by five success factors.

Illustration: Action fields of Innovation Excellence

Focus	• Which goals are pursued? • How is innovation measured? • How is the climate for innovation supported?
Process	• What must the innovation process deliver? • How is the innovation process controlled? • How is the process anchored in the company?
Tool	• Which methods are used in the course of innovation management? • How are old and new methodologies linked?
Skill	• Does the corporate culture promote innovation? • Do the employees have everything necessary for developing and implementing innovations?
Structure	• Have responsibilities and mechanisms for decision making been defined? • Have resources been assigned to the topic?

[5] Horváth & Partner: Benchmarking study: "F&E-CONTROLLING", reference period 2008

The company and/or the respective business unit, determines the innovative strategy by the **focus** applied. First, the basic decision for a generic strategy must be made. This should take into account their technology's life cycle, their competence as well as the company's competitive position.

Illustration: Alternative technological strategies

Generic Technology Strategy	Implication for the technology and competence portfolio
Technological Leadership	• Innovative leader in the industry (first mover) • Setup of a leading position in all technologies of the industry • Pronounced strength in all key and pacemaker technologies • If possible own developments in key and pacemaker technologies
Technological Presence Strategy	• Presence/strategy of the fast follower • Capacity to maintain at least a medium position in all technologies • If possible leadership position in one/few key technologies
Technological Niche Strategy	• Focus on limited number of industry critical technologies • Niche strategy in market segments, in which these technologies have a strong influence on the critical success factors
Technological Acquisition	• Strong market presence which is jeopardized by lack of technological strength • Acquisition of skills by acquiring companies/teams
Technological Joint Venture	• High degree of technological innovation but with limited implementation skills • Collaborative technology implementation with partners possessing complementary strengths
Technological Streamlining	• Medium to low innovation progression and defensive technology development • Maintenance of a presence only in the critical technologies of an industry
Technological Withdrawal	• To the greatest extent possible external procurement of know-how and reduction of own development resources

Only on the basis of the selected strategy can the specific development goals and respective key performance figures be derived. These provide the reference points for determining the specific development activities and budgets.

Especially for companies having a medium to weak competitive position with respect to a specific technology or business field, should the approach for finding a strategy be supported by the Blue Ocean Theory.[6)] The basic idea of this approach originates from the realization that during the past 20 years European companies focused primarily on optimizing the parameters of quality, cost and time. Subsequently, numerous industries have been characterized by excessive supply and falls in demand that then lead to prices wars. They represent the so-called "Red Oceans". Those who want to escape from this circle of competitive and cost pressure must create new markets, where they themselves make the rules, the so-called "Blue Oceans".

Illustration: Red vs. Blue Ocean

Strategy in the Red Ocean	Strategy in the Blue Ocean
• Competition in existing market	• Create new markets
• Beat and/or exceed competitors	• Avoid competitors
• Use existing demand	• Generate new demand
• Direct connection between benefit and costs	• Remove direct connection between benefit and costs
• Corporate activities are directed towards differentiation or low costs	• Corporate activities are directed towards differentiation and low costs

In order to open up a "Blue Ocean" a company must identify the factors characterizing the competition in its industry and decide when to leave the field to its' competitors. Resources freed up in this way can be used by the company to increase its performance with respect to other targeted factors and developing Unique Sell Propositions.

Once the focus has been set, suitable **processes** are required in order to, systematically, get from generating the idea, selecting and implementing the right innovation projects, to successfully marketing the new development. The DMADV

6 Kim, W. Chan and Mauborgne, Renée, 2005. Blue Ocean Strategy: How to Create Uncontested Market Space and Make the Competition Irrelevant. Boston: Harvard Business School Publications

phased approach (DMADV = DEFINE, MEASURE, ANALYZE, DESIGN, VERIFY) within the context of Design for Six Sigma^{+Lean} (DFSS^{+Lean}) offers a best-in-class approach for effectively and efficiently realizing ideas. The approach is an open framework into which new developments can be integrated. Thus, especially agile development methods such as SCRUM can be integrated into the ANALYZE and DESIGN phases in order to organize the development of a basic concept and the detailed design quickly and successfully. Prior to the start of the actual DMADV approach comes the so-called "Frontend of Innovation" which is also referred to as "Fuzzy Frontend" (FFE) due to its low degree of structure. During this, the relevant innovation key figures must be defined based on the innovation strategy and respective target values (e.g. target turnover in target market with new products) set. The generation of innovative ideas can take place during a period of information collection and evaluation (e.g. trends in the target market and in the respective industry). These must be evaluated with respect to their conformity with the strategy as well as their prospect of success. They are to be prioritized and integrated into the innovation project portfolio. The most promising ideas are designed and implemented systematically with the help of DFSS^{+Lean}. After a successful implementation benefit collection commences and the economic impact of the development can be assessed.

Illustration: Process of Innovation Excellence

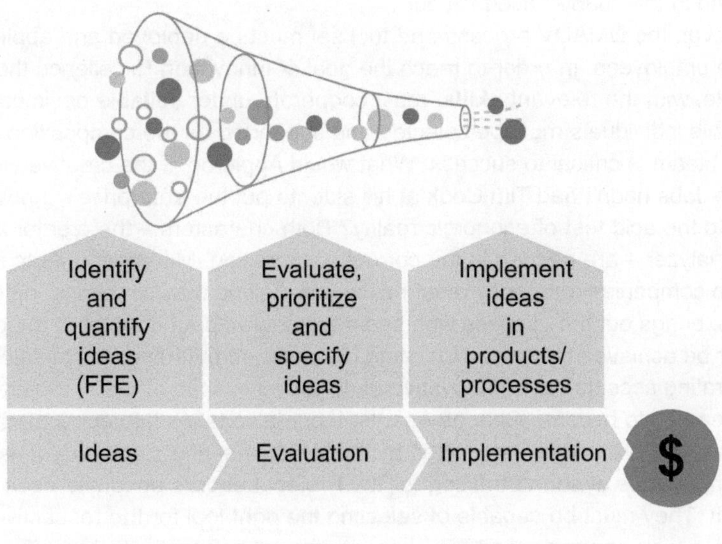

A coordinated procedural model provides an indispensable framework. This is enriched with a series of tools and the specific instructions necessary to make it work.

Illustration: Tools for Innovation Excellence

SCOUT			Imple-mentation
Define search fields	**Generate ideas**	**Develop business opportunities**	
Procedure: • Define idea Scorecard • Analyze patents • Analyze markets • Evaluate technologies • Analyze competitors (Benchmark) • Analyze trends	**Procedure:** • Look for jobs to be done • Identify unfulfilled needs • Analyze functions • Compare unfulfilled needs and functions • Derive ideas • Prioritize ideas	**Procedure:** • Display unfulfilled needs • Describe offer for the customer • Define business model • Identify value contribution • List assumptions and prerequisites which are essential for success • Evaluate risks • Prioritize business opportunity	**Design for Six Sigma^{+Lean}** • DEFINE • MEASURE • ANALYZE • DESIGN • VERIFY

DFSS^{+Lean} provides a comprehensive toolbox for the project team which is presented in detail in this book. A great number of these tools can be used to facilitate decisions and actions in the "Fuzzy Frontend" – i.e. the generation and evaluation of ideas – in order to effectively fill the innovation pipeline. Some of these will be considered in the chapter titled "Scout".
However, the DMADV process and tool set must be deployed and applied by capable employees. In order to reach the goal of Innovation Excellence the correct people, with the relevant **skills**, must cooperate under suitable basic conditions. Suitable individuals must be selected and supported as the composition of the innovation team is critical to success. What would Apple be, if the creative visionary Steve Jobs hadn't had Tim Cook at his side, to put the enterprise's innovative projects to the acid test of economic reality? Both characters – the creator as well as the analyzer – are needed in the correct combination. Moreover, efforts for innovation in companies frequently meet resistance. Taking new directions, off common paths, brings out the skeptics who seem to know without doubt that the goal will never be achieved that way. Not being discouraged by these critical voices and generating acceptance of innovative undertaking in spite of them, requires the Project Leaders to become Change Agents. If one also takes into account, that innovation projects are usually conducted by multifunctional and cross-departmental teams, the importance of sound training for the Project Leaders becomes even more apparent. They must be capable of selecting the right tool for the respective task in a targeted way and to apply it together with the team effectively and efficiently. For this purpose they need not only methodological competence but also leadership and facilitation know-how.

Finally, the fifth success factor is the consistent anchoring of the topic of innovation to the organization's **structure**. The way in which this takes place, sends a clear signal to the employees. Top Management must show that its' commitment to the increase of innovative strength is really to be taken seriously. If the respective roles and/or responsibilities are lacking within the upper management level, and the topic is not supplied with the necessary resources, success attributable to innovation will be left to chance. A real innovation culture, one that shapes the company permanently,cannot emerge in this way. If innovation activities are to be supported systematically and their strategic conformity made permanent role in the organizational structure must be provided. The task of such an "Innovation Leader" is to coordinate and promote innovation activities across the whole company. He/she supports the corporate divisions in the development of their own innovation strategy and derivation of respective actions. A Steering Committee consisting of representatives of Top Management ensures that the planned activities also match the company's strategic aspiration. A suitable organizational structure is shown in the illustration below.

Illustration: Organizational structure

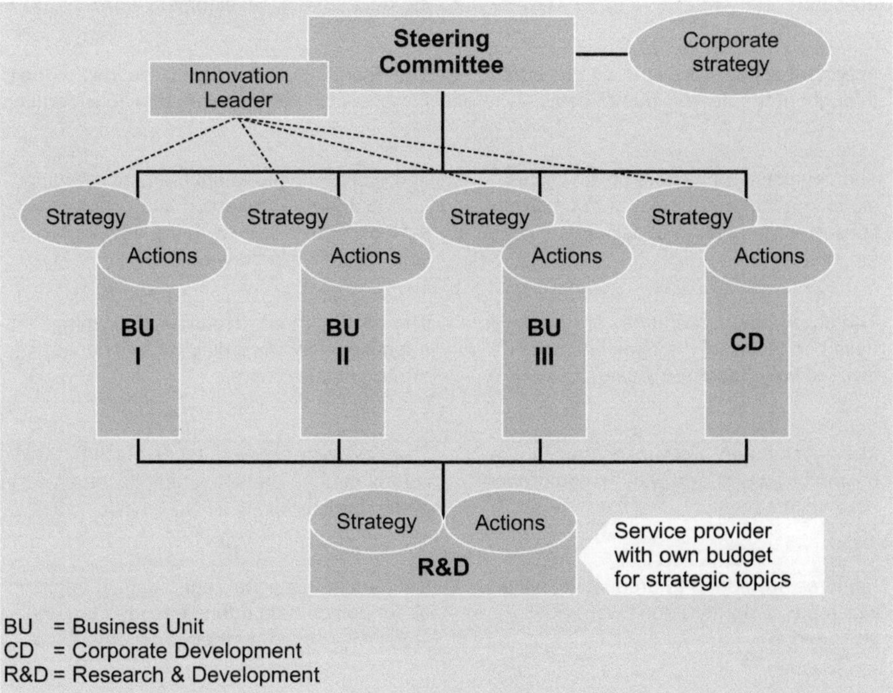

Irrespective of how mature a company is with respect to the above mentioned success factors, every company can advance further in the direction of Innovation Excellence with the help of the DFSS^{+Lean} approach presented in this book. The organization cannot avoid dealing with the success factors such as focus, skill and structure. Inevitably the following questions will emerge: Which ideas do we want to implement within the framework of a DFSS project? Which executive will function as the project's principal or Sponsor? Which employees are to form the project team and receive the necessary training? Decision processes which reliably lead to the correct answers must be institutionalized in the next step.

A further advantage of the DFSS^{+Lean} approach is that the phased, structured, tool-based procedure minimizes the typical risks experienced in development projects.

Illustration: Development risks vs. DFSS^{+Lean} approach

Development Risks	DFSS^{+Lean} Approach
Customer needs are not, or are incompletely, identified and products/services are developed without taking the customer into account.	The customer requirements for the product or process are studied intensively – the true needs of the target customers are the starting point for the development work.
Important aspects are ignored or forgotten throughout the development work.	The entire value chain from the idea to the final development is taken into consideration.
The development work is driven by R&D alone and important input from other functions is not taken into consideration or done too late.	In the core development team all relevant functions are represented and contribute their own point of view to the development work in a structured way.
The development resources are deployed based on incorrect priorities and not in a focused way (resource waste).	The development resources are deployed in a focused way according to the priorities of the target customers.
The development work takes place in an unstructured way and without comprehensible documentation.	The development work takes place through clearly defined phases and is documented in a standardized way.
Features are added to products/services which the customers don't want at all (Over engineering).	Customers are encouraged to give feedback at set points in time thus ensuring that the focus will remain on their requirements.

Moreover, DFSS^{+Lean} can be used not only for product but also for process development. The differences in application will be shown in the relevant sections of this book.

If a company already utilizes an established development process the DFSS^{+Lean} toolbox can nevertheless offer valuable inspiration and enhancement (cf. illustration 9). Although phase models often prescribe the basic steps of development work and define interim goals, they provide little specific assistance for the project team with respect to their realization. In such cases the existing procedural model can usually be combined, without problems, with the DMADV phase approach. Components and tools of DFSS^{+Lean} can be integrated where the existing development process displays gaps or weaknesses. These are all tools which have been tried and tested over many years and which are linked within the framework of DFSS^{+Lean} consistently. In this way the Development Team is provided with a Common Guideline for its next steps.

Illustration: Overview DMADV phase approach

	Tools	Goal
DEFINE	• Project Charter • Project Scope • Multi Generation Plan (MGP) • Gantt Chart • RACI Chart • Budget Calculation • Stakeholder Analysis Table • Communication Plan • Risk Analysis • SCRUM – Agile Project Management	• The project has been defined. • Problem and goal have been defined and a Multi Generation Plan has been created. • The project has been clearly scoped and the influence on other projects has been assessed. • Activities, Time and Resource Planning have been defined. • Potential project risks have been estimated.
MEASURE	• Portfolio Analysis • Customer Value Analysis • Customer Interaction Study • Questioning Techniques • Affinity Diagram • Tree Diagram • Kano Model • Benchmarking • House of Quality • Design Scorecard	• The relevant customers have been identified and segmented. • The customer needs have been collected, sorted and prioritized. • CTCs and measurements have been derived based on the customer needs. • Assign priorities to measurements; target values and quality key figures have been defined.

Continued on the following page

Illustration: Overview DMADV phase approach (continuation)

	Tools	Goal
ANALYZE	• Function Analysis • QFD 2 • Lead User Approach • Cross Innovation • Creativity Techniques • TRIZ • Benchmarking • Pugh Matrix • QFD 3 for target costs • FMEA • Early Prototyping (SCRUM) • Process Modeling • Advance Marketing	• The best concept has been selected from alternative high-level concepts. • Conflicts and contradictions in the selected concept have been resolved and requirements for necessary resources have been derived. • The residual risk has been defined, customer feedback has been collected and the concept has been finalized.
DESIGN	• Design Tree • Cost Breakdown Structure • Statistical Methods (Tolerancing, Hypothesis Tests, DOE) • Trimming • Design for X • Anticipatory Failure Determination • Process Design Principles • Lean Toolbox (Value Stream Map, 5S, SMED, TPM, Process Balancing) • Process Charts • Process Monitoring • Process Simulation • Piloting	• The detailed concept has been developed, optimized and evaluated. • The production process has been planned and optimized according to Lean guidelines. • The implementation of the process design has been prepared, involved employees have been informed and customer feedback was collected.
VERIFY	• Implementation Strategy • Transition Plan • Scale Up • Pilot Marketing • Process Documentation • Standard Operating Procedures (SOPs) • Implementation Teams • Process Performance • Project Documentation	• The transition to serial production has been conducted. • The production process has been implemented. • The market launch has been finalized. • A suitable process control mechanism has been developed. • Process and project documentation have been completed. • The process has been handed over completely to the Process Owner, the documentation was handed over and the project has been closed officially.

Even though the course of a project should always follow the same essential steps, it remains the team's responsibility to select suitable tools from the DFSS^{+Lean} Toolbox within these steps and, then, apply them in a targeted way based on the prevailing conditions. Industry and project specific aspects must be taken into consideration. The specific requirements for development activities e.g. of the pharmaceutical or the automotive industry must also be considered as well as the differences between the development of an IT application and a material. Therefore the methods listed in this book must not be seen as mandatory tools. Many of them can be applied within a different context or on a stand-alone basis. Methods and tools will be presented as we go through the phases DEFINE, MEASURE, ANALYZE, DESIGN and VERIFY. First, an overview will be given of the phase and then the tools which can be potentially used within it will be explained with the help of a uniform structure. Each essential tool will be exemplified with the help a practical example which will develop through each phase.

Moreover, the first chapter of this book, “Scout”, deals with the identification of suitable innovation projects. Before the realization of ideas can take place using the DMADV phase model, the product and/or service ideas must be generated and the most promising ones must be selected for implementation. Even though most companies don't lack creative ideas there should still be a systematic approach to the setup of an innovation portfolio in order to minimize the risk of false investments in the area of new product development.

DESIGN FOR SIX SIGMA+LEAN TOOLSET

SCOUT

Scout

Term
Scout, Scout phase

When
Before embarking on a DMADV project

Goal
- Find, evaluate and develop promising business opportunities
- Create prerequisites for a quick and successful implementation of innovation projects with DMADV

Overview Scout

Scout	DEFINE	MEASURE	ANALYZE	DESIGN	VERIFY
- Generate and describe business opportunity	- Set project scope - Start project - Manage project	- Identify and specify customer requirements	- Develop basic concept and prove feasibility	- Develop, test and implement detailed design	- Ensure commercial success - Close project

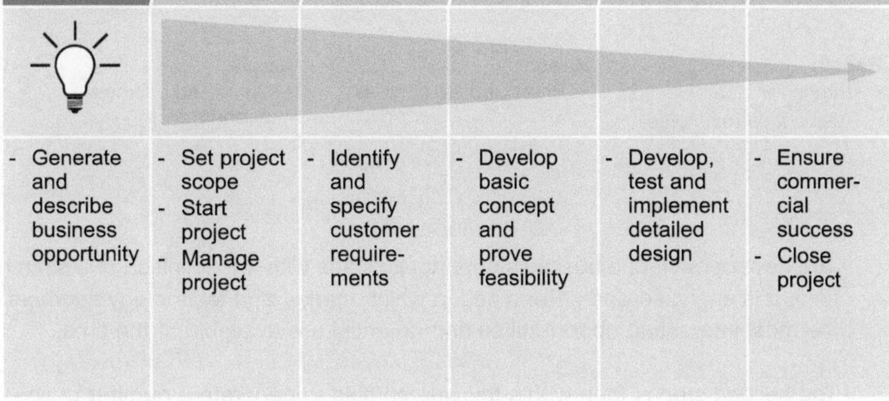

The Scout Phase serves the purpose of finding new and attractive business opportunities which are likely to lead to commercially successful products and services. In order to discover and evaluate them, the procedure in the Scout phase is adopted as follows:

Define search fields	Generate ideas	Develop business opportunities
Central question: In which search fields are ideas for business opportunities to be generated?	**Central question:** Which ideas are to be developed into business opportunities?	**Central question:** Which business opportunities are to be implemented with DMADV?
Procedure: • Define Idea Scorecard • Analyze patents • Analyze markets • Evaluate technologies • Analyze competitors (Benchmark) • Analyze trends	**Procedure:** • Search for jobs to be done, identify unfulfilled needs • Analyze functions • Compare unfulfilled needs and functions • Derive ideas • Prioritize ideas	**Procedure:** • Display unfulfilled needs • Describe offer for the customer • Define business model • Identify value contribution • List assumptions and prerequisites which are critical for success • Evaluate risks • Prioritize business opportunity
Result: • Criteria for idea selection • Relevant search fields	**Result:** • Prioritized list of ideas	**Result:** • Prioritized business opportunities

The development of a business opportunity starts with the definition of a search field. It is analyzed and determined in which market and technology segment the most interesting opportunities and potential are available at the time.

The second step is then within this search field to generate a number of specific ideas with the help of different approaches. These ideas are then developed into business opportunities in the third step.

In all three steps the results are developed systematically and prioritized repeatedly with the help of the same criteria so that the result of the Scout phase is an attractive and completely documented business opportunity. All required information necessary for the ensuing innovation project has been compiled in a compressed way.

1. **Define search fields**
   Search fields are scoped market and technology segments within which the customers expect the fulfillment of specific tasks. Where these requirements are insufficiently fulfilled and/or "unmet needs" exist, opportunity for innovations are available..

   In such segments the potential for commercial success is particularly high and above-average margins can be achieved. Ideally, search fields are already described by an innovation strategy derived from a business strategy.

   The identification of search fields can also take place in different ways e.g. systematic patent analyses can hint at new developments and technical trends with high innovation potential. Also, general or industry specific trend observations can indicate important developments in area of production, marketing, sales and supply chain. In the course of a competition analysis which looks at market positioning and fulfillment of functions, providers can also derive relevant search fields from existing product innovations.

   How promising individual search fields identified in this way are for a company can be evaluated by previously defined criteria. These may have been adjusted to the relevant market and industry environment.

Typical criteria for evaluating the attractiveness of search fields:

Category	**Evaluation Criterion**	**Evaluation (9,3,1,0)**
Strategy	• In line with corporate strategy • In line with innovation strategy	
Unfulfilled needs	• Strength of unfulfilled needs • Effort due to changeover on the customer's part • Unique Selling Proposition compared to competition	
Market attractiveness	• Turnover potential • Market growth • Intensity of competition	
Technology	• Availability of functions • Internally and externally available technologies	
Risk	• Technical risk • Market risk • Implementation risk (capability and capacity)	
Sum		

The category of unfulfilled needs is especially decisive for the prioritization of search fields with respect to chances for commercial success. Where tasks are only insufficiently fulfilled from the customer's perspective an appropriate offer to the customer is particularly attractive.

In order to be able to prioritize the search fields it helps to assign a value for each category between "very attractive" (9) and "very unattractive" (0). By summing up the values of various categories an attractiveness ranking of search fields can be set.

2. **Generate ideas**

In order to find an attractive business opportunity in the most promising search field, ideas for commercially successful products and services must now be generated. The basis for this promising and/or good idea is the customer requirements in a specific application situation – the so-called "job to be done". Where these needs are not fulfilled, i.e. where "unmet needs" exist, needs for new product offering and services exist.

In order to generate ideas for such new offers two, basically different, procedures can be applied: If the focus is on the analysis of insufficiently fulfilled customer needs a "Market Pull" approach could be adopted. Here, insights are gained with the help of different tools about which unfulfilled needs currently exist in the market and hence derive suitable potential offers.

If the search for ideas is primarily based on existing available technologies and functionalities for which suitable markets and applications are to be found a "Technology Push" approach could be used. The technology offer is also added to and/or optimized with respect to the customer needs in the new alternative market.

Mapping "Technology Push" and "Market Pull"

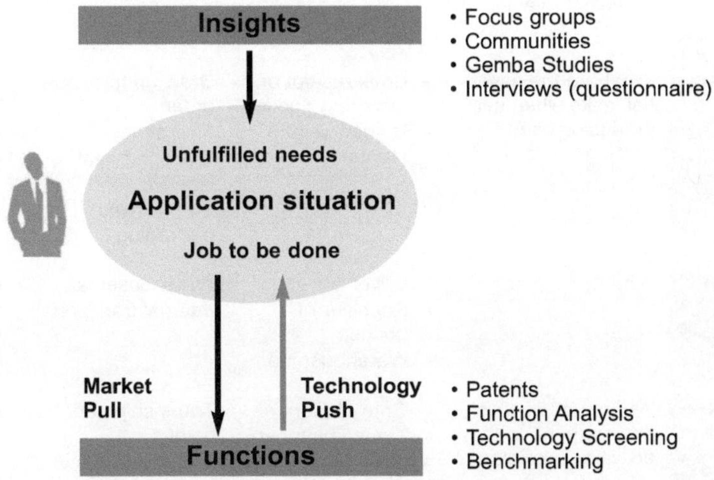

In both cases it must be examined first which "jobs to be done" the customer expects and in what specific situations. It is important that all application situations and the existing needs of customers and/or individuals involved in the process are taken into consideration. For this purpose the search field along the value or process chain is broken down into its essential steps as shown in the following example. The involvement of individuals who are engaged or working in the process, the environment as well as tools used will define the requirements in the different situations of application.

Below is an example of how this might look when looking at the delivery and assembly of a seat in a bus manufacturer's assembly plant:

Process steps	What are the essential steps in the process or value chain?	1. Deliver seat to plant	2. Transport seat internally	3. Assemble seat in bus
Sub-steps	What are the essential tasks when this product is used?	- Book receipt of incoming goods - Unload shipments	- Draw up transport order - Find seat - Transport seat - Unpack seat - Throw away packaging	- Prepare installation - Assemble seat
Persons	Who is involved?	- Driver - Receiver of goods - Warehouseman	- Warehouseman - Internal transport	- Work preparation - Assembly-man
Places	Where and in which environment do the activities take place?	- Gate of plant - Trans-shipment point - Location	- Trans-shipment point - Storage space in assembly hall	- Assembly hall
Tools	Which tools and products are currently being used?	- Cardboard boxes - Delivery note of paper	- Forklift	- Standard tool – spanner
Jobs to be done	Which task is to be fulfilled for the customer specifically?	- The warehouseman needs to recognize to which assembly place the packaged seat is to be delivered	- The right unpacked seats are to delivered to the assembly line in due time	- The seat is to be installed in a stable and reversible way within a short period of time

For a comprehensive identification of all jobs to be done surveys of all essential individuals involved in the process, Gemba Studies and customer interviews are absolutely essential. Only by doing this can reliable statements be made about how often and how urgently certain requirements for the product and/or the process occur and how high the dissatisfaction resulting from poorly fulfilled requirements is.
Based on this information the unfulfilled needs can be identified:

Job to be done	Importance	Frequency	Dissatis-faction	Strength of unfulfilled need
1 The warehouseman needs to recognize to which assembly place the packaged seat is to be delivered	3	9	3	81
2. The right unpacked seats are to be found at the assembly line in due time	9	3	3	81
3. The seat is to be installed in a stable and reversible way within a short period of time	9	9	9	729

The evaluation of the jobs needed takes place for each of the three categories, as in the above example, importance, frequency and dissatisfaction on a scale from 1 to 10. Next the three evaluations are multiplied. Based on the resultant value inferences on the strength of the unfulfilled needs can be made. Future generation of ideas using Brainstorming Workshops is driven by the most important unfulfilled needs.

The Brainstorming Workshops pursue the goal of finding and documenting as many ideas as possible for eliminating the unfulfilled needs.

In order to create a working environment which is as creative and as inspiring as possible, the idea generation team should be as heterogeneous and as unconventional as possible. This does not only concern the diverse knowledge areas but also age, sex, and background. Apart from involving creative and knowledgeable company employees external experts should also be invited to participate.

After a briefing has been conducted, all ideas are written down and collected e.g. on idea cards during the Workshop.

Category: Process step:	Name:
Title of idea:	
Relevant unmet needs:	Potential solution/offer:

The ideas which were captured in this way can now be further combined, summarized and visualized by the workshop participants before being prioritized. This prioritization takes place according to the same criteria that were used in assessing the search fields. Afterwards the most promising ideas can be developed further.

3. **Develop business opportunity**
   In order to be able to develop an accepted and promising business idea from the prioritized idea, a more detailed specification of the offer and an evaluation of the technical, organizational and economic feasibility are required. Moreover, the patent-related framework, the market and competitive situation and the customer acceptance of the offer should be analyzed, meaning that the interdisciplinary cooperation of marketing, development, production and customers is necessary.

   As soon as the favored idea is seen as feasible, a promising business opportunity can be formulated and presented taking into account the following:

   - Describe the identified and prioritized unfulfilled needs
   - Describe the offer for the customer
   - Estimate the value contribution (NPV) with target turnover and margin
   - Describe assumptions and prerequisites which are critical for success
   - Evaluate the risks
   - Draft the project Time Plan
   - Estimate the budget

   Several developed business opportunities can now be prioritized, once again, with the help of common criteria (e.g. strategy, unfulfilled needs, market attractiveness, technology and risk).

   Having developed a business idea in this way it is more likely to be received positively by potential sponsors.
   A promising and commercially successful product can be created from an innovative idea in the subsequent DMADV phases.

DESIGN FOR SIX SIGMA^{+LEAN} TOOLSET

DEFINE

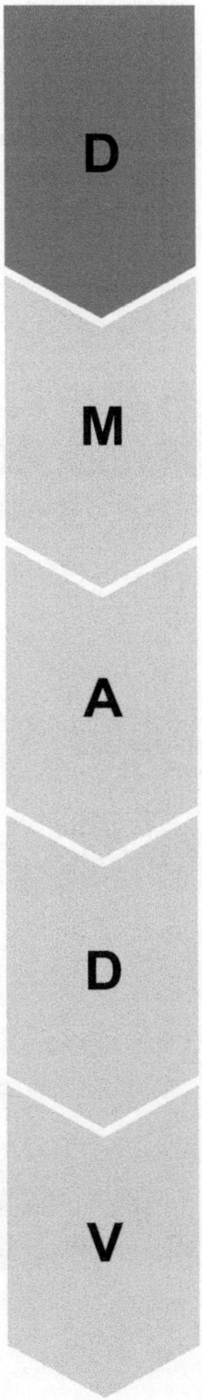

Central Question and Objective of the DEFINE Phase

- The DEFINE phase answers the question: "Why is this being done and what is the scope of this development activity?"
- Objective:
  - Define the project's starting point
  - Formulate and coordinate project goals
  - Determine project scope and management

The Meaning of the DEFINE Phase

- The product/process design is clearly scoped and defined in generations in order to keep development times short and to be able to focus on the most important customers
- The focus is on interdisciplinary teams and benefit innovation for customers and companies
- There are clear responsibilities in the team to ensure direct and effective communication
- Any potential project risks are monitored proactively and managed throughout the project duration

Procedure in the DEFINE Phase

- Develop the Business Case
- Define opportunities and goals
- Calculate monetary benefit
- Define roles
- Set project scope
- Develop a Multi Generation Plan
- Conduct a Best Practice Check
- Plan activities, times and resources
- Plan budget
- Set up Stakeholder Management
- Plan project communication
- Estimate project risks

DEFINE

MEASURE

ANALYZE

DESIGN

VERIFY

D M A D V

D.1 Initiate the Project

- How is the project aligned?

D.2 Scope the Project

- What is in the project scope?
- What are reasonable developmental generations?

D.3 Manage the Project

- What work packages must be completed in what time?
- Who is responsible for completing the work packages?
- Which Stakeholders must be addressed?
- What risks might occur and how might they be managed?

Result DEFINE

- Project Charter defined completely and clearly
- Project scope clearly outlined
- Project Management set up

DEFINE Gate Review/Phase Closure

D.1 Initiate the Project

- Project Charter
- Business Case
- Redesign and New Design
- Project Team

D.2 Scope the Project

- In/Out-Frame
- Multi Generation Plan (MGP)
- Best Practice Check

D.3 Manage the Project

- Project Structure Plan
- Network Plan
- Time Plan
- Agile Project Management
- Resource Planning
- RACI Chart
- Budget Planning
- Stakeholder Management
- Project Communication
- Risk Management
- Kick-off

D Gate Review

Project Charter

Term
Project Charter, Project Order

When
DEFINE: initiate the project; validation in MEASURE, ANALYZE, DESIGN and VERIFY

Goal
Summarize all necessary information for defining the project

Procedure
The Project Charter is the central document of the DEFINE phase. It summarizes all essential information at the beginning of the project. The information contained in it is updated throughout the project after each phase.
The Project Charter consists of the following elements:

Project Initiation	1. **Business Case** Explanation why the project should be carried out now 2.a **Redesign: Problems and Goals** Description of problems/opportunities as well as goals in clear, concise and measurable terms 2.b **New Design: Goal** Naming a new product/process and the respective guidelines and goals 3. **Project Benefit** Financial benefit of the project and if required unquantifiable soft savings 4. **Roles** Sponsor, Black Belt, Green Belt, team members, Master Black Belt
Project Scope	5. **Project Scope** Project Frame – what's "In" and "Out", Multi Generation Plan (MGP)
Project Management	6. **Project Management** Major steps and Milestones for reaching the goal, project risks

The Sponsor is responsible for drawing up a provisional Project Charter. The contents are coordinated with Management, Deployment Champion, Marketing, Sales and Black Belt/Green Belt.

Example Passenger Seat:

Project title:	Development of a passenger seat for Russia	Date:	01/18/2012	Version:	D 1
		Project number:			
		Project category/type:			
		Life cycle stage:			

Business Case:	Project Scope:	
Market studies have shown that transport companies in Russia will modernize their bus fleets by up to 80% over the next ten years. Our current seat model has been on the market for 5 years and fulfills the increased requirements with respect to quality and comfort insufficiently. Sales figures are decreasing due to increasing competition. In order to meet the market changes an innovative seat system for use in buses must be developed.	In:	Ergonomic seat design, variants for installation and removal, cushioning, easy maintenance, seat material
	Out:	Electronic features on seat, change of bus layout, measurements of seat system
	MGP:	(1) Generation I: 30% market share in Russia by 2014, modular standard fitting (2) (3)

Problem/Goal:	Roles:	
During the last 6 months the proportion of customer complaints in regard to processing and maintenance friendliness increased by 25%. Sales figures decreased by 20% in the same time period. In order to counteract this negative trend and to participate in the projected increase in demand a marketable seat system is to be developed by 10/01/2011. We intend to achieve a 30% share of the Russian market with this new system by 2014. The bus manufacturer and/or suppliers in Western Europe are to be involved actively in the development process.	Champion	Dr. Jacomo Franco
	Project Leader	Mr. Bernhard Fuchsberger
	Core team	Ms. M. (Marketing), Dr. Q. (Quality Management), Dr. F. (Product Development), Dr. Z. (Process Development), Mr. E. (Procurement), Ms. P. (Production), Mr. V. (Sales), Mr. C. (Controlling)

Monetary benefit:	Meilensteine:	
Output 2012: 20,000, output 2013: 30,000, output 2014: 50.000, profit per seat = 10 % of the net sale price, net sale price = $ 100, tax rate = 25 %, capital cost rate = 10 % = >EVA 2012 = $ 150,000, EVA 2013 = $ 225,000, EVA 2014 = $ 375,000, discounted EVA for 2012 = $ 664,500	DEFINE:	01/18/2012
	MEASURE:	02/10/2012
	ANALYZE:	04/05/2012
	DESIGN:	05/25/2012
	VERIFY:	07/26/2012
	Project handover:	08/28/2012

Potential risks:		
Increased political intervention in the Russian domestic market, potential liquidity problems among the trading partners, logistics constraints, silo thinking	Signature Champion	Signature Project Leader

Business Case

Term
Business Case, description of starting point

When
Before the project start, DEFINE: initiate the project

Goal
- Describe the business environment and the project's starting point
- Describe the project's meaning and impact for customers and companies
- Describe explicitly with facts & figures why the company should undertake the project now

Procedure
To define the Business Case and/or the description of the starting point the following central question must be answered:
- Why should the development project be conducted now and what is its monetary benefit?

To answer this question use information from:
- a preliminary market and competition analysis
- environment analysis within the company assessing how the new project might fit in it

EXTERNAL: Market and competition analysis	INTERNAL: Environment analysis
• Determine the target market and the target customers (size of market and/or potential customers in target market) • Knowledge of market (market entry barriers, price level etc.) • Existing customer requirements • Describe the benefit for the customer • Estimate the market and turnover potential and possible market growth. • Describe the competitive context (Benchmark)	• Is the project's goal in line with corporate strategy? • Which direct and indirect benefit does the company gain because of the project? • Are there any other projects in the company dealing with this or a similar topic? • Which other projects are being treated with the same priority?

The following individuals and/or areas should participate in the procurement and/or development of information:

- Deployment Champion (corporate strategic aspects)
- Marketing and Sales (forecasts on target market, customers, and competitive strategies)
- Management and executives
- Black Belt/Green Belt and potential team members, who contribute their specialist know-how

Comparison and evaluation of alternative project approaches should be conducted prior to the project and can be used as a basis for the Business Case in DEFINE. Evaluation of ideas and/or different project approaches should be done by taking into account relevant innovation KPIs.

1. Define innovation KPIs (as is and target values) and operationalize alternative project approaches
   In order to make the evaluation as valid as possible, KPIs should be directed towards internal and external factors and take into account market and/or customer oriented success potential not to mention internal implementation effort. A Tree Diagram enables the specification of potential KPIs; current and target values can be added

Illustration: KPIs on success potential

External perspective	Relevant criteria
Success potential	• Size of market/number of potential users • Level of, and variation in, demand • Knowledge of market • Market growth • Price level in target market • Market entry barriers • Intensity of competition • Unique Selling Proposition (USP) • Exclusiveness • Patent protection • Sustainability/life span • Influence on the company's other products • Potential cross-innovation • Future potential (leverage effect)

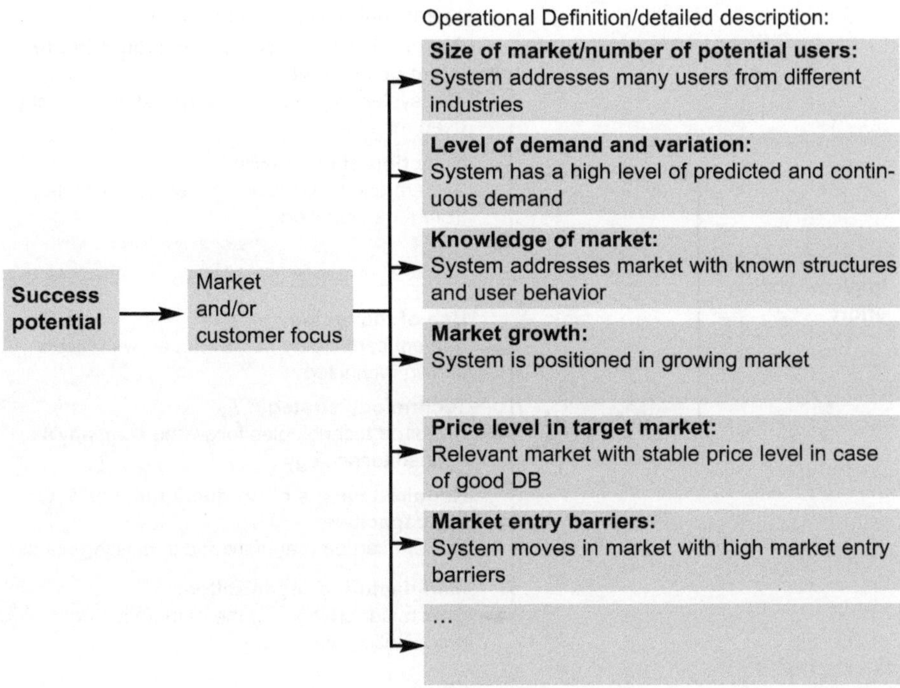

Mapping KPIs with respect to implementation effort

Internal perspective	Relevant criteria
Implementation effort	• Required inputs (primary material, semi-finished product, etc.) • Selection of suppliers • Required development personnel and know-how • Technology use • Technology strategy • Required means of production/manufacturing capacities • Manufacturing organization • Sales channels/outbound logistics • Required sales force • Customer interaction • Complexity reduction

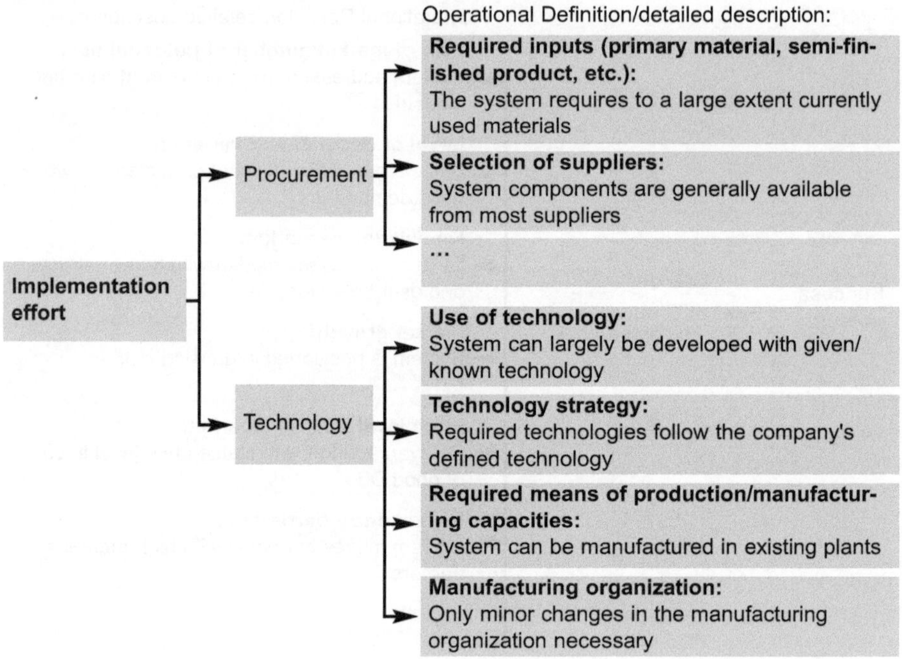

2. Quantify the gap between as is and target for each KPI (e.g. comparison of current market share with current product on the market with intended market share with new product in target market)

3. Weigh KPI goals according to strategic alignment

4. Benefit estimation based on alternative project approaches by quantification of expected contribution to "closing the gap"

5. Evaluate occurrence probability, i.e. probability that the respective product and/or process innovation can contribute the expected contribution to "closing the gap"

6. Conduct cost estimate for each project approach by estimating the project expenses up to initial revenue receipt (alternatively: up to first positive contribution to margin or until Break Even is achieved)

The above steps can be combined within the frame of a benefit comparison table as prepared below:

Mapping Benefit Comparison Table

KPI	KPI 1	KPI 2
Prioritization factor	2	3
As is	5	6
Target	8	12
Delta	3	6

	KPI 1	KPI 2	Average	Average	Sum	Project expen-ses	Benefit unit/ $
Impact[1)]	2	3					
Impact[1)] percentage	66%	50%	58%				
Occurrence probability	50%	75%		62.5%			
Score[2)]	0.66	1.13			1.79	300,000	167,597
Impact[1)]	1	4					
Impact[1)] percentage	33%	66%	50%				
Occurrence probability [2)]	90%	66%		78%			
Score[2)]	0.60%	1.33%			1.93	450,000	233,160

Efficiency factor/idea

Risk key figure/idea

Benefit unit/idea (by taking into account the priority, the efficiency factor and the risk)

Costs/benefit unit

1) Absolute change of KPI and change of KPI in percent of KPI delta
2) Percentage effect x occurrence probability x KPI prioritization factor

Comparison of project approaches on the basis of the respective efficiency factor, conducted by taking into account the expected project costs and costs/benefit unit is a necessary precursor to describing the Business Case.

Tip

- Show Management and important Stakeholders the background and context of the project in a comprehensible and fact driven way with the help of the Business Case
- Convey the necessary pressure to act at this point – make it compelling
- For more detailed information for drawing up a Business Case or a Business Plan, refer to specialist economic literature

Redesign

Term
Redesign Order, problem and target mapping

When
DEFINE: initiate the project

Goal
Describe the problem and the intended improvement

Procedure

1. Describe the problem clearly. The problem can be best described by answering the following questions:
   - What goes wrong or does not meet the requirements of the customers?
   - When and where do the problems occur?
   - How big are the problems?
   - What is the impact of the problems?
   - Why is a Redesign necessary to achieve the improvement?

2. Formulate the goal clearly, in a measurable way and set a final date. Should several problems be addressed by the project, several goals will have to be formulated accordingly

3. Include the problem and goal in the Project Charter

Tip

- When formulating the problem and goal you should adopt the SMART rule:
  - Specific
  - Measurable
  - Agreed to
  - Realistic
  - Time bound
- Ensure that your problem description contains no causes and does not apportion blame; the goal should not contain, nor describe, a solution

New Design

Term
New Design Order, mapping of product or service idea

When
DEFINE: initiating the project

Goal
Describe the idea for the new product or service and the opportunities it affords

Procedure
In case of a new design the starting point is usually not a problem but an idea, promising an entrepreneurial opportunity.

1. Describe the idea and the potential connected to it in a convincing way
   The following questions can help:
   - What is the idea for the product or service?
   - Which customer group(s) will profit from it?
   - Why is the pursuit of this idea advantageous from the business's perspective?
2. Formulate a clear and measurable goal on the basis of the described idea: It must articulate what the expected effect is on the essential key figures (e.g. yield or customer satisfaction) and by when if the idea is implemented
3. Include idea and goal in the Project Charter

Tip
- When formulating the problem and goal you should adopt the SMART rule:
  - Specific
  - Measurable
  - Agreed to
  - Realistic
  - Time bound
- Ensure that your problem description contains no causes and does not apportion blame; the goal should not contain, nor describe, a solution

Project Benefit

Term
Project Benefit, Hard and Soft Savings

When
DEFINE: initiating the project; validation in MEASURE, ANALYZE, DESIGN and VERIFY

Goal
Estimate the quantitative and qualitative benefit of the project for the company

Procedure
The objective of development projects is supporting and positively influencing the company's strategy and by implication the most important key performance indicators by which the company is controlled.

The benefit of a development project is evaluated by the Project Sponsor with the help of the Business Case which will have been verified with the assistance of Finance. The project benefit is the difference between what accrues under the as is state (problem/chance/idea) and what will accrue under the target state (goal) and expressed as a profitability variable e.g. margin, revenue, net profit etc.

The EVA® (Economic Value Added) is a financial performance indicator describing the profit generated over and above the capital costs. It is included in the following observation in order to exemplify the monetary effects of DFSS projects (Example of EVA® on the following page).

The depicted value drivers also influence other financial key figures, such as ROI (Return on Investments), Cash Flow, EBITDA (Earnings Before Interest, Taxes, Depreciation and Amortization), which are often used to steer and control operations in a company.
Apart from the monetary benefit the project often includes unquantifiable successes, the so-called Soft Savings. These should also be taken into consideration and described. (Note that sometimes these items can be measured but no financial benefit attributed.)

Classic Soft Savings are for example:

- Enhancement of prestige through excellent quality
- Increased brand recognition or company name awareness
- Increased customer satisfaction
- Increased employee motivation and retention of high potential people in the company
- Increased job security

Mapping EVA® – Value Driver Tree

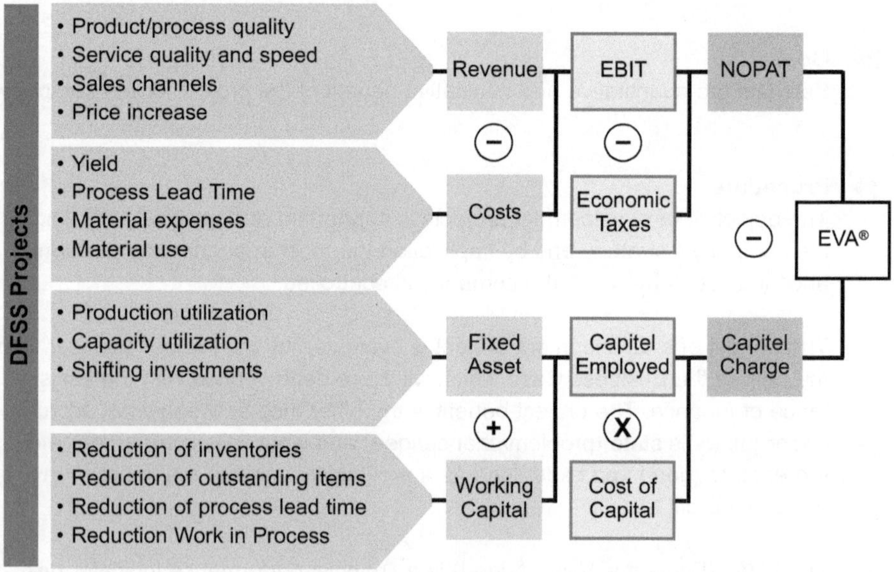

EVA® is a registered brand of the business consultancy Stern Stewart & Co.

Project Team

Term
Project Team, Core Team

When
DEFINE: initiating the project

Goal
Put a project team together that is capable of delivering the project

Procedure
A key success factor of DFSS projects is an interdisciplinary core team that consists of members from all relevant areas of the value chain:

- Marketing
- Sales
- Development
- Production
- Quality Management
- Customer Service
- Controlling

1. Before the project commences the Project Leader thinks about which specific areas require to be represented, in which phase(s) of the project this will be, and to what extent i.e. what level of commitment is required

2. In coordination with the Sponsor and the respective responsible people for individual divisions/departments, the core team members and their required level of commitment per project phase are determined

3. If required, experts and/or members of the extended team can be nominated and the required commitment level estimated

4. The resource plan is coordinated with the Sponsor who is responsible throughout the entire project for ensuring that the commitments that were agreed to,are actually made available to the project

Tip

It is important that you point out to team members, no later than the Kick-off Meeting, that responsibility shifts over the course of the project e.g. from Marketing/Sales to the Development Department and then to Production and Quality Management (cf. RACI Chart).

Project Scope

Term
In/Out-Frame, Project Scope, Project Frame, Project Focus

When
DEFINE: scope the project; validation in MEASURE, ANALYZE, DESIGN and VERIFY

Goal
- Focus the project work on coordinated contents
- Clearly determine and define the topics which are not part of the project, in order to keep the project scope realistic
- Ensure a common understanding of the product among the team and important Stakeholders

Procedure
The following questions can be helpful for scoping a project quickly and easily:
- Which product/which process should the team focus on?
- Which aspects, features and things are to be excluded from the scope of the project?
- How is the product to be developed going to be differentiated from other products or product families?
- How has the project been separated from other, ongoing activities/projects?

1. Visualize the Project Scope like a picture frame

2. All aspects affecting the project are positioned in relation to this frame:
   - The issues definitely to be considered in the project are placed inside the frame
   - The issues definitely not to be considered in the project, are placed outside the frame
   - The issues which cannot be clearly assigned to either "in" or "out" are positioned on the frame; the team can discuss/review final positions afterwards; if no clear decision can be made, the Sponsor is to be consulted

Example

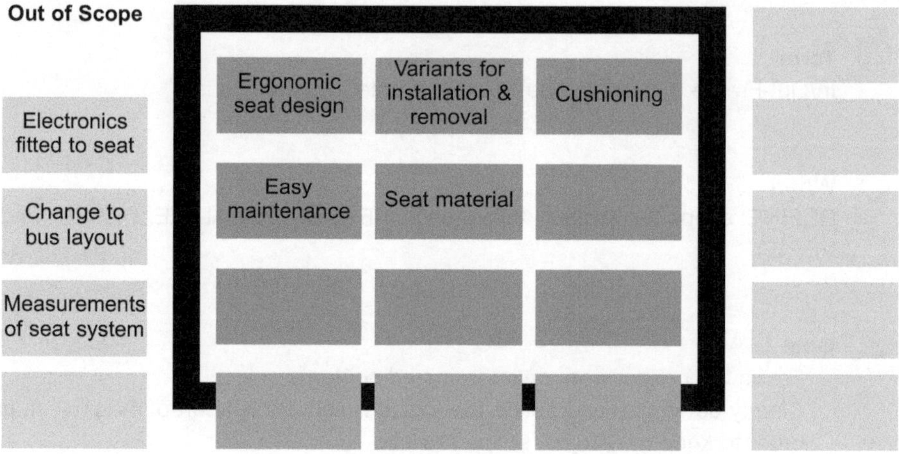
Out of Scope
Electronics fitted to seat
Change to bus layout
Measurements of seat system
Ergonomic seat design
Variants for installation & removal
Cushioning
Easy maintenance
Seat material

Multi Generation Plan

Term
Multi Generation Plan (MGP)

When
DEFINE: scoping the project

Goal
- Support long-term development planning
- Limit the scope of the current project
- Give a preview of potential future developments for the product or process

Procedure
The Multi Generation Plan describes the product development with the help of three successive and interrelated generations.
The contents of the three generations can be characterized generically as follows:

Generation I	Generation II	Generation III
Carry out the first step!	Improve the achieved position!	Take over the leadership position!
Generation I aims at cutting off urgent problems and closing gaps.	Generation II extends the secured product base and looks to aggressively develop new target markets.	Generation III strives for the realization of the final vision, e.g. "Be technological market leader".
Stop the bleeding!	Take the offensive!	Attain leadership!
Achieve market entry	Extend market share	Gain market leadership

Timeline

Each generation is described by
- its vision/its goal
- its characteristics
- required technologies and platforms

"How is the product development staged?"

	Generation 1	Generation 2	Generation 3
Vision	Passenger seat for the "rural market" in Russia, market share 30%	Passenger seat for the "rural market" in Eastern Europe, market share 40%	Passenger seat for the "rural market" worldwide, market share 50%
Description	Based on customer requirements in Russia, modular standard fitting	Extension of standard fitting based on new customer require-ments if required	Complexity increase in line with local condi-tions
Platform	Extension of the sales channels e.g. internet and print media	Extension of sales and services processes	Extension of production into other countries
Technology	Use of existing pro-cesses and technolo-gies, low investments in new technologies	Support by B2B e-Business Applications	
Beginning of development	2011	2013	2015
On the market/ in production	2014	2016	2020

➾ **Tip**

You can further characterize the generations taking into account the following aspects:

- Processes and systems which are important for achieving more efficient and quicker developments and bringing products onto the market
- Sales channels and structures which can place the product in its assigned target market
- Resources (e.g. employees, know-how etc.) required for realizing the respective vision (target of generation)
- KPIs specifying the goals of the respective generation specifically and in a measurable way

Best Practice Check

Term
Best Practice Check, Project Overview

When
DEFINE: scope the project; validation in MEASURE, ANALYZE, DESIGN and VERIFY

Goal
- Check the mutual influence between the DFSS project and other projects/ initiatives in the company
- Ensure an effective and efficient exchange of information between the project teams

Procedure
Review the interactions with other projects by answering the following questions:

Internal projects
- Are there other projects being conducted in the company at the same time, where there has been an increase in knowledge which might be used to help the DFSS project?
- Can other projects use the accumulated information and interim results of the DFSS project?
- Do other projects influence the DFSS project with respect to anticipated constraints in the test, production, marketing, procurement or sales environments?
- How is the information to be documented and communicated in order to ensure an effective and efficient exchange of information between the different project teams?

External projects
- What projects are being pursued by competitors?
- What initiatives or projects are being conducted by customers?
- Are regulatory changes to be expected during the project life cycle?

Example

Projects to be taken into consideration:

EXTERNAL	INTERNAL
New developments by competitors	Project Single Sourcing (Procurement)
Optimization initiatives by customers	Project Optimization of spray-painting process
EU initiative Fire protection in passenger buses	Project Restructuring

Project Management

Term
Project Management, Project Planning, Project Monitoring and Control

When
DEFINE, MEASURE, ANALYZE, DESIGN, VERIFY

Goal
Dynamically control project activities in order to reach the project goal with the available resources without exceeding the budget or the time frame (OTOBOS = on Time, on Budget, on Specification)

Procedure
Successful project management is characterized by some essential features:
- an interdisciplinary team and early integration of all involved areas and external partners (customers, suppliers, test institutes, universities, etc.)
- realistic and detailed planning of project activities, time and budget
- clear determination of competencies for project and line functions
- targeted and structured communication within the team as well as with Stakeholders
- continuous project monitoring and control with suitable tools
- structured documentation of project results

1. Determine all necessary work packages and activities and find out logical relations and dependencies

2. It should be taken into consideration that at this stage in the project not all work packages can be planned completely and that the use of agile methods can make sense

3. Plan derived work packages/activities with respect to the duration of the implementation and the implementation effort

4. Identify the budgets for the implementation of individual work packages and distribute the total budget

5. Put together the project team, find out resource availability, assign resources and responsibilities

6. Set up project communication, identify risks, develop action plans for prevention and reaction

7. Develop a Change Management concept, create a communication strategy

8. Monitor and control project progress continuously

Illustration: Project Management Elements

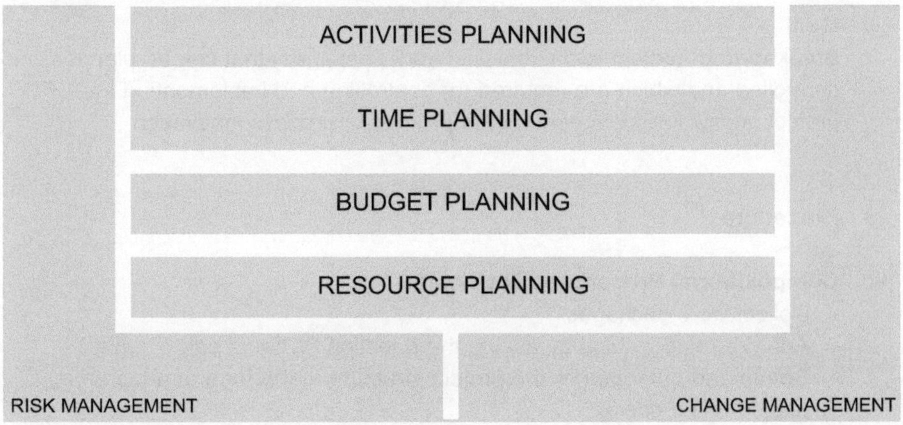

Work Breakdown Structure

Term
Work Breakdown Structure, Project Structure Plan, Activities Plan, Action Planning

When
DEFINE: manage project and tracking in MEASURE, ANALYZE, DESIGN and VERIFY

Goal
Breakdown of project into subtasks (work packages) that can be planned and controlled and which are required for controlling and implementing the DFSS project and/or for implementing the product or process innovation

Procedure

Compositional Procedure (Bottom-up)
- Collect work packages
- Analyze relationships by asking the question "What is part of what?"
- Set up and put together the project structure in the form of a hierarchy (cause-effect-chain)
- Add unmentioned but obviously required tasks and/or project parts

Decomposition Procedure (Top-down)
- Determine the project and/or implementation phases and break them down into main tasks
- Break down the main tasks into subtasks
- Describe sub tasks in finer and finer detail to create comprehensive work packages

Mapping activities planning, example MEASURE

	Activity:	
MEASURE	Identify customers and their requirements	Identify customers
		Formulate hypotheses about customer needs
		Formulate hypotheses about customer behavior in the process
		Plan customer observation and surveys
	Structure and prioritize customer needs	Observe how customers interact with the process
		Evaluate customer observations
		Interview customers
	Specify customer needs	Derive and evaluate customer needs
		Derive CTCs and output measurements
		Define target values and output measurements
		Estimate risks
		Review relevance of CTCs with the customer

Tip

- The number, content and detail of the work packages must be sufficient to enable you to draw up a robust and realistic Time Plan that will facilitate delivery of the project goal
- Bear in mind that each work package must be assigned to precisely one responsible person
- Ensure that the technical, material and organizational prerequisites for completing the work package are available
- To check for completeness ask yourself the question: "If I were to do all these things, can I successfully complete the phase?"

Network Plan

Term
Program Evaluation and Review Technique, Critical Path Method, Network Plan

When
DEFINE: managing the project and tracking it in MEASURE, ANALYZE, DESIGN and VERIFY

Goal
- Visualize logical relationships/dependencies between the activities and their temporal sequence in the course of the project especially when planning complex flows/projects
- Discover the critical path, i.e. the shortest period of time in which the project can be completed
- Minimize the project duration and make it more manageable

Procedure
1 Map work packages in a logical sequence; parallel steps are possible

2 Map the earliest beginning, the duration and the earliest end (Best Case) as well as the latest beginning and the latest end (Worst Case) for all work packages

3 Find out and mark the combined shortest period of time for completing the project by adding all work package durations (Best Case) – this is the critical path

Tip
- Take into consideration that differences can exist between the effort for carrying out an activity and the real time duration to complete
- In order to minimize the impact of task dependencies on project duration focus on the detail of relevant work packages to see what tasks can be carried out simultaneously
- Use different colors to keep an overview of finished and pending activities and highlight the critical path

Illustration: Network Plan

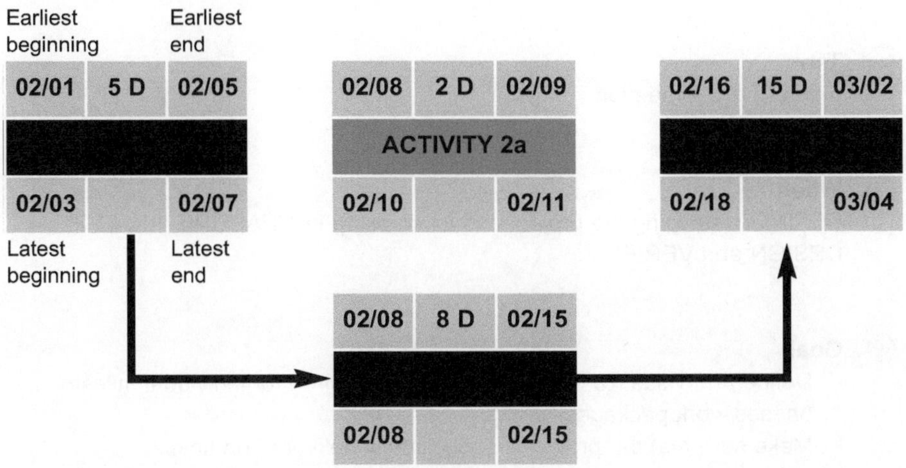

Time Plan

Term
Gantt Chart, Time plan

When
DEFINE: managing the project and its' tracking in MEASURE, ANALYZE, DESIGN and VERIFY

Goal
- Define and visualize duration and effort required for individual milestones/ phases/work packages
- Make sure that the project is successfully delivered on time

Procedure
1 Include work packages from the Activities Plan and the Network Plan

2 Fix start and end date of each activity

3 Additionally, implementation start and end dates and responsibilities must be set

4 If required, integrate further information such as implementation status, effort/benefit, dependencies etc.

5 Update the Time Plan continuously to reflect the real status so that comparison the original implementation plan* can be made so that the impact of any deviation can be assessed in terms of delay, effort or benefit loss etc. (*Base Time Line)

6 In case of deviations from the plan resources must be actively managed and necessary adjustments made to reduce the impact

The time required for each activity can be mapped graphically in a Gantt Chart in the form of a bar. Simultaneous and linked activities (dependencies) are better recognized this way. Software is available to facilitate this.

Illustration: Activities planning – MEASURE

	Activity:		Duration [days]	Start	End
MEASURE	Identify customers and their needs	Identify customers	2	07/06	07/07
		Formulate hypotheses about customer needs	2	07/07	07/08
		Formulate hypotheses about customer behavior in the process	2	07/08	07/09
		Plan customer observation and surveys	4	07/09	07/12
	Structure and prioritize customer needs	Observe how customers interact with the process	4	07/12	07/15
		Evaluate customer observations	6	07/15	07/20
		Interview customers	4	07/20	07/23
	Specify customer needs	Derive and evaluate customer needs	2	07/23	07/24
		Derive CTCs and output measurements	2	07/24	07/25
		Define target values and output measurements	2	07/25	07/26
		Estimate risks	2	07/26	07/27
		Review relevance of CTCs with the customer	2	07/27	07/28

Illustration: GANTT Chart – MEASURE

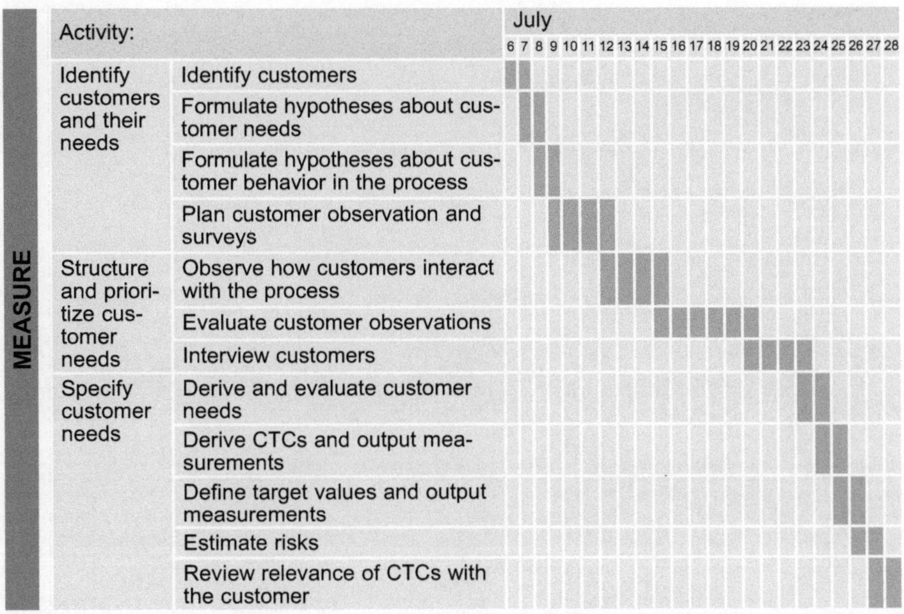

Tip

- If you are unsure about the realism of your milestone planning get support from your Coach
- Plan time buffers for activities to compensate for unpredictable influences or where there are big difference between the duration and the actual effort of a task
- Suitable software packages help track the current status of the activities and the Time Plan quickly and enable faster reaction in the event of deviations
- In order to guarantee a successful completion owners of individual activities often assign time reserves that are too generous during planning (often 50 - 200% more than required)
- You should agree a reduction of these time reserves with the owner and instead build an overall buffer at the end of the critical path; this way individual activities are conducted more quickly; the overall buffer serves the purpose of reaching the planned end date and is not just squandered away
- You should always visualize deviations from the Time Plan as “should be” – “as is” lines

Agile Project Management

Term
Agile Project Management, Agile Development, SCRUM, Agile Innovation

When
Through the entire project or in a targeted way in ANALYZE and DESIGN

Goal
Reduce complexity in development projects, to support implementation in a quick, inexpensive yet high quality way

Procedure
Agile Project Management & SCRUM were developed in the early 90's in the USA by software engineers and are strongly characterized by Lean approaches, and process thinking as propagated by Deming and the Toyota Production System. As a kind of "Lean" project management, Agile project management looks to address complexity of developments and/or late changes of customer specification as the project progresses.

Compared with DFSS, Agile Project Management & SCRUM includes neither tools nor clearly defined phases or steps and should thus be considered rather as a frame comprising the following elements:
- Principles
- Values
- Meetings
- Roles
- Artifacts

1. **Principles:**
   - Product vision:
     Every project begins with a product vision (project and development goals). This vision is formulated when the team meets for the first time

- Sprint:
  The project works in development cycles (so-called Sprints). A set number of requirements is developed and tested in each Sprint. The requirements are fulfilled in such a way that the system is "potentially shippable", i.e. it can actually be operated straight away. There is consequently no piloting version, rather piloting, as such, takes place throughout testing in the Sprint. A Sprint usually takes between one and four weeks. Sprints are always of equal duration (they are "time-boxed"). Time Boxing is an important principle in Agile and means, for instance, that for a Sprint Planning Meeting, four hours is set aside and that after this time the meeting is stopped. On one hand this contributes to stronger meeting discipline, and on the other, the team becomes better at estimating how much time is required for processing the tasks (the so-called Velocity)

2. Meetings:

- Product Planning Meeting:
  Product Owner, Team and SCRUM Master meet for the first discussion of the product vision; they go through relevant customer requirements (called "User Stories") in a central document (the Product Backlog); the Product Backlog is a simple list of customer requirements which are sorted according to priority

- Sprint Meeting:
  Before each new Sprint a preparation meeting will take place with the Product Owner, the team and the SCRUM Master; the team agrees on the number of User Stories which can be implemented during the Sprint

- Daily SCRUM:
  The team meets every day for 15 minutes (time-boxed). Every team member answers the following questions:
  – What did I do yesterday?
  – What am I going to do today?
  – Are there any obstacles/problems?

- Review and retrospective:
  After each Sprint the Product Owner and customer(s) will receive a short presentation of results (Review) with appropriate feedback being provided; after every Sprint the team meets with the SCRUM Master and conducts a lessons learned review(the so-called Retrospective) with a view to prioritize a few key activities for the next Sprint

3. **Roles:**
   - Product Owner:
     – Is in contact with the customer i.e. is their voice
     – Compiles the customer needs in the Product Backlog and updates it
     – Monitors results against objectives and goals
     – Attends the Reviews and is always available for questions by the team
     – Makes necessary decisions
   - SCRUM Master:
     – Ensures the adherence to the SCRUM rules and principles
     – Removes obstacles, facilitates, coaches, asks questions and negotiates
     – Is not the Project Leader
   - Team:
     – Drives the actual development ahead
     – Should cover versatile competence fields
     – Works in a self-organized way

4. **Artefacts** (the documents used in team work):
   - Product Backlog:
     – Prioritized list of customer needs and/or functionalities which are to be implemented during the development project
   - SCRUM Backlog:
     – Abstract from the Product Backlog, which is to be tackled in the next Sprint and is usually visualized on a Pin Board, the so-called Sprint Board
   - Product Burndown Chart:
     – A Chart, on which the completed "Story Points", i.e. the points of a completed task are deducted from the entirety (of Story Points) of the current product release or Sprint, showing a curve representing how much work remains to be done

➾ **Tip**

- SCRUM is perfectly suitable for managing early prototyping activities at the end of the Analyze phase
- Moreover, SCRUM is also suitable for controlling the entire detailed design

- Use SCRUM early on and in iterative optimization loops in order to combine, test and coordinate the defined design elements of the detailed concept in the Design Tree
- By doing this, potential weak points in the detailed concept can be recognized and corrected before the actual prototyping takes place
- You can also use SCRUM for accelerating DFSS IT projects, especially in the ANALYZE and DESIGN phases, in order to achieve quicker results
- A necessary prerequisite for the successful application of SCRUM is an interdisciplinary team with physical proximity and decision-making powers
- The tools and methods of the DMADV methodology remain relevant within Agile project management since they provide practical support to the approach

Illustration: SCRUM process

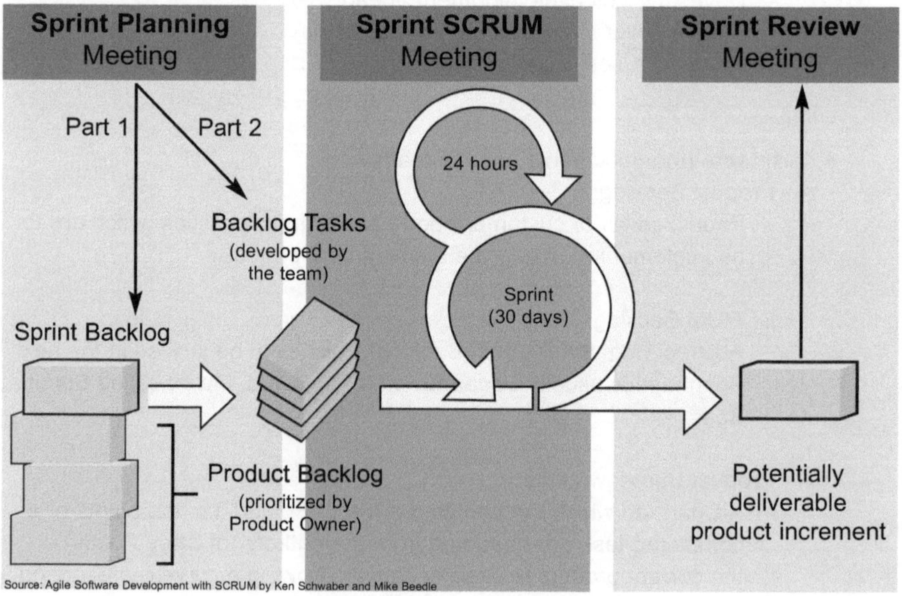

Source: Agile Software Development with SCRUM by Ken Schwaber and Mike Beedle

Resource Planning

Term
Resource Planning

When
DEFINE: managing the project and tracking it in MEASURE, ANALYZE, DESIGN and VERIFY

Goal
- Discover resource requirements
- Identify necessary resources and ensure availability for the project

Procedure

1. Determine and/or check team composition and answer the following questions:
   - Which sub-processes are involved? Who are the people involved?
   - Which areas are affected by the product and/or process innovation?
   - A SIPOC can help with these questions
   - Which specialists and methodological experts are required for processing the project?
   - How long should be allowed for team sessions and for the time between the sessions?
   - Who are the team members representing the Six Sigma organization (MBB, BB, GB)?
   - Are the right members in the team from a functional and hierarchical perspective?

2. Determine the resource need per team member based on the relevant activities and associated times. Ensure availability of the resources and actively manage them

3. Set realistic timeframes with the help of the following questions:
   - To what extent are the team members released from day-to-day business?
   - Apart from the core team, who else is needed and to what extent?
   - Who is the contact partner?

- Is external support required?
- When are vacation periods and/or who is on vacation when?
- Who is absent when (conference attendance, sales activities, etc.)?
- How can we obtain suitable replacement experts to cover for holiday absences etc?
- What is the estimated running time of the development project?

4. Define responsibilities (RACI Chart)

➾ **Tip**
Use a RACI Chart in order to determine the roles and responsibilities of those associated with the project work.

RACI Chart

Term
RACI Chart, definition of responsibilities

When
DEFINE: managing the project; whenever activities need to be assigned to individuals

Goal
- Define responsibilities for individual major tasks
- Avoid double work and inefficiencies in communication

Procedure
1. Identify/list all activities and all those involved

2. Clarify and determine the RACI roles for each activity:
   a. Responsible
   b. Accountable
   c. Consulted
   d. Informed

3. For each activity allocate names against R, A, C & I as appropriate, bearing in mind that only one person should be held Accountable per activity

4. Eliminate gaps and overlaps in responsibility

Mapping RACI roles on the following page

Mapping of RACI roles

RESPONSIBLE (R)	**"The Doer"** • Person(s) who complete(s) a task • Responsible for an action and/or for introducing an action • Responsibility is set by the person held accountable. (Accountable) • Activities/tasks can have several people who are responsible
ACCOUNTABLE (A)	**"Vouches for Others"** • Person who is ultimately responsible for completing a task and who can be held accountable • This individual possesses a veto • Activities/tasks can only have one person who is accountable
CONSULTED (C)	**"Consultant"** • The person(s) who is/are consulted in the run up to a final decision
INFORMED (I)	**"Information obligation"** • The person(s) who has/have to be informed about a decision or an action

Illustration: RACI Chart – MEASURE

R = Responsible
A = Accountable
C = Consulted
I = Informed

		Management	Marketing	R + D	Quality Management	Procurement	Production	Sales	Customer Service
		Ms. G.	Ms. M.	Dr. F.	Dr. Q.	Mr. E.	Ms. P.	Prof. Dr. V.	Ms. K.
MEASURE	Identification of target market	A	R					C	
	Collection of customer requirements	I	R		I				I
	Derivation of CTCs	I	R					R	C
	Benchmarking	I	R					R	C
	Identification of improvement potentials	I							
	Technical benchmarking		I	A	R		C	I	
	Derivation of target values			A	R				

Budget Planning

Term

Project Budgeting, Cost Planning, Project Budget Planning, Budget Planning

When

DEFINE: managing the project; tracking it in MEASURE, ANALYZE, DESIGN and VERIFY

Goal

- Find out the budget needed for the project
- Enable monitoring of budget adherence

Procedure

- List all budget and non-budget related cost categories
- Explain necessity for the respective effort
- List the point in time when any given cost will appear on the books
- Find out the costs per category
- Add up all costs to get the planned total budget

The required project budget is identified with the help of a cost statement.

Cost planning and monitoring, listing of items

					Planned (target)				Current (as is)				Deviation	
	Category	Detail What? For what? Who?	DFSS Phase (DMADV)	Project activity	Cost period	Net amount $	Before tax $	Total tax $	Cost period	Net amount $	Before tax $	Total tax $	As is – target	Ex-planation
Budget related (BR)	1. External services													
	2. Materials and tools													
	3. Travel expenses													
	4. Investments													
	5. Additional costs (e.g. rent, soft-ware licenses)													
Non BR	6. Internal costs (ac-cording to ICR*)													
					Sum:									

* ICR = Internal cost rate by employees taken from one department [$/h]

The adherence to the project budget is monitored throughout the entire project duration by the Project Leader who compares the actual costs with the original estimates. Potential differences must be explained.

➾ **Tip**

- Work together closely with the Sponsor when planning the budget! He/she usually ensures the availability of the project budget
- Consult a financial controller when calculating the internal cost rates
- In determining the project benefits you must consider the budgeted project costs
- Take into account the phases of a DFSS project and the costs related to these when listing the cost categories:
  - Delivery of a market research study
  - Delivery of trend analyses
  - Delivery of Gemba studies
  - Use of focus groups
  - Manufacturing of prototypes
  - Use of simulation tools etc.

Stakeholder Management

Term
Stakeholder Management

When
DEFINE: managing the project; update in MEASURE, ANALYZE, DESIGN and VERIFY

Goal
- Determine the attitude of leaders and other impacted parties, then take actions to win them over
- Generate support for the project
- Identify resistance and overcome it

Procedure
1. Create a list of individuals for the Stakeholder Analysis with the help of the following questions:
   - Who are the people impacted by the project?
   - Which departments and interfaces are involved?
   - Looking at individuals outside the project, who might be interested in its outcome?
   - Who feels positive about this project?
   - Who will profit by the success of the project?
   - How can the Stakeholders contribute to the project success?
   - Can some particular relationships be leveraged to help the project in a positive way?

2. Determine the attitude of each respective Stakeholder to the project

3. In case of resistance, determine the type of resistance (TPC Analysis: technical, political or cultural) which may cause a negative attitude. Do this for each Stakeholder

4. Assess the actual or expected behavior for each individual; note the perceived viewpoint of each person (o) as well as of the target area (x); visualize the gaps; find links between the players: Who influences whom?

5. Create an effective system for communication and strategy to influence others

Tip

- Conduct Stakeholder Analysis with the Project Sponsor – he/she will support you in identifying and analyzing the relevant Stakeholders and in identifying the appropriate actions
- Document the Stakeholder Analysis for future use in the project;
- However, the Stakeholders' names and the analysis of them, should be kept confidential – it could lead to misunderstandings and increased resistance.
- Update the Stakeholder Analysis frequently during the course of the project: Always check the success of planned actions and examine if "new" Stakeholders ought to be added
- Stakeholders can be identified both within and outside the organization

Illustration: Stakeholder Analysis

	Opinion towards project					Type of resistance		
Stake-holder	--	-	0	+	++	Technical (I can't do that!) Lack of abilities, lack of critical resources	Political (I'm not allowed to do that!) Concerns with regard to power or authority	Cultural (I don't like that!) Norms, mentality, habits, language

Illustration: Influencing Strategy

Stakeholder	Topics/concerns	Levers	Influence (by whom)
Mr. A	SAP introduction	Resources	Sponsor
Mr. B	Works council	Company agreement	Sponsor
Mr. C	Education	Training	Sponsor

Illustration: A Communication Plan

Content	Purpose	Recipient	Responsible person	Media	Times	Status
Message	Why is this message to be sent to the recipient?	Who is to receive the message? (RACI)	Who is responsible for the communication? (RACI)	What media will be used?	When or how often is the communication?	Was the message sent as planned?
				Examples: Email, Gate Review, posting, orally, newsletter, "Elevator Speech"		

By drawing up an "Elevator Speech" within the team it is ensured that any external communication is characterized by a uniform image of the project. Such an "Elevator Speech" should

- be short and precise (max. 90 seconds)
- contain only positive information
- be aligned towards increasing acceptance of the project

Additional means of communication and communication channels should also be used in a targeted way:

- Newsletter
- Information markets
- Corporate magazines
- Exchanges of experience
- Departmental sessions etc.

Tip

- Establish a relationship of trust with important contact partners and cultivate it
- Keep feedback loops short (avoid "Chinese Whispers Effect") and counteract rumors promptly
- Select formal uniform and simple communication methods; this increases understanding and therefore acceptance of the project
- Make sure information on the project is published regularly

Project Communication

Term
Project Communication

When
DEFINE: managing the project; update in MEASURE, ANALYZE, DESIGN and VERIFY

Goal
- Ensure regular and structured communication within and between the project team and individual sub-teams
- Ensure that the Sponsor is continuously informed of project's progress
- Regularly inform employees working in the project domain and create transparency

Procedure

1. **Internal:**
   a. Define necessary communication frequency for the project
   b. Plan the timing with the team members and the Sponsor during the Kick-off Meeting

2. **External to the project environment:**
   a. Set communication goal and determine recipient
   b. Set contents and media methods

Tip
- Use the Kick-off Meeting to set the communication pace for the project team
- Develop an "Elevator Speech" with your team. The so-called Elevator Speech will enable you and your team to describe the necessity and the goal of the project in a concise and clear way: contains the idea, benefit, current status and any desired support

Risk Management

Term

Risk Management, Risk Assessment, Project Risk Assessment

When

DEFINE: anticipating risks at the beginning or the planning stage of the project. Should continue for the duration of the project

Goal

- Identify risks and evaluate them with respect to the likelihood of occurrence and their potential influence on project success
- Continually manage risks
- Identify actions and implement to mitigate risks

Procedure

1. Collect potential project risks from the team (e.g. with the help of Brainstorming). The following risk categories should be taken into consideration:
   - Commercial risks (e.g. uncertainties with respect to investments required for closing the project successfully)
   - Economic risks (e.g. market uncertainties and activities of competitors)
   - Political risks (e.g. pending legislation changes)
   - Technical risks (e.g. uncertainties with respect to technical feasibility)
   - Change Management risks (e.g. lack of acceptance among important individuals or groups)

2. Estimate the probability of occurrence of the identified risks and their potential impact on the project success and map it using a matrix (cf. Risk Management Matrix on the following page)

3. Where required, take appropriate actions for reducing or eliminating risks; actions can target reducing the occurrence probability and/or the severity of impact

Illustration: Risk Management Matrix

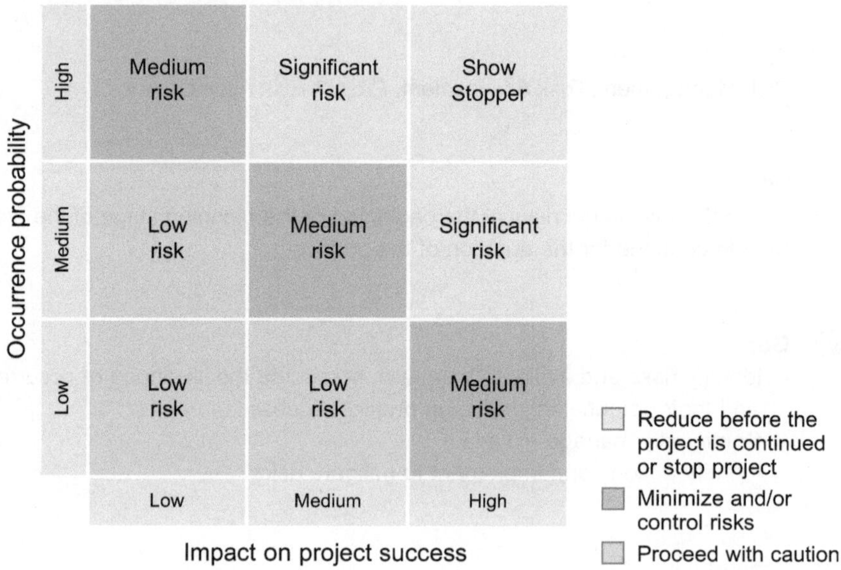

Tip

- When collecting potential risks you should also make use of the information collected in the Multi Generation Plan about required technologies/platforms, resources and sales channels
- It is the Project Leaders' responsibility to draw peoples' attention to the potential risks of a project and they should speak directly to the Sponsor about significant risks so that prompt and appropriate action can be taken to mitigate against them

"How will the risks to project success be minimized?"

Occurrence probability ↓ / Impact on project success →	Low: Risk	Low: Action	Medium: Risk	Medium: Action	High: Risk	High: Action
High	… …	… …	… …	… …	- Silo thinking obstructs cross-departmental cooperation	- Stake-holder-Analysis - Top Management Workshop - Communication Plan
Medium	… …	… …	- Liquidity problems by trade partners	- Targeted selection of partners and contractual protection	- Logistics constraints	- Clearly defined Supply Chain process with responsibilities - …
Low	… …	… …	- Demand volatility	- Consistent integration of sales - Setup of an early warning system	… …	… …

Occurrence probability (Low – Medium – High)

Impact on project success (Low – Medium – High)

Risk	Action	Who?	With whom?	Antil when?
Silo thinking obstructs cross-departmental cooperation	Set up steering committee, in which executives from all relevant areas are informed regularly about the project direction and make joint decisions about future procedures	Mr. X (Project Sponsor)	Mr. Y (member of the executive board)	02/10 (end of DEFINE phase)
Logistics constraints	...	...	...	...

Specifics for process development

When developing new processes the following questions are helpful in identifying potential threats:

- Are there centralized or corporate level processes that influence this project?
- Are there interface related projects or activities that might influence the project?
- Which data bases, software solutions, or process development tools are used in the affected areas and could they be adjusted?
- Are there market changes, economic trends, or strategy changes, which can have an influence on the process to be developed?
- Are there changes among customers that could have an influence on the process to be developed?

Kick-off Meeting

Term
Kick-off Meeting

When
DEFINE: managing the project; first project session for the entire team

Goal
- Introduce the project to the project team and make its purpose clear to everyone
- Inform the team members about their role within, and the nature of, a DMADV project
- Coordinate creation of the Project Charter and focus of the project team
- Introduce and start applying the tools of the DEFINE phase with the team
- Coordinate future activities and create an initial Time Plan

Procedure
1. Coordinate the dates with the Sponsor and ensure his/her participation
2. Develop the agenda with input from the Sponsor and the Master Black Belt
3. Invite all team and extended team members; the official invitation to the Kick-off event should contain the following:
   - Short email with general information about the project and the Kick-off event
   - Agenda (with times)
   - Goals of the session/day
   - List of participants
4. Book suitable rooms i.e. bright, comfortable, spacious and have sufficient amounts of facilitation material available (e.g. markers, meta plan boards, flip charts with paper)
5. Conduct the meeting according to the agenda
6. Take notes and publish them as soon as possible after the meeting

Gate Review

Term
Gate Review, Phase Check, Phase Transition, Phase Approval

When
At the end of each phase: DEFINE, MEASURE, ANALYZE, DESIGN and VERIFY

Goal
- Inform the Sponsor about the actions and results of each phase and answer any questions that he/she may have
- Present recommendations about the future direction of the project and pose any questions that need answering e.g. about scope issues
- After a full review of the information decide about the future course of the project e.g. Go/No Go

Procedure
1. The results are presented completely and comprehensibly

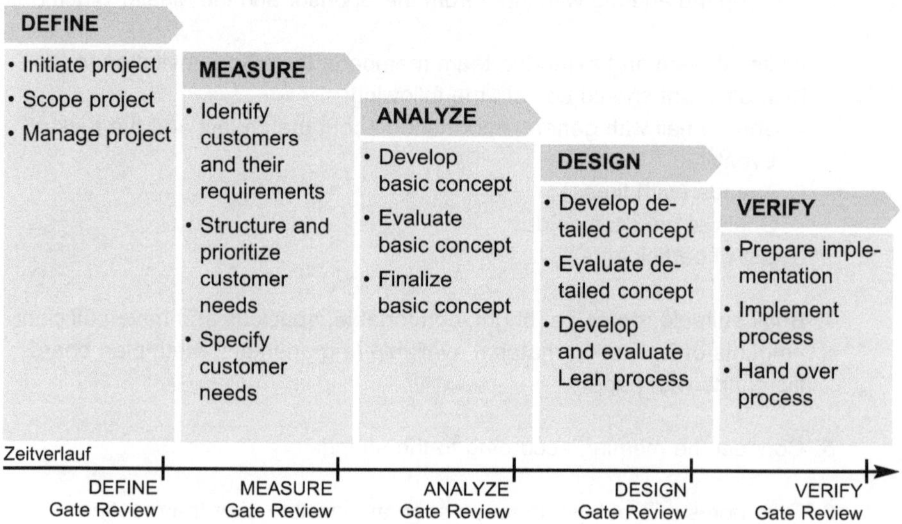

2. The Sponsor checks the current status of the project according to the following criteria:
   - Completeness of results
   - Probability of project success
   - Resource allocation in the project

3. The Sponsor decides if the project can enter the next phase

The following questions will help the Sponsor when carrying out the DEFINE Gate Review:

Initiate the project:

- What idea(s)/problem is (are) the reason for the project?
- What is the goal of the project?
- Who are the target customers?
- What is the estimated benefit of the project and how was it calculated?
- What does the market and competitive situation look like?

Scope the project:

- Which aspects are included in the project and which are not?
- How does this influence other projects?
- What is the long term development goal being pursued? What is the vision?

Manage the project:

- Who are the team members and why were they selected?
- How were roles and responsibilities defined in the core team?
- What activities and resources are required?
- How large is the time and budget requirement?
- On what basis has data been collected?
- How much acceptance for the project is there?
- What are the potential risks?
- What actions will be or have been introduced?

DESIGN FOR SIX SIGMA^{+LEAN} TOOLSET

MEASURE

D M A D V

Central Question and Objective of MEASURE Phase

- The MEASURE phase of the DMADV approach answers the question: "Who is the central customer of the project and what specific requirements does he/she require of the product or process?"
- Objective:
  - Identify customers and their needs
  - Interpret and translate customer needs into specific requirements of the product or process
  - Set relevant output measurements, target values and specifications
  - Reflect defined objective, alignment and scope of the project

Meaning of the MEASURE Phase

- Target and interested groups who are important to the product or process development are identified and addressed
- Mix of targeted methods used for collecting customer needs which will assist in the identification of product or process features which will enable the company to offer customers a Unique Selling Proposition
- Transfer the language of the customer into the language of the company, in order to create appropriate measurements, relevant target values and specifications, and to identify potential development conflicts early on

Procedure in the MEASURE Phase

- Identify customers
- Segment customers
- Select and conduct research methods
- Collect Voice of the Customer
- Derive customer needs
- Structure customer needs
- Categorize customer needs
- Prioritize customer needs
- Derive CTCs and measurements
- Conduct Benchmarking
- Prioritize measurements, determine target values and specifications
- Set quality key figures
- Estimate potential risk
- Collect customer feedback on the Design Scorecard

D
M
A
D
V

M.1 Identify Customers and Their Needs

- Who is the central customer of the project and how can her/his needs be reliably identified?

M.2 Structure and Prioritize Customer Needs

- Where do gaps exist?
- Where must the collection of needs be detailed further?
- How can needs be categorized into Basic, Satisfiers and Delighters?
- How can Satisfiers and Delighters be prioritized?

M.3 Specify Customer Needs

- What measurements, target values and specifications should be be assigned to individual customer needs?
- Are any potential development conflicts revealed based on the defined measurements?

Result MEASURE

- Target customer(s) defined in line with project scope (MGP)
- Customer needs collected using suitable methods
- Customer needs categorized, prioritized and specified

MEASURE Gate Review/Phase Closure

M.1 Identify Customers and Their Needs

SIPOC	Alternative Customer Categories	Market Research	Customer Segmentation	ABC Classification	Portfolio Analysis
Customer Scoring	Customer Value Measurement	System Value Chain	Life Cycle Roadmap	5 W + 1 H Table	Active and Passive Research Methods

M.2 Structure and Prioritize Customer Needs

Customer Needs Table	Affinity Diagram	Tree Diagram	Kano Model	N/3 Method	Nominal Group Technique
Analytic-Hierarchy Process	Simple Pair Comparison				

M.3 Specify Customer Needs

CTCs and Measure-ments	Bench-marking	QFD 1 (House of Quality)	Design Scorecard	Operational Definitions	Quality Key Figures

M Gate Review

Identify Customers

Term
Customer Selection

When
Before the project starts, DEFINE, MEASURE

Goal
Identify relevant target markets and target customers

Procedure
The entire market is segmented and a target market segment is determined. Relevant target customers in this target market segment are identified using suitable methods.

Mapping Market Segmentation

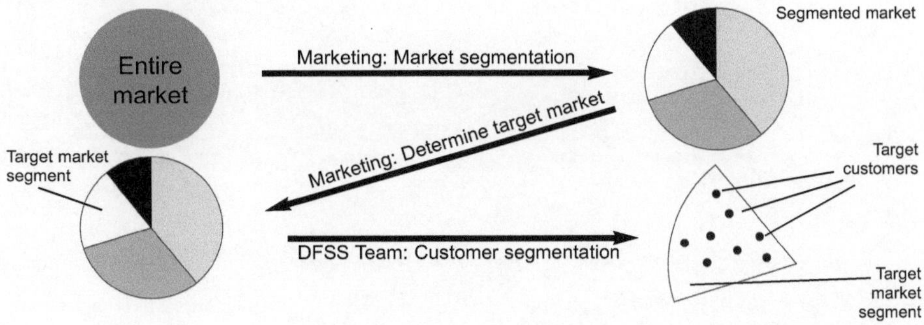

The target customers are characterized and/or segmented with the help of set criteria.

Even before a DFSS project is launched, ideas about relevant target customers and/or the target market could exist. This might include thoughts about target mix of new and existing customers. Marketing and Sales can prepare and provide the necessary information for consideration. A corresponding analysis of the market and the competitors is not conducted by the DFSS project team.

Prior to any further segmentation of customers the project team should think about relevant internal and alternative customer groups which it might be able to integrate into future development work respectively.

SIPOC

Term
SIPOC (Supplier, Input, Process, Output, Customer)

When
Before the project starts, DEFINE, MEASURE, identify customers and their needs

Goal
- Identify relevant internal interest groups, which must be considered for a process development
- Focus on internal suppliers and customers of a process

Procedure
1. Set start and end points of the process to be developed
2. Map the process to be developed in basic steps
3. Identify essential suppliers of required inputs and customers of outputs produced
4. Define group of interested parties, which will consist of identified suppliers and customers, internal and external, who are to be considered in process development

Illustration: SIPOC

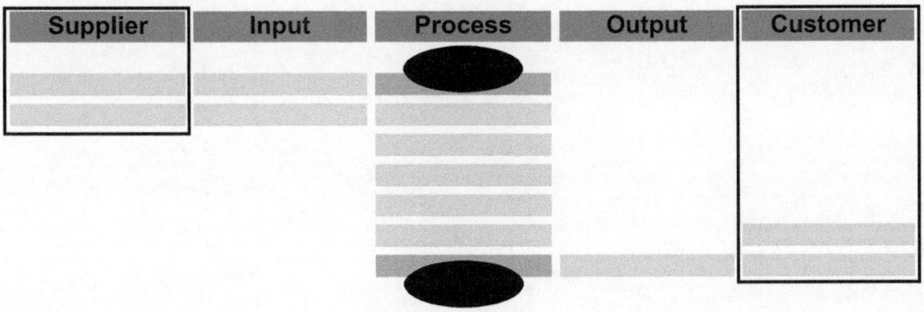

Tip
When putting the core team together you should make sure that the internal suppliers and customers defined in the SIPOC are represented.

Alternative Customer Categories

Term

Customer definition, customer categories

When

Before the project starts, DEFINE, MEASURE, identifying customers and their needs

Goal

- Extend the classic customer perspective in order to focus on further customer groups over and above the main revenue mainstay of sales resulting from an ABC classification
- Take into consideration allegedly uninteresting customer groups in order to derive new product or process ideas by broadening perspectives

Procedure

A distinction can be made between the following alternative customer categories:

1. Future Non-customers:
   - Prepare for pending defection as soon as a good alternative has been found
   - Relevant considerations:
     Why do customers want to leave and what are the most common reasons for this?

2. Denying Non-customers:
   - Do not use the offers of the industry since they do not appear to be attractive, affordable or useful from the customer's point of view
   - Relevant considerations:
     Why don't these customers use the products and services of the industry and which similarities are observed?

3. Undiscovered Non-customers:
   - Customers who haven't been considered in the industry as potential target customers so far since satisfaction of their needs has been attributed to other industries

- Relevant considerations:
  How have the needs of customers been structured and can common features be defined which make the customer interesting?

4. Groups of Buyers:
   - Purchasers, users and influencers are included as groups of buyers who are directly or indirectly involved in the purchase decision; however, a company usually focuses strongly on a special group of buyers
   - Example pharmaceutical industry:
     Companies from the pharmaceutical industry primarily focus on group of buyers with a strong influence (medical doctors); some companies also integrate groups of users (medical staff)

5. Heavy User (also Heavy Consumer or Buyer):
   - Customer groups which regularly buy above average quantities of a product and/or a service
   - Among this are hospitals, hotels, canteens and also families with several members
   - It should be noted, however, that Heavy Users are not necessarily loyal and/or brand-loyal; their behavior is rather characterized by a multi-brand strategy
   - Relevant considerations:
     How can the next generation of the product and/or service be designed in a way that these profitable customer groups develop into loyal customers?

6. Lead User:
   - Customer groups buying and using new products long before other customers recognize their need
   - They are innovative and visionary and function as leaders and trendsetters in their segment
   - In the automotive area these might be employees from a motorsport domain (technicians, constructors or mechanics); for the sporting goods industry these might include endurance and extreme athletes
   - They are outside the target market and are often confronted with more extreme conditions than the general public
   - They are technologically gifted and have an urgent need to improve their product or service
   - The Lead User expects that the product fulfills his specific requirements in the sense of a benefit innovation and solves his/her problem

- Information from Lead Users is very valuable throughout the product development since their behavior exceeds that of an "average customer"
- More innovative ideas can be generated from this which could provide the basis for recognizing and implementing the trends and requirements of tomorrow and the day after tomorrow into exciting new products

➾ **Tip**

- When Heavy Users are observed, the following procedure should be pursued:
  1. Describe the relevant Heavy User first
  2. Describe the respective Non-user afterwards
  3 How could a Non-user develop into a User?
- Apart from the presented alternative customer categories you should also think about potential alternative industries; a company doesn't only compete with direct competitors in its industry but also with companies in other industries offering alternative products and/or services
- Alternatives can be products and/or services having another form and function but serving the same purpose (e.g. cinema and restaurant visit); take into account the respective advantages of alternative industries and try to integrate and/or extend these further

Specifics for process development

- Transfer the identified customer groups onto internally oriented process developments; an example of future non-customers might reveal themselves as unmotivated users of an IT system
- In such cases you should also regularly question their motives in order to take these into account throughout the process development

Market Research

Term
Market Research

When
Before the project starts, MEASURE, ANALYZE, identifying customers and their needs

Goal
- Generate relevant project ideas in an early idea generation phase
- Make customer needs recognizable and measurable in the MEASURE phase
- Test product concepts and features in the ANALYZE phase and ask for impressions by customers and improvement ideas

The following applies in general:
First of all, qualitative market research methods should be used for a better understanding of customers who will later be the subject of quantitative market research methods. Qualitative market research collects subjective data and primarily deals with behavior, attitudes and values whilst quantitative market research primarily deals with objective data which can be measured and expressed by figures.
The reliable preparation, execution and evaluation of market research is essential for ensuring success.

Procedure

1 Set the examination goal

2 Determine the object of examination

3 Draw up the market research briefing
- Describe a product and its most important functions
- Describe a market segment and the customer group to be questioned
- Describe the requested information and the information goal precisely
- Compare the information goal with the already developed objective of the development project

4 Determine examination method and sample size

5 Analyze potential risks and sources for misinterpretation and distortions (Bias)

6 Decide if market research should be conducted with or without external support

7 Formulate a set of questions and an evaluation schema

8 Conduct test run of market research

9 Accompany market researcher, get an overview of test results and be prepared to adjust as necessary

10 Get an overview of the final results and evaluate them

Mapping:
Checklist for preparing, conducting, and selecting market research

1. General Matters	• Short description of development project (starting point, problem and objective) • Current phase of project • Peculiarities of market and/or customer environment • Client of market research
2. Objective	• Why is market research to be conducted within the scope of the project? • Important product information: - Unique Selling Proposition and/or differentiation potential - Core functions - Competitor products - Potential for cross innovations
3. Basic Conditions	• Start of market research • Planned closure of market research • Definition of target market • Definition of target group • Segmentation criteria • Incentives and salaries
4. Information Need	• What are the general information goals? • What information is to be collected? • Which information is to be tested and/or secured?

Tip

- When you plan conducting a market research study by yourself consider about an online approach which might offer:
  - Savings on printing and postage costs (postage, return postage, logistics handling)
  - Minimal or no cost for data collection since information has already been entered by the user directly into the system
  - An increase in data quality by reducing transmission errors, since no manual collection and/or transmission of answers is necessary
  - Early plausibility check of answers
  - Shorter response times of individuals surveyed
  - Opportunity to remind individuals via email in case surveys haven't been completed
  - Easy access to respondents using email
  - Reduction in duration of survey
  - Increased response rates
- Also use the advantages of online market research when you cannot contact the participants in the survey directly via email; use conventional channels to contact participants and inform them of the surveys, giving a web address and access code
- Participants who cannot participate in the online survey will be sent a copy of the survey used via the conventional channel; response data will be entered manually and can be combined with the data from the online survey

Customer Segmentation

Term
Customer Segmentation

When
Before the project starts, DEFINE, MEASURE, identify customers and their needs

Goal
- Segment customers with the help of selected criteria
- Derive strategies for the future integration of the customers within the scope of the development project

Procedure
Criteria for customer segmentation classify the target customers which are relevant for the project according to general and specific customer features.

When general criteria are applied, the segmentation is conducted on the basis of immediate customer features, i.e. irrespective of products or the purchasing situation and/or area of application.

Specific criteria, however, take into consideration existing relationships between customers, products and/or the purchasing situation and/or area of application.

See next page for example

Illustration: General and specific segmentation criteria

Measurement of Features		Customer Features	
		General (independent of product/situation)	Specific (product and/or situation-related customer features)
	Directly observable and/or measurable	• Geographic features (size of residence, region, city/rural region) • Demographic features (age, sex, personal status, household size etc.) • Socio economic features (income, occupation, education)	• Frequency of use • Purchase volume • Brand and supplier loyalty • Propensity to innovate • Application area
	Only indirectly measurable (derived)	• Personality features (inclination to take risk, accepted values, decision type) • Lifestyle (activities, interests, opinions)	• Purchasing motives, purchase expectations • Attitudes • Perceived benefit • Purchase intentions • Preferences • Elasticities

Directly observable criteria can be identified with little effort whereas indirectly observable criteria can usually only be derived via indicators and/or time-consuming and expensive market research.

The selection of segmentation criteria should be aligned towards the following requirements:

- Identifiable:
  The actual existence of market segments that differ significantly from each other should be verifiable with the selected segmentation approach

- Accessibility and responsiveness:
  The segments must be accessible with given communication and distribution methods; moreover, a continuous integration into the project, of selected customer groups, from the observed segment should be possible in order to obtain feedback on development work; the use of relevant marketing instruments should also be evaluated as a marketing mix throughout the implementation of the product on the market by customer feedback

- Stability:
  The segments should display a level of stability over time – they should at least remain stable over the time period from the segmentation, through the collection of customer needs, to the implementation of the product development

The following opportunities for customer segmentation can be used and developed as project support:

	Retrospective	Prospective
Quantitative	• ABC Classification • Portfolio Analysis • Scoring Method • Customer Value Measurement	• Portfolio Analysis • Scoring Method • Customer Value Measurement
Qualitative	• Scoring Method • Customer Value Measurement	• Scoring Method • Customer Value Measurement

ABC Classification

Term
ABC Classification

When
Before the project starts, DEFINE, MEASURE, identify customers and their needs

Goal
Focus on target customers generating the biggest turnover share

Procedure
Existing customers are assessed by considering their respective share of the company turnover. The basis is the turnover from the past year or the forecast turnover for the current year. If turnover fluctuates very strongly, the average turnover of the past three years can be calculated and taken as the base. Contribution margins can also be applied as a basis for comparison.

Illustration: ABC Classification

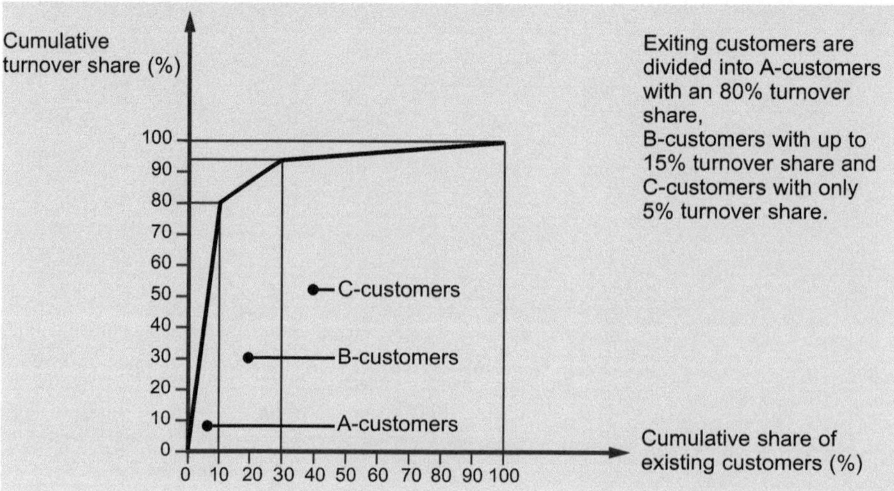

Specifics for process development

Use the ABC Classification also within the scope of process developments and replace the turnover component by relevant alternative benchmarks:

- Frequency of use and/or access
- Demand quantities and frequencies
- Extent of damage
- Scope of insurance
- Material values etc.

Portfolio Analysis

Term
Portfolio Analysis

When
Before the project starts, DEFINE, MEASURE, identify customers and their needs

Goal
Add and extend relevant information on potential market growth to the ABC Classification

Procedure
A third dimension is added to the two-dimensional ABC Classification, e.g. the market growth of the trade channel.

Mapping Portfolio Analysis

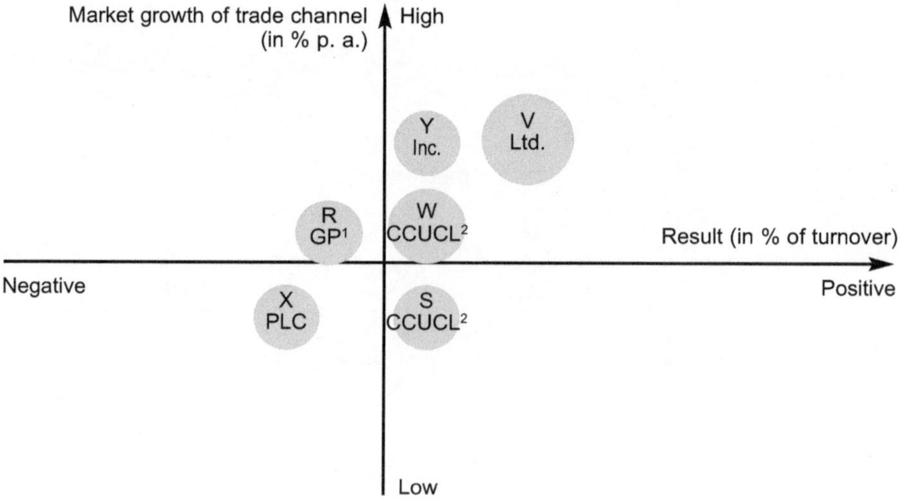

The size of the circle represents the magnitude of a third dimension.
[1] GP = General Partnership
[2] CCUCL = Company constituted under civil law

Customer Scoring

Term

Customer Scoring, Scoring Methodology, criteria based customer segmentation

When

Before the project starts, DEFINE, MEASURE, identify customers and their needs

Goal

- Prepare customer related information systematically
- Divide customers with the help of selected criteria into segments
- Derive strategies for individual segments in order to exploit and increase customer potential to a maximum

Procedure

The Scoring Methodology, a multi-dimensional procedure, makes use of a catalog of individually defined criteria, which will help the evaluation and segmentation of customers take place. By doing this not only can company and project specific influencing factors be assessed but also market and customer specific influencing factors. The weighing and evaluation of both, takes place in the context of past, current and future customer circumstances.

The Scoring Methodology can be conducted with the help of a criteria based selection procedure. This way individual customers or groups of customers can be compared with the help of weighted criteria. The selected criteria are of a qualitative nature and require a subjective estimate.

1 Define and specify relevant evaluation criteria for customer segmentation; align the perspective towards company and project specific as well as market and customer specific criteria

2 Prioritize defined evaluation criteria for preparing the selection procedure with the help of the pair comparison (cf. MEASURE "Analytic-Hierarchy Process")

3 Transfer the defined criteria, identified prioritizations and the customers to be observed into the criteria based selection procedure

4 Evaluate customers using the defined criteria:
- How strongly does the respective customer fulfill the defined criterion?
- Evaluate with 9 = strong, 3 = medium, 1 = weak, 0 = no contribution

5 After the evaluation has been conducted results can be summarized for comparison by summing weighted rankings i.e. multiplying the criteria score by the criteria weighting; from this the customers and/or customer groups which fulfill the defined criteria best are identified

		Customers			
Evaluation Criteria	Weighting	Customer 1	Customer 2	Customer 3	Customer 4
Demand volume	25%	9	0	1	1
Growth	10%	1	9	0	1
Market entry barriers	12%	0	0	9	1
Intensity of competition	10%	9	0	0	0
Loyalty	10%	3	3	0	1
Lead User Function	10%	3	0	0	0
Share of supplies	9%	9	0	1	1
...	8%	1	9	0	1
...	4%	0	0	9	1
...	2%	9	0	0	9
Sum:	100%	4.92	1.92	1.78	0.96
		51.56%	20.04%	18.58%	10.02%

Tip
- When needed add measurements and specifications (e.g. demand volume of x pieces/year) to the criteria in order to operationalize them and make them more specific to simplify the evaluation and selection procedure
- For the prioritization of evaluation criteria you should revert to a pair comparison (AHP) approach
- In this way you ensure that you compare external and/or customer related

criteria systematically and completely with internal corporate and/or project related criteria and observe the respective relevance

- Exemplary evaluation criteria:
  - Quantitative Parameter:
    - Customer turnover
    - Customer's contribution margin
    - Demand volume
    - Demand variation
    - Creditworthiness
  - Qualitative Parameter:
    - Customer loyalty
    - Cooperation potential
    - Innovation potential
    - Potential for further development
    - Development potential
- In addition to these the following dimensions of customer value measurement can be integrated albeit that they may need further specification:
  - Short-term success potential
  - Mid-term success potential
  - Strategic success potential

Customer Value Measurement

Term
Customer Value Measurement, Customer Value Management, 3 Dimensions Model, Customer Cube

When
Before the project starts, DEFINE, MEASURE, identify customers and their needs

Goal
- Collect and prepare historic and future information on individual customers, customer groups and/or the entire customer portfolio
- Position customers in a 3 Dimensions Model (Customer Cube)
- Derive areas of action in order to be able to exploit the identified customer value with the development project

Procedure
Before beginning with the Customer Value Analysis a decision has to be made about the scope of the analysis. There are two choices:

- Conduct a **partial analysis**, directed towards the company's specific objectives (e.g. customer satisfaction and turnover)
- Conduct a **complete analysis**, observing the customer value in a multi-dimensional way

Since the partial analysis represents a simplification of the complete analysis the procedure for conducting a complete analysis is explained here:

1. **Positioning of customers**

   The categorization of customer value takes place in three dimensions and is represented by a Customer Cube:

   1. **Short-term success potential**
   2. **Mid-term success potential**
   3. **Strategic success potential**

The focus is on whether a high or low potential of the individual dimensions exists. If a combination of several dimensions exists, these are weighted and the total value of the three dimensions is identified.

Mapping 3 Dimensions Model/Customer Cube

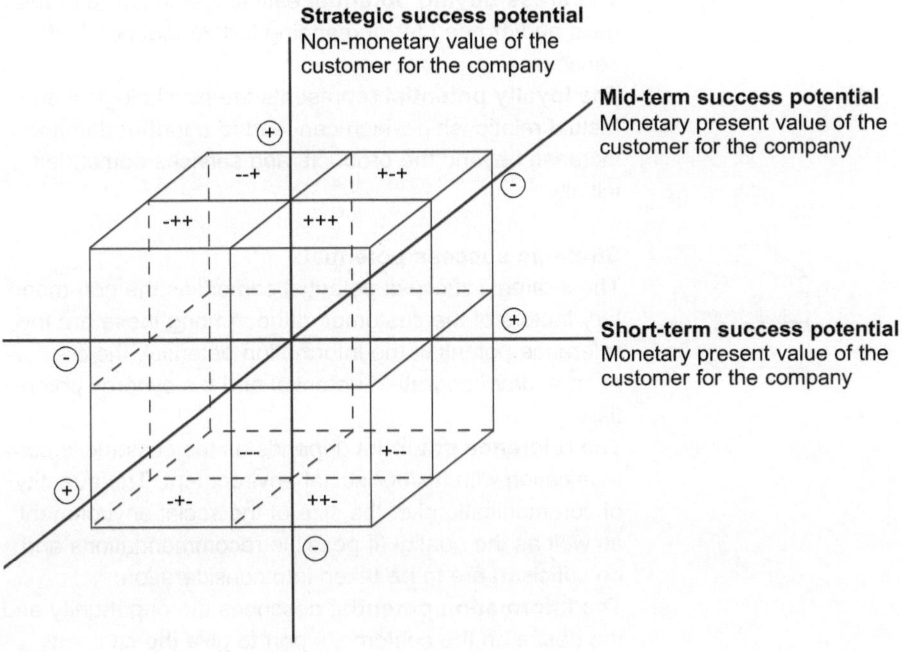

The highlighted dimensions in the Customer Cube can be specified as follows:

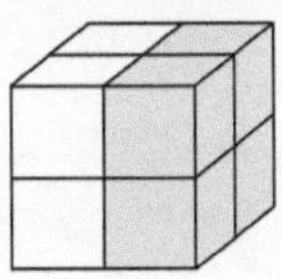

Short-term success potential:

The short-term success potential is a purely monetary variable depending on the yield potential, i.e. the customer's current monetary contribution to the company.

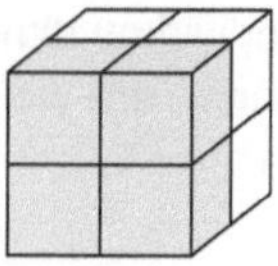

Mid-term success potential:
The mid-term success potential is also a purely monetary variable focusing on the customer's development. It depends on the development potential, the cross buying potential and the loyalty potential.
The **development potential** shows the expected contribution margin and the customer's potential growth.
The **cross buying potential** estimates the future readiness and/or need for demanding further independent services.
The **loyalty potential** represents the psychological and factual relationship which can lead to a further demand increase beyond the products and services demanded initially.

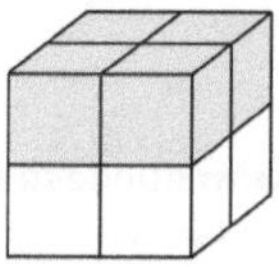

Strategic success potential:
The strategic success potential comprises the non-monetary factors of the customer value. Among these are the reference potential, the information potential, the cooperation and/or integration potential and the synergy potential.
The **reference potential** depends on the customer's communication with his/her social environment. The quantity of communication i.e. the size of the social environment as well as the quality (if possible recommendations and no criticism) are to be taken into consideration.
The **information potential** describes the opportunity and the desire on the customer's part to give the company constructive feedback on services and requests.
The **cooperation and/or integration potential** primarily refers to the B2B area indicating the potential to generate synergies and value increase by enhanced cooperation with providers and demanders within the value chain.
The **synergy potential** shows combination opportunities in the customer base.

Mapping of customer value dimensions according to market and resource potential

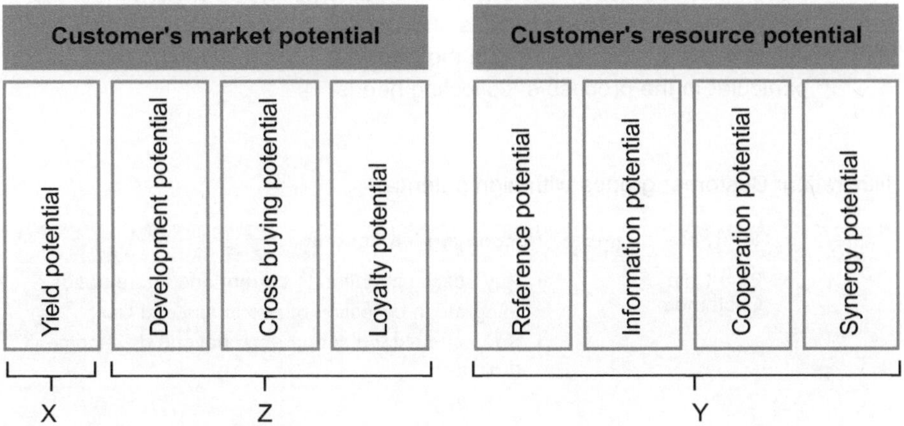

X: Short-term success potential
Z: Mid-term success potential
Y: Strategic success potential

2. **Segmentation**
After having evaluated individual customers, customer groups and/or the entire customer portfolio with respect to the three dimensions, the customers are accordingly positioned in the Customer Cube thereby segmenting customers into eight :
- Blue Chip Customers
- Opportunity Customers
- Prospective Customers
- Entertainment Customers
- Future Customers
- Running down Customers
- Selective Customers
- Disclaim Customers

The positioning in the cube enables a simple distinction into customers with respectively high and/or low potential. It must be taken into consideration, however, that the increased concentration on individuals of high potential is not recommended, since customers with a low potential are frequently more innovative and communicative e.g. with respect to efforts for optimization and development.

3. **Recommended actions**
   Recommended actions for each of the customer types are shown in the table that follows. These actions show how customers can be integrated according to their positioning in the development project in general and in particular in the process of collecting needs.

Illustration: Customer groups with high potential

x	y	z	Customer segment	Recommended actions
+	+	+	Blue Chip Customers	• Pay special attention to current and future needs • Integrate in collection of needs as Lead User • Keep and extend cooperation beyond development project
+	+	-	Opportunity Customers	• Classic Heavy User who purchases above-average quantities, however, is not necessarily loyal and/or brand loyal • Align collection of needs to how a higher degree of involvement and commitment can be promoted
-	+	+	Prospective Customers	• Pay special attention to future needs and integrate the customer in the collection of needs as Lead User • Consider more exact analysis of customer's positive development (e.g. cross buying) • Generally first priority for collection of needs
+	-	+	Entertainment Customers	• Align collection of needs in a targeted way towards the lack of growth and analyze reasons • Take insights into consideration throughout future development work in a targeted way • Increase Share of Wallet

Illustration: Customer groups with low potential

x	y	z	Customer segment	Recommended actions
-	+	-	Future Customers	• Pay special attention to current and future needs • Integrate in collection of needs as Lead User • Maintain and extend cooperation beyond the development project
+	-	-	Running down Customers	• Align collection of needs of future non-customer in a targeted way towards lack of reference and cooperation potential and analyze the reasons for this • Take insights into consideration in a targeted way throughout future development work
-	-	+	Selective Customers	• Align collection of needs in a targeted way as to how the positive indirect value contributions of the customer can also be transferred to other customers • More exact analysis of negative direct value contribution contributions in the course of collecting needs
-	-	-	Disclaim Customers	• More exact analysis if the customer is a denying non-customer who considers existing products and services as unattractive, unaffordable or useless • Direct collection of needs in a targeted way towards reasons for rejection or denial • Alternatively they can be undiscovered non-customers, customers who so far haven't been regarded as customers by the industry, since their assigned needs have been attributed to other industries. The collection of needs should be directed towards the customers' structure in a targeted way

➾ **Tip**

- Pay attention to the fact that the "soft" dimensions of customer value are assigned a higher importance in customer segmentation; only by doing this way can needs for new products and services be collected and derived
- In contrast to classic marketing strategies Running Down and Disclaim Customers need collection becomes interesting in this phase of the development project since they articulate valuable product optimizations by virtue of their position and thus provide indications of Unique Selling Propositions

System Value Chain

Term
System Value Chain

When
Before the project starts, DEFINE, MEASURE, identify customers and their needs

Goal
Extend the customer perspective to the entire value chain and integrate all relevant company internal and external suppliers and customers

Procedure

1. Place the product, the process and/or the service to be developed into the center of a simple value chain

2. Collect and list essential external and internal suppliers on the supplier's side of the product, process and/or service (input)

3. Collect and list essential external and internal customers on the demand side of the product, process and/or service (output)

4. Mark former target customers and reflect on them in the context of the now defined Multi Generation Plan

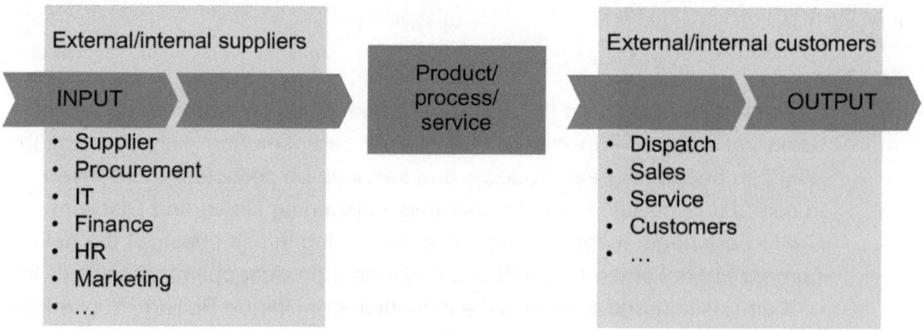

Tip

- Draw up a simple system value chain within an entire team Workshop
- Use facilitation cards or Post-its® for drawing it up; this allows for changes and enables a clear visualization of elements at all times
- Reflect on customer definition afterwards by taking into consideration the Multi Generation Plan; also take into consideration the necessary alignment of the development project with process specific requirements identified in the course of the MEASURE phase; particularly internal suppliers and customers should be taken into consideration when collecting additional needs information during the MEASURE phase

Lifecycle Roadmap

Term
Lifecycle Roadmap

When
MEASURE: identify customers and their needs

Goal
- Map a product's life cycle from the customer's point of view
- Formulate hypotheses on the interaction of target customers with the product
- Articulate exact starting points for designing the collection of customer needs

Procedure
1. Determine relevant target customers from whose perspective the Roadmap is to be generated

2. Determine relevant information on customers which could be relevant and useful for the Roadmap

3. Define the first and last phases in the life cycle of product utilization from the perspective of the selected customer (start/stop)

4. Complete the life cycle by adding interim steps

5. Evaluate individual life cycle phases according to their relevance for future product development

Fill out the following list for all relevant target customers:

Target customer	Life cycle	Relevance* (9/3/1/0)
Customer M.: • Existing customer • Demand behavior • Purchasing conditions • Product adjustments • Complaint behavior • ...	• Become aware of product	**9**
	• Select product	3
	• Purchase product	1
	• Transport product	0
	• Open packaging	**9**
	• Prepare product for use	**9**
	• Use product	**9**
	• Maintain product upon need, conduct repairs	3
	• Enhance product value	**9**
	• Throw product away	0

*Relevance for future product development: 9 = high, 3 = medium, 1 = low, 0 = none

➾ **Tip**

- As a team formulate hypotheses about the product life cycle from the first to last phase and translate them into process steps
- Only after having considered the customers' perspective look at the company perspective. Then look to evaluate individual elements for their relevance to the development of the product
- A potential guiding question for the evaluation should be "how possible is it to influence the respective phase of the life cycle through the product development?". Criteria such as potential for differentiation and/or Unique Selling Proposition and cross selling potential should be discussed for evaluation
- Reflect on the evaluation from the company's perspective bearing in mind what was written in the Multi Generation Plan
- Break phases down into further detail if you want to carry out a deeper evaluation; this way you can add extra information e.g. actual product utilization by a detailed process hypothesis and use it for preparing the Customer Interaction Study (Gemba)

Specifics for process development

- A similar approach is used for the life cycle of a process

- First define the internal customer from whose perspective the process is to be developed and mapped
- This way you ensure the process course to be developed e.g. from the perspective of the clerk in charge, from the initial to the final process step e.g. from the initial customer contact through the various input channels to final collection of the customer's order
- Now relevant process hypotheses about the interaction of target customers with the process can be formulated which will be verified and extended further later in the phase

5 W 1 H Table

Term
5 W 1 H-table

When
MEASURE: identify customers and their needs

Goal
- Structure available information
- Formulate hypotheses on the interaction of defined target customers with the system (product/process)

Procedure
Structure the existing information and hypotheses according to the five Ws and one H: Who? What? When? Where? Why? & How?

Example: Target Customer Table – Passenger Seat

Who?	What?	When?	Where?	How?	[redacted]
Commuters (80% share)	Use of seats to commute to work and/or home from work	Between 6 -10 a.m. and 3 - 7 p.m	Use of seats primarily in the front area and close to the doors	Use of seats wearing business clothes	Use of seats and yet be able to read and work
Students (20% share)	Use of seats to commute to school and/or home from school	Between 6 -10 a.m. and 12 - 3 p.m.	Use of seats primarily at rear of bus	Use of seats in bigger groups, strong desire to move	Use of seats to talk

Continued on the following page

Who?	What?	When?	Where?	How?	
Cleaning staff of transport company	Cleaning of seats	Three times a week, in the evenings after 10 p.m.	Complete cleaning of seat systems in bus line	Manual cleaning with commercially available cleaning agents	Ensure cleanliness for appearance and hygiene reasons
Maintenance staff of transport company	Installation and removal of all seats and/or individual parts of seat	As and when needed	Maintenance of seat systems in company workshops	Manual installation and removal of seat systems	Ensure functionality of seat systems
Bus manufacturer	Installation of seats in new buses	During production	Production site XY (in production line)	Manufacturing Procedure with standardized Tools	Efficient installation in line with set standards

➾ **Tip**

- The "Why" column contains initial assumptions on which to base customer need investigation
- As a team, draw up a simple system value chain and transfer it to the key interest groups from suppliers and customers into a Mindmap structure
- Add the described elements from the 5 W 1 H Table (e.g. what, how and why) and develop a comprehensive understanding, within the team, of the customer as the basis for the next steps

Illustration: Combining a System Value Chain and a 5 W 1 H Table

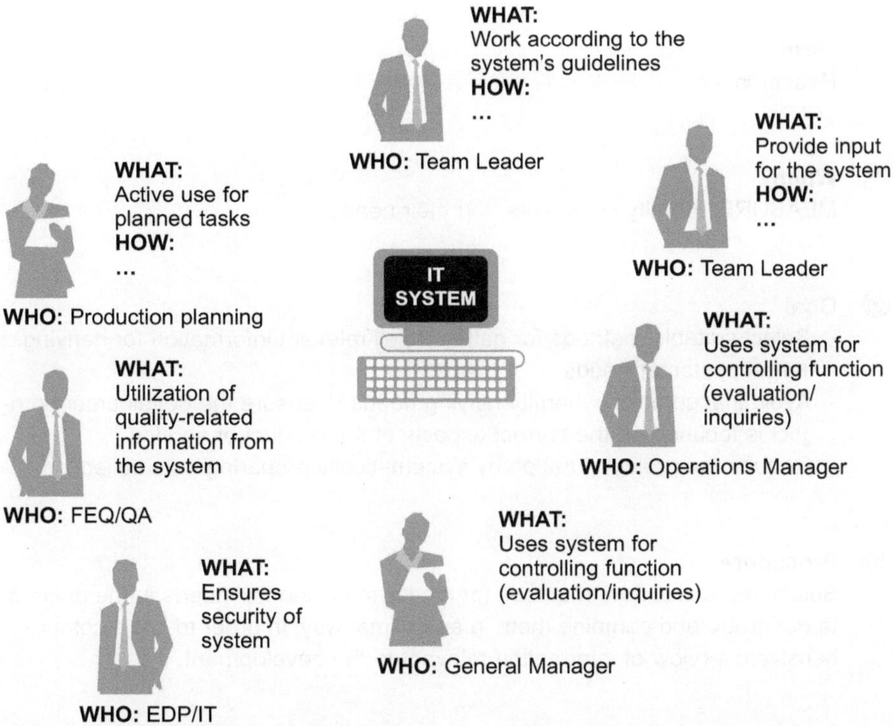

Select and Conduct Research Methods

Term
Research

When
MEASURE: identify customers and their needs

Goal
- Select suitable methods for gathering all relevant information for deriving target customer needs
- Avoid inaccuracies when identifying needs to ensure the development project is focused on the correct aspects of the product or service
- Reduce the project duration by systematically preparing data collection

Procedure
Select the appropriate methods for identifying customer needs in the defined target group and combine them in an optimal way, in order to get a comprehensive overview of information relevant to the development.

Mapping research methods on the following page

			Effort (1-5)	Benefit
Passive	Internal Research	Research in secondary sources on customer needs and requirements, customer values, possible product and service features, indicators for measuring success.	1	Quick overview of existing and/or easily accessible information and material
	Internal Research: web based	Web based collection and structuring of relevant information on the target customers, in order to generate first hypotheses about potential customer needs and prepare targeted external customer surveys.	1	Gain a quick overview of published information and material
Active	Customer Interaction Study	Observe the customer "at work" in order to gain a better understanding of his/her environment and activities. Identify needs. This provides information on unexpressed needs in particular.	5	Representative and very good source for unexpressed needs
	Participatory observation	Active observation of customer "at work", by actually participating in the work oneself i.e. doing the job.	4	
	1-to-1 Interview	Provides results on the needs and expectations of specific customers, the values of the customer, the customer's opinion about service aspects, requested product/service attributes and data for measuring success.	2	Generate detailed information and personal impressions
	Focus group interview	The focus group is suitable for identifying the opinion of a group of customers. The group should represent a specific customer segment and represent the segment's exact definition as well as the prioritization of the customer values.	3	Comprehensive need spectrum from heterogeneous groups/ clusters
	Survey	Serves the purpose of measuring customer needs and values as well as evaluating products and services by a large number of customers of one or several segments. Due to the big sample, it delivers "hard" facts for making decisions.	4	Big samples generate high statistical significance for the information gained

Effort: 1 = low, 5 = high

Tip

- Make sure that a combination of passive and active customer observation methods and surveys are used in order to get the most out of their relative pros and cons
- Follow a simple "P-A-P" principle*: Combine methods, carrying out advance research with passive methods and then using the results of this for driving subsequent active methods
- For subsequent evaluation and validation of results passive methods should be used e.g. in order to verify hypotheses on the observed customer situation; researching visit reports, corporate brochures and internet publications can prepare for a personal interview or a Customer Interaction Study on site, to be able to verify afterwards the insights from this
- In order to evaluate the quality and reliability of the process used for selecting research methods, you should categorize validity according to the Kano Model (see chapter "Kano Model"):
  - Passive methods primarily generate information on basic factors
  - Interviews, focus groups, and surveys additionally generate information on performance factors
  - Only Customer Interaction Studies and observation by participation enable the collection of information on Buzz Factors
- Since the identification of customer needs should provide a comprehensive overview within the scope of the development project, an appropriate combination of the presented methods should be undertaken

* PAP – The Principle of Alternate Possibilities, Harry Frankfurt, "Alternate Possibilities and Moral Responsibility" Journal of Philosophy, 66/23 (1969), 829-39

Internal Research

Term
Internal Research, Passive Research

When
MEASURE: identify customers and their needs

Goal
- Collect existing information on the target customers and structure them in a reasonable way
- Generate first hypotheses on potential customer needs
- Identify improvement gaps
- Prepare active customer surveys

Procedure
The following secondary sources contain relevant information and can provide indicators of potential customer needs:
- Information from Service or Sales
- Customer complaints
- Specialist journals
- Internet
- Patents etc.

Tip
- Internal (passive) research, as a cost-effective method, should generally be seen as preparation for the external (active) research (P-A-P Principle)
- Internal research will never be able to provide all the relevant information necessary for gaining a complete profile of customer needs; its main purpose is to help formulate hypotheses about customer behavior and needs which are then verified or discarded through active research

Web based Research

Term
Web based Research

When
MEASURE: identify customers and their needs

Goal
- Collect and prepare relevant information on the target customers in order to generate first hypotheses about potential customer needs
- Prepare active customer surveys

Procedure
The Internet can be used in various ways to collect the Voice of Customer. The difficulty is to detect it in an unfiltered way and/or identify it from more comprehensive contexts. The Internet allows access to customer and/or product specific platforms and forums in which manufacturers, end-users and decision makers exchange and discuss their opinions. From the project's perspective it is important to find these valuable, web based opportunities, and use them in a targeted way.

1. **Thematic forums**
   Members exchange their opinions about certain topics in thematic forums. Unfiltered Voice of the Customer can be found in forums due the informal nature of the exchange of opinions and ideas. A key task is to find the forum where your product or service is a topic of discussion. Locating the opinion of a specific target group can be realized very simply. Most forums have search functions facilitating the search for specific phrases. Every member has the opportunity of introducing new topics and put them up for discussion. It is thus possible to survey the other forum members directly for their own opinion.
   Registration in a forum is usually necessary if one intends to do research. Thematic groups exist for all key topics.
   Every forum has rules which need to be kept. The facilitators ensure that these rules are kept. Rules are mostly concerned with the placing of topics and the use of respectful language.

2. **Review sites**
   Review sites are platforms where customers can evaluate utilized products or services.
   They facilitate the buying decision for individuals who are interested in a product/service. By evaluating products, weak points are revealed and information on new ideas provided. Whilst it is often possible to read evaluations without registering on the portal prior registration is usually necessary if you want to make an evaluation. Remember though, that that such sites may also be partly be used by companies to promote their products. Review sites can be independent or integrated in Online stores.

3. **Twitter**
   Twitter enables members to send current status messages (Tweets or Updates) as text messages (140 signs). These can be received by all members and found using a search function.
   Twitter is a social network; because of the magnitude of information available however, it is also an attractive source of product information. Registration is required for using Twitter.

4. **Blogs**
   Blogs are mainly private web pages similar to a diary. The term "Blog" was derived from the English word "Web Log" which itself was derived from "World Wide Web" and "Logbook". The creators of these pages (bloggers) write so-called articles (posts or blogs), whenever they wish to do so, which can then be commented on by the reader.
   The wide range of blogs also includes some which specialize in product related topics. These can therefore provide Voice of Customer or subject specific comment, but always check for timeliness. In the thread (the index), posts are chronologically listed and simplified.
   There are many different search tools to help you locate the blog you seek if it actually exists.

5. **Search engines**
   Apart from standard search engines such as Google and Yahoo the so-called meta search engines are particularly appropriate for obtaining an overview of the search criterion after having completed just a few steps. Meta search engines have no data base of their own but take a user request and use other search engines to obtain the relevant information before consolidating it into a single response. Many high profile meta search engines are now available.

Tip

Use Boolean operators for targeted web based research in order to find product and/or process relevant data. Here are some examples and their uses:

- AND:
  All of the indicated search items must be included in the document
- OR:
  At least one term of the connected search items must be included in the document
- AND NOT (also NOT):
  The subsequent word mustn't be included in the document
- NEAR:
  The connected search items must be found in close proximity e.g. within 10 words of each other
- "Search item":
  This would search for and return any document containing the exact content of the inverted commas

Active Research

Term
Active Research

When
MEASURE: identify customers and their needs

Goal
- Actively collect relevant information on the target customers
- Verify formulated hypotheses about potential customer needs based on internal research

Procedure
- Direct collection of information from the customer

- Observation of customer/customers at the place of their interaction with the product or service

Customer Interaction Study

Term

Customer Relationship Modeling, Customer Interaction Study, Go to the Gemba, Gemba Study

When

MEASURE: identify customers and their needs

Goal

- Gain undistorted and complete information on the customer
- Identify actual customer needs without considering solutions at this stage

Procedure

The Customer Interaction Study comprises three steps:

1. Planning
2. Execution
3. Analysis

1. Planning

Determine the Customer Interaction Study' who, what, when, where, and how. The team formulates hypotheses on the customer's interaction with the system (product/process). It helps to visualize the interaction process when formulating the hypotheses.

A Customer Interaction Study is not limited to simply asking questions, the customer's environment and the customer themselves are also observed in great detail. A Customer Interaction Study is conducted where ever value is created for the customer(s).

Illustration: Customer values

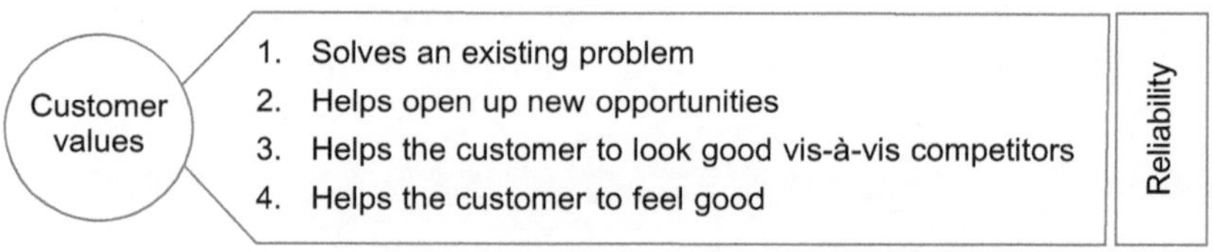

2. **Execution**
   The environment, situation and behavior of the customer is documented. The Voice of Customer is also noted. Needs identified during the Interaction Study are to be confirmed with the observed customer if possible.

   A standardized form should be used by all "observers" for documenting the information.

Example: Form for conducting a Customer Interaction Study – Passenger Seat

1. Preparation	Observer:	Mr. Peter Finger	Recording medium:
	Date and time:	10/23/2012 / 12:30 p.m.	Notepad
	Place of observation:	Moscow – parking lot AutoMoskov	
	Name/function of observed individual:	Mr. Sergej Abramovic	
	Contact details:	sergej.abramovic@automosk.ru	
	Further details about observed individual:	Mechanic, employed at AutoMoscov, responsible for bus interior	
	Situation and environment:	He is busy as he was late because of an accident in Moscow's inner city this morning where he had to go in order to talk to authorities	

2. Execution	**Process step**	#4: Uninstall old seat		
	Observation	He has to loosen 8 bolts with a wrench		
	Original VOC	"Shoot, this is taking too long"	"The bolts have rusted again"	"The bolts are quite dirty and I can't get a good grip with this wrench"
	Comment	He wants to finish the seat installation before going back to the city		

Continuation of presentation on following page

3. Analysis	**Identified need**	Quick assembly of seat	Simple assembly of seat	
	Validated by the observed individual	Yes	Yes	
	Further comments by the observed individual			

3. **Analysis**
   The hypotheses formulated during planning are now discarded, adjusted or verified. The following questions provide the focus of the analysis:
   - What do the customers really want?
   - What are their "actual" needs?
   - What is the value (benefit) for the customer/s?

Specifics for process development

- Use onsite inspections in the domain of the process development in order to observe similar process flows on site and to obtain as genuine and as complete information as possible
- You should focus on processes with similar inputs and/or generating similar outputs and should particularly focus on observing quality, availability and complexity criteria
- You should list and define primary materials, operating supplies, tools, data, information, signatures required, capturing information on when they have to be made available, their quality and who gets them
- Semi-finished goods, processed information and acknowledged forms should also be defined and evaluated as interim outputs in the current process

Tip

- The observing team should be interdisciplinary in order to make observations from as many different perspectives as possible
- Deploying multiple teams simultaneously is a reasonable option
- Observations should be recorded as rigorously as possible. Using videoing, audio recording, photos, drawings and comprehensive notes are all legitimate choices; the motto is "the more information the better!"

- The typical duration of a Customer Interaction Study is approximately two hours
- Customers often suggest solutions. However, the objective is to identify the actual needs underlying these proposals! The Customer Needs Table supports the distinction between solutions, specifications, complaints etc. and actual needs

Customer Interaction Study: Participatory Observation

Term
Participatory Observation

When
MEASURE: identify customers and their needs

Goal
- Gain undistorted and complete information on the customer/s by placing the observer within the situation to be observed and having them actively participating in it
- Identify the actual customer needs without considering solutions at this stage

Procedure
The participatory observation generally follows the steps of a Gemba study:
1. Planning
2. Execution
3. Analysis

In addition to the procedure outlined previously the following aspects should also be considered:

1. **Additions to planning:**
   Individuals conducting the observation:
   - Several people should participate in the active observation in order to collect a wider variety of perspectives and impressions
   - It is better if the observers are familiar with the environment (e.g. production hall, sales room etc.) as this eliminates the need for time consuming induction sessions
   - Together with the observers you should agree in advance when and how to document the observations

Documents and forms:

- The same form used in the Gemba study can be used for the participatory observation; however, it is best not to be too formal and work with a simple and quick form of notes
- Photos and video cameras can also be used to provide supporting material but it is important to get prior approval

Domain to be observed:

Only the most important individuals and/or groups of individuals in the intended observation area should be informed in advance about the planned data collection (e.g. the purchasing manager in the retail store)

2. **Additions to execution:**
   - The observed situation (e.g. the observation of the customer interview in a retail store) is to be neither disturbed nor influenced by the participating observer
   - In order to be able to record the situation as comprehensively as possible, it is best if several observers are deployed on site simultaneously
   - Impressions should be recorded as bullet points during the observation period and then formulated in more detail and precision later
   - As a rule the longer and the more frequently the participatory observation takes place, the better the data will be – potentially revealing patterns and exceptions more clearly because of the repetition

3. **Additions to the analysis:**
   - Final documentation should describe what observations were made, by whom and from what perspective
   - The evaluation team reads all observations and compares them; similarities and differences between the various observations are then interpreted

➾ **Tip**

- Pay attention to the fact that the observers should be as open and open-minded as possible so they have a greater chance to perceive the unexpected
- When conducting the observations you should not endeavor to record all events but to observe the things that can be seen precisely and attentively
- When you subsequently draw up notes and comments, you should compile descriptions with as much detail as possible

- When taking notes it is most important to focus on good descriptions of the situation rather than making evaluations or interpretations
- Make sure that you separate the following steps clearly:
  1. Perceptions by observation and participation
  2. Description (when drawing up notes afterwards)
  3. Evaluation and interpretation (for evaluation in the team)
- You should also pay attention to the fact that different observations of the same phenomenon are possible (e.g. depending on the perspective)
- Moreover, different interpretations of the same observation are possible (e.g. depending on the background and the individual evaluating the observation)

1-to-1 Interview

Term
One-to-One Interview

When
MEASURE: identify customers and their needs

Goal
Survey customers individually to gather further information on their needs and requirements

Procedure
An interview guideline is drawn up before talking to the customer.
It is important to formulate open questions which require a more detailed and informative response than just a "Yes" or "No" answer by the individual being surveyed.

Interview should follow a set structure:
- Explain the reason for the interview
- Explain the interview style
- Ask for permission to record the interview (audio)
- Ask open questions in accordance with the prepared interview guidelines
- Review gained insights and if required, use additional questions and comments to explore these insights and develop them further

Illustration: Advantages and disadvantages of a 1-to-1 Interview

Advantages of interview	Disadvantages of interview
• Flexibility when dealing individually with an interview partner • Opportunity to cover more complex issues • High response rates	• Cost intensive • Time intensive • Requires the interviewer to be a good listener • Risk that interviewer and interviewee fail to understand each other • Different results can be generated by different interviewers

Focus Group Interview

Term
Focus Group Interview

When
MEASURE: identify customers and their needs

Goal
Find out needs of a group of selected target customers

Procedure
A Focus Group Interview is conducted as an open discussion session where the goals, agenda, and timing have been set in advance. A focus group consists of approx. 7 to 13 participants is formed by people from the relevant customer segment. The discussion is led by a nominated facilitator, whose task it is, to ensure that the group remain focused on the topic. This type of interview should take between two to four hours.

Mapping advantages and disadvantages of a Focus Group Interview

Advantages of a Focus Group Interview	Disadvantages of a Focus Group Interview
• The participants encourage and inspire each other: One participant says something another individual hasn't thought of so allowing further ideas to develop • More cost-effective than many 1-to-1 interviews	• Difficult to manage when the individuals in the group don't agree with each other • People tend to follow others, or become passive; this can lead to a situation where only the needs of the "dominant" person are articulated • The time available belongs to the group as a whole; not every person receives the same "airtime"

Tip

- When preparing and conducting the Focus Group session the so-called "laddering procedure" can help lead to the actual key aspects very quickly
- The interviewer develops his/her line of questioning based on the responses provided by the interviewee(s) thus closing in on the fundamental requirements for the product or service

Survey

Term
Survey

When
MEASURE: identify customers and their needs

Goal
Find out representative information on target customers

Procedure
Determine and prepare the strategy, then conduct and analyze the survey. A structured questionnaire is designed which will be put to the customers. Given that any results should be statistically valid and representative of the target customer group, sample size and method of selecting interviewees should be carefully planned.
Depending on the customer segment and project scope, various kinds of surveys could be used and/or combined.

Mapping survey types

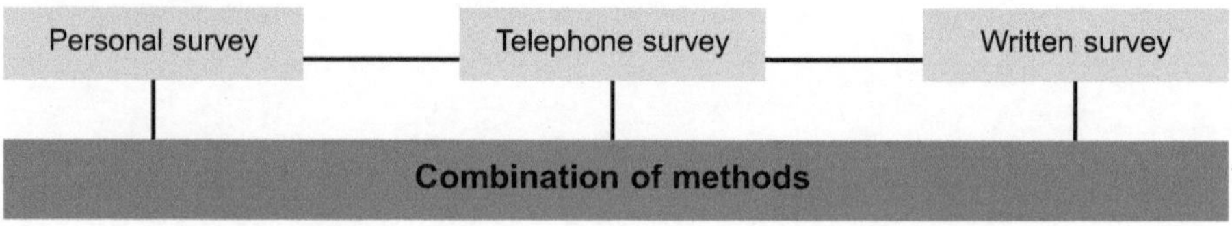

Personal Survey
Face to face talk through of the survey makes sense when complex issues need to be clarified. The required information can be generated in full:
- Personal surveys are generally more expensive than telephone or written surveys
- The training effort for interviewers is high because they need strong experience, a high level of social skill, and often a technological competence

Telephone surveys

A telephone approach makes sense when a large number of target customers have to be surveyed within a short space of time:

- They provide results quickly
- The interviewer's influence on the results is low
- However, limiting the scope of the survey to customers reachable only by phone could distort the survey results i.e. is not necessarily representative
- Possibly restricts the survey to simpler questions as more complex issues cannot be adequately explained

Written surveys

Postal or electronic surveys make sense when the budget and the resources are limited:

- Fewer personnel are required for sending a questionnaire via email or mail i.e. it provide economies of scale
- The cost of sending email or letter based surveys is not materially influenced by questionnaire size
- However, limiting the scope of the survey to only households with Internet connection may not be representative of the selected customer segment
- Response rates and level of information gained are generally lower than that of the other methods described

Questionnaire for surveys

Special care is needed when the questionnaire is compiled:

- All hypotheses formulated during internal research or the Customer Interaction Study should be covered
- The structure should be logical and clear
- Complex issues have to be constructed in comprehensibly; in some cases a picture is worth a thousand words!
- Professional layout and appearance are essential
- A personal covering letter should inform the recipient about the reason for the survey, guarantee that all information will be treated confidentially, and should list contact persons in case there are any questions
- Completion of the questionnaire should not take more than 30 minutes

Categories for questionnaires

Questionnaires can be divided into the following categories depending on their structure and directness:

- Structured and direct:
  This type of a questionnaire is used most frequently since it contains structured questions and answers are pre-defined i.e. a matter of selection.

Every question is clear and precise and allows very little room for interpretation. Moreover, pre-defined answers enable a rapid and balanced analysis of responses. The purpose and background of the survey are clearly communicated to the respondents and further explanatory detail can be added as necessary. If you are concerned about potential influences due to poorly trained or inexperienced interviewers which might lead to interviewer bias then, this type of a questionnaire is appropriate for you.

- Unstructured and direct:
  This type of a questionnaire contains questions of a broader nature which encourages more open answers. Nevertheless, the questions are still aligned towards the respective core topic. Here too, the purpose and the background of the survey are clearly communicated. Particularly in personal and telephone surveys this type of questionnaire is especially appropriate since the respondents are allowed additional room for maneuver. However, when choosing this type of survey it is important to use qualified interviewers in order to avoid potential interviewer introduced bias. A downside of this type of survey is the limited scope for analysis and evaluation of results.

- Structured and indirect:
  If it is considered that anonymity is important, for whatever reason, then an indirect type of the questionnaire is appropriate. Using a set of structured questions and a selection of predefined answers, neutral information can be generated which could otherwise have been influenced by naming the customer company. Typically a market research company would be used for this purpose.

Illustration: Advantages and Disadvantages of the Survey

Advantages of the survey	Disadvantages of the survey
• A significant number of market players can be surveyed • Statistically valid statements are possible • Evaluations and/or results emerging from up-to-date surveys are often mirrored by the strong interest shown by the respondents	• Answers are limited to the questions asked – for this reason the topics should be developed beforehand by using other survey and observation methods • Demands a great deal of effort and expense, coupled with generally lower response rates or refusal to participate • Negative attitude among the population towards question based surveys

Selection of respondents

Three methods are used for obtaining a representative selection of respondents:

Illustration: Methods for selecting respondents

Random Selection	Quota Selection	Haphazard Selection
For an unqualified random selection, each unit must have the same calculable chance of being selected. This method is applied for example in "random digit dialing" for telephone interviews, where the number is generated randomly and dialed automatically.	Here, rules are set for selecting the respondents. These rules are designed to gain as representative a selection as possible. However, only the quota attributes are sure to be representative	Haphazard Selection means selection occurs without any recognizable strategy. This can lead to a systematic distortion. This practice can be applied in probes or when the budget is limited and no other approach is possible. In general though, it is to be avoided.

Selection of respondents

When conducting a customer survey it would be ideal to have a complete sample. But in reality it is not usually possible to conduct a complete sample due to time and financial restrictions. In this case a survey with a representative sample size should be carried out.

Relevant quality criteria

For the assessment of a customer survey's quality the following quality criteria should be taken into consideration:

- Objectivity:
  Indicates to what extent the survey's result depends on who conducted the survey. Furthermore, a distinction can be made between the objectivity of conducting the survey (result of survey is not influenced by individual conducting the survey), the objectivity of analysis and of interpretation (subjective interpretations have no influence on the interpretation of the results).

- Reliability:
  As far as reliability is concerned, repeatability and in this way reliability of data collection within the frame of a survey is aimed at. A survey is referred to as reliable if the same result is achieved after a measurement of the same objects was repeated under the same conditions; it is assumed that the characteristics to be measured are stable over time.

- Validity:
  Validity describes the precision with which conclusions can be made from the collected data about the characteristic to be analyzed (e.g. from a survey's result about customer behavior to the actual customer behavior in the underlying market segment).
  Furthermore, a distinction between internal and external validity can be made, whereby internal validity refers to inner coherence and consistency during the survey questionnaires. Internal validity is given if the change of a dependent variable can be traced back clearly to the variation of the independent variable within the frame of the survey and is not influenced by further influencing variables.
  External validity is given if generalizations can be made from the results of a survey about the population and/or when the results can be transferred beyond the specific setting of the survey and to other situations and can thus be classified as universally valid.

→ **Tip**

- A Customer Interaction Study is absolutely essential for every development project; it reveals the potential for genuine breakthrough innovations!
- Against all odds, a study on site should be conducted by all means
- 1-to-1 Interviews and Focus Group Interviews are especially appropriate for target customer surveys in the sector of industrial goods markets
- Surveys are particularly appropriate for target customers in consumer goods markets
- A survey should be planned, conducted and analyzed together with experts from market research
- Nothing replaces direct contact and interaction with the target customers; the biggest mistake is assuming that one already knows everything about the customer(s) and identifying his (their) needs solely from meeting the customer in a conference room
- Systems (products/processes), developed on the basis of incomplete and unverified hypotheses, will most likely demand a great deal of rework and fail to highlight possible breakthrough innovations

Customer Needs Table

Term
Customer Needs Table

When
MEASURE: structure and prioritize customer needs

Goal
Derive actual customer needs from the identified customer information which were identified throughout the research

Illustration: Customer Needs Table – Passenger Seat

Voice of Customer (VOC)	Measure-ment	Possible cause	Possible Solution	Complaint	True Need
"The seats are bust all the time"				X	I want a robust seat
"I have to employ three people for maintenance"				X	I want a seat with no mainten-ance
"Passengers steal everything they can get their hands on"				X	I want a seat whose parts cannot be stolen
"Dirt can be seen too quickly"				X	I want a seat which is always clean
"Build a hard shell out of plastic"			X		I want a low cleaning effort
"The DIN standard for seat dimensions must be kept"	X				I want industry standards to be met

Tip

- Abstracting the information into solution-neutral formulations of needs is critical to success!
- This step supports the team in overcoming psychological barriers and designing genuine breakthrough innovations

Customer Need Structure

Term
Customer Need Structure

When
MEASURE: structure and prioritize customer needs

Goal
- Sort collected customer needs in order to make thought structures of customers transparent and to be able to recognize gaps
- Recognize need for action by collecting further needs by e.g. additional interviews

Procedure

See next page for Roadmap Of Customer Needs Structure

MEASURE **Structure customer needs**

PROCEDURE

Now that information has been collected on customer requirements, and probably a lot of it, it can be collated using a **Customer Needs Table ...**

— Customer need 1

— Customer need 2

— Customer need ...

Customer Needs Table

				x	
				x	
			x		
	x				

... the needs are grouped into an Affinity Diagram,

Affinity Diagram

SAFETY	DESIGN	...

... sorted into hierarchies in the Tree Diagram,

Tree Diagram

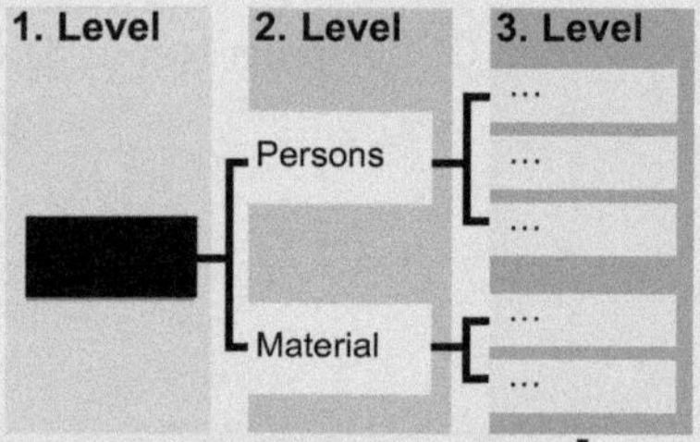

... categorized with the help of the Kano Model, → **... prioritized in the pair comparison (AHP) and** → **... provided with measurement, target values and specifications**

KANO-MODELL

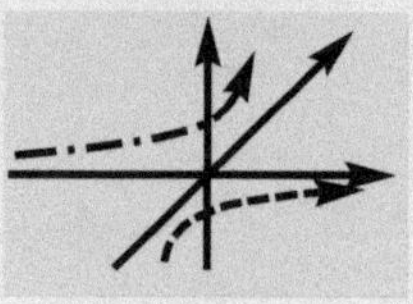

AHP

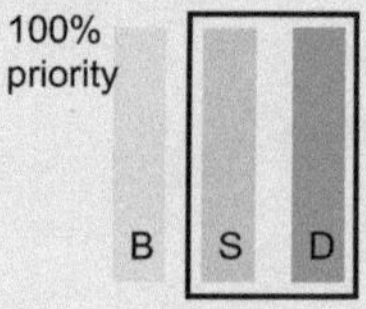

QFD

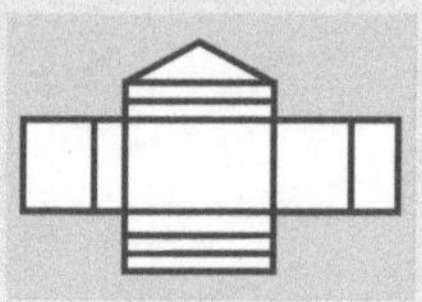

Affinity Diagram

Term
Affinity Diagram

When
MEASURE: : structure and prioritize customer needs

Goal
- Sort customer needs according to similar themes/content
- Recognize and understand customers' thought processes

Procedure
Write down the customer needs on cards or Post-its®, group them according to the "major need" and name the "major need".

Example: Affinity Diagram – Passenger Seat

Maintenance	Seat Robustness	Cleanliness
I want a seat requiring no maintenance	I want the seat to withstand vandalism	I want low cleaning effort
I want to be able to install and remove the seat quickly	I want a seat whose parts cannot be stolen	I want a seat which is always clean
I want the seat to be usable at any time	I want to be able to install and remove the seat quickly	

Tree Diagram

Term
Tree Diagram

When
MEASURE: structure and prioritize customer needs

Goal
- Group identified customer needs according to similar themes and sort into hierarchical levels
- Identify existing need gaps
- Create uniform and comparable levels of detail
- Recognize and understand customers' thought processes

Procedure
1. Transfer the information from an Affinity Diagram to a Tree Diagram

2. Split the Tree Diagram into different sublevels, in order to identify existing gaps and/or needs which haven't been identified yet

Illustration: Transfer of Affinity Diagram into a Tree Diagram

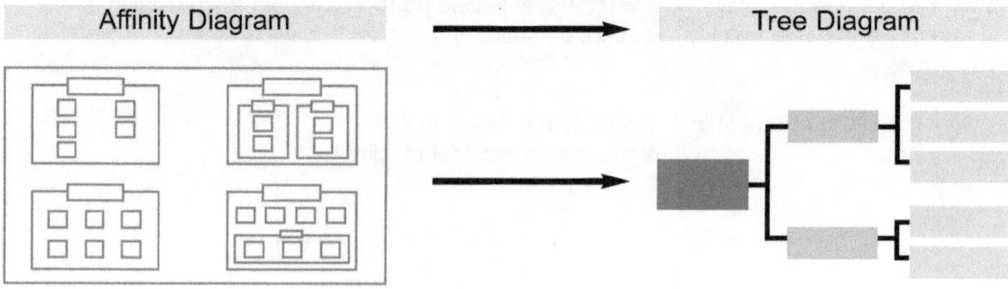

Example: Tree Diagram – Passenger Seat

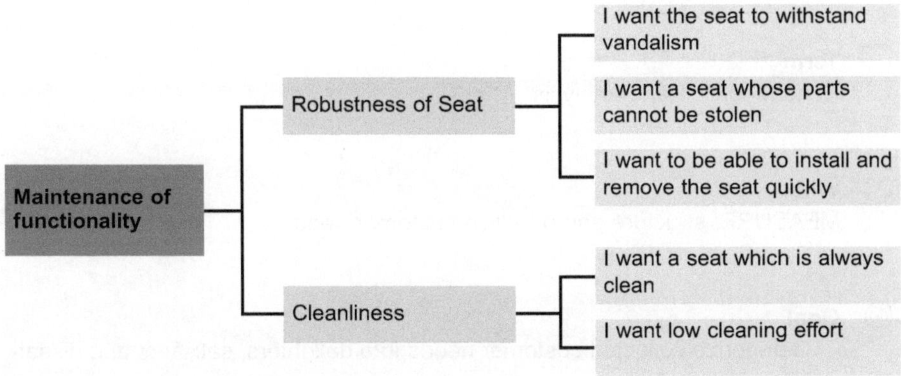

Tip

- There is no right or wrong when drawing up an Affinity Diagram; every person thinks in different structures; if there is no direct contact with the customer, the team should do the preliminary work and develop the Affinity and Tree Diagrams which the customer can check afterwards
- If differences emerge in clusters and thinking structures then it is up to the team to restructure them logically
- Tree Diagrams generate a common level of detail and support the identification and closure of thematic gaps; as the project evolves this leads to a significant reduction in misevaluations in the QFD 1 (House of Quality)

Kano Model

Term
Kano Model*, Kano Analysis

When
MEASURE: structure and prioritize customer needs

Goal
- Categorize collected customer needs into delighters, satisfiers and dissatisfiers
- Recognize needs that must or should be guaranteed

Procedure
- Every potential need is tested by asking the customer a positively and a negatively formulated question:
  – How do you feel if need x is not fulfilled?
  – How do you feel if need x is fulfilled?

- The customer is given the following response options:
  – I like that
  – That's normal
  – I don't care
  – I don't like that

- Based on the customer responses the needs can be classified into:
  – Basic factors (Dissatisfiers), i.e. features which are expected by the customer
  – Performance factors (Satisfiers), i.e. features which enable the customer to measure the quality of the process or product
  – Buzz factors (Delighters), i.e. features which exceed the customer's expectations

The following matrix helps with the assignment of needs:

* This classification is based on the model by Professor Dr Noriaki Kano (Rika University, Tokyo), developed in 1978.

Kano Table

Answer to a positively formulated question	Answer to a negatively formulated question →	I like that	Normal	I don't care	I don't like that
	I like that		Delighter	Delighter	Satisfier
	Normal				Dissatisfier
	I don't care				Dissatisfier
	I don't like that				

Kano Model

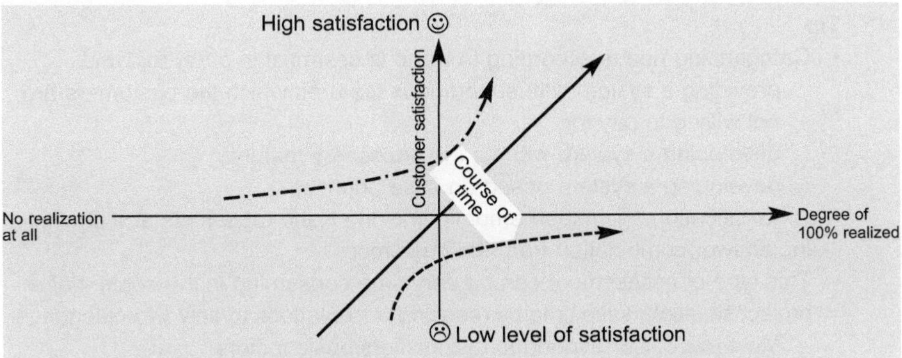

World-class in the market	Assert in the market	Enter the market
Delighters/buzz factors	Satisfiers/performance factors	Dissatisfiers/basic factors
• Not expected	• Specified	• Expected as given
• Not expressed	• Expressed	• Not expressed
• Not yet aware of	• Aware of	• Hardly aware of, taken for granted

Example passenger seat

Negative question:

- "What is your response if the rating of the seat system to withstand vandalism is not given?"
  I DON'T LIKE THAT

Positive question:

- "What is your response if the rating of the seat system to withstand vandalism is given?"
  I LIKE THAT

= **SATISFIER/PERFORMANCE FACTOR**
"The more resistant the seat system has been designed against vandalism, the more satisfied is the relevant target customer group"

Tip

- Categorizing needs according to Kano is essential in order to avoid:
  ... providing a system with superfluous features which the customers are not willing to pay for
  ... developing a system without the necessary features
  ... developing a system based on false priorities
- A classification against an empty cell in the Kano table hints at a conflicting answer combination from the customer
- This type of assessment can be very time-consuming in the context of a project situation; save time by reducing observations to only two categories:
  - Must Have: Correspond to dissatisfiers/basic factors
  - Nice to Have: Correspond to satisfiers and delighters
- Apart from the various factors identified following customer research you should also reflect, relevant laws and regulations, industry-specific regulations, safety requirements etc. as basic factors

Prioritize Customer Needs

Term
Customer Need Prioritization, Prioritize Customer Needs

When
MEASURE: structure and prioritize customer needs

Goal
Customers prioritize their identified satisfiers and delighters

Procedure
Methods for prioritizing satisfiers and delighters:

1. N/3 Method
2. Nominal Group Technique
3. Analytic Hierarchy Process
4. Simple Pair Comparison

The customers should weight the delighters and satisfiers which have been identified with the help of the Kano Model. Dissatisfiers are not evaluated since they always have the highest priority as basic factors and/or basic requirements. The system when it is developed must be capable of consistently fulfilling them 100% of the time.

Tip

- If you are confronted with a great number of defined customer needs in this phase of the project, which were collected in a bigger customer circle, you should conduct a preliminary prioritization with the help of the N/3 Method; looking at only the main themes that the surveyed customers mentioned most often during the evaluation
- If you are confronted with a great number of defined customer needs in this phase of the project, which were collected in a smaller customer circle, you

should conduct a preliminary prioritization of the needs on the second level of the Tree Diagram first; the preliminary prioritization should be conducted with using Pair Comparisons and you should observe only the most important thematic areas further

N/3 Method

Term
N over 3 Method, N/3 Method

When
MEASURE: structure and prioritize customer needs

Goal
Weighting priorities for delighters and satisfiers by a simple distribution of points

Procedure
1. Prepare the satisfiers and delighters which were categorized during the Kano Analysis so that all participants can see them
2. Every participant receives adhesive dots, the number of which corresponds to a third of the total number of needs identified as delighters and satisfiers
3. Participants clearly allocate their dots to those satisfiers and delighters that they think most important
4. After having distributed the adhesive dots and tallied them for each need discussions are allowed until a consensus has been reached (facilitator is challenged!) about the respective priorities
5. If needs receive an identical amount of adhesive dots they can be prioritized once again in another run
6. By distributing the adhesive dots and by the discussions a final ranking of delighters and satisfiers emerges

Mapping N/3 Method on the following page

Illustration: N/3 Method

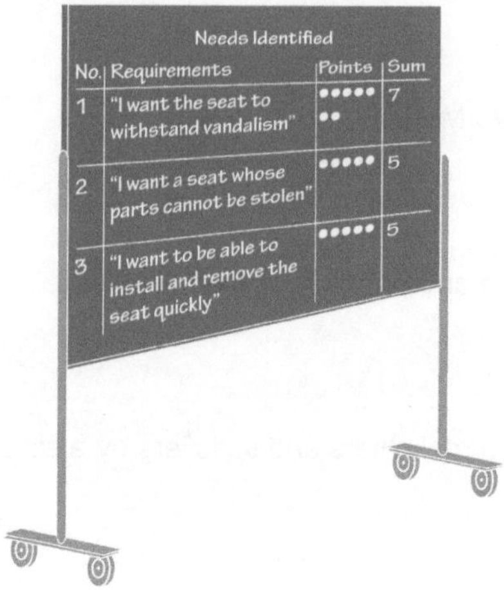

Tip

- It is best if the distribution of adhesive dots is conducted by the customers and/or together with the customers
- This should be facilitated by a member of the project team
- You should ensure in advance that the question formulation for evaluating the needs with the N/3 Method is clear and visible for everyone
- Remember that during the previous Kano categorization a clear separation into basics, satisfiers and delighters has already taken place so only consider the satisfiers and delighters; basics would always be more important in the evaluation
- The N/3 Method has the advantage that it is quick and uncomplicated; you should, however, take into account that the method can be influenced very easily by noise factors and that the result of the prioritization can be obstructed as well; it is possible for instance that participants are easily "influenced" by the opinions of others and distribute their points in line with the group "leaders" in order not to take a contrary position

Nominal Group Technique

Term
Nominal Group Technique

When
MEASURE: structure and prioritize customer needs

Goal
Identify weighting for delighters and satisfiers

Procedure

1. List the delighters and satisfiers and enter them into a table
2. Every team member/each customer determines the ranking of the delighters and satisfiers for him/herself; the delighter and/or satisfier which is considered to be most important receives the highest amount of points
3. Discuss the evaluation of outliers
4. Add up the values so that a ranking of delighters and satisfiers emerges

Illustration: Nominal Group Technique

Delighter/Satisfier		Person A	Person B	Person C	Sum
A	Satisfier 1	3	5	4	12
B	Delighter 1	7	3	3	13
C	Satisfier 2	4	2	2	8
D	Satisfier 3	5	6	6	17
E	Delighter 2	2	4	1	7
F	Satisfier 4	6	7	7	20
G	Delighter 3	1	1	5	7

Tip

- The distribution of rankings is ideally conducted by the customers or together with the customers
- If two or more delighters and/or satisfiers receive the same ranking have a discussion to determine why, and is it important. This should avoid any misunderstanding

Analytic Hierarchy Process*

Term
Analytic Hierarchy Process, AHP, Pair Comparison

When
MEASURE: structure and prioritize customer needs

Goal
- Compare and weight delighters and satisfiers in pairs
- Systematically derive a clear ranking

Procedure
The needs identified as delighters and satisfiers are structured in pairs to enable a comparative evaluation. The relative weighing of each need is calculated from weighing all pair combinations.

A scale from 1 to 9 is used for the evaluation:
1 = equally important
3 = a little more important
5 = more important
7 = much more important
9 = of extremely greater importance

A "less important" weighting in relation to "extremely less important" is evaluated with the help of the reciprocal fraction, e.g.:
1/7 = much less important
1/9 = extremely less important

Mapping AHP Contingency Table, example passenger seat on the following page

* Thomas L. Saaty (2000): Fundamentals of decision making and priority theory with the Analytic Hierarchy Process, Vol. VI of the AHP series, RWS Publications, Pittsburg/USA

Example: AHP Contingency Table Passenger Seat

Needs	I want the seat to with-stand vandalism	I want to be able to install and remove the seat quickly	I want a seat which is always clean	I want a low cleaning effort	Aggregate relativity	**Rel. importance [%]**
I want the seat to withstand vandalism	1.00	3.00	5.00	5.00		
	0.58	0.25	0.78	0.44	2.05	**51.0**
I want to be able to install and remove the seat quickly	0.33	1.00	0.20	0.33		
	0.19	0.08	0.03	0.03	0.33	**8.3**
I want a seat which is always clean	0.20	5.00	1.00	5.00		
	0.12	0.42	0.16	0.44	1.14	**28.5**
I want a low cleaning effort	0.20	3.00	0.20	1.00		
	0.12	0.25	0.03	0.09	0.49	**12.3**
Aggregate	1.73	12.00	6.40	11.33		
Aggregate relativity	1.00	1.00	1.00	1.00	4.00	100.0

The Pair Comparison proceeds from row to column, e.g.

- "I want the seat to withstand vandalism" is more important (3) than "I want to be able to install and remove the seat quickly"

  or

- "I want a seat which is always clean" is less important (1/5) than "I want the seat to withstand vandalism"

Then the aggregates of the columns are formed, e.g.

- Column:
  "I want the seat to withstand vandalism":
  1 + 0.33 + 0.2 + 0.2 = 1.73

I want the seat to withstand vandalism
1.00
0.58
0.33
0.19
0.20
0.12
0.20
0.12
1.73
1.00

- Spalte:
  I want a seat which is always clean":
  5 + 0.2 + 1 + 0.2 = 6.40

I want a seat which is always clean
5.00
0.78
0.20
0.03
1.00
0.16
0.20
0.03
6.40
1.00

The weighting of each defined cell is now compared to the respective column aggregate and a relative weighting calculated:

Needs	I want the seat to withstand vandalism
I want the seat to withstand vandalism	1.00
	0.58
I want to be able to install and remove the seat quickly	0.33
	0.19
I want a seat which is always clean	**0.20**
	0.12
I want a low cleaning effort	0.20
	0.12
Aggregate	**1.73**
Aggregate relativity	1.00

e.g. cell reference row "I want a seat which is always clean" and column "I want the seat to withstand vandalism" in relation to the column aggregate "I want the seat to withstand vandalism": 0.2/1.73 = 0.12

The total weighting of a single row is achieved by adding together the relative weightings of each cell in that row, e.g. the row

"I want the seat to withstand vandalism": 0.5 + 0.25 + 0.78 + 0.44 = 2.05.

Needs	I want the seat to withstand vandalism	I want to be able to install and remove the seat quickly	I want a seat which is always clean	I want a low cleaning effort	Aggregate rel.	**Rel. importance [%]**
I want the seat to withstand vandalism	1.00	3.00	5.00	5.00		
	0.58	**0.25**	**0.78**	**0.44**	**2.05**	**51.0**

The column aggregate of these values corresponds to the total weighting of all rows: 2.05 + 0.33 + 1.14 + 0.49 = 4.00.
Comparing the row aggregates with this total generates the final weighting of the individual needs in relation to the other needs (satisfier and delighter).

Tip

- An AHP Pair Comparison should ideally be carried out by the customers or together with the customers
- Facilitation of this should be done by a member of the project team
- It is best if the facilitator is supported by a time keeper and a scribe (e.g. for collecting individual evaluations)
- Make sure in advance that the question formulation for evaluating the needs is easily visible for everyone in the meeting room
- Use supporting software tools (e.g. Excel templates) for conducting the Pair Comparison and for gaining information about the individual evaluations
- If the evaluations given by the various customers are not consistent, the mean of the different comparisons can be calculated and entered into the corresponding cell of the Contingency Table
- It is not recommended to undertake an absolute weighting of needs using an ordinal scale (e.g. 1 = unimportant to 5 = very important) because this rarely leads to a meaningful prioritization!
- Experience shows that in this case all needs are evaluated as important and/or very important; it is therefore not possible to identify a reasonable prioritization or to ensure the targeted deployment of available resources
- In the Kano model it is very important to separate the needs into the three distinct categories – basic, satisfier and delighter. In the AHP you should only consider the relative importance of the satisfiers and the delighters as by definition the basics are the more important

Additional tip for using a simple Pair Comparison in a Workshop situation:

- When having to evaluate a great number of needs and having to cope with strong time restrictions you can apply a simple variant of the Pair Comparison
- It uses the strengths of the Pair Comparison but requires no evaluation of the relative importance of the needs

Mapping simple Pair Comparison, example passenger seat

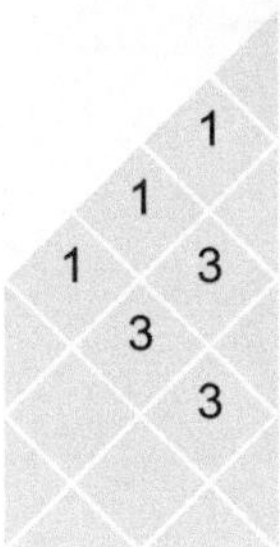

No.	Evaluation criterion	Number	Weighting
1	I want the seat to withstand vandalism	3	50.00%
2	I want to be able to install and remove the seat quickly	0	0.00%
3	I want a seat which is always clean	2	33.33%
4	I want a low cleaning effort	1	16.67%

Develop the simple Pair Comparison as follows:

- Basic question formulation: "Which defined need is more important?"
- Is the need "I want the seat to withstand vandalism" (no. 1) more important than the need "I want to be able to install and remove the seat quickly" (no. 2)?
- The need no. 1 "I want the seat to withstand vandalism" is considered to be more important by the respondents so the pairing is assigned a “1” in the evaluation pyramid
  – Every need can be compared with every other need and evaluated in this way
- The need "I want a seat which is always clean" (no. 3) is for example, considered to be more important by the respondents than the need "I want to be able to install and remove the seat quickly" (no. 2) so the pairing is assigned a “3”
- After having completed the evaluation of each possible pairing you count the number of responses in the evaluation pyramid; need no. 1 for example, has been evaluated 3 times as the more important need
- Next you summarize the scores in a column and compare the respective individual values with the column aggregate; the final weighting of the identified individual needs (i.e. Kano: satisfiers and delighters) is established

- In the Kano model it very important to separate the needs into the three distinct categories – basic, satisfier and delighter. In the AHP you should only consider the relative importance of the satisfiers and the delighters as by definition the basics are the more important

Derive CTCs and Key Output Metrics

Term

Derive CTCs and Key Output Metrics, derive CTCs and measurements

When

MEASURE: specify customer needs

Goal

Transform customer needs into specific and measurable customer requirements (Critical to Customer = CTC) with corresponding metrics

Procedure

Describe customer requirements rather than solutions

- Describe requirements in full sentences
- Formulate requirements which refer to a single item
- Formulate as specifically as possible
- Use concise descriptions
- Use positive statements
- Use measurable terms

(Test: requirement can be measured)

In the next step the CTCs are assigned corresponding metrics. The crucial factor is whether the respective measurement can give a clear indication of the specific requirement's fulfillment.

See next page for example

Example: Transformation Table – Passenger Seat

Need	CTC	Measurement
I want the seat to withstand vandalism	Each seat and its parts are resistant against improper treatment	Number of willful damages per seat
		Number of burn marks/stains per seat
		Number of slashes per seat
		Number of graffito per seat
		Number of graffito per seat after cleaning
I want a seat which is always clean	Each seat and its parts are clean at all times	Number of stains per seat/ per time
I want low cleaning effort	Each seat and its parts can be cleaned quickly	Cleaning time per time
I want to be able to install and remove the seat quickly	Each seat and its parts can be installed and removed quickly	Removal time per complete seat unit Assembly time per complete seat unit

Tip

- Note that CTCs don't have to be formulated for projects with low complexity
- Nevertheless, the benefit of changing from the customer's perspective "I want ... " to the system's perspective "Every item must have the feature…", which is linked with the CTC formulation, is often very helpful since CTCs support function analysis and the evaluation of alternative concepts in the ANALYZE phase

Conduct Benchmarking

Term
Benchmarking, comparison of system, comparison of product, comparison of process

When
MEASURE: specify customer needs

Goal
Evaluate competing systems with respect to fulfilling customer needs

Procedure
The customers are surveyed as to whether their needs are met by competing systems. For this purpose they assign single needs to a prescribed ordinal scale e.g.
1 = need is not met by own system/competing system
2 = need is weakly met by own system/competing system
3 = need is only adequately met by own system/competing system
4 = need is satisfactorily met by own system/competing system
5 = need is very well met by own system/competing system

The results are entered into a matrix for comparison purposes.

See next page for example

Example: Competitive Comparison – Passenger Seat

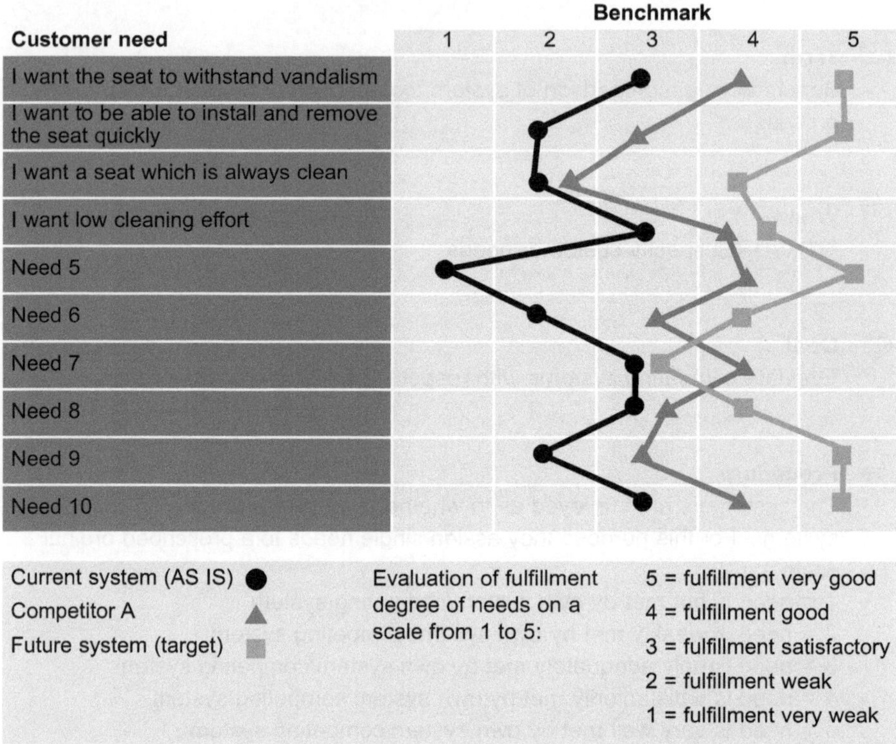

Tip

- Use the matrix for mapping the benchmark and define your own target values based on the positioning of competitors
- You should define the target values depending on the planned strategic positioning of the system since you probably intend to position yourself vis-à-vis competitors in the market using specific individual needs
- The needs on which you want to base a Unique Selling Proposition (USP), receive a higher priority compared to the AHP prioritization in order to be more competitive than the observed competitors

Quality Function Deployment

Term
Quality Function Deployment, QFD, House of Quality

When
MEASURE: specify customer needs

Goal
- Fulfill identified customer needs in a reliable way when developing new products and processes
- Focus on essential elements when products and processes are developed
- Define clear guidelines for the high-level and detailed design of product and process developments

Illustrating QFD Target System

Quality	Instrument for planning and developing quality functions based on customer requirements
+	
Function	Quality developments and quality improvements through the systematic and consistent collaboration of all areas of activity
+	
Deployment	Specification of the required quality into targets for individual company departments
=	"Stringent target system for converting the Voice of Customer into the language of the company."

Procedure

QFD pursues a derivation of goals in several stages, in clearly defined individual steps, by linking houses of quality. In each stage an interdisciplinary team ensures the correct and complete consideration of relevant information, which is prepared and documented, within the frame of the respective QFD Matrix.

Illustration of the procedural model of QFD in DMADV:

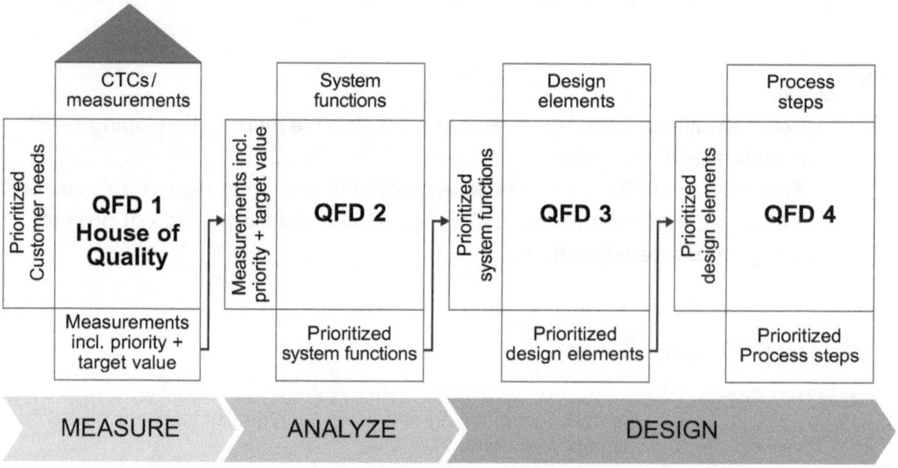

Tip

- Take into consideration that the "classic" QFD approach is a conceptual guiding idea rather than a standard procedure
- Furthermore the development of four houses of quality is rarely applicable in practice and rarely leads directly to success
- You should decide early on in each project if the development of a house of quality is reasonable in the respective project phase
- However, if target costs are to be derived in the ANALYZE and DESIGN phase for individual system components the development of QFD 1, QFD 2 and a modified QFD 3 is required as a combination of a component-functions matrix and elements of Target Costing

Quality Function Deployment 1

Term

Quality Function Deployment 1, QFD 1, House of Quality 1

When

MEASURE: specify customer needs, prioritize measurements, determine target values and specifications

Goal

- Translate identified customer needs into the language of the company as measurements, target values and specifications
- Prioritize measurements for evaluating future development work
- Identify potential design conflicts early on

Procedure

Illustration: Steps in QFD 1

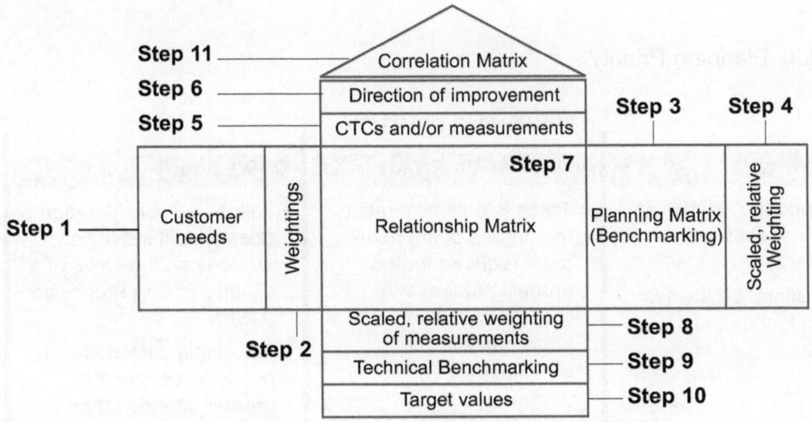

Step 1 (customer needs)
The customer needs which were identified with the help of the Tree Diagram and structured and categorized with the help of the Kano Model are inserted. It is important that all defined customer needs have a uniform hierarchy level.

Step 2 (weightings)
Fulfilling the dissatisfiers has a higher priority than fulfilling the delighters and satisfiers. Therefore, a priority of 100% is entered in the House of Quality for all dissatisfiers. For delighters and satisfiers the priority is entered as identified in the AHP.

Steps 3 and 4 (Planning Matrix and scaled, relative weightings)
The results from the "Customer Satisfaction Benchmarking" are summarized in the Planning Matrix. In this benchmarking the customers are asked how they evaluate the performance of existing systems: do they fulfill their needs? Their evaluation follows a scale from 1 (no fulfillment) to 5 (very good fulfillment).

The "improvement factor" is derived from this information and is complimented by the "need priority" of the AHP and a "USP factor". These three factors are corrected by a planning priority so as to obtain an adjusted scaled weighting of customer needs that serves as the basis for the Relationship Matrix.

Illustration: Planning Priority

PLANNING PRIORITY

AHP Priority	Improvement Factor	USP Factor
• Need priorities identified in AHP for satisfiers and delighters • Dissatisfiers are always given a high priority (100%)	• Takes into consideration the degree of improvement required for the product/process with respect to meeting customer needs (VOC Benchmark)	• Takes into consideration a possible differentiation strategy on the basis of a Unique Selling Proposition (USP). • Are single customer needs to be granted greater attention than USPs?
Comparison +	Comparison +	Comparison

= adjusted, scaled, relative weighting (evaluation basis for Relationship Matrix)

The Planning Priority defines the ranking between the single factors within the Planning Matrix:
- Need priority (derived from AHP)
- Improvement factor (VOC Benchmark)
- USP factor (differentiation based on Unique Selling Proposition)

This distinction is unavoidable because it is always a company-specific decision as to which factor is to be assigned the greatest importance in developing the system. The basis for determining the planning priority is an AHP consisting of three defined factors.

Illustration: AHP Matrix for weighing the factors of the planning priority

Planning priority	Need priority	Improvement factor	USP factor	Aggregate of rows	Aggregate scaled
Need priority	1.00	3.00	5.00	1.90	**63.33%**
Improvement factor	0.33	1.00	3.00	0.78	**26.00%**
USP factor	0.20	0.33	1.00	0.32	**10.60%**
Aggregate of columns	1.53	4.33	9.00	3.00	100.00%

The following scaling is used for the evaluation:
1 = equally important
3 = a little more important
5 = more important
7 = much more important
9 = extremely more important

Planning Priority		
Need priority [63.33%]	Improvement factor [26%]	USP factor [10.6%]

Tip

- The evaluation within the AHP Matrix for setting the planning priority is to be made by the DFSS Team in line with the company's strategic orientation

- The "improvement factor" results from the customer evaluation with respect to the fulfillment of single needs; this evaluation is used to define an improvement goal that is then assigned a corresponding improvement factor (see AHP Matrix); finally the improvement factor can be adjusted for the planning priority

Illustration: AHP Matrix for scaling the improvement goal

Improvement goal	5	4	3	2	1	Aggregate of rows	Scaled aggregate
5	1.00	2.00	3.00	4.00	5.00	1.9471	**38.94%**
4	0.50	1.00	3.00	5.00	5.00	1.5352	**30.70%**
3	0.33	0.33	1.00	3.00	4.00	0.8144	**16.29%**
2	0.25	0.20	0.33	1.00	3.00	0.4487	**8.97%**
1	0.20	0.20	0.25	0.33	1.00	0.2547	**5.09%**
Aggregate of columns	2.28	3.73	7.58	13.33	18.00	5	100.00%

Scaling for the definition of the improvement goal:
1 = need is not met by own system/competing system
2 = need is only weakly met by own system/competing system
3 = need is moderately met by own system/competing system
4 = need is well met by own system/competing system
5 = need is very well met by own system/competing system

The weighting corresponding to the improvement goal is entered into the Planning Matrix.

Illustration: Identification of Improvement Factor

Benchmark							
1 (no fulfillment)	2	3 (moderately met)	4	5 (very well met)	Improvement goal	Improvement goal scaled	Improvement factor total
26.0%							
A		B		C	0.389	35.0%	9.1%
A	B	C			0.163	14.6%	3.8%
A	B	C			0.163	14.6%	3.8%
A	C				0.090	8.1%	2.1%
	A	B	C		0.307	27.6%	7.2%
					0	0.0%	0.0%
					0	0.0%	0.0%
					0	0.0%	0.0%
					0	0.0%	0.0%
					0	0.0%	0.0%
					1.112	100.0%	26.0%

A = customer's viewpoint – current product
B = customer's viewpoint – competition
C = improvement goal

In addition to a need's priority and improvement factors, the "USP factor" (Unique Selling Proposition) evaluates its' potential in relation to competing systems. The USP evaluation can also be scaled using an AHP.

Illustration: Identification of the USP factor

USP	Strong	Moderate	Small	Aggregate of rows	Aggregate scaled
Strong	1.00	2.00	3.00	1.57	**52.47%**
Moderate	0.50	1.00	3.00	1.00	**33.40%**
Small	0.33	0.33	1.00	0.42	**14.20%**
Aggregate of columns	1.83	3.33	7.00	3.00	100.00%

For each customer need there is, as a result of the Planning Matrix, an adjusted scaled priority that reflects the impact of the correction factors, for example, the need's priority, its' improvement factor and its' USP factor. In turn, this priority provides the evaluation basis for future steps in the Relationship Matrix.

Example: Planning Matrix- Passenger Seat

			Benchmark								USP					
	Weighting (AHP)	Need priorities in total	1 (need not met)	2	3 (moderately met)	4	5 (very well met)	Improvement goal	Improvement goal scaled	Improvement factor in total	USP	Improvement goal	Improvement goal scaled	USP factor in total	Adjusted scaled priority	Ranking
Overall priority		63.33%	26.0%								10.6%					
I want the seat to withstand vandalism	29.19%	18.5%	A		B			0.307	23.1%	6.0%	Moderate	0.334	18.0%	1.9%	26.41%	2
I want a seat whose parts cannot be stolen	41.33%	26.2%		A	B			0.389	29.3%	7.6%	High	0.525	28.2%	3.0%	36.80%	
I want to be able to install and remove the seat quickly	17.72%	11.2%	A	B				0.307	23.1%	6.0%	High	0.525	28.2%	3.0%	20.24%	3
I want a seat which always looks clean.	4.64%	2.9%	A	B				0.163	12.3%	3.2%	Moderate	0.334	18.0%	1.9%	8.04%	5
I want a seat which can be cleaned quickly	7.12%	4.5%	A	B				0.163	12.3%	3.2%	Small	0.142	7.6%	0.8%	8.51%	4
		0.0%						0	0.0%	0.0%		0	0.0%	0.0%	0.0%	6
		0.0%						0	0.0%	0.0%		0	0.0%	0.0%	0.0%	6
		0.0%						0	0.0%	0.0%		0	0.0%	0.0%	0.0%	6
		0.0%						0	0.0%	0.0%		0	0.0%	0.0%	0.0%	6
		0.0%						0	0.0%	0.0%		0	0.0%	0.0%	0.0%	6
		63.3%						1.329	100.0%	26.0%		1.858	100.0%	10.6%	100.0%	

Customer needs with AHP weighting

Scaled priorities compared to overall priority

Adjusted, scaled priority with corresponding ranking

The customer need "I want to be able to install and remove the seat quickly" has been evaluated at 17.72% in the AHP Matrix. Combining the target improvement goal and the strategic necessity for fulfilling the need as a USP vis-à-vis competing systems, lead to an adjusted scaled priority of 20.24%. This priority provides the basis for evaluating the relevant measurements in the Relationship Matrix.

Step 5 (CTCs and measurements)
The CTCs and metrics identified using the transformation table are listed here.

Step 6 (improvement direction)
The improvement direction is fixed and given a symbol for each listed measurement.

Illustration: Improvement Direction

↑	1	Maximize
O	0	Fulfill this exactly
↓	-1	Minimize

Step 7 (Relationship Matrix)
The relationships between the prioritized customer needs and the measurements are formed in this matrix. Because the measurements are derived directly from the customer needs, each should strongly correlate with at least one need. The goal is to identify at least one strong relationship with a measurement for each need. The market expertise and technological know-how of the team members is critical for answering the following questions:

- To what extent do the measurements indicate the degree of fulfillment of the respective need?
- To what extent does a positive change of the measurement (based on the envisaged improvement direction) lead to a better fulfillment of the respective need?

A scale is used for the degree of the correlation.

Illustration: Correlation scale between measurement and customer need

Symbol	Meaning	Number
/	No correlation	0
△	Potential correlation	1
O	Moderate correlation	3
⦿	Strong correlation	9

Incorrect evaluations in the Relationship Matrix have far reaching repercussions on the project results. Uncertainties about evaluations must be taken seriously! It is very helpful to formulate an Operational Definition for the measurements (what is to be measured and how?). Strive for consensus in the ratings rather than objectivity!

Step 8 (scaled, relative weighting of measurements)
The scaled, relative weighting of the measurements enables the recognition of the focal points for the subsequent system development. The weighting of a measurement is calculated from the aggregate product of the correlation number and the overall priority.

Example: Relationship Matrix with scaled, relative weighting of measurements – Passenger Seat

Improvement direction		Min.	Min.	Min.	Min.	Min.	Min.	Min.	Min.	Min.	
CTCs and measurements		1.1 Number of replaced elements	1.2 Number of burn marks/stains per seat	1.3 Number of slashes per seat	1.4 Number of graffiti after clean-ing per seat	2.1 Number of missing elements	3.1 Assembly time for each single part	3.2 Assembly time for the whole seat	4.1 Number of complaints due to dirt	5.1 Time for cleaning	
	Weighting (AHP)	1. Each seat and parts are resistant against inproper treatment				2. Each seat and its components are theft-proof	3. Each seat and its individual parts can be installed and removed any time		4. Each seat and its individual parts look clean at all times	5. Each seat and its individual parts can be cleaned quickly	Adjusted, scaled priority from the Planning Matrix
I want the seat to withstand vandalism	29.19%	3	9	9	3	3	0	0	3	0	26.41%
I want a seat whose parts cannot be stolen	41.33%	0	0	0	0	9	1	1	0	0	36.80%
I want to be able to install and remove the seat quickly	17.72%	1	0	0	0	0	9	9	0	0	20.24%
I want the seat to look clean at all times	4.64%	0	0	0	9	0	0	0	9	1	8.04%
I want a seat which can be cleaned quickly	7.12%	0	0	0	0	0	3	1	1	9	8.51%
		0.99	2.38	2.38	1.52	4.10	2.44	2.27	1.60	0.85	
Scaled relative weighting		5.37%	12.82%	12.82%	8.18%	22.15%	13.19%	12.27%	8.64%	4.57%	

Step 9 and 10 (Technological Benchmarking and target values)
Technological benchmarking compares the current process performance capability of one's own system and competing systems with regard to the prioritized measurements.
For each measurement the present system in place (A) is compared and evaluated with the competing system (B). Depending on the target level (C) of the CTCs, specific target values and tolerances (USL/LSL) can be derived for the individual measurements.

Another possibility is to supplement technological benchmarking with the degree of difficulty involved in attaining a goal. The relevant question is: How great is the effort/input for attaining the set goal and/or how difficult is it to attain the goal?
The degree of difficulty is assessed on a scale of 1 (very simple goal attainment) to 5 (very difficult goal attainment). By multiplying the weighting of the respective measurement, a gage is generated for assessing the risk of realization. This shows which measurements are the most critical for implementing the project.

Example: Technological Benchmarking and target values with tolerances – Passenger Seat

	Measurements										
	Min.	Min.	Min.	Min.	Min.	Min.	Min.	Min.	Min.		
	1.1 Number of replaced elements	1.2 Number of burn marks/stains per seat	1.3 Number of slashes per seat	1.4 Number of graffiti after cleaning per seat	2.1 Number of missing elements	3.1 Time needed for each single part	3.2 Time needed for the whole seat	4.1 Number of complaints due to contamination	5.1 Time for cleaning		
	5.37%	12.82%	12.82%	8.18%	22.15%	13.19%	12.27%	8.64%	4.57%	0%	Scaled weighting
	No./month and seat	No. of new units/month	No. of new units/month	No. of new units/month	No. of new units/month	Minutes/parts	Minutes/seat	No./month	Minutes/seat		Unit
	Discrete	Discrete	Discrete	Discrete	Discrete	Continu.	Continu.	Discrete	Continu.		Data type
	0	0	0	0	0	5	10	1	2		Target value
	0	0	0	0	0	0	0	0	0		LSL
	0	0	0	0	0	8	12	0	4		USL
											Quality key figure
5		■	■		■	■		■			Technological Benchmark
4	■	B		■			■		■		
3	B		B	B	B	B		B			
2	A	A		A		A	B		B		
1			A		A		A	A	A		
Optional	2	4	5	3	5	5	4	■	4		Degree of difficulty
	10.73%	51.30%	64.12%	24.53%	110.73%	65.95%	49.08%	8.64%	18.26%	0.0%	Critical features for attaining goal
	8	4	3	6	1	2	5	9	7	10	Ranking

Step 11 (Correlation Matrix)
The Correlation Matrix helps to identify dependencies between measurements. In particular conflicts are pinpointed which must be solved innovatively in the subsequent system development. A simple example might be increasing the stability of an object, which in fact, might be in conflict with the reduction of its' weight. The measurements with their respective improvement direction are compared.

Illustration: Correlation Matrix

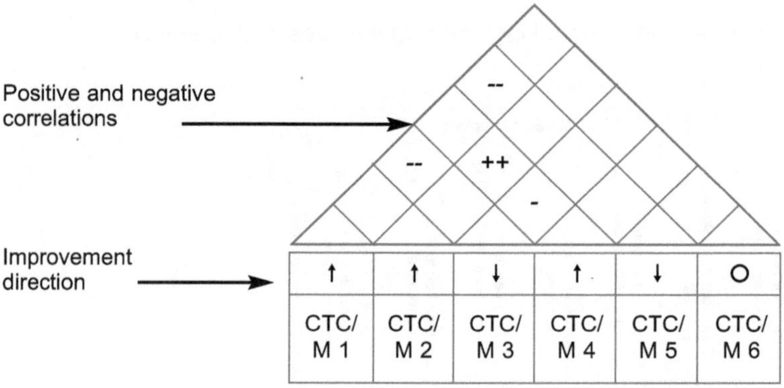

Correlation symbols:

- ++ Strongly positive effect
- \+ Slightly positive effect
- O No effect
- \- Slightly negative effect
- -- Strongly negative effect
- M Measurement

Improvement direction:

- ↑ Must be increased
- O Must remain constant
- ↓ Must be reduced

The correlation roof only represents positive and negative interactions between the measurements. The corresponding interaction can be considered and investigated, however, using the degree of difficulty (Step 9 "Technological Benchmarking"). The negative interactions shown in the correlation roof can be interpreted as technological conflicts/contradictions. Methods for solving such conflicts will be presented in the ANALYZE phase.

Example: Correlation Matrix – Passenger Seat

Correlation Matrix	Improvement direction	1.1 Number of - replaced elements	1.2 Number of burn marks/stains per seat	1.3 Number of slashes per seat	1.4 Number of graffito after cleaning per seat	2.1 Number of missing elements	3.1 Assembly time for each single part	3.2 Assembly time for the whole seat	4.1 Number of complaints due to dirt	5.1 Time for cleaning
Improvement direction		Min.	Min.	Min.	Min.	Min.	Min.	Min.	Min.	Min.
1.1 Number of replaced elements	Min.		++	++	+	++				
1.2 Number of burn marks/ stains per seat	Min.	++							+	
1.3 Number of slashes per seat	Min.	++								
1.4 Number of graffito after cleaning per seat	Min.	+								
2.1 Number of missing elements	Min.						–			
3.1 Assembly time for each single part	Min.					–				
3.2 Assembly time for the whole seat	Min.									
4.1 Number of complaints due to dirt	Min.	+								
5.1 Time for cleaning	Min.									

↑ Measurements ↑ Improvement direction

- The team identifies a negative interaction between the measurements for:
  - the assembly time for single parts and
  - number of missing parts
- If the assembly time is to be reduced, which is demanded by the improvement direction, theft-proof design can no longer be guaranteed
- This negative correlation has to be considered as the project proceeds; the team decides to look at the technological conflict with TRIZ

Design Scorecard

Term
Design Scorecard

When
MEASURE: specify customer needs, ANALYZE, DESIGN, VERIFY

Goal
- Summarize the most important results from the House of Quality
- Make need fulfillment measurable with the help of precisely defined criteria

Procedure
1. Insert and number the measurements defined in the House of Quality

2. Next, every measurement is to be assigned a relevant unit, an Operational Definition, a relevant specification and a target value

3. Define further quality key figures as appropriate

4. Estimate risks which may occur when important CTCs and/or measurements are not fulfilled

Example: Design Scorecard – Passenger Seat

#	1	2	3	4
Measurement	Number of willful damages per seat	Number of stains per seat/per time	Cleaning time per seat	Assembly time for whole seat unit
Unit	Number per seat	Number per seat	Minutes per seat	Minutes per seat
Priority	X %	X %	X %	X %
Operational Definition	Burn holes, stains (writing) and cuts in visible places of the seat and backrest are counted. A visual inspection of the seat system is conducted	Stains on the seat and backrest are counted. A visual inspection of the seat system is conducted	The time required for cleaning the seat and its backrest completely is measured in minutes. Start: First application of cleaning agent Stop: Restoration of operational capability	The time required for installing the seat system is measured. Start: Beginning of first working step for the installation of the seat system Stop: Seat system in fixed position completely installed and ready for use
Target value	0	0	5	4
LSL	--	--	0	0
USL	--	--	10	6
KPI	DPU	DPU	C_p, C_{pk}	C_p, C_{pk}
Conflict	Potential conflict between robustness, easy maintenance and seat comfort			
Effect in case of non-fulfillment	• Increased failure probability • Increased maintenance effort for customers • Bad reputation for suppliers			

Operational Definition

Term
Operational Definition

When
MEASURE: specify customer needs

Goal
- Create a uniform understanding of measurement and measurement method
- Avoid ambiguities in the data collection and interpretation of the results
- The measurement result is irrespective of the person conducting the measurement

Procedure
1. Formulate an Operational Definition for each measurement – what is to be measured and how?

2. If required sample cards with decision criteria can be defined

3. Check with the help of the Measurement System Analysis if the Operational Definition delivers reliable results

Illustration: Operational Definition

WHAT	Measurement	A detailed description of the object to be measured and of the measurement (what exactly is measured?)
HOW	Measurement instrument	A precise description of the measurement instrument (measure-ment device, mere visual inspection, etc.)
HOW	Measurement method	A description of the detailed measurement procedure i.e. position of object and measurement device, lighting, if required, handling of measurement device
HOW	Criteria for decision	In case of discrete measurements these are criteria for good parts or scrap, e.g. cards for comparison and sample cards etc.; continuous characteristics (e.g. the diameter of a pipe) can also be measured with the help of a gage in a discrete way

Example: Passenger Seat

Measurement	Number of stains per seat/per time
Measurement instrument	Visual inspection of seat and backrest
Measurement method	After having been removed the seat and the back of the seat are observed in full lighting. All visible surfaces are evaluated with the help of a comparison card for degree of dirt on seat and back surface
Criteria for decision	If the degree of dirt on the seat does not exceed A3 on the comparison card the seat is acceptable. If it exceeds A3 then the seat is to be reworked

Quality Key Figures

Term
Quality Key Figures, Process Performance

When
Closure MEASURE: specify customer needs

Goal
- Determine the performance capability of a system (product/process) with respect to the defined customer requirements
- Define the status quo in order to be able to conduct control measurements and comparisons later

Procedure
Among the most common quality key figures for establishing performance capability are:

DPMO	Defects Per Million Opportunities
ppm	Parts per Million
DPU	Defects per Unit
Yield	Yield
C_p und C_{pk}	Process capability indices
Process Sigma	Sigma value

Parts per Million (ppm)

Term
Parts per Million (ppm)

When
Closure MEASURE: specify customer needs

Goal
Evaluate defects from the customer's perspective since a unit with one or more defects must be evaluated as a defect from the customer's point of view

Procedure
- Determine the defect opportunities for a part/unit which result in its characterization as defective
- Determine the number of inspected parts/units and count the number of defective units
- Calculate the ppm value:

$$ppm = \frac{\text{Number of defective units}}{\text{Total number of units}} \cdot 1{,}000{,}000$$

Example: ppm – Passenger Seat Assembly Process: Parts per Million

- In 63 out of 80 back elements of the seat system to be assembled rework was necessary because of damages and imperfections:

$$ppm = \frac{63}{80} \cdot 1{,}000{,}000 = 787{,}500$$

- The ppm rate amounts to 787,500

Tip
- The DPMO value corresponds to the ppm value if there is only one defect opportunity

Defects per Unit (DPU)

Term
Defects per Unit (DPU)

When
Closure MEASURE: specify customer needs

Goal
Determine the average number of defects per unit

Procedure
- Define each defect opportunity that could occur in a unit
- Determine the number of inspected units and count the defects
- Calculate the DPU value:

$$DPU = \frac{\text{Total number of defects}}{\text{Total number of units}}$$

Example DPU – Passenger Seat Assembly Process: Defects per Unit

- In a total of 80 back elements assembled 108 defects were identified due to damages and imperfections:

$$DPU = \frac{108}{80} = 1.35$$

- The DPU rate amounts to 1.35. This means that one manufactured part has 1.35 defects on average

Tip
- Taken together, the three quality key figures DPMO, ppm and DPU provide a comprehensive image of the process performance – it is therefore strongly recommended to use all three key figures

Yield

Term
Yield

When
Closure MEASURE: specify customer needs

Goal
Determine the proportion of non-defective units and/or good parts in a process

Procedure
Yield: Indicates the proportion of good, non-defective units.

$$Y = \frac{\text{Number of non-defective units}}{\text{Total number of units}} \cdot 100$$

Rolled Throughput Yield: Calculates the probability that a unit will run through the entire process without becoming a defective. This total yield is calculated by multiplying the single sub process yields.

$$Y_{RTP} = Y_{Sub\ 1} \cdot Y_{Sub\ 2} \cdot \ldots \cdot Y_{Sub}$$

Normalized Yield: Determines the average yield per process step
Attention: This measurement can be misleading when the yields of the individual process steps differ strongly from one another.

$$Y_{Norm} = \sqrt[n]{Y_{RTP}}$$

Examples Yield – Example 1: Passenger Seat Assembly Process Yield

- Only 21 out of 80 back elements were non-defective
- We have a yield rate of 26.25%

$$\text{Yield} = \frac{21}{80} = 0.2625 = 26.25\%$$

Examples Yield –
Example 2 Passenger Seat Assembly Process Rolled Throughput Yield

- The following yields were calculated for the individual process steps of the assembly process:

$Y_1 = 92\%$ → $Y_2 = 82\%$ → $Y_3 = 84\%$ → $Y_4 = 82\%$ → $Y_5 = 95\%$

- The probability that a unit runs through the entire process without becoming defective is $Y_{RTP} = 0.92 \cdot 0.82 \cdot 0.84 \cdot 0.82 \cdot 0.95 \cong 0.494$

➾ **Tip**

- A distinction between two characteristics of yield can be made:
  1. The relationship between non-defective units and the total amount of units (yield in classic production)
  2. The ratio between the quantity of good product produced and the quantity of material used in making it. (yield in the chemical industry/pharmaceutical industry)
- The yield is usually determined before repair and/or rework occurs (First Pass Yield)

C_p- and C_{pk} Values

Term

C_p and C_{pk}

When

Closure MEASURE: specify customer needs

Goal

- Find out the relationship between the customer specification limits (tolerance limits) and the natural spread of the process (C_p value)
- Determine how centered the process is (C_{pk} value)

Procedure

C_p value:

- Determine the upper and lower specification limits
- Divide the distance between the upper and lower specification limits (tolerance) by 6 times the standard deviation of the process
- With non-normally distributed data: Divide the tolerance by the percentile distance of +/-3 standard deviations (corresponds to 99.73%)

With normal distribution	With non-normal distribution
$C_p = \frac{USL - LSL}{6s}$	$C_p = \frac{USL - LSL}{X_{0.99865} - X_{0.00135}}$

C_{pk} value:

- Divide the distance between the closest specification limit and the mean value by 3 times the standard deviation of the process; this now takes the position of the process into consideration
- With non-normally distributed data: divide the distance between the closest specification limit and the Median by half the percentile distance

With normal distribution	With non-normal distribution
$C_{pk} = \min\left[\frac{USL - \bar{x}}{3s}; \frac{\bar{x} - LSL}{3s}\right]$	$C_{pk} = \min\left[\frac{USL - \tilde{x}}{X_{0.99865} - \tilde{x}}; \frac{\tilde{x} - LSL}{\tilde{x} - X_{0.00135}}\right]$

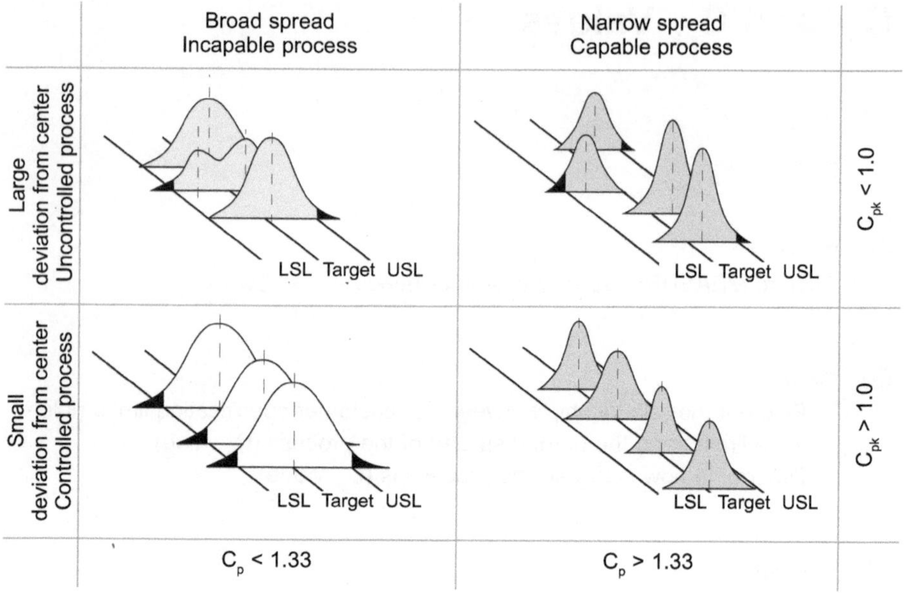

Example: C_p and C_{pk} in MINITAB®

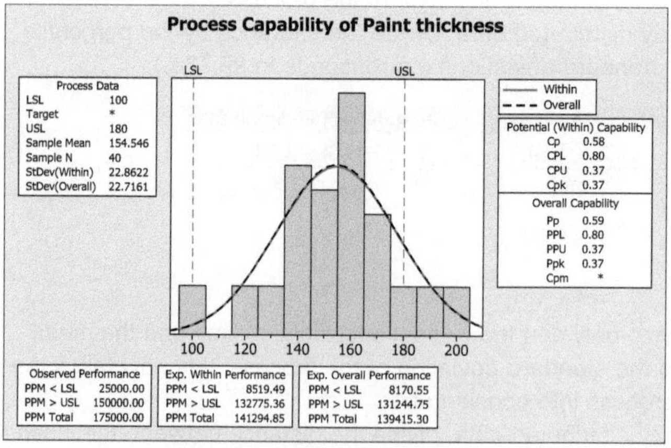

C_p = (USL-LSL)/(6s within) → >1 means: The process is narrower than the set limits

CPU= (USL-m)/(3s within) → <1 means: The upper limit is exceeded.

CPL = (m-LSL)/(3s within) → <1 means: The lower limit is exceeded.

C_{pk} = min{CPU, CPL} → <1 means: The process breaches at least one of the limits

m = midpoint between USL, LSL

Graphic result

The upper and lower specification limits of the sample:
A histogram shows the distribution of data in relation to the specification limits. The curve depicts the normal distribution. If both, short and long-term observations are provided then two curves will be shown.

The C_p and C_{pk} values:
The bigger these values are, the more capable the process. C_p = 2 Six Sigma level.

Since no subgroups were given, the short and long term process capability are identical.

Tip

- A high C_p value is a necessary but not a sufficient condition for a good process sigma value, it is also necessary to take into account process centering. A C_{pk} value allows us to do this, so only by having a high C_{pk} can a high process sigma be attained
- In order to reach a Sigma level of 6 (Six Sigma process), C_p and C_{pk} must assume the value of 2 (i.e. at least 6 standard deviations fit between the mean and the customer specification limits); due to the assumed process shift of 1.5 standard deviations Six Sigma+Lean corporations such as Motorola set C_p values of 2 and C_{pk} values of 1.5 as their goals
- When long term observations are used C_p and C_{pk} values are referred to as P_p and P_{pk}

Process Sigma

Term
Sigma Value, Process Sigma

When
Closure MEASURE: specify customer needs

Goal
- Map the performance capability of a process especially with respect to the specification limits
- Use as benchmark and/or Best Practice

Procedure
- Using **DPMO**:
  Identify from a Sigma conversion table (see appendix)

- Using **Yield**:
  Identify from a Sigma conversion table by using the First Pass Yield

Tip
- Do not enforce sigma values as the sole performance indicators in the company
- Use values, metrics and key figures which are easily understood and accepted by the customers and the employees alike

Gate Review

Term
Gate Review, Phase Check, Phase Approval

When
Closure MEASURE: when each phase is closed

Goal
- Inform the Sponsor about the results and actions of each phase
- Evaluate the results
- Decide about the future course of the project

Procedure
The results are presented in a complete and comprehensible way. The Sponsor checks the current status of the project according to the following criteria:
- Completeness of results
- Probability of project success
- Optimal resource allocation in the project

The Sponsor decides if the project can enter the next phase.

A variety of results from the MEASURE phase are presented in the MEASURE Gate Review to the Sponsor and the Stakeholders. The following questions must be answered during a complete and comprehensible presentation:

Identify customers and their needs
- How was the target market determined?
- How were the target customers of the product/process identified and what characterizes them?
- Were the target customers segmented? Which target customers have what priority?

Structure and prioritize customer needs

- Which survey techniques were selected? What are the results?
- How was the Customer Interaction Study conducted? What are the results?
- Have costs been estimated? Are they permissible?
- In what price range could the product/process be offered?

Specify customer requirements

- What are the customer needs? How are they related?
- Are there any contradictory customer needs and how have they been considered?
- Do competing systems exist and how do they fulfill the customer needs?
- Were the Voices of the Customer transformed into CTCs and measurements? With which priority?
- Has a technological Benchmarking been conducted? If so, how? What are the results?
- How were the target values and tolerances determined for the measurements?
- What defect rates might the measurements show?
- Are conflicts to be expected in the fulfillment of individual CTCs/measurements?
- What restrictions and obstacles were identified?
- What consequences should be expected if CTCs are not fulfilled?

On Project Management

- What are the "Lessons Learned" of the MEASURE phase and what next steps do they require?
- Why does it make sense to continue the project?
- Has the Business Case been adjusted in respect of new insights?
- Are any adjustments required to the previously defined Multi Generation Plan?

DESIGN FOR SIX SIGMA+LEAN TOOLSET

ANALYZE

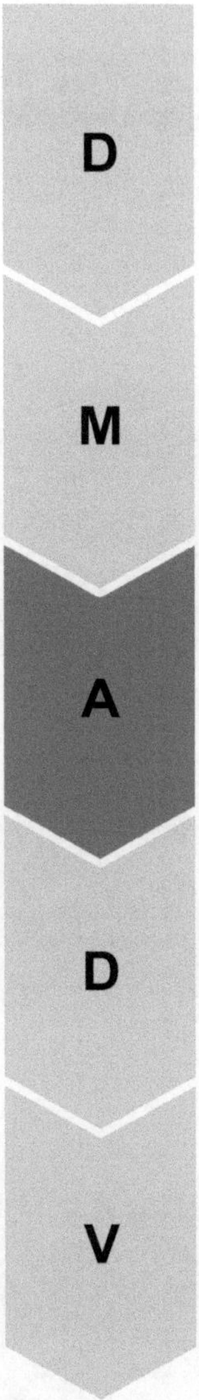

Central Question and Objective of the ANALYZE Phase

- The ANALYZE phase of the DMADV methodology answers the question: "What is the best high-level concept, that can be created, based on the defined customer requirements?"
- Objective:
  - Develop appropriate high-level concepts
  - Evaluate and optimize the high-level concepts
  - Finalize and coordinate the best high-level concept

The Meaning of the ANALYZE Phase

- Defining the basic system functionality i.e. what does the product or process have to do?
- Comparing product or process functions with customer requirements to reveal gaps or further development potential
- Comparing alternative concepts and eliminating potential weak points and conflicts
- Adopting the guiding principle that new systems should not be designed in a more complex way than the old systems
- Observing the relevant manufacturing flows and supporting processes, to evaluate whether they can accommodate, not only the new high level concept but also the specific customer requirements

Procedure in the ANALYZE phase

- Identify relevant required functionality
- Develop alternative high-level concepts
- Select the best detailed concept
- Determine target costs
- Resolve weak points and conflicts in the selected concept
- Collect customer and stakeholder feedback
- Define the basic process
- Prepare market launch
- Derive necessary resource requirements

D
M
A
D
V

A.1 Develop high-level concept

- What alternative high-level concepts are there, and how well do they match up to customer requirements?
- Which high-level concept fulfills the defined customer and business requirements best?

A.2 Evaluate high-level concept

- What might the individual system components cost taking into account the calculated target costs?
- Where are potential weak points and conflicts apparent in the concept?
- What feedback do we get from customers and Stake-holders?

A.3 Finalize high-level concept

- What would the manufacturing flows and the necessary supporting processes look like?
- How can the product's market launch be prepared?
- What resources are required for the concepts' implementation of the detailed design?

Result ANALYZE

- Best high-level concept design based on customer and business requirements
- Best high-level concept evaluated by taking into account the costs and the conflicts
- Relevant processes sketched out for the product domain
- Necessary resources defined and detailed design prepared

ANALYZE Gate Review/Phase Closure

A.1 Develop high-level concept

Functional Analysis	QFD 2	Concept Development	Methods for External Search for Solutions	Methods for Internal Search for Solutions
Morphological Box	Conjoint-Analysis	Methods for Concept Evaluation	Concept Evaluation Matrix	Pugh Matrix

A.2 Evaluate high-level concept

QFD 3 for Calculation of Target Costs	TRIZ	FMEA	Risk Analysis	Storytelling
Early Prototyping	Customer/ Stakeholder Feedback			

A.3 Finalize high-level concept

SIPOC	Market Launch	Resources

A Gate Review

Develop High-level Concept

Term
High-level Concept Identification

When
ANALYZE: develop high-level concept

Goal
- Develop alternative high-level concepts based on prioritized system functionality
- Identify the best concept by taking into account the defined customer and business requirements

Procedure
The requirements which were identified within the scope of the MEASURE phase can potentially be fulfilled in different ways. Therefore in the ANALYZE phase it is important to identify promising solution options and to combine them to create alternative high-level concepts.
The best concept is selected from the various competing concepts. Selection criteria include the identified customer requirements, the degree of innovation and the implementation effort required to deliver the concept. Afterwards, the best concept is further optimized and its' design enhanced. The level of detail increases continuously. The nature of the project will determine what level of detail is required before a high-level concept can be frozen (closure ANALYZE); it is project specific and should be agreed by the project team upon entering the ANALYZE phase.

From high-level to detailed concept

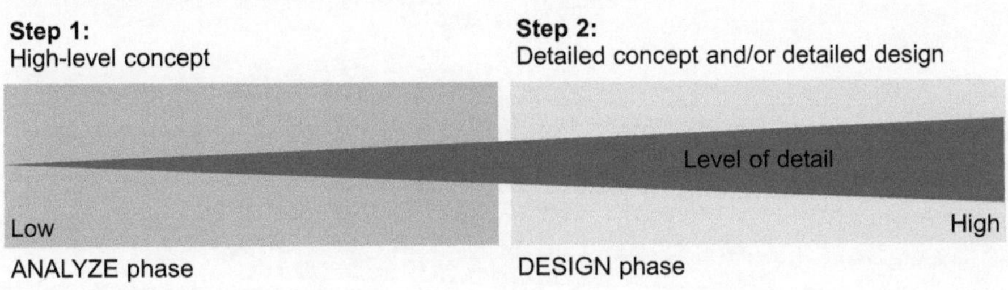

The basis for developing alternative concepts is initially an analysis of the relevant required system functions. The abstract and solution-free formulation of system functionality supports the extension of the solution space and thus the development of innovative ideas.

Functional Analysis

Term
Functional Analysis

When
ANALYZE: develop high-level concept

Goal
- Describe the system without solutions as an impact system of interacting functions
- Identify and prioritize all relevant sub-functions within this impact system

Procedure
Based on the question "WHAT must the system do, or be able to do, in order to fulfill the defined requirements?" the necessary functions are formulated by using a noun and a verb. This can be achieved by Brainstorming.

Example: Functional Analysis – Passenger Seat

No.	CTC	Measurement	Opt. Direction	Functions
1	Each seat and its parts are resistant against improper treatment	Number of willful damages per seat	Minimize	• Protect seat • Absorb power
2	Each seat and its parts are clean at all times	Amount of dirt per seat and time unit	Minimize	• Repel dirt
3	Each seat and its parts can be cleaned quickly	Cleaning time per seat	Minimize	• Remove dirt particles • Ensure accessibility
4	Each seat and its parts can be installed and removed quickly	Removal time per complete seat unit	Minimize	• Ensure accessibility • Undo fixings • Fasten seat
...	...	...	...	...

Specifics for process development

The functions are connected to the high-level process steps i.e. at a SIPOC level during the process development. Starting from the CTCs and/or measurements the question is asked: "WHAT do we have to do to fulfill these requirements?" The resulting activities should be listed sequentially and, by using appropriate verbs and nouns, become your new high level process steps.

CTC	Measurement	Functions
Every area delivers the data completely	Proportion of incomplete data deliveries (per division)	Define required data
		Communicate required data to divisions
		Set responsibilities in divisions
		Control completeness of data

Prioritize Functions

Term
Quality Function Deployment 2, QFD 2, House of Quality 2

When
ANALYZE: develop high-level concept

Goal
- Check the connection between the identified system functions and the measurements
- Weight the system functions by taking into consideration the measurement prioritization

Procedure
The relationships between the identified functions and the weighted measurements are identified in QFD 2.

Mapping QFD 2

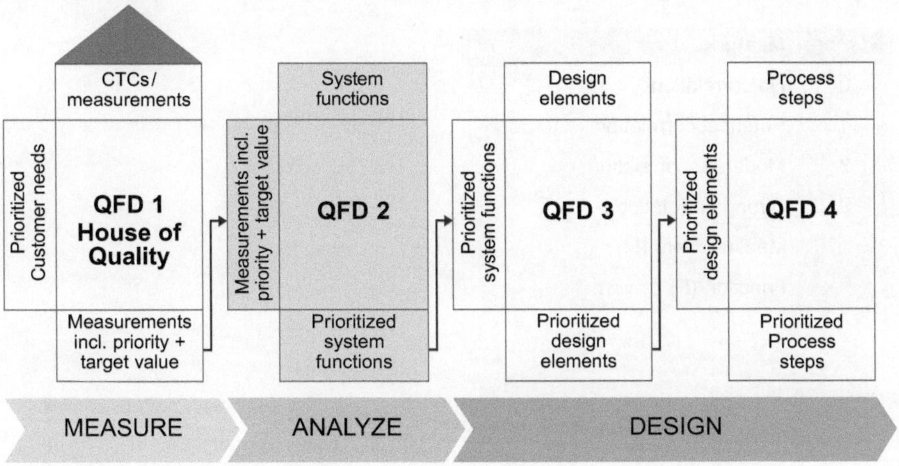

The influence of the system functions on the individual measurements is analyzed.

Thus, the following questions can be answered:
1. Can all relevant measurements be realized with the help of the identified system functions or do "blind spots" remain?
2. How strong is the influence of an individual system function on the respective measurement? Are specific system functions particularly important or unimportant or even superfluous?

Illustration: QFD 2 – Generic example

	System functions				
	F 1	**F 2**	**F 3**		**Scaled, relative weighting M**
Measurement 1	9				25%
Measurement 2	3	9			45%
Measurement 3		1	9		30%
Weighting (absolute)	3.6	4.35	2.7		10.65 —— 100%
Scaled, relative Weighting	33.8%	40.8%	25.4%		100%

Legend for evaluating the cause-effect relation

Number	Meaning
0	No correlation
1	Potential correlation
3	Moderate correlation
9	Strong correlation
	Measurement (M)
	Function (F)

		Function					
Measurement	Prio	Keep body	Absorb impact	Repel dirt	Resist violence	Ensure accessibilities	Prevent theft
Number of willful damages per seat	39%				9		
Number of stains per seat	28%			9			
Cleaning time per seat	22%	3		9			
Installation time per seat unit	11%					9	
	Σ	...	...	...	...	...	...
	Prio	...	...	...	...	...	...

The measurements and their priorities derived in the MEASURE phase are now used to arrive at a weighting for the system functions. With base on this weighting the priorities in the concept development can be set.

Tip

- When a QFD 2 is drawn up it might be sensible to integrate a correlation matrix; by using a correlation matrix you can make potential conflicts between product and/or process functions transparent
- Generally, the Ideality Principle should be taken into consideration, which, loosely translated, means maximizing the sum of useful functions and minimizing the sum of harmful functions

$$\text{Ideality} = \frac{\sum \text{useful functions} \uparrow}{\sum \text{harmful functions} \downarrow}$$

- Apart from QFD 2 you can also use other methods for deriving transfer functions

Methods for deriving transfer functions

Mathematical	Mathematical models
Physical	Empirical experiments through trials (prototyping)
Analogous	Experiments with similar systems

Develop Alternative Concepts

Term
Concept Generation

When
ANALYZE: develop high-level concept

Goal
Generate alternative concept ideas for realizing the system functions

Procedure

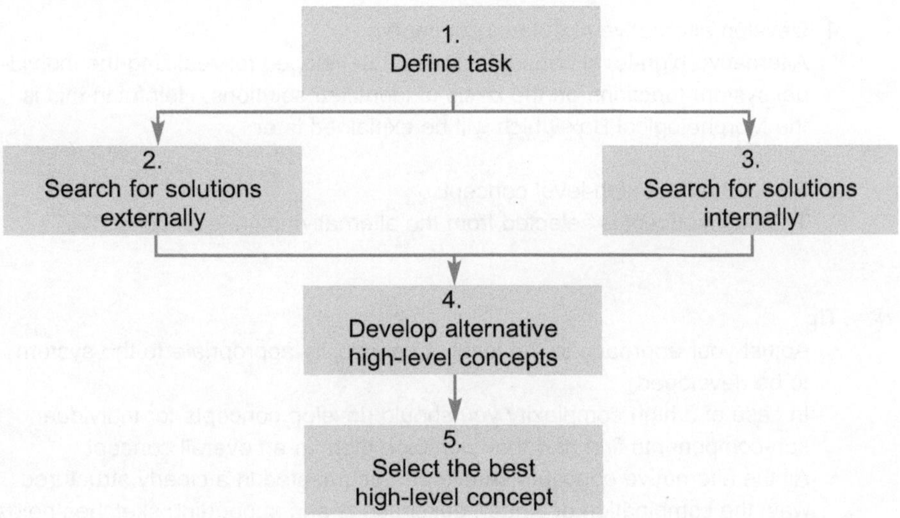

1. Define task:
   The more clearly the task is formulated the easier the search for innovative solutions becomes. The identified relevant system functions serve as a starting point. The tasks become the basis for the question e.g.: "How can we design the seat in such a way that it repels dirt?"

2. Search for solutions externally:
   Good solutions are sought which others have identified for the same or similarly structured tasks ("How do others provide this function?"). This takes place for example, with the help of a best practice and/or benchmarking approach or with the help of trend and patent analyses.

3. Search for solutions internally:
   Innovative solutions are sought with the help of one's own creativity and/or specialist know-how. A great number of creativity techniques are available for use. We will present the following techniques:
   - Brainstorming
   - Brainwriting
   - Anti Solution Brainstorming
   - Mind-mapping
   - SCAMPER
   - Analogy Brainstorming

4. Develop alternative high-level concepts:
   Alternative high-level concepts must be developed for realizing the individual system functions on the basis of identified solutions. Helpful in this is the Morphological Box which will be explained later.

5. Select the best high-level concept:
   The best concept is selected from the alternative high-level concepts.

Tip

- Adjust your approach to the level of complexity appropriate to the system to be developed
- In case of a high complexity you should develop concepts for individual sub-components first and then combine them in an overall concept
- All the alternative concepts should be documented in a clearly structured way; the combination of written descriptions and supporting sketches helps you, your team, the Champion and the customers involved in the evaluation to select the optimal concept later on

Search for Solutions Externally

Term
External Search, External Search for Solutions

When
ANALYZE: develop high-level concept

Goal
Identify alternative solutions for realizing the system functions with the help of existing approaches

Procedure
The following approaches can be used individually or in combination to search for external solutions:

- Benchmarking/Best Practice
- Patent Analysis
- Trend Analysis
- Inspect VOCs (customer solutions)
- Lead User Approach
- Cross Innovation
- Function databases
- Evolution of Technological Systems

Tip

- The external search for solutions is not absolutely necessary since you can also obtain good solutions with the help of your teams' own creativity and their specialist know-how; however, you reduce the risk of unsuccessful developments and the loss of the associated development effort if you succeed in identifying an already tried and tested approach
- If you want to use other peoples' solutions, then "Intellectual Property" becomes a consideration; a Patent Analysis would then be obligatory

Specifics for the process development

During process development the following methods are appropriate in searching externally for solutions:

- Benchmarking/Best Practice
- Patent Analysis
- Trend Analysis
- Inspect VOCs (customer solutions)
- Lead User Approach
- Cross Innovation
- Function databases (only in case of technological questions)
- Evolution of technological systems (only in case of technological question)

Benchmarking

Term
Benchmarking, Targeted Comparison of Performance

When
DEFINE, MEASURE, ANALYZE, develop high-level concept

Goal
- Derive solution ideas from existing "Best-in-Class" systems (products/processes)
- Targeted comparison of performance in relevant fields/functions

Procedure
Benchmarking compares similar systems which are regarded as exemplary. This allows strengths and weaknesses to be identified and relevant hints about the current positioning of the products and/or processes vis-à-vis competitors are obtained. A distinction is made between performance and process benchmarking. From the insights gained about the relative strengths of the compared systems, solution approaches for designing one's own system can be developed.

	MEASURE phase	ANALYZE phase	DESIGN phase
Performance Benchmarking	• Comparison with competitors from the customer's point of view	• Best Practice	• Best Practice
Process Benchmarking	• Technological Benchmarking	• General product or service concepts • Function elements	• Alternatives for the detailed design

Benchmarking is conducted in three phases

1. Planning	2. Execution	3. Analysis
• Determine the scope of benchmarking	• Prepare Data Collection Plan	• Interpret data
• Identify benchmarking partner(s)	• Conduct data collection	• Derive concept ideas
• Determine relevant information		

The level of competition is considered when selecting suitable benchmarking systems.

Direct competitor	Same industry	Latent competitor	Business partner	Other industry	Internal partner

High ← Level of competition → Low

The higher the level of competition, the greater the effort needed for benchmarking and the more difficult benchmarking becomes.
It is therefore recommended to search for suitable internal benchmarking partners in one's own company. Other internal processes, methods or company departments are appropriate for comparison.

Tip

- Internal benchmarking demands a strong willingness to be transparent and an open error culture
- In external benchmarking attention should be paid especially to the comparability of data and key performance indicators; companies often define these differently; you should therefore clarify the background and operational definition of figures, data and facts that you want to use when preparing and executing the performance comparison

Patent Analysis

Term
Patent Research, Patent Analysis

When
ANALYZE: develop high-level concept

Goal
- Generate solution approaches inspired by innovative ideas of others
- Ensure that "Intellectual Property" is respected

Procedure
By the targeted analysis of patents that were handed in and/or are protected, technological trends and new developments are detected. For one thing, the topics and contents of patents can provide important hints, as can the number of patents found on a specific topic.

For strategic patent analyses the patents can be evaluated according to certain indicators. Reasonable criteria for analysis could be:
- Activity (Number of registered patents)
- Recency (How old are the patents?)
- Dominance (How strong is the influence on society, politics and the economy?)
- Range (For which regions have the patents been registered?)
- Concentration (Who has handed in patents – many or few individuals and companies?)
- Complexity (What kinds of patents are being handed in?)
- Involvement (Which companies are involved?)
- Universality (Does the patent have a unique characteristic?)
- Differences (Are there differences among the patents on specific topics?)
- Growth (What about the growth in the patent markets?)

A target oriented search can be conducted based on scientific and technological issues, novelty or infringement issues or state of the art issues; all with the help of the International Patent Classification (IPC).

The International Patent Classification is a hierarchically structured system for sorting and classifying patents and is used in more than 100 countries. This enables international comparison. When patents are registered, they are assigned to one or several groups and receive a code consisting of two capital letters and some Arabic figures.

Structural sections by IPC:	
Section A	Human necessities
Section B	Performing operations, transportation
Section C	Chemistry, metallurgy
Section D	Textiles, paper
Section E	Building industry, earth boring, mining
Section F	Mechanical engineering, lighting, heating, weapons, blasting
Section G	Physics
Section H	Electrotechnology

These categories and sub categories enable searches to be structured and targeted in particular areas based on your needs. Whether it be by country or type of interest. The uniform design of patents enables a simple analysis of patent specifications since relevant information can be filtered easily by targeted search.

Illustration: Structure of Patent

High-level structure of patents

1. "First Page" bibliographical data and abstract	2. Description	3. Claims	4. Drawings (if applicable)

First Page: Bibliographical data and abstract	Description	Claims
• Features of patent • Date of patent • International Patent Classification (IPC) • Name and address of applicant • Name of all inventors involved • Name and address of legal representative (patent attorney) • Patent citations • Territorial protection • Title of inventor and summarized description (abstract) of innovation	• Technological field of invention • State of the art • Criticism of state of the art • The technological task underlying the invention • Explanation of task's solution approach • Advantages of solution • Description of at least one example	The patent claim consists of one sentence. This sentence frequently displays a dichotomy: • Generic term - The generic term includes the title of the invention and the known features of the invention • Designating part - The designating part lists the new features of the invention • The first patent claim is the principal claim comprising all essential features of the invention • Dependent claims contain elaborations of the invention as protected in the principal claim; dependent claims depend on the principal claim • Independent claims contain further solutions of the technological task, which are irrespective of the solution in the principal claim

Trend Analysis

Term
Trend Analysis

When
ANALYZE: develop high-level concept

Goal
- Derive solution approaches from technology or industry trends
- Ensure that important trends from technology, marketing, sales and the supply chain are taken into consideration in the development

Procedure
Specialized institutes and/or providers analyze current global innovations and detect technological as well as service and marketing trends. They are classified into three categories according to their expected temporal and social impact:

1. **Megatrends:**
   Describe the biggest social currents with a time horizon of more than ten years

2. **Macro trends:**
   Refer to a time horizon of five to ten years and describe sub-trends/themes in the respective megatrends and form groups under which micro trends can be subsumed

3. **Micro trends:**
   Best practice cases from the areas marketing, technology and lifestyle which are considered to be a specific characteristic or proof of the existence of a mega and/or macro trend

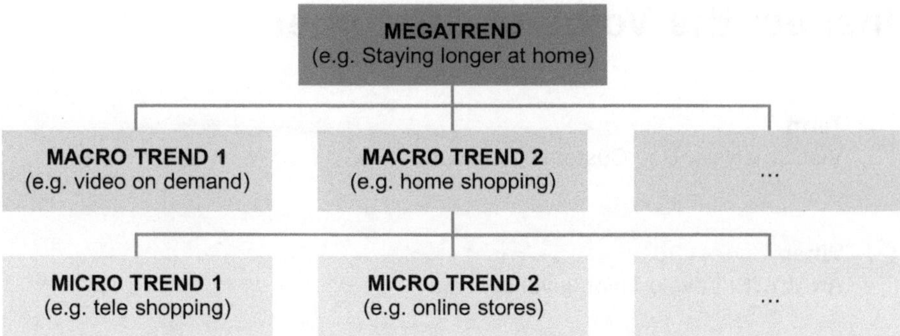

The trends which are relevant for the respective industry or the technology are identified and macro as well as micro trends are used as input for generating ideas.

➾ **Tip**

- Not all trends are directly and immediately relevant for the respective development project
- Trends can often be used as inspiration for targeted brainstorming workshops
- The following questions should be taken into consideration:
  - "Which products and services could support and reinforce a specific trend?"
  - "How would an existing product and/or service have to be developed so as to follow a particular trend?"

Inspect the Voice of Customer

Term
VOC Analysis, Use Customer Solutions

When
ANALYZE: develop high-level concept

Goal
Create solution approaches from the solution ideas of customers

Procedure
Customers frequently utter their own ideas about how their problems could be solved or their requirements fulfilled when contributing to the surveys conducted in the MEASURE phase.
When solution approaches are developed, the survey content should be inspected again in order to discover any promising ideas.
For this purpose the logic chain of the functions is traced back via the measurements and the CTCs to the needs. All VOC which contains a solution is included in the respective category of the Customer Needs Table from the MEASURE phase.

Function → Measurement → CTC → Need → VOC ("Solution")

Customer Needs Table

Complaint	Solution	Specification	Request	Other	"True" need
	It is best not to have fabric covers but hard shells				I want a robust seat
					...
	Quick locks would be good				I want to be able to install and remove the seat quickly
					...

➾ **Tip**

- Note that it might be harder to get the customer excited about solutions that he has suggested himself
- The customer solutions often refer to sub components of the entire process due to their limited knowledge of all its' aspects
- Keep the proportion of customer solutions in your concept as low as possible and/or add further innovative elements

Lead User Approach

Term
Lead User Approach

When
ANALYZE: develop high-level concept

Goal
Derive solutions from ideas/approaches from especially innovative users who are dealing with the system intensively

Procedure
The term lead user refers to a small group of users whose needs anticipate trends of the market and who believe that fulfilling their special needs has a great benefit for them.

This group is frequently inventive itself, in order to satisfy their own needs before commercial products become available.

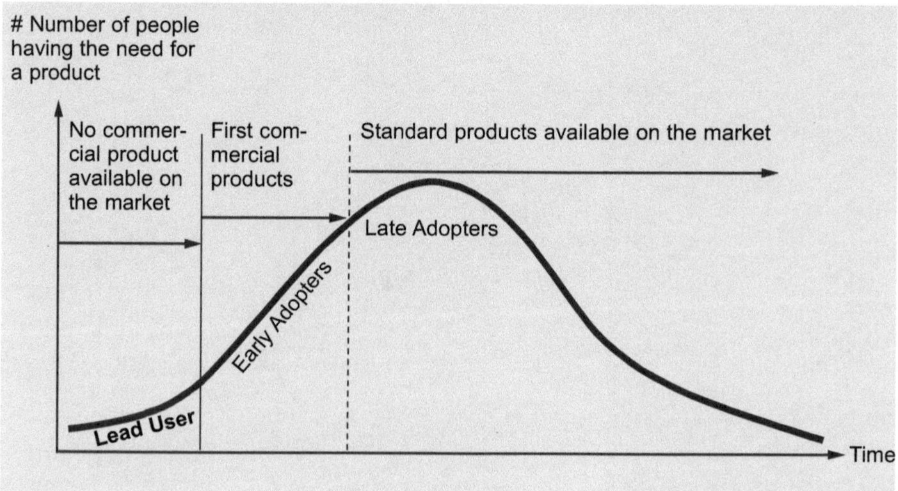

If a company wants to profit from lead user solutions during the product development, the following procedure is recommended:

1. Identify lead user
   - Search the Internet (blogs, forums, etc.)
   - Search in the material of appropriate events (conventions, fairs, etc.)
   - Provide toolkits (e.g. CAD Module on an Internet platform), which enables potential lead users to develop solutions on their own

2. Analyze solutions of lead users

3. Adopt relevant/appropriate approaches in to your own activities

The concept of the lead user is most useful for the development of products where quick innovation cycles are essential or complex technologies are applied (e.g. information technology, bio-technology, medical technology mechanical engineering, etc.).
Moreover, lead users play a role when markets are strongly characterized by trends and trend breaks (e.g. automotive industry, food industry or fashion industry).

➾ **Tip**

- Develop the search for lead users further; it can lead to particularly innovative solutions for your target market in analogous fields.
  Example: Develop a new, high-performance braking system for a motorcar
  → lead user from same industry (automotive): Formula 1, Dragster
  → lead user from more advanced analogous industry (aerospace): NASA (space shuttle), Airbus (A380)
- If you want to use solutions created by lead users it is necessary to check if the solution is patented

Cross Innovation

Term
Cross Innovation

When
ANALYZE: develop high-level concept

Goal
Solutions can be developed by a cross-industry combination of successful approaches

Procedure
The approach of cross innovation is based on the conviction that breakthrough ideas for new business fields or products rarely come into being by focusing on one's own industry. The area of "crossing" with other industries and trends is rather more inspiring (e.g. the combination of a coffee shop and an attorney's office has resulted in the successful concept of "Legal Grind", a café in the USA, where customers can get low-priced legal advice). The big advantage of this approach is that it connects two areas of life. A company can offer its' customers not only a completely new product but a completely new concept by using Cross Innovation. As a result buyers will always perceive the product as more innovative than those of the competitors who focus on the further development of their existing core products.

1. Identify the industries which are interesting for "Crossing" (which industries, who interact with our target market or customers where the new product or service will be offered, act in an especially innovative or successful way?)

2. Identify successful products and current trends in the selected industries (What trends are apparent in the selected industries? Which companies and products are particularly successful?)

3. Collect associations for combining this "successful model" with your own product idea (which aspects/solution approaches can reasonably be combined with your own product?)

➾ **Tip**

- By reverting to already tried and tested know-how in other industries you can significantly reduce your development costs while minimizing your development risks and shortening the development time
- Companies using the Cross Innovation approach see the following advantages:

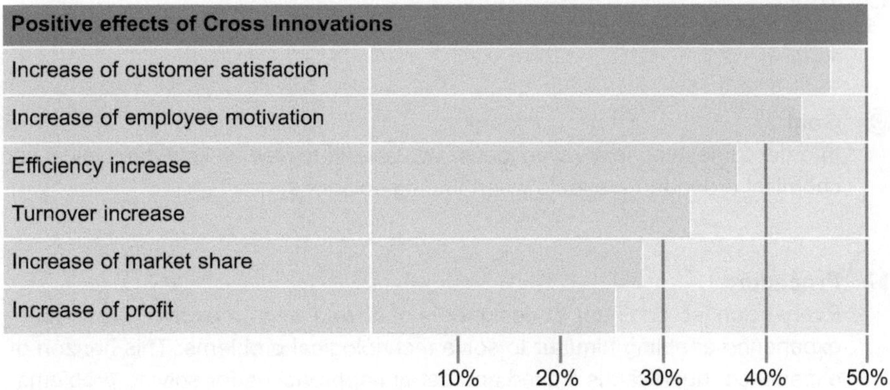

Question: Which positive effects can you think of that have been achieved by the networked procedure of Cross Innovation?

Source: Muckenschnabl/Zukunftsinstitut, 2009

DEFINE
MEASURE
ANALYZE
DESIGN
VERIFY

Function Database

Term
Function Database

When
ANALYZE: develop high-level concept

Goal
In order to develop innovative ideas, you should review all known physical and chemical procedures/effects when finding solutions

Procedure
Every scientist, engineer or designer is endowed with an evolved horizon of experience enabling him/her to solve technological problems. This horizon of experience, however, is limited and certain approaches for solving problems can dominate.
Resorting to a function database aims at broadening the horizon for solving problem in a focused way by detecting innovative alternatives. It represents an "encyclopedia" of opportunities for the fulfillment of functions.
Apart from licensed databases (e.g. TechOptimizer), a free database is accessible on the Internet (http://function.creax.com).
It is used as follows:

1. Indicate (aggregate) state of object for which a specific impact is to be achieved (solid; liquid; gas; field)

2. Select the function in the drop-down list which is to be fulfilled by the object (e.g. protect, fixate, heat, etc.)

3. The database provides appropriate procedures/effects including examples of application

4. Inspect the indicated procedures/effects and integrate them as required in a solution concept

Example: Function Database – Passenger Seat

Function	Alternatives	Example
Protect seat ("solid" object)	Coating	Fiber coating with invisible glass film
	Additive	Polybrominated diphenyl ether as flame inhibitor
	Lamination	Coating with synthetic foil
	Hydrophobical material	Material with lotus effect
Fixate seat ("solid" object)	Ferromagnetism	Magnetic locks
	Hooks	Snap-in hooks
	Mechanic connector	Bolt with nut
	Shape-Memory Effect	Cryo-Fit connections

Specifics for the process development

Since function data bases contain chemical and physical (thermodynamic, mechanic, electromagnetic etc.) effects, they are only helpful for process development if one is solving technological problems (e.g. cooling or heating of materials).

Evolution of Technological Systems

Term
Evolution of Technological Systems

When
ANALYZE: develop high-level concept

Goal
Predict the future evolutionary steps of a technology to push the system development forward in a targeted way

Procedure
Like biological systems technological systems move through four typical evolution phases:

1. **Infancy**
   Phase prior to entering the market, system evolves rather slowly
2. **Rapid growth**
   Entering the market, the speed of evolution accelerates quickly
3. **Maturity**
   The system is established, the system's evolution slows down and ceases
4. **Decline**
   A new system takes the place of the old one

The relationship between the costs and benefits of a system is subject to corresponding changes and in a chart takes the form of an S over the four phases.

Mapping s-curve analysis

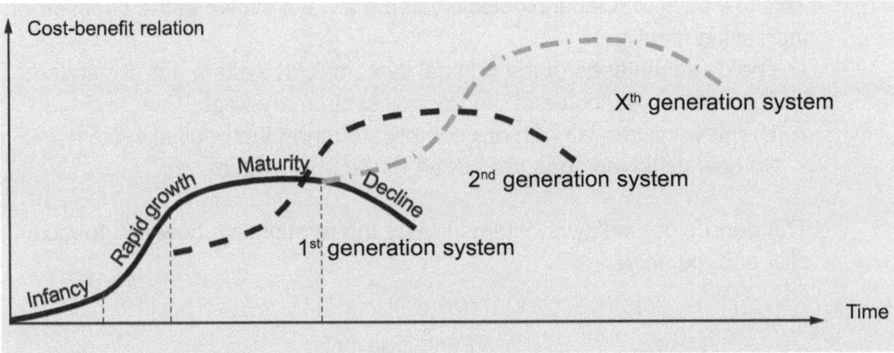

A timely entry into a new system has to be found if constant growth in the cost-benefit relationship is to be achieved.

According to the system's positioning on the s-curve, the following two basic questions arise:

1. Which changes need to be made to the system to advance its position on the s-curve?
2. What could the new generation of the system/the technology look like?

The nine laws of evolution for technological systems can provide the answers to these questions.

Nine laws of evolution for technological systems

1. **Law of increasing ideality of systems**
2. **Law of non-uniform development of subsystems**
3. **Law of transition to super systems**
4. **Law of increasing flexibility of systems**
5. **Law of transition from the macro to the micro level**
6. **Law of shortening energy flow in systems**
7. **Law of harmonizing the rhythm in systems**
8. **Law of increasing automation of systems**
9. **Law of increasing controllability of systems**

1. **Law of increasing ideality of systems**
   This law says that technological systems always evolve in the direction of increasing ideality:
   - The disadvantages of the original system (old) system are eliminated.
   - The positive attributes of the original system are kept
   - The new system is not more complicated than the original system
   - No new disadvantages are added to the new system

   The degree of ideality is determined by the relationship between functionality and expense.

$$\text{Degree of ideality} = \frac{\text{Functionality}}{\text{Effort (e.g. \$, energy, weight, etc.)}}$$

Illustration: Development of Home Appliances

Home appliance	Price in USD 1947	Price in USD 1997	Improved functionality
Refrigerator	1,470	700	Twice as large; ice machine
Washing machine	1,770	380	Quieter; improved energy efficiency
Television	3,180 (black & white)	300	Color image; stereo sound; remote control

2. **Law of non-uniform development of subsystems**
   This law says that the different components of a technical system are always in different stages of development.
   Hence there are always components which are behind other subsystems (their position on the s-curve) in terms of their development level.

   The more complex the system, the less uniform are the stages of development of its components. The resulting contradictions emerging in the system as a result have to be resolved if the development process is to be pushed forward.

Illustration: Development Stages in Computer Technology

The miniaturization of electronic components in computer technology (from conductor to micro-electronics) has enabled a massive reduction in the size of appliances. However, this potential to decrease size can only be exploited if the control and presentation concepts (keyboard and monitor) are also developed.

3. **Law of transition to super systems**
   This law says that technological systems generally evolve from mono to bi or poly systems. The integration of two independent mono systems generates a more complex bi system.

Illustration: Development of Hi-Fi Systems

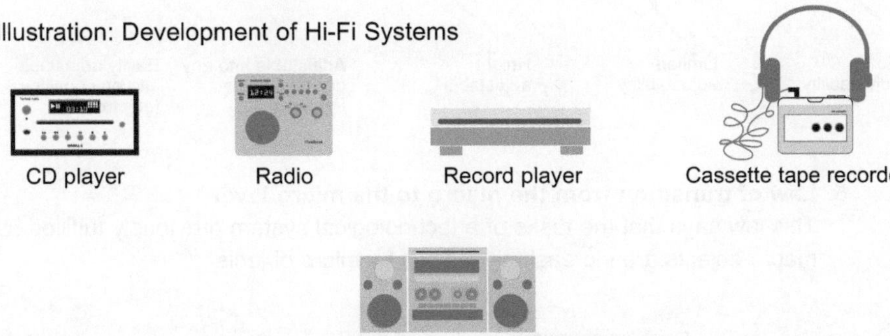

Hi-fi system with all 4 components

Especially in the mature phase of system life cycles additional functions are added to enhance their attractiveness. If a system is able to fulfill a large number of additional functions this is a good indicator that it has already exceeded its evolutionary peak and will soon be replaced by a new technology.

4. **Law of increasing flexibility of systems**
   This law says that technological systems develop increasingly flexible structures and as a consequence always become increasingly adaptable. Increasing flexibility can be achieved in two different ways:

1. Increasing the flexibility of the function in a system

Rigid system → Modular system → Programmable system → Self-adaptive system

2. Increasing the flexibility of the structure in a system

Rigid system → System with one flexible point → System with many flexible points → Elastic system → Liquid-based system → Field-based system

Illustration: Car Steering Wheels*

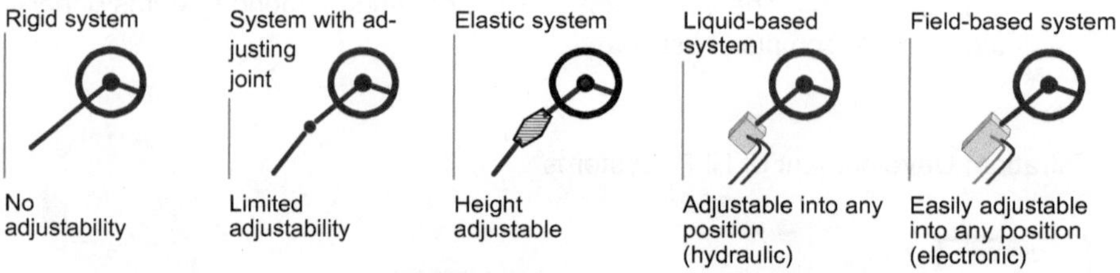

5. **Law of transition from the macro to the micro level**
   This law says that the tasks of a technological system previously fulfilled by macro objects are increasingly fulfilled by micro objects.

Monolithic structure → Crystal or aggregated molecules → Single molecules → Atoms or ions → Elementary particles

The advantage is enhanced controllability, often in conjunction with increased functionality (e.g. evident in a comparison between mechanical and digital watches).

* Cf. Bernd Gimpel, et. al. (2000): Ideen finden. Produkte entwickeln mit TRIZ. Hanser Verlag, Munich/ Vienna, p. 105.

The micro structures of a system can fulfill different tasks:

- Micro structures can carry out functions previously performed by macro structures
  Example: By replacing traditional mechanical cutting tools with photons delivered by a laser
- Micro structures control the physical attributes and behavior of macro structures
  Example: Spectacle lenses with photo-chrome particles which alter their translucence depending on the level of solar radiation

6. Law of shortening energy flow in systems

This law says that the distance to be covered by the energy in a technological system reduces through the generations.

In the most effective system the energy source directly impacts the working appliance:

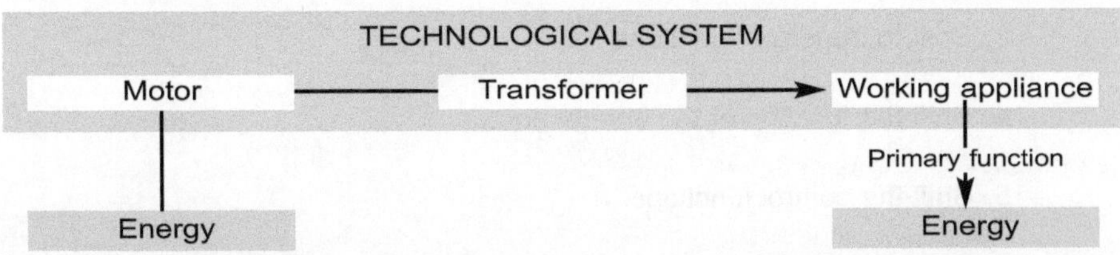

The trend to ever shorter energy flows within systems is evident in the development of industrial machines. Around 1920 energy was still directed from a central motor via drive shafts and belts to the individual machines but from the 1930s onwards a separate electric motor was installed in the individual machines, shortening the distance the energy had to cover through the system.

7. Law of harmonizing the rhythm in systems

This law says that the effectiveness of a technological system is raised through the increasing harmonization/synchronization of the movements of all of its parts.

Harmonization can be achieved in three different ways:

1. Coordinate the temporal sequences of moving system components
2. Utilize resonance
3. Eliminate undesired temporal sequences

Examples:
- An aircraft needs a coordinated control of movement (adjusting and aligning the different control vanes); while this used to be done manually, coordination in modern machines has been improved by using computers ("fly-by-wire")
- Airbrush pistols need a coordinated opening and closing of the air and paint valves to ensure that no paint drips onto the target surface.

8. Law of increasing automation of systems

This law says that the degree of automation increases as a technological system evolves.

Functions originally performed by man are taken over by a system that develops as follows:

1. Man performs the function
2. Shift to tool functions
3. Shift to transition functions
4. Shift the function of the energy source
5. Shift the control functions

Illustration: Movement

9. Law of increasing controllability of systems

This law says the controllability of a technological system constantly improves because the Sufield interactions within and outside the system continually increase.

Illustration: Food Preparation

Improving the interaction is achieved in the following ways:

1. Replace an uncontrollable field or one which is difficult to control with a controllable one (e.g. a field based on gravity with a mechanical field, or a mechanical one with an electromagnetic field)
2. Raise the degree of flexibility of the elements in a interaction
3. Adjust the field frequency to the natural frequency of the object or tool

Internal Search for Solutions

Term
Internal Search

When
ANALYZE: develop high-level concept

Goal
Identify alternative solutions for realizing the system functions with the help of one's own creativity

Procedure
The following approaches are appropriate individually or in combination for finding solutions internally.

- Brainstorming
- Brainwriting
- Anti Solution Brainstorming
- Mindmapping
- SCAMPER
- Analogy Brainstorming
- Morphological Box

Brainstorming

Term
Brainstorming, creativity technique, structured collection and evaluation of ideas

When
ANALYZE: develop high-level concept

Goal
Develop concept ideas on the basis of the system functions

Procedure
Brainstorming leads to a collection of ideas which are as diverse and extensive as possible. Strictly enforcing the following Brainstorming rules will guarantee the widest number of ideas:

Rules for Brainstorming:
1. Every proposal counts
2. No discussion about the proposals
3. No "killer phrases" allowed
4. No explanations when collecting ideas
5. All participants in the session are involved
6. Let others finish and listen

A Brainstorming session should proceed as follows:
1. Formulate and document the rules, hang them up in the room
2. Determine and write down the topic and/or goal
3. Collect ideas (Brainstorming): three to five minutes
4. Explain and structure the ideas (Affinity Diagram)
5. Derive and visualize criteria
6. Evaluate the collected ideas on the basis of the criteria
7. Select the "good" ideas and pursue them further

Brainstorming can be structured with the help of an Ishikawa Diagram.

Example: Ishikawa Diagram – Passenger Seat

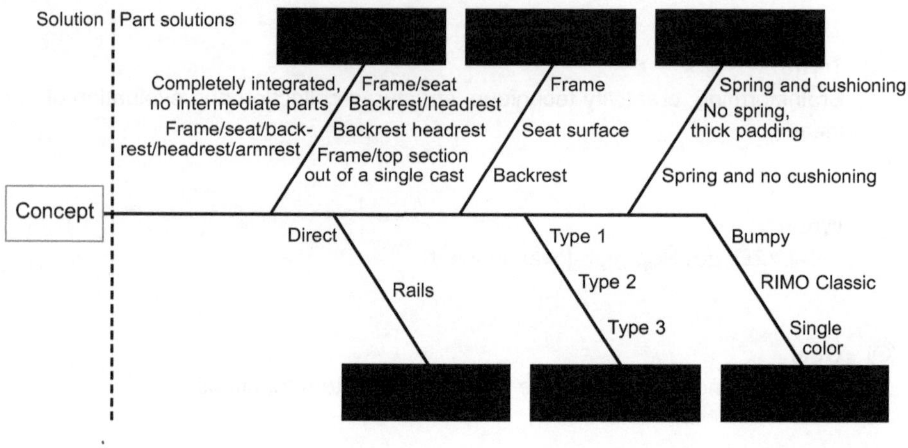

Tip

- The labels given to the "fish bones" (design elements of the system) can be adjusted individually in the Ishikawa Diagram; it is important that the whole system is considered
- The diagram provides a structure for the Brainstorming and makes it easier to generate ideas and serves as an ideal basis for pursuing concept ideas further
  - Separate the phase during which ideas are generated from the subsequent evaluation when the ideas are summarized and prioritized
  - Consolidate the ideas by eliminating duplicates, find (cluster) connected ideas/thoughts, formulate cluster headlines and combine ideas
  - Prioritize the ideas afterwards with appropriate methods; adjust the selection of methods to timeframes
  - When selecting and applying creativity techniques you should bear in mind that in our everyday life one brain hemisphere is preferred and thus better trained; the left brain hemisphere is all about figures, data, facts, rational logic and analytical thinking whereas the right brain hemisphere drives imagination and intuition; it is the latter which should be targeted
  - Individuals who are used to forming analogies and to thinking laterally may feel restricted when suddenly having to solve a problem in a structured way; conversely, a person thinking rather more rationally cannot produce new ideas at the push of a button; this concerns the team as well as the facilitator; both have to feel at ease with the technique

- Not every technique is appropriate for each situation; the selection of the creativity technique must ensure a balance between context, team and facilitator

Brainwriting

Term
Brainwriting, creativity technique, structured collection and evaluation of ideas

When
ANALYZE: develop high-level concept

Goal
Develop detailed ideas on selected topics

Procedure
Brainwriting is used when – due to the complex topic – it seems to make sense to collect ideas in a calmer, more concentrated atmosphere or when individuals involved are not available at the same location.
A Brainwriting session is structured as follows.

1. Determine the topic together
2. Note individual solution idea(s) on a piece of paper or in an email
3. Pass on the solution ideas to the next person:
   - in a room clockwise
   - in emails according to a pre-established order
4. Examine the idea of the person before you, elaborate on it or develop a completely new idea
5. At the end of an agreed time the ideas are collected centrally
6. The result is presented and discussed
7. Select solution proposals afterwards

Brainwriting 6-3-5

The Brainwriting 6-3-5 follows the classic Brainwriting. The name stands for the idea that 6 persons write down and pass on 3 ideas in 5 minutes.

Idea Card Method

This method is another variation of the classic Brainwriting. The ideas are collected on cards or Post-its®. The advantage is that there is no need to copy the ideas later.

Brainwriting Pool

Brainwriting Pool means that the ideas are not passed systematically in a clockwise direction but are collected in the middle of the table. Each team member can draw one or more ideas from the pool and develop it/them further.

Notebook Method

Instead of collecting the ideas during a session, each team member can write down his own ideas irrespective of time and place. The ideas are passed on to other team members via email. After a certain period of time they are collected and discussed during a team meeting. A typical time frame for this kind of Brainwriting is two to four weeks.

Tip

- Brainwriting often develops unusual and creative relationships between combinations of ideas, including chains of ideas – sometimes this can exceed what Brainstorming generates due to its more rigid structure and orientation
- Brainwriting is especially suitable for diverse teams with dominant and less dominant members. People with less dominant personalities are given more room to move and express their ideas

Anti-Solution Brainstorming

Term
Anti-Solution Brainstorming

When
ANALYZE: develop high-level concept

Goal
- Collect ideas for finding solutions by thinking about the question: What could make the situation even worse?
- Resolve blockages preventing the identification of solutions

Procedure

1. Formulate rules
2. Identify and write down the topic: What could make the situation even worse?
3. Develop ideas
4. Collect ideas
5. Explain ideas and structure and/or cluster them
6. Transform ideas into improvement proposals
7. Dig deeper into individual ideas through renewed Brainstorming

The Brainstorming rules apply.

➾ **Tip**

- Use red cards for collecting anti-solutions and green cards for positive solutions
- Having generated anti-solutions, further Brainstorming should be conducted in order to derive positive solutions within the team

Mind Mapping

Term
Mind Mapping, creativity technique, structured collection and evaluation of ideas

When
ANALYZE: develop high-level concept

Goal
Structure the collection of ideas by visualizing connections between alternative solution ideas

Procedure
Mind Mapping is a technique which shows semantic relationships between the ideas. Synergy effects are used by the parallel utilization of linguistic and visual skills. A mind map revolving around the key topic emerges.

Example: Mind Map – Passenger Seat

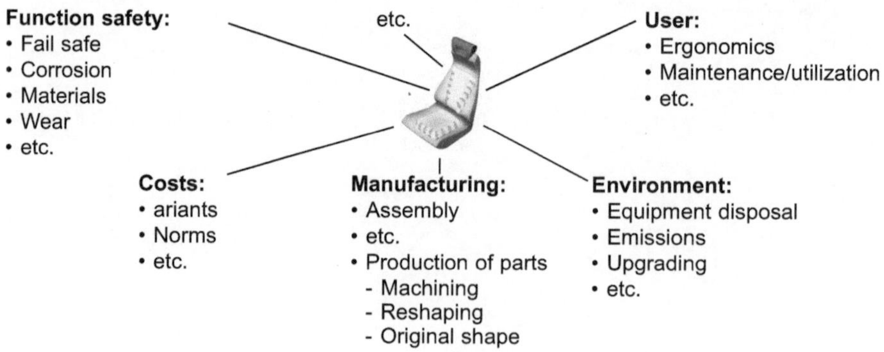

Depicting complex structures in a clear way triggers further supporting or modifying ideas. It also allows single aspects to be pursued in greater detail.

SCAMPER

Term
SCAMPER, creativity technique, structured collection and evaluation of ideas

When
ANALYZE: develop high-level concept

Goal
Supplement developed concept ideas by structured questioning of the concept environment

Procedure
Structured scrutinizing of the concept environment generates and channels creative ideas.
This analysis is undertaken with the help of the SCAMPER checklist.

SCAMPER checklist

Substitute	• Which concept elements are replaceable and by what? • Are there comparable concepts from other companies, areas, etc.?
Combine	• What can be combined in the concept? • Can this be combined with other ideas, broken down into modules and/or transformed into a different image
Adapt	• How can concept elements be adapted? • Can parallels be identified?
Modify	• How can concept elements be changed, for instance be enlarged or scaled down, strengths, dimensions and/or distances etc.?
Put to other uses	• How can concept elements be separated from their designated function? • Can they be put to other uses?
Eliminate/erase	• How can concept elements be eliminated and/or erased?
Reverse/rearrange	• What effect does reversing concept elements have? • Can the order be rearranged? • Can the elements be substituted?

➾ **Tip**

- Use MTC Analysis in addition to Modify:
  - M: Measurements
  - T: Time
  - C: Costs
- This approach assumes that development projects are always characterized by restrictions; financial means are limited, a certain experimental scope must be maintained and the procedure must be completed within a certain period of time
- Taking these aspects into account the approach tries to drive innovative thinking by suggesting that measurements, time and costs are either, unlimited, or are not available at all:

1. The measurements of the system are:
   - increased ad infinitum
   - reduced to zero

   i.e. how might the system and/or individual components look like if:
   - the system was infinitely big?
   - the system was infinitely small?

2. The time available:
   - is increased ad infinitum
   - is reduced to zero

   i.e. how would the system and/or individual components look like if:
   - infinite time was available?
   - no time was available?

3. The costs connected with the development:
   - are increased ad infinitum
   - are reduced to zero

   i.e. how would the system and/or individual components look like if:
   - costs could be increased infinitely?
   - no costs could be incurred at all?

Analogy Brainstorming

Term
Best Practice Sharing, Analogy Brainstorming

When
ANALYZE: develop high-level concept

Goal
Support creativity within the frame of finding solutions by leaving the familiar solution space and searching for creative approaches in entirely new areas

Procedure

1. Select the feature of the original system or find a feature with the help of the 8-P-feature checklist
2. Find another system which has this feature
3. Formulate question: How did that system solve this problem or how would it solve this problem?
4. Transfer the solution to the original system

Analogy Brainstorming goes beyond mere Best Practice Sharing in as much as the latter focuses on the adaptation of an especially high-performing solution found elsewhere.

See feature categories/questions on the next page

Feature category	Questions
People	Who is involved?
Processes	What procedures exist?
Places	Where does something happen?
Purposes	What goals/purposes are pursued?
Parameters	What features does the system have?
Policies	What rules/traditions exist?
Problems	What problems exist?
Parts	What parts exist?

Tip

- Involve all Stakeholders early on and create room for your own variants and adjustments; the implementation of insights gained elsewhere frequently meets resistance even if the solution is sophisticated; this is known as the "Not invented here" syndrome
- Bear in mind that the effort to relate back to the original problem is not easy; this can cause frustration

Morphological Box

Term
Morphological Box, Zwicky Box

When
ANALYZE: develop high-level concept

Goal
Combine the individual solution ideas into alternative high-level concepts

Procedure
The Morphological Box brings together all concept development elements and their possible features and characteristics, in order to highlight all conceivable combination possibilities. This takes place based on the defined and prioritized system functions.
Features are assigned to the prioritized functions. They should be independent of one another if possible ("decoupled"). Relevant characteristics are now defined for each feature.

Example: Morphological Box – Passenger Seat

	HOW CAN THIS BE REALIZED?			
	Alternative 1	Alternative 2	Alternative 3	Alternative 4
Repel dirt	Removable plastic foil	Material with lotus effect	Polybrominated diphenyl ether as flame inhibitor	Fiber coating with glass film
Absorb impact	Honeycomb structure	Elastic underlayer	Air cushioning	Counter weight
Keep body	Wooden body	Plastic container	Coated metal mold	…
Avoid theft	Snap-in hooks	Mechanic connector	Ferro-magnetism	Shape Memory Metal
	Low Budget		High End	

After having compiled the alternatives, promising combinations of individual features are identified with the help of a matrix i.e. that make economic and technological sense.

Tip

- It is best if you restrict the number of high level concepts selected for more detailed investigation to no more than 5
- It can help if titles for high-level concepts are chosen in advance (e.g. Low Cost Concept, Quick Hit Concept, Concept, Super Innovative Concept, etc.)
- You should ideally have a short description, draft or a model of each concept in order to facilitate their comparison and the final selection

Conjoint Analysis

Term
Conjoint Measurement, Trade-off Analysis, Conjoint Analysis, Decompositional Procedure

When
ANALYZE: develop high-level concept

Goal
- Collect customer preferences systematically with respect to individual system features
- Identify the contribution of individual system features/components to the overall benefit of a system (de-compositional procedure)
- Identify the system which delivers maximum benefit from the customer's perspective

Procedure
The Conjoint Analysis examines which preferences the customers have with respect to the characteristics of a system's features and attributes. It thus supports the system's configuration for maximum benefit. The Conjoint Analysis functions according to the de-compositional approach, i.e. the customer's overall judgment (evaluation of overall benefit and/or overall attractiveness of system) is broken down into single judgments (part worth) with respect to the system's components. It is then identified which combinations of characteristics of a system's features and attributes maximize the system's overall benefit from the customer's perspective.

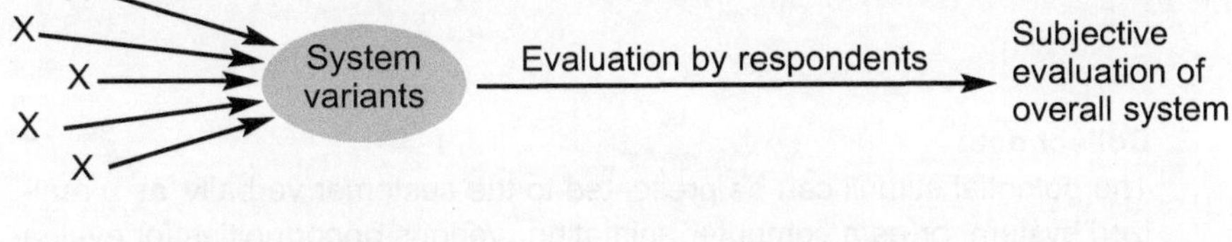

The Conjoint Analysis evaluates system variants which have been derived from a Morphological Box. The following procedure is recommended for conducting the Conjoint Analysis:

1. **Select attributes and characteristics**
   The following aspects should be considered when selecting attributes and their characteristics:
   - The attributes must be relevant, i.e. they should have an influence on the buying decision
   - The producer must be able to influence the attributes
   - The attributes should be independent of one another
   - Characteristics which are "musts" (exclusion criteria), must not be utilized
   - It must be feasible to realize these characteristics
   - The individual characteristics must compensate one another (e.g. a more complicated assembly can be compensated by improved seat comfort)
   - The number of attributes must be limited – the effort and expense of examining increase exponentially

2. **Determine the survey design**
   Fictitious systems (so-called stimuli) are formed out of the different attribute characteristics. They are listed in the following chart

Example: Survey Design – Passenger Seat

Stimuli no.	Assembly	Cushioning	Cover	Seat comfort
1	Simple (< 1 min.)	Firm	Fabric/leather	Armrest
2	Complicated (> 5 min.)	Firm	Fabric/leather	No armrest
3	Simple (< 1 min.)	Soft	Fabric/leather	No armrest
4	Complicated (> 5 min.)	Soft	Fabric/leather	Armrest
5	Simple (< 1 min.)	Firm	Synthetic	No armrest
6	Complicated (> 5 min.)	Firm	Synthetic	Armrest
7	Simple (< 1 min.)	Soft	Synthetic	Armrest
8	Complicated (> 5 min.)	Soft	Synthetic	No armrest

3. **Collect data**
   The potential stimuli can be presented to the customer verbally, as a realized system, or as a computer animation. Various opportunities for evaluation exist:

- Ranking
- Evaluation using an ordinal scale or AHP (Analytic Hierarchy Process)
- One or more customers are asked to evaluate the fictitious systems (stimuli)

4. Estimate the benefit values

The relationship between the identified preferences and the part worth of individual characteristics can be formulated as follows:

$$y_k = \sum_{j=1}^{J} \sum_{m=1}^{M_j} \beta_{jm} \cdot x_{jm} = \beta_{11} \cdot x_{11} + \beta_{12} \cdot x_{12} + \ldots + \beta_{JM_j} \cdot x_{JM_j}$$

The valuation of the stimulus k matches its overall benefit y:
y_k : Overall benefit value for stimulus k
The overall benefit consists of the aggregate of all part worth components

The benefit value for a user β of an individual attribute j depends on its special characteristic m:
β_{jm}: part worth values for the characteristic m of attribute j

Possible attribute characteristics which do not possess the evaluated stimulus k are removed by the binary variable x from the calculation:
x_{jm}: Binary variable with the value 1 for existing attribute characteristics and the value 0 for non-existing attribute characteristics

The mathematical specification of the overall benefit can also be obtained using an algorithm for Design of Experiments, e.g. with the help of Minitab®.

Conjoint Analysis with Minitab®

▶▶ Procedure

A Conjoint Analysis using the software program Minitab® (Factorial Design) derives the optimal combination of possible attribute characteristics from the identified overall benefit of the individual stimuli and thus shows which fictitious system (stimuli) has the highest overall user value for the customer. For example the team wishes to find out which characteristics enable a passenger seat to be best marketed to the target customer.

The four selected key attributes and their characteristics are:

1. Installing the seat: simple (< 1 min) vs. complicated (> 5 min)
2. Cushioning: firm vs. soft
3. Cover: fabric/leather vs. synthetic
4. Seat comfort: armrest vs. no armrest

Because the number of possible stimuli rises exponentially with the number of variable attributes, it often makes sense to use a reduced experimental design. Designs such as a fractional factorial or part factorial design can be drawn up in Minitab®

Example: Conjoint Analysis with Minitab® – Passenger Seat

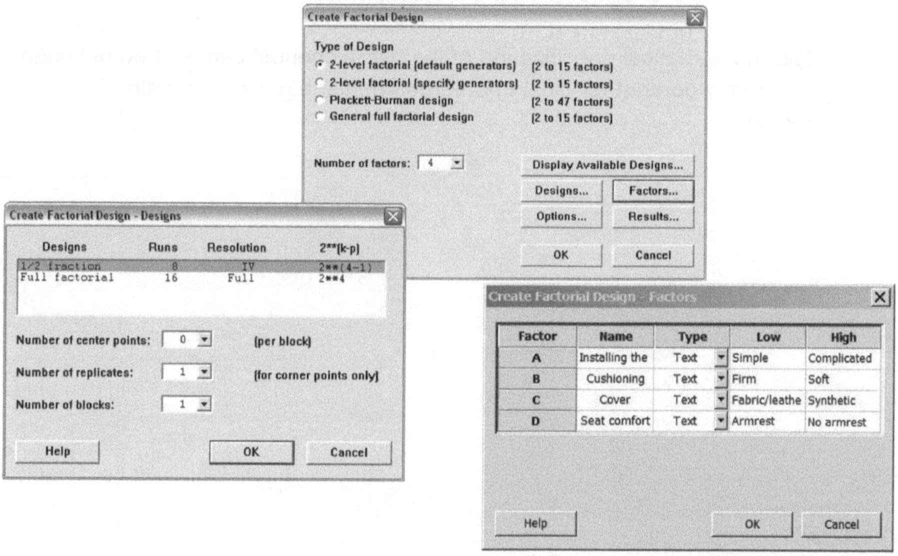

The number of observed attributes (number of factors) and their characteristics (factor levels) are entered into Minitab®.
The fictitious systems can be evaluated in a pair comparison using an AHP or can be ranked.
The resulting priority is captured as the overall benefit value in a Minitab® Worksheet. The scaling of the overall benefit value ranges from 1 (worst overall value) to 8 (best overall value).

Example: Minitab® Worksheet for the Conjoint Analysis – Passenger Seat

↓	C1	C2	C3	C4	C5-T	C6-T	C7-T	C8-T	C9
	StdOrder	RunOrder	CenterPT	Blocks	Installing	Cushioning	Cover	Seat comfort	Ranking
1	1	1	1	1	Simple (< 1 min.)	Firm	Fabric/leather	Armrest	6
2	2	2	1	1	Complicated (> 5 min.)	Firm	Fabric/leather	No armrest	4
3	3	3	1	1	Simple (< 1 min.)	Soft	Fabric/leather	No armrest	8
4	4	4	1	1	Complicated (> 5 min.)	Soft	Fabric/leather	Armrest	7
5	5	5	1	1	Simple (< 1 min.)	Firm	Synthetic	No armrest	1
6	6	6	1	1	Complicated (> 5 min.)	Firm	Synthetic	Armrest	3
7	7	7	1	1	Simple (< 1 min.)	Soft	Synthetic	Armrest	5
8	8	8	1	1	Complicated (> 5 min.)	Soft	Synthetic	No armrest	2

The first graph for analyzing data is generated.

Example: Minitab® Pareto Chart from DOE – Passenger Seat

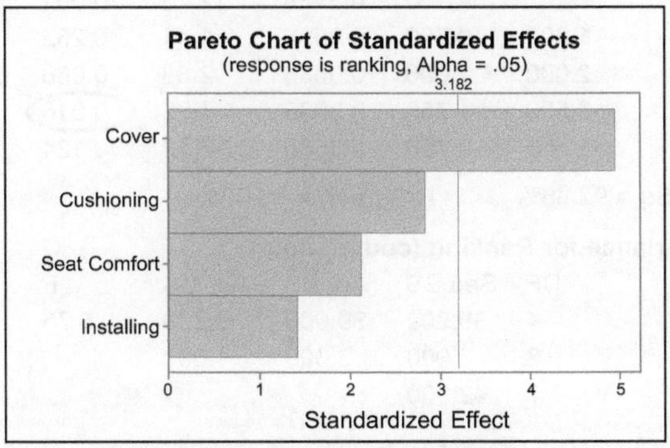

In this case the Pareto Chart shows that the seat cover is the most important attribute. The other attributes do not appear to be statistically significant.
A Main Effects Plot can also reveal which characteristics are preferred.

Example: Minitab® Main Effects Plot from DOE Passenger Seat

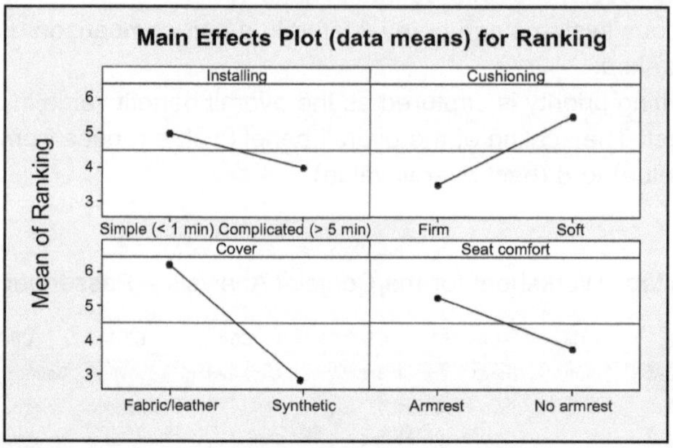

This example shows that a simple installation, soft cushioning, a fabric/leather cover, and a seat with armrests are preferred by the target customers.

The result in the Minitab® session window confirms this.

Example: Minitab® Session Window from DOE – Passenger Seat

Estimated Effects and Coefficients for Ranking (coded units)

Term	Effect	CoefSE	Coef	T	P
Constant		4.500	0.3536	12.73	0.001
Installing	-1.000	-0.500	0.3536	-1.41	0.252
Cushioning	2.000	1.000	0.3536	2.83	0.066
Cover	-3.500	-1.750	0.3536	-4.95	0.016
Seat comfort	-1.500	-0.750	0.3536	-2.12	0.124

< 0.05

S = 1 R-Sq = 92.86% R-Sq(adj) = 83.33%

Analysis of Variance for Ranking (coded units)

Source	DF	Seq SS	Adj SS	Adj MS	F	P
Main Effects	4	39.000	39.000	9.750	9.75	0.046
Residual Error	3	3.000	3.000	1.000		
Total	7	42.000				

The p-values indicate which attributes are statistically significant. Only the value generated for the cover is below the significance level of 0.05.

By maximizing the ranking the Minitab® "Response Optimizer" indicates the attribute characteristics which the system with the highest overall benefit value should have for the customer.

Example: Minitab® Response Optimizer from DOE

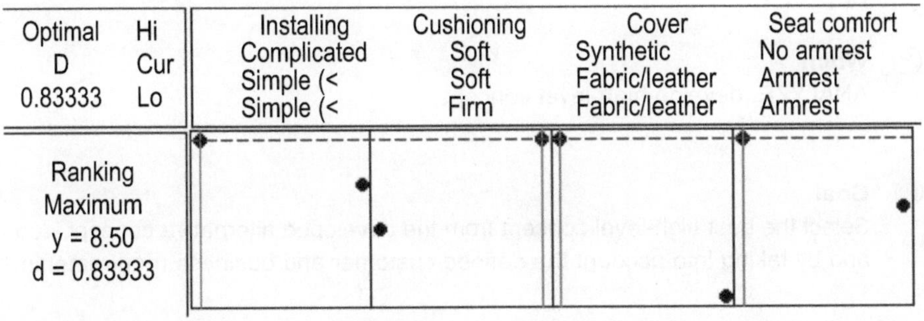

Result of the Conjoint Analysis:
The optimal overall benefit value of y = 8.5 is achieved by a seat that is simple to install, has soft cushioning with a fabric/leather cover and armrests. The optimal overall benefit value exceeds in this case even the highest ranking/the highest overall benefit of the evaluated stimuli.

Tip

- The Conjoint Analysis is a complex procedure for which statistical knowledge is required; you should therefore get support from an expert in the field of market research
- Using Conjoint Analysis, especially in the B2B area, to develop a concept is good for building customer relationships since you actively involve them in a structured way

Identify the Best High-level Concept

Term
High-Level Concept Selection

When
ANALYZE: develop high-level concept

Goal
Select the best high-level concept from the developed alternative concept ideas and by taking into account the defined customer and business requirements

Procedure
For the evaluation of the alternative high-level concepts, within the scope of the selection procedure, the following should be considered:

1. **How capable are the alternative high-level concepts of fulfilling the customer requirements?**
   After ensuring that the concepts do not violate must criteria (e.g. legal or corporate guidelines) it must be evaluated how well each is capable of fulfilling the customer requirements specified in the Design Scorecard. Even if a concept has already been identified using Conjoint Analysis that provides the maximum benefit from the customer's perspective, it is reasonable to systematically check the alternative concepts once again as to their ability to fulfill the CTCs. This way any remaining weaknesses and/or optimization potential can be revealed.

2. **How high is the innovation power of the concepts?**
   With the help of appropriate criteria the novelty of each concepts can be evaluated and tested as to how differentiated it is from the competition and so worth patenting or not.

3. **How appropriate/feasible are the concepts from the company's point of view?**
   The capability of the company to implement the concepts and/or provide the effort connected to them is taken into consideration throughout the selection procedure.

The results of the above three perspectives of evaluation are entered into the concept evaluation matrix which makes the best concept recognizable. In the following illustration concept C should be preferred since it displays the best combination of concept, innovative capability and implementation effort.

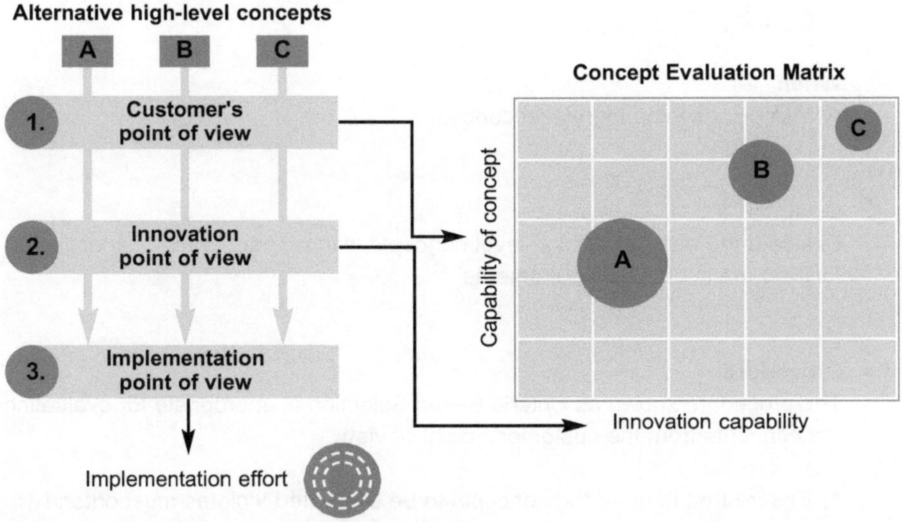

➾ **Tip**

- You should focus your attention on no more than 5 alternative high-level concepts when selecting the best high level concept
- It can help to title each of the high-level concepts to be considered in advance (e.g. Low Cost Concept, Quick Hit Concept, Concept, Super Innovative Concept, etc.)
- Ideally, you should have a short description, draft or a model for each concept in order to facilitate the comparison

Evaluation of Concept from the Customer's Point of View

Term
Concept Capability Check

When
ANALYZE: develop high-level concept

Goal
Evaluate the alternative high-level concepts with respect to their capability of fulfilling the customer requirements

Procedure
The procedure known as Criteria Based Selection is appropriate for evaluating the concepts from the customer's point of view:

1. Ensure that none of the concepts to be evaluated violates must criteria (e.g. legal guidelines, corporate strategy, technical restrictions, etc.)

2. Enter the identified customer requirements (CTCs) including their priorities into the vertical rows of a matrix

3. Enter the alternative high-level concepts into the horizontal fields of the matrix (best done by giving it a telling name and/or title e.g. concept "Super safe", concept "Super comfortable", or the like)

4. Evaluate the degree of fulfillment per requirement provided by the alternative concepts on a scale from 1 to 5 ("how well does the concept fulfill the requirement?") and enter it into the respective cell of the matrix

   Scale: 1 = no fulfillment at all
   2 = poor fulfillment
   3 = medium fulfillment
   4 = good fulfillment
   5 = perfect fulfillment

5. Multiply the requirement's degree of fulfillment with its priority to give a weighted value

6. Add up the weighted values of each concept

7. Use aggregate of weighted values for comparing the concepts

Exampel: Criteria Based Selection – Passenger Seat

Requirement CTC	Prio	Concept "Super installation"		Concept "Super clean"		Concept "Super robust"	
		Evaluation	Weighted	Evaluation	Weighted	Evaluation	Weighted
Each seat and its components resist improper treatment	0.35	2	0.7	4	1.4	5	1.75
Each seat and its parts are clean at all times	0.225	3	0.675	5	1.125	4	0.9
Each seat and its individual parts can be cleaned quickly	0.125	3	0.375	5	0.625	3	0.375
Each seat and its parts can be installed and removed quickly	0.30	5	1.5	4	1.2	3	0.9
	Aggregate	13	3.25	18	4.35	15	3.925

Concept Evaluation from Innovation Point of View

Term
Check of Innovative Strength, evaluation of innovation degree

When
ANALYZE: develop high-level concept

Goal
Evaluate the alternative high-level concepts with respect to their novelty and their ability to differentiate from the competition and existing patents

Procedure
Evaluate the concepts with the help of the criteria based selection with respect to the following criteria:

1. **Singularity of concept**
   To what degree is the concept suitable for positioning oneself as an innovative provider in the market place?
   Scale:
   1 = "Me-too product" to 5 = "entirely new approach"

2. **Chances of addressed market/market segment**
   To what extent can a new market and/or a new market segment be opened up ("Blue Ocean")?
   Scale:
   1 = "a saturated market is served" to 5 = "a new market can be opened up"

3. **Chances of the used technology**
   Is the technology used in the concept already in its' mature phase or is it new?
   Scale:
   1 = "mature technologies are used" to 5 = "the concept is based on a technology with a promising future"

4. **Patentability**
   To what extent is the concept appropriate for patenting and is it possible to provide protection against plagiarism?
   Scale:
   1 = "Concept cannot be patented and can be copied easily" to
   5 = "Concept can be completely patented and can be protected very well"

   For each concept calculate the mean value of the individual evaluations and enter it in the evaluation matrix as the degree of innovation capability.

Example: Criteria based selection for the evaluation of innovation capability – Passenger Seat

Criteria for innovation capability	**Concept "Super installation"**	**Concept "Super clean"**	**Concept "Super robust"**
	Evaluation	Evaluation	Evaluation
Singularity of concept	2	4	4
Chances of addressed market segment	3	3	3
Chances of the used technology	3	5	3
Patentability	3	4	3
Mean Value	2.75	4	3.25

➾ **Tip**
- Adjust the criteria used for evaluating innovation capability to the target market and to industry specific requirements
- If required you can weight the individual criteria and follow the procedure for criteria based selection

Concept Evaluation from the Company's Point of View

Term
Evaluation of Realization Effort

When
ANALYZE: develop high-level concept

Goal
Evaluate the alternative high-level concepts with respect to the expected realization effort connected to them

Procedure
Evaluate the concepts with the help of the criteria based selection with respect to the following criteria:

1. **Availability of required development resources**
   Are the resources required for realizing the project available (time, money, employees)? Is the necessary know-how available for realizing the concept?
   Scale:
   1 = "required resources are currently not available" to 5 = "resources are available for the project without problems"

2. **Opportunities for procurement**
   Can the required (primary) materials be procured with the existing supplier structure? What does the availability of the required materials look like?
   Scale:
   1 = "Required materials cannot be procured with the existing supplier structure" to 5 = "Required materials can be procured without problems"

3. **Prerequisites for production**
   To what extent are the required manufacturing plants and/or capacities available? Is the required manufacturing know-how available?

Scale:
1 = "Required plants and know-how are currently not available" to 5 = "Production is possible with the existing plants and without problems"

4. **Prerequisites for sales**
   To what extent are the required sales channels and the sales logistics available? Is the required (technical) know-how available among the sales representatives?
   Scale:
   1 = "Required sales infrastructure not yet available" to 5 = "Sales infrastructure entirely sufficient"

   Calculate the mean value of the individual evaluations and enter in the evaluation matrix as an indicator of the realization effort required from the company's point of view. This will be used as the circle size variable.

Example: Criteria based selection for the evaluation of the realization effort from the company's point of view – Passenger Seat

Criteria for evaluating the realization effort	**Concept "Super installation"**	**Concept "Super clean"**	**Concept "Super robust"**
	Evaluation	Evaluation	Evaluation
Availability of required development resources	4	4	4
Opportunities for procurement	4	4	2
Prerequisites for production	4	3	3
Prerequisites for sales	4	3	3
Mean Value	4	3.5	3

➾ **Tip**

- Adjust the criteria used for evaluating innovation capability to the target market and to industry specific requirements
- You can weight individual criteria based on their priorities and then follow the procedure for criteria based selection

Example: Concept Evaluation Matrix – Passenger Seat on the following page

Example: Concept Evaluation Matrix – Passenger Seat

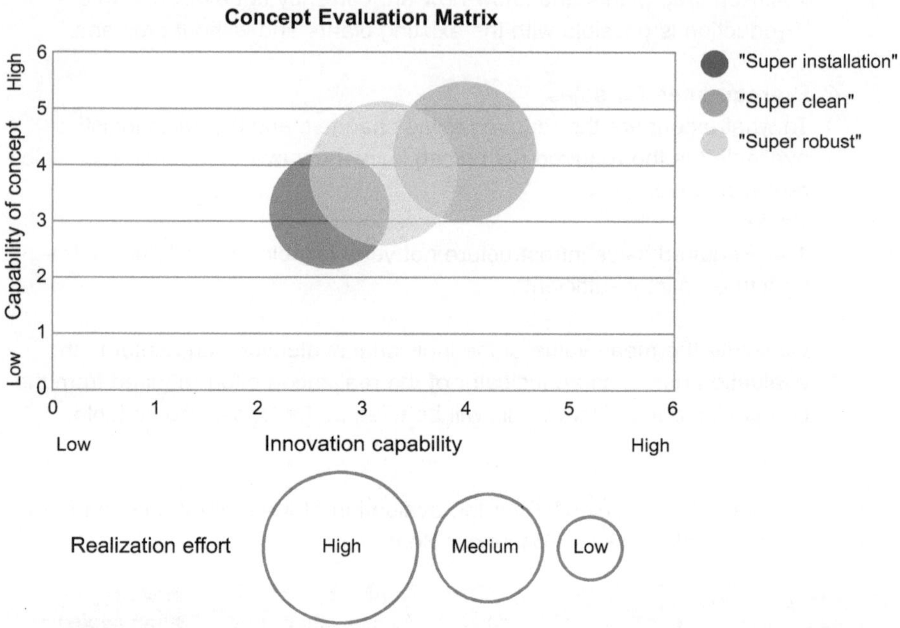

In the above example the Concept Evaluation Matrix shows that the concept "Super clean" should be favored. Not only does it score well as a concept but also for its' level of innovation capability. The slightly higher realization effort compared to the concept "Super installation" should be accepted and the concept “Super clean” should be pursued further.

Concept Evaluation with the Help of the Pugh

Term

Pugh Analysis*, Pugh-Matrix, selection procedure based on Pugh

When

ANALYZE: develop high-level concept

Goal

1. Conduct a simplified concept evaluation by taking into account the following aspects
   - Concept capability
   - Innovation capability
   - Realization effort

   in a single evaluation table
2. Identify opportunities for concept optimization

Procedure

1. Ensure that none of the concepts to be evaluated violate must criteria (e.g. legal guidelines, corporate philosophy, technical restrictions, etc.)

2. Enter the evaluation criteria (CTCs, degree of innovation and realization effort) including their priorities (if required) into the vertical fields of the matrix

3. Enter the alternative high-level concepts into the horizontal fields of the matrix (best done by giving concepts a telling name and/or title, e.g. concept "super safe", concept "super comfortable", or the like)

4. A concept (usually the existing one or that of a competitor) is set as a reference standard; every criterion is given the value 0 for the standard

---

* Stuart Pugh (1991): Total Design – Integrated Methods for Successful Product Engineering, Pearson Education, Peachpit Press, Berkeley, CA, USA

5. The alternative concept ideas are compared with the standard with respect to fulfilling the individual criteria; a better evaluation vis-à-vis the standard concept is tagged with a plus sign (+) and a poorer one with a minus (-), if no difference (0)

Illustration: Pugh Matrix. Comparison of alternative concept ideas

Alternative / Criterion	Standard	Concept 1	Concept 2	Prioritization
Criterion 1	0	+	-	3
Criterion 2	0	+	-	4
Criterion 3	0	0	+	2
Criterion 4	0	-	0	1
Aggregate +	0	2	1	
Aggregate -	0	1	2	
Aggregate 0	4	1	1	
Weighted aggregate +	0	7	2	
Weighted aggregate -	0	1	7	

6. For each concept the number of same valuations is added together and weighed according to the prioritization of the evaluated criteria (e.g. concept 1: prioritization of criterion 1 (= 3) + prioritization of criterion 2 (= 4) corresponds to the weighted aggregate + (= 7))

Illustration: Pugh Matrix, valuation of alternative concept ideas

Alternative / Criterion	Standard	Concept 1	Concept 2	Prioritization
Criterion 1	0	+	-	3
Criterion 2	0	+	-	4
Criterion 3	0	0	+	2
Criterion 4	0	-	0	1
Aggregate +	0	2	1	
Aggregate -	0	1	2	
Aggregate 0	4	1	1	
Weighted aggregate +	0	7	2	
Weighted aggregate -	0	1	7	

7. An analysis of the strengths and weaknesses of the alternative concept ideas is now possible by considering the following questions:
   - Does a concept exist that dominates the others?
   - Why is it dominant?
   - What are its weaknesses?

   Can these weaknesses be compensated for by characteristics taken from other concept ideas? (optimizing combination)

8. Develop an optimized (hybrid) solution approach based on this analysis by replacing the weaknesses of the best concept with suitable elements from other concepts

Illustration: Pugh Matrix optimizing weak characteristics of a best concept

Alternative / Criterion	Standard	Concept 1	Concept 2	Prioritization
Criterion 1	0	+	-	3
Criterion 2	0	+	-	4
Criterion 3	0	0	+	2
Criterion 4	0	-	0	1
Aggregate +	0	2	1	
Aggregate -	0	1	2	
Aggregate 0	4	1	1	
Weighted aggregate +	0	7	2	
Weighted aggregate -	0	1	7	

9. Adopting an iterative procedure the optimized concept in the Pugh Matrix is compared and evaluated once again with the standard

Tip

- To differentiate concepts further with respect to a specific criterion you can also assign several plus and minus signs per concept
- A column-oriented focus on the strongest concept within the Pugh Matrix reveals possible conflicts/contradictions in the degree of fulfillment of single criteria
- These contradictions can be described and solved with help of the TRIZ methodology

Evaluate High-level Concept

Term

High-level Concept Evaluation

When

ANALYZE: evaluate high-level concept

Goal

- Identify target costs for the product and distribute them across the components
- Identify and eliminate weaknesses of the concept
- Obtain the approval of the concept by customers and Stakeholders

Procedure

1. The estimated production costs are identified as part of a target cost calculation and are then distributed according to their priority across the product components
2. Any necessary cost adjustments can be realized with optimization approaches from the TRIZ toolbox
3. Conflicts and/or contradictions between different elements/parameters of the concept can also be solved using TRIZ methods
4. Further weaknesses of the concept can also be identified and addressed with the help of a Failure Mode and Effect Analysis
5. Further implementation and/or project risks are identified and evaluated with a Risk Management Matrix
6. Based on application examples, computer simulations or prototypes, precise feedback can be obtained from customers and Stakeholders

Determine Target Costs

Term
Target Costing

When
ANALYZE: evaluate high-level concept

Goal
- Ensure the profitability of a new product/service
- Distribute allowable overall costs among the sub-components according to their priority

Procedure
Based on the equation

Target costs = accepted price - target margin

the goal is to ensure profitability. This is achieved by the following procedure:

1. **Identify accepted price**
   The price that the customer accepts is determined by observing the market (what prices are paid for existing or comparable solutions on the market?) and/or information from customers (gained e.g. by market research, in the course of sales meetings or via specific target customer surveys). The Price Sensitivity Meter (PSM)* developed in 1976 by Peter van Westendorp is highly appropriate for a targeted customer survey.
   The customers are asked four questions:
   a From what price onwards do you consider the product to be cheap?
   b From what price onwards do you consider the product to be expensive?
   c From what price onwards do you consider the product to be so expensive that you wouldn't buy it any longer?
   d From what price onwards do you consider the product to be so cheap that you would mistrust its quality?

* WESTENDORP, PETER H. VAN (1976): NSS Price Sensitivity Meter (PSM) – A new Approach to study Consumer Perceptions of Prices, Paper for the 25th Esomar Congress

You are given a scale, ranging in approximately 20 steps from the lowest price to the highest price found on the market. From the answers the price area which is acceptable for the majority of customers can be easily identified.

2. **Determine target margin**
   The company must determine what contribution margin it expects from the new product/service. In order to support corporate growth with innovative products the target margin should be significantly above the average margin of existing products. Strategically it might also be reasonable to accept a margin which is lower than usual, in order to correctly position the new product/service in the market. This would be the case if it was hoped to gain positive effects on other business activities (Cross and Up-Selling) or a long-term competitive advantage (increase in prestige, increase of brand value, etc.). Thus, the target margin is not determined by the project team but by responsible management who can take into account the overall economic benefit of the innovation and the market situation.

3. **Identify target costs**
   Having an acceptable price and the desired margin allows target costs to be calculated. Apart from procurement and manufacturing costs, these also include the costs for marketing as well as after sales costs. The last two types of cost should initially be approximated based on experience and estimates by experts, and subsequently should be further specified step by step as the development work continues. These are to be deducted from the allowable total costs.

4. **Distribute target costs across system components**
   The remaining amount is now distributed, in line with the identified product function priorities from QFD 2, with the help of QFD 3, to the individual system components in the following way:
   a **WHAT**: **Insert** functions and priorities from QFD 2 (derive priorities if required via AHP) into the vertical fields
   b HOW: Add the components of the best high-level concept from the Morphological Box into the horizontal fields
   c Evaluation Matrix: "To what degree does the component contribute to fulfilling the respective function?"
   d Technical weight: Form aggregate of columns per component
   e Distribute allowable costs in line with their priority across the components
   f Identify expected costs (costs for development and manufacturing) with

the help of values based on experience and/or estimates by experts and compare them to the allowable costs

g Ascertain any need for adjusting the costs:
- Increase in component value, if expected costs are lower than estimated costs
- Reduction of component value, if expected costs are higher than estimated costs

5. **Conduct adjustment of component value**
Based on the calculation of the QFD 3 the need for reducing the component value can be supported by TRIZ or Design for X.

6. **Monitor development of costs**
The allowable costs of the individual components are continuously monitored in the course of the project in order to ensure that the target costs are reached.

Example: Target Cost Matrix (QFD 3 for target costs calculation) - Passenger Seat

		System components		
		Seat cover	**Suspension system**	**Holder**
Functions	**Priority**	Cover with fiber coating of glass	Air cushioning	Ferro-magnetism
Protect seat	**40%**	30%	10%	0
Absorb force	**30%**	5%	25%	0
Fixate seat	**30%**	0	5%	25%
Technical weight of component		35%	40%	25%
[redacted]		$ 106.57	$ 121.79	$ 76.12
Estimated costs		$ 97.43	$ 119.35	$ 82.82
Cost adjustment (component value increase/component value reduction)		+ $ 9.13	+ $ 2.44	**-$ 6.70**
		+9%	+2%	**-8%**
		Value increase	Value increase	**Reduction**

Specifics for the process development

The approach described for calculating the target costs can only partly be applied to process development. This is in part due to the fact that allowable costs for a process are usually given by management and the process is not supposed to generate a margin. It is also in part due to the fact that it is often difficult to assign a clearly quantifiable contribution in fulfilling the functions to individual process steps.

However, the application of target costing is appropriate using QFD 3, if certain technical components (e.g. IT applications) are included. In these circumstances it is usually possible to apportion a cost contribution for fulfilling the functions.

Eliminate Weaknesses and Conflicts in the Selected Concept

Term

Optimize Design Concept

When

ANALYZE: evaluate high-level concept

Goal

- Resolve potential contradictions in the selected concept
- Eliminate weaknesses of the selected concept
- Realize required cost savings

Procedure

Compile conflicts identified so far:

- Compile a list of the contradictions identified in the correlation roof of the House of Quality (MEASURE phase) or while the concept was being selected (ANALYZE phase) and address with TRIZ
- Realize calculated necessary cost reductions (if required with the help of TRIZ) from Target Costing
- Detect and eliminate further weaknesses of the concept using a Failure Mode and Effect Analysis (FMEA)

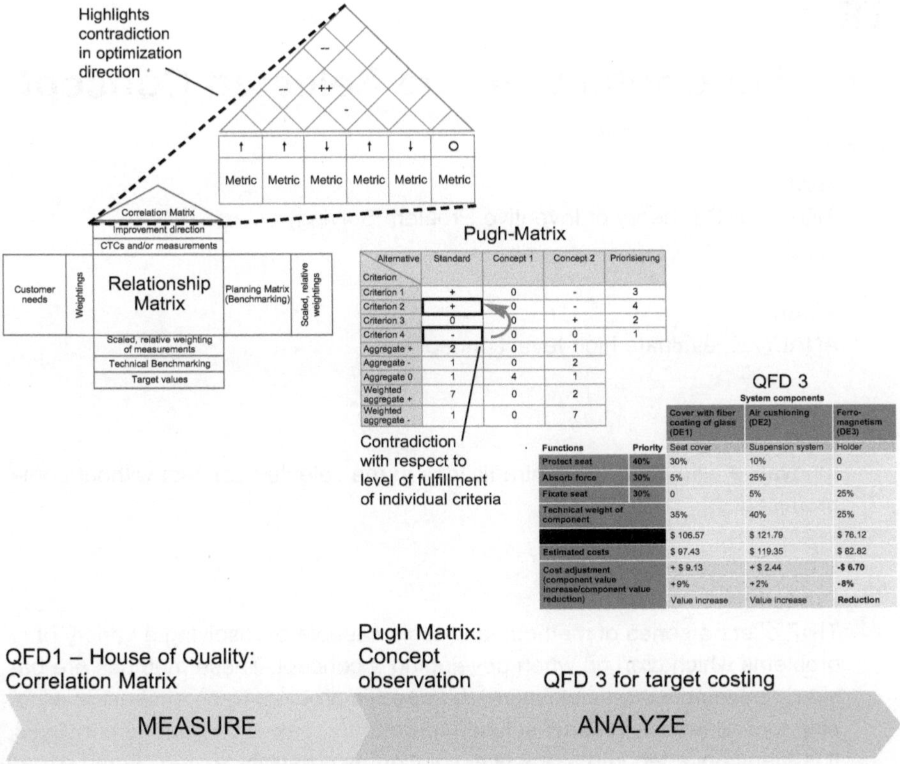

Technical conflicts between measurements can be identified in the correlation matrix of the House of Quality, but, if the weight and size are reduced, a potential conflict with the requested strength arises. Here, TRIZ innovation principles can be used for solving the conflict.

While the conflicts highlight negative interactions between measurements, they actually represent negative relationships between the collected customer requirements. These become apparent when comparing concepts in a Pugh Matrix e.g. the customer wants the lightest possible product (i.e. easier to carry) but at the same time, it should also be strong (i.e. to guarantee stability). Here, the TRIZ separation principles can be used for finding a solution.

.

TRIZ
Resolve Conflicts in the Selected Concept

Term
TRIZ*, TIPS (Theory of Inventive Problem Solving)

When
ANALYZE: evaluate high-level concept

Goal
Innovative elimination of contradictions in the selected concept without compromises

Procedure
TRIZ offers a series of methods and tools capable of resolving a variety of problems which crop up when developing a concept. These methods are basically oriented towards elevating the specific problem to an abstract level, a step that allows an abstract solution to be found based on general principles. It is then converted into a practical solution by creativity, expertise and experience.

Illustration: TRIZ – general procedure

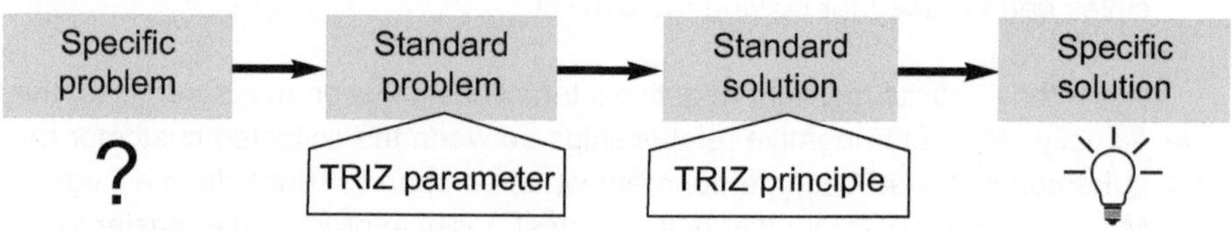

Every potential problem belongs to one of the following five main groups for which the various tools and methods of TRIZ provide solutions.

* Altshuller, Genrich S. (10/15/1926–09/24/1998)

Main groups in TRIZ

Engineering/ technical conflicts	Physical conflicts	Incomplete functional structures	Escalating complexity	System optimizations
Improving the operation of one object leads to a deterioration of another operation	Useful and harmful actions impact on the same object	There are insufficient, useful functions or the required useful functions are missing	The system is too complex and expensive	Although the current system functions, improvement is necessary for attaining competitive advantage

Tip

- You can use TRIZ approaches to get ideas even if the solution or concept is already developed
- The origin and the main focus of TRIZ is solving technical problems: You should therefore be aware that transferring TRIZ methods and principles to non-technical or process related topics may not always be possible

Technical Contradictions

Term
Engineering Contradictions, Technical Contradictions

When
ANALYZE: evaluate high-level concept, eliminate weaknesses and conflicts in the selected concept

Goal
Find innovative ideas for eliminating technical contradictions in the selected concept without making compromises

Procedure
An engineering or technical contradiction exists in a system when improving one parameter results in the deterioration of another.

Illustration: Technical Contradiction

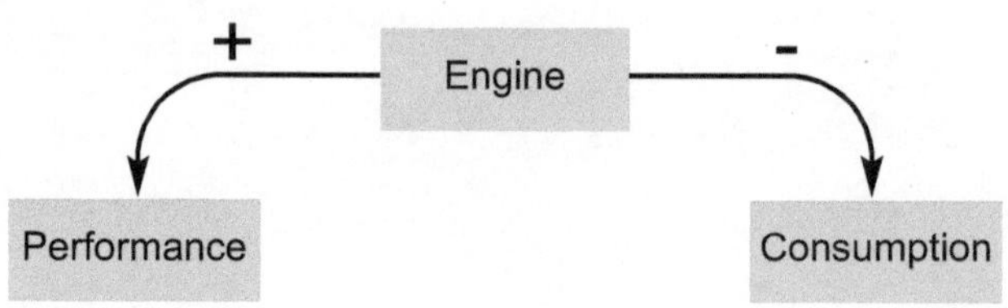

The TRIZ method describes 39 general engineering parameters which may contradict one another.

Mapping "The 39 Engineering Parameters in TRIZ"

1. Weight of a moving object	15. Durability of a moving object	29. Production precision
2. Weight of a stationary object	16. Durability of a stationary object	30. External effects harming an object
3. Length of a moving object	17. Temperature	31. Harmful factors generated by the object moving
4. Length of a stationary object	18. Illumination intensity	
5. Area of a moving object	19. Energy use by a moving object	32. Ease of production
6. Area of a stationary object	20 Energy use by a stationary object	33. Ease of operation
7. Volume of a moving object	21. Power	34. Ease of repair
8. Volume of a stationary object	22. Waste of energy	35. Adaptability or versatility
9. Speed	23. Waste of substance	36. Complexity in structure
10. Force	24. Loss of information	37. Complexity in measuring and monitoring
11. Stress or pressure	25. Waste of time	
12. Shape	26. Quantity of substance/matter	38. Extent of automation
13. Stability of an object's composition	27. Reliability	39. Productivity
14. Strength	28. Measurement accuracy	

These defined engineering parameters can be divided into physical-technical factors (e.g. weight, length, volume) and system-technical factors (e.g. reliability, productivity).

The 39 engineering parameters at a glance:

1. Weight of a moving object
   The effects generated by a moving object's own weight on an area (guiding element). Moving objects are those that can change their position on their own or through external forces.

2. Weight of a stationary object
   The effects generated by a stationary object's own weight on an area (fundamental). Objects are stationary when they cannot change their position on their own or through external forces.

3. Length of a moving object
   Dimensions – length, breadth, height or depth of a moving object.

4. Length of a stationary object
   Dimensions – length, breadth, height or depth of a stationary object.

5. Area of a moving object
   Area of an object that can change its position in space as a result of internal or external force.

6. Area of a stationary object
   Area of an object that cannot change its position in space as a result of internal or external force.

7. Volume of a moving object
   Volume of an object that can change its position in space as a result of internal or external force.

8. Volume of a stationary object
   Volume of an object that cannot change its position in space as a result of internal or external force.

9. Speed
   Working or processing speed with which an operation or a process can be executed.

10. Force, intensity
    Force necessary to effect physical changes on an object or system. These changes can be whole or partial, permanent or temporary.

11. Stress or pressure
    Amount of force that generates stress or pressure within the impact sphere of an object.

12. Shape
    Shape of an object or system. The shape can change entirely or partially, permanently or temporarily while a force impacts on it.

13. Stability of an object's composition
    Stability of the system as a result of internal or external effects on its' parts or subsystems.

14. Strength
    Ability of an object to withstand external forces before succumbing to them in some way.

15. Durability of a moving object
    Duration or time span in which a moving object can fully fulfill its function.

16. Durability of a stationary object
    Duration or time span in which a stationary object can fully fulfill its function.

17. Temperature
    Rise or fall in temperature of an object or a system while fulfilling its function.

18. Illumination intensity
    Intensity of light in, around or through the system, including the light quality and other characteristics of light.

19. Energy use by a moving object
    Energy needed by a moving object or system.

20. Energy use by a stationary object
    Energy needed by a stationary object or system.

21. Power
    Power (work/unit of time) that is necessary to actually perform a function.

22. Waste of energy
    Increased incapacity of an object or system to incorporate energy, especially when production is not taking place.

23. Waste of substance
    Reduction or loss of substance of an object or system, especially when production is not taking place.

24. Loss of information
    Reduction or loss of data or the input to a system.

25. Waste of time
    Necessary increase in time to carry out an operation.

26. Quantity of substance/matter
    Number or quantity of elements comprising an object or system.

27. Reliability
    The capacity of an object or system to fulfill its function over a specific time span or cycle.

28. Measurement accuracy
    The accuracy of a measurement in relation to the real value.

29. Production precision
    The precision of production in relation to the construction specifications.

30. External effects harming the object
    External factors which reduce the efficiency or quality of the object or system.

31. Harmful factors generated by the object moving
    Internal factors which reduce the efficiency or quality of the object or system.

32. Ease of production
    Objects or systems can be produced easily.

33. Ease of operation
    Objects or systems can be easily operated.

34. Ease of repair
    Objects or systems can be easily repaired after intensive use or damage.

35. Adaptability or versatility
    The capacity of an object or system to adapt to changed conditions.

36. Complexity in structure
    Number and diversity of elements comprising individual objects or systems as well as their interactions.

37. Complexity in measuring and monitoring
    Number and diversity of elements to measure and monitor objects and systems as well as the costs for an acceptable error contribution.

38. Extent of automation
    The possibilities to operate objects or systems without the help of people.

39. Productivity
    The relationship between operating time and total time.

The paired comparison of technical parameters in a matrix form facilitates their application and supports both the transference of the real conflict as well as establishing the relevant innovative principles that should be applied. This matrix is known as the "contradiction matrix".
Within the frame of TRIZ, 40 innovative principles are described as generally applicable solution approaches for conflicts defined using the technical parameters.

Illustration: "The 40 Innovative Principles of TRIZ"

1. Segmentation and dissection	16. Partial or excessive actions ("less is more")	29. Pneumatics and hydraulics
2. Taking out and spinning off		30. Flexible shells and thin films
3. Local quality	17. Another dimension	31. Porous materials
4. Asymmetry	18. Mechanical vibration	32. Color changes
5. Merging	19. Periodic action	33. Homogeneity
6. Multifunctionality and Universality	20. Continuity of useful action	34. Discarding and recovering
7. Nested doll	21. Skipping	35. Parameter changes
8. Anti-weight and counterbalance	22. "Blessing in disguise" or "Turn lemons into lemonade"	36. Phase transitions
9. Preliminary anti-action		37. Thermal expansion
10. Preliminary action	23. Feedback	38. Strong oxidants
11. Preventive activities/cushioning	24. "Intermediary"	39. Inert atmosphere
12. Equipotentiality	25. Self-service	40. Composite materials
13. The other way around/inversion	26. Copying	
14. Spheroidality – curvature	27. Cheap short-living objects	
15. Dynamics	28. Mechanics substitution	

The 40 innovative principles at a glance

1. Segmentation and dissection
   a. Divide an object into independent component parts, e.g.
      - Build a PC out of modular components
      - Replace a large truck with a small truck with trailer

b. Simplify the decomposition and/or assembly of an object e.g.
   - Module system
   - Quick locks with tubes
c. Increase the degree of segmentation of an object e.g.
   - Freely extendable garden hose

2. Taking out and spinning off
   a. Eliminate interfering functions, components, or qualities of an object e.g.
      - Relocate a noisy compressor outside the work area or building
      - Relocate loud units of air-conditioning system outside the living area
   b. Single out and restrict to, only necessary elements or functions of objects e.g.
      - Recording and playing barking dogs as part of a security alarm
      - Playing animal noises to frighten off birds at airports

3. Local quality
   a. Change the uniform (constant) structure of an object or its environment to a non-uniform structure e.g.
      - To combat coal dust in a mine a fine spray of water is spread across the working area; however, this hampers work in the drill zone
      - The working areas in the mine are separated by a further layer of water, this time larger drops are used, which keep the fine spray localized
   b. Distribute different functions to different components of an object e.g.
      - Pencil with eraser
      - Swiss army knife
   c. Create optimal conditions for each sub-function of an object e.g.
      - Lunchbox with special compartments for hot and cold beverages, and food

4. Asymmetry
   a. Replace symmetrical shapes and forms with asymmetrical ones e.g.
      - Asymmetrical vessels or asymmetrical mixing forms improve the mixing of materials (blenders, cement mixers)
      - Reinforce the exterior of tyres to minimize the harmful impact of curbs
   b. Increase existing asymmetrical effects

5. Merging
   a. Concentrate identical or similar objects and operations in the same room or space e.g.

- Merge PCs into a network
- Electronic chips on both sides of a plug-in card

b. Operations carried out simultaneously or sequentially (bringing them together in time) e.g.
   - Medical diagnostic instrument analyzing different parameters of blood simultaneously
   - Lawnmower fitted with mulch cover

6. Multifunctionality and Universality
   Multifunctional design of objects reduces the quantity of other parts e.g.
   - Sofa that can be converted into a bed
   - A stroller that can be converted into a car safety seat
   - Minivan seats which can be used to sit, sleep, or transport goods

7. Nested Doll (Matryoshka)
   a. Space saving placement of identical objects into one another e.g.
      - Matryoshka Russian wooden dolls that fit inside each other
      - Stackable seats to save space when storing
      - Telescopic antenna
      - Refillable drop-action pencil
      - Camera lens with zoom function
   b. Space saving placement of different objects into each other e.g.
      - Storage function of automatic seatbelts
      - Extendable aircraft landing gear

8. Anti-weight and counterbalance
   a. Reduce the weight of an object by generating buoyancy e.g.
      - Air tanks in the hull of a ship or in a submarine
      - Sandwich construction in planes, surfboards, etc.
   b. Utilize dynamic forces e.g.
      - Wing shape of aircraft creates lift
      - Generate traction for sports cars through a rear wing

9. Preliminary anti-action
   a. Take steps to counter or control harmful effects or undesirable stress e.g.
      - Spoke of a wheel
      - Bolted connection to absorb force

10. Preliminary action
    a. Take into account predictable actions early e.g.

- Craft knife with blade segments which can be broken off
- Tool change system

b. Sensible spatial arrangement of necessary objects which need to be readily available e.g.
   - Replenishment pull system in production

11. Preventive activities/cushioning
    Take into account the possible consequences of an object which has relatively low reliability with counter activities e.g.
- Back-up parachute
- Magnetized anti-theft strip on consumer goods

12. Equipotentiality
    Create a constant spatial level e.g.
    - Canal locks for regulating the level of water and/or elevating or lowering ships
    - Work on the underbody or engine (from below) of a car

13. The other way around/Inversion
    a. Invert the actions used to solve a problem e.g.
       - "Bring the mountain to Mohammed"
    b. Invert the moving and non-moving qualities e.g.
       - Rotate the part instead of the tool
       - Ergometer, treadmill
    c. Reverse the object or process
       - Emptying pivotally mounted containers (rail, ship)

14. Spheroidality – Curvature (in all dimensions)
    a. Curvature of straight lines and flat surfaces e.g.
       - Parabolic mirror
    b. Extend two-dimensional movements e.g.
       - Computer mouse
       - Trackball
       - Ballpoint pen
    c. Use centrifugal forces e.g.
       - Centrifugal casting
       - Clothes dryer

15. Dynamics
    a. Variable design of an object or its environment e.g.
       - Automatically adjustable car seat, rearview mirror, steering wheel

b. Divide an object into parts or segments which move relative to one another e.g.
   - "Gooseneck" for car radios, flashlights, lamps
   - Clockworks
   - Gears

c. Turn fixed objects or rigid processes into flexible or replaceable ones

16. Partial or excessive actions ("less is more")
   a. Extend or limit individual object functionality e.g.
      - Rotate freshly painted cylinder to remove excess paint

17. Another dimension
   a. Avoid obstacles by adding further dimensions e.g.
      - Move an infrared computer mouse in a room instead of across a surface
      - 3D chess
   b. Utilize storage possibilities e.g.
      - CD jukebox
      - Tool changing systems
   c. Change of position e.g.
      - Tipper truck
   d. Project objects into neighboring areas e.g.
      - Concave reflector on the north side to illuminate a glasshouse

18. Mechanical vibration
   a. Utilize the vibrations made by objects e.g.
      - Vibrating knife to remove a plaster cast as a way of preventing damage to the skin
      - Vibration of funnels to optimize the flow of abrasives
   b. Increase the frequency of vibrating objects
   c. Exploit an object's own resonant frequency
   d. Transition from mechanical to piezo-electric vibrators e.g.
      - Cleaning of lab equipment in ultrasonic bath
   e. Functional combination of ultrasonic vibrations with electromagnetic fields

19. Periodic action
   a. Transition from continuous to periodic actions e.g.
      - Flashing of warning lights to improve visibility
      - Remove rusted screws with pulsating instead of continuous exertion of force
   b. Change the frequency of periodic actions

c. Use recurrent pauses e.g.
   - Do something else whilst waiting

20. Continuity of useful action
    - All components work continuously at full load
    - Eliminate all idle times or intermittent actions

21. Skipping
    Raise the speed of damaging or dangerous actions; work areas that are harmful but indispensable for the process have to be finished or exited as quickly as possible

22. "Blessing in disguise", converting harmful factors into positive effects
    a. Positive use of harmful factors or effects – especially from the environment
    b. Eliminate harmful factors by combining them with one another
    c. Eliminate a harmful factor by amplifying it so that it is useful

23. Feedback
    a. Introduce feedback
    b. Vary or reverse feedback

24. Intermediary
    a. Integrate a sub-object to transfer an effect e.g.
       Chilled electrodes with another liquid metal, with a lower melting point, lying between them to avoid energy loss when electric energy is applied.
    b. Temporarily connect one object to another but in an easily detachable way

25. Self-service
    a. Enable an object to “help itself” in performing auxiliary functions e.g. abrasive surface design of a bottling apparatus for abrasive materials enabling it to repair itself.
    b. Use waste resources or "bi-product analogy" (energy, materials)

26. Copying
    a. Replace complex, expensive, fragile objects or objects which are difficult to manage with cheap, simple copies
    b. Visually represent and object or system scaling as necessary

c. Replace optical copies with infrared or ultraviolet copies e.g. Measure high objects using their shadows

27. Cheap short living objects
Replace a sophisticated, expensive object with a disposable, cheap object:
- Disposable diapers
- Single use scalpels
- Single use syringes

28. Machine substitution
a. Replace a mechanical system with an optical, acoustic or odorized system
b. Use interactions of electrical, magnetic and/or electromagnetic fields with the object
c. Move from static to movable fields, from constant to alterable, and from unstructured to structured fields
d. Use ferromagnetic particles e.g.
- Improve the connection between a metal and a thermoplastic by attaching an electromagnetic field to the metal

29. Pneumatics and hydraulics
Use gas and liquid components instead of solid parts; apply parts that are inflatable or filled with liquids, an air cushion, or hydro-static and hydro-reactive parts; packaging of fragile goods in padded envelopes or air cushions during transport

30. Flexible shells and thin films
a. Deploy malleable, flexible wrap and thin sheets
b. Isolate an object from its environment using flexible wrappings and thin sheets e.g.
- Spray leaves with a PE film to protect a plant from evaporation

31. Porous materials
a. Porous design of an object or its elements (inserts, coatings, etc.)
b. Fill an already porous object

32. Color changes
a. Change the color of an object or its external environment
b. Change the transparency of an object or its external environment

c. Make visible an object difficult to recognize by adding color
d. Use fluorescent colors

33. Homogeneity
Apply identical or very similar materials e.g.
- Abrasive surface design of a bottling apparatus for abrasive materials to achieve a continuous "healing of itself"

34. Discarding and recovering
a. Discard object parts which are no longer necessary or have fulfilled their function e.g.
   - Rocket or missile stages
   - Bullet casings
b. Modify used parts during an operation
c. Restore used parts during an operation

35. Parameter changes
Do not only make use of simple changes of an object's aggregate state (solid, liquid, into gas) but also transitions into "pseudo" or "quasi" states and interim states (elastic, solid bodies, thixotropy substances)

36. Phase transitions
Exploit the effects during a phase transition of a substance e.g.
- Exploit the evaporation energy of water
- Fill hollow bodies with water in order to measure the expansion of the bodies or to split them once the water freezes

37. Thermal expansion
a. Use changes in the volume of materials caused by warmth
b. Combine materials with different rates of thermal expansion e.g. Bimetal switches

38. Strong oxidants
a. Enrich air with oxygen
b. Replace enriched air with pure oxygen
c. Expose air or oxygen to ionizing radiation
d. Deploy ozonized oxygen
e. Replace ozonized (or ionized) oxygen with ozone

39. Inert atmosphere
    a. Replace the normal environment with an inert one
    b. Carry out the process in a vacuum e.g. process groceries in a protective atmosphere (e.g. nitrogen)

40. Composite materials
    Apply composite materials e.g. aircraft construction (carbon fiber composites)

TRIZ Contradiction Matrix

Term
Contradiction Matrix

When
ANALYZE: evaluate high-level concept, eliminate weaknesses and conflicts in the selected concept

Goal
Translate specific technical problems into a conflict between two technical parameters and identify relevant innovative principles for generating solution approaches

Procedure
The problem is articulated as a contradiction between two general technical parameters. The following approach is recommended:
- Which technical parameters need to be improved ("improving feature")?
- Which technical parameters would this improvement negatively influence ("worsening feature")?

The innovative principles suitable for solving the formulated contradiction can be obtained from the contradiction matrix based on the paired parameters.

Utilizing the innovative principles suggested by the matrix the team can use its' creativity, expertise and experience to create an innovative solution which resolves the contradiction without compromise.

Illustration: Application of the TRIZ contradiction matrix

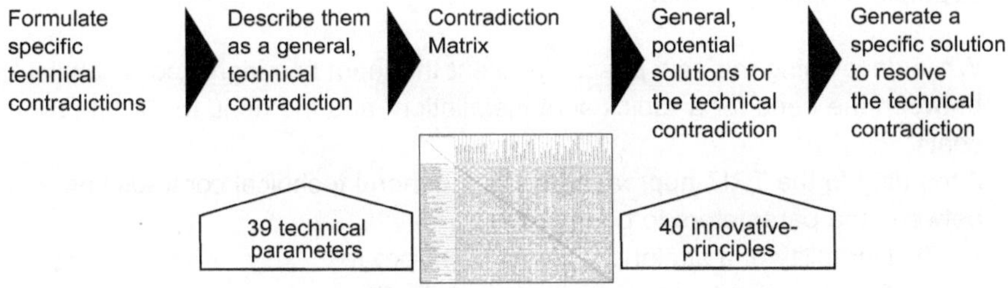

Illustration: TRIZ Contradiction Matrix

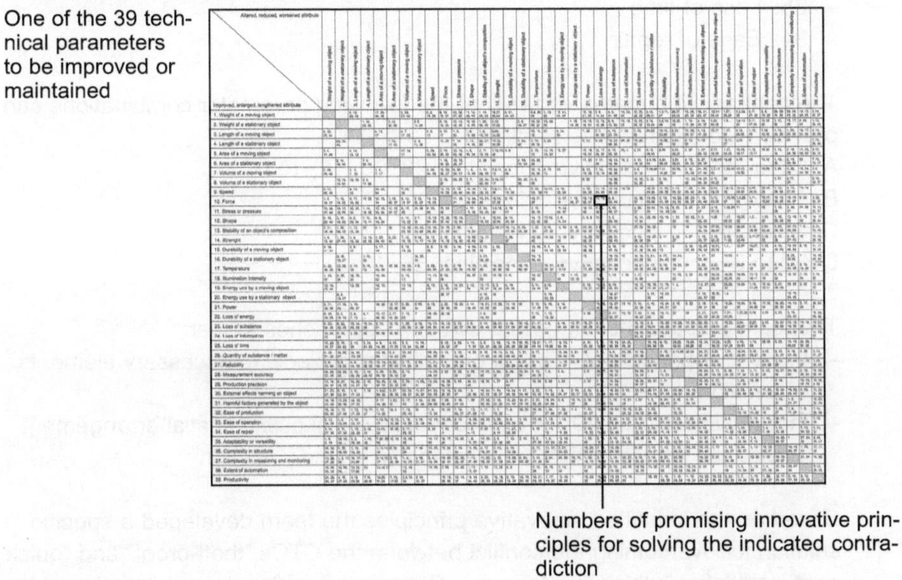

An enlarged TRIZ contradiction matrix is included in the appendix.

➾ **Tip**

The contradiction matrix reduces the effort by focusing on just a few promising innovative principles for solving your conflict; however, in case of doubt, each of the 40 innovative principles should be examined to see if they are applicable; the goal is to generate as many good solution ideas as possible.

Applying a TRIZ contradiction matrix
Example: Passenger Seat

While developing the new passenger seat the team has identified a conflict between the need for a "quick seat installation" and the need for "theft-proof seats".
According to the TRIZ approach, this is a general technical contradiction between the parameters to be improved:
- 16: Durability of a stationary object (theft-proof)
- 30: External effects harming the object (theft)

and the parameters which will be negatively impacted because of this improvement:
- 25: Loss of time
- 34: Ease of repair

The contradiction matrix shows that the following parameter combinations can provide a solution for this conflict:

A: 16 to 25	innovative principles: 28, 20, 10, 16
B: 16 to 34	innovative principles: 1
C: 30 to 25	innovative principles: 35, 18, 34
D: 30 to 34	innovative principles: 35, 10, 2

The team pursues the suggested solution approaches further:
- Innovative principle 2: "Taking out – limiting objects to necessary elements or functions"
- Innovative principle 10: "Preliminary action – sensible spatial arrangement of required objects which are available in a timely fashion"

Based on the identified innovative principles the team developed a specific solution for overcoming the conflict between the CTCs "theft-proof" and "quick seat installation"; this was done in a Brainstorming Session and produced the following ideas:
- A continuous fixing channel rail separate from the seat construction in which all of the seats can be installed one behind the other
- This channel is fixed to the floor with attachments that can be closed or opened from a central switch
- This reduces installation time while raising security against theft because the seat can no longer be removed with conventional tools

Tip

- The identified TRIZ innovative principles are suggestions for changing the technical system – they shouldn't be taken literally; imagination and creativity are needed when developing specific solutions!
- Combinations and reversals (e.g. "merging" instead of "segmenting") of the suggested innovative principles can also lead to reasonable solutions

Physical Contradictions

Term
Physical Contradictions

When
ANALYZE: evaluate high-level concept, eliminate weaknesses and conflicts in the selected concept

Goal
Innovative elimination of physical contradictions in the selected concept without compromises

Procedure
A physical contradiction exists in a system when the whole system or one of its components has to take on two contradictory states to fulfill a parameter.

Mapping physical contradictions according to TRIZ – Example:

Parameter xy

- sweet – sour
- open – closed
- short – long
- hot – cold
- gaseous – solid
- big – small
- inflammable – non-flammable

In principle, there are three possibilities for solving physical contradictions
- Separate the contradictory requirements
- Fulfill the contradictory requirements
- Avoid the contradiction

Mapping possible solutions for physical contradictions according to TRIZ

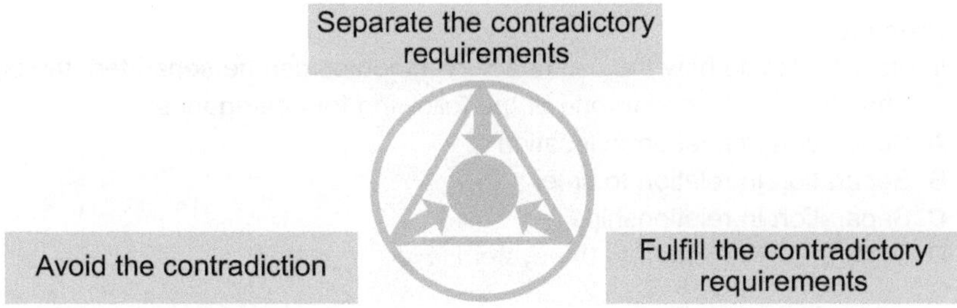

Examples

The production of machine parts from of a specific type of steel requires that the steel be heated to 2,192°F before it can be molded. It turns out that the surface of the material is damaged by the reaction with air when the temperature is heated above 1,472°F.

On the one hand, the steel must have a temperature of 2,192°F if it is to be moldable, while on the other hand, it cannot become hotter than 1,472°F, otherwise damage occurs.

A company produces oval glass elements with a thickness of 1 mm. In the first working step rectangular parts are cut and their edges ground off. These parts break however, due to the thinness. On the one hand, the parts have to be thin to match customer requirements, but on the other hand, they also need to be thick enough so they don't break when they are ground.

Mapping physical contradictions according to TRIZ

Example:

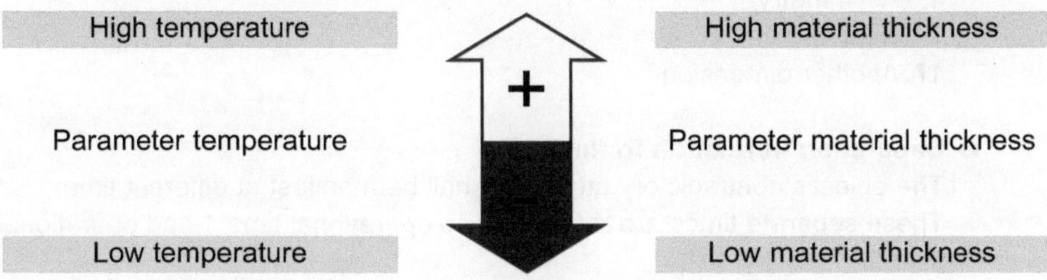

Separate the contradictory requirements

Procedure:
In order to decide how the contradictory attributes can be separated, the problem has to be assigned to one of the following four categories:

A Separation in relation to location
B Separation in relation to time
C Separation in relationships
D Separation in relation to the system level

A Separation in relation to location
The object's contradictory attributes shall be manifest at different places. These separate locations are known as operational zone 1 and operational zone 2.

Mapping separation in relation to location

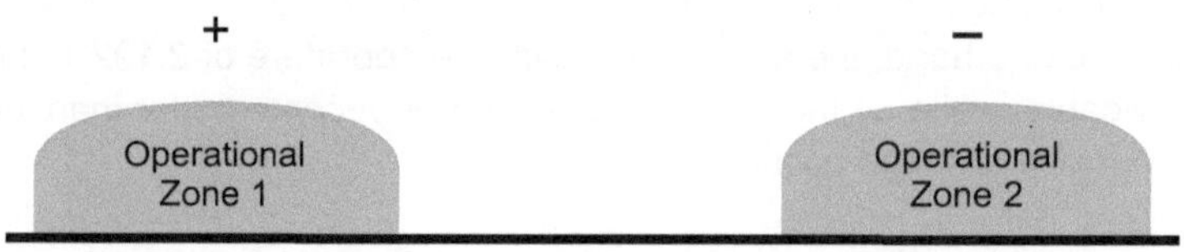

Example:
The steel parts must be heated to 2,192°F internally, but their surfaces are not to exceed 1,472°F.

Suitable innovative principles in this case might be:
1. Segmentation
2. Taking out
3. Local quality
4. Asymmetry
7. Nested doll
17. Another dimension

B Separation in relation to time
The object's contradictory attributes shall be manifest at different times. These separate times are referred to as operational time 1 and operational time 2.

Illustration: Separation in relation to time

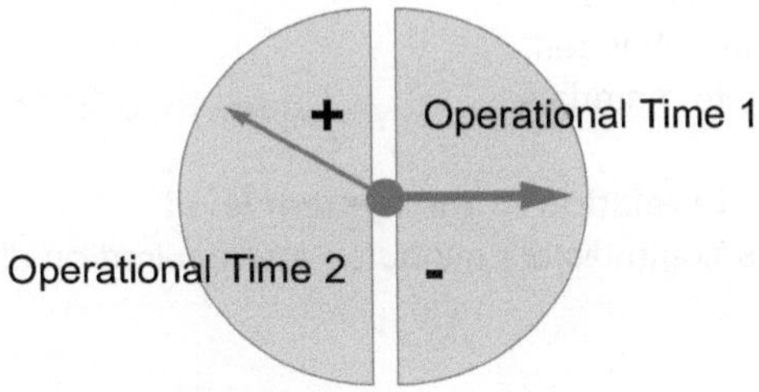

Example:
An umbrella must be as big as possible when it is raining. When it isn't raining the umbrella must be as compact as possible.

Suitable innovative principles in this case are among others:
9. Preliminary anti-action
10. Preliminary action
11. Preventive activities/cushioning
15. Dynamics
34. Discarding and recovering

C Separation in relationships
Die widersprüchlichen Eigenschaften des Objekts sind in der Relation zu unterschiedlichen anderen Objekten erforderlich.

Mapping Separation in relationships

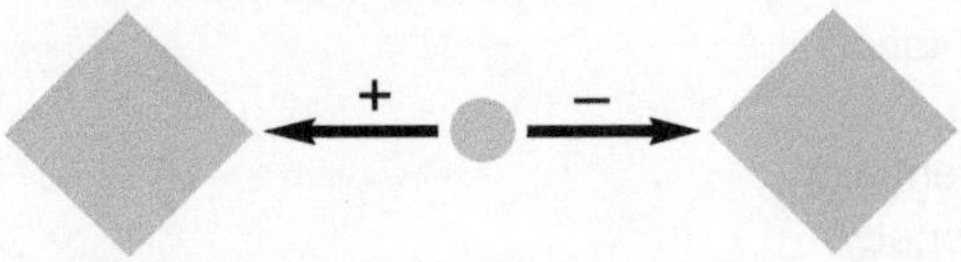

Example:
With regard to weight aircraft wings must be as small as possible; for propulsion and uplift, however, they must be as big as possible.

Suitable innovative principles in this case are among others:
3. Local quality
17. Another dimension

19. Periodic action
31. Porous materials
35. Parameter changes
40. Composite materials

D Separation in relation to the system level
The object's contradictory attributes are required on different system levels.

Mapping separation in relation to the system level

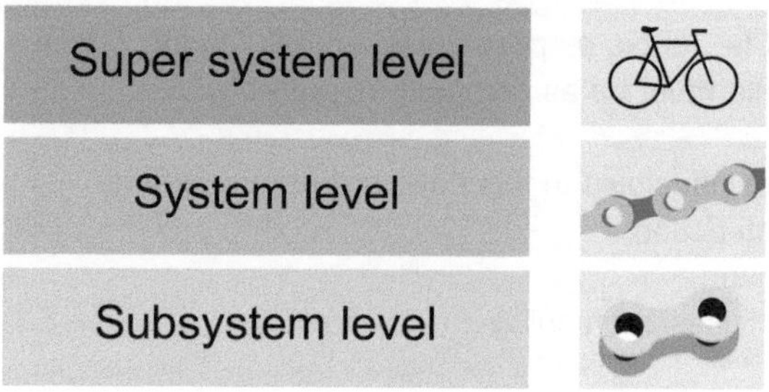

Example:
The system bicycle chain must be flexible on the super system level (bicycle) in order to enable transmission of power through pedals and sprockets. On the subsystem level (links) it must be solid to guarantee the necessary stability.

Suitable innovative principles in this case might be:
1. Segmentation
5. Merging
12. Equipotentiality
33. Homogeneity

Fulfill the contradictory requirements

Procedure:
So-called "smart materials" can fulfill contradictory requirements in some cases.

These include for example:

- **Shape-Memory Alloys** (SMA); metals which have a "memory" and can take on different shapes when they are heated or cooled
- **Electrorheological or magnetorheological fluids;** these fluids change their viscosity within milliseconds (fluid – solid) when an electrical and/or magnetic field affects them; one example is how the fluid in a hydraulic cycle can take on the function of a medium for pressure transmission at the same time as it acts as a steering medium; electorheological flow resistance allows valves to be developed which have no need for movable parts and so operate almost wear-free

Avoid the contradiction

Procedure:
New approaches can make the resolution of rigid conflicts superfluous.

Example:
When rain is accompanied by strong winds we need a large umbrella to keep us dry. On the other hand, the umbrella must be very small in order not to offer the wind much contact surface. This contradiction becomes unimportant as soon as one starts using a suitably designed raincoat covering the whole body.

➾ **Tip**
The existing system often provides the starting point for the problem definition and the direction selected is unconsciously very narrow in its dimensions, so that only improvement approaches become visible for the basic problem rather than genuine solutions. You can prevent this by abstracting the problem with the help of TRIZ and extending the solution space in a targeted way.

Sufield Analysis – Incomplete Functional Structures

Term
Sufield Analysis, WEPOL Analysis

When
ANALYZE: evaluate high-level concept, eliminate weaknesses and conflicts in the selected concept

Goal
Eliminate suboptimal, functional structures in the system

Procedure
In the Sufield Analysis a technical system is defined as the combination of at least two substances (S_1 and S_2) which interact with the help of a field (F). Each system can be depicted in this way.

Mapping Sufield Analysis

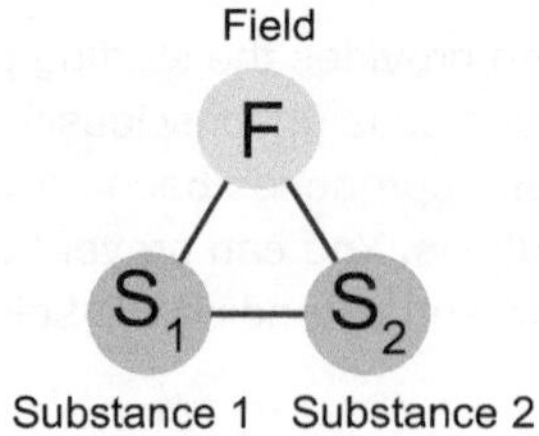

S_1 is the substance that is to be changed, processed, transformed and/or controlled.
The substance S_2 serves as the tool, instrument, or medium.
The field F represents the force or energy with which S_2 affects S_1.

Illustration: Sufield Analysis

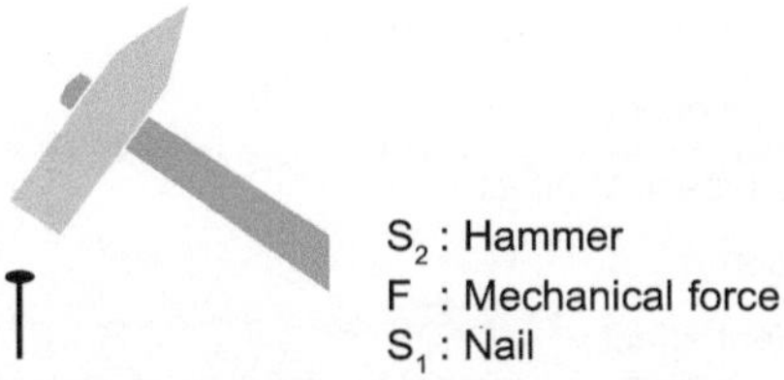

With the help of a Suffield model a real problem can be abstracted and solved with the help of "standard solutions".

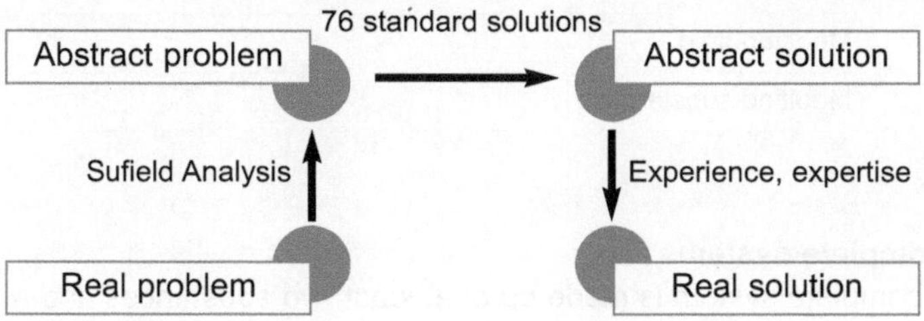

The Sufield Analysis distinguishes between four basic models of technical systems:

1. Complete systems
2. Incomplete systems
3. Complete but inefficient systems
4. Complete but harmful systems

The basic models are visualized with defined symbols.

See next page for Sufield Analysis Symbols

Mapping Symbols for the Sufield Analysis

Δ	Symbolic form of a Sufield model
——	Unspecific effect
——→	Desired (specific) effect
←——→	Interaction
- - - - - -→	Insufficient effect
∿∿→	Harmful effect
⇨	Indicates the direction from given to desired Sufield model
F ——→	Field affects a substance
——→ F	Field generated by a substance
F'	Modified field
S'	Modified substance

1. **Complete systems**
   A complete system is made up of at least two substances and a field that affects them in a desired way.

Mapping Sufield Analysis with complete systems

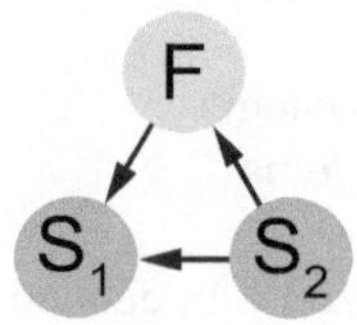

Example:
Using a compactor (S 2) which deploys mechanical force (F), metal parts (S1) are beaten into the desired shape.

2. **Incomplete systems**
   Systems are incomplete when one or more components are missing. The desired effect cannot be generated.

Example:
The condition of a refrigerator's cooling unit is to be examined. A fluorescent substance is mixed into the cooling liquid and the unit radiated with ultra-violet light in a dark room.
By doing this Leaks will become visible.

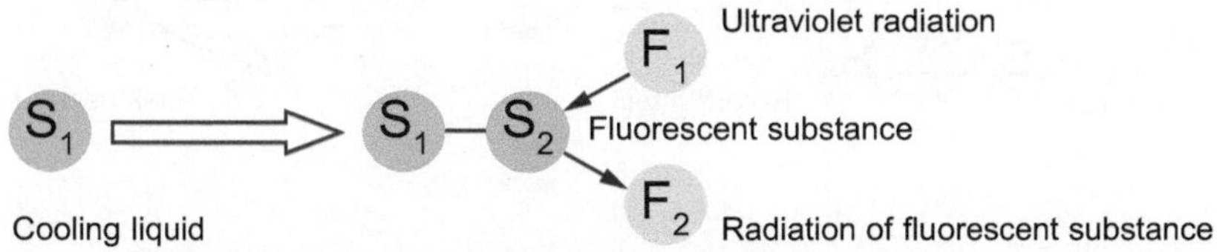

Incomplete systems must be completed to achieve the desired effect.

3. **Complete but inefficient systems**
Complete systems are inefficient when their intended effect does not reach the desired degree.

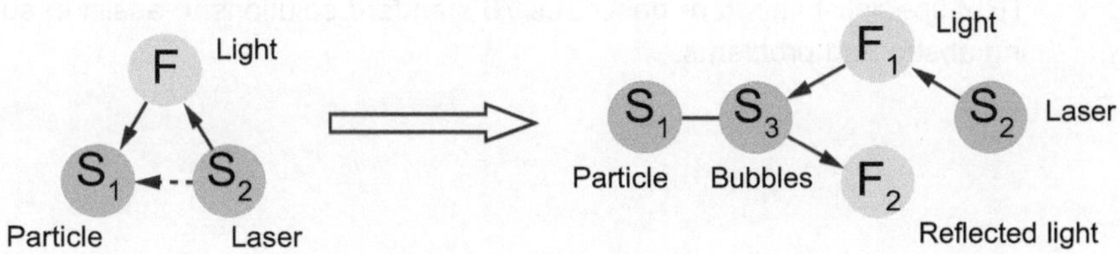

Example:
Optically complete clear fluids are to be examined to see if there are any impuities. Because these impurities are non-magnetic a laser is used for scanning the fluid. However, some of these particles are so small that they are unable to provide a good reflection of the light. By heating the fluid the surrounding particles begin to boil. The resulting bubbles can be detected easily.

Inefficient systems must be improved.

4. Complete but harmful systems
A harmful interaction takes place between the components in these systems.

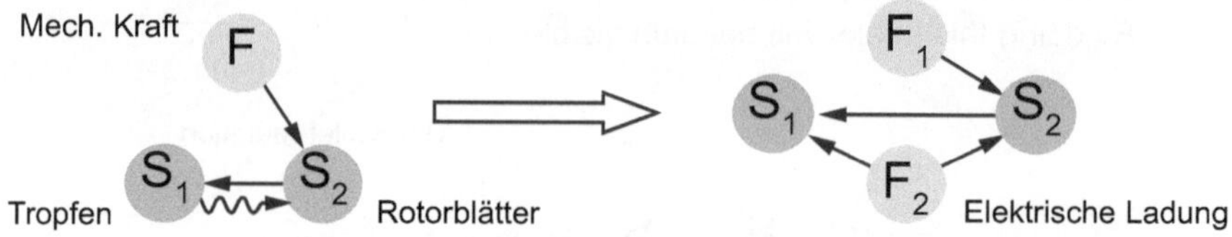

Example:
The rotor blades of steam turbines are surrounded by a mixture of steam and water drops. Relatively large drops (typically between 0 and 800 µm in diameter) collide with the rapidly rotating damaging their surfaces. To avoid this he rotor blades and the water drops are charged with the same electrical charge. They therefore repel one another.

The harmful effects of a system should be eliminated.

TRIZ specialist literature describes 76 standard solutions to assist in solving abstracted problems.

76 Standard Solutions

Term
76 Standard Solutionsn

When
ANALYZE: evaluate high-level concept, eliminate weaknesses and conflicts in the selected concept

Goal
Eliminate suboptimal, functional structures in the system

Procedure
The 76 standard solutions can be arranged thematically into the following five groups:
1. Composition and decomposition of complete Sufield Models
2. Improvement of Sufield Models
3. Transition to super and subsystems (macro and micro levels)
4. Detection and measurement
5. Assistance for applying the standard solutions

The 76 standard solutions at a glance

1. Composition and decomposition of complete Sufield Models

1.1 Composition of Sufield Models (SFM)

1.1.1 Complete an incomplete SFM
1.1.2 Complete when an internal additive can be used
1.1.3 Complete when an external additive can be used
1.1.4 Use local/existing resources for completion
1.1.5 Generate further resources by changing the system environment
1.1.6 Use an excess (material) to complete a task and then eliminate the surplus
1.1.7 If a surplus of an action is harmful, attempt to direct it to other components in the system
1.1.8 Introduce local protective substances for completion

1.2 Decomposition of Sufield Models

1.2.1 Eliminate harmful interactions by introducing a third substance, S_3

1.2.2 Eliminate harmful interactions by introducing a third substance, S_3, whereby S_3 can be a modification of the two existing substances, S_1 and/or S_2

1.2.3 Direct the effect onto a less important substance, S_3

1.2.4 Introduce a new field to compensate harmful effects

1.2.5 Turn a magnetic on or off in line with need

2. Improvement of Sufield Models

2.1 Transition to complex Sufield Models

2.1.1 Interlink several SFMs

2.1.2 Duplicate a SFM

2.2 Further development of a Sufield Model

2.2.1 Employ fields which can be controlled more easily

2.2.2 Fragment S_2

2.2.3 Deploy capillaries and porous substances

2.2.4 Increase the level of dynamics

2.2.5 Structured fields (e.g. stationary waves)

2.2.6 Structured substances (e.g. ferroconcrete)

2.3 Rhythm coordination

2.3.1 Create harmony between the rhythm (frequency) of the affecting field with one of the two substances (or controlled non-agreement)

2.3.2 Synchronize the rhythm, the frequency of fields

2.3.3 Bring independent actions into rhythmic coordination

2.4 Complex improved Sufield Models

2.4.1 Use ferromagnetic substances and magnetic fields

2.4.2 Use ferromagnetic particles, granules, powders

2.4.3 Use ferromagnetic liquids

2.4.4 Use capillary structures in combination with ferromagnetism

2.4.5 Use complex ferromagnetic SFMs, e.g. external magnetic fields, ferromagnetic additives, etc.

2.4.6 Introduce ferromagnetic material into the system environment if the system is unable to magnetize itself

2.4.7 Use natural effects (e.g. Curie point of metals)

2.4.8 Use dynamic, variable or self-adapting magnetic fields

2.4.9 Change the structure of a material by incorporating ferromagnetic particles and applying a magnetic field to move the particles

2.4.10 Calibrate the rhythms
2.4.11 Use electrical current instead of ferromagnetic particles to generate magnetic fields
2.4.12 Use electrorheology

3. Transition to super and subsystems (macro and micro levels)

3.1 Transition into bi and poly systems

3.1.1 Combine systems to bi and poly systems
3.1.2 Create or intensify the connections between the individual elements in bi and poly systems
3.1.3 Improve the efficiency of bi and poly systems by enlarging the difference between individual components
3.1.4 Simplify bi and poly systems by eliminating superfluous, redundant or similar components
3.1.5 Balance negative attributes of the whole system with positive attributes of individual components (or vice versa)

3.2 Transition to micro systems

3.2.1 Miniaturized components or whole systems

4. Detection and measurement

4.1 Indirect methods
4.1.1 Avoid detection and measurement
4.1.2 Carry out detection and measurement on a copy
4.1.3 Replace measurement with two successive detection procedures

4.2 Composition of Sufield Models for measuring

4.2.1 Detect or measure using an additional field
4.2.2 Introduce additives, substances that can be easily detected and measured
4.2.3 Introduce fields into the system environment which can be easily detected and measured if nothing can be added to the system
4.2.4 If additives cannot be introduced to the system environment change the state of something that is already present in the system environment and measure the effect of the system on this changed substance/object

4.3 Improvement of measurement systems

4.3.1 Use natural effects to improve measurement systems
4.3.2 Use resonance phenomena for measurement
4.3.3 Use objects linked by resonance phenomena for (indirect) measurement

4.4 Transition to ferromagnetic measurement systems (this was a popular method before the introduction of microprocessors and fiber optics, etc.)

4.4.1 Deploy ferromagnetic substances and magnetic fields
4.4.2 Replace substances with ferromagnetic substances and detect or measure via the magnetic field
4.4.3 Generate complex, linked SFMs with ferromagnetic components
4.4.4 Introduce ferromagnetic materials to the system environment
4.4.5 Use the impact of naturally magnetic effects for measurement

4.5 Evolution of detection and measurement

4.5.1 Generate bi and poly systems
4.5.2 Detect and measure the first and second derivations in time and space instead of the original functions (e.g. change in frequency instead of speed [Doppler Effect])

5. Assistance for applying standards

5.1 Introduce substances

5.1.1 Indirect methods (e.g. introduce vacuums or cavities as substances)
5.1.2 Divide the elements into smaller units
5.1.3 Use the self-elimination of substances
5.1.4 Use expandable substances e.g. air bags, foam

5.2 Introduce fields

5.2.1 Use a field to trigger the generation of another field
5.2.2 Use fields from the system environment
5.2.3 Use substances of generating fields (e.g. magnetic substances)

5.3 Phase transitions

5.3.1 Change the aggregate state or the phase of substances
5.3.2 Use two aggregate states or phases of a substance
5.3.3 Use the physical effects accompanying a phase transition
5.3.4 Use effects which result from the simultaneous presence of two phases (e.g. use "phase transition" metals)
5.3.5 Improve the interaction between the elements or phases of a system

5.4 Use natural phenomena

5.4.1 Use self-controlled, reversible physical transformations
5.4.2 Use storage or amplifying effects

5.5 Substance particles

5.5.1 Generate substance particles (e.g. ions) by decomposing more complex structured substances (e.g. molecules)

5.5.2 Generate substance particles (e.g. atoms) by combining less complex structured substances (elementary particles)

5.5.3 If a substance cannot be decomposed, begin to decompose on the second highest substance level; if it is not possible to combine substance particles, begin to combine on the next highest substance level

Tip

- TRIZ is largely used in the domain of technological and technical scientific problems
- However, a successful deployment is also possible for non-technical problem formulations; the areas of Business Management, architecture, education, administration and politics are particularly appropriate for applying the innovative principles amongst others
- However, you should note that individual principles are evaluated differently here:
  - Principle 1 "Segmentation" is universal in character
  - Principles 31 "Porous materials" and 38 "Strong oxidants" should be classified as technical principles
- Only some of the 40 innovative principles can be applied to non-technical problem situations, the so called universal principles
- The following universal principles should be applied for solving non-technical problems:

Illustration: Universal Principles

1. Segmentation	10. Preliminary action	21. Skipping
2. Taking out	11. Preventive activities/cushioning	25. Self-service
3. Local quality	12. Equipotentiality	26. Copying
5. Merging	13. The other way around/Inversion	28. Mechanics substitution
6. Universality	15. Dynamics	
9. Preliminary anti-action	20. Continuity of useful action	

- In order to enable pragmatic use of TRIZ, the innovation checklist should be used for a detailed analysis of the problem

- It contains relevant questions on available resources, solution opportunities, restrictions and problems, to clarify the situation, determine the task and set the scope or constraints
- Questions on the innovation checklist do not exclusively focus on technological issues rather they are intended as a template for processing any sorts of problem in an innovative fashion
- The list describes a systematic approach to dealing with the specific problem situation
- The innovation checklist contains the following:

Aspect		Comments	Idea bank
1. **The system to be improved and its environment** "What am I dealing with?"	• Name of system		
	• Primary useful functions of system		
	• Operating principles of system		
	• System environment		
2. **Available resources and their potential** "Which resources can I use?"	• Substance-based resources		
	• Resources in the shape of a field		
	• Functional resources		
	• Information resources		
	• Temporal resources		
	• Spatial resources		
3. **Problem situation** "Why does a problem exist at all?"	• Aimed at improvement of system		
	• Mechanism or effect of disadvantage		
	• Background of problem		
	• Further problems to be solved		

Aspekt		Comments	Parking lot
4. **Change of system** "Within which boundaries may I move when solving the problem?"	• Which changes?		
	• What boundaries are allowed for changes to the system?		
5. **Selection criteria of solution concepts** "What do I want to achieve with the new product?"	• Targeted technical features		
	• Targeted economic features		
	• Targeted Time Plan		
	• Novelty in line with expectations		
	• Other selection criteria		
6. **History of solution attempts** "Has the problem been solved elsewhere?"	• Previous attempts for problem solving		
	• Other systems containing a similar problem		

Failure Mode and Effect Analysis (FMEA)

Term
Failure Mode and Effect Analysis FMEA

When
ANALYZE: evaluate high-level concept

Goal
Identify weaknesses in the high-level concept, evaluate them and if required, define actions to eliminate them, since with the design concept's increasing level of detail the costs of errors rises exponentially in the event of undiscovered weaknesses in the design

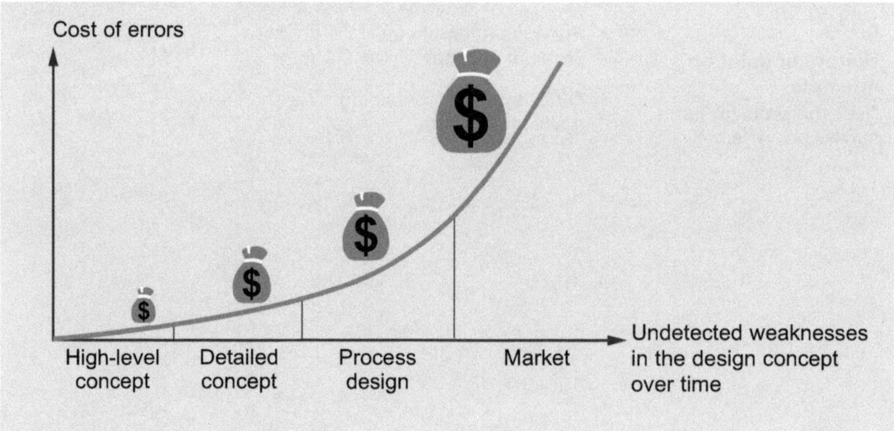

▶▶ Procedure

Conducting an FMEA comprises 5 steps.

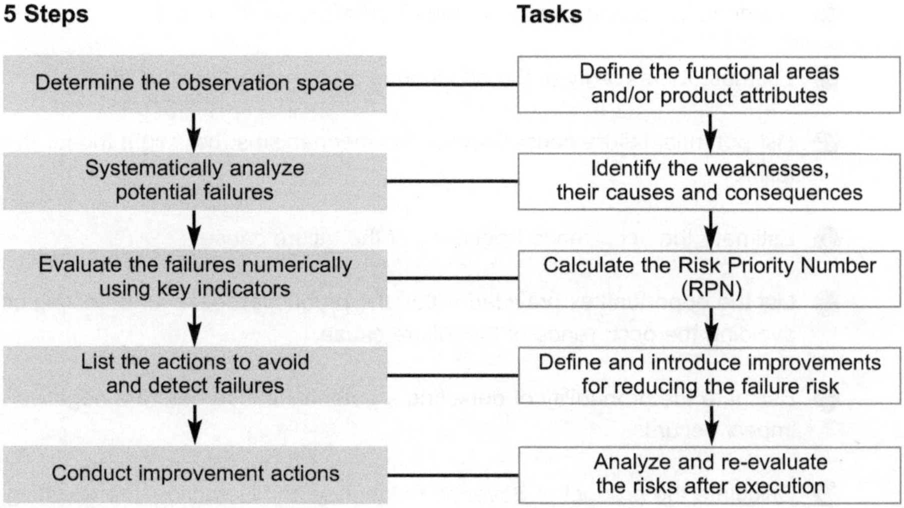

Process/product: ●								FMEA date: (Original)				
FMEA Team:								(changed)				
Black Belt:								Page:		out of:		
FMEA Process											Action results	
Position Function Process step	Potential failure mode	Potential effects of failure	Severity	Potential causes/ mechan-isms of failure	Frequency	Current control/ monitoring	Detection	RPN	Recom-mended actions	Responsi-bilities and closure date	Realized actions	Revised "RPN"
●	●	●	●	●	●	●	●	●	●	●	●	●

● Record general information on the project in the documentation first

● Describe precisely the analyzed process function and/or the analyzed process step (product function)

- Describe the potential failure modes: Why might the product/process not meet the requirements of a specific operation?
- Describe the consequences of the error/failure on the result
- Estimate the severity of the effect generated by the potential failure
- List potential failure causes and/or the mechanisms that might trigger this failure
- Estimate the occurrence frequency of the failure cause
- List the opportunities provided within the product/process for detecting or avoiding the occurrence of the failure cause
- Estimate the probability of detecting a potential cause before a negative impact occurs
- Calculate the product of Severity, Frequency and Detection; the resulting rating scale is the RPN (Risk Priority Number) which prioritizes the fields requiring action; in depth analysis will be required if the RPN is high
- Define actions which reduce the ratings of Severity, Frequency, and/or Detection in the event of high RPNs (usually >125)
- Name responsible people and set closure date
- Describe completed actions and document implementation date
- Estimate the effect of completed actions by revising estimates of Severity, Frequency and Detection and recalculate the RPN

Rating scale: Severity	
1	Remains unnoticed and has no effect
2	Remains unnoticed and has only an insignificant effect
3	Causes only minor inconvenience
4	Causes a minor loss of performance
5	Causes a loss of performance which results in a customer complaint
6	Causes a loss of performance which results in breakdown of functionality
7	Defective functionality results in enormous customer dissatisfaction
8	Product or service becomes unusable
9	Product or service is illegal
10	Customer or employee is injured or killed

Rating scale: Frequency	
1	Every 100 years
2	Every 5 to 100 years
3	Every 3 to 5 years
4	Every 1 to 3 years
5	Every year
6	Every 6 months
7	Once a month
8	Once a week
9	Once a day
10	Several times per day

DEFINE
MEASURE
ANALYZE
DESIGN
VERIFY

Rating scale: Detection Probability	
1	Failure cause is obvious and can be easily prevented
2	All units are inspected automatically
3	Statistical process control with systematic examination and prevention measures for failure causes
4	Statistical process control is conducted with systematic examination of failure causes
5	Statistical process control is conducted
6	All units are checked manually and prevention actions for failure causes are installed
7	All units are checked manually
8	Frequent manual examination of failure causes
9	Occasional manual examination of failure causes
10	The failure cannot be detected

Example: Concept FMEA – Passenger Seat

Concept-FMEA											Results of action	
Position Function Process step	Potential failure mode	Potential effects of failure	Severity	Potential causes/ mechan-isms of failure	Frequency	Current control/ monitoring	Detection	RPN	Recom-mended actions	Responsi-bilities and closure date	Realized actions	Revised "RPN"
Holding rail	Dirt	Difficult installation	6	Stone, sand, etc.	10		9	540				
	Accident	Seats are pulled from attachment	10		7	Training of drivers	10	700				**Extract**
Frame	Liquids	Corrosion	8	Water, acids	8		8	512				
Join frame/seat	Simple dis-assembly	Theft	5		10		8	400				
	Accident	Seats are pulled from attachment	10		7	Training of drivers	10	700				
Synthetic shell of seat	Vandalism	Sharp edges	10	Knifes, breakage	8		8	640				
	Spilled liquids	Damage	5	Water, acids	8		8	320				

➾ **Tip**

- The evaluation of the RPN is always industry and company specific
- Nevertheless, it usually makes sense to define relevant actions for reducing the risk if the RPN is > 125
- Reducing the RPN and/or the corresponding failure potential of a function can be achieved primarily by conducting activities which influence the frequency and/or the detection probability
- An FMEA can be used for a variety of purposes and can take place at various times during the development work:
  - Concept FMEA
  - Process FMEA
  - Design FMEA
  - System FMEA
  - Subsystem FMEA
  - Components FMEA
  - Assembly FMEA
  - Production FMEA
  - Machine FMEA

Risk Analysis

Term
Risk Evaluation, Risk Matrix, Risk Management Matrix

When
ANALYZE: evaluate high-level concept

Goal
Analyze and manage the (project) risks connected with implementing the concept

Procedure
As done in the DEFINE phase, the risks connected with implementing the concept should be evaluated systematically with the help of a Risk Management Matrix in the ANALYZE phase as well.
It is sensible to collect commercial, economic, political, technical as well as Change Management risks and to classify these into the Matrix according to their likelihood and their impact on the project success (procedure described in the DEFINE phase). If big risks or "show stoppers" exist, counter measures should be developed and appropriate communication with the clients should take place.

Illustration: Risk Management Matrix

Occurrence probability	Low	Medium	High
High	Medium Risk	Big Risk	Show Stopper
Medium	Low Risk	Medium Risk	Big Risk
Low	Low Risk	Low Risk	Medium Risk
	Impact on project success		

- Reduce before the project is continued or stop the project
- Minimize and/or control risks
- Proceed with caution

Storytelling

Term
Storytelling, Use Case, describe an application case

When
ANALYZE: evaluate high-level concept

Goal
Exemplify the product/service concept in order to enable customers as well as Stakeholders to form an opinion and give clear feedback

Procedure
1. Think of examples
2. Describe examples verbally
3. Make drawings of the examples (e.g. in the style of a comic or the storyline of a film)

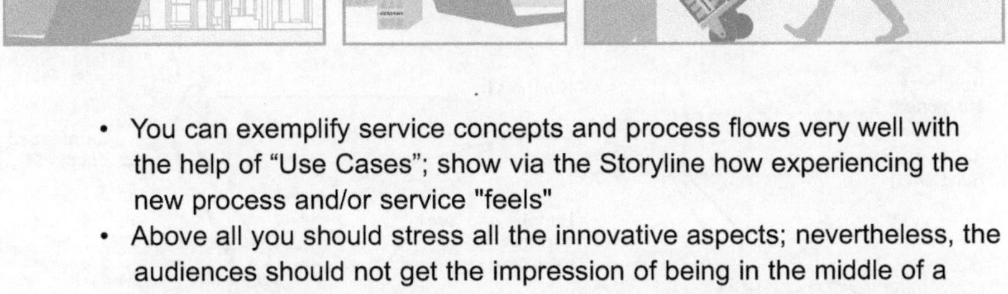

- You can exemplify service concepts and process flows very well with the help of "Use Cases"; show via the Storyline how experiencing the new process and/or service "feels"
- Above all you should stress all the innovative aspects; nevertheless, the audiences should not get the impression of being in the middle of a sales presentation; the focus for everyone involved should be the critical assessment of the concept
- Storytelling can also be helpful at an earlier stage when alternative concepts are discussed

Early Prototyping

Term
Prototyping, Just Enough Prototyping, Early Prototyping

When
ANALYZE: evaluate high-level concept

Goal
Demonstrate the product concept in order to enable customers and the Stakeholders to form their own opinion and give clear feedback

Procedure
A model is made of the new product using simple tools in order to provide a basis for feedback. Depending on the product type materials such as clay, plastic material, modeling compound, cardboard, etc. can be used. Apart from the "Use Cases" mentioned above, sketches can be appropriate for virtual products. Simple design templates are available in MS Office e.g.to make screens for input into IT applications.

If a product already exists, it can be used as basis for making the model and adjusted with the above mentioned tools.

Example: Drawing of Draft – Passenger Seat

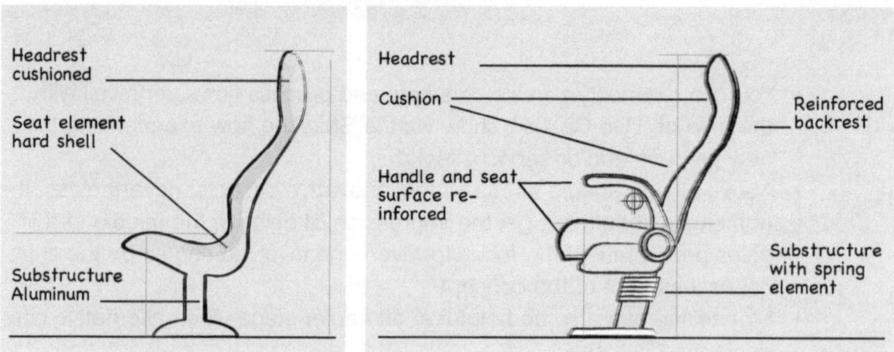

➾ **Tip**

- The decisive aspect of Early Prototyping is to promote the provisional character of the prototype at all times; otherwise it is possible that the customers' feedback focuses on detailed aspects (color, font size, etc.) instead of on the product's essential characteristics
- Based on the prototype you can also enter into an interactive process of further development with the customers, by having the customer draw up variants of the prototype themselves (co-creation)
- Use agile methods such as SCRUM in order to control not only Early Prototyping but also the interaction with the customers, receiving their feedback and to making use of the customer feedback

Collect Customer/Stakeholder Feedback

Term
Receive Customer and Stakeholder Feedback

When
ANALYZE: evaluate high-level concept

Goal
- Agree on the best optimized high-level concept with customers and stakeholders in order to get a confirmation of the concept's attractiveness and/or necessary need for adjustment before the detailed development starts
- Release of budget for detailed development, infrastructure, resources and start of production

Procedure
Representative target customers (e.g. within the frame of focus groups) and important stakeholders are invited to judge the developed concept.
The team presents the status of the development work:
- Short presentation of the project's previous approach (e.g. with the help of a Power Point presentation)
- Presentation of the final high-level concept in the form of a "Use Case" and if possible, a computer-supported visualization (CAD/CAM) of a prototype/demonstrator

Each of the people invited is asked to give his/her subjective feedback. The feedback is analyzed and necessary adjustments are made to the high-level concept.

Finalize High-level Concept

Term
High-level Concept Finalization

When
ANALYZE: finalize high-level concept

Goal
- Draw up a high-level process matching the high-level concept
- Roughly plan the product's market launch
- Define and organize the resources which are necessary for future detailing and implementation

Procedure
1. Roughly draw up the production process and/or the process and its delivery, based on the high-level concept which has been coordinated with customers and stakeholders

2. Identify an appropriate strategy for positioning the product on the market based on the competitive situation

3. Define the necessary requirements for detailing and implementing resources

Define the High-level Process

Term
High-level Process Development

When
ANALYZE: finalize high-level concept

Goal
Determine the production process and/or the process and its delivery in accordance with the high-level concept

Procedure
If the elements of the product concept have been set, the rough definition of the production process and/or the process and its delivery should take place. This ensures that process related questions can be discussed before entering the DESIGN phase (who/what is required in which form for producing/delivering the process? Where in production/delivery is it possible to leverage things that already exist and where is it necessary to take a completely new direction?). This way resource requirements and challenges will be revealed.
A SIPOC diagram is appropriate for visualizing the high-level process.

Supplier	Input	Process	Output	Customer

The SIPOC diagram, however, should be created in reverse, starting from the relevant customer i.e. in the sequence COPIS.

1. List relevant customers under the section "Customer" (take into account internal and external customers!)

2. Indicate outputs which these customers are supposed to receive from the process i.e. the actual product and other things such as information in the form of data, reports, invoices, etc.

3. Determine limits (stop and start point) of the process

4. Describe the process between its' start and stop points in rough steps (maximum 10); describe the individual steps using a combination of noun and verb
5. Compile the inputs which are essential for the process
6. If possible, you should list the suppliers of these inputs (if necessary use general terms such as "Supplier fabrics", "Supplier coating material", etc.)

Example: High evel process definition using SIPOC – Passenger Seat

Supplier	Input	Process	Output	Customer
Manufacturer of bus (e.g. Auto Moscov)	Order	Commission production order		
	Metal	Build frame		
Suppliers (available)	Cushioning material	Make cushioning		
	Fabric	Coat cover material	No experience yet	
Supplier (not available)	Coating material			
Supplier must be found		Assemble seat	Change of inventory	Production planning
		Order Commission		
		Draw up invoice	Invoice	Collection system
		Deliver order	Bus seat ready for assembly	Manufacturer of bus (e.g. Auto Moscov)

Tip

- If you consider the provided level of detail in the SIPOC as too generic and therefore do not regard it as added value, you can also use a process flow chart for the visualization
- You can add further information on critical inputs and outputs for individual process steps within the Flow Chart

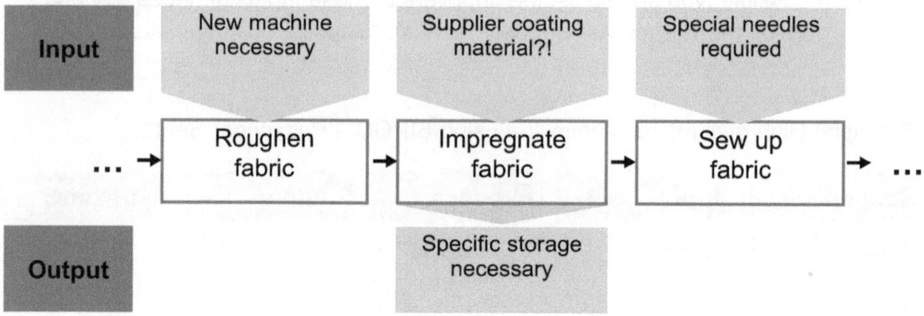

- Mark current weaknesses and open questions in the process mapping; this makes the definition of actions and work packages easier for the detailed development later
- You should pay attention to the following basic facts throughout the process design:
  1. Do not outsource value adding activities, especially if they are part of your core business
  2. Minimize value enabling activities; even if they are required by the company in order to render value addition, they are not perceived by the customer as direct added value (e.g. process steps for control or internal information processing)
  3. Put activities which are identical for all customers and/or product variants to the beginning of the process and put the customer and/or product specific activities to the end of the process
  4. Where possible, you should decouple support activities and the information flow from the value adding process steps
  5. Install Poka Yoke mechanisms (for defect prevention) where necessary
  6. Minimize complexity by avoiding interfaces and/or handovers
  7. Develop the process first and the necessary technical support (IT tools etc.), afterwards, not vice versa!

Prepare Market Launch

Term
Market Launch, Market Positioning

When
ANALYZE: finalize high-level concept

Goal
- Initiate internal and external marketing activities early on in order to enable successful marketing in the defined target market
- Convince target customers of the development's advantages

Procedure
Innovation marketing does not just start with a product launch but requires more advance planning. By this stage product marketing will already have been established alongside the project and alignment to the future Market phase will be finalized in the VERIFY phase. The following illustration shows how a marketing approach could run alongside the development project:

Illustration: Actions for innovation marketing

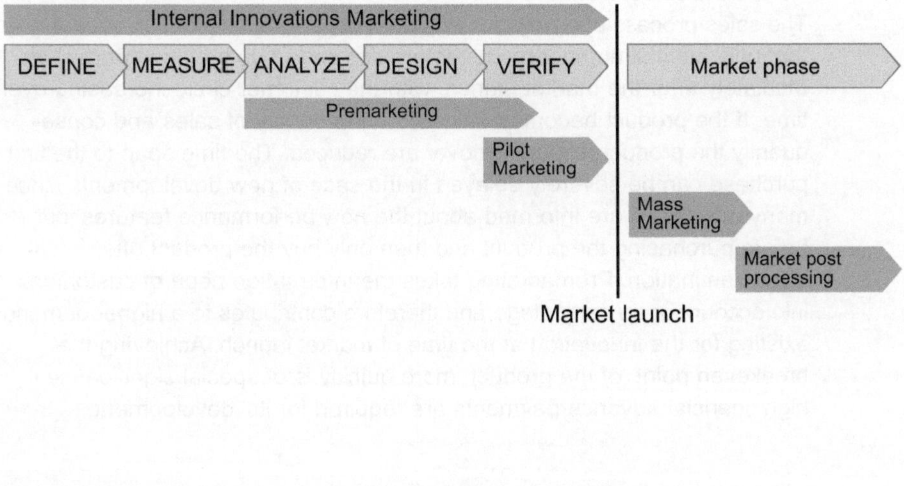

Internal innovation marketing

Innovations must be "marketed" in one's own company early on in order to recognize internal resistance and eliminate it. For affected employees innovations often mean a change of former working procedures and practices, and, they are often perceived as disturbance, nuisance or senseless turmoil. Internal resistance should be reduced in a targeted way and must be taken into consideration for innovation projects by:

- Early integration of different departments in the development project (interdisciplinary teams)
- Project related internal communication
- Internal training and workshops for presenting the project work

Internal innovation marketing should begin with the project launch and play a central role in each of the subsequent phases of the DMADV procedure. Before the market launch of the product innovation takes place, it is important that all individuals involved back the project result and support it in front of the market and customers.

Premarketing

Premarketing comprises all marketing actions which are conducted before the market launch of the product innovation. Customers, sales agents and public disseminators are informed about the market launch in a targeted way in order to accelerate the sales process after the market launch, reduce buying resistance and set standards early on.

1. Accelerate sales processes:
   The sales process of a new development usually takes place in such a way that only a small number of customers decide to buy the new product immediately after the market launch, with the customer circle increasing over time. If the product becomes outdated, the quantity of sales and consequently the product related turnover are reduced. The time span to the first purchase can be severely delayed in the case of new developments since many customers are informed about the new performance features just before purchasing the product and then only buy the product after intensive examination. Premarketing takes the information need of customers into account at an early stage and therefore contributes to a higher demand existing for the innovation at the time of market launch. Achieving the breakeven point, of the product, more quickly is of special significance if high financial advance payments are required for its' development.

Mapping effect of Premarketing

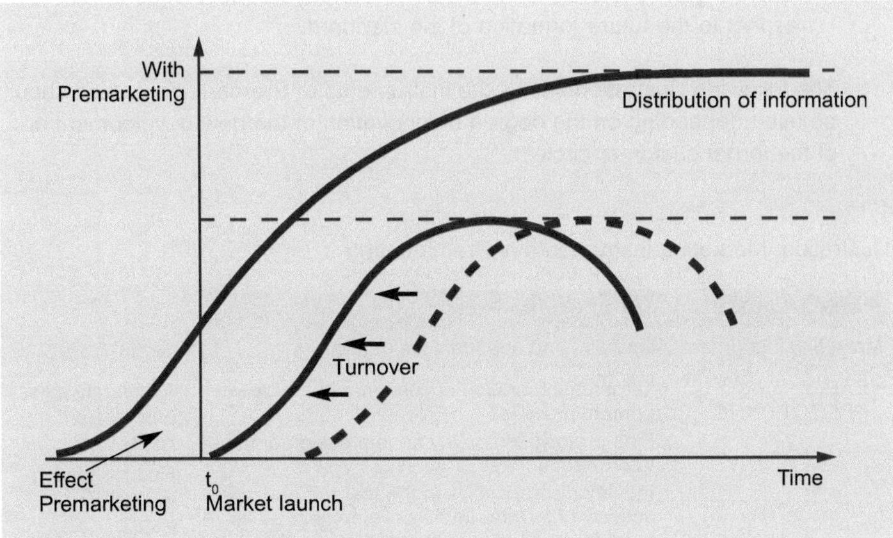

2. Reduce buying resistance:
   New attributes and product characteristics frequently cause restraint on the customer's part which has a negative impact on sales; some time passes until large quantities can be sold. This type of restraint can be frequently observed in the telecommunication industry since it is very difficult for many customers to evaluate the performance capability of new products before purchasing them. If resistance of this type is not tackled proactively with appropriate communication measures, the new development can become a flop.

3. Set standards:
   Standards represent product or performance features of the new development which are accepted not only by the providers but also by the customers and market partners who orient themselves towards these products. Standards emerge not only by legal agreement (norms) but also by respective market acceptance (de facto standard). The provider who succeeds in establishing a standard on the market first, has great advantages when it comes to further marketing of the innovation. Competitors are forced to change over to this standard or face intense competition. The major effect of premarketing is that the early announcement of the innovation pushes

one's own standard. This can ideally lead to the fact that the demand for competing products is obstructed since the buyers feel insecure with respect to the future formation of the standard.

The following illustration shows the instruments of Premarketing which should be used depending on the degree of innovation of the new development and of the target customer circle.

Illustration: Marketing instruments of Premarketing

Marketing instruments of Premarketing		
Marketing instrument	Alignment and opportunities of use	Effect(s)
• EARLY PROTOTYPING	• Targeted integration of customer in the development process • First prototypes convey an impression of the future product • Identify discrepancies in the test run and actively "interfere" in the development so as to be better able to compare the new development with the customers' specifications • Test prototypes in the customer's working environment in order to receive further information which is relevant for the product development	• Accelerate sales processes: HIGH • Reduce buying resistance: HIGH • Set standards: MEDIUM
• AGENDA SETTING	• Take advantage of the high media attention on new developments • Provide media with information in a targeted way before the market launch, in order to gain attention from the potential customer circle • Agenda Setting can be completed with the help of publications in specialist magazines, press conferences and press releases • Targeted use of general media should only be considered, if the object of innovation and its' sales potential justify it. • A new development serving a very specialized market should also use respective specialist magazines and/or specialist fairs for information transfer	• Accelerate sales processes: HIGH • Reduce buying resistance: HIGH • Set standards: HIGH

Marketing instruments of Premarketing		
Marketing instrument	Alignment and opportunities of use	Effect(s)
• DIRECT MARKETING	• The customers can be directly addressed about the planned market launch via mailing campaigns • After evaluating the responses the target customers can be further segmented and targeted with specific information material, invitations to agents or fairs, trade samples, etc.	• Accelerate sales processes: MEDIUM • Reduce buying resistance: LOW • Set standards: LOW
• INTEGRATION OF AGENTS	• In particular if product and service innovations require further explanations • Final customer receives advance information on the planned product by personal consultation with agent • Respective advertising materials and advance announcements underline/reinforce the announcements of the agent and create curiosity and interest • However, if market offers competitive products, there is the risk that the agents prefer the direct and immediate sale to a product which hasn't been introduced to, and established on the market in order to reach a business transaction more quickly	• Accelerate sales processes: MEDIUM • Reduce buying resistance: HIGH • Set standards: MEDIUM

Identification of Necessary Resources

Term
Identification of Necessary Resources

When
ANALYZE: finalize high-level concept

Goal
Clarify the necessary resource need for the detailed further development of the concept

Procedure
The assessment of the required resources takes place on the basis of the concept draft which is now available. With the help of descriptions, sketches, computer visualizations and prototypes which have already been created, the necessary resources can be systematically identified and requested on time.

The following resources are necessary for the future development of the detailed design:
- Time
- Money
- Manpower (skills and number)
- Equipment, materials, machines, etc.

The activities, time and resource planning must be continued in detail at this point(see DEFINE).

Gate Review

Term
Gate Review, Phase Check, Phase Approval

When
At the end of each phase

Goal
- Inform the Sponsor about the results and actions of the respective phase
- Evaluate the results
- Decide about the further course of the project

Procedure
- Present the results of the phase completely and in a comprehensible way

- The Sponsor checks the current status of the project according to the following criteria:
  – Completeness of results
  – Probability of project success
  – Optimal resource allocation for the project
  – Release of development budget

- The Sponsor decides if the project can enter the next phase of the project

The following list of questions can be of assistance for the project team when preparing the presentation of results as well as for the Sponsor when conducting the phase approval:

Develop high-level concept:
– Were the functions of the product/process formulated clearly and completely?
  What are they?
– Were solutions sought internally and externally?
– What alternative high-level concepts were developed?

- How was the best concept selected?
- What strengths and weaknesses does the selected concept have?

Evaluate high-level concept

- Were target costs derived for the product?
  What was the approach?
- Was the concept checked for weaknesses and conflicts?
- Were identified weaknesses and conflicts eliminated?
- Was customer and stakeholder feedback obtained?
- What was the approach and how was the feedback?

Finalize high-level concept

- Has the high-level process supporting the product concept been developed?
  What does it look like?
- Has a market launch been prepared?
  What is the approach?
- What does the market positioning of the concept look like?
- Have the required resources been identified for the detailed planning?
- Are they available?

On Project Management

- Is it necessary to adjust the Business Case?
- Does the project schedule have to be adjusted?
  Can design activities be accelerated if required?
- Has the Multi Generation Plan been taken reviewed?
  Does it have to be adjusted?
- What risks exist if the project is continued?
  Can they be managed actively?
- Should the project be continued?
- What are the Lessons Learned of the ANALYZE phase?
- What are the next steps in the project?

DESIGN FOR SIX SIGMA^{+LEAN} TOOLSET

DESIGN

Central Question and Objective of the DESIGN Phase

- The DESIGN phase of the DMADV procedure answers the question: "Which detailed system design is the best design of the high-level concept?"
- Objective:
  - Describe the system design in detail with respect to the complete and consistent fulfillment of the defined specifications and functions
  - Develop and prepare target flows by taking into consideration Lean principles

The Meaning of the DESIGN Phase

- Definition and evaluation of relevant design elements for implementing the detailed design
- Detailed development of the product and detailed process design on the basis of the best developed high-level concept
- Definition of the detailed process design using design principles for the further detailed description of the high-level concepts
- Process Management defines the most important KPIs for controlling the product and/or process performance and ensuring sustainability
- Prototypes, based on the detailed product and process design, are the basis for further customer feedback

Procedure in the DESIGN Phase

- Map alternative design characteristics of the system
- Define design elements
- Draw up a Cost Breakdown Structure
- Apply Tolerance Design
- Compare alternative design elements
- Develop a Scorecard for detailed concept
- Apply prototyping methodologies and create prototypes
- Conduct simplification of the system with the help of Trimming and Design for X
- Estimate risks and avoid them
- Evaluate current process capability
- Draw up process design
- Draw up work and operating instructions
- Plan facilities, buildings, equipment and resources

349

D.1 Develop detailed concept

- What does a detailed design, which is capable of implementation, look like?
- Do the design elements of the detailed design fulfill the defined customer requirements?

D.2 Evaluate detailed concept

- How can the detailed design be further simplified and/or which reliable, more cost effective and ecofriendly design elements can be integrated in the detailed design?

D.3 Develop and evaluate Lean process

- What does an efficient and effective process look like after taking into account the detailed design capable of implementation?
- What are the components for appropriate process control and how can the new process be piloted?

Result DESIGN

- Best detailed design identified for the fulfillment of the defined customer requirements
- Detailed design implemented in prototypes
- Relevant production flows and supporting processes defined and tested

DESIGN Gate Review/Phase closure

D.1 Develop detailed concept

Design Tree	Cost Breakdown Structure	Tolerance Design	Hypothesis Tests	Design of Experiments (DoE)

D.2 Evaluate detailed concept

Design Scorecard	Prototyping	Trimming	Design for X	Anticipatory Failure Determination
Poka Yoke				

D.3 Develop and evaluate Lean process

Evaluation of Process Performance	Design Principles	Process Diagrams (SLD, VSM)	Minimize Process Lead Times	5 S Concept
SMED	TPM	KPI Systems	Process Monitoring	Process Simulation
Work and operating instructions	Procurement Strategies	Prince 2, ITIL, CMMI, V Model	Pilot Processes	

D Gate Review

DEFINE
MEASURE
ANALYZE
DESIGN
VERIFY

Develop Detailed Design

Term
Develop Detailed Design

When
Closure ANALYZE, DESIGN: develop detailed concept

Goal
- Develop a detailed concept which is capable of implementation
- Develop and map alternative characteristics of the design elements and check their suitability with respect to fulfilling the defined customer requirements

Procedure
The customer requirements which were supplemented by measurements, target values and specifications in the MEASURE phase, are transferred into the best high-level concept identified in the ANALYZE phase.

Illustration: From the customer needs to the detailed concept on the following page

From ANALYZE to DESIGN – Overview

Alignment MEASURE ➡ ANALYZE ➡ DESIGN

MEASURE phase
Who is the project's central customer and what specific requirements do they expect from the system?

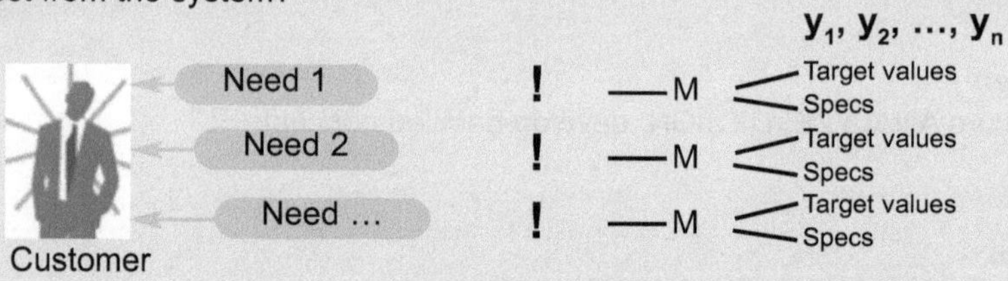

ANALYZE phase
What is the best high-level concept on the basis of the defined customer requirements?

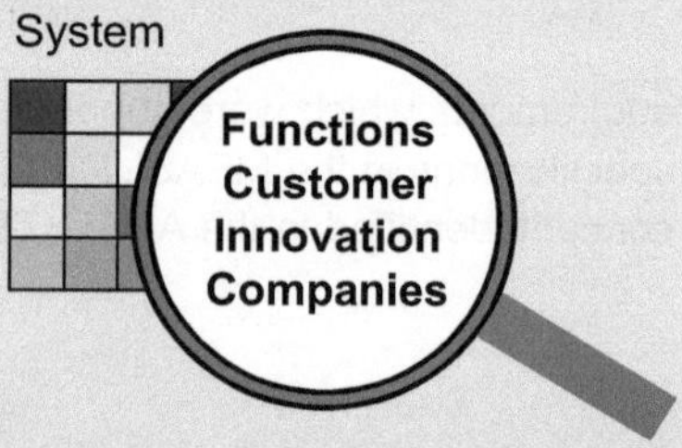

DESIGN phase
Which detailed system design is the best for delivering the high-level concept?

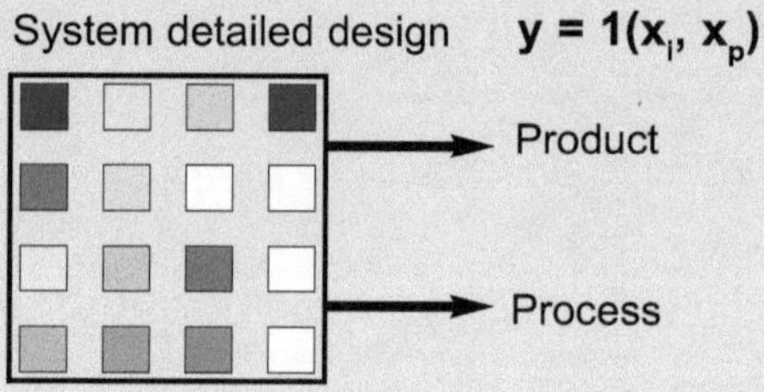

Having selected and developed the best high-level concept in ANALYZE, the resulting detailed concept is now further developed and evaluated.

To begin with alternative design elements are drafted for the system components, which were identified in ANALYZE, in order to evaluate them later with respect to costs, complexity, feasibility, fulfillment of set specifications and residual risks. Characteristics and design elements are reworked until all the various system functions can be fulfilled without contradictions.

Example:
System design across the ANALYZE and DESIGN phases – Passenger Seat

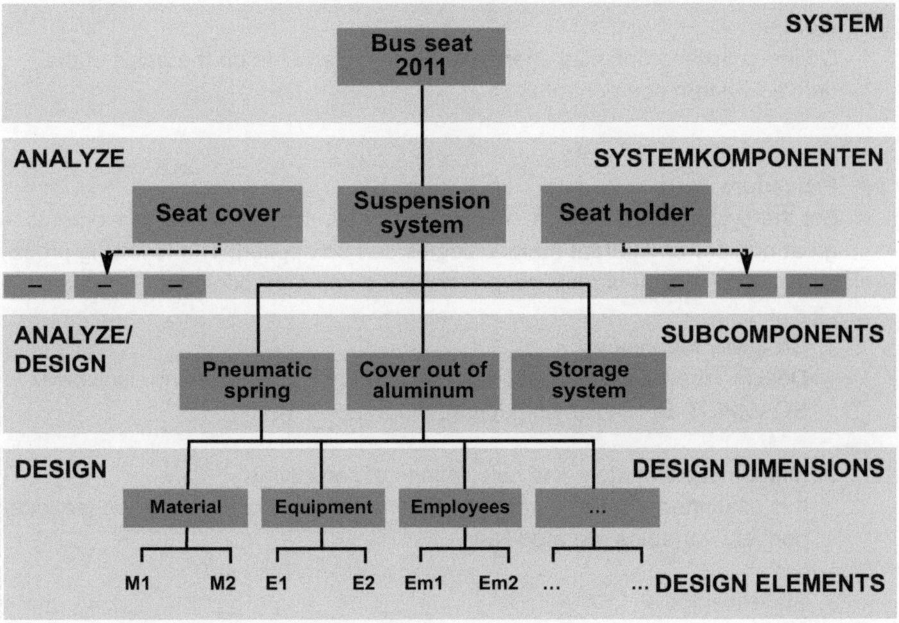

Map Alternative Design Characteristics of the System

Term

Develop Detailed Design, Develop Detailed Concept

When

ANALYZE, DESIGN: develop detailed concept

Goal

Define, map and optimize alternative design elements on the basis of the derived system components

Procedure

For the systematic definition, mapping and optimization of design elements, a great number of different methodologies can be applied. The following procedural model indicates respective techniques and methodologies:

1. Creativity techniques
   Design Tree, Morphological Box, Mind Mapping, Brainwriting, Ishikawa, SCAMPER, ERCI, Benchmarking, TRIZ

2. Mathematical models and calculations of sensitivity
   It is examined to what extent dependent variables changes when the independent variables are modified

3. Statistical tools
   Further relationships can be recognized through data analysis (e.g. hypothesis tests, ANOVA, Regression)

4. Design of Experiments
   Design of Experiments is a further opportunity of examining the sensitivity of systems; the data for statistical analysis can be generated through Monte Carlo Simulations, CAD-CAM (Computer Aided Design/Manufacturing) or other simulations methods

The listed techniques and methods are presented here. In project practice, however, the sequence can deviate and tools can be applied to different design elements depending on the question presented.

TIP
Disturbance variables also influence the cause-effect relationship because they can lead to deviations (noise) from the ideal function; they can be identified e.g. with the help of an FMEA and should be fully taken into consideration.

Define Design Elements

Term
Define Detailed Design, Define Design Elements

When
DESIGN: develop detailed concept

Goal
Describe system components of the best high-level concept systematically and in detail through the relevant design dimensions

Procedure
In order to transfer the system components of the best high-level concept as defined in the ANALYZE phase into a detailed concept it is important to identify all relevant subcomponents and to describe them in detail including their own individual design elements.

Either a Design Tree or a Morphological Box provides an excellent structure to assist in sequential mapping and evaluation of alternative design elements.

Steps for drawing up a Design Tree:

1. List the system components defined in Analyze and add identified subcomponents

2. Add the relevant information to the system components from the QFD 3 for target costing:
   - The technical weight of the system component (contribution to fulfilling function)
   - Allowable proportion of costs (target costs of system component)
   - Need for action regarding increase of component value and/or reduction of component value (comparison target costs and expected costs)

3. Describe the system and/or the subcomponents in greater detail along the relevant dimensions of the product and process design:

- Material procurement:
  - Material substitution
  - Change of supplier
  - Change of procurement strategy etc.

- Equipment:
  - Need for new machines and facilities
  - Production procedures etc. to be modified

- Employees:
  - Qualification for new production procedures etc.

- Architecture and design:
  - Further specification of subcomponents in design elements (e.g. measurement, color, etc.)

- Service:
  - Altered service requirements
  - Dealing with components etc. that are prone to failures

Depending on the project a Design Tree can be extended to include further design dimensions e.g. process, facilities and buildings, IT as well as marketing and sales. The Design Tree ensures comprehensive preparation of the detailed design and provides the basis for further evaluation.

As the project progresses, and especially after observing the Cost Breakdown Structure, the Design Tree should be continuously added to, checked and updated. Also any unsolved problematic areas and/or fields requiring action should be visualized in the Design Tree so that it represents a Roadmap for the status of the development.

Example: Design Tree – Passenger Seat on the following page

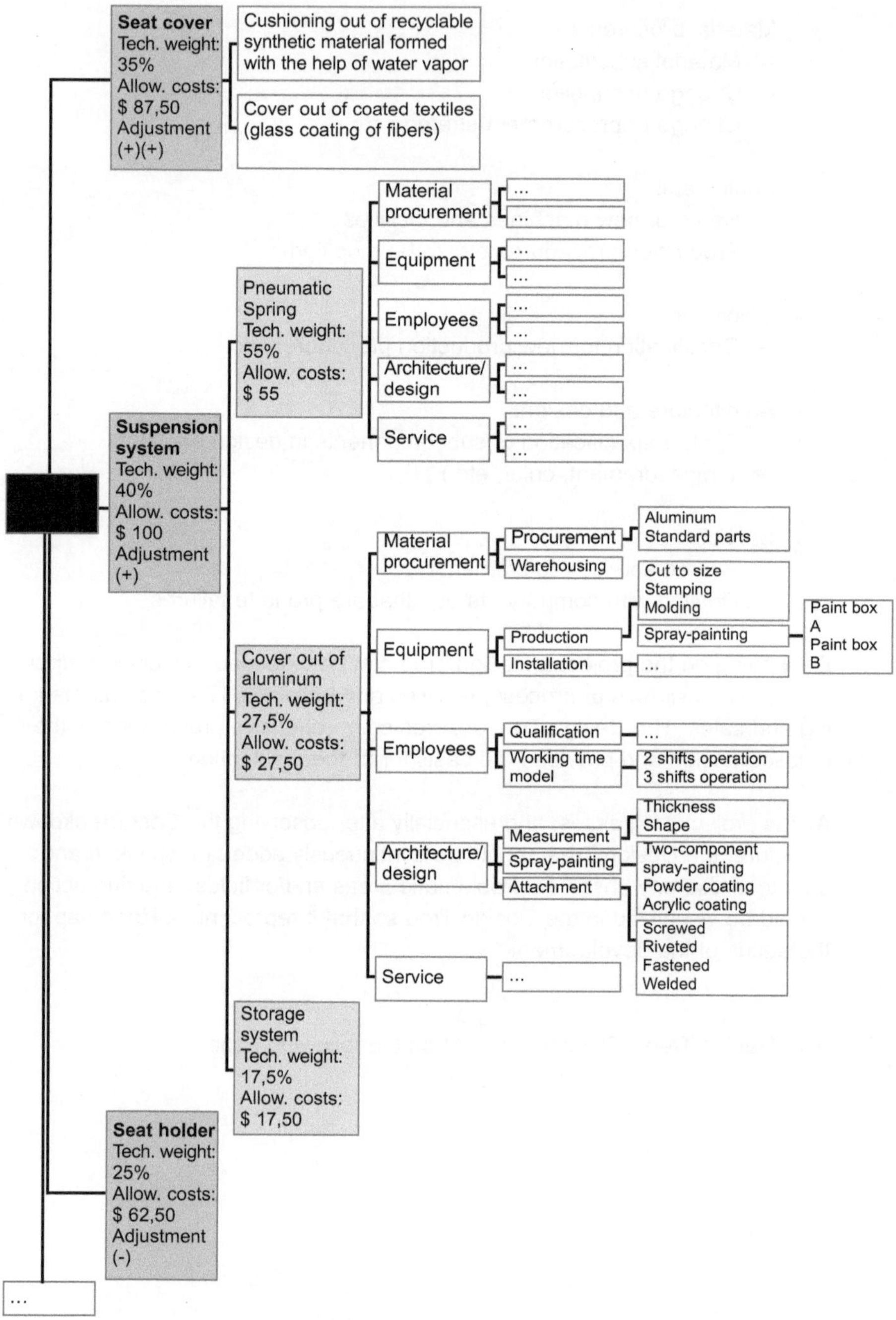
Seat cover
Tech. weight: 35%
Allow. costs: $ 87,50
Adjustment (+)(+)
Cushioning out of recyclable synthetic material formed with the help of water vapor
Cover out of coated textiles (glass coating of fibers)
Suspension system
Tech. weight: 40%
Allow. costs: $ 100
Adjustment (+)
Pneumatic Spring
Tech. weight: 55%
Allow. costs: $ 55
Material procurement
Equipment
Employees
Architecture/ design
Service
...
Cover out of aluminum
Tech. weight: 27,5%
Allow. costs: $ 27,50
Material procurement
Procurement
Warehousing
Aluminum
Standard parts
Equipment
Production
Installation
Cut to size
Stamping
Molding
Spray-painting
Paint box A
Paint box B
Employees
Qualification
Working time model
2 shifts operation
3 shifts operation
Architecture/ design
Measurement
Spray-painting
Attachment
Thickness
Shape
Two-component spray-painting
Powder coating
Acrylic coating
Screwed
Riveted
Fastened
Welded
Service
Storage system
Tech. weight: 17,5%
Allow. costs: $ 17,50
Seat holder
Tech. weight: 25%
Allow. costs: $ 62,50
Adjustment (-)
...

➾ **Tip**

- As an alternative to the Design Tree you can use a Morphological Box
- Insert the mapped design dimensions (material procurement, equipment, employees, architecture as well as design and service) into the Morphological Box and compare the alternative design elements to the respective subcomponents (e.g. pneumatic spring)

Cost Breakdown Structure

Term
Cost Breakdown Structure

When
DESIGN: develop detailed concept

Goal
- Ensure the profitability of the new product/service
- Assign the allowable total costs of a system component to its subcomponents
- Identify financial flexibility for developing design elements

Procedure
In addition to the calculation of target costs in QFD 3 the Cost Breakdown Structure assigns the proportion of allowable costs of a system component to its' subcomponents.

1. Insert functions with the identified weighting for the system component from QFD 3 for Target Costing
(steps 1 and 2)

2. Insert technical weight, allowable proportion of costs and any need for adjusting the component value of the system component
(Step 3)

3. Add the respective subcomponents
(Step 4)

4. Assign function weighting (from step 2) to the subcomponents according to their contribution to fulfilling the function
(Step 5)

5. Standardize the weight of the subcomponents
(Step 6)

6. Assign allowable costs, according to their weights, to the subcomponents (steps 7 and 8)

Example: Cost Breakdown Structure – Passenger Seat

		Suspension system (air suspension) **Tech. weight: 40%** **Allow. costs: $ 100** **Adjustment: +**		
Function	Weight from QFD 3	Pneumatic Spring	Cover out of aluminum	Storage system
Protect seat	0.1	0 **0%**	0.1 **25%**	0 **0%**
Absorb force	0.25	0.2 **50%**	0 **0%**	0.05 **12.5%**
Fixate seat	0.05	0.02 **5%**	0.01 **2.5%**	0.02 **5%**
Aggregate	0.4 **100%**	0.22 **55%**	0.11 **27.5%**	0.07 **17.5%**
Allowable costs per subcomponent		**$ 55.00**	**$ 27.50**	**$ 17.50**

➾ **Tip**

- Draw up a matrix for cost components for each system component and its subcomponents
- Transfer the information on the subcomponents back to a Design Tree or a Morphological Box in order to estimate any further financial flexibility for developing the design elements and to be able to indicate the need for action, if required
- You can influence the respective component values using the indicated design dimensions such as material procurement, equipment, employees, architecture as well as design and service

Tolerance Design

Term
Tolerancing, Tolerance Design

When
DESIGN: develop detailed concept

Goal
Identify production tolerances from the mechanical components of the design elements

Procedure
Examine the interaction between the tolerances of the design elements in respect to their effect on the output variable and determine specifications.

First step: Determine the relevant requirements
Deviations were already noticeable when the measurements and specifications were determined; they can now to be verified based on transfer functions. The requirements for the Tolerance Design are as follows:
- Fulfillment of CTCs
- Tolerances that can be produced
- Profitability, i.e. avoidance of excessive production costs that are driven by strict tolerances

Tolerance Design follows the guiding principle:
"As precisely as necessary and as roughly as possible."

Second step: Select a suitable calculation methodology
The following methods have been established in practice depending on the existing transfer function of the task:
1. Worst Case Analysis (WCA)
2. Root Sum Square Method (RSS)
3. Monte Carlo Analysis (MCA)

Illustration: Assistance for identifying a suitable calculation procedure

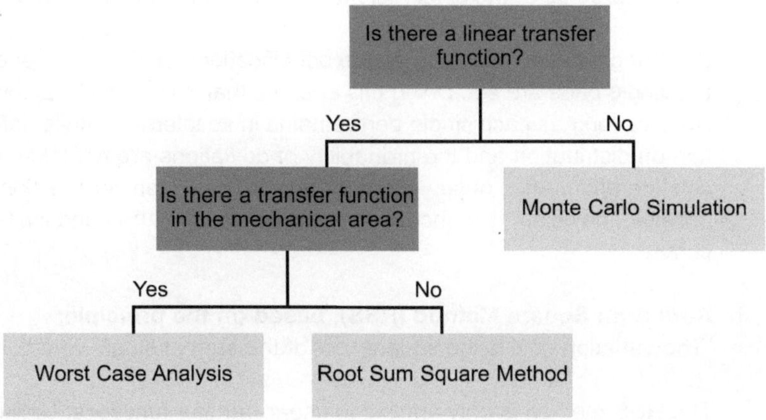

Third step: Determine the tolerance of the output variable with the help of the selected calculation methodology

Example: Setting tolerances

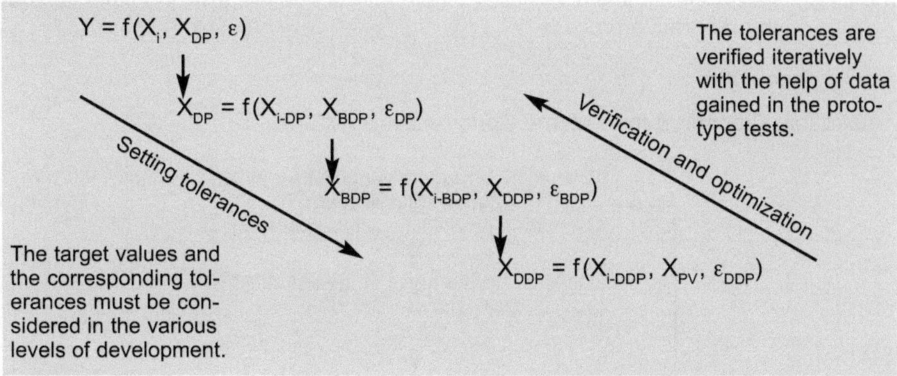

Legend:

Y = output	$X_{i\text{-}BDP}$ = input/signal for the broad design parameter
X_i = input/signal for the output	X_{DDP} = detailed design parameter
X_{DP} = design parameter	$X_{i\text{-}DDP}$ = input/signal for the detailed design parameter
$X_{i\text{-}DP}$ = input/signal for the design parameter	ε = noise, disturbing variables
X_{BDP} = broad design parameter	

a. **Worst Case Analysis (WCA), based on the principle:**
"The aggregate of the single tolerances results in the tolerance of Y."

By taking into consideration all the specifications of Y, the tolerances of the single parts are set. Doing this ensures that Y remains in its tolerance limits as long as each single part remains in its tolerance limits. Information on distribution and the probability of deviations are not taken into consideration. Also, observation of the non-linear transfer function is possible, however, it is uncommon due to the high effort and limited accuracy.

b. **Root Sum Square Method (RSS), based on the principle:**
"The variation of Y is the square root of the sum of single variations."

The RSS method is only applied to linear transfer functions. Unlike the WCA method, the RSS method considers the variation in tolerances of design elements.

c. **Monte Carlo Analysis (MCA)**
The Monte Carlo Analysis is a simulation method which is conducted with the help of special software such as Crystal Ball® and SigmaFlow® with which distribution assumptions are made enabling conclusions about the variation of Y. The single tolerances are given optimal settings in an iterative process.

Illustration: Procedure for a Monte Carlo Analysis (MCA)

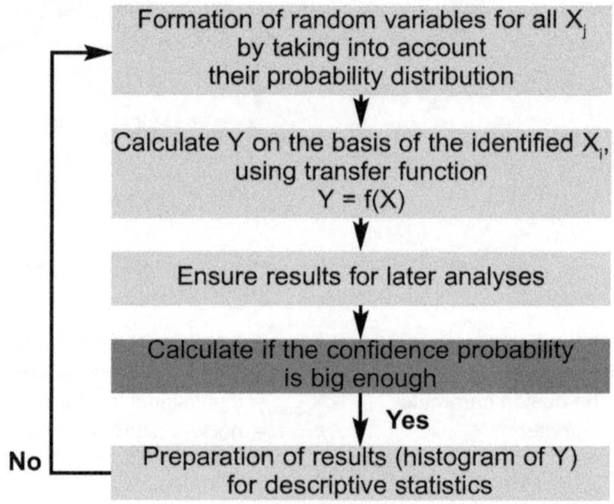

➾ **Tip**

- Pay attention to the fact that statistical procedures (RSS Method and Monte Carlo Analysis) are to be preferred to WCA, since they contain additional information on the spread not just the location
- Nevertheless, the application of the WCA can make particular sense in case of simple linear transfer tasks, especially in the construction area; it is an effective means to make a quick and conservative estimation as to whether the selected tolerances of Y are appropriate; take into consideration, however, that it doesn't provide the best solution at optimal costs

Illustration: Worst Case Analysis and statistic analysis methods in comparison

	Advantages	**Disadvantages**	**Applicability**
WCA	Guarantees the complete interchangeability of assembly and operations	• The assumption of the biggest tolerances leads to an unnecessarily high total tolerance	• Linear transfer tasks especially for mechanic transfer functions • Single production
Statistical procedure	Take into account information on spread and location	• Only partial interchangeability of assembly parts • Computer support required	• Mass and large batch production

Compare Alternative Design Elements

Term
Alternative Design Comparison

When
ANALYZE, DESIGN: develop detailed concept

Goal
- Compare alternative design elements and their characteristics with respect to their performance capability
- Optimize detailed concept further

Procedure
Depending on the question formulation and the data type, different statistical tools can be used to compare and optimize the identified alternative design elements and their characteristics.

Illustration: Use of different statistical tools according to data type

		Output (Y)	
		Continuous	**Discrete**
Input (X)	**Continuous**	• Correlation • Simple and multiple linear regression • DOE	• Logistic regression • DOE
	Discrete	• Tests for the mean value • Tests for variances • Non-parametric tests • Analysis of Variance • Design of Experiments	• Tests for proportions (One and Two Proportion Test, χ^2 Test) • DOE

Hypothesis Tests

Term
Alternative Design Comparison

When
ANALYZE, DESIGN: develop detailed concept

Goal
- Compare alternative design elements based on data
- Determine relevant influencing factors

Procedure
When working with data from design projects, inferences on the relationships in the population should be drawn from a small number of samples.
It is therefore necessary to identify the scope a sample must have to enable a valid statement about the true parameters (e.g. mean value, median, proportions, variance, etc.) of the population to be made.

Illustration: Population and Sample

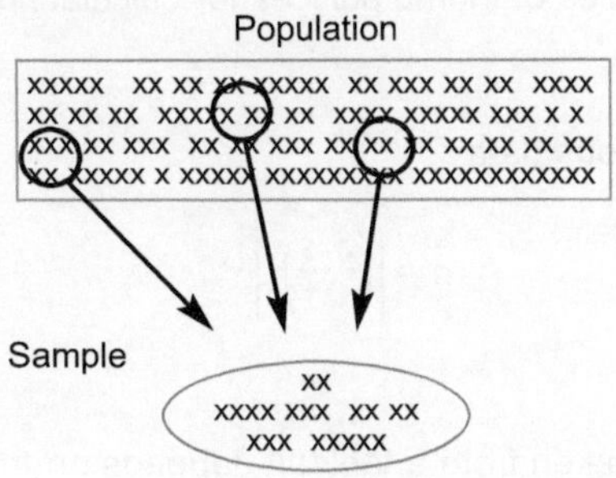

Confidence intervals are formed to compare the parameters calculated from the samples with equivalents from the population. The confidence intervals assert that, with a confidence level of x, which is determined by the testing person in advance (usually 95% or 99% or a significance level of 5% or 1% respectively), the true value (parameters) from the population lie within these intervals.

Confidence interval for mean value

$$\left[\bar{x} - z\left[\frac{s}{\sqrt{n}}\right];\ \bar{x} + z\left[\frac{s}{\sqrt{n}}\right]\right]$$

The breadth of the confidence interval is influenced by the spread of the sample (s), the certainty (z), and the sample size (n). The sample size n is determined in turn by the temporal effort and the resulting costs for the examination, but also depends on the desired meaningfulness of the values which are to be generated:

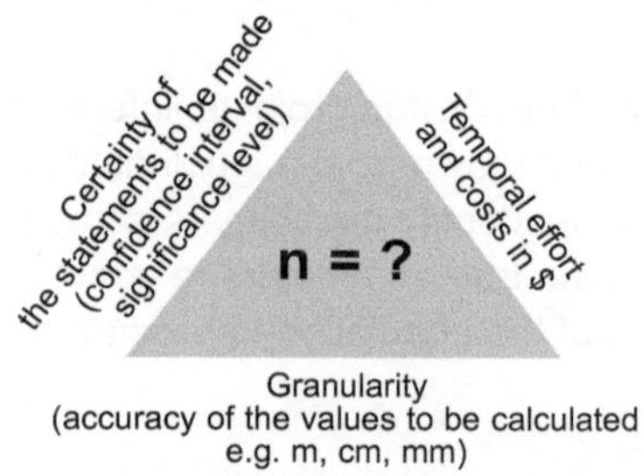

In general, using rules of thumb suffices for calculating a suitable sample size.

Sample size for continuous data

$$n = \left\lceil\left(\frac{z \cdot s}{\Delta}\right)^2\right\rceil$$

- n = sample size
- z = this value is taken from a table; it depends on the selected confidence interval; here z 95% = 1.96 or z 99% = 2.575
- s = standard deviation of the sample
- Δ = granularity (accuracy of the calculated values in the unit of s, e.g. m, cm, mm)
- ⌈x⌉ = this symbol means that the number x is rounded up to the next whole number

Sample size for discrete data

$$n = \left\lceil \left(\frac{z}{\Delta}\right)^2 \hat{p} \cdot (1 - \hat{p}) \right\rceil$$

- n = sample size
- z = this value is taken from a table; it depends on the selected confidence interval; here z 95% = 1.96 or z 99% = 2.575
- $\hat{p}$ = proportion of defective units in the sample, e.g. 32% defective units correspond to $p = 0.32$. If the proportion p is unknown, it is initially calculated with $p = 0.5$
- Δ = granularity (accuracy of the calculated values in the unit of p, e.g. 10% = 0.1, 1% = 0.01, etc.)
- $\lceil x \rceil$ = this symbol means that the number x is rounded up to the next whole number

Side condition: $n \cdot p \geq 5$ or $n \cdot (1 - p) \geq 5$

Hypothesis tests can be conducted based on the taken samples to compare concepts or determine the influencing factors. A statistical test is a procedure that uses a test statistic to verify the statistical validity of a hypothesis about the population based on a sample and a given level of confidence.

Such a hypothesis test is based on the formulation of two complimentary assertions: the null hypothesis H_0 and the alternative hypothesis H_A.

- The **null hypothesis H_0** asserts:
  Equality exists; there's no difference!
- The **alternative hypothesis H_A** asserts:
  There is no equality; a difference exists!

Statistical tests can only ascertain differences, not accordance. For this reason, the null hypothesis is usually established in order to be refuted.

Making decisions on the basis of statistical tests entails a certain degree of uncertainty: one cannot be 100% certain that the decision is correct. At the same time, however, statistical tests are designed in such a way that the probability of making an incorrect decision is minimized.

A null hypothesis is rejected when the result of a sample shows that the validity of the established null hypothesis is improbable. What is ultimately considered to be improbable is determined in advance by the so-called significance level and/or confidence interval. The most frequently used significance levels are 0.05 (5%) and 0.01 (1%), and/or 95% and 99% confidence intervals. The significance level is connected with the potential for wrong decisions.

There are basically two types of wrong decisions or errors in statistical tests: the α-error and the β-error:

Illustration: α- and β-errors

		Reality	
		H_0	H_A
Decision	H_0	Right decision	Error of the second type (ß-error)
	H_A	Error of the first type (α-error)	Right decision

The null hypothesis is not rejected although it is not valid in reality. A difference in the population is not identified.

The null hypothesis is rejected although it is valid in reality.

The statistical decision is made by comparing the significance level (α) with the p-value (p ≡ probability). The p-value indicates the actual probability from the present samples that the null hypothesis is falsely rejected. The p-value therefore equates to the residual risk when the null hypothesis is rejected. For this reason it is also known as error probability. The p-value is calculated with statistics software, e.g. Minitab®.

- If the p-value is small, e.g. smaller than the set α (significance level), the null hypothesis must be rejected; the following phrase sums up what is to be done: "If P is low, H_0 must go!"
- If the p-value is bigger than the α-level, this means that possible emerging differences are not statistically significant

A statistical test comprises the following steps:

1. Define problem and goal (what is to be investigated for what purpose?)
2. Formulate the hypotheses (H_0: condition of equality)
3. Set the significance level α (as a rule $\alpha = 0.05$ or $\alpha = 0.01$)
4. Select a suitable statistical test (e.g. Two-Sample t-Test)
5. Conduct the test statistic with the help of a statistics program (e.g. Minitab®)
6. Interpret the test statistic and/or p-value
7. Make a decision
8. Verify your decision: If H_0 is not rejected, verify β with the help of a statistics program!

A great number of statistical hypothesis tests exist. Some practice-relevant tests are described on the following pages.

Discrete Data – testing proportions

Test	When/for what	Hypotheses	Prerequisites
Binomial test One Proportion Test	Compare a proportion with a theoretical and/or given proportion with binominally distributed data, e.g.: good (non-defective)/ bad (defective) test	$H_0: p = p_{Target}$ $H_A: p \neq p_{Target}$	Binominally distributed data $n \geq 100$ and/or $n \cdot p \geq 5$ and $n \cdot (1 - p) \geq 5$
Binomial test Two Proportion Test	Compare proportions of a characteristic in two samples	$H_0: p_1 = p_2$ $H_A: p_1 \neq p_2$	Binominally distributed data $n \geq 100$ and/or $n \cdot p \geq 5$ and $n \cdot (1 - p) \geq 5$
χ^2 (Homogeneity) Test Chi-Square Test	1. Compare proportions of a characteristic in two or more samples 2. Compare proportions in two or more populations	$H_0: p_1^1 = p_1^2 = \ldots = p_1^j$ $p_2^1 = p_2^2 = \ldots = p_1^2$ $\vdots$ $p_j^1 = p_j^2 = \ldots = p_j^j$ H_A: at least one proportion is different	Nominal data $n \geq 100$ and/or $n \cdot p \geq 5$ and $n \cdot (1 - p) \geq 5$

Continuous Data – testing the mean value

Test	When/for what	Hypotheses	Prerequisites
One Sample t Test	Compare the mean value of a sample with a target value	$H_0: \mu = \mu_{Target}$ $H_A: \mu \neq \mu_{Target}$	n ≥ 30 and/or normally distributed data
Two Sample t Test	Compare the mean values of two independent samples	$H_0: \mu_1 = \mu_2$ $H_A: \mu_1 \neq \mu_2$	n ≥ 30 and/or normally distributed data, independent samples
Two Sample paired t Test	Compare mean values of two dependent samples	$H_0: \mu_1 = \mu_2$ $H_A: \mu_1 \neq \mu_2$	n ≥ 30 and/or normally distributed data, paired, dependent samples
One Way ANOVA	Compare mean values of several independent samples	$H_0: \mu_1 = \mu_2 = \ldots = \mu_i$ H_A: at least one mean value is different	Equal variances or equal samples, independent samples

Continuous Data – testing variances

Test	When/for what	Hypotheses	Prerequisites
F-Test/Levene's Test Two Variances	Compare variances of two independent samples	$H_0: \sigma_1^2 = \sigma_2^2$ $H_A: \sigma_1^2 \neq \sigma_2^2$	F-Test: Normally distributed data. Levene's Test: no distribution assumption, independent samples
Bartlett's Levene's Test Test For Equal Variances	Compare variances of several independent samples	$H_0: \sigma_1^2 = \sigma_2^2 = \ldots = \sigma_i^2$ H_A: at least one variance is different	Bartlett's Test: Normally distributed data. Levene's Test: no distribution assumption, independent samples

Example: One Sample t Test – Minitab© Output

```
One Sample T: Paint thickness

Test of mu = 140 vs not = 140


Variable      N     Mean    StDev  SE Mean                95% CI       T       P
Paint
Thickness    80  153.859   35.654    3.986  (145.925; 161.793)   3.48  0.001
```

Result:
Here $p < 0.05$. There is a statistically significant difference. The hypothesis H_0 can be rejected.

Example: One Sample t Test

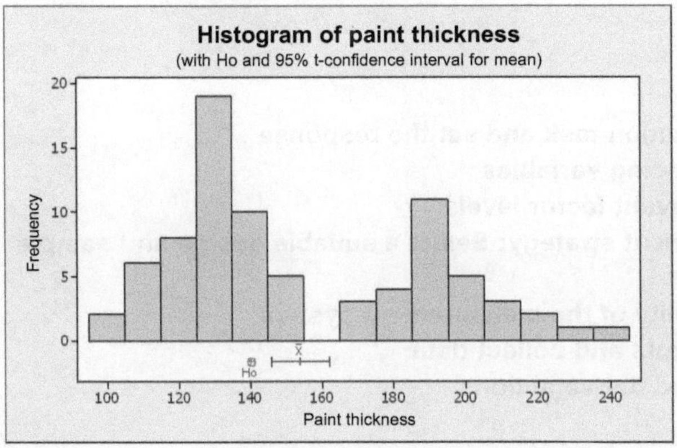

Graphic result:
the difference between the target value and mean value of the sample is statistically significant.
The hypothesis H_0 can be rejected.

Design of Experiments (DoE)

Term
Design of Experiments, DOE

When
ANALYZE, DESIGN: develop detailed concept, evaluate detailed concept

Goal
- Compare alternative design elements based on data
- Determine significant factors as well as their effects and interactions
- Draw up and/or supplement the transfer functions
- Determine the optimal characteristics of the design elements
- Determine the optimal equipment settings (process variables)

Procedure
1. **Define the optimization task and set the response**
2. **Identify the influencing variables**
3. **Determine the relevant factor levels**
4. **Derive the experiment strategy: Select a suitable design and sample size**
5. **Ensure the capability of the measurement system**
6. **Conduct experiments and collect data**
7. **Analyze results and derive actions**

1. Define the optimization task and set the responses
- Select the product and/or process to be analyzed
- Set the goals
- Set the responses for measuring whether the goals are achieved
- Make sure that the responses have the following characteristics:
  - Completeness: All key process and product characteristics have been covered
  - Dissimilarity: Each response describes a different relationship
  - Relevance: each response bears a clear relation to the goal of the analysis

- Linearity: If there are several similar possible responses, select the response that depends linearly on the influencing variables
- Quantification: The responses should be as continuous and/or metric as possible

2. Identify the influencing variables

- Identify and document decisive influencing variables with the help of structured Brainstorming.
  Important tools are:
  - Cause-Effect Diagram
  - Tool 3 (testing the relation between output measurements and input/ process measurements)
  - FMEA

- Moreover, results gained in process and data analysis can be taken into consideration:
  - Data Stratification
  - Hypothesis tests
  - Analysis of Variance
  - Regression analysis

- The final evaluation should be based on the following criteria:
  - Importance of a factor
  - Accuracy of the potential setting
  - Reproducibility of the setting
  - Effort and expense for changing the levels

3. Determine the relevant factor levels

- A maximum and minimum are set as factor levels. Two factor levels are selected initially:
  - Continuous variables: The maximum and the minimum should be located in a reasonable area, so that the response is still quantifiable
  - Discrete variables: If the factor levels are discrete, e.g. there are five producers, initially at least, use the two most important as factor levels

4. Derive the experiment strategy

- Set the sample size (plan the experiment scope)
- Determine the number of blocks

- Decide on randomization and/or take into consideration restrictions in randomization (e.g. due to the costs of the experiment setup)
- Determine the factor level combinations: full-factorial or fractional factorial DOE

Full factorial DOEs

- All factor settings are combined in a full-factorial DOE

Illustration: Full Factorial DOE – fuel consumption

Speed (km/h)	Tire pressure (Bar)	Fuel (octane)	Consumption (l/100km)
100	2	91	10
150	2	91	15
100	3	91	9
150	3	91	7
100	2	98	9
150	2	98	14
100	3	98	6.5
150	3	98	13

- In this way, the effects generated by the factors and their interactions can be identified completely
- The number of characteristic combinations to be examined depends exponentially on the number of factors:

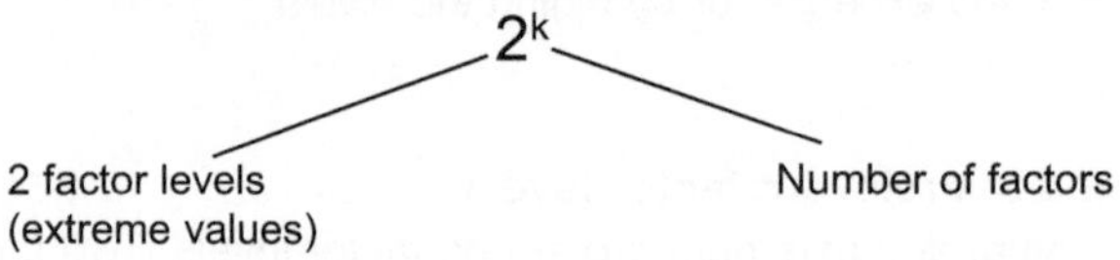

Fractional factorial DOEs

- Fractional factorial (or part-factorial) DOEs reduce the number of single tests

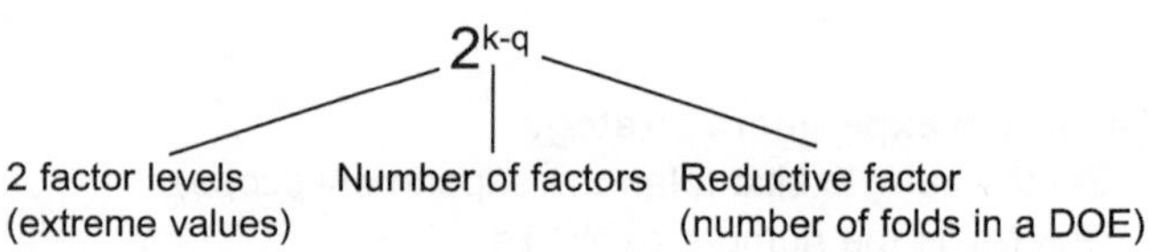

- Testing the significance of the factors is still possible even with a loss of some information; the loss of information refers to the confounding of specific effects, e.g. the effects of main factors and the effects of interactions are indistinguishable from one another; which kind of confounding is present depends on the corresponding resolution type

Illustration: Solution Types of the Fractional Factorial DOE

Number of experiments	Number of factors													
	2	3	4	5	6	7	8	9	10	11	12	13	14	15
4	Full	III												
8		Full	IV	III	III	III								
16			Full	V	IV	IV	IV	III	III	III	III	III	III	III
32				Full	VI	IV	IV	IV	IV	IV	IV	IV	IV	IV
64					Full	VII	V	IV	IV	IV	IV	IV	IV	IV
128						Full	VIII	VI	V	V	IV	IV	IV	IV

Resolution type	Confounding	Evaluation
III	Main factors are confounded with two-factor interactions	Critical
IV	Main factors with three-factor interactions/ two-factor interactions with two-factor interactions	Less critical
V	Main factors with four-factor interactions/ two-factor interactions with three-factor interactions	Uncritical

The analysis is conducted in the same way as a full-factorial design.

As a rule, a full-factorial DOE is too expensive and elaborate; the following procedure is recommended (block procedure) if the experiments can be conducted successively:

- *Block 0:* Good-Bad Trials
  - There are two different settings for each factor which lead to clearly different values of the observed response; all factors are set in such a way that a "good" result can be expected (e.g. low defect rates, high levels of agent concentration) according to expert opinion; next, all factors are set in such a way that a "bad" result can be expected (e.g. high defect rate, low concentration of agents)
  - The goal is to find out whether effects exist; if no effects are located this can be due to the fact that the selected factors are not relevant or the signal-noise relationship is too weak, i.e. the noise is too "loud"
  - At this point the experiments should be stopped and, if required, further factors should be determined or the noise should be eliminated

- *Block 1:* Screening Experiments
  - It is not unusual that 10 or even up to 15 factors are selected
  - If effects exist in principle, experiments with resolution III or IV should be conducted first
  - The important question is: do effects of a sufficient dimension exist?
  - The goal is to locate the relevant factors in this phase ("separate the wheat from the chaff"); it is often possible to significantly reduce the number of relevant factors and conduct further DOEs with far fewer experiments
  - When deciding to leave out factors attention must be paid to possible interactions; in practice reducing factors in a resolution III is therefore to be avoided

- *Block 2:* Fold Over Experiments

  Fold-over experiments supplement screening DOEs, i.e. supplemented by the missing experiments to achieve a better resolution type. This is a reversal of the signs deployed in the starting DOE.
  - The goal is to reduce the number of factors to the really important ones; in this way it is possible to estimate the interactions
  - The statistical analysis can provide first approaches for the optimal settings (Response Optimizer)

- *Block 3:* Completion Experiments
  - If there is reason to assume that the relationships are non-linear, i.e. squared effects or effects of a higher order of the relevant factors exist, additional experiments are carried out which, besides the minimum and maximum settings, take into consideration additional mean values

  - This is known as the Response Surface Methodology (e.g. Central Composite Design)

- *Block 4:* Optimization Experiments
  - Optimal settings were proposed throughout the statistics analysis generated by the preceding experiments
  - The goal is to test the optimal settings of the factors

Estimate the costs:
Make sure that the costs are reasonable in relation to the expected results. If the effort appears to be too big, examine whether the costs can be reduced by doing without factors and/or factor levels, by block building and/or randomization, or carrying out a smaller number of experiments – without, of course, endangering the goal.
The goal should be reconsidered, if required.

5. **Ensure the capability of the measurement system**
   - Develop the Operational Definition and conduct a Measurement System Analysis
   - A Measurement System Analysis verifies if the measurement system is appropriate; improve the system if required

6. **Conduct experiments and collect data**
   - Prior to actually conducting the experiments it is recommended to conduct a couple of preliminary tests and/or pilot experiments; the goal is to verify whether the estimated expense and effort is realistic, and if the result is consistent, i.e. the noise has been eliminated
   - When conducting the experiments make sure that everything runs according to plan; this means that each experiment has to be monitored individually

7. **Analyze results and derive actions**
   - The statistical analysis of the results proceeds according to the methodologies of the regression (smallest square method) and variance analysis
   - The graphic and analytic results are reviewed after each block in order to determine the next steps; in this respect, conducting a DOE is an iterative process

- When analyzing the results and deciding on the next steps, one or several experts involved in the process should be consulted in order to avoid false conclusions; such conclusions can conceal the true relationships, e.g. through measurement errors or noise; the results should be checked at all times to see if they make sense

Tip

- Along with the mean values as responses, a classic factorial DOE is also suitable for observing the variance in a response; in this case the factors responsible for the variation can be recognized and a meaningful reduction of variation can be conducted
- In order to stabilize variation it is necessary to transform the variance s^2; this is done either through the root transformation (in this case the result is the standard deviation s) or a logarithmic transformation ($\ln [s^2]$)
- In order to consider the standard deviation as a response, several measurements of an experiment, repeats, are necessary

Example: DOE applications in production

Product	Response (Y)	Factors (X)
Baking mix	• Weight 1cm^3	• Amount of flour • Amount of baking powder • Granularity of cocoa
Oven	• Degree of browning • Uniformity of browning	• Ventilation speed • Form of heating spiral • Sealing
Can	• Width • Depth	• Color aluminum • Amount of oil • Construction A/B • Tools H/Z

Examples of DOE applications in the service industry

Industry	Response (Y)	Factors (X)
Logistics	• Inventory costs	• Supplier • Delivery conditions • Payment conditions
Market research	• Willingness to acquire the product (ranking scale)	• Description of product attributes • Packaging • Placement
Financial services	• Process Lead Time for processing the request	• Processing a form (manual or electronic) • Approval • Processing (sequential or parallel) • Processor (industry specialist or general education)

Example: Design of Experiments – Passenger Seat

The DFSS team must decide on a paint type for spray-painting the seat frame.

- In the high-level design the decision was made to use a standard alloy for the frame material
- The CTCs corrosion and wear resistance can only be achieved with this alloy through spray-painting
- The paint of the manufacturer Xylosud and the manufacturer Miller can be chosen; both produce a corrosion-resistant paint which is effective in case one of the common liquids comes into contact with the paint (cleaning agents, sulfuric acids, coke, juice)
- The team decides to examine the suitability of the paint in terms of its thickness using DOE

The question to be answered is which paint and which pretreatment achieve the greatest wear resistance.

Output measurement is
Y1 : paint thickness

Input and Process measurements Factor levels [- ; +]

X1 : temperature	[20; 25]
X2 : pressure	[15; 30]
X3 : thinner	[10; 20]
X4 : pretreatment	[A; B]

Determining the relevant factors and interactions is supported by a statistics program (such as Minitab®).

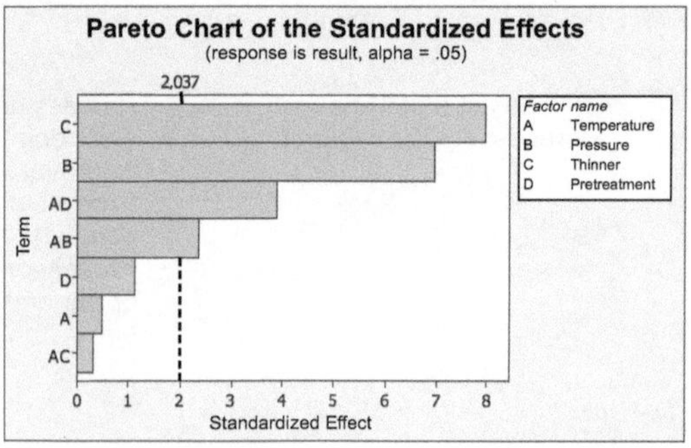

The dotted line shows a Confidence Level of 5%, i.e. α-value = 0.05.

The graphic result of the analysis is a Pareto Chart. The statistically significant factors have longer bars which pass the red line (significance level 5%).

The optimal settings can be identified with the Response Optimizer.

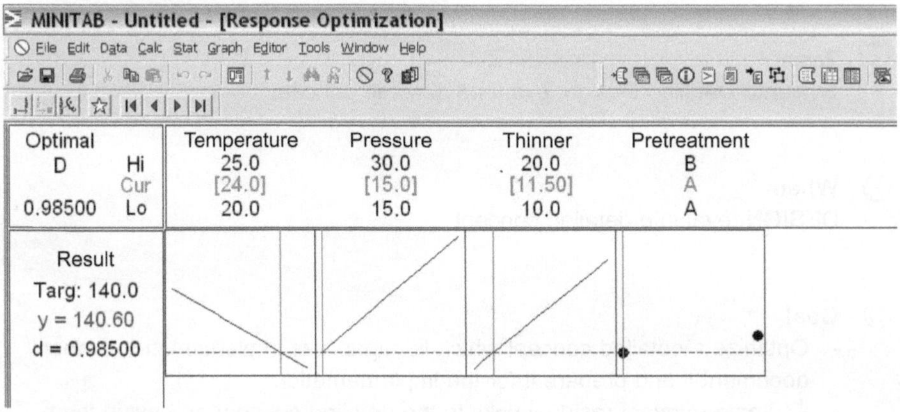

The DOE helps to identify the optimal settings for the continuous factors such as temperature, pressure and thinner as well as the best pretreatment for the spray-painting.

DEFINE MEASURE ANALYZE DESIGN VERIFY

Evaluate Detailed Concept

Term
Evaluate Detailed Design, evaluate detailed concept

When
DESIGN: evaluate detailed concept

Goal
- Optimize a detailed concept which is capable of implementation further, document it and prepare it for the implementation
- Estimate relevant residual risks in the detailed concept and avoid them early on

Procedure
After having developed individual design elements, the entire detailed concept is finally tested and questioned. For this purpose, the final status of the detailed concept is documented in a Design Scorecard in order to be able to conduct test runs and/or field tests with respective prototypes. Moreover, experience gained from test runs provides the basis for a further simplification of the system.

Illustration: Evaluate detailed concept

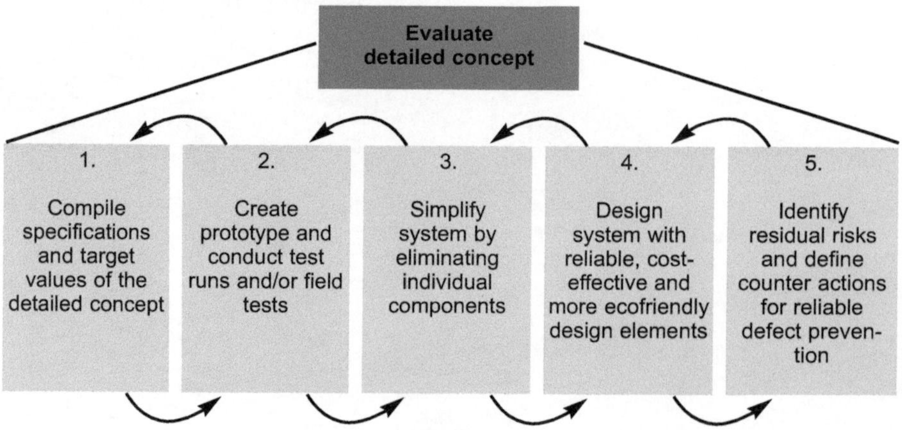

Develop Design Scorecard for the Detailed Concept

Term
Design Scorecard

When
DESIGN: evaluate detailed concept

Goal
- Document relevant design parameters, operational definitions, target values, specifications and further quality key figures for each hierarchy level of the detailed concept
- Create a basis of decision making for the selection of individual design elements

Procedure
Once the best final design elements and their specifications have been set, their measurements, specifications and target values are documented in a Design Scorecard. A Design Scorecard is drawn up for each hierarchy level.

A Data Collection Plan is drawn up to generate a baseline (sample strategy, sample size and responsibilities etc.).

Causes of measurement system variation can be identified with the help of a Gage R&R and various graphical tools depicting the results (Run Charts and Control Charts etc).

The documentation of the design elements in the Scorecard is the basis for building relevant prototypes.

Tip
- You should also document the specifications of alternative design elements in the Design Scorecard in order to understand the result of the process for developing the detailed concept

- This way decisions for or against a design element can be made in a logical way which then allows further iterative optimization loops after the test phase if required

Illustration: Design Scorecards on different hierarchy levels

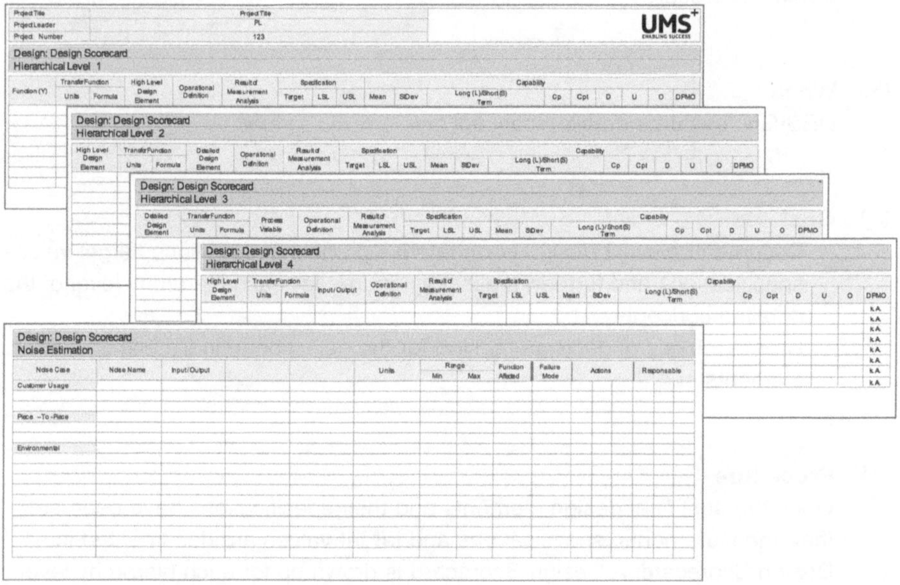

Test Detailed Concept

Term
Testing Detailed System Design

When
ANALYZE: develop high-level concept; DESIGN: evaluate detailed concept

Goal
Develop the detailed concept further step by step from the function pattern to the pilot series and adjust it to additional internal and external requirements in iterative optimization loops

Procedure
The detailed concept which has largely been developed based on its' conceptual and theoretic aspects, is now progressed in a development and optimization process to the final pilotable product. This process takes place in reasonable sub-steps in order to monitor the respective outcomes produced which are then reviewed in small Gate Reviews.

A distinction can be made between the following stages of delivering the detailed concept:

- Function pattern (Demonstrator):
  The defined core functions have largely been realized and correspond to the future product; shape, measurements and performance can still vary; the procedures for the necessary production plants are selected based on the function pattern
- Function prototype:
  Apart from the core functions all further important functions have been realized; the function prototype corresponds to the future product with respect to shape, measurements, functions and performance; with respective patterns the customer acceptance is tested; necessary supplies and tools have been defined and their availability has been ensured

- Prototype:
  All functions of the product and necessary equipment have been implemented; customer feedback from the function prototype has been taken into consideration; the required number of prototypes for further validation are available; necessary supplies and tools are largely available

- Pilot series:
  The first series of products have been manufactured under near realistic manufacturing conditions, in statistically sufficient quantities in order to be able to draw conclusions about quality, efficiency and qualification; necessary optimizations of the production flow took place and feedback from further field tests with customers was taken into consideration in order to prepare for the introduction of full production

Illustration: Test detailed concept

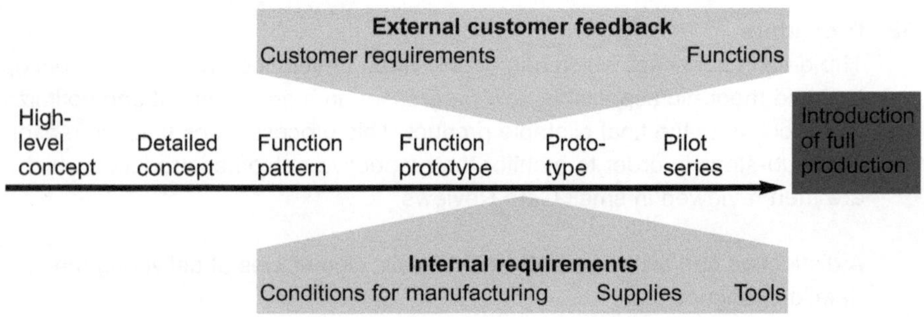

➾ **Tip**

- A function prototype already enables a more precise analysis of existing production flows, means of production and supplies, internal and external logistic processes, relevant procurement processes and the complete material flow chain
- It forms the basis for a parallel development and/or optimization of existing production procedures and flows, and the design of the necessary infrastructure

Prototyping Methods

Term
Prototyping, Just Enough Prototyping, Early Prototyping, Reverse Engineering

When
DESIGN: evaluate detailed concept

Goal
- Create a system early on in order to test the fulfillment of the defined customer requirements and functions
- Develop the detailed concept in iterative optimization loops from the function pattern to the pilot series
- Recognize and correct weaknesses in the selected detailed concept
- Recognize starting points for system simplification

Procedure
A prototype can be realized in the described development stages from the function pattern to the pilot series not only in a physical but also in a simulated model of the system. The following methods are appropriate for this:

1. **Traditional Method (Tool and Die):**
   The product is realized physically. The Tool-and-Die Method is often applied to analyze the peculiarities of repairs and maintenance (e.g. constructing a suspension system to test wear and conduct maintenance work).

   In this context, the Just Enough Prototyping is an accelerated type of prototyping and enables a first orientation with regard to it fulfilling function, form, measurements and performance requirements with minimal effort:

   a. Simple drawings of the system support the exemplification and visualization. A sketch should be maintained in order to focus the discussion of the team on the actual functionality

   b. In the next step an existing product is converted by simple makeshift modification to a prototype. In this way for instance, the office of the future can be imitated with any means available in a conference room, or, the next generation of the passenger seat can be exemplified with components of previous seat systems

c. The prototypes in the ongoing development process of the Just Enough Prototyping become more and more specific and realistic. User friendliness and handling are tested. Manually created within the team, clay models and “clickable dummies” become the first steps towards completely functioning prototypes that can fully test the new product

2. **CAD Prototyping:**
With the CAD Method (Computer Aided Design) products and their transfer functions can be mapped in an increasingly realistic way. In contrast to the traditional Tool-and-Die, this method is more cost-effective and time-saving.

In this context, Reverse Engineering creates digital models from real data and generates the necessary starting data for the CAD Prototyping. The object needn’t be a functioning finished part, but can be for example, a clay model which was manually constructed in a "Just Enough Prototyping" Session, or it can be just a two-dimensional sketch. This is enough to describe the characteristic form of a part to be modeled. For the realization, a digital three-dimensional model is scanned as a parent part, read into an appropriate CAD Software and positioned in the virtual room. Afterwards, curve and area functions are used to develop the virtual model further by adding the appropriate reference data.

Trimming

Term

Trimming, Value Engineering, Complexity Reduction

When

DESIGN: evaluate detailed concept

Goal

- Evaluate system components with the help of the functionality-costs-relationship
- Simplify the complete system by eliminating individual components

Procedure

Experience shows that technical systems with a high degree of complexity are generally less reliable than simpler ones. It therefore makes sense to reduce the complexity of a system. This is achieved by trimming, as individual system components are made superfluous and then eliminated.

Illustration: Reducing parts and therefore complexity by Trimming

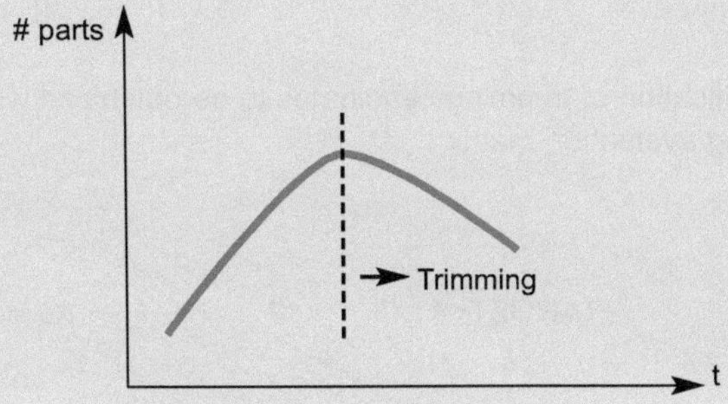

Components whose value to the system is low already are considered appropriate for this. Functional Analysis and Value Engineering, an approach developed by Lawrence Miles, help to identify such trimming candidates.

According to this approach the value of a component and/or a sub-system is determined by the relation of its functionality to its costs:

$$\text{Value} = \frac{\text{Functionality}}{\text{Costs}}$$

The functionality of a system component is defined by the proportion of its contribution to the overall function as well as its relationship to the system's other components.

Illustration: Identification of trimming candidates to be eliminated with the help of a Value Engineering evaluation matrix

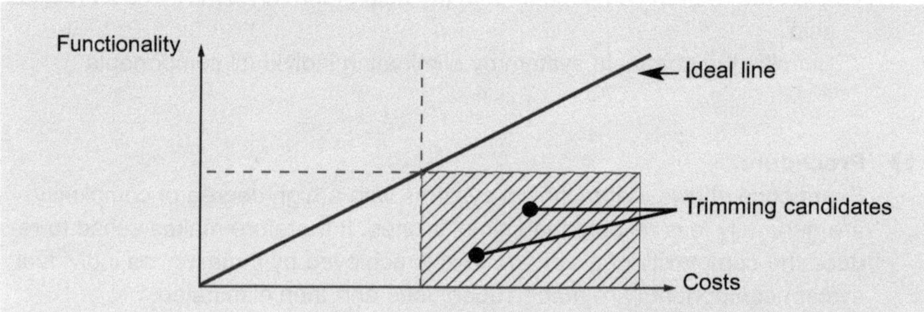

Components placed in the bottom right quadrant of the matrix have the lowest value for the system. The aim is to make these trimming candidates superfluous and to remove them from the system.

Illustration: Identification of trimming candidates to be optimized with the help of a Value Engineering evaluation matrix

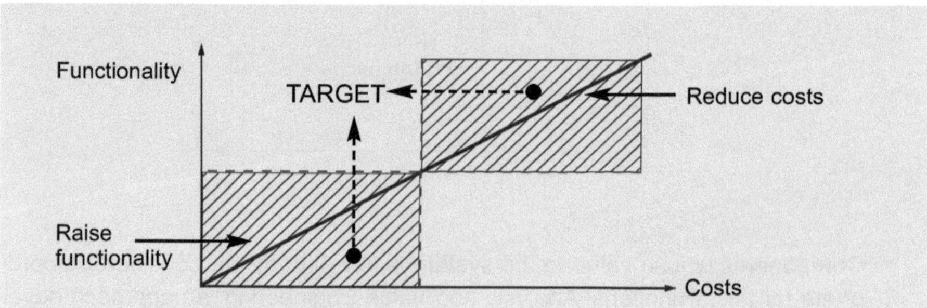

If a component is in the upper right quadrant of the matrix its costs must be cut.
If a component is in the lower left quadrant its functionality should be increased.

To determine the relative functionality of a system component the following steps should be undertaken:

1. Determine the main function of the system
2. Draw up a functional model of the system
3. The components are sequenced according to their distance from the main function. The component which is most distant from the main function is given the lowest functional ranking 1
4. The functional ranking of the respective components is multiplied by the number of their functions (= absolute functionality)
5. The relative functionality is gained by dividing the absolute functionality of the individual components through the aggregate of all absolute functionalities

Illustration: Determining the relative functionality – Functionality of the components of a toothbrush

❶ The main function of a toothbrush is to remove plaque from teeth

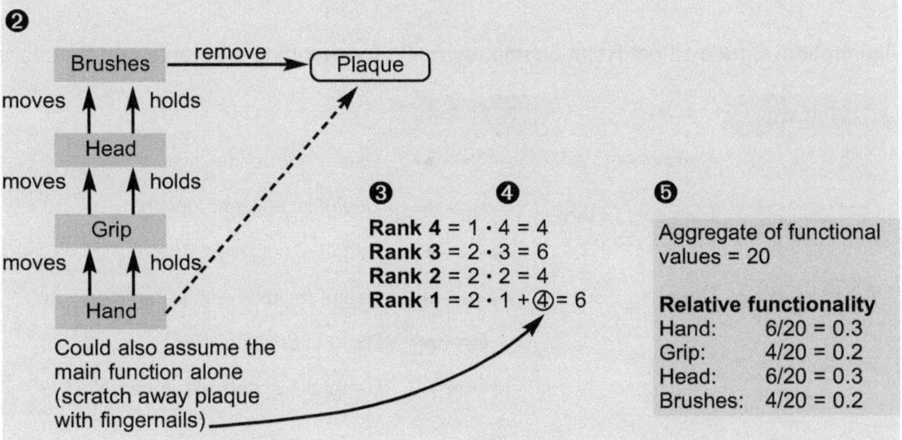

These relative functionalities allow the system components to be sorted into an evaluation matrix. The components suitable for trimming actions can be determined due their unfavorable value (functionality-cost relationship).

The question about these trimming candidates is:
How can they be made superfluous to the system and be eliminated from it and/or how can the remaining system components take over these individual functions without obstructing the total function?

Tip

- Use the procedure and symbolism for mapping the functions and drawing up the function model, in order to analyze the cause-effect relationships between the system components concerned and to identify useful and/or harmful relationships (conflicts and/or contradictions) between the components
- Break down the system and its operating environment into individual elements and describe them by appropriate nouns; their respective impacts and/or functions are described with the help of verbs
- A distinction should be made between the following functions:
  - "Useful functions" (UF)
  - "Harmful functions" (HF)
- In material development functions can very often be equated with the material's characteristics, e.g.:
  - Electrical resistance (conduct electricity)
  - Firmness (withstand mechanical tension)

Illustration: Cause-Effect Relationship, symbols for mapping elements and functions

Elements	
	Component
Super system	Elements of the super system
Product	Product

Functions	
	Useful normal function
	Useful insufficient function
	Useful excessive function
	Useful function with parameters
	Harmful function
	Harmful function with parameters

Example: Function mapping motor vehicle – super system "motor vehicle"

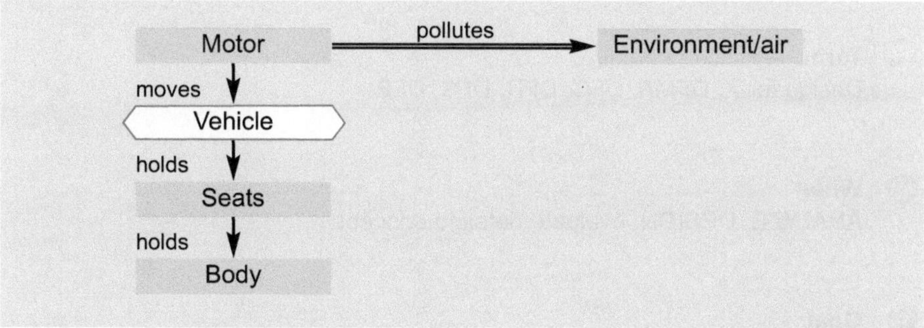

Example: Function mapping tooth brush – super system "tooth brush"

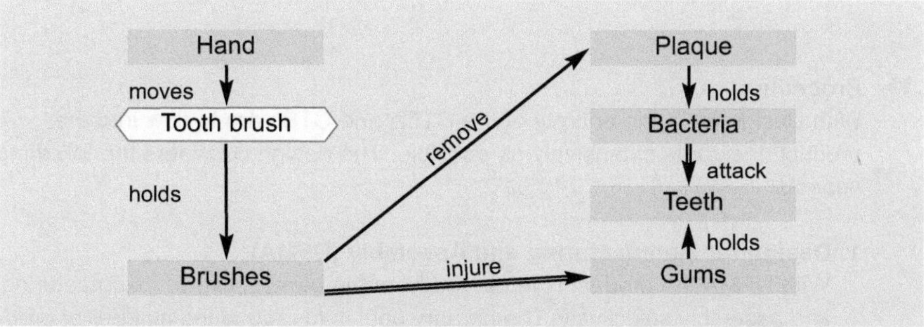

Design for X

Term
Design for X, DFMA, DFC, DFR, DFS, DFE

When
ANALYZE, DESIGN: evaluate detailed concept

Goal
- Develop alternative, reliable, cost-effective, environmentally friendly design elements
- Question existing systems and their configuration and simplify and/or reduce their complexity in a structured way

Procedure
With Design for X the entirety of the CTCs and CTBs are to flow into the product design as extensively as possible. The design considers the following aspects:

1. **Design for Manufacturing and Assembly (DFMA)**
   With DFMA the product is improved from the perspective of manufacturing and assembly selectively. The primary goal is to reduce the number of product parts as much as possible and to analyze their influence on production. The alternative solutions are evaluated in terms of costs and failure resistance.

2. **Design for Configuration (DFC)**
   DFC is used to realize the requested external variant diversity with as few components and processes as possible. The interfaces and dependencies between the components are defined and their contribution to satisfying customer wishes is examined.

3. **Design for Reliability (DFR)**
   DFR anticipates failure opportunities and improves the reliability of the design over a longer period of time. Besides reducing complexity, this is achieved by standardizing the parts and materials. The design elements

should withstand and/or counter environment influences. Weaknesses which could lead to packing damaging transport and repairs should also be considered. Stress resistance, influence of environmental factors and inappropriate handling by end customers are also tested and taken into consideration.

4. **Design for Maintainability (DFM)**
   DFM aims at reducing maintenance times and necessary material consumption within the scope of maintenance activities. Work accidents are also supposed to be reduced systematically when maintenance works are conducted. DFM aims at a comprehensive analysis of maintenance activities in order to initiate respective actions, for improving reaction capability in the event of machine stoppage, for the increased use of standard components primarily for parts which are susceptible to failure, for the reduction of required storage space in the maintenance area and for general simplification of maintenance activities.

5. **Design for Services (DFS)**
   Design for Services is applied to determine and optimize future service tasks, raise customer satisfaction, reduce life cycle costs, and improve the life span so that it is environmentally compatible and sustainable. A consistent Design for Services ensures that parts are easily identifiable and accessible and by utilizing modular systems reduces service demand.

   The following steps are recommended for realizing a Design for Services:
   1 Define the service actions
   2 Simplify the diagnosis
   3 Evaluate and optimize the costs for parts
   4 Set and optimize the working costs
   5 Simplify the entire execution

6. **Design for Environment (DFE)**
   Design for Environment looks at the ecological and economic consequences at the end of a products' life cycle. It helps to reduce the environmental burden, boosts ability to recycle and thus reduces waste disposal costs.

   The following steps reduce the impact of the product on the environment and the resulting follow-up costs:
   1 Define the environmentally harmful materials/procedures

2 Define the consumption
3 Evaluate the costs (protection measures for employees and environment, disposal and consumption)
4 Search for alternative materials/procedures

Illustration: Design for X – Passenger Seat

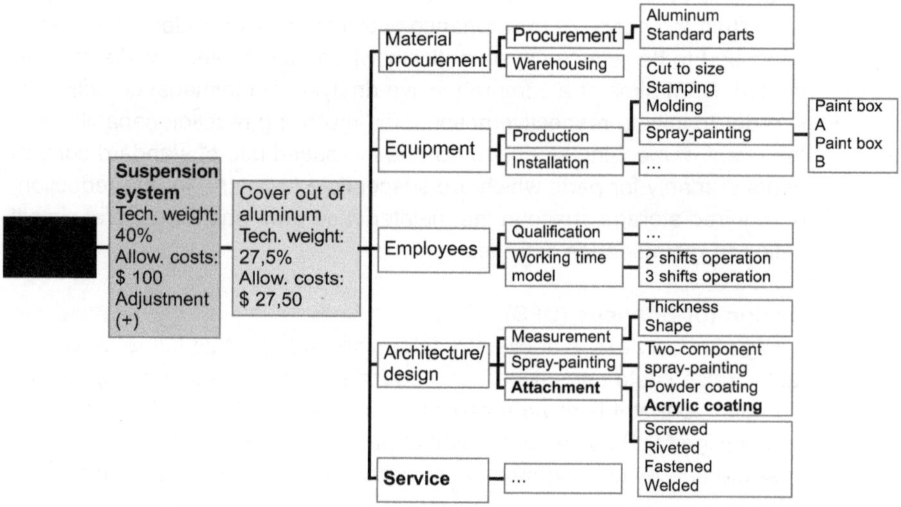

Design for Environment:
- Use of an acrylic coating to reduce the proportion of solvents
- ...

Design for Maintainability:
- Quick locks for parts that are prone to failure in order to simplify maintenance activities
- ...

Design for Service:
- Optimize accessibility of the suspension system
- Modular structure of the suspension system
- ...

Tip
- Use and supplement the following suggested checklist for the application of different DFX approaches
- Use and add the Design Tree and start with the question formulation on the last level

	Questions	Answers
1	How many subcomponents does the complete system have?	
2	Are all subcomponents required for fulfilling the functions?	
3	Are all necessary interfaces between the necessary subcomponents error-proof?	
4	Can the necessary parts be further standardized?	
5	Can alternative materials be used?	
6	Are the necessary parts exposed to environmental influences which could have a harmful impact?	
7	Are there weaknesses and/or problems caused by existing packaging materials and transport systems?	
8	Are there any apparent optimization approaches in regard to maintenance with respect to reaction capability, using standard components for parts that are prone to failures or reducing storage space for maintenance parts?	
9	Are all parts of the system that are prone to failures easily accessible and reachable?	
10	Can approaches for reducing environmental pollution and for enhancing the reusability of the system be derived?	

Anticipatory Failure Determination (AFE)

Term
Anticipatory Failure Analysis, Anticipatory Failure Determination, Subversive Failure Analysis

When
ANALYZE, DESIGN: evaluate high-level concept; DESIGN: evaluate detailed concept

Goal
Identify potential sources of failure in an anticipatory way and with reference to the product as well as the process

Procedure
Anticipatory Failure Determination (AFD) provokes failures in a targeted way with the help of the following questions:

- What is the requested target function of the system and what could happen to bring about the system's breakdown?
- Which options in the process and which resources available to the system and the system environment could be used to sabotage the system?

The reason for the breakdown is transformed into a desired function i.e. how could this failure be created? Individual TRIZ and Poka Yoke methods can now be applied to the inverted problem. The defective function – is inverted into the primary, beneficial function of the system. Subsequently, an attempt is undertaken to generate this "beneficial function" through the given system and environment conditions.

An Anticipatory Failure Determination comprises four steps.

Mapping Steps of the Anticipatory Failure Determination

1. **Define the target function:**
   Define the target functions per component with subsequent inverting, i.e. what has to happen or be done to disable the target function? The inverted problem has to be reinforced further.

2. **Define resources:**
   Define the circumstances which can contribute to the breakdown of the complete system and/or individual system components. Among these are: production flows, required production and manufacturing equipment, tools, internal and external logistic processes, relevant procurement processes and qualities of primary material. Also further system and/or environment resources e.g. line speed, vibration frequency and the speed of equipment, can also lead to sabotage of the target function.

3. **Overcome contradictions:**
   Search for solutions with the help of TRIZ methods: formulate technological and physical contradictions, 40 innovative principles, ARIZ, etc.

4. **Avoid failures:**
   Selected failure opportunities are now converted back to the actual problem formulation. Actions for avoiding these failures (Poka Yoke) are defined.

➾ **Tip**

- Anticipatory Failure Determination is also suitable as preparation for an FMEA and to analyze connected failures and chains of failures
- Thinking barriers are broken through and subjective thinking is reduced
- Inverting the problem generates useful information because the inverted perspective reveals aspects about the affected system and its environment which would not emerge in a conventional approach, which is problem-oriented

Avoid Risks

Term
Poka Yoke, Avoid Defects, Avoid Risks

When
DESIGN: evaluate detailed concept

Goal
Take actions to ensure total elimination of errors i.e. by 100%

Procedure
Potential errors are analyzed and prevented before they can occur as a result of the identified causes.

A distinction is made between the different error types:
- **Operating errors:** Twisting, changing or mistaking parts
- **Forgetfulness:** Important working steps are forgotten
- **Misunderstandings:** Human beings see apparent solutions before they are familiar with the situation
- **Overlooking:** Errors occur because people do not look at an object long enough or from a far enough distance
- **Beginners:** Errors due to lack of experience
- **Inadvertent errors:** Errors due to lack of attention
- **Slowness:** Errors when workflows unexpectedly stop and/or slow down
- **Lack of standards:** Errors due to missing and/or incomplete work or process instructions
- **Surprises:** Errors when workflows run differently than expected
- **Willful errors:** Errors due to intentional action or consciously ignoring rules and regulations
- **Intentional errors:** errors made deliberately e.g. sabotage or theft

The following steps are taken to prevent (Yoke, Japanese for "prevent") these errors (Poka, Japanese for "error"):

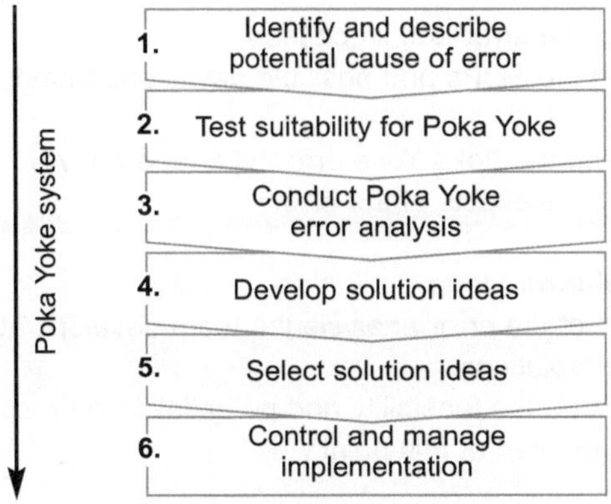

1. **Identify and describe potential cause of error**
   The error-relevant data is analyzed from different perspectives:
   - Location and frequency of error pattern
   - Type of error (stochastic or systematic)
   - Point of time when the error was detected
   - Significance and impact of the error

   The goal is a detailed and measurable description of the error pattern and of the error environment.

2. **Test suitability for Poka Yoke**
   In the Poka Yoke system sufficient specification of the error pattern is a prerequisite for a successful elimination of errors.
   The following questions should therefore be answered with "yes":
   - Is the point of origin of the error pattern known?
   - Is the part causing the error known?
   - Is the activity causing the error known?

   If more than one question is answered with no, the error must be further investigated and specified.

3. **Conduct Poka Yoke error analysis**
   An analysis of the error and the process in which the error originates proceeds as follows:

- Observing the error and its causes
- Reviewing the SOPs and possible deviations from the standard procedure
- Determining the Poka Yoke error type (see above)
- Observing the effects of the error

4. Develop solution ideas

On the basis of the error analysis the team develops at least three alternative ideas for solutions.

First remarks on the feasibility and potential of the respective solution idea can be documented at this point.

5. Select solution ideas

With the help of a Pugh Matrix the solution alternatives are evaluated and prioritized in terms of the following aspects:

- Feasibility/implementation
- Costs/benefits
- Potential for avoiding errors
- Effects on the process and/or follow-up process

The best Poka Yoke system solution can now be identified. If no solutions can be identified for preventing errors at origin, then the error should be detected as early as possible. See mapping examination methods on the following page.

6. Control and manage implementation

The following activities should be carried out to ensure a stable implementation of the selected Poka Yoke solution:

- Plan and document resources and activities required for implementation
- Initiate, accompany and monitor implementation
- Define Reaction Plans
- Control the error pattern and if required take counter actions (PDCA)

Illustration: Examination Methods

Examination Methods		
Traditional Examination	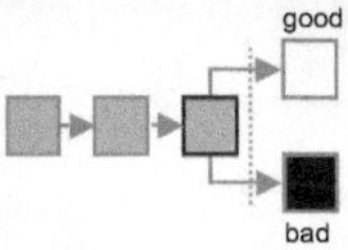	• Discerns good parts and scrap and/or rework • Reduces the defective parts delivered to the customer • Does not prevent error production • Slow feedback via scrap and/or rework
Statistical Examination	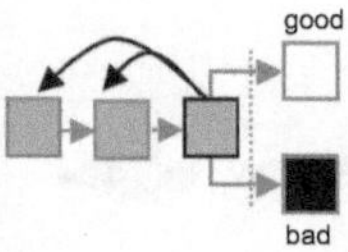	• System for reducing examination costs • Does not prevent error production, does not ensure non-defective parts • Errors can be passed through due to examination of samples rather than all • Slow feedback via scrap and/or rework
Continuous Examination	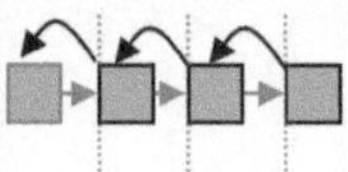	• Every process step examines the quality of the previous process • 100% of the parts are examined • Does not prevent error production • High effort/expenses of examination – efficient only for small amounts
Self-examination	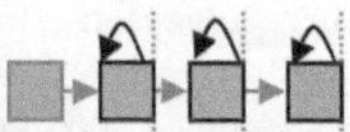	• Every process step examines its own quality • Immediate feedback and corrective actions • Stops the further processing of a defective part • High effort/expenses of examination – 100% of the parts are examined
Complete Examination	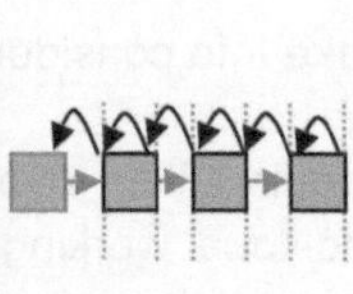	• Every process examines its own quality and that of its supplier • Problems are identified before the process step is finished • Immediate feedback and corrective actions • Stops the further processing of a defective part • High effort/expenses of examination – 100% of the parts are examined

Poka Yoke, examples

Error prevention: Using clearly dimensioned locking systems directly prevents errors.

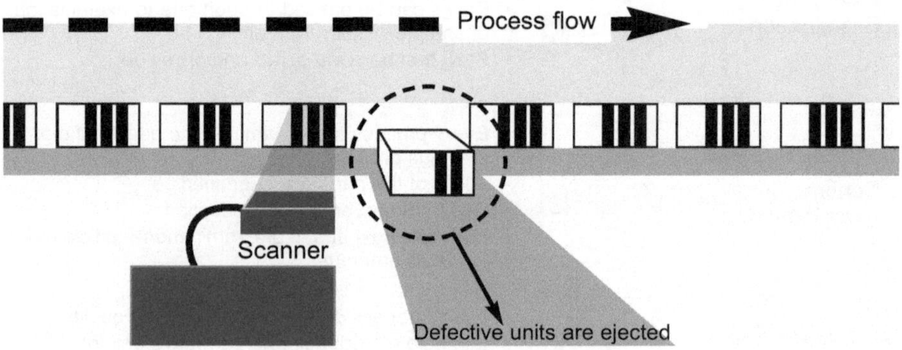

Early identification of errors: A barcode system identifies defective units.

➾ **Tip**

Throughout the implementation you should take into consideration the following criteria for a good Poka Yoke mechanism:

- It can be realized easily and quickly and requires little investments
- It is part of the process and requires no additional working step
- Its effect is immediately recognizable at the potential defect source and thus enables immediate correction by the employee
- It integrates the employee directly and does not serve the purpose of controlling the employee, but supports him/her with respect to generating quality
- Combined with further actions it can make control at the end of the process superfluous

Develop and Evaluate Lean Process

Term

Lean Process Development and Evaluation, Develop and Evaluate Lean Process

When

DESIGN: develop and evaluate lean process

Goal

- Develop efficient and effective processes which support the production and service of the detailed concept
- Define appropriate process control and pilot the new process

Procedure

After finalizing the detailed product concept the necessary production and supporting processes surrounding the system are redesigned and/or existing flows optimized.

1. Evaluate the current process capability
2. Draw up the process design
3. Minimize the Process Lead Times
4. Plan facilities and buildings
5. Optimize equipment
6. Provide employees
7. Set up KPI systems
8. Set up process monitoring
9. Simulate processes

10. Draw up work and operating instructions

11. Plan material procurement

12. Provide IT

13. Pilot processes

Tip

- Think about using process simulations on a step by step basis to develop the draft production and administrative processes into an optimal detailed process design
- Take into account internal and external requirements e.g. regulatory requirements, during the process design, which the processes need to comply with

Evaluate Current Process Capability

Term
Evaluation of Process Performance/Capability, Evaluate Process Capability

When
DESIGN: develop and evaluate lean process

Goal
- Identify relevant process and input variables on the basis of the product specific detailed concept and evaluate them with respect to quality, costs and capacity
- Show development and optimization needs of process and input variables

Procedure
The process and input variables defined in the Design Tree are now tested for their availability and reliability. In the Design Scorecard product specific design parameters, target values and specifications have also been defined, which should be taken into consideration in the development and evaluation of the process flows.

This step is necessary in order to ensure that requirements with respect to quality, costs and capacity for the implementation of the detailed concept are fulfilled.

Limited performance capability of any individual process or input variables must be targeted for improvement and optimization.

The following illustration shows relevant questions for analyzing process and input variables on the basis of the Design Tree and the Design Scorecard. The analysis is focused on quality, capacity and costs:

See next page for example

Questions for Process and Input Variable Analysis

Process 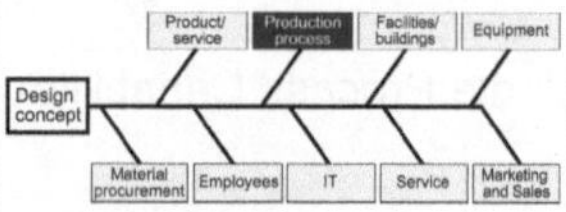	• Is today's existing process capable of producing the desired quality? • Is the existing process capable of delivering sufficient quantities to meet customer need? • Can the existing process adhere to the planned production costs?
Facilities/buildings 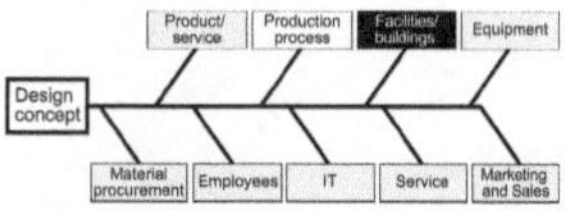	• Do the facilities match current environmental, health and safety standards (OSH/HSE)? • Are the facilities capable of delivering 5 S standards? • Is there enough storage space? • Is the operating concept suitable for a good production management – are the distances small?
Equipment 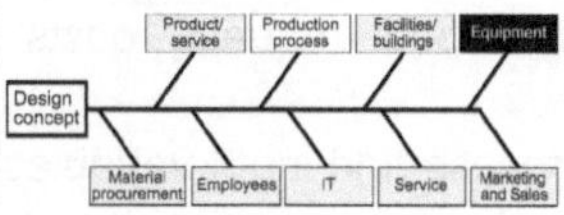	• Is the existing equipment (machines/tools) capable of delivering the desired quality? • Are the right tools available in the right quantity? • Are the costs for tools and facilities affordable (operating costs, wear, etc.)?
Material procurement 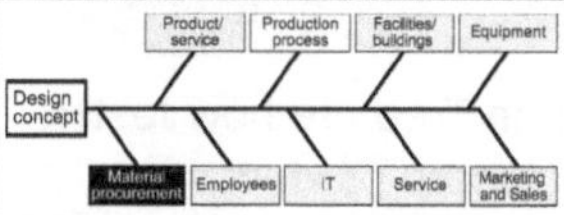	• Is the quality of the standard material sufficient? • Can the material be procured in sufficient amounts at the requested times? • Are the procurement costs of the material in line with planned costs?
Employees 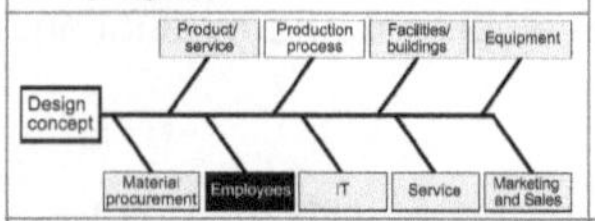	• Are the employees sufficiently trained to manufacture the product adequately? • Are enough employees available? • Are labor costs in line with the planned costs?
IT 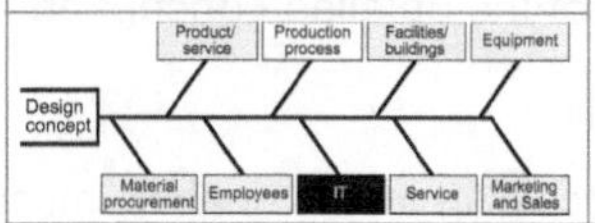	• Is IT support for the process ensured e.g. order and inventory management, quality management? • Are all materials used recorded in the system? • Are IT costs in line with the planned costs?
Service 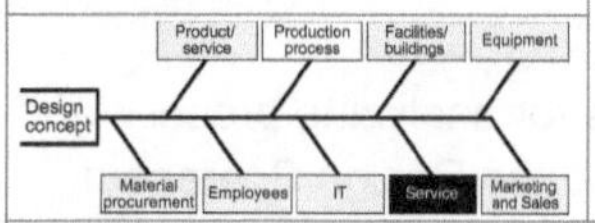	• Can existing service structures sufficiently fulfill potential service tasks for improving the lifespan? • Is the necessary know-how available to support the product? • Are service costs in line with the planned costs?
Marketing and Sales 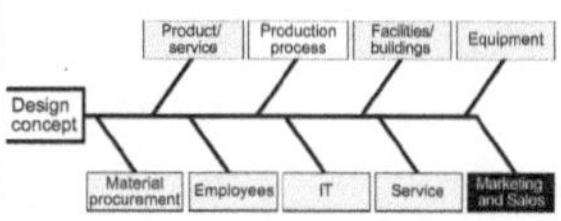	• Are existing marketing and sales concepts sufficiently in order, to successfully launch the system on the market? • Are sales familiar with the new Unique Selling Propositions of the system and can they be communicated appropri-ately? • Are costs for marketing and sales in line with planned costs?

The individual evaluations should be as quantitative as possible, so should be based on key performance indicators or critical figures (e.g. C_p, C_{pk}, costs, capacities) and Control Charts.
The results of the evaluation are presented systematically in a matrix. The following question is decisive for the appraisal of the evaluated production components:

- To what extent are the necessary requirements of the individual components fulfilled?

The following question is decisive for the appraisal of the evaluated process and input variables:

- To what extent do the design elements fulfill the necessary requirements for implementing the defined detailed concept?

The individual results of the as is analysis can be evaluated and visualized respectively on a scale from 1 to 5 within an evaluation matrix.

Illustration: Evaluation Matrix

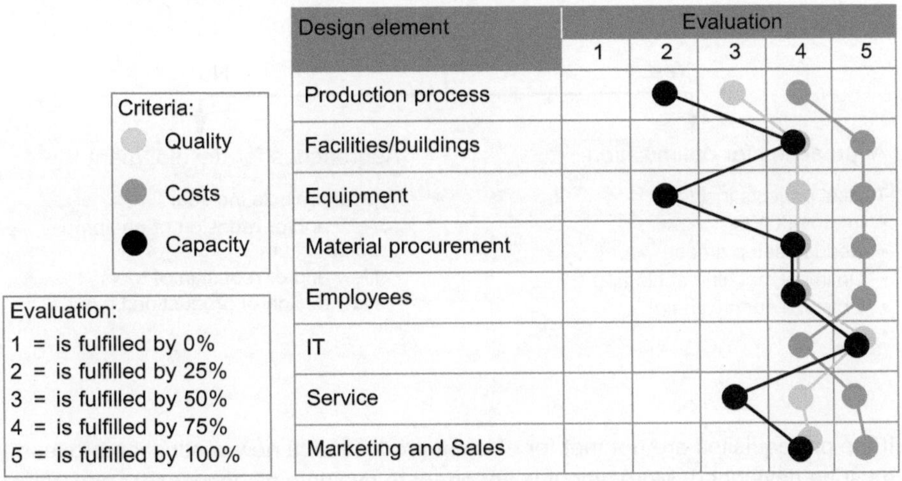

A decision on how to proceed will to be made on the basis of this overview. The target state on the individual criteria can be derived based on the as is analysis. A comparison of as is and target state shows any additional need for development and/or optimization.

Illustration: Optimization Direction on the basis of the evaluation results

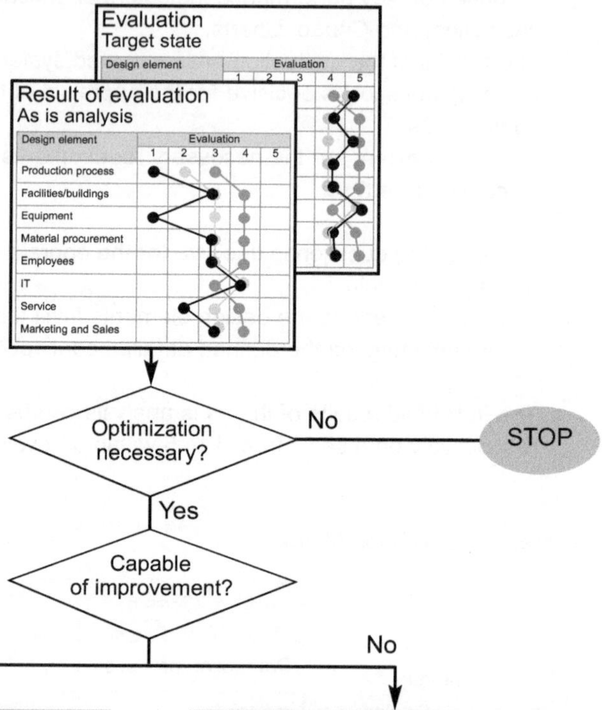

Approaches for optimization

Typical projects include:

- Improve OEE
- Reduce setup times
- Optimize machine settings
- Optimize supply chain
- …

Approaches for development

Typical projects include:

- New and/or redesign of equipment components
- New and/or redesign of tools
- New design of production layout
- …

If the prerequisites are not met for efficient production, a new production process must be developed. Moreover, it is important to evaluate all necessary supporting processes (procurement, marketing, sales, service, etc.) and to indicate respective needs for development and/or optimization.

Draw up Process Design

Term
Process Design Principles, Draw up Process Design

When
DESIGN: develop and evaluate lean process

Goal
Create basics for an efficient and effective process which is designed in line with basic design principles

Procedure
In order to optimize and/or redesign production and supporting processes for the product specific detailed concept, design principles must be taken into consideration to ensure a lean process.

A lean process requires only a minimum amount of working space, resources and time. A lean process should focus on its value addition and be directed towards customer activities.

Requirements of a Lean Process:

1. **Focus:**
   The focus of a lean process is on value-adding elements, especially in core business areas

2. **Analyze value:**
   Value-enabling elements are questioned in a lean process and tested for their necessity in the process environment

3. **Organize:**
   In a lean process similar customer activities are located at the start of the process and customer specific activities are located at the end

4. **Separate:**
   Information, communication and transport streams are clearly separated in a lean process

5. **Prioritize:**
   Process design in a lean process is always given the highest priority, especially if IT systems and/or methodologies are deployed

6. **Avoid:**
   Simple mechanisms for early defect prevention (Poka Yoke) are deployed in a lean process; they are particularly important where activities are continuously repeated

7. **Simplify:**
   Unnecessary complexity in a process is reduced and/or eliminated in a lean process, especially non-essential interfaces and authorization and/or release loops

Draw up Flow Chart

Term
Flow Chart, Cross Functional Diagram, Swim Lane Diagram

When
- DESIGN: develop and evaluate lean process

Goal
- Visualize process steps
- Display complexity and optimization potential in the overall process (number of interfaces, different media, etc.)
- Clarify responsibilities

Procedure
1. Insert start and stop points of the process described in the SIPOC drawn up in Analyze
2. Define individuals involved in, and responsible for, the process and assign detailed process steps between the start and stop points and respective responsibilities
3. Map all process steps in the way they actually occur, in order to recognize optimization potential

Illustration: Swim Lane Diagram

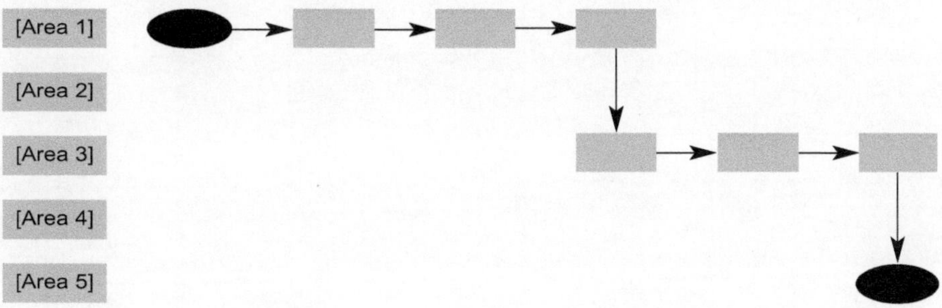

Tip

- Extend each process step by adding required input and output parameters to the chart so as to observe the quality, availability and complexity
- You should list, for example, primary material, operating supplies, tools, data, information, signatures, etc. as required input parameters and define the required level of quality and to whom they have to be made available
- Also add, semi-finished goods, processed information and confirmed forms as interim outputs of the current process and evaluate on the same basis

Illustration: Extension input-process-output

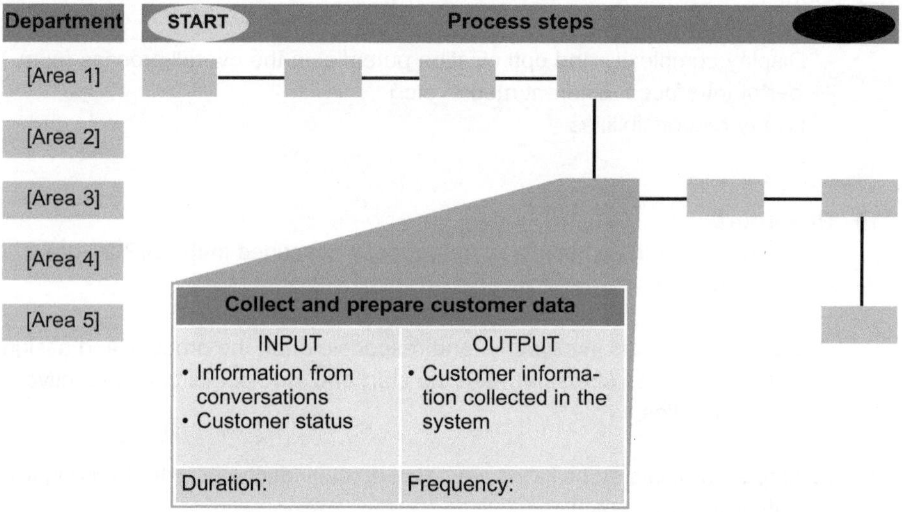

Draw up Value Stream Map

Term
Value Stream Map, VSM,

When
DESIGN: develop and evaluate lean process

Goal
- Map and evaluate relevant material and information flows from supplier to the final customer
- Identify sources of waste and their causes in the process and reveal optimization potential

Procedure

1. **Define processes and products:**
   Draw up an overview of all relevant process information on a high level by using a SIPOC diagram; define the process frame by marking the start and stop events.

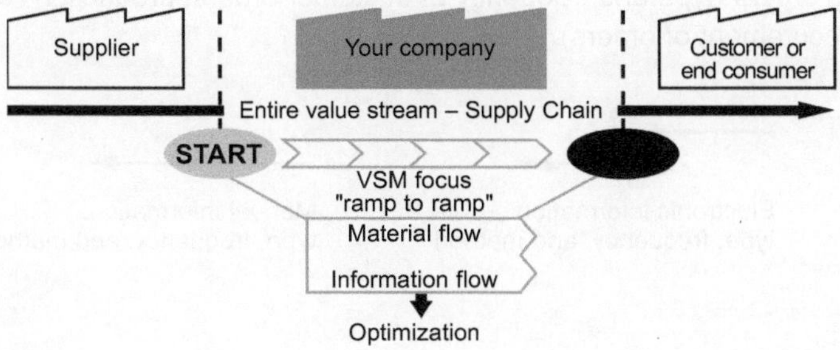

   Document the as is state with the help of a top-down flow chart: vertical division of core processes into individual sub-processes, afterwards identify the right observation level for sketching the value stream; identify customer relevant product groups or families in order to be able to focus on the areas in the Value Stream Map which have the highest impact on the customer (ABC/XYZ Analysis).

2. **Visualize flow chart:**
   Visualize the entire value stream: The observation works "upstream" starting from dispatch.

Process step | Customers/ suppliers | Transport | Product to customer | 300 units 7 days Inventory

3. **Define material and information flow:**
   Make the distinction between push and pull movements when recording the direction in which materials flow within the process; include all inspections of goods received and any quality checks.

Push Material flow | First-in-first-out sequence | Scrap | Rework | Pull Material flow

Document the direction of information flow afterwards, starting with incoming orders (type and frequency of customer orders, production releases, procurement of orders).

Electronic information: type, frequency, and method | Manual information: type, frequency, and method

4. **Define process data boxes and times:**
   Document all process relevant data (processing time, setup time, scrap rate, yield, machine availability etc.); define the process related lead times and proportion value adding and non-value adding time; derive the indicators for determining Process Efficiency.

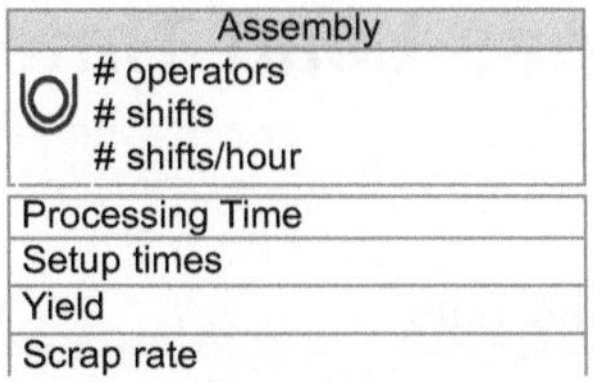

Process data

5. **Validate "as is" state:**
   Has everyone involved in the process checked the Value Stream Map drawn up in order to validate all interfaces, material flow and information flow?

➾ **Tip**

- Check the measurements recorded in the data boxes fulfill the respective Operational Definitions
- Check the information (process steps, information and material flow) on site
- Pay attention to uniform measurement units e.g. minutes/hours

Example:

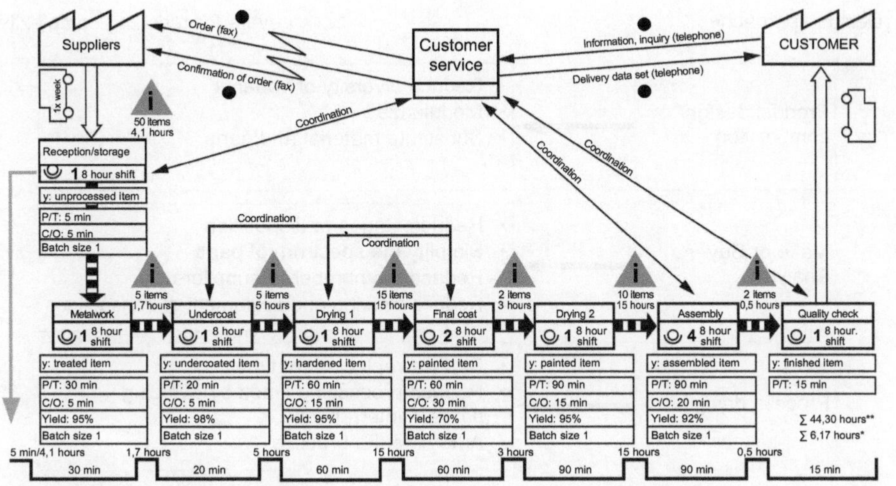

* value adding, * non-value adding, C/O: changeover time, P/T: processing time

Minimize Process Lead Times

Term
Minimizing Process Lead Time

When
DESIGN: develop and evaluate lean process

Goal
- Eliminate complexity and waste in the process
- Enhance the flexibility of the process
- Increase the capacity at constraints in the process

Procedure
1. Reduce complexity

Mapping possibilities to reduce complexity

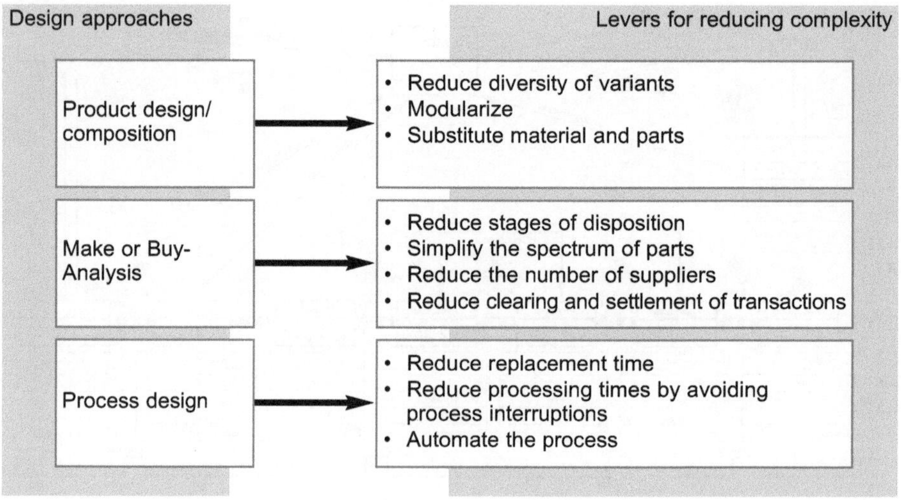

2. **Eliminate non-value adding activities**
   The Process Lead Time is often extended significantly by non-value adding activities.

Illustration: TIMWOOD

Seven types of waste in PRODUKTION		
T	**Transport**	• Movement of materials/products from one place to another • Repacking, transport on conveyor belts and bands, etc. unless paid by the customer
I	**Inventory**	• Material/product waits to be processed • Warehouse, buffer, temporary storage and "illegal" stock
M	**Motion**	• Superfluous movements or bad ergonomics • Workplaces are far away from one another, search for materials, etc.
W	**Waiting**	• Delays within work flow • Waiting for material, releases, downtimes, etc.
O	**Overproduction**	• More is produced than necessary • By the avoidance of setup procedures etc. • Exploitation of productivity as key control parameter
O	**Overprocessing**	• More value addition than the customer is willing to pay for • By misunderstood and unknown customer needs etc.
D	**Defects**	• Defects which must be eliminated and/or scrap • Due to false machine settings, materials, etc.

Seven types of waste in ADMINISTRATION		
T	Transport	• Unnecessary transport of information • Movement of documents, passing through hierarchies
I	Inventory	• Unnecessary inventories • Documentation of concluded projects, unused working tools and data inventory, multiple filing
M	Motion	• Unnecessary routes • Distance covered when searching for documents, consulting colleagues, ergonomic obstacles
W	Waiting	• Waiting times/idle periods • Waiting for decisions, returns, handovers, warm-up time of office equipment
O	Overproduction	• Superfluous information • More information than the customer, the following processes or the current process phases require (emails, copies, memos, etc.)
O	Overprocessing	• Useless activities • Unread reports, statistics and protocols, unnecessary data entries and copies
D	Defects	• Defects (possibly due to media disruptions) • Illegible faxes and notes, incomplete information

This approach is a key aspect when increasing Process Efficiency.

$$\text{Process Efficiency [\%]} = \frac{\text{Value adding time [t]}}{\text{Process Lead Time [t]}} \cdot 100\ [\%]$$

3. Reduce inventories and raise capacity at constraints

Basic definitions:

Capacity

The maximum product amount (output) that a process produces within a specific period of time.

Bottleneck (Time Trap)
The process step that causes the greatest delay in a process – there can only be one time trap in a process.

Constraint
A constraint is a process step that is incapable of producing the exit rate demanded by the customer (internal or external) i.e. production is below the takt rate driven by the customer need. A constraint is always a bottleneck, but a bottleneck is not always a constraint!

Work in Process (WIP)
Stock within the process: Things associated with started but not yet completed activities e.g. materials, waiting customers, orders, part assembled items, emails etc.

Exit Rate
The output of a process per unit of time e.g. 50 units per hour.

Takt Rate [quantity/time]
The quantity of a product (output) that the customer needs over a specific period of time.
Example: our customer demand equates to a takt rate of 100 parts per day.

Takt Time [time/quantity]
The time span in which the process has to produce a unit to satisfy customer demand
Example: The takt time is 45 seconds per part.

Existing constraints can be identified by using a Task Time Chart. The processing times of the respective process steps are collected in a chart and related to the calculated takt time.

See next page for example

Illustration: Task Time Chart starting situation – Processing times by process step

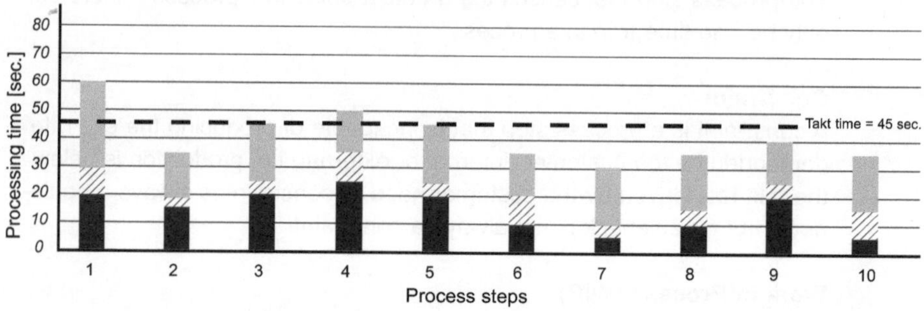

Constraints have an impact on the performance capability of a process because they require larger inventories, more machines, more personnel, more materials and more time to fulfill customer requirements. Constraints can be avoided by reducing non-value adding activities, minimizing waste and merging single process steps.

The relationship between Process Lead Time (PLT), Work In Process (WIP) and process capability (Exit Rate) is described in Little's Law.

Illustration: Little's Law

$$PLT = \frac{WIP}{Exit\ Rate}\ [time] \quad \rangle \ Min\ PLT\downarrow \quad \rangle \ \frac{Min\ WIP\downarrow}{Max\ Exit\ Rate\uparrow}$$

PLT = Process Lead Time [time]
WIP = Work In Process [units]
[inventory of units currently in the process]
Exit Rate = Process output [unit/time]
[production capacity]

Example: Reducing the Process Lead Time – Production process with three steps. Recording the individual step cycle times, results in the following:

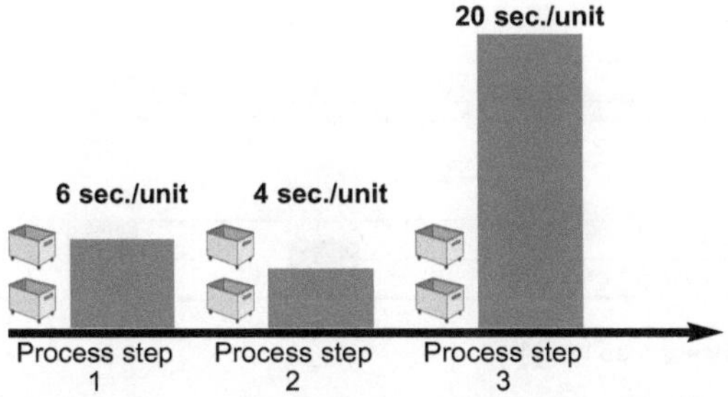

❶ **Status Quo**
Assumed WIP:
6 units
Bottleneck:
20 seconds/unit
Capacity at bottleneck:
3 units/min. (1 unit every 20 sec.)

$$\text{DLZ} = \frac{6 \text{ units}}{3 \text{ units/minute}} = 2 \text{ minutes}$$

How does reducing the WIP affect the Process Lead Time in the production process while maintaining the same capacity?

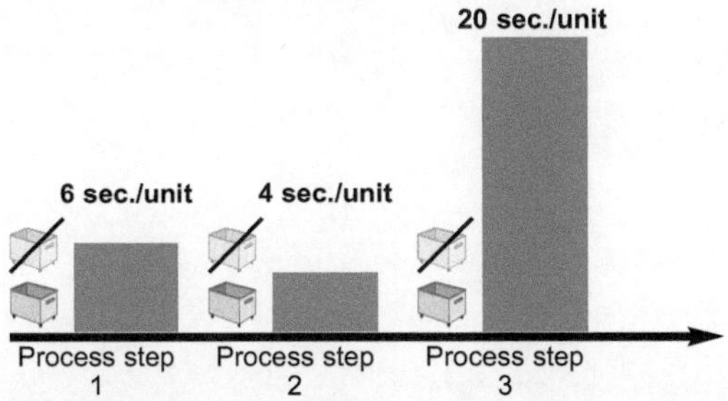

❷ **Reducing WIP**
Assumed WIP:
3 units
Bottleneck:
20 seconds/unit
Capacity at bottleneck:
3 units/min. (1 unit every 20 sec.)

$$\text{PLT} = \frac{3 \text{ units}}{3 \text{ units/ minute}} = 1 \text{ minute}$$

Reduction of PLT by 50%

How does adding additional capacity at the constraint affect the Process Lead Time in the production process?

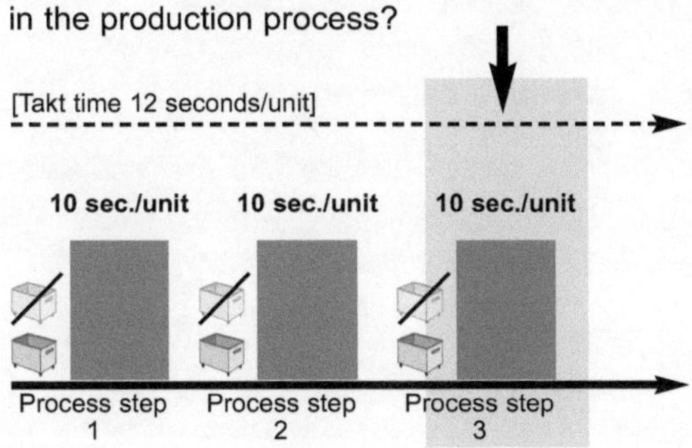

❸ **Capacity increase**
Assumed WIP:
3 units
Bottleneck:
10 seconds/unit
Capacity at bottleneck:
6 units/min. (1 unit every 10 sec.)

$$\text{PLT} = \frac{3 \text{ units}}{6 \text{ units/ minute}} = 30 \text{ seconds}$$

Reduction of PLT by a further 50%

Illustration: Process balancing – Real optimization approach

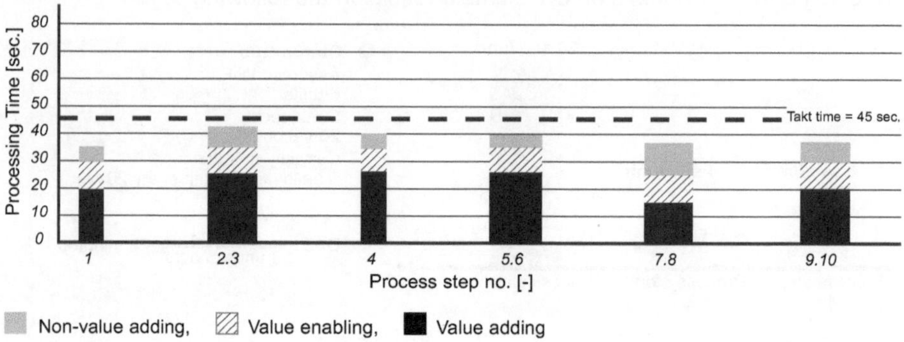

Plan Facilities and Buildings

Term

Plant Layout, Facility Layout Planning, Plan Production Layout, Planning of Facilities and Buildings

When

DESIGN: develop and evaluate lean process

Goal

- Optimize the internal distances to be covered
- Create the basics for a clean, safe, and highly efficient working environment
- Take into consideration safety related aspects throughout planning
- Ensure the performance capability of the process

Procedure

Planning facilities and buildings usually occurs on a smaller scale, if only individual production steps need to be modified. However, attention must also be paid to the optimal design of the logistic environment for dispatching and transporting parts, for basic concepts for order and cleanliness and relevant safety requirements especially in the environment of a plant.

The following approaches should be taken into consideration:

- Develop a layout for production
- Draw up plans for manufacturing/offices
- Define and optimize the distances employees and materials have to travel
- Define the equipment of manufacturing/offices
- Integrate and ensure 5 S concept accommodated
- Review and budget OSH/HSE capability (e.g. ventilation, separate routes, etc.)
- Analyze the material and information flow within the individual process variables in order to identify optimization opportunities in the actual working environment

Draw up a Spaghetti Diagram

Term
Spaghetti Diagram

When
DESIGN: develop and evaluate lean process

Goal
- Plan and visualize material and information flows
- Identify sources of waste and their causes
- Show optimization potential

Procedure
The routes necessary for conventional production and/or forwarding of documents and information are visualized. Required employees, tools, materials and information can be highlighted by simple symbols or by different colors. Weaknesses (waste) in the process are marked as well as paths which are critical for efficient processing e.g. crossings of footways and driveways, "traffic jams" or dangers.

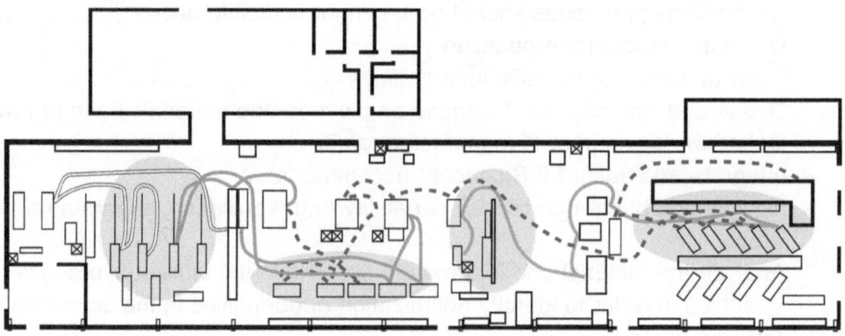

➾ **Tip**

- Add process related times such as processing times, transport and waiting times in order to be able to support further analysis based on the Spaghetti Diagram
- Combine a Value Stream Map with a Spaghetti Diagram in the work area and/or plant environment where you suspect the existence of a bottleneck This way you can visualize material supply and removal, necessary setup processes and maintenance activities etc. Add this information to the Value Stream data boxes and/or the other supplementary comments
- You should also include transport routes (e.g. conveyor belts, forklifts lanes, cranes, walkways etc) in the graphic in order to identify areas with a high risk of accident; classify these areas, according to the risk of accident, into primary and secondary
- You should bear in mind that the Spaghetti Diagram can quickly become confusing for complex processes with many variables (e.g. different materials and people)

5 S Concept

Term

5 S, Sort – Seiri, Set in Order – Seiton, Shine – Seiso, Standardize – Seiketsu, Sustain – Shitsuke, 5 A

When

DESIGN: develop and evaluate lean process

Goal

Create the basics for a clean, safe and highly effective working environment and ensure it is sustainable

Procedure

1. Sort – Seiri

- In general: Sort and mark all materials and/or objects in the working environment; sort out material which is not required
- Attach red adhesive stickers or tags to materials which are not directly required ("Red Tagging"); keep the materials in a marked zone for a defined period of time
- After a certain time period the marked objects should be …
  … removed if they have been tagged as "unnecessary", i.e. they are sold or thrown away
  … kept if they have been tagged as "required"
  ... thrown away, in case of doubt, follow the rule: "When in doubt, throw it out!"

Example: "Red Tagging"

RED TAG			
No.			
Date			
Department			
Category	1	Inventory	
	2	Machines/other equipment	
	3	Devices	
	4	Tools and provisions	
	5	Other (description)	
Description			
Qantity			
Total value			
			Date
Availability/ evaluation	a	Taken to "red storage"	
	b	Thrown away	
	c	Taken to more suitable environment	
	d	Left at original place	

2. Set in Order – Seiton

- Arrange and mark required objects in such a way that anyone can find them and return them to the correct place
- Objects have been sorted and have "their permanent place" which has been labeled accordingly
- Work with tools: colored markings for defined areas
  - "Home addresses" in order to be able to assign objects
  - Labels and color or shadow marks designating the object and its storage place

Example: "Shadow Board"

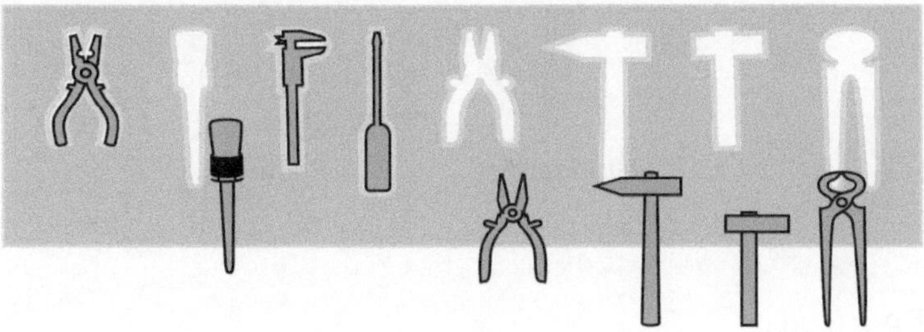

- Use the principles of Motion Economy

3. Shine – Seiso

- Remove dirt and other impurities as well as scraps and waste materials in the immediate working environment
- Define the cleaning goals and the areas to be cleaned
- Draw up and distribute a Schedule and define responsibilities
- Define procedures for regular cleaning activities
- Schedule periodic audits and set objectives for audit

4. Standardize – Seiketsu

- Create a consistent procedure for topics that have to be carried out on a daily basis within the frame of Sort, Set in Order and Shine
- Proceed according to the motto: "Do things correctly, at all times!"

5. Sustain – Shitsuke

- Integrate 5 S mentality into daily business and all processes; 5 S should not degenerate into a one-off event ("Flavor of the Month")

- Integrate 5 S audit forms for collecting and presenting results (Radar Chart)
- Establish a regular audit cycle to ensure sustainability

The current state can be represented by a Spider diagram:

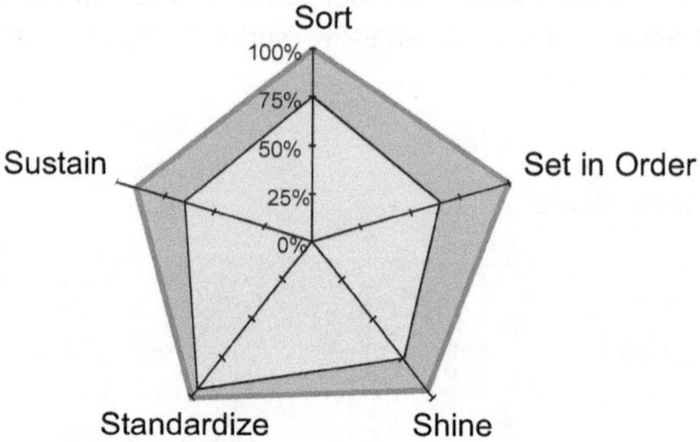

Tip

- Take photos of the area before and after the 5 S action; this is a simple means to visualize results and create acceptance
- Integrate all employees into the 5 S action
- Use the advantages of the 5 S methodology in order to significantly reduce search times as well as potential sources of accident in the working environment
- Make sure that Sustain is lived, otherwise the majority of all previous efforts will have been in vain
- Take into consideration that 5 S can be applied not only in production but also in administrative environments

Optimize Equipment

Term
Operational Equipment Planning, Equipment Planning, Plan Equipment

When
DESIGN: develop and evaluate lean process

Goal
- Identify necessary machines, equipment, tools and production facilities
- Ensure and optimize the availability of required machines and facilities

Procedure
Based on previous steps the respective requirements for equipment, tools and machine modifications can now be defined in readiness to change the production process. If machines, equipment and tools are to be procured and/or developed, attention should be paid to creating specific and measurable definitions of requirements well in advance in order to be able to evaluate their performance capability and reliability early on.

Apart from the maintenance friendliness of new machines and equipment the need for replacement parts should be coordinated early on with the manufacturer in order to be able to guarantee that replacement parts are available during operation.
New manufacturing techniques with new primary materials and tools usually include new and/or modified setup procedures which need to be planned and tested respectively.

SMED (Single Minute Exchange of Die) as a systematic procedure for reducing the setup time and TPM (Total Productive Maintenance) as a concept for preventive maintenance provide supporting approaches:

1. Reduce setup times

The setup time is defined as the interval between the last good part of a batch to the first good part of the following batch (produced at planned process speed).

Illustration: Setup time

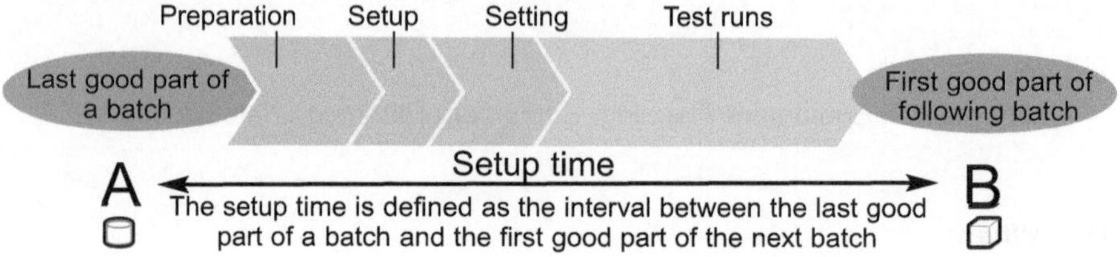

When the setup time is to be reduced, the focus is especially on internal setup activities:

Setup time reduction, also referred to as SMED (Single Minute Exchange of Die) comprises four steps:

1. Document setup process and divide individual activities into internal and external activities
   a. Internal setup activities can only be conducted when the equipment stands still (e.g. replacement of tools)
   b. External setup activities can be conducted while the equipment is operating (e.g. prepare material, invoice batch)
2. Convert internal activities into external ones
3. Streamline remaining internal activities
4. Eliminate adjustments and test runs

Illustration: Four steps for reducing the setup time in batch production

Short setup times are the central prerequisite for economic production with small batch sizes

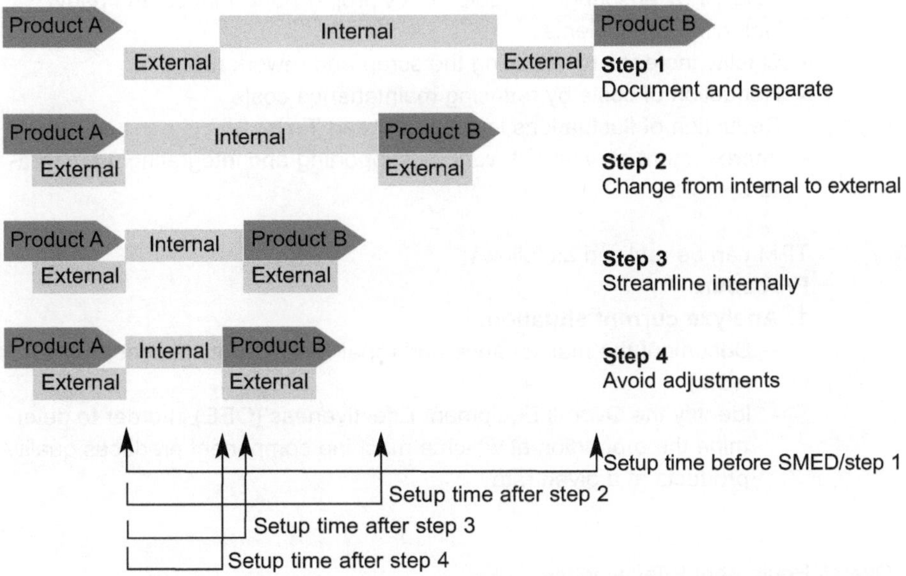

A setup time reduction with SMED can also be applied in the administrative environment if:

- information must be traced back in order to complete reports,
- required information is not available upon demand,
- different tasks are processed in turns.

Each of the mapped steps contains waste and leads to the fact that value adding work is prevented and/or obstructed. Setup time reduction in the administrative environment requires observation of all activities which obstruct the value-adding work process. Reasons for failure are systematically questioned and solutions are developed. Disrupting activities deemed necessary are simplified, e.g. electronic notes and reminders are used, in order to accelerate complex working steps.

2. TPM (Total Productive Maintenance)

Preventive maintenance adopting the TPM approach is directed towards the following objectives:

- Productivity increase through a reduction in production disruptions and failures by ensuring the equipment's proper performance and early detection of problems
- Quality increase by reducing the scrap and rework rate
- Reduction of costs by reducing maintenance costs
- Reduction of fluctuations in Process Lead Time
- Improvement of safety at work by supporting and integrating the operators

TPM can be realized as follows:

Procedure

1. Analyze current situation:

- Document the maintenance and repair costs (replacement parts and work)
- Identify the Overall Equipment Effectiveness (OEE) in order to determine the proportion at which a machine component produces quality products at a given rate

Overall Equipment Effectiveness

Availability level	Performance level	Quality level
Equipment downtimes Mechanic, pneumatic, hydraulic or electric defects which lead to production downtimes. **Losses due to setup and settings** Equipment is changed over for the production of a new part	**Idle time and short downtimes** equipment is operating without producing anything, as well as short disruptions (e.g. due to employees' breaks, search for material, etc.) or short failures of equipment which can quickly be solved (e.g. confirm an error message). **Reduced takt speed** The equipment is not operated at the required takt speed.	**Quality losses** Activities which are connected with the quality assurance of a product (scrap and rework). Rework frequently means that the whole production step must be run through once again. Equally, scrap means that the machine time used for producing a scrap part hasn't been used in a value adding way.

Calculate OEE as a product of availability, performance and quality:

Scheduled busy time		T = Total no. of hours for plant operation [h]	
Availability level (AL)	$AL = \frac{A}{T} \cdot 100\%$	A = Availability [h]	Plant breakdown; Setup and setting
Performance level (PL)	$PL = \frac{S}{A} \cdot 100\%$	S = Max. takt speed [h]	Speed loss; Idle time, standstill
Quality level (QL)	$QL = \frac{Q}{S} \cdot 100\%$	Q = Good Quality [h]	Rework, scrap

OEE	$OEE = AL \cdot PL \cdot QL$

Availability level (AL)	$AL = \frac{T_{Run}}{T_B} \cdot 100\%$
Performance level (PL)	$PL = \frac{T_{Cons} \cdot N_{Produced}}{T_{Run}} \cdot 100\%$
Quality level (QL)	$QL = \frac{N_{Good}}{N_{Produced}} \cdot 100\%$

T_B = Scheduled busy time [h]

T_{Run} = Machine running time [h]

T_{Cons} = Constraint takt time [h/unit]

N_{Good} = Number of finished goods [unit]

$= N_{Produced} - N_{Scrap} - N_{Rework}$

Example	$OEE = 73\% \cdot 91\% \cdot 80\%$ $= 53\%$

85% World Class

2. **Restore equipment into reliable state:**
   - Inspect and clean machine, identify necessary repairs, and mark defects which need to be eliminated
   - Machine is cleaned thoroughly by team
   - Mark areas in need of repair
   - Document all necessary repairs and draw up Time Plan to fix
   - Conduct repairs
   - Improve access opportunities to the area and/or the respective equipment so actions such as regular cleaning, lubricating, setting and inspection can be conducted

3. **Implement concept for "planned maintenance":** Structure maintenance process in order to ensure a stable production process and to monitor the equipment with planned ("preventive") actions in such a way that no unscheduled downtimes occur:

   1. Define the maintenance priorities:
      Identify the areas (equipment and/or production areas) which occupy the maintenance department most in terms of breakdowns, replacement parts, etc. Introduce logbooks to record all incidents affecting the equipment.

   2. Create a stable starting basis:
      Undertake a detailed analysis of weaknesses for parts and components susceptible to breakdown, using logbooks and other relevant documents.

   3. Introduce information, planning and steering system (IPS):
      Document and analyze the equipment specific information about the equipment park and the contingent maintenance activities. On this basis, maintenance actions can be planned, controlled and coordinated by computer.

   4. Introduce process related maintenance:
      Undertake regular inspection and maintenance of the equipment. Maintenance plans listing the tasks to be undertaken annually, monthly or weekly help to coordinate these inspection and maintenance actions.

5. Optimize internal operations:
   Optimize existing internal maintenance flows when conducting maintenance actions. Potential optimization approaches:
   - Shorten the diagnosis of defects – failure analysis is often very time-consuming
   - Optimize the storage of replacement parts – optimal storage directly influences the repair time
   - Optimize the exchange of replacement parts – reduce the time needed to exchange parts

6. Ensure sustainability – continuously improve planned maintenance program:
   - Guarantee the early detection of problems by providing good training to employees with respect to maintenance methods for prevention and forecast
   - Install visual controls
   - Implement 5 S
   - Regularly control and improve the machine performance
   - Has the equipment causing most problems been identified?
   - Have all weaknesses been effectively eliminated?
   - Can the IPS system be further improved?
   - Can the maintenance plans and standards be further improved?
   - How effective are the maintenance department's actions and workflows?
   - Can repair and service times be reduced further?
   - What are the key TPM indicators showing?
     – Overall Equipment Effectiveness (OEE)
     – Mean Time Between Failure (MTBF) and
     – Mean Time To Repair (MTTR)

See example on next page

Example Maintenance Plan

Machine No.				Machine Name		
Chart No.				Manufacturer		
No.	Machine parts to be lubricated		Lubricant	Quantity	Method	By
1	Pneumatic control panel	□	ESSO ZD5	50 ml	Orange oil can	PRD
2	Storage 1 and 2	△	SHELL T32	2 l	Blue tank	RS
3	Shaft	○	ESSO ZD5	0.5 l	Control oil level	PRD
4	Spindle	△	SHELL T16	1 l	Violet jerry can	RS
5	Gear unit	□	ESSO ZD5	10 ml	Control oil level	PRD
6	Track 1 and 2	△	SHELL T3	/	Green container	RS
7	Hinge joints	△	SHELL T11	/	Blue oil can	RS
8	Power unit 3	△	ESSO ZD5	0.5 l	Red jerry can	PRD
9	Power unit 4	△	SHELL T11	1.5 l	Blue oil can	PRD
10	Power unit 5	△	ESSO ZD5	1 l	Red jerry can	PRD

□ Daily
○ Weekly
△ Monthly

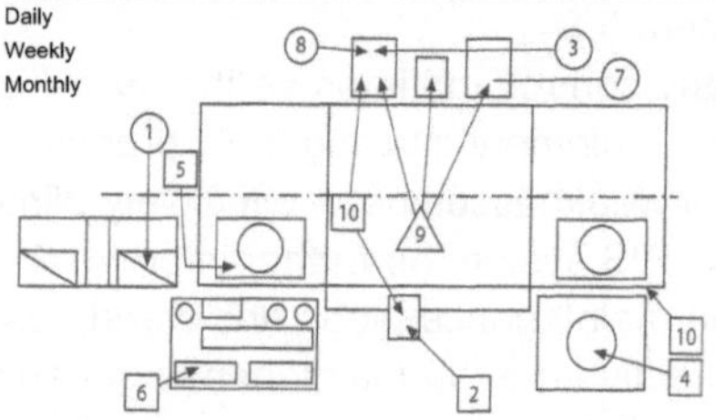

RS = Repairs & Servicing	PRD = Production	Creation date

Tip

- Plant/machine operators must be instructed in basic maintenance activities such as cleaning, lubricating, identifying damage, etc.
- Basic familiarity with, and know-how of, the equipment contributes to raising the productivity and reducing downtime
- TPM is only successful if all employees are actively involved

Provide Employees

Term

Work Plan, Personnel Planning, Work Organization, Staff Planning

When

DESIGN: develop and evaluate lean process

Goal

Plan personnel deployment for implementing the modified and/or new processes in line with demand

Procedure

- Define workplaces and activities
- Define work organization
- Determine the necessary skills
- Determine how many people are required
- Set a training plan
- Test workplace layout, test workload of employees and ergonomics, OHS, HSE
- Define remuneration and incentive systems

Tip

A suitable training concept should address the following questions:

- Which work steps are primarily influenced by the change?
- Who is responsible for these work steps and who does them?
- Who are the internal suppliers and customers of the process step?
- How can these individuals be prepared for the change?

- How can it be guaranteed that training content will be optimally transferred to everyday work?
- How can the change be communicated to the external world (Elevator Speech)?

Implementation teams are formed in the training. These teams support the change in a variety of ways, acting as change agents:
- They ensure the communication about the change activities in their respective departments
- They secure the complete changeover to the new process by acting as contact partners on site
- They report implementation risks and problems to the DFSS team at an early stage

Set up a KPI System

Term
KPI (Key Performance Indicator), Key Figure System

When
DESIGN: develop and evaluate lean process

Goal
- Increase transparency of Supply Chain, especially with respect to quality and performance (time and costs) of the new process/product
- Monitor the new process in an efficient way
- Control the process in real time, especially at the beginning of the implementation
- Ensure reasonable effort for collecting the parameters

Procedure
1. The entire supply chain is divided into the areas of procurement, production and distribution

Illustration: Control of value added chain with KPI

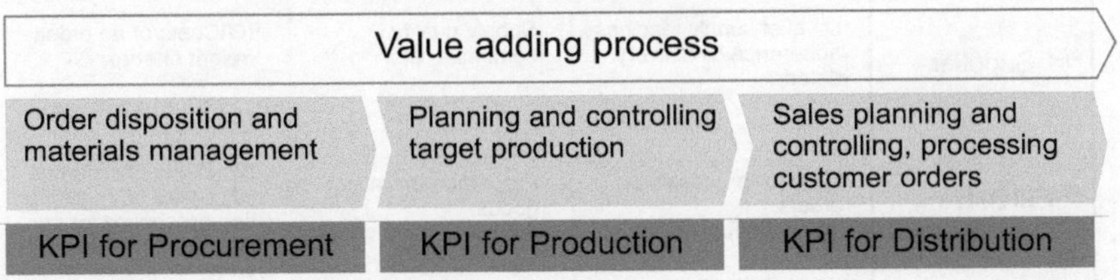

2. Determine relevant key figures (KPI) for each area oriented towards the individual value adding stages and control/optimization criteria such as time, quality, costs

See next page for example

Illustration: KPI levels for the areas of the Supply Chain in the individual stages of value addition

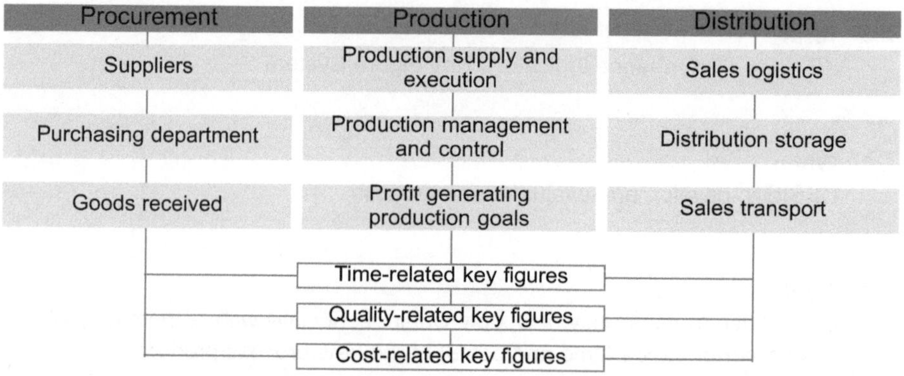

Relevant KPIs for Procurement

It is recommended to define the following KPIs for the procurement area:

- Influencing factors which affect the performance capability of procurement
- Demand, the quantity and the point in time of procurement (short and long-term)
- Procurement cost of materials
- Purchasing payments and securing of material supply

Example: Procurement Matrix – Passenger Seat

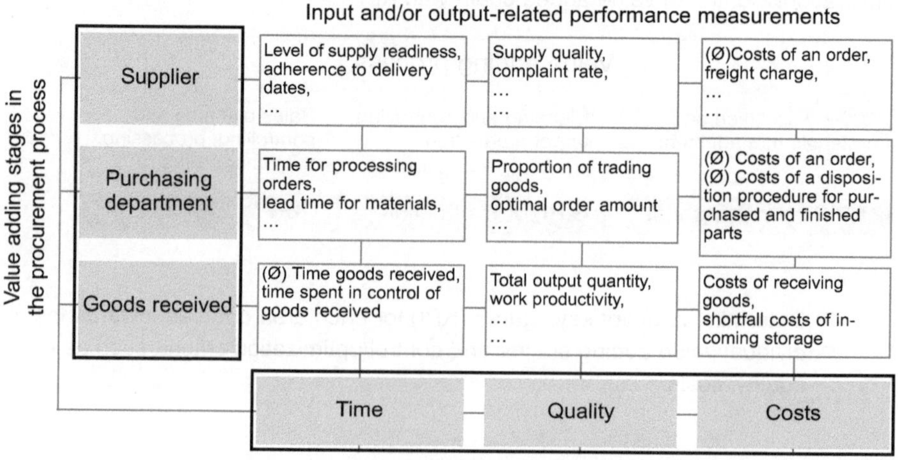

Relevant KPIs for production

A KPI system in the production area should provide information about the following things:

- The development trends of the process
- Efficiency, reliability and accuracy of production activities
- Relationships between the individual activities
- Optimization potentials in the production area

Example: Production Matrix – Passenger Seat

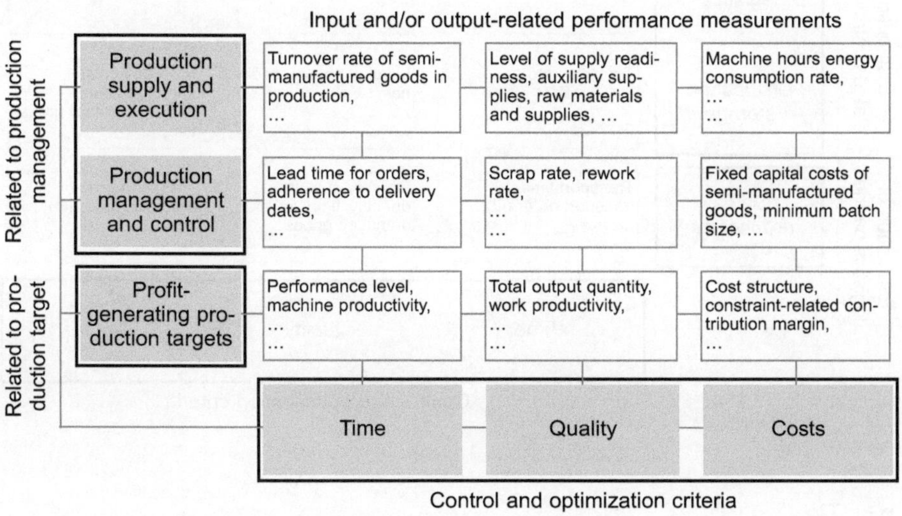

Relevant KPIs for distribution

Suitable key figures for the KPI system in distribution enable:

- Transparency in relation to quality and costs
- Identification of optimization potential
- Securing the sustainability of improvements

Example: Distribution Matrix – Passenger Seat

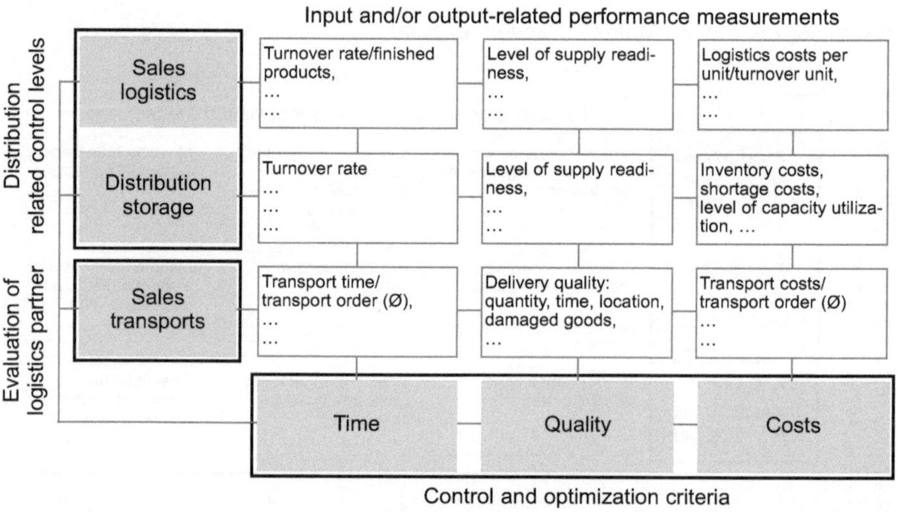

➾ **Tip**

- If procedural models for implementing new processes already exist in your company e.g. SCOR (Supply Chain Operations Reference model), you should check if they also include the derivation of key figures
- Define key figures which enable a total evaluation of the supply chain
- The benefit (controllability of process) should justify the effort

Set up Process Monitoring

Term
Monitoring, Process Monitoring, Process Control

When
DESIGN: develop and evaluate lean process

Goal
- Monitor process capability
- Create prerequisites in order to be able to compare product and process specific target values with as is values and to be able to intervene early in case of deviations
- Enable targeted reaction and introduction of action

Procedure
Having selected the KPIs regular recording, monitoring and control of process performance should be introduced with the help of the key figures selected. The following procedure applies:

1. Ensure standardized collection of key figures:
   Amongst this are: clear definition of the key figure, the measurement period, the collection interval, the data source and the responsibilities (What, How, When, How often, Where and Who?). All data should be collected within the scope of a key figure profile, in a consistent way and then evaluated.

See example on next page

Example: Key Figure Profile – Passenger Seat

Key figure	Number of defective seat covers
Definition	Proportion of defective seat covers in relation to the total number of seats produced
Dimension	%
Target value	Target value 1% per year
Measurement period	Weekly
Repetition	Permanent measurement
Data collector	Quality control
Data recipient	Process Owner
Evaluation/ reporting	Head of Quality Control
Responsible	Process Owner

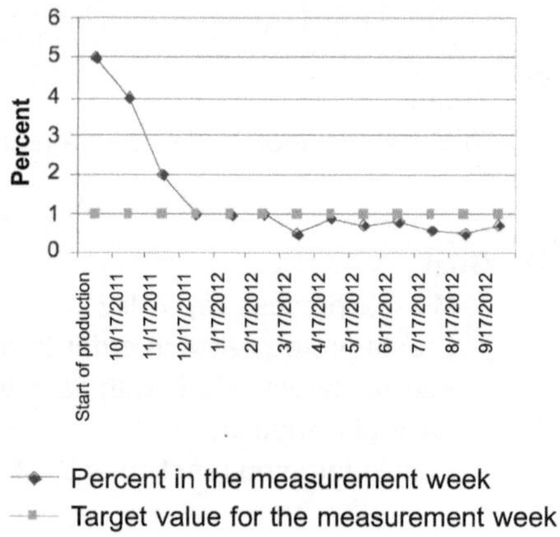

2. Conduct continuous Monitoring:
   It is recommended to map the key control and optimization criteria in a "Cockpit Chart". The Cockpit Chart is usually aligned to the following parameters:
- Time (e.g. takt time, Process Lead Time)
- Quality (e.g. FTQ rate, scrap, rework,)
- Costs

Example: Cockpit Chart – Passenger Seat

Part no.	Seat type	SOP	Plan 2010 units	FTQ	Cycle time Target	-as is	OEE	FTQ	Inventory level TC	RW T.	Value Analysis Pot.	CSt. real.	Post calc. Mat.	Prod.
4711 007	PS 007	2009	15,000	96.8	60	61	87.5	96.8	29.4	5	12.50	8.00	-6.00	4.00
4711	PS	2010	5.000	[redacted]	55	[redacted]								

3. Set specification limits (given by the customer):
   LSL = Lower Specification Limit
   USL = Upper Specification Limit

4. Determine statistical Control Limits with the help of Control Charts :
   LCL = Lower Control Limit
   (≈ - 3 standard deviations the from mean value);

   UCL = Upper Control Limit
   (≈ + 3 standard deviations the from mean value).

Example: Control Chart (Xbar-R)

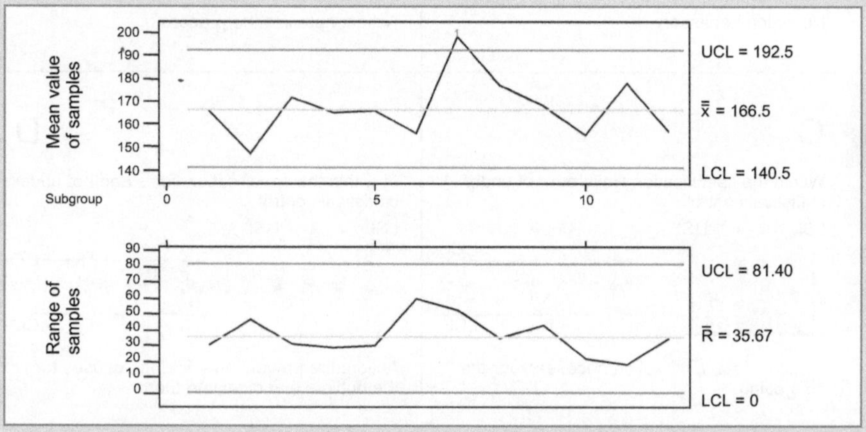

5. Monitor the process capability by taking into account the following process performance scenarios:

   A. The process is within the specifications and under control statistically: no action required

   B. The process is not within the specifications but is under control statistically: search for common causes and optimize the process

   C. The process is within the specifications but not under control statistically: monitor the process closely

   D. The process is not within the specifications and is not under control statistically: search for special causes and conduct actions for "fire prevention"

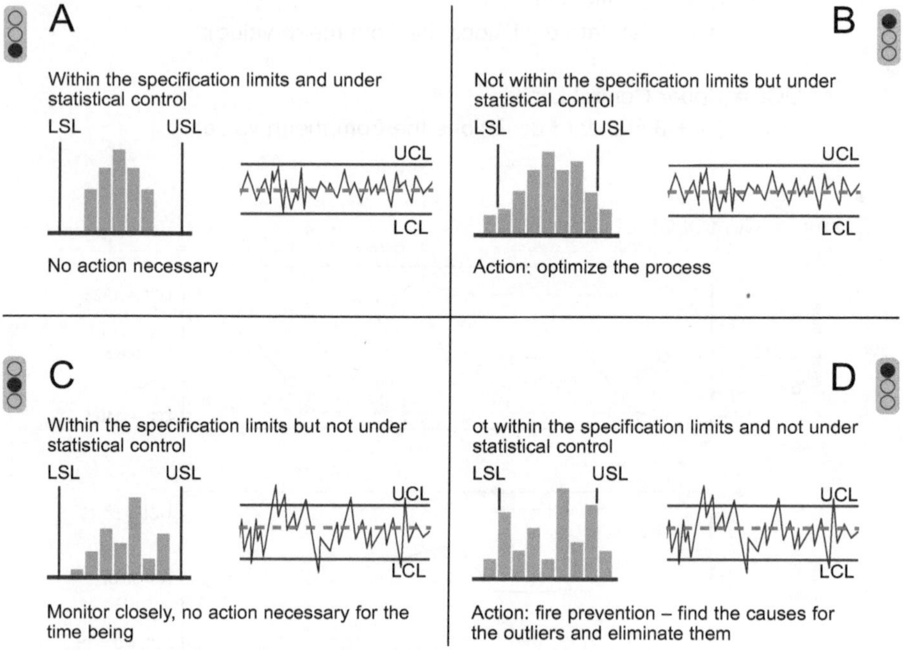

Illustration: Monitoring of Process Capability

LSL = Lower Specification Limit:
the lower limits specified by the customer
USL = Upper Specification Limit:
the upper limits specified by the customer
LCL = Lower Control Limit:
the lower control limits calculated statistically
UCL = Upper Control Limit:
the upper control limits calculated statistically

6. Draw up a Process Management Diagram:
Document the procedure of the newly set up and/or partly optimized process within the frame of a Process Management Diagram (process mapping, Monitoring with KPIs and Reaction Plan). Monitor the targeted process performance with the help of the defined KPIs continuously and resort to the respective Reaction Plan in case of deviations.

Process Management Diagram			
Target process documentation	Monitoring		Reaction Plan
Target process from mm/dd/yy 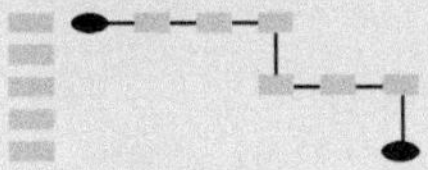	Product KPIs	Process KPIs	Actions in case of reduced process performance ... ...

Illustration: Process Management Diagram

➾ **Tip**

- Monitoring supported visually by images, colors and markings ensures sustainability
- The Process Owners should be trained in dealing with process monitoring
- In the context of Process Management Diagrams you should pay special attention to using introduced and/or known quality key figures and graphic mapping methods to avoid unnecessary problems of comprehension

Simulate Processes

Term
Process Simulation, Process Prototyping, Simulate Processes

When
DESIGN: develop and evaluate lean process

Goal
- Test a complete process or subareas of a process in a test environment with the objective of optimizing it
- Further develop the process detailed design in iterative optimization loops with respect to increasing Process Efficiency, reducing Process Lead Times and a lowering level of material use, optimizing procurement channels and availability of material
- Recognize risks in the defined process detailed design and introduce counter measures

Procedure
After having developed necessary process and input variables in order to be able to manufacture the defined detailed concept with the right equipment and in the right working environment it is now important to develop the designed production and administrative processes by taking into account the defined key figures step by step to an optimal process detailed design with the help of simulations.

The simulation of processes contributes to mapping planned flows in a realistic way. Processes can be modeled, simulated, tested and finally optimized in a targeted way. The optimization of processes aside of the system prevents interference in production. With simulation projects not only can planned standard scenarios be explored but the impact of special cases can be played out e.g. failure of manufacturing capacities due to machine constraint, the operation of additional equipment or the change of a shift system.

The following steps describe the typical procedure for setting up a process related simulation environment:

1. **Map the process as a Flow Chart in the simulation software:**
   This can be done in the form of a simple flow chart, a Swim Lane Diagram or a Value Stream Map.

2. **Collect data on individual process steps (activities) and materials (transactions) and define expenses when decisions are made:**
   Information on the duration of the delivery, respective expenses and the value observation can be added to the process steps. The value observation of the activity refers to the question "does the respective activity generate added value from the customer's perspective or is to be classified merely as waste?" The Process Lead Time for delivery can be distributed normally or in a uniform way e.g. based on a defined period of time (e.g. between 5 and 10 minutes). For a uniform distribution the occurrence probability is the same for all values. A normal distribution follows a normal distribution curve.

3. **Define process relevant resources (machines, employees, IT, etc.) and their criteria (quantity, degree of availability, etc.):**
   Each element required for executing the process is recorded as a process resource. In each process step it is determined how many resources are required for the respective step. If possible dependencies between two resources emerge these must be taken into consideration. Hence, an employee is marked as busy as soon as specific equipment produces or – as in the case of a maintenance activity – the equipment is blocked as soon as maintenance personnel commence work on the equipment.

4. **Draw up Time Plans and assign resources (equipment times, shift systems, etc.):**
   Define the general time frame for the complete process since start and stop points are required for executing the process simulation. A test run of the process lasting one month can be chosen as Time Frame. Relevant information on shift systems for equipment operators and available production time of required equipment and machines etc. will also need to be defined.

5. **Test, control and improve simulation:**
   If the simulation model has largely been completed, a first test run can be conducted. The test checks if the transactions pursue the right path in the model and lead to an outcome i.e. run through the complete process and don't remain stuck in the process. For this purpose the process is simulated on the basis of the defined Time Frame across all process steps and resources and is evaluated respectively.

6. **Define and evaluate process relevant reports:**
   When the simulation has been completed the simulated data can be analyzed. Reports on the following topics can be generated using the reporting function:
   - Key figures on process times (e.g. Process Lead Time, processing time and waiting times)
   - Value Analysis on the process (e.g. process efficiencies)
   - Costs of the process and/or of individual process steps and their resources
   - Degree of utilization and availability of resources
   - Information on critical waiting times and downtimes in the process

   User-defined reports enable the compilation of tailored information on process performance and can be aligned to the defined key figures of the process.

Tip

- A simulation could also make sense to save money within the frame of product development
- Therefore, a simple model is often set up in practice; this in turn means that the model used often represents a rough simplification of reality; note then, that these simplifications obstruct the accuracy of the simulation results
- You should also bear in mind that a model only provides results in the areas which were modeled. It is possible therefore that the results you observed could be significantly influenced by parameters you did not model or vice versa; you should therefore validate models to ensure the results you expect are indeed the results in reality
- Also pay attention to the significance of incomplete or insufficient data when studying process parameters and ensure an appropriate data quality and quantity – in advance

Draw up Work and Standard Procedures

Term
Standard Operating Procedures (SOPs), Work Instruction

When
DESIGN: develop and evaluate lean process

Goal
Describe the specific tasks of the individuals involved in the process in detailed work and standard operating procedures

Procedure

1. The following basic steps need to be conducted in order to draw up work instructions and/or Standard Operating Procedures for the optimized or newly developed process:
   - Define process settings and material specifications
   - Coordinate Service Level Agreements (SLAs) with internal and external suppliers
   - Describe the process steps in detail
   - Test SOPs with a second person using defined test criteria and get it signed
   - Inform and train employees
   - Instigate sustainable Change Management
   - Define responsibilities

2. When drawing up comprehensible and appropriate SOPs take the following into consideration:
   - Use a level of detail which sufficiently clarifies what activities take place, when and where
   - Use graphic elements such as Process Flow Diagrams, Flow Charts and Value Stream Maps etc. to further support the description of activities and expected results
   - Clarify in clear and intelligible language the reasons for the activity so that even people not involved in the process can understand it
   - Provide sufficient hints and advice to restrict variation e.g. about known Cause-Effect relationships

- Formulate explanations in keeping with target audience
- Pay attention to simple and central access opportunities i.e.:
  – The SOP should be available to everyone both online and as hard-copy
  – The responsibilities for documenting the activities must be clearly defined
  – Internal ability to link documents must exist
  – Mechanisms for updating and optimizing documents should exist to support continuous development of them

Example: Work and Standard Operating Procedures – Passenger Seat

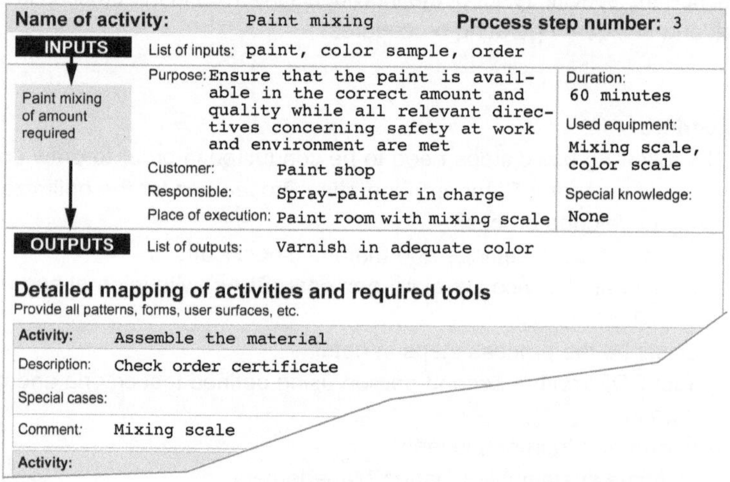
Name of activity: Paint mixing **Process step number:** 3

INPUTS List of inputs: paint, color sample, order

Paint mixing of amount required

Purpose: Ensure that the paint is available in the correct amount and quality while all relevant directives concerning safety at work and environment are met

Customer: Paint shop

Responsible: Spray-painter in charge

Place of execution: Paint room with mixing scale

Duration: 60 minutes

Used equipment: Mixing scale, color scale

Special knowledge: None

OUTPUTS List of outputs: Varnish in adequate color

Detailed mapping of activities and required tools

Provide all patterns, forms, user surfaces, etc.

Activity: Assemble the material

Description: Check order certificate

Special cases:

Comment: Mixing scale

Activity:

Plan Material Procurement

Term
Material Procurement Planning

When
DESIGN: develop and evaluate lean process

Goal
Ensure material provision at the right time, in the required quantity, at the required quality and at minimal costs

Procedure
In addition to machines, equipment and tools, it is important to ensure an integrated and reliable delivery chain in order to be able to determine the material need depending on quantity and time as precisely as possible. Unnecessary storage costs and spatial constraints occur in the event of premature material deliveries, whereas production downtimes and/or delays can be the consequence of delayed material deliveries of unreliable suppliers.

Expected consumption rates are calculated precisely; the required materials are classified according to respective value limits (ABC/XYZ Classification). Selection of suppliers is conducted by taking into account the defined value limits since different procurement strategies will result from these.

Illustration: Planning and control key figures in material procurement

Input	Process	Output
Number of suppliers according to value limits (ABC/XYZ Classification)	Quantity • Orders • Suppliers • Items per employee in purchasing	Proportion of deliveries on time
Average purchasing volume per active supplier	Proportion of items with available general agreements and/or proportion of items without general agreements	Proportion of deliveries in correct quantity
Total change number of suppliers compared to previous year and impact on expenses due to change of suppliers	Average costs of an order (in defined value and quantity limit)	Cash Out rate (proportion of purchasing volume compared to total incoming financial volume)
Proportion of items with more than one supplying source	Average processing time of an order (in defined value and quantity limit)	Delivery readiness (number of on time delivered need requisitions/total number of requisitions)
Proportion of items according to value limits (ABC/XYZ Classification)		

Illustration: ABC/XYZ Classification

	A	B	C
X	High value share Constant need High accuracy of predictability	Medium value share Constant need High accuracy of predictability	Low value share Constant need High accuracy of predictability
Y	High value share Varying need Medium accuracy of predictability	Medium value share Varying need Medium accuracy of predictability	Low value share Varying need Medium accuracy of predictability
Z	High value share Irregular need Low accuracy of predictability	Medium value share Irregular need Low accuracy of predictability	Low value share Irregular need Low accuracy of predictability

By combining the value of an item and the accuracy of its predictability and/or planning certainty, relevant procurement strategies are reached. For example, primary material, classified as AX, is addressed by taking into account the following:

- Supplier policy:
  - Single and/or Dual Sourcing
  - Motivation of suppliers

- Contract design:
  - Lang-term partnership agreements
  - Close and reliable partnership with the supplier which is controlled with appropriate key figures

- Procurement channels:
  - Procurement directly with manufacturer
  - Consignment warehouse

- Supply policy:
  - Bundling of quantities
  - Low level of safety stocks

Single Sourcing

Single Sourcing for procurement of required materials i.e. provision by a single source, requires high delivery reliability, quality and flexibility. There are also further prerequisites for Single Sourcing:

- Early integration of supplier especially during product development (take into account relevant requirements in Measure and the high-level concept development)
- Supplier's contractual commitment for the entire product life cycle
- Detailed evaluation of supplier's performance (audit)

The following advantages and disadvantages should be taken into consideration:

Advantages	Disadvantages
• Close cooperation • Lower prices through higher order volume • Lower costs for orders and transactions • Less suppliers and contacts • Lower logistic complexity and better control	• Constraints in case supplier is not available • Short-term change difficult and expensive • Dependence on supplier (quality, prices) • Exposure of company know-how • Low level of flexibility

Dual Sourcing

Dual Sourcing i.e. provision from two sources of competing suppliers, is a security strategy which, on the one hand, is supposed to guarantee the company's security of supply and on the other hand, supports competition between the suppliers.

Strategic raw materials are especially appropriate for this strategy e.g. aluminum, parts with long delivery times and other constraint items, all of which, if in short supply, could cause significant financial loss to the company.

Advantages	Disadvantages
• Lower prices through higher order volume (70/30 distribution) • Improved security of supply • Lower costs for orders and transactions • Less suppliers/contacts • Lower logistic complexity	• Exposure of company know-how • Partly dependent on suppliers (quality, prices) • Medium flexibility

Purchase Pull Systems

Purchase Pull Systems are based on need-based dimensioning of buffer stocks for primary material in order to avoid delivery constraints especially for AX materials. The inventory level is regulated by triggering an order with the supplier when a specific level, calculated on the basis of consumption data (reorder level), has been breached. When reorder levels are calculated, the supplier's reliability must be taken into consideration.

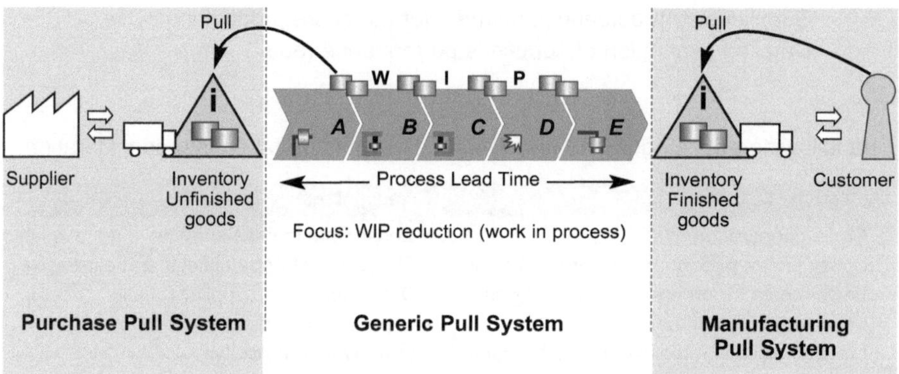

The following procedure should be considered when a Purchase Pull System is implemented:

1. Test prerequisites for Purchase Pull:
   - The Replenishment Time (delivery time) is higher than the time allowed by the customer; otherwise consider contract manufacturing a possibility
   - The demand (of production) varies only to a limited extent and displays no strong seasonal relationship; specific value: $s/\overline{x} < 1$
   - The variation of the Replenishment Time (delivery time) is low

2. Test the Pull System capability:
   - Pull capability is possible if the coefficient of variation ($s/\overline{x}$) from the standard deviation of the demanded quantity per time unit (s) and the mean value of the production amount ($\overline{x}$) is smaller than 1

Provide IT

Term
Supply and Demand for Information Technology (IT), Provide IT

When
DESIGN: develop and evaluate lean process

Goal
- Avoid conflicts between incompatible IT methods and procedures
- Use synergy effects between DFSS and existing IT methods
- Increase acceptance among important Stakeholders by integrating existing procedures

Procedure
Within the scope of the process optimization and/or redesign existing project and process management methods should be taken into consideration and harmonized within IT.
The development of hardware and software, systems and processes can be divided into four categories:

1. Project management standards such as: PMI (Project Management Institute), PRINCE2 (Projects in Controlled Environments) which is often used in IT

2. Process Frameworks based on best practice approaches, such as: ITIL Framework (Information Technology Infrastructure Library), MOF (Microsoft Operations Framework), COBIT (The Control Objectives for Information and related Technology), ISO (International Standard Organization) or TQM (Total Quality Management)

3. Process development and evaluation approaches such as: CMMI (Capability Maturity Model Integration) for services or acquisition and/or EDEN (German acronym for “Successful, Continuous, Efficient, Sustainable”)

4. Methods for developing software, hardware and systems. A distinction can be made between two groups:
   - Agile Software development methods, such as SCRUM and,
   - Waterfall models, e.g. the V-model

For the harmonization of the approaches within the DFSS procedure the objective and/or alignment of the necessary process optimization and/or new development must be defined. Depending on the objective and the project type, different methods, approaches and procedures are reasonable. A distinction between four essential objectives can be made:

1. IT as an environment and frame for new processes:
   Design one or several new processes in IT, e.g. the development of a process for processing IT service tickets

2. IT as the basis for integrating new hardware, software and systems:
   develop new systems, software or hardware in IT, e.g. the development of new services or software programs and/or IT hardware components

3. IT as support for new products and services:
   Develop new marketable products and services outside IT, e.g. the development of a new product or a new service

4. IT as support for new processes:
   Develop new processes outside IT, e.g. the development of a process for regularly collecting customer satisfaction data

See following page for example

Categories	New development of IT processes	Development of IT systems, HW, SW	Development of processes (not IT)	Development of products and services (not IT)
Approaches for project management, such as e.g. PMI or PRINCE 2	X	X	X	X
Process Frameworks, such as e.g. ITIL or MOF	X		X	
Approaches for process development, such as e.g. CMMI	X	X	X	X
IT specific development methods, such as e.g. SCRUM		X		X

Illustration: Alignment and categories of IT depending on the objective

The most frequent methods and approaches in IT are presented and classified in the following overview:

Methods / Criteria	PRINCE 2	ITIL Framework	CMMI	V-Model "XT"	SCRUM
Category	Project management methodology	Process Framework	Approach for process/ product development	Development approach for SW, HW & systems	Development approach for SW, HW & systems
Objective	Best practice model for structured execution of projects, especially in IT	Guideline of best practice processes for IT departments for the optimization of IT processes	Guideline of a reference model for the optimization of the organization and work efficiency in IT	Procedural model for planning and conducting projects for system development	Approach for the quick definition and achievement of partial results of SW, HW and system development
Strengths	Focus on project benefit, link with management and project management, structured "hands-on" approach, project subdivision into four phases, methodology which can be acquired quickly	Tried and tested and widely spread best practice model for IT processes; ITIL V3 integrates the continuous process improvement and Six Sigma	Best Practice approach also appropriate for measuring the process maturity, generally applicable to all organizations, focus on lived process	Collection of customer requirements at the beginning, structured phase model by linking the phases, XT version possible as for project specific needs, project management and quality assurance taken into consideration	Quick partial results, team oriented procedure, changes possible any time, low bureaucratic effort, consideration of human aspects
Weaknesses	Document orientation poses danger of excess bureaucracy, no analysis of project requirements in advance, no consideration of the human factor	Very comprehensive approach, can be realized only partly in small IT departments if required; transformation effort with respect to company specific requirements	Clarifies "What is to be done", but not "How can this be done"	Product results only visible at the end of the project, long process running times, no phase approvals, no strict chronological order, danger that individual results do not harmonize	Danger that partial solutions do not harmonize, overload of team with respect to self-organization, homogeneous teams

Illustration: Selected approaches and methods in IT

Methods / Criteria	PRINCE 2	ITIL Framework	CMMI	V-Model "XT"	SCRUM
Impact on DFSS projects	• Competing project management approaches • Documents of the same kind in both methods • DFSS projects must be integrated into the PRINCE 2 environment	• Clear guide-lines which IT processes are neces-sary • ITIL processes must be taken into consider-ation in the high-level concept and in the detailed concept of process development (ANALYZE, DESIGN)	• Require-ments of the pro-cess need to be com-pared to require-ments of the CMMI approach • Maturity level approach can be used generally as basis for DFSS	• Decision necessary which phase model is to be the dominating one • Supplemen-tation of the respective, leading model with tools of the other models • V-Model in DFSS: use of test phase to evaluate con-cept • DFSS in V-Model: Consideration of Gate Reviews in the phases of the V Model	• Acceleration of DFSS especially in ANALYZE and DESIGN • Generate quick partial results

Illustration: Impact on DFSS projects

➾ **Tip**

In order to ensure high acceptance for the approach, the DFSS project should be actively communicated and marketed as supporting rather than competing with existing approaches.

In your line of argument you should always focus on the fact that the DFSS procedure is based on answering questions rather than executing tools.

Pilot Processes

Term
Pilot, test the new process in a limited environment

When
DESIGN: develop and evaluate lean process

Goal
- Test the performance capability of the developed process and improve it should it deviate from expectation
- Create basis for the Roll-out

Procedure
Piloting the new process takes place in four steps, in the sequence Plan-Do-Check-Act (PDCA):

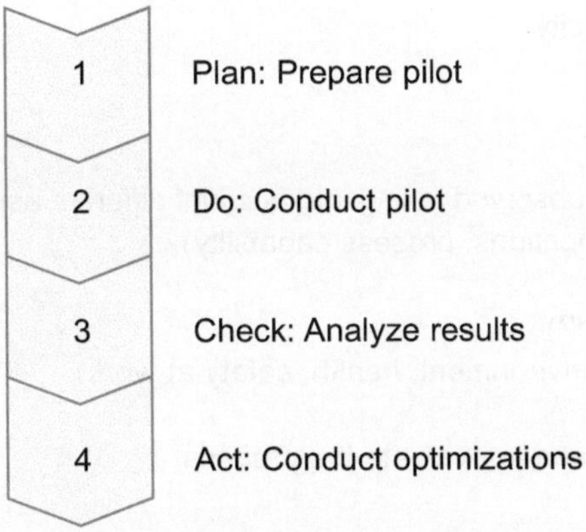

1. **Plan: Prepare pilot**
   Necessary activities for preparing the pilot include:
   - Determine pilot team
   - Define scope of pilot
   - Draw up Implementation Plan
   - Estimate risks

- Provide facilities and buildings
- Provide and set up machines and equipment
- Set up IT infrastructure
- Provide a sufficient amount of material by selected suppliers
- Train employees "on the job"

Potential failures in the new process can be identified and analyzed in advance with the help of a process FMEA (see Analyze). Strategies for the early detection and avoidance of any identified failure modes can be developed.

The FMEA can therefore, serve as the basis for a Reaction Plan.

2. Do: Conduct pilot

In order to enable early rectification of problems, conducting the pilot should take place step by step:

- Prepare test runs
- Conduct test runs
- Optimize machine settings (DOE)
- Increase quantities step by step
- Run production at full capacity

3. Check: Analyze results

Each relevant process step is observed taking into account different aspects:

- Quality (fulfillment of specifications, process capability)
- Capacity (exit rate, speed)
- Costs (target costs reached?)
- Regulatory requirements (environment, health, safety at work)

Different key figures can be used to evaluate the process.

Illustration: Key figures for validating a process

Validate process steps		
Quality	Specifications	LSL, USL
	Process capability	C_p, C_{pk}, LCL, UCL
Capacity	Exit rate	Quantity/time
	Speed	Process Lead Time
Costs	Production costs	Material, personnel costs, etc.
	Circulating assets	Inventories in dollar (work in process and inventory of finished products)
	Capital assets	Depreciation for machines, buildings, etc.
Regulatory requirements	Environment	Emissions
	Health	Ergonomics and stress
	Safety at work	Risk of injury

4. **Act: Conduct optimizations**
   If the results do not match the planned goals, the weaknesses should be eliminated. After eliminating the weaknesses the PDCA cycle starts again from the beginning – until the desired results have been achieved. Afterwards, the approval for Roll-out can take place.

Example: Pilot program and PDCA – Passenger Seat

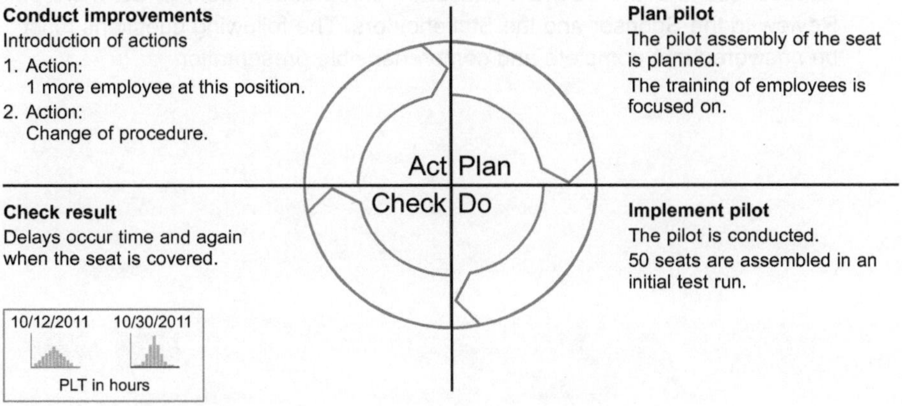

Gate Review

Term
Gate Review, Phase Check, Phase Approval

When
Closure DESIGN: at the closure of each phase

Goal
- Inform the Sponsor about the results and actions of the respective phase
- Evaluate the results
- Decide about the further course of the project

Procedure
The results are presented in a complete and comprehensible way. The Sponsor checks the current status of the project according to the following criteria:
- Completeness of results
- Probability of project success
- Optimal resource allocation in the project

The Sponsor decides if the project can enter the next phase.

Various results of the DESIGN phase are presented in the final DESIGN Gate Review to the Sponsor and the Stakeholders. The following questions must be answered in a complete and comprehensible presentation:

Develop detailed concept

- Have transfer functions been developed for evaluating the design parameters and what are these transfer functions?
- Have alternative characteristics of design elements been determined and compared?
- What statistically significant differences have been identified?
- Has Tolerance Design been applied?
- What were the results?
- Has Design of Experiments been applied?
- What were the results?

Evaluate detailed concept

- Has a simplification of the system been conducted with the help of Trimming?
- Has Design for X been applied?
  What were the results?
- Has a Design Scorecard been developed for the detailed concept?
- Has a prototype been implemented?
  What insights were derived from this?
- How were the test results integrated in the final design?
- What risks have been identified and what actions were derived to avoid these risks?
- Has the market launch been prepared?
  In what way?

Develop and evaluate Lean process

- Have the relevant process and input variables been identified?
- Have the relevant process and input variables been tested?
  What were the results?
- Has the current process capability of the process been evaluated?
  How was it evaluated?
- Has the process design been drawn up?
- Have all regulatory requirements been taken into consideration?
- Has all process related date been mapped in a VSM?
- Has the Process Efficiency been tested and optimized?
  With which tools?
- Has a production layout been developed?
- How and was it tested?

- What were the results?
- Will the process be monitored with the help of a reasonable key figure system?
- Which tests could be conducted to ensure that the Reaction Plan functions in the event of a deviation?
- Has a process simulation been carried out?
  What were the results?
  How has the material procurement been planned?
  How has consistent quality of deliveries been ensured?
- How has the availability of employees been ensured?
  How about their capability and motivation?
- Which IT resources are supposed to support the lean process?
- Has a comparison with existing IT approaches been conducted?

Cost calculation and marketing

- How have the detailed product costs been broken down?
- What manufacturing costs (MC 1) have been calculated?
- What quality costs are expected (rework, scrap etc.)?
- Are the target costs met?
- What does the detailed marketing and sales plan look like?
  What costs have been calculated for this?
- What contribution margins (CM 1 to CM 4) have been calculated?

DESIGN FOR SIX SIGMA+LEAN TOOLSET

VERIFY

Central Question and Objective of the VERIFY Phase

- The VERIFY phase of the DMADV procedure answers the question: How can the detailed design be implemented successfully and how can responsibility for the new process be handed over?
- Objective:
  - Plan and oversee full scale production
  - Finalize market launch
  - Develop appropriate process control
  - Complete process and project documentation
  - Hand over process responsibility to Operations
  - Conduct official project closure

The Meaning of the VERIFY Phase

- Design and oversee the realization of the process design in order to guarantee a smooth commencement of full scale manufacturing
- Finalize market launch and support with appropriate marketing actions
- Inform everyone involved about the new flows, responsibilities and interfaces
- The project and the responsibility for the newly developed process are officially handed over to the Process Owner

Procedure in the VERIFY Phase

- Derive implementation strategies
- Draw up a detailed Implementation Plan
- Finalize market launch
- Draw up process documentation and standard operating procedures (SOPs)
- Define the implementation team
- Visualize and control process performance
- Finalize project documentation
- Hand over process responsibility
- Conduct project closure

D M A D V

V.1 Prepare implementation

- How is the change to full scale manufacturing ensured without significant delays and downtime?
- How can targeted marketing actions accompany the market launch?

V.2 Implement process

- How are new tasks and working equipment documented and how will this information be made available?
- Is the implementation supported by appropriate implementation teams?
- How will process performance be controlled?

V.3 Hand over process

- What does the final process and project documentation look like?
- How will the process responsibility be handed over to Operations?
- How will the official project closure be conducted?

Result VERIFY

- Start of series manufacturing realized
- Market launch closed and market accompaniment prepared
- Process and project documentation completed and process responsibility handed over to Operations
- Project closure conducted

VERIFY Gate Review/Phase closure

V.1 Prepare implementation

- Implementation Strategy
- Transition Plan
- Scale up
- Implementation Plan
- Pilot Marketing
- Prepare Mass Marketing
- Prepare Market Follow-up

V.2 Implement process

- Process Documentation
- Standard Operating Procedures (SOPs)
- Implementation Teams
- Process Performance

V.3 Hand over process

- Project Documentation
- Lessons Learned
- ERIC Square

V Gate Review

Prepare Implementation

Term
Implementation, Implementation Planning

When
VERIFY: prepare implementation

Goal
- Derive an implementation strategy for the new process design
- Draw up a detailed Implementation Plan

Procedure
Upon finalizing and piloting the detailed concept, the implementation of the new and/or modified manufacturing process is designed in order to guarantee a smooth start to ongoing manufacturing.

In a step-by-step approach a manufactured amount is selected under nearly realistic manufacturing conditions in order to be able to conduct optimizations and/or adjustments in the process. The requested reliability and stability of series manufacturing can be ensured in this way. Moreover, the product's market launch is accompanied by appropriate marketing activities and further activities are prepared during and/or at the end of the market phase.

1. Derive implementation strategies

2. Draw up Implementation Plan in detail

3. Finalize market launch

Derive Implementation Strategies

Term
Implementation Strategy

When
VERIFY: prepare implementation

Goal
Plan the implementation of the process design in such a way that the change-over of live manufacturing can take place without delays and downtime

Procedure
The implementation of the defined process design can take place in various ways. There can be phases, sequences, and fully integrated. The chosen implementation strategy should be supported continuously by a transition plan and/or a detailed scale-up.

Implementation Strategies

In phases	The process design is implemented and realized according to defined implementation phases at one site
In sequences	The defined process design is initially implemented and realized at one site before it is implemented at another site
Integrated	Integrated implementation of a defined process design at all sites at the same time

It makes sense to combine these implementation strategies. The following questions are helpful:

- How much time is available for implementing the process?
- Which projects and/or initiatives are affected by the implementation? What is the scope?
- Which projects and initiatives can be integrated as support?

- What resources are required for the respective implementation strategy?
- What is the implementation's impact on operational business?

In order to keep the impact on the process changeover as low as possible on other serial processes, the changeover should first take place at a single pilot site. Ideally, the old and the new process should be operated in parallel to each other until sufficient quantities can be manufactured with the new process under normal manufacturing conditions.

Illustration: Transition Plan

Move	Parallel operations	Step-by-step transition
Shut down and move existing process operation to an alternative site until changeover to the new process has been completed	Parallel operations of old and new process until new one is stable	Utilization of production times with low machine capacity utilization in order to conduct changeover to the new process step-by-step

For all mapped variants of a transition plan a more detailed Scale-up should be conducted. A Scale-up Plan avoids errors for more complex changes by introducing them step by step and adjusting them in successive optimization loops. The structure of a Multi Generation Plan supports the creation of a Scale-up Plan.

Mapping Scale-up steps

Step 1	Step 2	Step 3
Carry out the first step! e.g.: Test production with two machines	Improve the achieved position! e.g.: Full utilization of available capacity	Quantum leap with effective success e.g.: Setup of further production lines

The implementation of the process design should be enhanced by further relevant aspects which are documented in an Implementation Plan:

- Set scope for implementation
- Identify all areas affected by the new and/or modified process design
- Define important Stakeholders, contact partners and individuals responsible for the implementation

Draw up Implementation Plan in Detail

Term
Implementation Planning

When
VERIFY: prepare implementation

Goal
- Prepare the implementation of the process design in such a way that full manufacturing starts without delays and downtime
- Provide infrastructure and instruct employees
- Ensure ongoing process control

Procedure
Previous project work has primarily been focused on the development and verification of the product and the respective manufacturing process and/or necessary, supporting processes, but now the focus is on ensuring a defect-free transition to operations.

The development project now becomes an implementation project. The following elements of an Implementation Plan should be understood at this point as a final checklist, i.e. they may already be included in detailed process design contained within the DESIGN phase.

1. Test the Project Charter as to whether it is up to date and feasible:

- **Business Case, problem and target definition:**
  The starting situation of the development project should be tested for communicability and should be revised if required. Check goals with respect to result, time and resources with the project champion (client) by taking into account current company specific or market related conditions and decisions as for feasibility and conduct adjustments if required.

See next page for Project Charter example

Example: Project Charter

Project title: Development of a passenger seat for Russia	Date: 01/18/2012	Version: D 1
	Project number:	
	Project category/type:	
	Life cycle stage:	
Business Case:	Project scope:	
Market studies have shown that transport companies in Russia will modernize their bus fleets by up to 80% over the next ten years. Our current model has been on the market for 5 years and fulfills the increased requirements with respect to quality and comfort only insufficiently. The sales figures decrease due to increasing competition. In order to meet the market changes an innovative seat system for the use in buses must be developed within the frame of the project.	In:	Ergonomic seat design, variants for installation and removal, cushioning, easy maintenance, seat material
	Out:	Electronic features on seat, change of bus layout, measurements of seat system
	MGP:	(1) Generation I: 30% market share in Russia by 2014, modular standard fitting (2) (3)
Problem/Goal:	Roles:	
During the last 6 months the proportion of customer complaints with respect to processing and maintenance friendliness increased by 25%. The sales figures decreased by 20% in the same time period. In order to counteract this negative trend and to participate in the projected increase in demand a marketable seat system is to be developed by 10/01/2011. With this system we intend to achieve a market share of 30% on the Russian market by 2014. The bus manufacturer and/or suppliers in Western Europe are to be involved actively in the development process.	Champion	Dr. Jacomo Franco
	Project Leader	Bernhard Fuchsberger
	Core team	Ms. M. (Marketing) Dr. Q. (Qality Management) Dr. F. (Product Development) Dr. Z. (Process Development) Mr. E. (Procurement) Ms. P. (Production) Mr. V. (Sales) Mr. C. (Controlling)
Monetary benefit:	Milestones:	
Output 2012: 20,000, output 2013: 30,000, output 2014: 50.000, profit per seat = 10% of the net sale price, net sale price = $ 100, tax rate = 25%, capital cost rate = 10% = >EVA 2012 = $ 150,000, EVA 2013 = $ 225,000, EVA 2014 = $ 375,000, discounted EVA for 2012 = $ 664,500	DEFINE:	01/18/2012
	MEASURE:	02/10/2012
	ANALYZE:	04/05/2012
	DESIGN:	05/25/2012
	VERIFY:	07/26/2012
	Project handover:	08/28/2012
Potential risks:		
Increased political intervention in the Russian domestic market, potential liquidity problems among the trading partners, logistics constraints, silo thinking	Signature Champion	Signature Project Leader

- **Test Multi Generation Plan:**
  Clarify if the first generation of the originally defined Multi Generation Plan has been fulfilled or if rework is necessary due to the pending project

closure; the start of the next generation should be prepared and initiated (e.g. consideration of further customer and market segments, extension of production volume, new features in product design, etc.).

Illustration: Multi Generation Plan

Generation I	Generation II	Generation III
Carry out the first step! Generation I aims at cutting off urgent problems and at filling gaps.	Improve the achieved position! Generation II extends the secured product base and devotes itself offensively to developing new target markets.	Take over the leadership position! Generation III strives for realizing the final vision, e.g. "Be technological market leader".
Stop the bleeding!	Take the offensive!	Attain Leadership!
Achieve market entry	Extend market share	Achieve market leadership

– **Determine the implementation team:**
Define role descriptions with subject-specific and temporal requirements on team selection and guarantee availability of resources. Define project management and steering committee with respective competence and rules regarding communication and escalation. Plan support and monitoring of the introduction and subsequent growth phase.

– **Conduct risk assessment** for the implementation and market launch with client and team.

Potential risk categories:

Result-oriented:
- Degree of target achievement with respect to customer requirements
- Insufficient production features/functions
- Performance capability of the process

Temporal:
- Postponements of dates

Resources:
- Budget restrictions

- Limited availability of machines, supplies, tools, personnel, etc.

Employees:

- Lacking expertise on application/implementation of the product/the process

Tools for risk assessment:

Process FMEA, Anticipatory Failure Determination (AFD), Risk Management Matrix

– **Escalation management:**

Define rules about how to deal with short or mid-term deviations of the defined target state (product and/or process performance according to customer requirement):

- Define a clear escalation process with rules
- Define temporal "Frozen Zones" in which no further product can be manufactured and/or a process is operated; within this zone a targeted cause analysis can be driven forward
- Limitations of budget/effort

– **Change Management:**

Update and add Stakeholder Analysis and define a strategy and respective actions in order to guarantee the project success

Tools:

Stakeholder Analysis, Communication Plan, Elevator Speech, RACI Charts

Illustration: Stakeholder Analysis

	Attitude towards project					
Stakeholder	**--**	**-**		**+**	**++**	**Maßnahmen**
Mr. A			O		+	
Mr. B	O		+			
Mrs. C		O		+		

-- strongly against, - moderately against, O neutral, + moderately in favor, ++ strongly in favor

2. Operational Project Management:

The implementation should be accompanied by other elements of operational project control.

- **Implementation Frame:**
  An Implementation Frame visually illustrates the relevant topics included (IN) and/or topics which have been explicitly excluded (OUT); topics which still need to be clarified can be placed on the frame. The Project Champion can be consulted and their position clarified. The implementation Frame therefore clearly limits the implementation topics once again and generates a common understanding among everyone involved in the implementation

Illustration: Implementation Frame

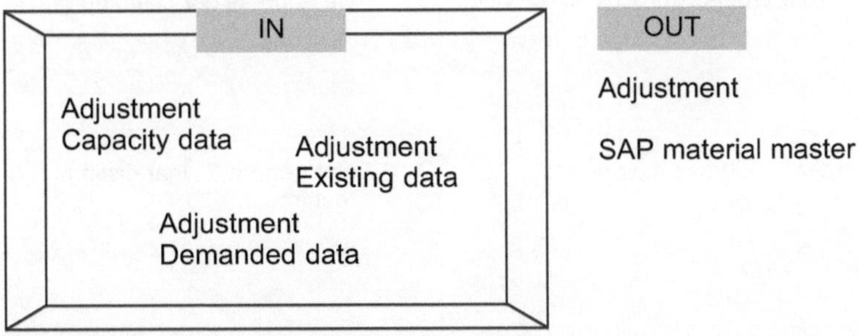

- **Work Breakdown Structure:**
  Work Breakdown Structure Draw up and/or refine Work Breakdown Structure in order to break down activities to individual persons/groups and to be able to delegate work packages

- **Scheduling:**
  Chronological organization of work packages with dependencies. Allocation of resources (necessary materials, deliveries, persons) by taking into account the availability; update of resource planning (when is what required); optimization of scheduling along the critical path

  Tools:
  Gantt Chart, Milestone Plan, Network Technique, critical path

– **Project communication:**
Market partial success in the implementation in order to ensure acceptance for the entire project; stress the respective benefits in communication:
- "We are creating something new ... "
- "The customer profits from ..."
- "We will profit from ... "

Show decisive Unique Selling Propositions of the developed product/process in contrast to existing products in the portfolio and/or vis-à-vis competing products on the market.

Illustration: Project Report

With new products/processes/services ...	we address our customer(s) ...
in order to address their needs ...	and generate clear distinguishing features ...

A compilation of decisive advantages for the customer and the company can take place also in the form of the following table:

Illustration: Advantages for the customer and the company at a glance

Target market, target customer(s):	Explanation for target market and target customers:	Relevant customer needs:	Need fulfillment:	Benefit contribution for companies:
• What is the target of the developed system?	• Why was the system developed and/or further developed for this circle?	• Which needs are characteristic of this group?	• How can the developed system contribute to fulfilling the needs?	• What is the benefit from the system development?

In addition to that single sided project reports are appropriate which are intended to deliver the decisive elements of product and/or process development; moreover, clients should be informed regularly about the project status (e.g. via an Elevator Speech and the level of finalization).

Mapping project report on project communication

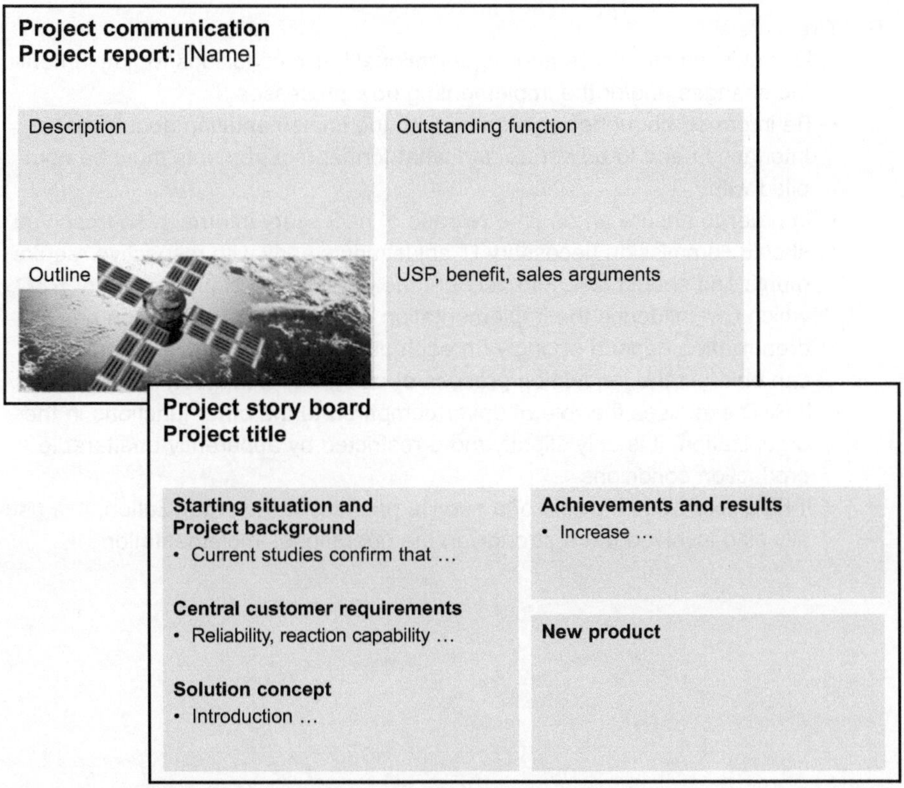

- **Update the schedule and the resource planning:**
  The schedule and the resource planning is continuously adjusted due to unplanned delays, emerging resource constraints and deviations from the budget (procurement costs, training costs, installation, etc.); procedures which need to take place at the same time and/or changes of the critical path are discussed taking this into account

- **Active risk management:**
  Continuous check if the implementation generates new insights about market or process requirements and if they can give rise to potential risks; does the benefit calculation of the development project have to be adjusted because for example the planned production costs will increase due to fluctuations of the raw material or labor market?

Tip

- Use existing structures and organizational basic conditions for implementing changes and/or the implementing new processes
- Be informed about before starting with the implementation about what interfaces need to be set up and what formal requirements must be complied with
- In order to ensure an on time release of necessary internal resources you should commission necessary IT adjustments early with respective requirements and should take into account the organizational integration of R&D which can influence the implementation of the project; innovation and organization depend strongly on each other; an unfavorable organization can obstruct the innovation process or can make it (almost) impossible
- If R&D exercises the role of driver compared to the other functions in the organization, it is only slightly more restricted by apparently unalterable production conditions
- If R&D exercises the role of a service provider in the organization, it is usually also involved more strongly in the operational implementation

Mapping "Organizational Integration R&D"

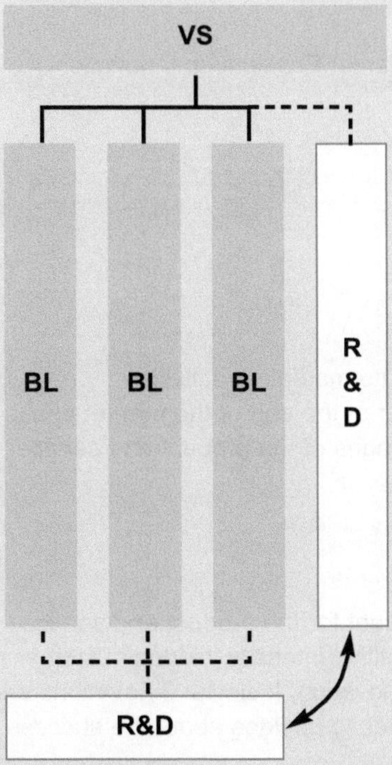

R&D – Driver or service provider of the Business Line?

Advantages driver role:

+ Stronger focus on strategic basic research/redesign
+ Lower limitations by apparently unalterable basic conditions of production

Advantages service provider role:

+ Closeness to operative implementation (Feasibility)
+ Financing by the Business Lines (BL responsible as client for the $ result)
+ R&D's role comprehension reduces the danger of "Hidden R&D" in the Business Lines

Finalize Market Launch

Term
Market Launch, Market Launch Strategy, Market Positioning, Innovation Marketing

When
VERIFY: prepare implementation

Goal
- Accompany market launch with appropriate marketing activities
- Prepare marketing activities during and/or at the end of the market phase
- Convince target customers of the advantages of the product and service innovation

Procedure
A targeted marketing concept is highly relevant for the product and service innovation especially in case of high competitive intensity, hidden customer benefit and high strategic significance of innovation. If similar innovations are launched nearly at the same time, the marketing decides about the success or failure of the innovation.

An early initiation of accompanying marketing actions ensures a target group specific alignment of communication and arouses the customer's interest before the market launch in order to achieve a clear advantage vis-à-vis innovation followers. Targeted actions of a pilot marketing should begin in the immediate environment of the market launch. Additional marketing actions after the product launch on the market should already be prepared at this point. Among these are marketing activities of mass marketing and market follow-up cultivation.

Mapping actions of innovation marketing in the development and market phase

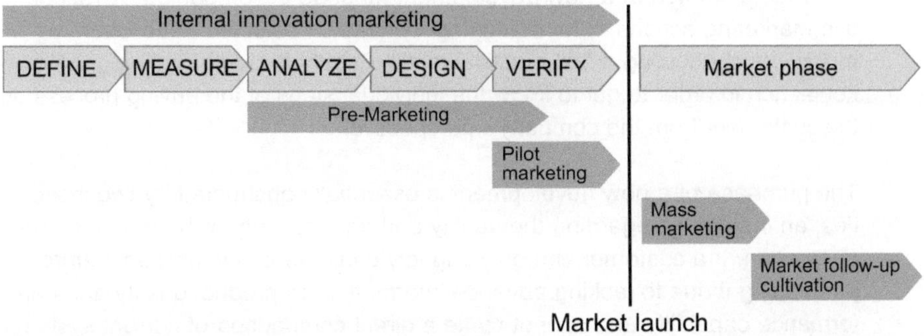

A reasonable combination of actions from pre-, pilot and mass marketing must take into consideration that as users of demanding systems the customers of today often act and communicate on three levels:

- Demand-oriented:
  The level of demand determines the day-to-day business and is determined by the direct price-performance comparisons between the providers; customers attach great importance on performance factors which can lead to a stronger focus on price competition and to continuous further development loops

- Problem oriented:
  Problem formulations and/or respective product complaints provide a good basis for the targeted further development of products and their performance capability; tackling the problem situation of the user can lead to the fact that a tangible price performance advantage is achieved in the long run, however, only little innovative potential can be increased; the customer focuses primarily on basic and performance factors

- Need oriented:
  Recognizing obvious and hidden needs of customers which cause the customer to demand a product, is elementary for successful benefit innovation and includes respective buzz factors
  Actions for better communication of innovation marketing must be aligned internally and externally to three levels and must differ clearly with respect to their design. Marketing actions must not confound customer requests (demand, problem, need) and send the wrong signals

Pilot marketing
The market launch of a new development must be accompanied by respective marketing actions. Pilot marketing focuses on selected Lead Users as multipliers conducting a first practical market test before and/or during the market launch in order to get to know the individual steps of the buying process of the customers from the company's perspective.

The purchase of a new development is essentially obstructed by two insecurities: an insecurity regarding the quality and an insecurity with respect to the generation. If a customer can only vaguely evaluate a new product before purchasing it due to lacking advance information on product quality and performance capability and cannot make a direct comparison of current systems which are being used, he usually perceives this as a qualitative insecurity. The customer is inclined not to buy the product directly but try and reduce his uncertainty first.

Moreover, the shorter and shorter development cycles and the more short-lived markets lead to a temporal insecurity on behalf of the customer. Products run through their life cycles partly only within a couple of years and the customer gets into a "waiting position" for the following generation. This can be observed above all in the telecommunication industry.

Mapping insecurity as for quality and generations in the buying process

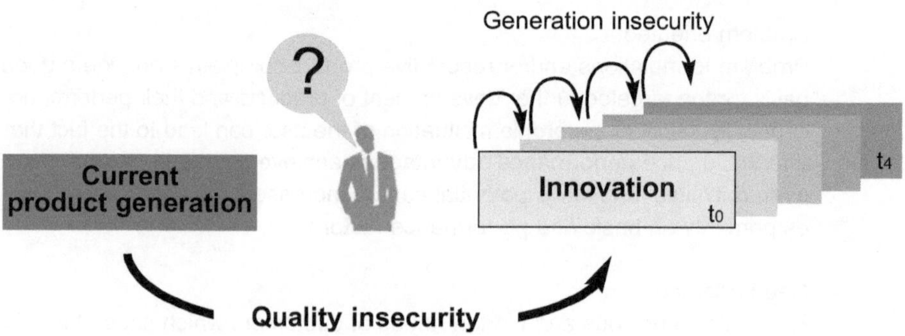

Four categories become apparent on this basis, which are to be taken into consideration when actions of pilot marketing are designed in the environment of market launch.

Mapping Customer's reaction in the environment of market launch

Customers' reactions in the environment of market launch		
Category	Characteristic	Result
• Buy current products IMMEDIATE BUYER	• Customer decides to buy a product from a provider who is already known to him • Alternatively, the customer makes a conscious decides in favor of a competitive product • The decisive aspects for the buying decision include the qualitative level, the price-quality ratio and the necessity to purchase the product	• Due to the urgency, new developments are not of prime interest • Innovation interest: LOW
• Wait for price reductions PRICE SENSITIVE	• Even though the customer is satisfied with the performance capability of the current product, he doesn't consider the price to be appropriate • The buying process is delayed since the customers regularly have to get current information on the price at regular intervals	• Search for information causes additional effort vis-à-vis the direct purchase • Innovation interest: LOW
• Skip product generations SKIPPER	• Customer is consciously waiting for future product generations • He is satisfied with the performance capability of the current product and is ready to spend more money for a better performance and to wait • The current product is rejected and skipped and/or not purchased • Particularly affected are information technology, electrical engineering, engineering as well as the car industry	• Cannibalization effects between the different product generations • If required, customer defection to competitors • Creation of incentives for the market entry of further providers • Innovation interest: VERY HIGH
• Decide against purchase REJECTOR	• Customer cancels the entire buying process consciously without intending to reconsider purchase at a later stage • Customer's explanation is that the product cannot solve his problem • Customer thinks that future product generations are not capable of doing this either	• Customer looks for product for best possible problem solution • Innovation interest: MEDIUM - HIGH

For pilot marketing the importance of customers must be stressed who skip a product generation and consciously resort to a "waiting position". They are future-oriented and usually open towards innovative products and performance features.

Mapping actions of pilot marketing depending on customer category

Targeted actions of pilot marketing	
Category	Actions
• Buy current products IMMEDIATE BUYER	• Provide hints at innovative performance features if the idea for the new product is launched immediately after purchasing the current product • Information is supposed to avoid disappointment about early purchase of a product which is perceived as outdated if required • Clear, comprehensible and comprehensive information of advantages of innovation
• Wait for price reductions PRICE SENSITIVE	• Communicate waiting time • Inform about the performance disadvantages of current products • Emphasize performance features of the new development and connect with clear benefit arguments
• Skip product generations SKIPPER	• Intensify actions of pre-marketing • Point out performance features of the new development
• Decide against purchase REJECTOR	• Show and/or offer alternative products and/or solution opportunities

Mass marketing
Mass marketing aims at the best possible marketing of an innovation which has been launched. In contrast to pre-marketing and/or pilot marketing the focus is not on preparing the market launch but on promoting the ongoing business. Actions for mass marketing should already be prepared at this point and coordinated with pilot marketing.

Thus the central task of mass marketing is the correct positioning and control of the product on the market. For this purpose the positioning of the competitors and the evaluation of the competitive situation during the market phase must be taken into consideration and evaluated continuously.

- Compare own launched product with products of competitors by target customers
- Evaluate specific product features and sort according to the mapped competitive advantage matrix

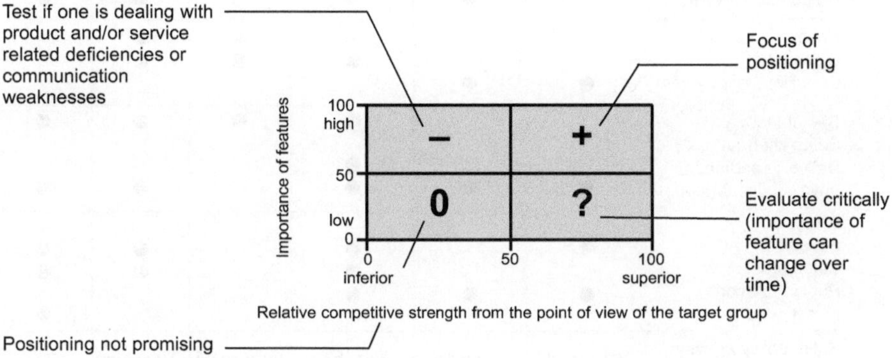

Reflect on previous major focus of the marketing strategy and adjust if required by taking this positioning as basis.

A promising marketing strategy usually contains the four aspects:

Product	Price
• Equipment/packaging • Labeling • Service/customer service	• List) prices • Discounts/conditions • Pricing policies over time
Place	**Promotion**
• Marketing area • Sales channels • Sales organs • Logistics	• Advertising, public relations • Sales promotion • Personal sale

Within these areas, depending on the economic sector, different Marketing instruments are used for mass marketing.

Mapping marketing strategy according to economic sector*

	Sales policy instruments	Investment goods		Consumer goods		Services	
		Raw material producing enterprises	Production companies of finished products	Brand manufacturers	Manufacturer of trademarks	Trade	Other
Product	• Product quality	●	●	●			●
	• List of services		●	●		●	●
	• Warranties		●	●			
	• Customer Service		●	●		●	●
Price	• Price			●	●	●	
	• Discounts			●	●	●	
	• Conditions of payment	●	●				
Place	• Site of last sale			●	●	●	●
	• Sales channel			●			
	• Delivery readiness, physical distribution	●	●	●		●	●
Promotion	• "Classic advertisement"	●		●		●	●
	• Sales promotion			●		●	●
	• Public Relations	●	●	●			
	• Direct advertising						●
	• Sales policy activity level	Very small	Small	Very big	Very small	Very big	Big

Innovation pricing must be pointed out, since a wrong pricing strategy from the beginning can have fatal consequences for the marketing phase of the innovation. A distinction can be made between two ideal-typical strategies of innovation pricing:

- Skimming strategy:
  Innovation is launched with a relatively high price which is reduced in the further course of time if required; the skimming strategy is recommended for innovations with strong strategic competitive advantages and low price elasticity

- Penetration strategy:
  The innovation is placed on the market with a low introductory price in order to achieve high sales figures and/or market shares within a short period of time; and enforcement of standards can be introduced and promoted

* Cf. Meffert, Herbert (1998): Marketing – Grundlagen marktorientierter Unternehmensführung, 8th edition, Munich: Gabler Verlag, p. 891

Market follow-up cultivation

The market follow-up cultivations aims at a comprehensive analysis of marketing actions during and at the end of the market phase. Based on this analysis market reactions can be recognized and used for follow-up generations. Also actions for the market follow-up cultivation should already be prepared at this point and coordinated with pilot and mass marketing.

- During the Marketing phase:
  - The goal of marketing this reflection is to check key figures to get hints at specific aspects of the provider-customer-relationship; for example, with complex specialty products with a correspondingly higher degree of innovation a review of the and performance required by the customer is needed in the real world situation; in contrast to this, for more standardized products with a lower degree ofinnovation a straightforward market review is appropriate
  - Actions for analyzing the direct customer interaction with the product, the company and the competitors include:
    - Customer satisfaction analyses
    - Acceptance Tests
    - Analysis of performance capability of product in application situation
    - Company's competence from the customers' point of view (service etc.)
    - Comparison of performance features of one's own product and the competitors
    - Analysis of competitor's communication and distribution channels
    - Analysis of customers' behavior when changing provider

- At the end of the product life cycle:
  - A comparison of marketing actions and marketing reactions during the marketing phase provides hints at the marketing of a new product generations as defined by a Multi Generation Plan
  - Advance marketing for a new product helps transition from the old to the new products
  - The advantages of the next generation product rather than those of the expiring product should be conveyed to the customers; the goal is to maintain the demand for the product category and to direct that towards the future product
  - Limited editions or a targeted launch of media information can be used e.g. as specific actions for preparing a follow-up generation or to arouse interest among the customer

– Replacement marketing pursues the "AIDA" principle:

- **A**: Attention
  Generate attention for the next generation
- **I:** Interest
  Arouse interest among the customers
- **D**: Desire
  Reinforce and incite the customer's desire to buy
- **A**: Action
  The customer decides to purchase the next generation

Implement Process

Term
Process Implementation

When
VERIFY: implement process

Goal
- Create intuitively clear description of the newly designed process for the individuals involved in the process
- Ensure quick start-up phase
- Ensure robust defect free process

Procedure
Upon completing the Implementation Plan and the final alignment of relevant marketing activities for market launch, everyone involved should be informed about the new procedures.

Changing responsibilities and interfaces are identified which must be aligned to new and/or existing process standards.

1. Draw up new process documentation and standard operating procedures (SOPs)

2. Define implementation team

3. Visualize and control process performance

Draw up Process Documentation and Standard Operating Procedures (SOPs)

Term
Standard Operating Procedures (SOPs), process documentation, work instructions

When
VERIFY: implement process

Goal
- Describe new and/or modified tasks and working resources in the changed process in a clear and simple way
- Clearly define process responsibilities and interfaces
- Enable autonomous decisions of the individuals involved in the process to avoid process loops
- Enable fulfillment of the set process standards
- Simplify initial training of new employees
- Ensure uniform documentation of the new and/or modified process

Procedure
Upon the successful delivery of the pilot all required changes are drawn up and/or integrated in existing process documentation and SOPs. The following steps should be taken into consideration for this:

- Prepare process documentation:
  – Go through newly developed and/or modified process and survey employees about the implemented innovations and/or changes
  – Determine the necessary level of detail for the documentation together with the people involved in the process; keep an eye on the suitability for everyday use

- Visualize process:
  – Visualize sequence of process steps, responsibilities and interfaces
  – Take inputs and outputs into consideration as well as their sources and customers (internal/external)

– Document and visualize working steps

- Define new and/or changed working steps in the standard operating procedures

- Set up workplaces according to the new standard operating procedure

- Train employees for the delivery of the new process

- Secure process documentation
  Take corporate standards into consideration:
  - QM guidelines
  - EDP/OSH standards
  - Formats for documentation, etc.

- Ensure existence of an update process
  Define contents, dates and responsibilities for continuous updates

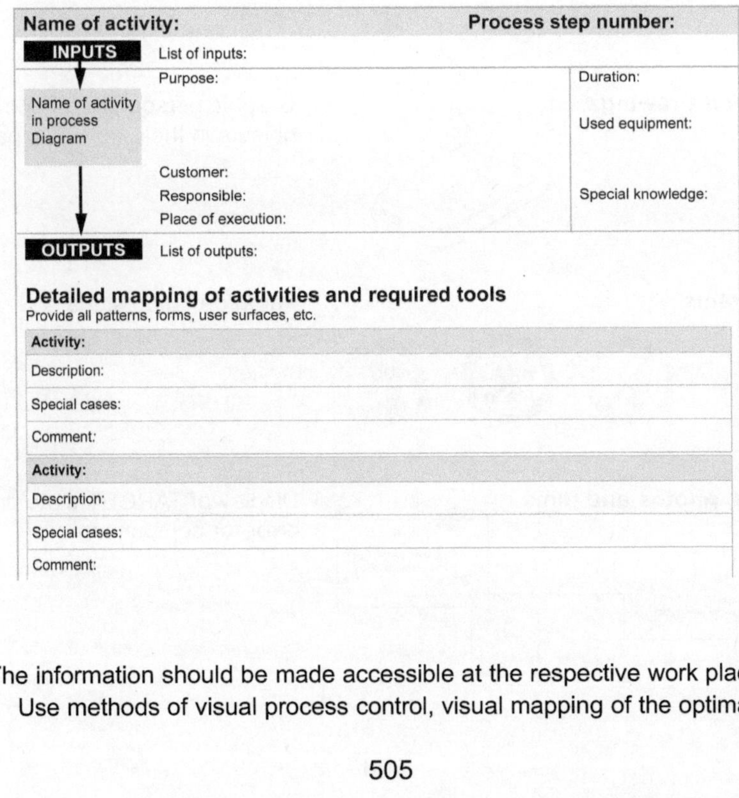

The information should be made accessible at the respective work places:

- Use methods of visual process control, visual mapping of the optimal workplace according to 5 S, visual description of the most important movements and of the correctly manufactured product

- Visual mapping of the most important production parameters such as machine settings, takt, WIP, process lead times, etc.

Depending on the set goal, different tools can be used:

Tool	Application
Graphic illustrations and Charts 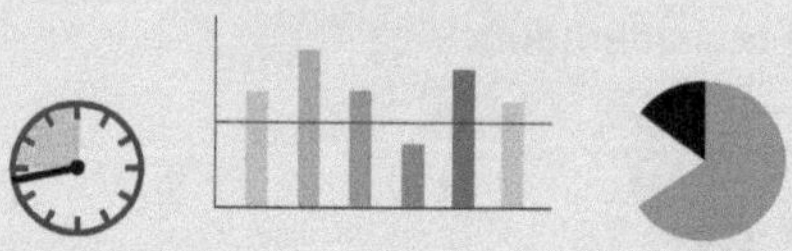	Visualization of figures and data, if required in conjunction with overviews of figures and tables
Flow Charts 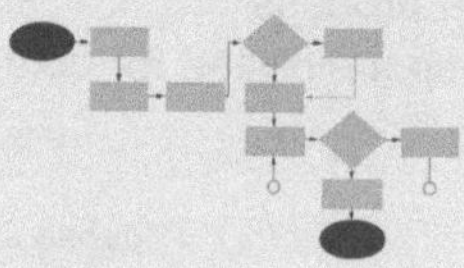	Visualization of processes, i.e. of process steps and their sequence
Exploded Drawings 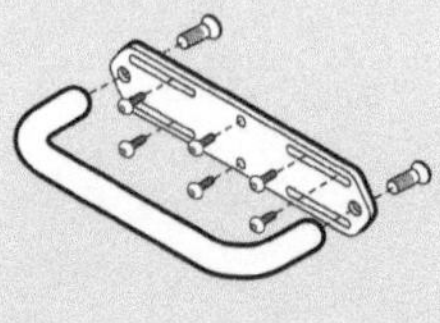	Graphic perspective of complex objects in their individual parts
Pictograms 	For information signs
Images, photos and films 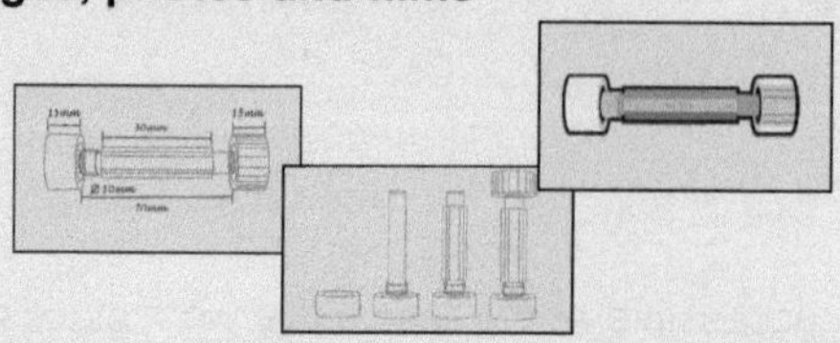	Display of TARGET state or error for comparison

Tool	Application
Reference parts 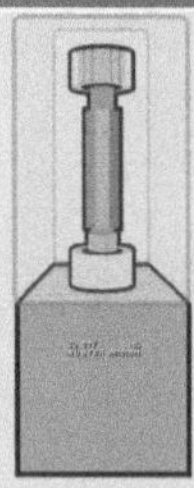	Three-dimensional visualization of TARGET state or error for comparison

Visual Management in production

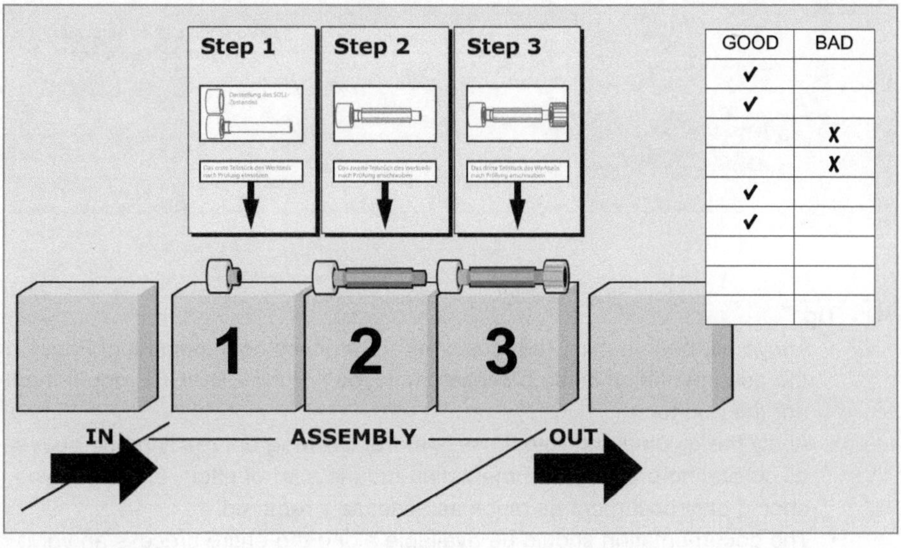

Visual Management in the service process on the following page

Visual management in the service process

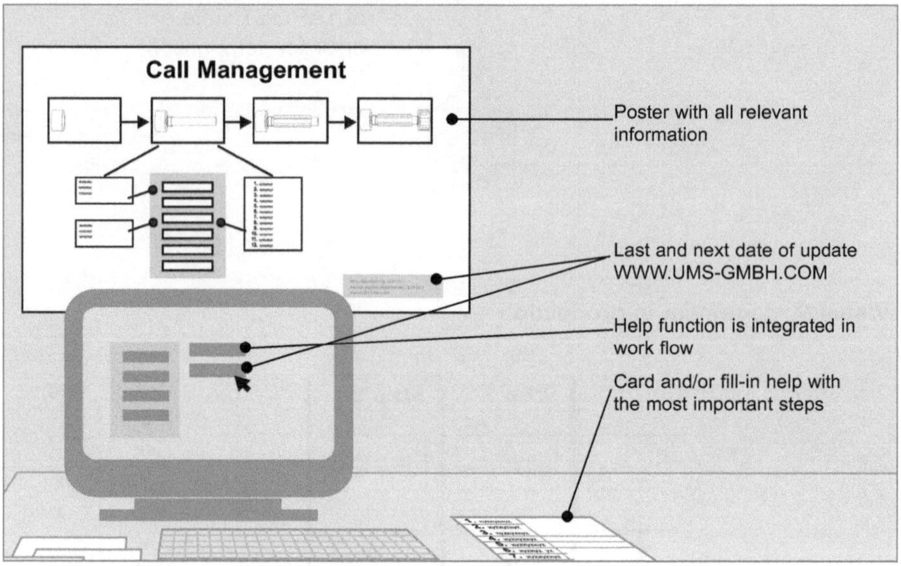

Tip

- Above all, bear in mind the "customer" of process documentation: Focus of the documentation is the process employee not the EDP/OSH department nor the auditor
- Apply the guiding principle: Less is more! Drawing up, reading and above all cultivating process documentation means a lot of effort; that's why you should only document as much as absolutely required
- The documentation should be available along the entire process and/or it should be possible to have a look at the documentation quickly and easily, best done by appropriate Visual Management
- Consider Visual Management as a tool with which the individuals involved in the process identify even more strongly with the new product and/or process; this enhances comprehension and motivation
- Also use Visual Management in administrative processes

Define Implementation Team

Term
Implementation Team

When
VERIFY: implement process

Goal
- Control the transition to full scale manufacturing actively
- Improve probability of implementation success by distributing improvement and process control tasks to a wide group of people

Procedure
1. Nominate the implementation team and define responsibilities with the help of RACI; the Process Owner can be supported by a small team consisting of e.g. Marketing and Sales, Purchase and Procurement, Product Management and Product and Process Development, in order to assist in the following phases:
   - Implementation phase up to full scale manufacturing
   - Product introduction phase and the subsequent growth phase

   The team should consist of members of the core project team.

2. Determine frequency of meetings and dates; these should be conducted at regular intervals

3. The implementation status should be discussed in the meetings and if required actions should be instigated; the following topics would usually be discussed:
   a. How was process performance yesterday, what problems were encountered?
   b. What improvement ideas are there?
   c. Are there any work backlogs and what are the goals for today?
   d. What tasks and priorities are there? How will these be allocated?
   e. Is there any customer feedback about the product or process that needs to be taken into consideration?

4. The team decides about the implementation of necessary actions and monitors results of these continuously

Tip

- Take the different responsibility areas for product management, production and marketing into consideration when putting the team together
- Conduct the monitoring and daily control of the process at the place of action, since sustainable process control lives by acceptance
- Regularly check the availability of supplies and tools required for the process
- Use standard charts to visualize process performance and interpret them jointly with the Process Owner, process managers and shift foremen etc.
- Identify deviations and define improvement actions to be taken; management should request and support them

Visualize and Control Process Performance

Term
Glass Wall Management*

When
VERIFY: implement process

Goal
Control processes by greater transparency

Procedure

1. Define work units (groups, teams, departments, etc as 'microenterprises' which can be controlled with the help of key figures; these units live a pronounced customer-supplier-customer relationship with respect to internal and external interfaces which will eventually lead to a waste-free value chain

2. The most important key figures are visualized continuously for everyone to see, irrespective of the results

3. The units use the figures in order to control and improve their processes continuously

4. Progress and success are communicated and visualized; competitive thinking is promoted this way

Example on the following page

* The concept of Glass Wall Management was developed and presented by Kiyoshi Suzaki in his book "Results from the Heart: How Mini-Company Management Captures Everyone's Talents and Helps Them Find Meaning and Purpose at Work" (Free Press, 2002).

Example:

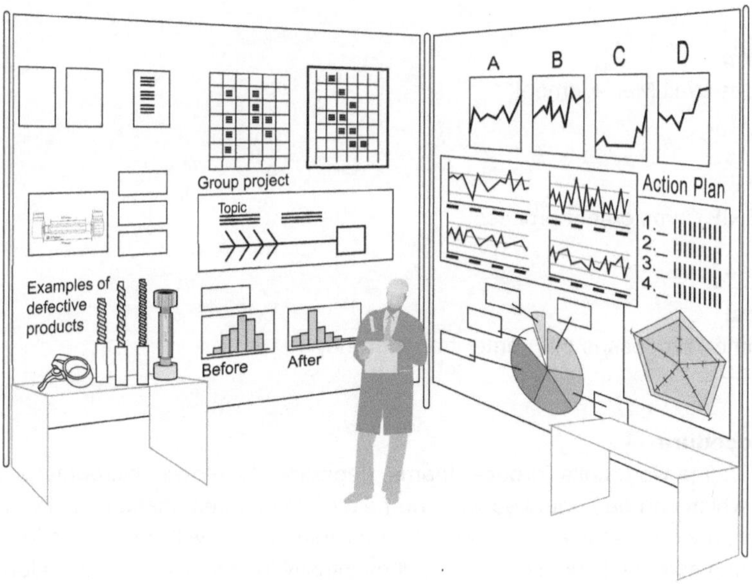

Tip

Check in advance if a suitable working environment and corresponding culture (openness and error culture) for implementing Glass Wall Management exists

Hand over Process

Term
Handover, process handover

When
VERIFY: hand over process

Goal
- Document project results in a comprehensible way
- Document experience, knowledge and insights gained from project work
- Hand over process responsibility to the Process Owner
- Close project work officially

Procedure
The project and the responsibility for the developed process are officially handed over to the Process Owner. Complete and comprehensible project documentation provides the basis for this, since it shows key connections between the requirements and the product and process features, and is the basis for further follow-up activities.

1. Finalize project documentation

2. Hand over process responsibility and conduct project closure officially

Illustration: Process Owner involvement

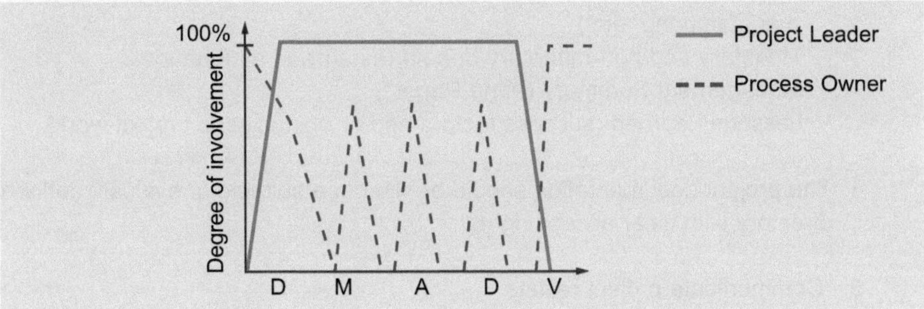

Finalize Project Documentation

Term
Project Documentation

When
VERIFY: hand over process

Goal
- Summarize the project's content in a generally intelligible way
- Ensure exchange of experience beyond the DFSS project team
- Communicate project results

Procedure
1. The documentation of the development project should be collated and updated continuously; responsibilities for project documentation should be clarified at the very beginning of the DFSS project in the team with the help of a RACI

2. A uniform format for documenting the tools and methods should have been defined (e.g. Workbook in MS Excel, MS Power Point, etc.)

3. The project documentation is supposed to show the procedure in the project in a reproducible way; thus, the team should take the following minimum requirements for project documentation into account:
   - Project Charter
   - Roadmap and decisive results of each phase
   - Project result (displaying prototype, results of field tests, results of live manufacturing, etc.)
   - Monetary and non-monetary benefit (as agreed by Finance)
   - Management Summary ("One Pager")
   - Lessons Learned (success factors and/or obstacles to project work)

4. The project documentation should be filed in a suitable, previously defined, directory with clear access rights

5. Communicate project results

Illustration: Management Summary (One Pager) and Workbook for project documentation

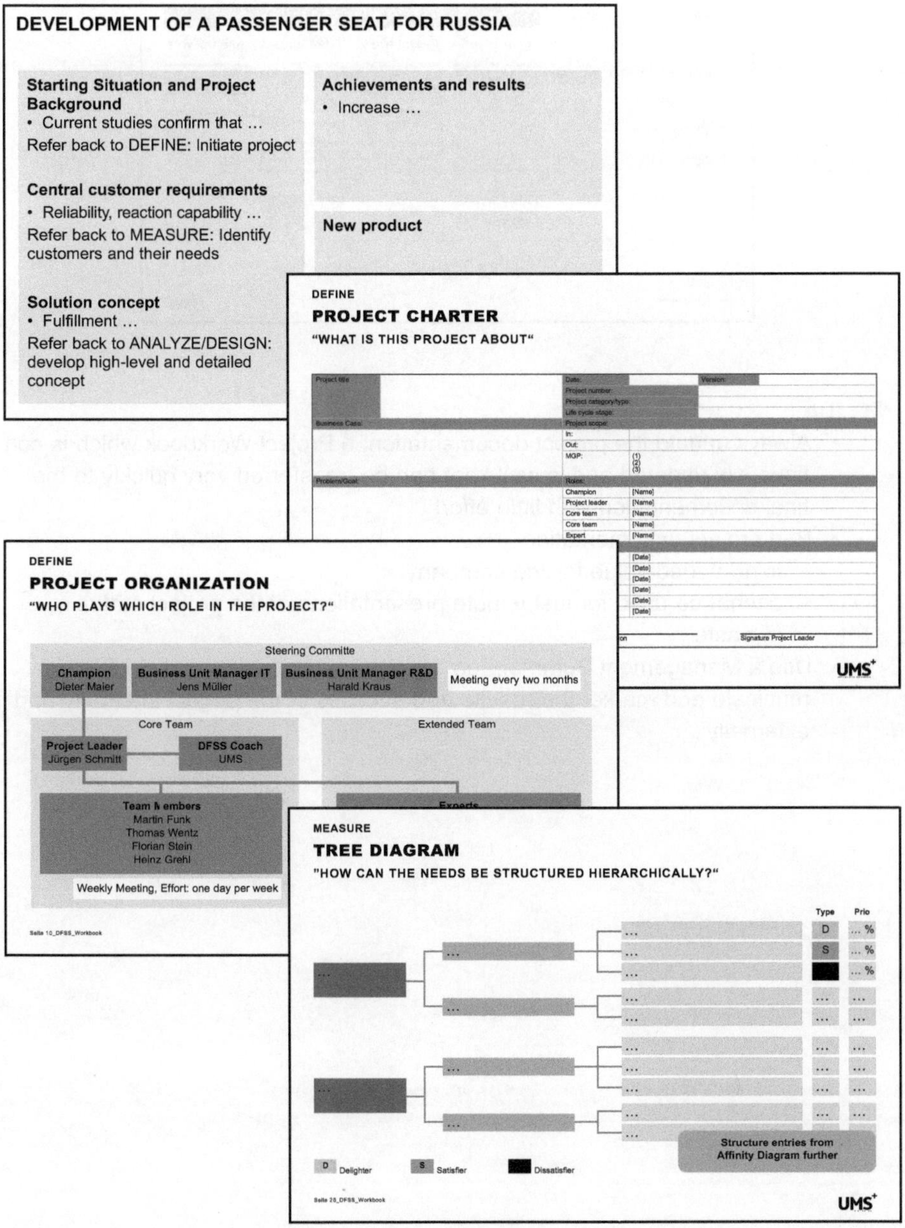

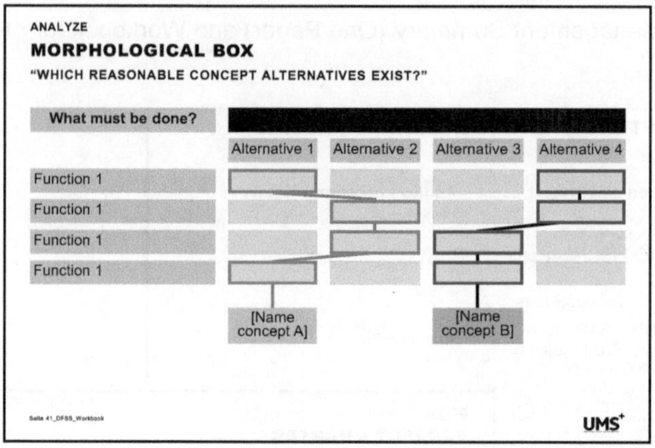

Tip

- Always update the project documentation; a Project Workbook which is continuously updated and is well kept can be transferred very quickly to the final documentation with little effort
- Bad project documentation
  - is no added value for the company
  - cannot be used for last minute presentations of the project status
  - is waste
- Use a Management Summary on one to two pages to successfully communicate and market the results and success of the project internally and externally

Hand over Process Responsibility and Conduct Project Closure

Term
Handover, Project Closure

When
VERIFY: hand over process

Goal
- Hand over process responsibility to the Process Owner
- Hand over final project documentation officially
- Close project and project work officially

Procedure
When the project is handed over to the Process Owner, she/he takes over the responsibility for the process development officially. Separate areas of responsibility should be taken into consideration by paying particular attention to the product and/or process development and should be assigned respectively to product management and/or production (manufacturing, assembly). The design of the responsibility areas should have already been taken into consideration when the implementation team was put together and for supporting the introduction and growth phase of the product.

Efficient process management is based on the following elements:
- Complete and current process documentation and SOPs
- Clearly defined responsibilities
- Relevant KPIs, control parameters and target values
- Regular and correct Monitoring
- Regular Performance Reviews
- Audits

Apart from the content-related preparation of the project work, the final project documentation should also focus on experience, knowledge and insights gained and answer the following questions:
- What has been learned within the scope of the project work?

- Where did weaknesses and/or constraints become apparent in project work?
- Which response measures are successful?
- What further projects and/or follow-up activities would be useful?
- What could be done better in the next projects/the following project generations?

Illustration: Lessons Learned

Team / Resources	Time plan
• Were the team and the resources available? • Was the planning adhered to? • What had a positive impact on keeping to plan and should be done exactly the same way next time? • What had a negative impact and how could it be avoided next time?	• Was the Time Plan kept? • What had a positive impact on keeping the Time Plan and should be made exactly in this way next time? • What had a negative impact and how could it be avoided next time?
Goals / Results	**Further important points**
• Was the goal reached? • What had a positive impact on keeping the goal and should be done exactly this way next time? • What had a negative impact and how could it be avoided next time?	• What else was advantageous for the project and should be done exactly the same way next time? • What was obstructive and should be avoided next time?

In addition to gathering of experiences and insights from project work the ERIC square methodology can be applied:

- **Eliminate**:
  Which elements (experiences, insights, occurrences) would the project team want to eliminate in future development projects and/or follow-up project generations?

- **Reduce**:
  Which elements (experiences, insights, occurrences) would the project team want to reduce in future development projects and/or follow-up project generations?

- **Increase:**
  Which elements (experiences, insights, occurrences) would the project team want to increase in future development projects and/or follow-up project generations?

- **Create:**
  Which elements (experiences, insights, occurrences) would the project team want to create in future development projects and/or follow-up project generations?

Illustration: ERIC square

Eliminate	Reduce
• Poor availability of important experts ...	• Size of team – limitation on key areas, such as ...
Increase	**Create**
• Integration of Marketing for customer segmentation ...	• Use of focus groups for the collection of customer needs in the MEASURE phase

➾ **Tip**

- Ensure official recognition of the team's performance in the form of a certificate, an official publication or credits etc.
- An official project closure meeting pursues the goal of conveying to the team a sense of achievement and the feeling of successfully completed work

Gate Review

Term
Gate Review, Phase Check, Phase Approval

When
Completion of VERIFY, at the end of each phase

Goal
- Inform the Sponsor about the results and actions of each phase
- Evaluate the results
- Decide about the further course of the project

Procedure
The results are presented in a complete and comprehensible way. The Sponsor checks the current status of the project according to the following criteria:
- Completeness of results,
- Probability of project success,
- Optimal resource allocation in the project.

Various results of the VERIFY phase are presented to the Sponsor and the Stakeholders in this the final Gate Review.

The following questions can be of assistance to the project team when preparing the presentation of the results as well as to the Sponsor when conducting the phase approval:

Prepare implementation

- Is it ensured that the serial production can be changed without significant delays and downtimes?
- How is the transition phase designed?
- What do the detailed Scale-up steps look like?
- Has the Project Charter been tested for updates and feasibility?
- Which elements, if any,have been adjusted as a result of this?
- Which activities must be pushed forward for the preparation of the second generation of the MGP?
- Has an implementation team been defined?
- Have activities been conducted successfully which are necessary for the market launch?
- How can it be determined if the activities undertaken are sufficient?

Implement process

- How has the final process been documented?
- Are new and/or changed tasks and the equipment used in the process clearly and simply described?
- Is information accessible at the respective work places?
- Is the transition phase supported by an implementation team?
- How is it composed?
- Is the process performance controlled?
  How is it implemented in the respective areas?

Hand over process

- How is the final project documentation structured?
- Is the project documentation sufficiently informative about the project content and project success?
- How will it become apparent that the Process Owner has completely taken over the responsibility (for the development) and relieved the development team of it?
- What Lessons Learned are derived from the project work?

DESIGN FOR SIX SIGMA+LEAN TOOLSET

Process DESIGN Workout

Process Design Workout

Term
Process Design Workout, Rapid Process Design

When
- In addition to product development during the phases ANALYZE, DESIGN
- Internal process development

Goal
- Manage increasing complexity of support processes
- Quickly align support processes to critical requirements and design and implement them

Procedure
Support processes (control and support processes) are often more complex than necessary due to historical changes in the framework. Emergency measures often lead to increased complexity of processes and to a corresponding impairment of performance and process capability of the affected process(es).

The Process Design Workout counteracts an increase of complexity, costs and defect opportunities in support processes while fulfilling critical success requirements. In terms of sales the relative proportion of employee capacity spent in optional processes rises continuously. The total cost e.g. for providing information and processing information increases.

Complexity of supporting processes can be managed in a simple way, if they are redesigned according to the same procedural model as operative processes. Also in the context of process development the question about the decisive individuals involved in the process arises, who have certain expectations, opinions and requests for the process performance from their customer perspective. It is important to identify and specify these different requirements, in order to be able to use them as a basis for the further design of process related requirements.

Similar to the procedure of the product development, the development of the process can be aligned to the collected requirements in order to design process relevant outputs, individual process steps and relevant inputs in such a way that the requested process capability is achieved and an efficient and effective detailed process design can be created.

Illustration: Procedure for process design

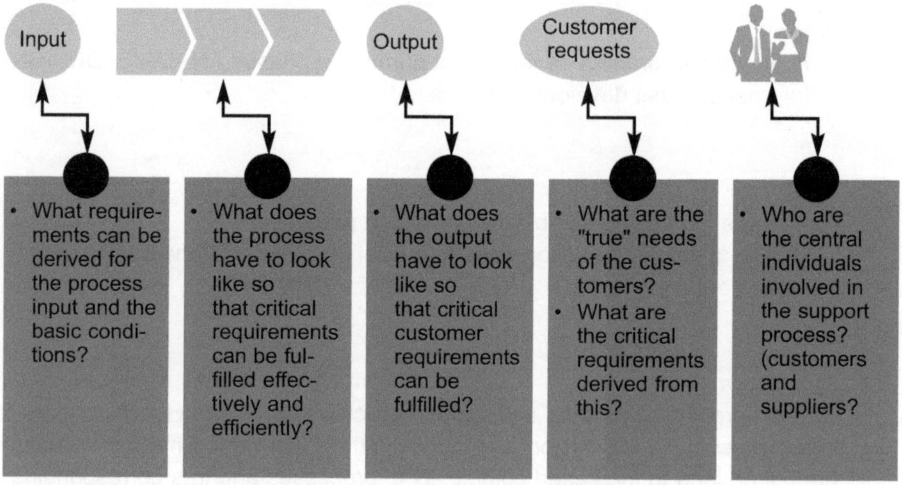

The Process Design Workout follows the DMADV procedure and comprises several stages from identifying customers and their needs to the specific process design. However, the DMADV phases should be restricted to phases for Preparation, Execution and Follow-up, because an accelerated procedure is required.

The relatively comprehensive preparation phase thus provides the basis for efficient workshops, requiring relevant competence, presence and the immediate capability to make decisions.

Illustration: Comparison DMADV and Process Design Workout

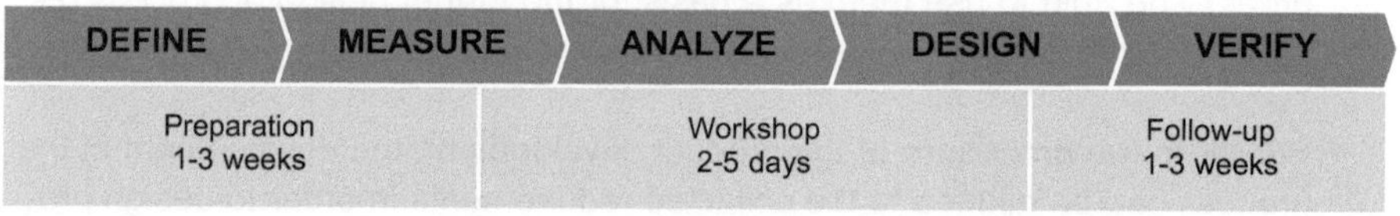

The following prerequisites should be taken into consideration when delivering of a Process Design Workout and should be checked in advance:

- Focus on a clearly scoped process:
  - No external customers
  - Implementation in one's own area
  - Few interfaces and internal customers (≤ 5)

- Employees of the affected area can be reached during the workshop, so that onsite inspections and verifications can be conducted

- Larger IT adjustments or changes of facilities are not to be expected

- Customers/individuals involved in the process and the Sponsor can be reached throughout the entire workshop and are capable of making decisions

- Team members have enough time to work on the project

- The project leader is very experienced in facilitation and conflict resolution, capable of realizing projects and is experienced with DFSS approach to process design

- The team is motivated, willing to realize innovative ideas and ready to implement and live changes

The project can be designed in such a way, and, depending on the Project Charter, that it takes into consideration the critical availability of individual core team members since they are primarily required in the workshop phase and can give their input then. The project leader can control the deployment of the core team and/or individual core team members upon need, in order to relieve the team members whose availability is limited in the Preparation and Follow up phases.

The Preparation phase provides the necessary basis for the subsequent attendance phase in the form of a Workshop. The preparation aims at clearly defining and scoping the targeted process development, identifying potential risks of project work early on, determining relevant target customers and preparing and largely conducting the collection of needs.

The following mapping shows the relevant guiding questions, tools and methods and respective peculiarities in the application for conducting the preparation phase within the frame of a Process Design Workout.

Illustration: Guiding questions, tools and methods as well as special considerations in the preparation phase

Process Design Workout, phase 1: PREPARATION		
Guiding questions	Tools and methodologies	Peculiarities for the application
• How is the Process Design Workout aligned? • How can the process to be developed be scoped? • What is to be observed within the frame of a Workout? • What are reasonable follow-up generations of the project? • What existing information can be used? • Which Stakeholders must be addressed? • Which risks can occur and how are they managed? • What are the responsibilities for completing the work packages? • Who is the target customer of the process and how can his needs be identified in a reliable way?	**1.1 Initiate the project:** • Project Charter (starting situation, goal, scoping, benefit, Time Plan, Team) **1.2 Scope the project:** • SIPOC • IN/OUT & MGP **1.3 Manage the project:** • Analysis of process environment • Stakeholder Management • Risk Analysis • Team definition (SIPOC) • Project planning **1.4 Identify customers and their needs:** • Identify customers & individuals involved in the process (SIPOC and alternative customers) • Collect customer needs (e.g. interviews, process inspection)	• Scope the process (start and stop) with the help of a SIPOC and add process output and/or necessary input parameters for the process delivery • Develop observation frame with the help of the In/Out Frame based on the I-P-O perspective of the SIPOC and define follow-up project generations if required • Align the analysis of the process environment to the following aspects: - Are there interface related projects/activities which can influence the project? - Do comparable internal or external processes exist today? - Which IT based implications need to be taken into consideration during the process design? - Are there economic trends, market or strategy changes which can have an influence on the process to be developed? - Are there changes among the customers, having an influence on the process to be developed? • Take the SIPOC into consideration within the frame of **defining the team** as well as when selecting **relevant target customers** and internal suppliers and customers of the process when the team is selected and the customer definition is taken into consideration

The contents of the preparation phase should be reviewed with the Sponsor in order to be able to introduce and conduct corrective steps early on and before the Workshop phase starts. A Gate Review should be conducted no later than the closure of the preparation phase based on the customer needs which have been developed. Moreover, the project leader should aim at an early integration of the Sponsor in creating and/or coordinating the Project Charter so as to be able him or the team to communicate a coordinated project focus.

The customer definition should take into consideration alternative customer categories such as Heavy User, Lead User and soon to be non-customers in order be able to consider their need structure in a structured way when needs are collected. This would allow users of an IT system, for example, to be categorized as soon to be non-customers, or as heavy users good for a focus group based on their user behavior.

In the Workshop phase attention should be paid to the fact that if possible all defined target customers (internal suppliers and customers of the process) and/or their representatives from the respective areas are available and can participate in the Workshop.

Illustration: Guiding questions, tools and methods as well as special considerations in the Workshop phase

Process Design Workout, Phase 2: WORKSHOP		
Guiding questions	Tools and methodologies	Peculiarities for the application
• What are the needs of the defined individuals involved in the process (internal suppliers and customers of the process)? • Where do gaps in the collection of needs exist? • Where must the collection of needs be detailed further?	**2.1 Structure and prioritize customer needs:** • Affinity Diagram • Tree Diagram • Define CTCs/CTBs • KANO Model • N/3 and/or simple pair comparison	• Actively involve process customers in the structuring and detailed collation of the needs with the help of the **Affinity and Tree Diagram** • Filter great numbers of needs using a **prioritization** on second level of the Tree Diagram (e.g. with the help of a simple pair comparison) and only observe the most important topics resulting from this further

Matrix is continued on the following page

Mapping guiding questions, tools and methods as well as special considerations for the Workshop phase – continued

Process Design Workout, Phase 2: WORKSHOP		
Guiding questions	Tools and methodologies	Peculiarities for the application
• Which CTCs/CTBs can be derived? • How can the CTCs/ CTBs be categorized and respectively priori-tized into Basics, Satisfier and Delighter? • Which measurements, target values and specifications can be assigned to the CTCs/CTBs? • Do contradictory requirements become apparent? • Which process high-level concept fulfills the defined CTCs/CTBs best and how can it be further optimized if required? • Which restrictions in the process must be taken into consider-ation? • How can an efficient and effective process in the detailed design be designed by taking into account the neces-sary process parame-ters?	**2.2 Specify customer needs** • Process Benchmarking • Process Design Score-card (requirement, CTC/CTB, if required measurement, opera-tional definition, target values, specifications, conflicts) • Budget coordination **2.3 Develop and evaluate process high-level concepts** • SIPOC • Morphological Box • Criteria based compari-son and/or Pugh Matrix • Principles of Lean Process Design • Optimization of neces-sary and non-value adding and value enabling activities • Process FMEA and/or AFD for processes **2.4 Develop and evaluate process detailed design** • Draw up process design (working and standard operating pro-cedures, interfaces IT, resource planning, piloting/simulation)	• Use categorization of CTCs/CTBs according to **KANO to reduce** to only two categories: - Must-have: Dissatisfiers/basic factors - Nice-to-have: Satisfier and Delighter • Apart from the collected basic factors you should also check and add further relevant require-ments with respect to conformity with laws and regulations, indus-try-specific regulations and safety requirements • Document **conflicts/contradic-tions** between CTCs and CTBs and integrate in the Morphological Box • Align high-level and detailed design to the **principles of Lean Process Design**, in order to require a minimum level of room and work space, resources and time • Test process if possible in a **simulated environment** in order to gain insights about the real process flow and to be able to develop it further in iterative optimization loops

Illustration: Content-related structure of a five-day workshop phase

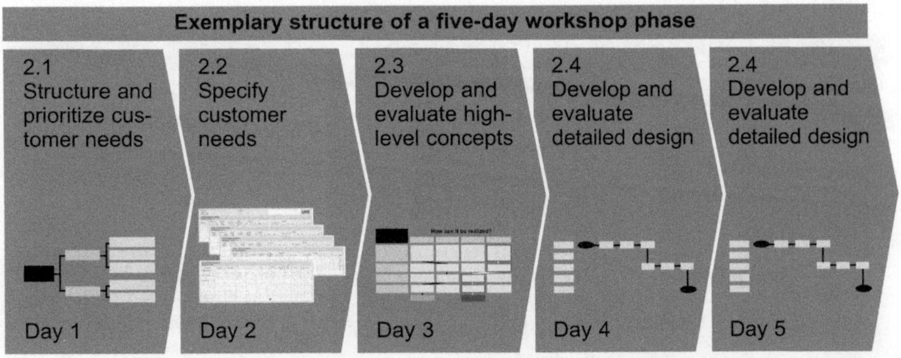

At the end of the workshop a process which is capable being piloted should have been defined in the detailed design phase. The detailed design should then be further developed in additional iterative optimization loops in the follow-up.

When the project closure takes place, at least the following results should exist:
- Documentation of project work
- Documentation of process flows
- Preparation and delivery of necessary training measures
- Implementation of a system for process control

Mapping guiding questions, tools and methods as well as peculiarities for the follow-up on the following page

Mapping: Guiding questions, tools and methodologies as well as special considerations in the follow-up

Process Design Workout, Phase 3: FOLLOW-UP		
Guiding questions	Tools and methodologies	Peculiarities for the application
• How can the key elements of project work be documented and made available for follow-up projects? • Which target group should be trained internally for process design? • What are the components of an appropriate process control? • How can the Rollout be designed? • How can the realization of the process development be continuously monitored?	**3.1 Finalize process detailed design:** • Project documentation (incl. Management Summary, One/Three Pager) • Training concept (incl. documents/ handout, simulations, etc.) • Process KPI system • Process monitoring **3.2 Implementation** • Launch and realization of process • Process audits	• Use **project documentation** to secure project work and also for the purpose of follow-up and as a basis for training actions • Align key **control and optimization criteria (KPIs)** in a Cockpit Chart to time related parameters (e.g. process lead time, process efficiency, etc.), quality (e.g. scrap, rework, etc.) and costs and visualize them

5S	Sort / Set in Order / Shine / Standardize / Sustain
AFD	Anticipatory Failure Determination
AHP	Analytic Hierarchy Process
ANOVA	Analysis of Variances
BB	Black Belt
C/O	Changeover (Setup Time)
CAD	Computer-aided Design
CAPS	Computer-aided Process Simulation
CIT	Change Implementation Tools
CTB	Critical to Business
CTQ	Critical to Quality
DFC	Design for Configuration
DFE	Design for Environment
DFMA	Design for Manufacturing and Assembly
DFR	Design for Reliability
DFS	Design for Services
DFSS	Design for Six Sigma
DMADV	Define, Measure, Analyze, Design, Verify
DMAIC	Define, Measure, Analyze, Improve, Control
DOE	Design of Experiments
DPMO	Defects per Million Opportunities
DPU	Defects per Unit
EBIT	Earnings before Interest and Taxes
EBITDA	Earnings before Interest, Taxes, Depreciation and Amortization
EHS	Environment / Health / Safety
etc.	et cetera
EVA®	Economic Value Added
FMEA	Failure Mode and Effect Analysis
FTA	Fault Tree Analysis
GB	Green Belt
HR	Human Resources
IT	Information Technology
KPI	Key Performance Indicator

LCL	Lower Control Limit
LSL	Lower Specification Limit
Max	Maximum
MBB	Master Black Belt
MCA	Monte Carlo Analysis
MGP	Multi Generation Plan
min	Minimum
NOPAT	Net Operating Profit after Taxes
PLT	Process Lead Time
P/T	Processing Time
PDCA	Plan, Do, Check, Act
ppm	Parts per million
QFD	Quality Function Diagram
R&D	Research & Development
R&R	Repeatability & Reproducibility
RACI	Responsible / Accountable / Consulted / Informed
RPN / RPZ	Risk Priority Number
RSS	Root Sum Square Method
RTY	Rolled Throughput Yield
SCAMPER	Substitute / Combine / Adapt / Modify / Put to other uses / Eliminate
SIPOC	Supplier / Input / Prozess / Output / Customer
SLA	Service Level Agreements
SMA	Shape-Memory Alloys
SMED	Single Minute Exchange of Die
SU	Setup
Sufield Analysis	Substance-Field Analysis
TIMWOOD	Transport / Inventory / Motion / Waiting / Overproduction / Overprocessing / Defects
TIPS	Theory of Inventive Problem Solving
TRIZ	Teoriya Reshemiya Izobretatelskikh Zadach (Russian acronym for Theory of Inventive Problem Solving)

UCL	Upper Control Limit
USL	Upper Specification Limit
USP	Unique Selling Proposition / Point
VOC	Voice of Customer
VSM	Value Stream Map
WCA	Worst Case Analysis
WIP	Work in Process

PROCESS SIGMA TABLE INCLUDING 1.5 SIGMA SHIFT

Yield	Process Sigma	Defects per 1,000,000	Defects per 100,000	Defects per 10,000	Defects per 1,000	Defects per 100
99.99966%	6	3.4	0.34	0.034	0.0034	0.00034
99.99946%	5.9	5	0.54	0.054	0.0054	0.00054
99.99915%	5.8	9	0.85	0.085	0.0085	0.00085
99.99867%	5.7	13	1	0.1	0.01	0.001
99.99793%	5.6	21	2	0.2	0.02	0.002
99.99683%	5.5	32	3	0.3	0.03	0.003
99.99519%	5.4	48	5	0.5	0.05	0.005
99.99277%	5.3	72	7	0.7	0.07	0.007
99.98922%	5.2	108	11	1.1	0.11	0.011
99.98409%	5.1	159	16	1.6	0.16	0.016
99.97674%	5	233	23	2.3	0.23	0.023
99.96631%	4.9	337	34	3.4	0.34	0.034
99.95166%	4.8	483	48	4.8	0.48	0.048
99.93129%	4.7	687	69	6.9	0.69	0.069
99.90324%	4.6	968	97	9.7	0.97	0.097
99.86501%	4.5	1350	135	13.5	1.35	0.135
99.81342%	4.4	1866	187	18.7	1.87	0.187
99.74449%	4.3	2555	256	25.6	2.56	0.256
99.65330%	4.2	3467	347	34.7	3.47	0.347
99.53388%	4.1	4661	466	46.6	4.66	0.466
99.37903%	4	6210	621	62.1	6.21	0.621
99.18025%	3.9	8198	820	82.0	8.20	0.820
98.92759%	3.8	10724	1072	107.2	10.72	1.072
98.60966%	3.7	13903	1390	139.0	13.90	1.390
98.21356%	3.6	17864	1786	178.6	17.86	1.786
97.72499%	3.5	22750	2275	227.5	22.75	2.275
97.12834%	3.4	28717	2872	287.2	28.72	2.872
96.40697%	3.3	35930	3593	359.3	35.93	3.593
95.54345%	3.2	44565	4457	445.7	44.57	4.457
94.52007%	3.1	54799	5480	548.0	54.80	5.480
93.31928%	3	66807	6681	668.1	66.81	6.681
91.92433%	2.9	80757	8076	807.6	80.76	8.076
90.31995%	2.8	96800	9680	968.0	96.80	9.680
88.49303%	2.7	115070	11507	1150.7	115.07	11.507
86.43339%	2.6	135666	13567	1356.7	135.67	13.567
84.13447%	2.5	158655	15866	1586.6	158.66	15.866
81.59399%	2.4	184060	18406	1840.6	184.06	18.406
78.81446%	2.3	211855	21186	2118.6	211.86	21.186
75.80363%	2.2	241964	24196	2419.6	241.96	24.196
72.57469%	2.1	274253	27425	2742.5	274.25	27.425
69.14625%	2	308538	30854	3085.4	308.54	30.854
65.54217%	1.9	344578	34458	3445.8	344.58	34.458
61.79114%	1.8	382089	38209	3820.9	382.09	38.209
57.92597%	1.7	420740	42074	4207.4	420.74	42.074
53.98278%	1.6	460172	46017	4601.7	460.17	46.017
50.00000%	1.5	500000	50000	5000.0	500.00	50.000
46.01722%	1.4	539828	53983	5398.3	539.83	53.983
42.07403%	1.3	579260	57926	5792.6	579.26	57.926
38.20886%	1.2	617911	61791	6179.1	617.91	61.791
34.45783%	1.1	655422	65542	6554.2	655.42	65.542
30.85375%	1	691462	69146	6914.6	691.46	69.146
27.42531%	0.9	725747	72575	7257.5	725.75	72.575
24.19637%	0.8	758036	75804	7580.4	758.04	75.804
21.18554%	0.7	788145	78814	7881.4	788.14	78.814
18.40601%	0.6	815940	81594	8159.4	815.94	81.594
15.86553%	0.5	841345	84134	8413.4	841.34	84.134
13.56661%	0.4	864334	86433	8643.3	864.33	86.433
11.50697%	0.3	884930	88493	8849.3	884.93	88.493
9.68005%	0.2	903200	90320	9032.0	903.20	90.320
8.07567%	0.1	919243	91924	9192.4	919.24	91.924

DEFINE	MEASURE	ANALYZE	DESIGN	VERIFY
• Develop business case • Define problems and objectives • Calculate financial bene-fits • Define roles • Set project framework • Develop Multi Generation Plan • Check impact on other projects • Plan and schedule resource and activities • Prepare budget • Change management plan • Assess project risks	• Identify customers • Customer segmentation • Select and perform research methods • Collect customer testimonials • Derive customer needs • Restructure customer needs • Categorize customer needs • Prioritize customer needs • CTCs and derived variables • Perform benchmarking • Prioritize measures that determine target values and specifications • Assess risk • Set quality indicators • Update Risk Management	• Identify relevant functions • Develop alternative rough concepts • Select best rough concept • Establish target costs • Eliminate weaknesses and conflicts in selected concept • Obtain customer and sta-keholder feedback • Prepare draft process • Prepare launch • Determine resource requirements • Finalize concept • Update Risk Management	• Map alternative design characteristics of the system • Draft Cost Structure • Compare alternative design elements • Develop Scorecard for detailed concept • Conduct system simplification (Design X, Trimming) • Identify and mitigate risks • Create detail process design • Draft facilities, building, equipment and materials procurement plan • Establish KPI systems and process monitoring • Simulate processes • Provide IT • Pilot processes • Update Risk Management	• Implementation strategies derived • Detail Implementation planning • Finalize launch • Prepare market monito-ring • Create process documen-tation and work instruc-tions (SOPs) • Define implementation team • visualize and control pro-cess performance • Finish project documenta-tion • Handover process res-ponsibility • Complete project

Additional material from , Design for Six Sigma^{+Lean} Toolset
ISBN 978-3-642-41454-1 (978-3-642-41454-1_OSFO1),
is available at http://extras.springer.com

EXTRAS ONLINE